Advances in Social Bots

Advances in Social Bots

Guest Editors

Yangyang Li
Yangzhao Yang
Hu Huang

MDPI

Basel • Beijing • Wuhan • Barcelona • Belgrade • Novi Sad • Cluj • Manchester

Guest Editors
Yangyang Li
Key Laboratory of Cyberculture Content Cognition and Detection
University of Science and Technology of China
Hefei
China

Yangzhao Yang
Chinese Institute of Command and Control
Beijing
China

Hu Huang
Shenzhen Graduate School
Peking University
Shenzhen
China

Editorial Office
MDPI AG
Grosspeteranlage 5
4052 Basel, Switzerland

This is a reprint of the Special Issue, published open access by the journal *Electronics* (ISSN 2079-9292), freely accessible at: https://www.mdpi.com/journal/electronics/special_issues/8TF9JML95C.

For citation purposes, cite each article independently as indicated on the article page online and as indicated below:

Lastname, A.A.; Lastname, B.B. Article Title. *Journal Name* **Year**, *Volume Number*, Page Range.

ISBN 978-3-7258-6418-8 (Hbk)
ISBN 978-3-7258-6419-5 (PDF)
https://doi.org/10.3390/books978-3-7258-6419-5

Contents

Yangyang Li, Yangzhao Yang and Hu Huang
Advances in Social Bots
Reprinted from: *Electronics* **2025**, *14*, 4885, https://doi.org/10.3390/electronics14244885 . . . . . **1**

Zhiyu Xie, Fuqiang Niu, Genan Dai and Bowen Zhang
From Claims to Stance: Zero-Shot Detection with Pragmatic- Aware Multi-Agent Reasoning
Reprinted from: *Electronics* **2025**, *14*, 4298, https://doi.org/10.3390/electronics14214298 . . . . . **5**

Yu Song, Pei Liu and Yunpeng Wu
Enhancing the Sustained Capability of Continual Test-Time Adaptation with Dual Constraints
Reprinted from: *Electronics* **2025**, *14*, 3891, https://doi.org/10.3390/electronics14193891 . . . . . **21**

Xi Zeng, Guangchun Luo and Ke Qin
Joint Event Detection with Dynamic Adaptation and Semantic Relevance
Reprinted from: *Electronics* **2025**, *14*, 234, https://doi.org/10.3390/electronics14020234 . . . . . . **44**

Yash Sharma, Pedro Ferreira and Laura Justham
Hardness Classification Using Cost-Effective Off-the-Shelf Tactile Sensors Inspired by Mechanoreceptors
Reprinted from: *Electronics* **2024**, *13*, 2450, https://doi.org/10.3390/electronics13132450 . . . . . **60**

Yi Qu, Sheng Wang, Hui Feng and Qiang Liu
Geometry Optimization of Stratospheric Pseudolite Network for Navigation Applications
Reprinted from: *Electronics* **2024**, *13*, 2397, https://doi.org/10.3390/electronics13122397 . . . . . **81**

Weiyi Chen, Lingjuan Miao, Jinchao Gui, Yuhao Wang and Yiran Li
FLsM: Fuzzy Localization of Image Scenes Based on Large Models
Reprinted from: *Electronics* **2024**, *13*, 2106, https://doi.org/10.3390/electronics13112106 . . . . . **99**

Yi Qu, Sheng Wang, Tianshi Pan and Hui Feng
IPCB: Intelligent Pseudolite Constellation Based on High-Altitude Balloons
Reprinted from: *Electronics* **2024**, *13*, 2095, https://doi.org/10.3390/electronics13112095 . . . . . **115**

Erfeng Xu, Junwu Zhu, Luchen Zhang, Yi Wang and Wei Lin
Research on Aspect-Level Sentiment Analysis Based on Adversarial Training and Dependency Parsing
Reprinted from: *Electronics* **2024**, *13*, 1993, https://doi.org/10.3390/electronics13101993 . . . . . **140**

electronics

MDPI

Editorial

Advances in Social Bots

Yangyang Li [1], Yangzhao Yang [1] and Hu Huang [2,*]

[1] Institute of Social Computing, Academy of Cyber, Beijing 100041, China
[2] School of Cyber Science and Technology, University of Science and Technology of China, Hefei 230026, China
* Correspondence: huanghu@ustc.edu.cn

1. Introduction

The digital landscape of the 21st century has been irrevocably shaped by the rise of automated actors. "Social bots"—algorithms designed to generate content and mimic human interaction—have evolved from simple, script-based novelties into sophisticated entities capable of influencing global discourse [1]. While early research primarily viewed these bots as deterministic tools for automated customer service or news aggregation, their capabilities have expanded alongside their potential for misuse. Today, malicious actors are implicated in amplifying low-credibility content, manipulating financial markets, and exacerbating political polarization by infiltrating echo chambers [2,3].

Recent years have witnessed a critical inflection point: the transition from "scripted automation" to "cognitive autonomy". With the advent of Large Language Models (LLMs) and multimodal foundation models, the distinction between human and machine behavior has blurred significantly. Reports indicate that automated traffic surpassed human traffic for the first time in 2024, a shift driven largely by AI-powered agents [4]. Unlike their predecessors, these modern actors possess "emergent abilities"—such as reasoning, planning, and emotional mimicry—allowing them to navigate complex social dynamics with unprecedented fluidity [5,6].

This evolution has fundamentally altered the nature of online interaction. On an individual level, modern agents can now engage in long-term strategic planning and exhibit persuasive capabilities that rival human interlocutors. Studies suggest that LLM-driven agents can tailor rhetoric to specific user demographics, rendering them potent tools for computational propaganda [7]. On a collective level, these agents are increasingly deployed in "social sandboxes" to simulate human community dynamics. While offering a new lens for computational sociology, this also raises concerns about the scalability of synthetic misinformation [8].

Consequently, authenticating these interactions is becoming increasingly difficult as the "detection boundary" shifts. Traditional models often fail to capture the long-range dependencies inherent in multi-party, multi-turn discussions, a challenge exacerbated by the scarcity of realistic conversational datasets [9]. Furthermore, the ability of agents to utilize external tools (e.g., search engines, APIs) allows them to ground responses in real-time data. This capability enables them to bypass detection methods that rely on identifying factual hallucinations or static knowledge cutoffs [10].

We are thus witnessing a paradigm shift from disembodied software scripts to "Social Agents"—autonomous systems capable of constructing internal world models, maintaining long-term memory, and interacting with the physical world through Embodied AI [11]. However, realizing this vision requires a holistic approach. To build agents that can effectively persuade without hallucinating, or navigate physical spaces without failing, we must address fundamental challenges in intent understanding, environmental grounding,

https://doi.org/10.3390/electronics14244885

and resilient connectivity. This Special Issue, "Advances in Social Bots," was curated to bridge this gap, providing the architectural blueprints that span from cognitive algorithms to the necessary physical infrastructure.

2. Thematic Overview: From Cognition to Infrastructure

The articles in this Special Issue illustrate that the evolution from "scripted bots" to "autonomous agents" is not a single leap, but a layered evolution. To support the high-level emergent behaviors described in the Introduction—such as strategic persuasion and adaptive social interaction—advancements are required across three fundamental layers: cognitive processing, perceptual adaptation, and physical infrastructure.

2.1. Cognitive Intelligence: Understanding Stance and Sentiment

For a social agent to interact meaningfully, it must understand not just what is said, but the stance and sentiment behind it.

Xie et al. (Contribution 1) address the challenge of zero-shot stance detection. They propose the PAMR (Pragmatic-Aware Multi-Agent Reasoning) framework, which utilizes LLMs in a multi-agent architecture. By explicitly modeling pragmatic cues like sarcasm, their work demonstrates how agents can "think" collaboratively to decipher implicit stances without task-specific training. Similarly focusing on linguistic nuance, Xu et al. (Contribution 2) explore the granularity of emotion. Their research on aspect-level sentiment analysis allows machines to disentangle complex sentence structures, a critical capability for agents engaging in nuanced human–machine dialogue.

Furthermore, Zeng et al. (Contribution 3) tackle the temporal dimension. They propose the DASR framework for joint event detection, utilizing incremental learning to mitigate "catastrophic forgetting." This ensures that agents can continuously adapt to emerging hot topics, mitigating the risk of knowledge obsolescence in long-term social simulations.

2.2. Perceptual Robustness: Vision and Adaptation

Modern agents operate in a multimodal world. As the boundary between real and synthetic content blurs, perceptual robustness becomes critical—both for agents to ground themselves in reality and for systems to distinguish authentic interactions from fabricated ones.

Chen et al. (Contribution 4) explore the intersection of vision and geography. Their FLsM model utilizes large-scale visual models for the fuzzy localization of image scenes. This capability is vital for verifying the authenticity of user-generated content and distinguishing between real-world activity and fabricated bot personas. Addressing the stability of AI models in changing environments, Song et al. (Contribution 5) introduce a "Dual Constraints" method for Continual Test-Time Adaptation (CTTA). This research ensures that detection algorithms remain robust even as data distributions shift over time, providing a defense against bots that constantly alter their behavioral patterns to evade detection.

2.3. Embodied Intelligence and Network Infrastructure

Finally, as agents transition from digital chatbots to embodied robots interacting with the physical world, they require sensitive perception and resilient communication networks to maintain autonomy.

Sharma et al. (Contribution 6) provide a glimpse into embodied interaction. Their work on hardness classification using cost-effective tactile sensors mimics human mechanoreceptors, paving the way for service robots that can socially and physically interact with humans.

Supporting these distributed agents requires robust connectivity. Qu et al. contribute two pivotal studies on the network layer. In their first paper (Contribution 7), they optimize the geometry of Stratospheric Pseudolite Networks (SPNs). In their second paper (Contribution 8), they propose an Intelligent Pseudolite Constellation (IPCB) based on high-altitude balloons. These studies lay the foundation for a resilient, wide-coverage communication infrastructure. Such networks are essential for coordinating swarms of autonomous agents in remote areas where terrestrial networks fail, ensuring that the "social" connection remains unbroken.

3. Future Outlook

The research presented in this Special Issue highlights an inevitable trajectory: the convergence of Generative AI, robotics, and social computing. We are witnessing a paradigm shift from simple "Social Bots" to sophisticated "Social Agents"—entities that can think, see, and touch. Unlike their text-centric predecessors, these future agents will operate across three interconnected dimensions, as evidenced by the contributions in this volume.

- Cognitive Autonomy: Future agents must move beyond script adherence to exhibit emergent behaviors and reasoning. They will simulate complex human societal dynamics in computational sandboxes, a direction supported by recent surveys on LLM-based multi-agent systems [12].
- Multimodal Perception: Agents will seamlessly transition between digital platforms and physical forms. They will require visual adaptability to verify reality and tactile sensitivity to interact with the physical world.
- Resilient Infrastructure: The sustainability of these agents will depend on robust network architectures capable of supporting distributed, autonomous swarms in even the most remote environments.

As these technologies mature, the challenge for the scientific community shifts from merely detecting these actors to understanding their complex interactions with human society. The boundary between human and machine is blurring, necessitating new governance frameworks to ensure that these powerful agents align with human values.

4. Conclusions

We extend our gratitude to all the authors, reviewers, and the editorial team who made this Special Issue possible. "Advances in Social Bots" stands as a testament to the field's diversity, successfully bridging the gap between abstract software algorithms and tangible hardware engineering.

By integrating cognitive intelligence, perceptual robustness, and physical infrastructure, this collection provides the foundational blueprints for the next generation of autonomous systems. We hope this Special Issue inspires further inquiry into the symbiotic future of human and machine intelligence.

Funding: The work was partly supported by the National Natural Science Foundation of China (U25B2042).

Conflicts of Interest: The authors declare no conflicts of interest.

List of Contributions

1. Xie, Z.; Niu, F.; Dai, G.; Zhang, B. From Claims to Stance: Zero-Shot Detection with Pragmatic-Aware Multi-Agent Reasoning. *Electronics* **2025**, *14*, 4298. https://doi.org/10.3390/electronics14214298.
2. Xu, E.; Zhu, J.; Zhang, L.; Wang, Y.; Lin, W. Research on Aspect-Level Sentiment Analysis Based on Adversarial Training and Dependency Parsing. *Electronics* **2024**, *13*, 1993. https://doi.org/10.3390/electronics13101993.

3. Zeng, X.; Luo, G.; Qin, K. Joint Event Detection with Dynamic Adaptation and Semantic Relevance. *Electronics* **2025**, *14*, 234. https://doi.org/10.3390/electronics14020234.
4. Chen, W.; Miao, L.; Gui, J.; Wang, Y.; Li, Y. FLsM: Fuzzy Localization of Image Scenes Based on Large Models. *Electronics* **2024**, *13*, 2106. https://doi.org/10.3390/electronics13112106.
5. Song, Y.; Liu, P.; Wu, Y. Enhancing the Sustained Capability of Continual Test-Time Adaptation with Dual Constraints. *Electronics* **2025**, *14*, 3891. https://doi.org/10.3390/electronics14193891.
6. Sharma, Y.; Ferreira, P.; Justham, L. Hardness Classification Using Cost-Effective Off-the-Shelf Tactile Sensors Inspired by Mechanoreceptors. *Electronics* **2024**, *13*, 2450. https://doi.org/10.3390/electronics13132450.
7. Qu, Y.; Wang, S.; Feng, H.; Liu, Q. Geometry Optimization of Stratospheric Pseudolite Network for Navigation Applications. *Electronics* **2024**, *13*, 2397. https://doi.org/10.3390/electronics13122397.
8. Qu, Y.; Wang, S.; Pan, T.; Feng, H. IPCB: Intelligent Pseudolite Constellation Based on High-Altitude Balloons. *Electronics* **2024**, *13*, 2095. https://doi.org/10.3390/electronics13112095.

References

1. Ferrara, E.; Varol, O.; Davis, C.; Menczer, F.; Flammini, A. The rise of social bots. *Commun. ACM* **2016**, *59*, 96–104. [CrossRef]
2. Shao, C.; Ciampaglia, G.L.; Varol, O.; Yang, K.C.; Flammini, A.; Menczer, F. The spread of low-credibility content by social bots. *Nat. Commun.* **2018**, *9*, 4787. [CrossRef] [PubMed]
3. Cresci, S. A decade of social bot detection. *Commun. ACM* **2020**, *63*, 72–83. [CrossRef]
4. Thales Group. 2025 Bad Bot Report. 2025. Available online: https://www.imperva.com/resources/resource-library/reports/2025-bad-bot-report/ (accessed on 24 November 2025).
5. Park, J.S.; O'Brien, J.; Cai, C.J.; Morris, M.R.; Liang, P.; Bernstein, M.S. Generative Agents: Interactive Simulacra of Human Behavior. In Proceedings of the 36th Annual ACM Symposium on User Interface Software and Technology (UIST'23), New York, NY, USA, 29 October–1 November 2023. [CrossRef]
6. Xi, Z.; Chen, W.; Guo, X.; He, W.; Ding, Y.; Hong, B.; Zhang, M.; Wang, J.; Jin, S.; Zhou, E.; et al. The rise and potential of large language model based agents: A survey. *Sci. China Inf. Sci.* **2025**, *68*, 121101. [CrossRef]
7. Salvi, F.; Ribeiro, M.H.; Gallotti, R.; West, R. On the conversational persuasiveness of large language models: A randomized controlled trial. *arXiv* **2024**, arXiv:2403.14380. [CrossRef]
8. Gao, C.; Lan, X.; Lu, Z.; Mao, J.; Piao, J.; Wang, H.; Jin, D.; Li, Y. S3: Social-network simulation system with large language model-empowered agents. *arXiv* **2023**, arXiv:2307.14984. [CrossRef]
9. Niu, F.; Yang, M.; Li, A.; Zhang, B.; Peng, X.; Zhang, B. A Challenge Dataset and Effective Models for Conversational Stance Detection. In Proceedings of the 2024 Joint International Conference on Computational Linguistics, Language Resources and Evaluation (LREC-COLING 2024), Torino, Italy, 20–25 May 2024. [CrossRef]
10. Wang, G.; Xie, Y.; Jiang, Y.; Mandlekar, A.; Xiao, C.; Zhu, Y.; Fan, L.; Anandkumar, A. Voyager: An Open-Ended Embodied Agent with Large Language Models. *arXiv* **2023**, arXiv:2305.16291. [CrossRef]
11. Fung, P.; Bachrach, Y.; Celikyilmaz, A.; Chaudhuri, K.; Chen, D.; Chung, W.; Dupoux, E.; Gong, H.; Jégou, H.; Lazaric, A.; et al. Embodied ai agents: Modeling the world. *arXiv* **2025**, arXiv:2506.22355. [CrossRef]
12. Guo, T.; Chen, X.; Wang, Y.; Chang, R.; Pei, S.; Chawla, N.V.; Wiest, O.; Zhang, X. Large Language Model Based Multi-agents: A Survey of Progress and Challenges. In Proceedings of the Thirty-Third International Joint Conference on Artificial Intelligence, Jeju, Republic of Korea, 3–9 August 2024; pp. 8048–8057.

electronics

MDPI

Article

From Claims to Stance: Zero-Shot Detection with Pragmatic-Aware Multi-Agent Reasoning

Zhiyu Xie [1,†], Fuqiang Niu [2,†], Genan Dai [1] and Bowen Zhang [1,*]

[1] School of Artificial Intelligence, Shenzhen Technology University, Shenzhen 518118, China
[2] School of Cyber Science and Technology, University of Science and Technology of China, Hefei 230026, China
* Correspondence: zhangbowen@sztu.edu.cn
† These authors contributed equally to this work.

Abstract: Stance detection aims to identify whether a text expresses a favorable, opposing, or neutral attitude toward a given target and has become increasingly important for analyzing public discourse on social media. Existing approaches, ranging from supervised neural models to prompt-based large language models (LLMs), face two persistent challenges: the scarcity of annotated stance data across diverse targets and the difficulty of generalizing to unseen targets under pragmatic and rhetorical variation. To address these issues, we propose PAMR (Pragmatic-Aware Multi-Agent Reasoning), a zero-shot stance detection framework that decomposes stance inference into structured reasoning steps. PAMR orchestrates three LLM-driven agents—a linguistic parser that extracts pragmatic markers and canonicalizes claims, an NLI-based estimator that produces calibrated stance probabilities through consensus voting, and a counterfactual and view-switching auditor that probes robustness under controlled rewrites. A stability-aware fusion integrates these signals, conservatively abstaining when evidence is uncertain or inconsistent. Experiments on SemEval-2016 and COVID-19-Stance show that PAMR achieves macro-F1 scores of 71.9% and 73.0%, surpassing strong zero-shot baselines (FOLAR and LogiMDF) by +2.0% and +3.1%. Ablation results confirm that pragmatic cues and counterfactual reasoning substantially enhance robustness and interpretability, underscoring the value of explicit reasoning and pragmatic awareness for reliable zero-shot stance detection on social media.

Keywords: zero-shot stance detection; multi-agent framework; large language model

1. Introduction

Social media has become a central arena for public discourse, where individuals express their opinions on controversial issues such as politics, social movements, healthcare, and environmental policies. Understanding public stance toward such topics is crucial for applications including opinion mining, misinformation detection, crisis monitoring, and policy-making support. Stance detection—the task of classifying a text as favorable, against, or neutral with respect to a target—has therefore emerged as a fundamental problem in natural language processing (NLP) and social computing [1]. Early studies treated stance detection as a supervised classification task on social media platforms such as Twitter, focusing on political debates and event-specific targets [2,3]. Subsequent research leveraged neural architectures and pre-trained language models to improve contextual understanding [4,5], yet most methods still depend on large annotated datasets and struggle to generalize across unseen targets. These limitations have motivated recent exploration into cross-target and zero-shot stance detection paradigms [6,7], which aim to infer stance for new topics without task-specific training data. In practice, however, stance on social media is rarely expressed in a straightforward manner. Instead, it is often conveyed implicitly through pragmatic cues such as sarcasm, negation, rhetorical questions, or figurative language [8]. These linguistic devices can obscure or even invert the intended polarity, making it difficult for systems to distinguish stance from surface sentiment [9]. This suggests that effective stance detection should not treat the input merely as a text–target pair, but rather as a reasoning task that accounts for the author's underlying claim and its pragmatic framing.

Traditional stance detection methods initially focused on supervised neural architectures, including recurrent networks, attention mechanisms, and graph-based models, which rely on large quantities of annotated stance data [3,4,10,11]. With the rise of pre-trained language models (PLMs), fine-tuning strategies achieved stronger performance by leveraging contextual embeddings [5,12]. However, these methods still face two fundamental challenges. First, annotated stance data remain scarce and unevenly distributed across targets, making supervised training impractical for new domains. Second, generalization to unseen targets remains difficult, even with powerful PLMs, as stance expressions often vary in subtle ways across topics [13].

To alleviate these issues, recent research has turned to cross-target [6] and zero-shot stance detection (ZSSD) [7]. Cross-target methods attempt to transfer stance knowledge from source domains to unseen targets using contrastive learning, knowledge injection, or graph reasoning. Meanwhile, zero-shot approaches leverage PLMs and large language models (LLMs) directly via prompting or inference reformulation [14]. Despite recent progress, two persistent gaps remain. First, many approaches conflate stance with sentiment, especially in emotionally charged or figurative language. Second, existing models lack explicit mechanisms to handle pragmatic phenomena—such as sarcasm, hedging, and negation—or to assess whether stance predictions are stable under slight linguistic variations. These challenges suggest that monolithic architectures may struggle to disentangle the complex and often interacting cues that shape stance expression.

These observations suggest several concrete desiderata for zero-shot stance detection systems: (1) extract explicit, target-linked canonical claims rather than relying solely on surface expressions; (2) remain robust to pragmatic confounds such as sarcasm and negation; (3) ensure prediction stability through counterfactual and perspective-based probing; (4) adopt calibrated decision strategies that abstain to neutral when evidence is thin or unstable.

Inspired by recent advances in multi-agent reasoning—particularly division of labor, coordination, and self-verification capabilities [15,16], we introduce **PAMR**—Pragmatic-Aware Multi-Agent Reasoning—a zero-shot stance detection framework that decomposes the task into interpretable subtasks and re-assembles their signals through stability-aware decision making. PAMR orchestrates three LLM-driven agents: (1) a Linguistic Parser that distills the input into a canonical claim while extracting pragmatic markers; (2) an NLI-based Estimator that produces a calibrated distribution {favor, against, neutral}; and (3) a Counterfactual and View-Switching module that probes robustness by re-evaluating stance under meaning-preserving rewrites (e.g., removing sarcasm, switching voice). A lightweight stability-aware fusion integrates these signals, conservatively assigning neutral when confidence is low, predictions are unstable, or top classes are tied. PAMR requires no task-specific fine-tuning, yields auditable intermediate outputs (claims, pragmatic markers, robustness flips), and explicitly mitigates common failure modes such as sarcasm-induced polarity errors.

The main contributions of this paper are summarized as follows:

1. We propose PAMR, a pragmatic-aware multi-agent framework for zero-shot stance detection that factors inference into claim normalization, probabilistic NLI, and robustness probing, enabling interpretable and modular reasoning.
2. We introduce a counterfactual and view-switching probe that quantifies stability of stance under meaning-preserving edits and perspective shifts; this signal directly informs a stability-aware fusion rule that curbs over-confident polarity errors and disentangles stance from sentiment.
3. We conduct extensive experiments on benchmark datasets, demonstrating that PAMR matches or surpasses strong zero-shot baselines. Ablations reveal the contributions of pragmatic cues and robustness probing, while additional analyses show that PAMR produces interpretable intermediate artifacts that enable fine-grained audits.

Overall, this study bridges computational modeling and social media discourse analysis by explicitly integrating pragmatic reasoning into stance inference. We aim to provide a framework that not only improves zero-shot prediction accuracy but also deepens interpretability for real-world applications such as misinformation monitoring and public opinion tracking. The remainder of this paper is organized as follows. Section 2 introduces related work. Section 3 describes the proposed method. Section 4 presents the experimental setup. Section 5 reports the experimental results and analysis. Section 6 discusses key findings and limitations. Section 7 concludes the paper.

2. Related Works

2.1. Traditional Stance Detection Methods

Stance detection, closely related to sentiment analysis, argument mining, and fact verification, has been studied extensively in natural language processing. Early research primarily relied on supervised learning with handcrafted features and classical classifiers. With the development of deep learning, neural architectures such as CNNs [17,18] and LSTMs [10] became prevalent, enabling models to capture contextual and sequential dependencies in text. Graph neural networks were later introduced to encode relations among posts, users, and targets, further enriching stance representations. The emergence of PLMs significantly advanced the field. Fine-tuning strategies reformulated stance detection as a text–target classification problem by concatenating target and input sequences [19]. To reduce computational cost, lightweight adaptation methods froze most PLM parameters while tuning small modules [14]. Beyond fine-tuning, prompt-based techniques framed stance detection as masked language modeling. By filling in templates such as "The attitude toward <Target> is [MASK]," PLMs could predict stance more effectively. Recent studies improved this paradigm by designing adaptive prompts tailored to different targets [20]. In addition to general PLMs, social-media-oriented variants such as BERTweet [21] and CT-BERT [22] have been developed to better capture the linguistic characteristics of Twitter and COVID-19 discourse. While these models enhance representation learning on noisy text, they remain limited in addressing pragmatic phenomena such as sarcasm or negation and lack interpretability or stability validation mechanisms. Multi-modal variants have also been proposed, combining textual and visual cues through prompt-based mechanisms [23]. Despite these innovations, most traditional approaches remain reliant on annotated supervision and show limited robustness when confronted with pragmatic language use.

2.2. Zero-Shot Stance Detection

To overcome the dependence on labeled data, zero-shot stance detection has been widely explored. In this setting, models must transfer knowledge from source targets to entirely new ones without direct supervision. Early work employed contrastive learning to align stance features across domains, as in JointCL [24], while other approaches leveraged external knowledge bases to bridge semantic gaps between known and unseen targets, such as TarBK. Graph-based methods further advanced the field by constructing heterogeneous or multi-view graphs that capture transferable stance signals among tweets, claims, and targets. Recently, researchers have investigated reasoning-based enhancements to improve zero-shot generalization. LogiMDF [25] integrates first-order logic constraints into multi-decision fusion, ensuring consistency across LLM outputs while leveraging hypergraph propagation. These methods highlight a shift from purely data-driven transfer to structured reasoning, which improves both robustness and explainability. Nevertheless, pragmatic challenges such as sarcasm and rhetorical framing remain largely unresolved in ZSSD.

2.3. LLM-Based Stance Detection Methods

The remarkable zero-shot and few-shot capabilities of large language models have reshaped stance detection research. One line of work directly treats LLMs as stance predictors: with carefully designed prompts, LLMs can classify stance without additional training [14,26]. Strategies include direct zero-shot prompting, chain-of-thought reasoning, and prompt designs enriched with background knowledge. While flexible, such direct use often leads to unstable outputs and inconsistent predictions. Another line of research employs LLMs as knowledge providers that augment smaller, trainable models. For example, domain-relevant background information can be elicited from LLMs and injected into stance classifiers [27], combining the interpretive capacity of LLMs with the adaptability of supervised models. More recent efforts emphasize structured reasoning with interpretable intermediate steps. Enhanced CoT methods [28–30] decompose stance prediction into step-wise inferences (e.g., from factual entailment to subjective alignment), but rarely define rule-based constraints or auditable outputs. Some systems, such as FOLAR, integrate symbolic representations like sentiment trajectories or discourse role tags, while LogiMDF incorporates logical rules over hypergraphs for traceable reasoning. However, the internal reasoning paths often remain implicit or non-verifiable.

2.4. Summary and Research Gap

Across these paradigms, most prior studies emphasize semantic transfer or large-scale contextual modeling but seldom incorporate explicit pragmatic reasoning or stability verification. For instance, social-media-oriented models such as BERTweet, CT-BERT, and RoBERTa-Twitter capture linguistic variations in tweets but remain limited in handling pragmatic phenomena like sarcasm or negation. Reasoning-based approaches (e.g., LogiMDF, CIRF) introduce logical or cognitive structures, yet their reasoning traces are often implicit and lack auditable intermediate outputs. Overall, existing frameworks can be contrasted along three dimensions—(1) whether pragmatic cues are explicitly modeled, (2) whether prediction stability is verified, and (3) whether interpretable intermediate reasoning is provided. PAMR addresses these gaps by integrating all three: it explicitly encodes pragmatic markers, employs counterfactual probing for stability, and yields interpretable intermediate artifacts for transparent stance reasoning.

3. Methods

3.1. Task Definition

Given a short text x and a target t, the goal of zero-shot stance detection is to predict

$$y \in \{\texttt{favor}, \texttt{against}, \texttt{neutral}\}$$

without using any annotated stance data for t, i.e., $\mathcal{T}_{\text{train}} \cap \mathcal{T}_{\text{test}} = \emptyset$. Our approach decomposes this prediction into multi-agent reasoning steps with interpretable intermediate outputs and evaluates prediction stability via agent-level sampling.

3.2. PAMR Overview

As shown in Figure 1, PAMR (*Pragmatic-Aware Multi-Agent Reasoning*) decomposes zero-shot stance detection into four interpretable stages, each implemented as an explicit agent that transforms inputs into structured intermediate representations. These agents interact in a modular pipeline to progressively refine stance decisions.

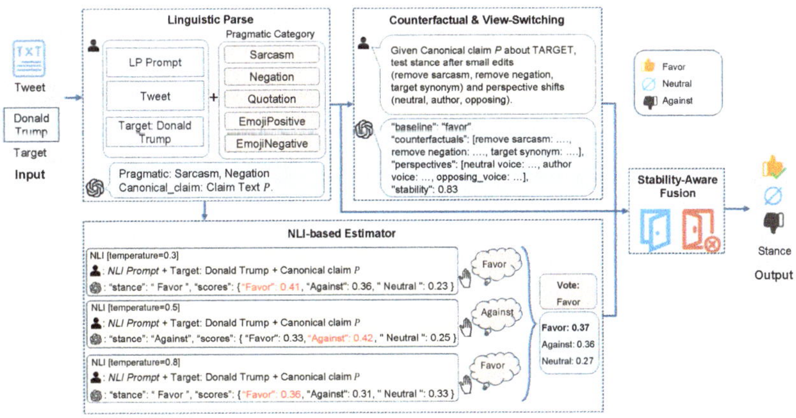

Figure 1. PAMR Framework overall.

(1) **Linguistic Parse Agent**: takes the raw social media post x and target t as input, and outputs a *target-linked canonical claim* along with pragmatic markers such as sarcasm, hedging, or negation. This ensures that subsequent reasoning is grounded in normalized propositions instead of noisy surface forms.

(2) **NLI Estimation Agent**: Reformulates stance classification as a natural language inference task. Given the canonical claim and target, it runs multiple inference passes using diverse prompts to produce both a probability distribution over stance labels and a consensus vote count, mitigating randomness.

(3) **Counterfactual and View-Switching Agent**: Evaluates the stability of the predicted stance by applying minimal, meaning-preserving perturbations (e.g., changing tone, perspective). It outputs a scalar *stability score* reflecting how robust the original decision is under linguistic variation.

(4) **Stability-Aware Fusion Agent**: Receives all signals—canonical claim, pragmatic tags, NLI probabilities and consensus, stability score—and integrates them under a conservative policy. It abstains to "neutral" if confidence is low or predictions are unstable, ensuring robustness.

Each agent has a well-defined role, input–output interface, and interpretable output, and the full pipeline corresponds directly to the annotated modules in Figure 1. In particular, the Linguistic Parse Agent is responsible for the canonicalization and pragmatic enrichment mentioned in the figure.

3.3. Linguistic Parse

The Linguistic Parse Agent aims to normalize noisy social media text into a target-linked canonical claim, while detecting pragmatic markers (e.g., sarcasm, negation, quotation, emoji) that may invert or obscure stance polarity. Given a tweet x and target t, the LLM generates a structured output containing pragmatic tags and a canonical claim c.

Canonical Claim Extraction: The model distills the author's underlying proposition regarding the target into a clear canonical claim, ensuring that the subsequent stance reasoning operates on an explicit statement rather than noisy surface text.

Pragmatic Tagging: The parser identifies pragmatic markers such as sarcasm, negation, quotation, and emoji usage, which are later incorporated in fusion to calibrate stance decisions.

Input Data: Tweet x, Target t
Prompt: Identify pragmatic markers (sarcasm, negation, quotation, emoji) and rewrite the tweet into a canonical claim about the target.
Expected Output: Pragmatic tags and a canonical claim c.

3.4. NLI-Based Estimator

The NLI-based Estimator stage reframes stance detection as a natural language inference (NLI) problem. The canonical claim c and target t are paired as premise–hypothesis inputs to the LLM. To mitigate stochasticity and improve reliability, the model performs n independent inference runs under different decoding parameters.

Multi-run Voting: Each run outputs a stance label $\{favor, against, neutral\}$ and a probability distribution $\mathbf{p}$. A majority vote and averaged probability vector $\bar{\mathbf{p}}$ are then aggregated for stability.

Input Data: Canonical claim c, Target t
Prompt: Given c and t, determine whether the author's stance is favor, against, or neutral. Output both the label and probability scores.
Expected Output: $\{stance, scores\}$ per run, aggregated into majority label $y^\star$, vote ratio r, and averaged scores $\bar{\mathbf{p}}$.

3.5. Counterfactual and View-Switching

The Counterfactual and View-Switching (CVS) stage evaluates robustness of stance predictions by rephrasing the canonical claim under minimal edits and perspective shifts. This stage probes whether stance remains consistent across semantically faithful variations.

Counterfactual Edits: The claim is modified by (i) removing sarcasm, (ii) removing negation, and (iii) replacing the target with a synonym. Perspective Shifts: The claim is paraphrased into three perspectives: neutral voice, author voice, and opposing voice. Stability Score: The proportion of paraphrases consistent with the baseline stance is reported as the stability score S. By definition, the stability score $S \in [0,1]$ quantifies the proportion of paraphrases whose predicted stance remains consistent with the baseline. Higher values of S therefore indicate greater prediction stability under controlled linguistic rewrites, and thus higher confidence in the stance.

Input Data: Canonical claim c, Target t
Prompt: Given c, re-evaluate stance after small edits (remove sarcasm, remove negation, target synonym) and perspective shifts (neutral, author, opposing).
Expected Output: JSON with baseline stance, stance for each edit/perspective, and stability score S.

3.6. Stability-Aware Fusion

Thresholds such as $\tau_{\text{unstable}} = 0.40$ control model behavior; for example, if $S < \tau_{\text{unstable}}$, the model abstains by predicting "neutral" to avoid over-confident errors. Finally, we aggregate signals with a gate-by-gate cascade that mirrors Figure 1. Let $\bar{\mathbf{p}}$ be the averaged scores, $y^\star/r$ the consensus, S the stability, and Π pragmatic tags. Algorithm 1 lists the fusion procedure.

Filtering Policy

(1) Validity: If c is empty or $\bar{\mathbf{p}}$ missing, return `neutral`. (2) Stability: If $S < \tau_{\text{unstable}}$, return `neutral`. (3) Pragmatics-aware confidence: Let $y = \arg\max \bar{\mathbf{p}}$ and $p_{\max} = \max \bar{\mathbf{p}}$; if `Sarcasm` or `Negation` $\in \Pi$, subtract a penalty λ_{prag} from $p_{\max}$. (4) Consensus override under low confidence: If adjusted $p_{\max} < \tau_{\text{prob}}$ and $r \geq \tau_{\text{cons}}$, output $y^\star$; else `neutral`. (5) Prefer consensus: If $r \geq \tau_{\text{cons}}$, output $y^\star$. (6) Small-margin flip: When $y =$ `favor` and lead $= p_{\text{favor}} - \max(p_{\text{against}}, p_{\text{neutral}}) \leq \tau_{\text{flip}}$, flip to `against` if `Sarcasm` or `Negation` $\in \Pi$. (7) Tie-to-neutral: If the top-2 gap $< \epsilon$, output `neutral`. (8) Fallback: Otherwise output y.

Algorithm 1 Stability-Aware Fusion.

Require: Π, $\bar{\mathbf{p}}$, $y^\star$, r, S; thresholds $\tau_{\text{unstable}}, \tau_{\text{prob}}, \tau_{\text{cons}}, \tau_{\text{flip}}, \epsilon$; penalty λ_{prag}
if c empty or $\bar{\mathbf{p}}$ missing **then return** `neutral`
if $S < \tau_{\text{unstable}}$ **then return** `neutral`
$y \leftarrow \arg\max_i p_i,\ \ p_{\max} \leftarrow \max_i p_i$ ▷ $\bar{\mathbf{p}} = (p_{\text{favor}}, p_{\text{against}}, p_{\text{neutral}})$
if `Sarcasm`$\in \Pi$ **or** `Negation`$\in \Pi$ **then**
    $\tilde{p}_{\max} \leftarrow \max(0, p_{\max} - \lambda_{\text{prag}})$
else
    $\tilde{p}_{\max} \leftarrow p_{\max}$
end if
if $\tilde{p}_{\max} < \tau_{\text{prob}}$ **then**
    if $r \geq \tau_{\text{cons}}$ **then**
        return $y^\star$
    else
        return `neutral`
    end if
end if
if $r \geq \tau_{\text{cons}}$ **then**
    return $y^\star$
end if
if $y =$ `favor` **then**
    lead $\leftarrow p_{\text{favor}} - \max(p_{\text{against}}, p_{\text{neutral}})$
    if lead $\leq \tau_{\text{flip}}$ **and** (`Sarcasm`$\in \Pi$ **or** `Negation`$\in \Pi$) **then**
        return `against`
    end if
end if
Let $p_{(1)} \geq p_{(2)}$ be the top-2 of $\bar{\mathbf{p}}$
if $|p_{(1)} - p_{(2)}| < \epsilon$ **then**
    return `neutral`
end if
return y

4. Experiments

4.1. Experimental Data

To evaluate the effectiveness of our approach, we run thorough experiments on two datasets: SemEval-2016 Task 6 (SEM16) [2] and COVID-19-Stance (COVID19) [31].

- SemEval-2016 (SEM16) [2]: The SEM16 dataset contains 4870 tweets, each targeting various subjects and annotated with one of three stance labels: "favor", "against", or "neutral". SEM16 provides six targets (Donald Trump (DT), Hillary Clinton (HC), Feminist Movement (FM), Legalization of Abortion (LA), Atheism (AT), and Climate Change (CC)). Following [25] for zero-shot evaluation, we exclude Atheism (AT) and

Climate Change (CC) and use only the official test split for fair comparison with prior zero-shot settings. Per-target class counts for the four retained targets are reported in Table 1.

- COVID-19-Stance (COVID-19) [31]: We also use the COVID-19 stance dataset, which assesses public attitudes toward pandemic-related policies and figures across four targets: *Wearing a Face Mask* (WA), *Keeping Schools Closed* (SC), *Anthony S. Fauci, M.D.* (AF), and *Stay at Home Orders* (SH). Each tweet is labeled with *Favor/Against/Neutral*. As with SEM16, we report results using the test split only; class distributions are shown in Table 1.

Table 1. Statistics of the datasets used in our experiments.

Dataset	Target	Favor	Against	Neutral	Total
SEM16	DT	148	299	260	707
	HC	163	565	251	979
	FM	268	511	170	949
	LA	167	544	222	933
COVID-19	AF	492	610	762	1864
	SH	615	250	325	1190
	WA	693	190	668	1551
	SC	400	782	346	1528

Both datasets were chosen as they are widely adopted benchmarks for stance detection and provide diverse, topic-specific challenges. SEM16 offers classic political and social targets that test generalization across ideological domains, while COVID-19-Stance introduces a contemporary and highly pragmatic context involving public health discourse. We acknowledge that both datasets are relatively small in size (a few thousand tweets each), which may limit the statistical power of comparisons. To mitigate this, all evaluations are conducted in a strict zero-shot setting using the official test splits only, ensuring comparability with prior studies. Furthermore, our stability-aware fusion mechanism and counterfactual probing explicitly reduce the sensitivity of results to sampling variance, partially alleviating dataset-size constraints.

4.2. Evaluation Metrics

We adopt the Macro-F1 score F_{avg} computed as the average of the F1 scores of the *Favor* and *Against* categories (the *Neutral* class is excluded from the average), following standard practice in stance detection [2,7,26]. Macro-F1 equally weights each class and is less affected by label imbalance, which is critical for social media datasets where "neutral" or "against" instances often dominate. In addition, we report per-target Macro-F1 averages to ensure comparability with prior zero-shot studies.Let P_y and R_y denote precision and recall for class $y \in \{\text{favor}, \text{against}\}$ The per-class F1 and the macro average are defined as

$$F_{\text{favor}} = \frac{2 \cdot P_{\text{favor}} \cdot R_{\text{favor}}}{P_{\text{favor}} + R_{\text{favor}}}, \tag{1}$$

$$F_{\text{against}} = \frac{2 \cdot P_{\text{against}} \cdot R_{\text{against}}}{P_{\text{against}} + R_{\text{against}}}, \tag{2}$$

$$F_{\text{avg}} = \frac{F_{\text{favor}} + F_{\text{against}}}{2}. \tag{3}$$

4.3. Baseline Methods

To ensure a comprehensive evaluation, we compare PAMR with a broad range of existing stance detection approaches, which can be grouped into traditional DNNs, fine-tuning-based methods, and LLM-based frameworks.

(1) TraditionalDNNs.

- **BiLSTM and Bicond [32]**: These two approaches utilize separate BiLSTM encoders, where one captures sentence-level semantics and the other encodes the given target, thereby enabling the model to jointly represent stance-related information.
- **CrossNet [33]**: This model leverages BiLSTM architectures to encode both the input text and its corresponding target, while introducing a target-specific attention mechanism before classification, which enhances the model's ability to generalize across unseen targets.
- **TPDG [34]**: This method automatically identifies stance-bearing words and distinguishes target-dependent from target-independent terms, adjusting them adaptively to better capture the relationship between text and target.
- **TOAD [35]**: To improve generalization in zero-shot scenarios, TOAD adopts an adversarial learning strategy that allows the model to resist overfitting to specific targets while transferring stance knowledge.

(2) Fine-tuning methods.

- **TGA-Net [36]**: This approach establishes associations between training and evaluation topics in an unsupervised way, using BERT as the encoder and fully connected layers for classification, thereby linking topic domains without annotated supervision.
- **Bert-Joint [37]**: It combines bidirectional encoder representations from transformers that have been pre-trained on large-scale unlabeled corpora, producing dense contextual embeddings for both tokens and full sentences.
- **Bert-GCN [38]**: This method enhances stance detection with common-sense knowledge by integrating both structural and semantic graph relations, which makes it more effective in generalizing to zero and few-shot target scenarios.
- **JointCL [24]**: It unifies stance-oriented contrastive learning with target-aware prototypical graph contrastive learning, allowing the model to transfer stance-relevant features learned from seen topics to unseen targets.
- **TarBK [39]**: By leveraging Wikipedia-derived background knowledge, TarBK reduces the semantic gap between training and evaluation targets, thereby improving the reasoning capability of stance classifiers.
- **PT-HCL [7]**: This contrastive learning approach utilizes both semantic and sentiment features to improve cross-domain stance transferability, enabling robust generalization beyond source data. Its combination of semantics and sentiment proves critical for disambiguating stance from emotion.
- **KEPrompt [40]**: This method proposes an automatic verbalizer to generate label words dynamically, while simultaneously injecting external background knowledge to guide stance recognition. It reduces reliance on manually designed verbalizers and improves flexibility.

(3) LLM-based methods.

- **COLA [14]**: This approach employs a three-stage framework where different LLM roles are orchestrated for multidimensional text understanding and reasoning, resulting in state-of-the-art zero-shot stance performance. It demonstrates the effectiveness of task decomposition within LLM pipelines.

- **Ts-CoT [41]**: It introduces a chain-of-thought prompting mechanism for stance detection with LLMs, upgrading the base model to GPT-3.5 in order to take advantage of improved reasoning capacity. The CoT design encourages step-by-step reasoning rather than direct label prediction.
- **EDDA [27]**: This method exploits LLMs to automatically generate rationales and substitute stance-bearing expressions, thereby increasing semantic relevance and expression diversity for stance detection. By focusing on rationales, EDDA improves both performance and interpretability.
- **FOLAR [42]**: A reasoning framework that augments stance detection with factual knowledge and chain-of-thought logical reasoning, aiming to improve interpretability and robustness in zero-shot settings.
- **LogiMDF [25]**: A logic-augmented multi-decision fusion framework that extracts first-order logic rules from multiple LLMs, constructs a logical fusion schema, and employs a multi-view hypergraph neural network to integrate diverse reasoning processes for consistent and accurate stance detection.

4.4. Implementation Details

We implement PAMR using `GPT-3.5-turbo` accessed via the OpenAI API. For Linguistic Parse and CVS, the maximum generation length is set to 256 and 512 tokens, respectively, with temperature 0.3. For NLI-based estimation, we run the model three times with temperatures $\{0.3, 0.5, 0.8\}$ and aggregate results by majority vote and averaged probabilities. Fusion thresholds are fixed across datasets: $\tau_{\text{unstable}} = 0.40$, $\tau_{\text{prob}} = 0.45$, $\tau_{\text{cons}} = 0.60$, $\tau_{\text{flip}} = 0.15$, $\epsilon = 0.02$, and pragmatic penalty $\lambda_{\text{prag}} = 0.05$. All evaluations are conducted in a strict zero-shot setting using the official test splits, and performance is reported in terms of macro-F1 over `favor` and `against`.

5. Overall Performance

5.1. Analysis of Main Results

Table 2 reports the results on the SEM16 dataset. Early neural baselines such as BiLSTM, Bicond, and CrossNet achieve relatively low performance, with average scores around 34–38. With the introduction of pre-trained models, methods like JointCL, PT-HCL, and NPS4SD raise the average performance to around 52–55. More recent LLM-based approaches, such as COLA, FOLAR, and LogiMDF, achieve competitive results on specific targets (e.g., COLA and FOLAR both exceed 81 on HC). However, their performance tends to fluctuate across different targets. In contrast, PAMR achieves strong and consistent results across all targets, with best scores on FM (75.9) and LA (71.8), and a balanced overall average of 71.9, surpassing all baselines. This demonstrates the effectiveness of pragmatic-aware reasoning and stability fusion in achieving robust zero-shot stance detection. Here, Avg denotes the arithmetic mean over all targets.

Table 3 shows the results on the COVID-19 dataset. Traditional baselines (CrossNet, BERT, TPDG) perform poorly, with average scores between 40 and 45. Stronger models like TOAD and JointCL improve performance, while LogiMDF achieves the best results on AF (70.4) and WA (75.4), and FOLAR performs strongly on WA (73.1). Nevertheless, PAMR consistently achieves competitive or best performance on nearly all targets, obtaining the top scores on SC (72.0), SH (72.2), and WA (78.8). Its overall average reaches 73.0, outperforming all comparison methods. This indicates that PAMR is better able to handle pragmatic confounds such as sarcasm and negation in COVID-19 tweets, yielding more stable and reliable stance predictions. Again, Avg is the mean over all targets in the dataset.

Table 2. Results on the SEM16 dataset. F_{avg} is reported for each target. The Avg is the arithmetic mean over all listed targets. The best results are in bold. ‡ indicates the first-best result.

Method	HC	FM	LA	DT	Avg
BiLSTM	31.6	40.3	33.6	30.8	34.1
Bicond	32.7	40.6	34.4	30.5	34.6
CrossNet	38.3	41.7	38.5	35.6	38.5
TPDG	50.9	53.6	46.5	47.3	49.6
TOAD	51.2	54.1	46.2	49.5	50.3
TGA-Net	49.3	46.6	45.2	40.7	45.5
Bert-Joint	50.1	42.1	44.8	41.0	44.5
Bert-GCN	50.0	44.3	44.2	42.3	45.2
JointCL	54.4	54.0	50.0	50.5	52.2
TarBK	55.1	53.8	48.7	50.8	52.1
PT-HCL	54.5	54.6	50.9	50.1	52.5
KEPROMPT	57.0	53.6	53.0	41.8	51.3
NPS4SD	60.1	56.7	51.0	51.4	54.8
COLA	**81.7**	63.4	71.0	68.5	71.2
Ts-CoT$_{GPT}$	78.9	68.3	62.3	68.6	69.5
EDDA	77.4	69.7	62.7	**69.8**	69.9
FOLAR	81.9	71.2	69.9	–	–
LogiMDF	75.1	67.9	68.0	67.6	69.7
PAMR (Ours)	73.7	**75.9** ‡	**71.8** ‡	66.1	**71.9**

Table 3. Results on the COVID-19 dataset. We report per-target Macro-F1 and the overall Avg, which is the mean across all targets. The best numbers are in bold. ‡ indicates the first-best result.

Method	AF	SC	SH	WA	Avg
CrossNet	41.3	40.0	40.4	38.2	40.0
BERT	47.3	45.0	39.9	44.3	44.1
TPDG	46.0	51.5	37.2	48.0	45.7
TOAD	53.0	68.3	62.9	41.1	56.3
JointCL	57.6	49.3	43.5	63.1	53.4
Ts-CoT	69.2	43.5	66.5	57.8	59.3
COLA	65.7	46.6	53.5	73.9	59.9
FOLAR	69.5	67.2	65.4	73.1	68.8
LogiMDF	**70.4**	68.8	64.9	75.4	69.9
PAMR	68.6	**72.0** ‡	**72.2** ‡	**78.8** ‡	**73.0**

To further validate these improvements, we conduct paired t-tests between PAMR and the strongest baseline across all targets on both datasets. The results show that PAMR's performance gains are statistically significant ($p < 0.05$), confirming that the observed advantages are unlikely due to random variation.

5.2. Ablation Study

To better understand the contributions of different components in PAMR, we conduct ablation experiments by removing the Linguistic Parser (w/o LP) and the Counterfactual and View-Switching module (w/o CVS). The results are shown in Figure 2.

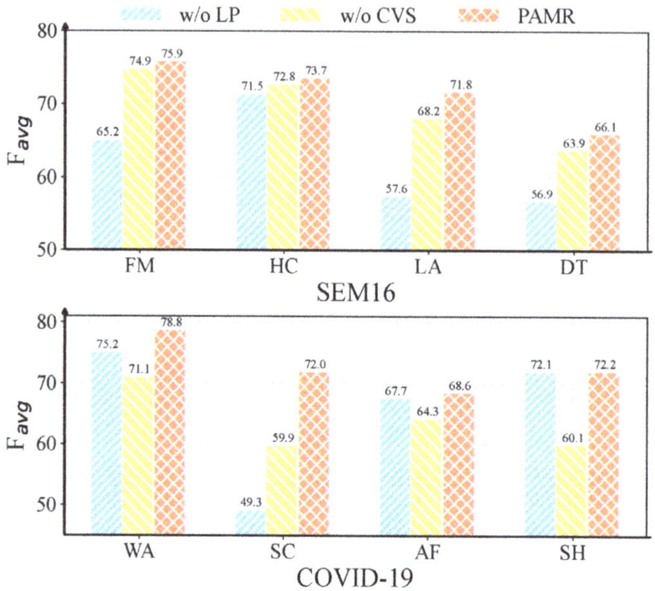

Figure 2. Ablation study of PAMR on SEM16 and COVID-19.

We observe that removing either component consistently degrades performance across both COVID-19 and SemEval-2016 datasets. Specifically, eliminating the Linguistic Parser leads to a substantial drop, e.g., from 72.0 to 49.3 on SC and from 66.1 to 56.9 on DT, confirming the importance of extracting canonical claims and pragmatic cues for stance reasoning. Similarly, removing the Counterfactual and View-Switching module also results in performance declines, particularly on WA (from 78.8 to 71.1) and FM (from 75.9 to 74.9), highlighting the necessity of stability probing to mitigate pragmatic confounds and prevent over-confident errors.

Overall, the full PAMR framework consistently outperforms its ablated variants, demonstrating that both pragmatic parsing and counterfactual stability probing are complementary and essential for robust zero-shot stance detection.

5.3. Case Study

To further illustrate the behavior of PAMR, we analyze representative examples from the evaluation set, as shown in Table 4.

Case 1. In this example, the canonical claim and the original tweet exhibit strong semantic alignment. The NLI-based estimation assigns a dominant probability to `favor` ($p = 0.65$), clearly surpassing both `against` and `neutral`. The CVS stability score reaches $S = 0.83$, well above the instability threshold, indicating robust consistency under counterfactual and perspective shifts. Moreover, majority voting yields complete unanimity (3/3 runs), reinforcing the reliability of the prediction. Since no pragmatic markers (e.g., sarcasm or negation) are present, the system directly outputs `favor`, which matches the gold annotation. This case shows that, under standard conditions, PAMR can capture entailment relations both stably and accurately.

Table 4. Case studies on **SemEval-2016** Trump target showing intermediate outputs.

Tweet	Claim	Pragmatic Tags	NLI Scores	Stability	Prediction
People are saying that kids will not have a safe place to go to if the schools are closed. Large gatherings are not safe with coronavirus. The coronavirus is not safe.	Keeping schools closed may leave kids without a safe place to go, but large gatherings are unsafe due to coronavirus.	[Quotation]	{favor: 0.65, against: 0.15, none: 0.20}	0.833	Favor
"2 years ago Hillary never answered whether she used private email. Liberal media passed on reporting." #equality	Questioning Hillary's email while excusing others reflects the sexist double standards women in politics face.	[Sarcasm, Quotation]	{favor: 0.50, against: 0.46, none: 0.04}	0.666	Against

Case 2. Here, the canonical claim and initial NLI probabilities suggest an approval stance (`favor`) toward the target, with the aggregated vote leaning in the same direction. However, the CVS stability score is only $S = 0.67$, indicating borderline robustness. Importantly, pragmatic markers such as `Sarcasm` are detected. Within the fusion stage, these cues trigger a polarity flip, shifting the final decision from `favor` to `against`, which aligns with the gold annotation. This case demonstrates PAMR's ability to exploit pragmatic signals to correct initial misclassifications in sarcastic contexts, thereby disentangling stance polarity from sentiment polarity.

6. Discussion

Our results highlight key insights into zero-shot stance detection. PAMR consistently outperforms baselines on both SEM16 and COVID-19 datasets, demonstrating that explicit modeling of pragmatics and stability improves generalization. Unlike prior models that conflate sentiment with stance, PAMR leverages pragmatic cues like sarcasm and negation to reduce polarity errors. Its modular design—with components for claim extraction, counterfactual probing, and fusion—enables interpretable outputs and fine-grained analysis. While PAMR currently relies on GPT-3.5, all reasoning steps are prompt-based and generalizable, making the framework compatible with future open-source or symbolic alternatives. Fusion strategies further enhance robustness under figurative or domain-specific inputs. Future work may explore lightweight replacements to improve accessibility and stability.

7. Conclusions

In this paper, we introduced PAMR, a pragmatic-aware multi-agent reasoning framework for zero-shot stance detection on social media. The framework enables explicit and interpretable reasoning by integrating linguistic, inferential, and counterfactual analyses without requiring target-specific training. Experiments on the SemEval-2016 and COVID-19-Stance datasets show that PAMR achieves stable improvements over strong zero-shot baselines, confirming the value of incorporating pragmatic cues and stability probing. While our results demonstrate promising robustness and interpretability, the model has been evaluated only on English social media datasets; its applicability to other domains or high-stakes contexts should be considered exploratory. Future work will extend

PAMR to multilingual and multi-modal settings, investigate domain adaptation under distribution shifts, and explore hybrid integration with symbolic reasoning to further enhance stability and transparency in zero-shot stance detection. Beyond these directions, PAMR offers a flexible foundation for future research on pragmatic reasoning and stance analysis. Its modular components can be adapted to evaluate or enhance other stance models, while its interpretable outputs provide useful tools for analyzing how pragmatic cues shape stance, supporting broader advancements in social media understandings.

Author Contributions: Conceptualization, Z.X. and B.Z.; Methodology, Z.X. and F.N.; Investigation, G.D.; Writing—original draft, Z.X. and F.N.; Writing—review and editing, B.Z. All authors have read and agreed to the published version of the manuscript.

Funding: This research is supported by the Natural Science Foundation of Top Talent of SZTU (grant no. GDRC202320) and the Research Promotion Project of Key Construction Discipline in Guangdong Province (2022ZDJS112).

Data Availability Statement: No new data were created in this study. The datasets analyzed are publicly available benchmark datasets (SemEval-2016 and COVID-19), which can be obtained from their respective sources. Processed data and code are available from the corresponding author upon reasonable request.

Conflicts of Interest: The authors declare no conflicts of interest.

References

1. Küçük, D.; Can, F. Stance detection: A survey. *ACM Comput. Surv. (CSUR)* **2020**, *53*, 12. [CrossRef]
2. Mohammad, S.; Kiritchenko, S.; Sobhani, P.; Zhu, X.; Cherry, C. Semeval-2016 task 6: Detecting stance in tweets. In Proceedings of the 10th International Workshop On Semantic Evaluation (SemEval-2016), San Diego, CA, USA, 16–17 June 2016; pp. 31–41.
3. Addawood, A.; Schneider, J.; Bashir, M. Stance classification of twitter debates: The encryption debate as a use case. In Proceedings of the 8th International Conference on Social Media & Society, Toronto, ON, Canada, 28–30 July 2017; pp. 1–10.
4. Sun, Q.; Wang, Z.; Zhu, Q.; Zhou, G. Stance detection with hierarchical attention network. In Proceedings of the 27th International Conference on Computational Linguistics, Santa Fe, NM, USA, 21–24 August 2018; pp. 2399–2409.
5. Li, Y.; Caragea, C. Target-Aware Data Augmentation for Stance Detection. In Proceedings of the 2021 Conference of the North American Chapter of the Association for Computational Linguistics: Human Language Technologies, Online, 6–11 June 2021; pp. 1850–1860.
6. Xu, C.; Paris, C.; Nepal, S.; Sparks, R. Cross-target stance classification with self-attention networks. *arXiv* **2018**, arXiv:1805.06593.
7. Liang, B.; Chen, Z.; Gui, L.; He, Y.; Yang, M.; Xu, R. Zero-Shot Stance Detection via Contrastive Learning. In Proceedings of the ACM Web Conference 2022, Virtual, 25–29 April 2022; pp. 2738–2747.
8. Hong, G.N.S.Y.; Gauch, S. Sarcasm Detection as a Catalyst: Improving Stance Detection with Cross-Target Capabilities. *arXiv* **2025**, arXiv:2503.03787. [CrossRef]
9. Maynard, D.; Greenwood, M. Who cares about Sarcastic Tweets? Investigating the Impact of Sarcasm on Sentiment Analysis. In *Proceedings of the Ninth International Conference on Language Resources and Evaluation (LREC'14)*; Calzolari, N., Choukri, K., Declerck, T., Loftsson, H., Maegaard, B., Mariani, J., Moreno, A., Odijk, J., Piperidis, S., Eds.; European Language Resources Association (ELRA): Reykjavik, Iceland, 2014; pp. 4238–4243.
10. Du, J.; Xu, R.; He, Y.; Gui, L. Stance classification with target-specific neural attention networks. In Proceedings of the International Joint Conferences on Artificial Intelligence, Melbourne, Australia, 19–25 August 2017.
11. Wei, P.; Lin, J.; Mao, W. Multi-target stance detection via a dynamic memory-augmented network. In Proceedings of the The 41st International ACM SIGIR Conference on Research & Development in Information Retrieval, Ann Arbor, MI, USA, 8–12 July 2018; pp. 1229–1232.
12. Li, Y.; Sosea, T.; Sawant, A.; Nair, A.J.; Inkpen, D.; Caragea, C. P-stance: A large dataset for stance detection in political domain. In Proceedings of the Findings of the Association for Computational Linguistics: ACL-IJCNLP 2021, Online Event, 1–6 August 2021; pp. 2355–2365.
13. Conforti, C.; Berndt, J.; Pilehvar, M.T.; Giannitsarou, C.; Toxvaerd, F.; Collier, N. Synthetic Examples Improve Cross-Target Generalization: A Study on Stance Detection on a Twitter corpus. In Proceedings of the Eleventh Workshop on Computational Approaches to Subjectivity, Sentiment and Social Media Analysis, WASSA@EACL 2021, Online, 19 April 2021; pp. 181–187.

14. Lan, X.; Gao, C.; Jin, D.; Li, Y. Stance detection with collaborative role-infused llm-based agents. In Proceedings of the International AAAI Conference on Web and Social Media, Buffalo, NY, USA, 3–6 June 2024; Volume 18, pp. 891–903.
15. Guo, T.; Chen, X.; Wang, Y.; Chang, R.; Pei, S.; Chawla, N.V.; Wiest, O.; Zhang, X. Large Language Model based Multi-Agents: A Survey of Progress and Challenges. *arXiv* **2024**, arXiv:2402.01680. [CrossRef]
16. Tran, K.T.; Dao, D.; Nguyen, M.D.; Pham, Q.V.; O'Sullivan, B.; Nguyen, H.D. Multi-Agent Collaboration Mechanisms: A Survey of LLMs. *arXiv* **2025**, arXiv:2501.06322. [CrossRef]
17. Zarrella, G.; Marsh, A. MITRE at SemEval-2016 Task 6: Transfer Learning for Stance Detection. In Proceedings of the 10th International Workshop on Semantic Evaluation (SemEval-2016), San Diego, CA, USA, 16–17 June 2016; pp. 458–463.
18. Sobhani, P.; Inkpen, D.; Zhu, X. A dataset for multi-target stance detection. In Proceedings of the 15th Conference of the European Chapter of the Association for Computational Linguistics: Volume 2, Short Papers, Valencia, Spain, 3–7 April 2017; pp. 551–557.
19. He, Z.; Mokhberian, N.; Lerman, K. Infusing Wikipedia Knowledge to Enhance Stance Detection. *arXiv* **2022**, arXiv:2204.03839.
20. Wang, S.; Pan, L. Target-Adaptive Consistency Enhanced Prompt-Tuning for Multi-Domain Stance Detection. In Proceedings of the 2024 Joint International Conference on Computational Linguistics, Language Resources and Evaluation (LREC-COLING 2024), Torino, Italia, 20–25 May 2024; pp. 15585–15594.
21. Nguyen, D.Q.; Vu, T.; Nguyen, A.T. BERTweet: A Pre-trained Language Model for English Tweets. In Proceedings of the 2020 Conference on Empirical Methods in Natural Language Processing: System Demonstrations, Online, 16–20 November 2020; pp. 9–14. [CrossRef]
22. Müller, M.; Salathé, M.; Kummervold, P.E. COVID-Twitter-BERT: A Natural Language Processing Model to Analyse COVID-19 Content on Twitter. *arXiv* **2020**, arXiv:2005.07503. [CrossRef] [PubMed]
23. Liang, B.; Li, A.; Zhao, J.; Gui, L.; Yang, M.; Yu, Y.; Wong, K.F.; Xu, R. Multi-modal Stance Detection: New Datasets and Model. *arXiv* **2024**, arXiv:2402.14298. [CrossRef]
24. Liang, B.; Zhu, Q.; Li, X.; Yang, M.; Gui, L.; He, Y.; Xu, R. Jointcl: A joint contrastive learning framework for zero-shot stance detection. In Proceedings of the 60th Annual Meeting of the Association for Computational Linguistics (Volume 1: Long Papers), Dublin, Ireland, 22–27 May 2022; Volume 1, pp. 81–91.
25. Zhang, B.; Ma, J.; Fu, X.; Dai, G. Logic Augmented Multi-Decision Fusion Framework for Stance Detection on Social Media. *Inf. Fusion* **2025**, *122*, 103214. [CrossRef]
26. Li, A.; Liang, B.; Zhao, J.; Zhang, B.; Yang, M.; Xu, R. Stance Detection on Social Media with Background Knowledge. In Proceedings of the 2023 Conference on Empirical Methods in Natural Language Processing, Singapore, 6–10 December 2023; pp. 15703–15717.
27. Ding, D.; Dong, L.; Huang, Z.; Xu, G.; Huang, X.; Liu, B.; Jing, L.; Zhang, B. EDDA: An Encoder-Decoder Data Augmentation Framework for Zero-Shot Stance Detection. In Proceedings of the 2024 Joint International Conference on Computational Linguistics, Language Resources and Evaluation (LREC-COLING 2024), Torino, Italia, 20–25 May 2024; pp. 5484–5494.
28. Fei, H.; Li, B.; Liu, Q.; Bing, L.; Li, F.; Chua, T.S. Reasoning Implicit Sentiment with Chain-of-Thought Prompting. *arXiv* **2023**, arXiv:2305.11255.
29. Ling, Z.; Fang, Y.; Li, X.; Huang, Z.; Lee, M.; Memisevic, R.; Su, H. Deductive Verification of Chain-of-Thought Reasoning. *arXiv* **2023**, arXiv:2306.03872.
30. Cai, Z.; Chang, B.; Han, W. Human-in-the-Loop through Chain-of-Thought. *arXiv* **2023**, arXiv:2306.07932.
31. Glandt, K.; Khanal, S.; Li, Y.; Caragea, D.; Caragea, C. Stance detection in COVID-19 tweets. In Proceedings of the 59th Annual Meeting of the Association for Computational Linguistics and the 11th International Joint Conference on Natural Language Processing (Long Papers), Online, 1–6 August 2021; Volume 1.
32. Augenstein, I.; Rocktaeschel, T.; Vlachos, A.; Bontcheva, K. Stance Detection with Bidirectional Conditional Encoding. In Proceedings of the Conference on Empirical Methods in Natural Language Processing, Sheffield, Austin, TX, USA, 1–5 November 2016.
33. Du, J.; Xu, R.; He, Y.; Gui, L. Stance Classification with Target-specific Neural Attention. In Proceedings of the Twenty-Sixth International Joint Conference on Artificial Intelligence, IJCAI-17, Melbourne, Australia, 19–25 August 2017; pp. 3988–3994. [CrossRef]
34. Liang, B.; Fu, Y.; Gui, L.; Yang, M.; Du, J.; He, Y.; Xu, R. Target-Adaptive Graph for Cross-Target Stance Detection. In Proceedings of the WWW '21: The Web Conference 2021, Virtual Event/Ljubljana, Slovenia, 19–23 April 2021; pp. 3453–3464.
35. Allaway, E.; Srikanth, M.; Mckeown, K. Adversarial Learning for Zero-Shot Stance Detection on Social Media. In Proceedings of the 2021 Conference of the North American Chapter of the Association for Computational Linguistics: Human Language Technologies, Online, 6–11 June 2021; pp. 4756–4767.
36. Allaway, E.; Mckeown, K. Zero-Shot Stance Detection: A Dataset and Model using Generalized Topic Representations. In Proceedings of the 2020 Conference on Empirical Methods in Natural Language Processing (EMNLP), Online, 16–20 November 2020; pp. 8913–8931.

37. Devlin, J.; Chang, M.W.; Lee, K.; Toutanova, K. BERT: Pre-Training of Deep Bidirectional Transformers for Language Understanding. In Proceedings of the 2019 Conference of the North American Chapter of the Association for Computational Linguistics: Human Language Technologies, Volume 1 (Long and Short Papers), Minneapolis, MN, USA, 2–7 June 2019; pp. 4171–4186.
38. Liu, R.; Lin, Z.; Tan, Y.; Wang, W. Enhancing zero-shot and few-shot stance detection with commonsense knowledge graph. In Proceedings of the Findings of the Association for Computational Linguistics: ACL-IJCNLP 2021, Online Event, 1–6 August 2021, pp. 3152–3157.
39. Zhu, Q.; Liang, B.; Sun, J.; Du, J.; Zhou, L.; Xu, R. Enhancing Zero-Shot Stance Detection via Targeted Background Knowledge. In Proceedings of the 45th International ACM SIGIR Conference on Research and Development in Information Retrieval, Madrid, Spain, 11–15 July 2022; pp. 2070–2075.
40. Huang, H.; Zhang, B.; Li, Y.; Zhang, B.; Sun, Y.; Luo, C.; Peng, C. Knowledge-Enhanced Prompt-Tuning for Stance Detection. *ACM Trans. Asian-Low-Resour. Lang. Inf. Process.* **2023**, *22*, 159. [CrossRef]
41. Zhang, B.; Fu, X.; Ding, D.; Huang, H.; Li, Y.; Jing, L. Investigating Chain-of-thought with ChatGPT for Stance Detection on Social Media. *arXiv* **2023**, arXiv:2304.03087.
42. Dai, G.; Liao, J.; Zhao, S.; Fu, X.; Peng, X.; Huang, H.; Zhang, B. Large Language Model Enhanced Logic Tensor Network for Stance Detection. *Neural Netw.* **2025**, *183*, 106956. [CrossRef] [PubMed]

electronics

MDPI

Article

Enhancing the Sustained Capability of Continual Test-Time Adaptation with Dual Constraints

Yu Song, Pei Liu and Yunpeng Wu *

School of Computer Science and Artificial Intelligence, Zhengzhou University, Zhengzhou 450001, China; ieysong@zzu.edu.cn (Y.S.); peiliu123@gs.zzu.edu.cn (P.L.)
* Correspondence: ieypwu@zzu.edu.cn

Abstract: Continuous Test-Time Adaptation aims to adapt a source model to continuously and dynamically changing target domains. However, previous studies focus on adapting to each target domain independently, treating them as isolated, while ignoring the interplay of interference and promotion between domains, which limits the model's sustained capability, often causing it to become trapped in local optima. This study highlights this critical issue and identifies two key factors that limit the model's sustained capability: (1) The update of parameters lacks constraints, where domain-sensitive parameters capture domain-specific knowledge, leading to unstable channel representations and interference from old domain knowledge and hindering the learning of domain-invariant knowledge. (2) The decision boundary lacks constraints, and distribution shifts, which carry significant domain-specific knowledge, cause features to become dispersed and prone to clustering near the decision boundary. This is particularly problematic during the early stages of domain shifts, where features are more likely to cross the boundary. To tackle the two challenges, we propose a Dual Constraints method: First, we constrain updates to domain-sensitive parameters by minimizing the representation changes in domain-sensitive channels, alleviating the interference among domain-specific knowledge and promoting the learning of domain-invariant knowledge. Second, we introduce a constrained virtual decision boundary, which forces features to move away from the original boundary, and with a virtual margin to prevent features from crossing the decision boundary due to domain-specific knowledge interference caused by domain shifts. Extensive benchmark experiments show our framework outperforms competing methods.

Keywords: continual test-time adaption; test-time adaption; domain adaption; domain generalization; continual learning

1. Introduction

Continual Test-Time Adaptation (CTTA) focuses on enhancing machine learning models' ability to adapt continuously in dynamic environments where input data distributions shift over time. This capability is crucial in various real-world scenarios. For instance, in autonomous driving [1–4], a vehicle may encounter changing conditions such as transitioning from daylight to nighttime or from sunny to rainy weather. To maintain high performance, models must adapt effectively to these evolving data distributions through Continual Test-Time Adaptation.

Numerous methods have been developed to tackle the challenges of Continual Test-Time Adaptation, including the use of teacher–student models [5–7], data augmentation techniques [5], semi-supervised learning [8], Low-Rank Learning [9], sample replay [10], and Masked Autoencoders [11].

Although previous studies have achieved significant success, a key limitation must be pointed out: These studies still follow the approach of Test-Time Adaptation (TTA) [12], treating the problem of Continual Test-Time Adaptation in isolation and focusing solely on adapting to each domain independently. This limits the model's sustained capability, similar to a greedy algorithm [13], causing the model's performance to become trapped in local optima [14] and failing to achieve the ideal global optimum [15]. This is because these methods overlook the essential nature of Continual Test-Time Adaptation: in this setting, domain adaptation is dynamic, ongoing, and continuously evolving, resulting in different domains either interfering with or promoting each other.

Therefore, addressing the following two challenges is crucial for enhancing the sustained capability of Continual Test-Time Adaptation: (1) How to effectively alleviate the mutual interference of domain-specific knowledge. (2) How to effectively learn domain-invariant knowledge across domains.

This study revisits the differences between Continual Test-Time Adaptation and Test-Time Adaptation, focusing on the long-neglected issue of sustained capability, and identifies two key factors that severely limit the model's sustained capability and prevent it from getting trapped in local optima:

1. Lack of constraints in parameter updates: The model contains a large number of domain-sensitive parameters, which tend to learn domain-specific knowledge. When the model encounters a new domain, the distribution difference between the old and new domains causes these domain-sensitive parameters to behave abnormally, leading to unstable channel representations that incorporate substantial domain-specific knowledge from previous domains, especially in the early stages of adapting to the new domain. This results in severe interference between domain-specific knowledge. More importantly, relying on domain-specific knowledge for classification significantly hinders the learning of domain-invariant knowledge.
2. Lack of constraints in decision boundary: Domain shifts carry a significant amount of domain-specific knowledge, causing the features generated by the model to spread out more. The lack of constraints makes these features prone to clustering near or crossing the decision boundary. Under the interference of domain-specific knowledge, features near the decision boundary are more likely to cross, particularly in the early stages of the new domain.

As shown in Figure 1, we analyze the channel representations under three different corruption levels (level-5, level-3, and level-1) of CIFAR10C (The method for calculating unstable channel representations is provided in Section 4.4.3). In Figure 1a, we observe that different levels of corruption have varying impacts on the stability of channel representations. The higher the level of corruption, the more unstable the channel representations become. The domain-sensitive parameters are highly sensitive to this domain-specific knowledge, resulting in anomalies and instability in the channel representations.

In Figure 1b, we observe that domain-sensitive channels are much more responsive to domain shifts than domain-robust channels. When encountering different domains, the representation of these channels tends to exhibit instability and are prone to significant variations. The cause of this abnormal behavior lies in the domain-sensitive parameters, which tend to overfit domain-specific knowledge such as background, lighting, and texture. When a domain shift occurs, this domain-specific knowledge undergoes drastic changes, leading to significant variations in the channel representations. The instability of channel representations and the presence of a large amount of domain-specific knowledge can cause significant interference in domain adaptation, especially in the early stages of domain shift. In contrast, domain-robust parameters focus on learning domain-invariant knowledge, such as contours and shapes. This domain-invariant knowledge remains unaffected by domain shifts, resulting in more stable channel representations.

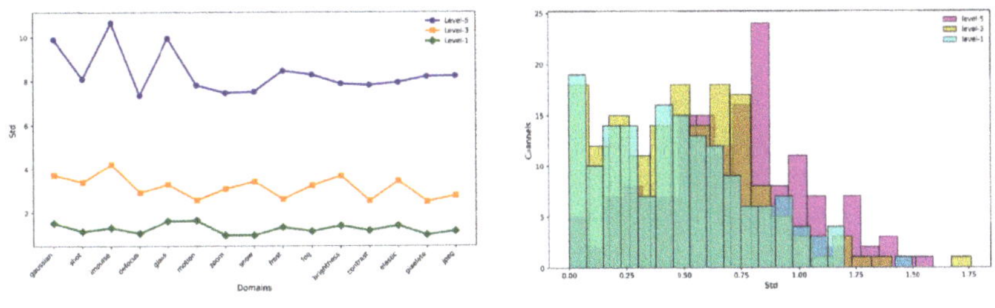

(**a**) (**b**)

Figure 1. An illustration of the relationship between channels and unstable representations: (**a**) The instability of channel representations increases with the level of corruption. This indicates that domain shifts impact the channel outputs, and the stronger the domain shift, the greater the instability of the channel outputs. (**b**) The instability of representations varies across different channels. Domain-robust channels exhibit stable representations with smaller variance, typically concentrated on the left, suggesting that these channels have learned more domain-invariant knowledge, making them resilient to data distribution shifts and less prone to changes. In contrast, domain-sensitive channels show unstable representations with larger variance, typically concentrated on the right, as they have learned domain-specific knowledge, making them vulnerable to data distribution shifts and more susceptible to changes.

Based on our observation in Figure 1, we can see that domain-specific channels exhibit instability and are prone to change. Therefore, we propose a parameter update constraint method that estimates the relationship between changes in channel representations and the loss increases caused by parameter updates, suppressing domain-sensitive parameters by minimizing the changes in domain-sensitive channel representations. This constraint not only effectively alleviates the interference caused by domain-specific knowledge but also promotes the learning of domain-invariant knowledge by reducing the sensitivity of the parameters. Additionally, we provide theoretical evidence that our method can effectively enhance the model's generalization ability and promote the learning of domain-invariant knowledge.

Furthermore, as shown in Figure 2, domain shifts carry substantial domain-specific knowledge, and when the model has not yet adapted to the new domain, features are more susceptible to the effects of the domain shift, making them more likely to cross the original decision boundary. We introduce a virtual decision boundary that constrains the features generated by the model to move away from the original decision boundary, preventing them from clustering near it. This constraint also creates a virtual margin between two decision boundaries. During domain shifts, this virtual margin provides sufficient buffering to prevent features from crossing the original decision boundary. The strongly constrained virtual decision boundary effectively mitigates the interference caused by domain-specific knowledge in the early stages of domain adaptation.

Overall, we propose a Dual Constraints method that combines channel-based parameter constraints and feature-based virtual decision boundary constraints, effectively addressing the two major challenges of domain knowledge interference and learning domain-invariant knowledge, thereby enhancing the model's sustained capability(More motivations and details will be elaborated in Section 3.1). Our contributions can be summarized as follows:

1. We propose a novel parameter constraint method that minimizes the representation changes in domain-sensitive channels, which, respectively, enhance and suppress the learning of domain-invariant and domain-specific knowledge. In addition, we theoretically prove that it can effectively enhance the model's generalization ability.
2. We introduce a strongly constrained virtual decision boundary that creates a virtual margin, forcing features away from the original decision boundary, effectively mitigating the problem of features crossing the boundary during domain shifts.
3. Dual Constraints enhance the model's sustained capability and achieve excellent performance, surpassing all existing state-of-the-art methods.

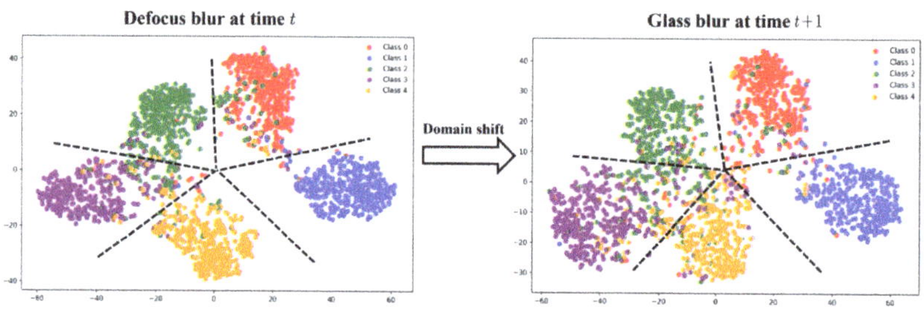

Figure 2. Decision boundary lacks constraints. Consider a domain shift occurring between two adjacent time points, t and $t+1$. Features that are clustered near the decision boundary at time t are subjected to a stronger domain shift, causing them to cross the decision boundary and manifest as features at time $t+1$.

2. Related Work

2.1. Unsupervised Domain Adaptation

Unsupervised Domain Adaptation (UDA) [16] assumes that there is a domain shift between the source domain and the target domain, with labeled data available in the source domain but no labels in the target domain [17]. The goal is for the model to perform well on the unlabeled target domain data. UDA methods often use distribution distance metrics to align the feature distributions between the source and target domains during training. For example, Maximum Mean Discrepancy (MMD) [18–20] is a statistical test that assesses whether two distributions are equal based on observed samples from those distributions. Adversarial training [21–23] is another common approach, which involves aligning distributions using two adversarial roles: a domain classifier and a feature generator. Unlike traditional Unsupervised Domain Adaptation methods, Test-Time Adaptation (TTA) aims to adapt a model trained on a source domain to a new target domain without accessing the original source data during inference, and many methods have been proposed to solve TTA, such as TENT [12] and SHOT [24]. TENT [12] updates the trainable batch normalization parameters from a pretrained model by minimizing the entropy of the model's predictions during testing. SHOT [24] combines entropy minimization and diversity regularization with label smoothing techniques to train a general feature extractor from the pretrained source model.

2.2. Domain Generalization

Domain generalization (DG) aims to extract knowledge from the source domain that can generalize well to unseen target domains. Many methods learn domain-invariant representations by aligning the distributions of the target and source domains. These methods include adversarial learning [25], causal learning [26], and meta-learning [27]. Another approach is to enhance the model's generalization capability by generating more source domain data, specifically by augmenting the diversity of the source data through data augmentation [28]. These data augmentation techniques primarily involve style transfer [28,29] and pixel-level augmentation [30]. Although these methods have demonstrated promising results, they may still learn excessive domain-specific features, as they rely on implicit assumptions to remove domain-specific characteristics via image-level augmentation or model-level constraints. Some studies have pointed out that Convolutional Neural Networks (CNNs) tend to classify objects based on local texture features that contain domain-specific characteristics [31–33]. To address this, they propose using penalty loss functions to suppress the model from learning local features such as texture, background, and lighting. This penalty encourages the model to rely on global features for classification. In contrast to local features, global features, including shape and contours, remain more stable and less prone to changes across domains. Therefore, this penalty on local features forces the model to learn domain-invariant properties, thereby enhancing its generalization ability.

2.3. Continual Learning

Continual learning (CL) aims to enable models to learn new tasks from a continuously evolving data stream while retaining previously acquired knowledge. One of the main challenges is catastrophic forgetting, where models tend to forget prior knowledge when learning new tasks. To address this issue, several approaches have been proposed, including regularization methods [34], which protect important weights from excessive updates, architecture expansion methods [35], which adapt the model by expanding its structure to accommodate new tasks, and memory replay methods [36–39], which store and replay data from previous tasks to mitigate forgetting. Despite significant progress, continual learning still faces challenges related to computational and memory overhead and maintaining strong generalization performance under non-stationary data distributions.

2.4. Continual Test-Time Adaptation

Unlike traditional Test-Time Adaptation (TTA) [40–42], which assumes a fixed target domain and works in a source-free online manner, CTTA accounts for the dynamic nature of real-world data distributions. This means CTTA requires models to adapt online and source-free across evolving domains. CoTTA [5] is a leading approach in continual domain learning, using average teachers and data augmentation for reliable pseudo-labels and robust model updates. DSS [8], inspired by semi-supervised learning, employs FreeMatch [43] to generate label thresholds for filtering pseudo-labels. RoTTA [44] stabilizes Batch Normalization updates to mitigate batch size variations during adaptation. RMT [6] addresses asymmetry in cross-entropy in teacher–student models, proposing symmetric cross-entropy for better gradients. Reshaping [10] addresses the catastrophic forgetting problem using sample replay. MAE [11] leverages mask autoencoders to learn domain-invariant knowledge. Unlike previous work, this study focuses on the sustainability of continuous domain adaptation, aiming to find the global optimum rather than settling for local optima.

3. Method

3.1. Problem Setting

Given a pretrained model M^0 with initial parameters θ_0, originally trained on source domain data (X^S, Y^S), the objective of CTTA is to iteratively adapt model M to a sequence of target domain datasets. During this process, the source domain data (X^S, Y^S) is not accessible, and the target domain datasets $\{X_0^T, X_1^T, \ldots, X_n^T\}$ are unlabeled. At each time step t, the parameters θ_t are updated to θ_{t+1} to better align the model M^{t+1} with the current target domain X_t^T. For the sake of clarity, we define the model output as:

$\hat{y} = M(x) = G(Z) = G_\ell(F_\ell(x))$. $F(x)$ is the feature extractor, and the feature map at the ℓth layer is $Z_\ell = \{Z_{\ell,1}, Z_{\ell,2}, \ldots, Z_{\ell,C_\ell}\} = F_{\ell-1}(Z_{\ell-1})$, where $Z_\ell \in \mathbb{R}^{C_\ell \times H_\ell \times W_\ell}$. $G(\cdot)$ represents the classifier, and $W = \{W_1, W_2 \ldots W_C\}$ denotes the weights of the classifier.

3.2. *Domain-Sensitive Parameter Suppression*

3.2.1. Motivation of Domain-Sensitive Parameter Suppression

Existing methods typically use the following objective function to optimize model parameters for domain adaptation during testing:

$$\min_{M^{t+1}} \mathcal{L}(M^{t+1}(X_t^T), \hat{Y}_t^T) \tag{1}$$

where at time t, the target domain data X_t^T is encountered, and the model parameters are updated based on the loss function $\mathcal{L}(\cdot)$ to form the new model M^{t+1}. It is important to note that before any parameter update at time t, $M^{t+1} = M^t$.

Existing methods naively use Equation (1) as the objective function, leading to an unconstrained parameter update process. A large amount of domain-sensitive parameters are highly sensitive to domain-specific knowledge and change rapidly to fit such knowledge [31,45], such as background, lighting, and texture. Although existing methods introduce EMA [46], where $M^t = \beta M^{t-1} + (1-\beta)M^t$, and set $\beta = 0.999$ to alleviate performance degradation caused by rapid parameter changes, this parameter weighting approach has several issues: historical parameters contain a large amount of knowledge from past domains, which interferes with the current domain adaptation process, and hyperparameters cannot be adjusted adaptively. Moreover, this method is often used as a performance-boosting technique: people know that using it improves performance, but not using it worsens performance, without exploring the essence of the problem.

3.2.2. Implementation of Domain-Sensitive Parameter Suppression

To tackle this challenge, we propose a novel parameter constraint method that minimizes the representation changes in domain-sensitive channels, effectively mitigating the rapid updates and overfitting of domain-sensitive parameters to domain-specific knowledge.

Now, we consider two consecutive models, M^t and M^{t-1}. Theoretically, the parameter difference between two adjacent models is very small, but CTTA is a continuous process where the model continuously adapts to the target domain, which will eventually result in a huge cumulative effect. To analyze the feature maps change caused by the model parameter changes between adjacent moments, let Z'_ℓ and Z_ℓ are the feature maps generated by the $\ell - 1^{th}$ layer of $M^t(\cdot)$ and $M^{t-1}(\cdot)$, respectively. We can compute the loss given by Z'_ℓ via the first-order Taylor approximation as follows:

$$\begin{aligned} \mathcal{L}(G_\ell(Z'_\ell), \hat{y}) &\approx \mathcal{L}(G_\ell(Z_\ell), \hat{y}) \\ &+ \sum_{c=1}^{C_\ell} \left\langle \nabla_{Z_{\ell,c}} \mathcal{L}(G_\ell(Z_\ell), \hat{y}), Z'_{\ell,c} - Z_{\ell,c} \right\rangle_F \end{aligned} \tag{2}$$

where $\langle \cdot, \cdot \rangle_F$ is the Frobenius inner product, i.e., $\langle A, B \rangle_F = \mathrm{tr}(A^\top B)$, and $\nabla_{Z_{\ell,c}} \mathcal{L}(G_\ell(Z_\ell), \hat{y})$ is the gradient in the backpropagation of $M^{t-1}(\cdot)$. Using Equation (2), we define the loss increment $\triangle\mathcal{L}(Z'_{\ell,c})$ for the cth channel at the ℓth layer caused by parameter changes as follows:

$$\triangle\mathcal{L}(Z'_{\ell,c}) := \left\langle \nabla_{Z_{\ell,c}} \mathcal{L}(G_\ell(Z_\ell), \hat{y}), Z'_{\ell,c} - Z_{\ell,c} \right\rangle_F \tag{3}$$

Thus, we aim to minimize the loss change caused by parameter updates between two adjacent models at consecutive moments, as follows:

$$\min_{M^{t+1}} \sum_{\ell=1}^{L} \sum_{c=1}^{C_\ell} \mathbb{E}_{(x)\sim X_t^T} \left[\triangle \mathcal{L}(Z'_{\ell,c}) \right]^2 \tag{4}$$

We need to update $M_t(\cdot)$ to $M_{t+1}(\cdot)$ at time t using data X_t^T. If we minimize Equation (4) using standard stochastic gradient descent, in addition to calculating the gradients of the feature maps produced by $M^t(\cdot)$ with respect to the samples, we also need to calculate and store the gradients of the feature maps produced by $M^{t-1}(\cdot)$ with respect to the samples, which significantly increases computational cost or memory overhead.

Using the Cauchy-Schwarz inequality, we derive the upper bound for the objective function of Equation (4) as follows:

$$\begin{aligned} &\mathbb{E}\left[\triangle \mathcal{L}(Z'_{\ell,c})\right] \\ &= \mathbb{E}\left[\langle \nabla_{Z_{\ell,c}} \mathcal{L}(G_\ell(Z_\ell), \hat{y}), Z'_{\ell,c} - Z_{\ell,c} \rangle_F\right] \\ &\leq \mathbb{E}\left[\|\nabla_{Z_{\ell,c}} \mathcal{L}(G_\ell(Z_\ell), \hat{y})\|_F \cdot \|Z'_{\ell,c} - Z_{\ell,c}\|_F\right] \\ &\leq \sqrt{\mathbb{E}\left[\|\nabla_{Z_{\ell,c}} \mathcal{L}(G_\ell(Z_\ell), \hat{y})\|_F^2\right] \cdot \mathbb{E}\left[\|Z'_{\ell,c} - Z_{\ell,c}\|_F^2\right]} \end{aligned} \tag{5}$$

By optimizing the upper bound of the objective function, we avoid storing large amounts of the gradients of the feature maps or performing additional backpropagation. Substituting Equation (5) into the objective function to minimize Equation (4), we obtain

$$\min_{M^{t+1}} \sum_{\ell=1}^{L} \sum_{c=1}^{C_\ell} \mathbb{E}_{x\sim X_t^T} \|\nabla_{Z_{\ell,c}} \mathcal{L}(G_\ell(Z_\ell), \hat{y})\|_F^2 \cdot \|Z'_{\ell,c} - Z_{\ell,c}\|_F^2 \tag{6}$$

According to Figure 3, we observe an interesting phenomenon: The degree of unstable representation of the channels is positively correlated with the magnitude of the gradient values. Larger gradients cause the parameters to change rapidly, indicating that domain-sensitive parameters are sensitive to domain-specific knowledge and fit such knowledge through rapid changes, leading to unstable channel representations that are prone to variations and contain a large amount of domain-specific knowledge. This is consistent with the conclusion derived from the Equation (6): the larger the gradient value, the higher the degree of suppression of the parameters.

Therefore, we define a channel sensitivity importance weight as $I_{\ell,c}^t$:

$$I_{\ell,c}^t = \|\nabla_{Z_{\ell,c}} \mathcal{L}(G_\ell(Z_\ell), \hat{y})\|_F^2 \tag{7}$$

However, we cannot directly optimize using this equation. There are significant differences in the magnitude of channel gradients across layers, so it is necessary to balance the scale of importance across layers. Finally, we construct the channel sensitivity importance weight $I_{\ell,c}^t$ as follows:

$$I_{\ell,c}^t = \begin{cases} I_{\ell,c}^t = 0 & t = 0 \\ I_{\ell,c}^t = \dfrac{I_{\ell,c}^{t-1} + \|\nabla_{Z_{\ell,c}} \mathcal{L}(G_\ell(Z_\ell), \hat{y})\|_F^2}{\frac{1}{C_{\ell,c}} \sum_{c=1}^{C_\ell} \|\nabla_{Z_{\ell,c}} \mathcal{L}(G_\ell(Z_\ell), \hat{y})\|_F^2} & t > 0 \end{cases} \tag{8}$$

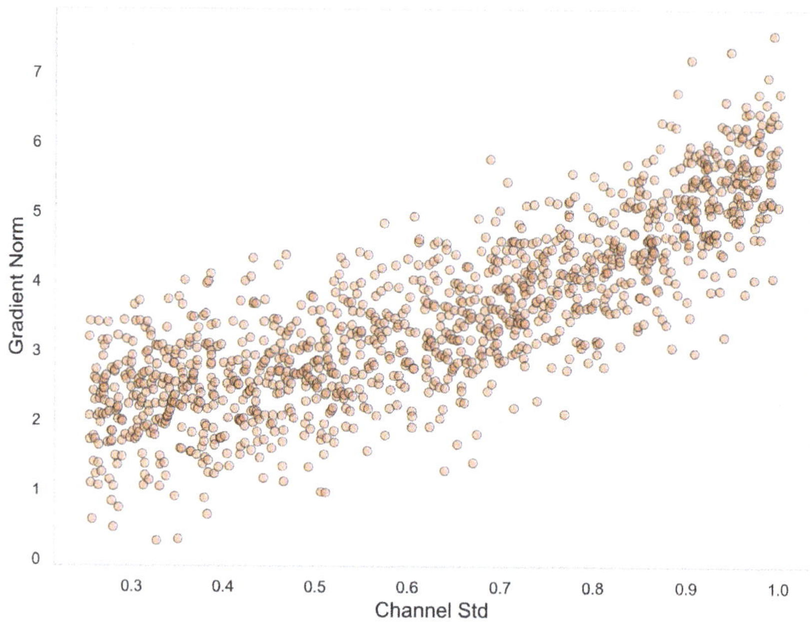

Figure 3. Gradient norm and channel-unstable representation relationship.

Therefore, the final parameter constraint objective function can be expressed as

$$\mathcal{L}_{DSP} = \min_{M^{t+1}} \sum_{\ell=1}^{L} \sum_{c=1}^{C_\ell} \mathbb{E}_{x \sim X_t^T} I_{\ell,c}^t \cdot \|Z'_{\ell,c} - Z_{\ell,c}\|_F^2 \tag{9}$$

3.2.3. Theoretical Analysis of Domain-Sensitive Parameter Suppression

In this section, we prove from the perspective of Lipschitz continuity that our method can effectively suppress domain-sensitive parameters, promote the learning of domain-invariant knowledge, and enhance the generalization ability of the model.

We first provide the definition of Lipschitz continuity. Given $\Omega \subset \mathbb{R}^n$, let $\theta_1 \in \Omega$ and $\theta_2 \in \Omega$. For a function $h : \Omega \to \mathbb{R}^m$, if there exists a constant K such that the following holds:

$$\|h(\theta_1) - h(\theta_2)\|_2 \leq K\|\theta_1 - \theta_2\|_2, \quad \forall \theta_1, \theta_2 \in \Omega \tag{10}$$

then, h is called Lipschitz continuous.

According to existing studies [33,47], if the loss function has a smaller Lipschitz constant K, it indicates that the loss function landscape is flatter, which consequently leads to better model generalization. On the contrary, if the Lipschitz constant K is very large, it indicates that the model parameters are highly sensitive to input variations, with even minor changes leading to drastic shifts in the model's output, which means that the model lacks generalization ability.

Now, consider the model parameters θ_{t-1} and θ_t at two consecutive moments during continual domain adaptation. The change in the loss function can be expressed as:

$$\|\mathcal{L}(\theta_t) - \mathcal{L}(\theta_{t-1})\|_2 = \|\mathcal{L}(\zeta)(\theta_t - \theta_{t-1})\|_2 \tag{11}$$

where $\zeta = c\theta_t + (1-c)\theta_{t-1}$, and $c \in [0,1]$. By applying the Cauchy-Schwarz inequality, we have

$$\|\mathcal{L}(\theta_t) - \mathcal{L}(\theta_{t-1})\|_2 \le \|\nabla L(\zeta)\|_2 \|\theta_t - \theta_{t-1}\|_2 \quad (12)$$

Considering $\theta_t = \theta_{t-1} - \eta\nabla\theta_{t-1}$, we have $\theta_t \to \theta_{t-1}$, so $\nabla L(\zeta) \approx \nabla\mathcal{L}(\theta_{t-1})$, and the above can be rewritten as

$$\|\mathcal{L}(\theta_t) - \mathcal{L}(\theta_{t-1})\|_2 \lesssim \|\nabla\mathcal{L}(\theta_{t-1})\|_2 \|\theta_t - \theta_{t-1}\|_2 \quad (13)$$

From Equation (13), it can be seen that minimizing $\|\nabla\mathcal{L}(\theta_{t-1})\|_2$ is equivalent to minimizing the Lipschitz constant K. According to Equation (9), we have

$$\mathcal{L}_{DSP} \propto I^t_{\ell,c} = \|\nabla_{Z_{\ell,c}}\mathcal{L}(G_\ell(Z_\ell), \hat{y})\|_F^2 \quad (14)$$

Thus, based on Equation (14), penalizing the gradient norm forces the model parameters to generate smaller gradient norms, which is equivalent to reducing the Lipschitz constant of the model. As a result, the model achieves better generalization performance.

3.3. *Virtual Decision Boundary*

3.3.1. Motivation of Virtual Decision Boundary

In Figure 4, we clarify the reasons behind the lack of robustness in the decision boundary and present our proposed solutions. In Figure 4a, we observe that during domain adaptation, the features extracted by the feature extractor $F(\cdot)$ tend to cluster near or slightly cross the decision boundary, which is indicated by the red boxes. In Figure 4b, we examine domain shifts, where we assume that two successive adapted domains have deviations ε_k and ε_{k+1} from the source domain. The deviation between them is $\Delta\varepsilon = \varepsilon_{k+1} - \varepsilon_k$. When $\Delta\varepsilon > 0$, the feature bias increases, leading to the possibility that features clustered near the decision boundary may cross over, resulting in errors. To address these challenges, we propose the virtual decision boundary. As shown in Figure 4c, this method introduces a virtual margin between the original decision boundary and a newly created virtual decision boundary. This virtual margin pushes features away from the original boundary, reducing the likelihood of clustering near it. Additionally, the virtual margin provides sufficient buffer space to help prevent features from crossing the decision boundary during domain shifts.

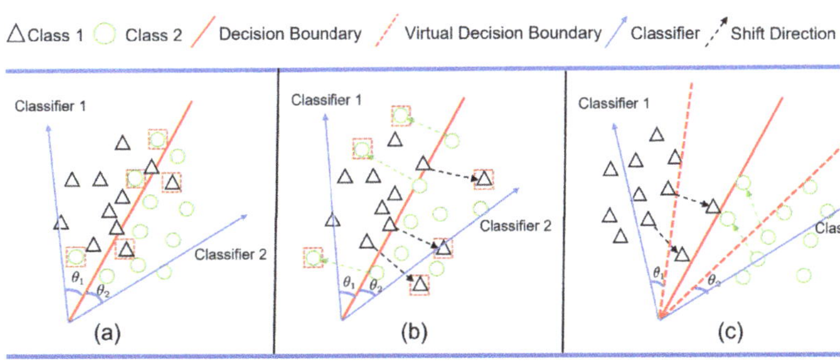

Figure 4. An illustrative example showcasing our motivation and solution: (**a**) During continuous domain adaptation, features are prone to cluster around or cross the decision boundary. (**b**) During domain shifts, features clustered at the decision boundary are prone to crossing it. (**c**) The virtual decision boundary forces the generated features away from the original decision boundary, creating a virtual margin that prevents features from crossing the decision boundary during domain shifts.

3.3.2. Implementation of Virtual Decision Boundary

Before addressing this issue, we start with a simple binary classification problem. Consider a binary classification problem where we have a sample from class 1 and use the feature extractor function $Z = F(x)$ to obtain the features of the sample x. We introduce a parameter m that is used to scale the inequality, forming a stricter virtual decision boundary. This can be mathematically expressed as $\|W_1\|_2\|Z\|_2\cos(m\theta_1) > \|W_2\|_2\|Z\|_2\cos(\theta_2)$ $(0 \leq \theta_1 \leq \frac{\pi}{m})$, where θ is the angle between the classifier vector and the feature vector, and $m \in [1, +\infty)$, as the following inequality holds:

$$\|W_1\|_2\|Z\|_2\cos(\theta_1) \geq \|W_1\|_2\|Z\|_2\cos(m\theta_1) > \|W_2\|_2\|Z\|_2\cos(\theta_2) \tag{15}$$

Therefore, $\|W_1\|_2\|Z\|_2\cos(\theta_1) > \|W_2\|_2\|Z\|_2\cos(\theta_2)$ has to hold. So the new classification criteria is a stronger requirement to correctly classify x, producing a more rigorous virtual decision boundary for class 1. Thus, the loss function for binary classification of class 1 can be written as

$$\mathcal{L} = \frac{e^{\|W_1\|_2\|Z\|_2\cos(m\theta_1)}}{e^{\|W_1\|_2\|Z\|_2\cos(m\theta_1)} + e^{\|W_2\|_2\|Z\|_2\cos(\theta_2)}} \tag{16}$$

We can also apply the same constraint to class 2 as we did for class 1. In CTTA, we can transform a binary classification problem into a multiclass classification problem.

$$\mathcal{L} = \frac{1}{|N|}\sum_{i=1}^{N}\frac{e^{\|W_c\|_2\|Z_i\|_2\cos(m\theta_c)}}{e^{\|W_c\|_2\|Z_i\|_2\cos(m\theta_c)} + \sum_{j\neq c}^{C} e^{\|W_j\|_2\|Z_i\|_2\cos(\theta_j)}} \tag{17}$$

In addition, in order to make Equation (17) hold for all $\theta \in \pi$, we construct $\psi(\theta)$ as follows:

$$\psi(\theta) = \begin{cases} \cos(m\theta) & 0 < \theta < \dfrac{\pi}{m} \\ \mathcal{D}(\theta) & \dfrac{\pi}{m} < \theta < \pi \end{cases} \tag{18}$$

where m is a non-negative number that is closely related to the classification margin. With larger m, the classification margin becomes larger, and the learning objective also becomes harder. Meanwhile, $\mathcal{D}(\theta)$ is required to be a monotonically decreasing function and $\mathcal{D}(\frac{\pi}{m})$ should equal $\cos(\frac{\pi}{m})$. We construct a specific $\psi(\theta)$ as follows:

$$\psi(\theta) = (-1)^k\cos(m\theta) - 2k, \theta \in [\frac{k\pi}{m}, \frac{(k+1)\pi}{m}] \tag{19}$$

where $k \in [0, m-1]$ and k is an integer. So, the final virtual decision boundary loss can be represented as follows:

$$\mathcal{L}_{VDB} = \frac{1}{|N|}\sum_{i=1}^{N}\frac{e^{\|W_c\|_2\|Z_i\|_2\psi(\theta_c)}}{e^{\|W_c\|_2\|Z_i\|_2\psi(\theta_c)} + \sum_{j\neq c}^{C} e^{\|W_j\|_2\|Z_i\|_2\cos(\theta_j)}} \tag{20}$$

3.3.3. Dynamic Virtual Margin of Virtual Decision Boundary

In the virtual decision boundary method, the width of the virtual margin m is not uniform across different domains and classes at any given time. This variability arises because different domains experience varying degrees of domain shift, and each class is affected differently, leading to varying densities of sample features near the decision boundary. To address this, we propose a dynamic margin that adjusts the width of the virtual margin according to the shift degree of each domain and class. A practical way to implement this is by using the classifier from the source model as a reference point. The deviation of each class from its corresponding class in the source domain can be quantified using cosine distance. Additionally, the overall domain shift can be assessed by calculating the average cosine distance of all samples from their positions in the source domain. Specifically, we first measure the shift degree $\bar{D}$ of the current domain relative to the source domain:

$$\bar{D} = \frac{1}{2}\left(1 - \frac{1}{|N|}\sum_{i}^{N}\frac{W_c \cdot Z_i}{||W_c||_2||Z_i||_2}\right) \tag{21}$$

Then, we measure the shift degree D_c of each class relative to the corresponding class in the source domain:

$$D_c = \frac{1}{2}\left(1 - \frac{1}{|N_c|}\sum_{j}^{N_c}\frac{W_c \cdot Z_j}{||W_c||_2||Z_j||_2}\right), c \in C \tag{22}$$

Thus, at any given time t, the dynamic margin for each class in different domains can be expressed as

$$m_c^t = \begin{cases} m & t = 0 \\ \beta \cdot m_c^{t-1} + (1-\beta) \cdot \left(\lceil \frac{D_c}{\max(D)} \cdot e^{\bar{D}} + 1 \rceil\right), & t > 0 \end{cases} \tag{23}$$

where $D = \{D_1, D_2, \ldots, D_C\}$, and β is set to 0.999, used to robustly update, preventing its value from undergoing drastic changes. This dynamic adjustment allows the model to more effectively adapt to the specific shifts encountered across different domains and classes, enhancing its robustness and accuracy. The m is a hyperparameter used to initialize the virtual margin, and the specific setting is discussed in detail in the experimental section.

3.4. Loss Function

Based on Equations (1), (8) and (20), we can derive the following overall loss function:

$$\mathcal{L}_{all} = \mathcal{L}(M(X^T), \hat{Y}^T) + \mathcal{L}_{DSP} + \mathcal{L}_{VDB} \tag{24}$$

We constrain the model using this function to enhance its sustained capability.

4. Experiment and Results

4.1. Experimental Setup

4.1.1. Datasets and Task Setting

Building on the previous works [5,6,44], our method undergoes evaluation on three classification CTTA benchmarks, which encompass CIFAR10-to-CIFAR10C, CIFAR100-to-CIFAR100C, and ImageNet-to-ImageNetC. In the segmentation CTTA, we conduct assessments on the Cityscapes-to-ACDC, using the Cityscapes [48] as the source domain and the ACDC [49] as the target domain.

4.1.2. Compared Methods and Implementation Details

We compare our method with the original model (Source) and multiple CTTA methods, including BN [50], TENT [12], CoTTA [5], RoTTA [44], SATA [51], RMT [6], PETAL [7], DSS [8], Reshaping [10]. For the classification task, all methods are implemented using the same backbone architecture and pretrained model as used in our approach. Specifically, we utilize the pretrained WideResNet-28 [52] for CIFAR10C, ResNeXt-29 [53] for CIFAR100C, and ResNet-50 [54] for ImageNetC, and use the largest corruption severity (level 5). For the segmentation CTTA task, we use the ACDC dataset as the target domain, which includes images captured under four distinct, unobserved visual conditions: Fog, Night, Rain, and Snow. To simulate continuous environmental changes akin to real-world scenarios, we cyclically iterate through the same sequence of target domains (Fog → Night → Rain → Snow) multiple times.

4.2. Classification CTTA Tasks

4.2.1. CIFAR10-to-CIFAR10C Gradual and Continual

For the CIFAR10-to-CIFAR10C task, we evaluate our methods under two distinct settings. The first setting is a gradual task; the model sequentially adapts to fifteen target domains where the corruption severity level gradually changes between the lowest and highest extremes. The corruption type only changes when the severity reaches its lowest point. As shown in Table 1, our method achieves the lowest error rate of 8.3%, representing 2.1% improvement over the CoTTA method.

The second setting is a standard continual task where the model sequentially adapts to fifteen target domains, each with a corruption severity level of 5. The results, shown in Table 2, indicate that directly applying the source domain model yields an average error rate of 43.5%. The BN [50] method improves this performance by 23.1% compared with the source-only baseline. Among all compared methods, DSS achieves the lowest error rates of 12.2% on motion. SATA [51] achieves the lowest error rates of 10.2%, 14.1%, 13.2%, 10.3% on zoom, snow, frost, and contrast, respectively. Reshaping [10] achieves the lowest error rates of 17.1%, 12.7%, 15.9% on elastic, pixelate, and jpeg. In other scenarios, our proposed method either outperforms or is on par with the other approaches, ultimately achieving the lowest overall average error rate, reduced to 14.5%.

Table 1. Classification error rate (%) for the gradual CIFAR10-to-CIFAR10C task. The best results in each column are highlighted in **bold**.

Dataset	Source	BN [50]	TENT [12]	CoTTA [5]	Ours
CIFAR10C (Error %)	24.8	13.7	30.7	10.4	**8.3**

Table 2. Classification error rate (%) for the standard CIFAR10-to-CIFAR10C Continual Test-Time Adaptation task. The best results in each column are highlighted in **bold**.

	Gaussian	Shot	Impulse	Defocus	Glass	Motion	Zoom	Snow	Frost	Fog	Brightness	Contrast	Elastic	Pixelate	Jpeg	Mean
Source	72.3	65.7	72.9	46.9	54.3	34.8	42.0	25.1	41.3	26.0	9.3	46.7	26.6	58.5	30.3	43.5
BN [50]	28.1	26.1	36.3	12.8	35.3	14.2	12.1	17.3	17.4	15.3	8.4	12.6	23.8	19.7	27.3	20.4
TENT [12]	24.8	20.6	28.6	14.4	31.1	16.5	14.1	19.1	18.6	18.6	12.2	20.3	25.7	20.8	24.9	20.7
CoTTA [5]	24.3	21.3	26.6	11.6	27.6	**12.2**	10.3	14.8	14.1	12.4	7.5	10.6	18.3	13.4	17.3	16.2
RoTTA [44]	30.3	25.4	34.6	18.3	34.0	14.7	11.0	16.4	14.6	14.0	8.0	12.4	20.3	16.8	19.4	19.3
RMT [6]	24.1	20.2	25.7	13.2	25.5	14.7	12.8	16.2	15.4	14.6	10.8	14.0	18.0	14.1	16.6	17.0
PETAL [7]	23.7	21.4	26.3	11.8	28.8	12.4	10.4	14.8	13.9	12.6	7.4	10.6	18.3	13.1	17.1	16.2
SATA [51]	23.9	20.1	28.0	11.6	27.4	12.6	**10.2**	**14.1**	**13.2**	12.2	7.4	**10.3**	19.1	13.3	18.5	16.1
DSS [8]	24.1	21.3	25.4	11.7	26.9	**12.2**	10.5	14.5	14.1	12.5	7.8	10.8	18.0	13.1	17.3	16.0
Reshaping [10]	23.6	19.9	26.0	11.8	25.3	13.2	10.9	14.3	13.5	12.7	9.0	11.9	**17.1**	**12.7**	**15.9**	15.8
Ours	**20.1**	**16.5**	**23.4**	**11.2**	**24.1**	12.6	10.3	14.3	13.6	**12.1**	**7.3**	10.9	18.2	12.9	17.0	**14.5**

4.2.2. CIFAR100-to-CIFAR100C

The results for the CIFAR100-to-CIFAR100C continual task, as shown in Table 3, further demonstrate the effectiveness of our method. Our approach achieves the lowest error rates across the Gaussian, shot, impulse, glass, motion, zoom, snow, frost, elastic, and jpeg, and it also obtains the lowest overall average error rate. Compared with the source-only baseline, our method improves performance by 17.9%, and it surpasses the Reshaping [10] method with a further 1.2% reduction in error rate.

Table 3. Classification error rate (%) for the standard CIFAR100-to-CIFAR100C Continual Test-Time Adaptation task. The best results in each column are highlighted in **bold**.

	Gaussian	Shot	Impulse	Defocus	Glass	Motion	Zoom	Snow	Frost	Fog	Brightness	Contrast	Elastic	Pixelate	Jpeg	Mean
Source	73.0	68.0	39.4	29.3	54.1	30.8	28.8	39.5	45.8	50.3	29.5	55.1	37.2	74.7	41.2	46.4
BN [50]	42.1	40.7	42.7	27.6	41.9	29.7	27.9	34.9	35.0	41.5	26.5	30.3	35.7	32.9	41.2	35.4
TENT [12]	37.2	35.8	41.7	37.9	51.2	48.3	48.5	58.4	63.7	71.1	70.4	82.3	88.0	88.5	90.4	60.9
CoTTA [5]	40.1	37.7	39.7	26.9	38.0	27.9	26.4	32.8	31.8	40.3	24.7	26.9	32.5	28.3	33.5	32.5
RoTTA [44]	49.1	44.9	45.5	30.2	42.7	29.5	26.1	32.2	30.7	37.5	24.7	29.1	32.6	30.4	36.7	34.8
RMT [6]	40.2	36.2	36.0	27.9	33.9	28.4	26.4	28.7	28.8	31.1	25.5	27.1	28.0	26.6	29.0	30.2
PETAL [7]	38.3	36.4	38.6	25.9	36.8	27.3	25.4	32.0	30.8	38.7	24.4	26.4	31.5	26.9	32.5	31.5
SATA [51]	36.5	33.1	35.1	25.9	34.9	27.7	25.4	29.5	29.9	33.1	24.1	26.7	31.9	27.5	35.2	30.3
DSS [8]	39.7	36.0	37.2	26.3	35.6	27.5	25.1	31.4	30.0	37.8	24.2	26.0	30.0	26.3	31.1	30.9
Reshaping [10]	38.8	35.0	35.4	26.7	33.2	27.4	25.0	**27.4**	26.8	29.8	24.1	**25.1**	26.9	24.9	**28.0**	29.0
Ours	**33.9**	**32.6**	**31.8**	**25.4**	**30.2**	**26.7**	**24.8**	28.9	**24.3**	**29.7**	**23.4**	**25.1**	**26.5**	**23.3**	30.6	**27.8**

4.2.3. ImageNet-to-ImageNetC

Table 4 presents the performance comparison for various methods on the challenging ImageNet-to-ImageNetC continual task. Our method stands out by achieving the lowest average error rate among all the methods evaluated. Notably, it significantly outperforms the recently proposed Reshaping method across several difficult corruption types, including Gaussian (72.2% vs. 78.5%), shot (70.7% vs. 75.3%), impulse (68.3% vs. 73.0%), and glass (71.3% vs. 73.1%).

Table 4. Classification error rate (%) for the standard ImageNet-to-ImageNetC Continual Test-Time Adaptation task. The best results in each column are highlighted in **bold**.

	Gaussian	Shot	Impulse	Defocus	Glass	Motion	Zoom	Snow	Frost	Fog	Brightness	Contrast	Elastic	Pixelate	Jpeg	Mean
Source	95.3	95.0	95.3	86.1	91.9	87.4	77.9	85.1	79.9	79.0	45.4	96.2	86.6	77.5	66.1	83.0
BN [50]	87.7	87.4	87.8	88.0	87.7	78.3	63.9	67.4	70.3	54.7	36.4	88.7	58.0	56.6	67.0	72.0
TENT [12]	81.6	74.6	72.7	77.6	73.8	65.5	55.3	61.6	63.0	51.7	38.2	72.1	50.8	47.4	53.3	62.6
CoTTA [5]	84.7	82.1	80.6	81.3	79.0	68.6	57.5	60.3	60.5	48.3	36.6	66.1	47.2	**41.2**	46.0	62.7
RoTTA [44]	88.3	82.8	82.1	91.3	83.7	72.9	59.4	66.2	64.3	53.3	**35.6**	74.5	54.3	48.2	52.6	67.3
RMT [6]	79.9	76.3	73.1	**75.7**	72.9	64.7	56.8	56.4	58.3	49.0	40.6	**58.2**	47.8	43.7	44.8	59.9
PETAL [7]	87.4	85.8	84.4	85.0	83.9	74.4	63.1	63.5	64.0	52.4	40.0	74.0	51.7	45.2	51.0	67.1
SATA [51]	74.1	72.9	71.6	**75.7**	74.1	64.2	55.5	55.6	62.9	46.6	36.1	69.9	50.6	44.3	48.5	60.1
DSS [8]	84.6	80.4	78.7	83.9	79.8	74.9	62.9	62.8	62.9	49.7	37.4	71.0	49.5	42.9	48.2	64.6
Reshaping [10]	78.5	75.3	73.0	**75.7**	73.1	64.5	56.0	55.8	**58.1**	47.6	38.5	58.5	**46.1**	42.0	43.4	59.0
Ours	**72.2**	**70.7**	**68.3**	75.9	**71.3**	**62.0**	**54.8**	**55.2**	61.3	**43.3**	40.5	61.2	48.6	42.1	**41.7**	**57.9**

4.3. Semantic Segmentation CTTA Task

Cityscapes-to-ACDC

We validate the effectiveness of our approach in the more challenging segmentation CTTA task by adapting the pretrained Segformer model from the Cityscapes dataset to the ACDC dataset, as shown in Table 5. Our method outperforms the previous entropy minimization method (TENT [12]), teacher–student method (CoTTA [5]), and sample reshape method (Reshaping [10] by 9.7%, 4.4%, and 0.4%, respectively). Notably, our method demonstrates better stability compared with others, with performance continuously improving throughout the adaptation process. This is attributed to the effective suppression of domain-sensitive parameters, which forces the model to learn more domain-invariant knowledge while mitigating the interference from domain-specific knowledge.

Table 5. Semantic segmentation results (mIoU in %) on the Cityscapes-to-ACDC CTTA task. The four test conditions are repeated ten times to evaluate the long-term adaptation performance. The best results in each column are highlighted in **bold**.

Time	t →																				
Round			**1**					**4**					**7**					**10**			**All**
Condition	*Fog*	*Night*	*Rain*	*Snow*	*Mean*	*Fog*	*Night*	*Rain*	*Snow*	*Mean*	*Fog*	*Night*	*Rain*	*Snow*	*Mean*	*Fog*	*Night*	*Rain*	*Snow*	*Mean*	**Mean ↑**
Source	69.1	40.3	59.7	57.8	56.7	69.1	40.3	59.7	57.8	56.7	69.1	40.3	59.7	57.8	56.7	69.1	40.3	59.7	57.8	56.7	56.7
BN	62.3	38.0	54.6	53.0	52.0	62.3	38.0	54.6	53.0	52.0	62.3	38.0	54.6	53.0	52.0	62.3	38.0	54.6	53.0	52.0	52.0
TENT [12]	69.0	40.2	60.1	57.3	56.7	66.5	36.3	58.7	54.0	53.9	64.2	32.8	55.3	50.9	50.8	61.8	29.8	51.9	47.8	47.8	52.3
CoTTA [5]	70.9	41.2	62.4	59.7	58.6	70.9	41.2	62.4	59.7	58.6	70.9	41.2	62.4	59.7	58.6	70.9	41.2	62.4	59.7	58.6	58.6
Reshaping [10]	71.2	42.3	65.0	62.0	60.1	**72.8**	43.6	66.7	63.3	61.6	72.5	42.5	66.8	63.3	61.3	72.5	42.9	66.7	63.0	61.3	61.3
ours	**71.8**	**43.1**	**65.2**	**62.3**	**60.6**	72.6	**44.6**	**66.7**	**63.5**	**61.9**	**72.8**	**43.5**	**67.8**	**63.4**	**61.9**	**73.1**	**43.9**	**67.3**	**64.0**	**62.1**	**61.7**

4.4. Ablation Study and Further Analysis

4.4.1. Ablation Study

We conducted an ablation study to assess the effectiveness of the key components of our approach across three benchmarks. For clarity, we refer to the virtual decision boundary as VDB and Domain Sensitivity Parameter Suppression as DSP. As shown in Table 6, incorporating VDB and DSP results in reduced error rates across all benchmarks. The combination of VDB and DSP leads to an even greater reduction in error rates, highlighting the synergistic effect of these components when used together.

Table 6. Ablation: Contribution of our proposed VDB and DSP. The best results in each column are highlighted in **bold**.

	VDB	DSP	CIFAR10C	CIFAR100C	ImageNetC
0			16.2%	32.5%	62.7%
1	✓		15.4%	30.1%	60.3%
2		✓	15.1%	28.3%	59.2%
3	✓	✓	**14.5%**	**27.8%**	**57.9%**

4.4.2. Integration with Existing Methods

Next, we integrate our method with existing methods, namely TENT [12], CoTTA [5], RMT [44], and DSS [8]. The experiments are conducted on the CIFAR10C and CIFAR100C datasets. Utilizing the official code of each method, we enhance the accuracy of all methods, as shown in Table 7. For example, our method reduces the error rate of CoTTA from 16.2% to 14.2% on CIFAR10C, from 32.5% to 26.9% on CIFAR100C, and from 62.6% to 57.8% on ImageNetC. These experiments and results demonstrate that our method can be seamlessly integrated with other CTTA methods to enhance performance.

Table 7. Integration with existing methods. Our method can be seamlessly integrated with other CTTA methods to boost performance.

Method	CIFAR10C	CIFAR100C	ImageNetC
TENT+ours	18.9% (+1.8%)	56.6% (+4.3%)	65.3% (+6.7%)
CoTTA+ours	14.2% (+2.0%)	26.9% (+5.6%)	57.8% (+4.8%)
RMT+ours	15.4% (+1.6%)	28.7% (+1.5%)	57.4% (+2.5%)
DSS+ours	14.1% (+1.7%)	26.1% (+4.8%)	58.5% (+6.1%)

4.4.3. Analysis of Domain Sensitivity Parameter Suppression

In this section, we focus on evaluating the effectiveness of Domain Sensitivity Parameter Suppression (DSP). The dataset used is CIFAR10C, and the pretrained network is

WideResNet-28. The methods compared include CoTTA [5], DSS [8], and RMT [6]. First, we measure the model's sensitivity to the domain by calculating the unstable representations of all channels in the network. The calculation formula is $Std_{channels} = \sum_{l=0}^{L} \sum_{i=0}^{C^l} \sqrt{\|\tilde{C}_i^l - C_i^l\|_2^2}$ where $\tilde{C}_i^l$ represent the channel features generated by the source domain data in the target domain network, and C_i^l represent the channel features generated by the target domain data in the target domain network. As shown in Figure 5, our method consistently achieves the lowest unstable activation values across all domains. This indicates that DSP effectively suppresses the learning of domain-specific knowledge by regulating the updates of domain-sensitive parameters. This, in turn, enhances the learning of domain-invariant knowledge and prevents the model from overfitting to current domain knowledge during continuous domain shifts, thereby mitigating interference with future domains that may be encountered. Furthermore, we examined the relationship between the number of channels in the first layer of the network and their corresponding unstable activation values. As shown in Figure 6, the unstable representations of DSP exhibit a smaller variance, with activation values concentrating around 0.35 across all channels. This demonstrates that the introduction of DSP can effectively control channel instability, enabling the model to maintain strong robustness and effectively counteract the effects of domain shifts.

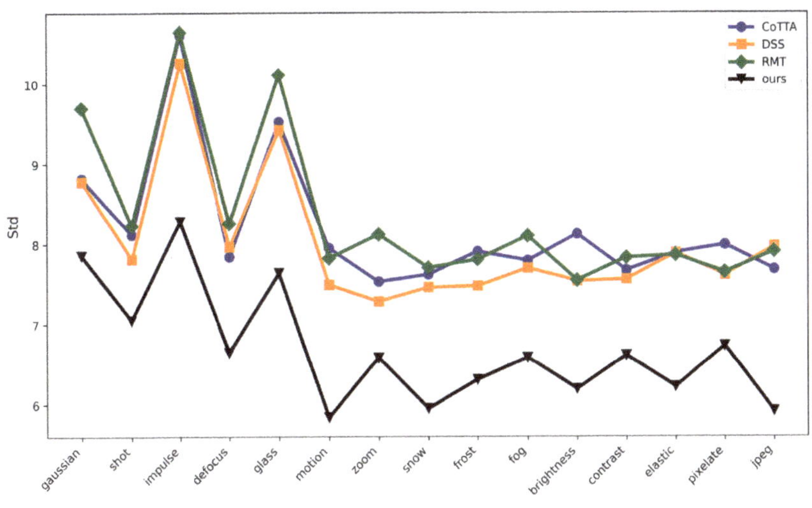

Figure 5. Channels std in all domains.

Additionally, inspired by adversarial training, we incorporate an auxiliary discriminator network to evaluate the effectiveness of DSP in promoting the learning of domain-invariant representations. Specifically, we add a source domain discriminator to the first layer of the WideResNet-28 network and train this discriminator on the CIFAR-10 dataset. The source domain discriminator takes the output of the first layer as input and generates a score indicating the likelihood that the input belongs to the source domain. Intuitively, if the domain-sensitive parameters are effectively suppressed, the model's output should remain highly robust and well aligned with the source domain, regardless of the domain shifts. As shown in Figure 7, compared with other methods on CIFAR10C, our approach consistently maintains high robustness to the source domain across all domains, indicating that the domain-sensitivity parameters have been effectively suppressed.

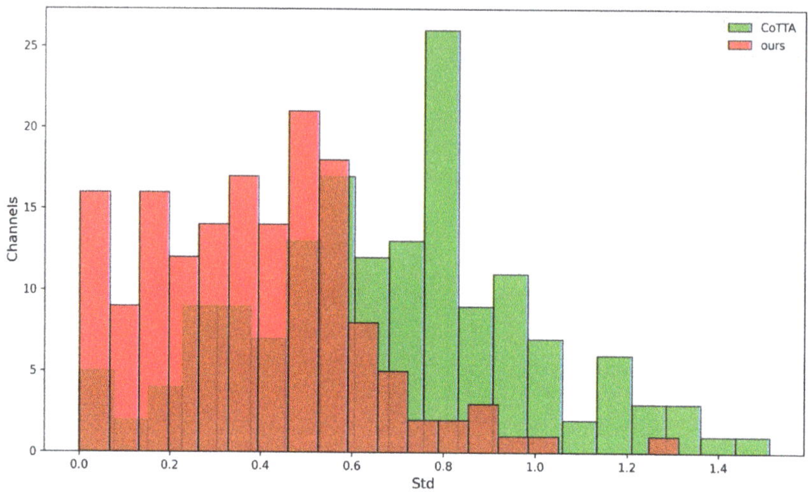

Figure 6. Channels std in first layer.

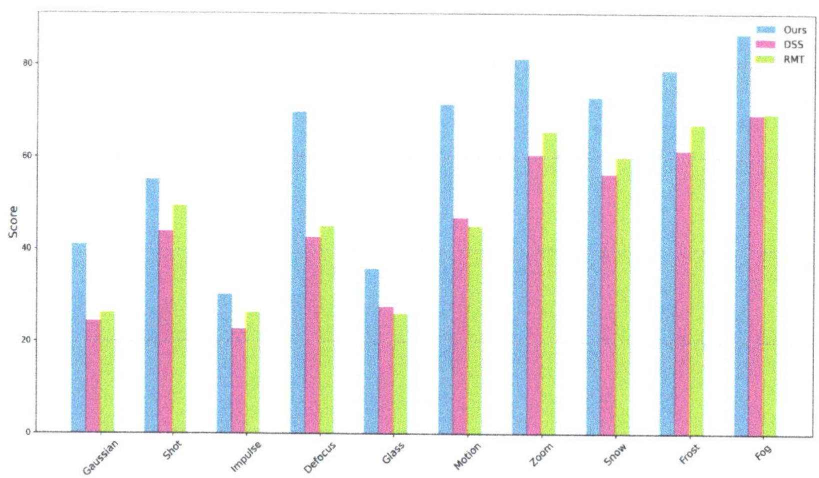

Figure 7. Domain robustness score.

4.5. Analysis of Dynamic Virtual Margin

In this section, we focus on discussing the parameter initialization of the virtual decision margin m and the superiority of dynamic virtual margins. First, we explore the effects of different initialization values for the parameters on three datasets (CIFAR10C, CIFAR100C, ImageNetC). As shown in Table 8, the initialization of the virtual decision margin parameters is highly robust. When m is set between 1 and 6, it performs well across all three datasets. However, when m exceeds 6, the performance begins to gradually decline. This is because a larger virtual margin increases the difficulty of model convergence during the initial phase of domain adaptation, leading to a drop in performance. We recommend initializing with a smaller virtual margin, allowing the margin to adaptively adjust to an appropriate value to avoid convergence difficulties caused by a large margin in the early stages of adaptation.

Overall, the dynamic virtual margin can adaptively adjust the margin width based on domain and class difficulty, and is not sensitive to initialization parameters, showing good robustness with excellent performance across a broad initialization range.

Next, we investigate the performance of fixed and dynamic virtual margins. As shown in Table 9, dynamic virtual margins outperform fixed virtual margins on all datasets. This is due to the dynamic margin's ability to adaptively adjust the margin size based on domain and class difficulty, achieving optimal performance.

Table 8. Performance results (error rate %) with different values of m on CIFAR10C, CIFAR100C, and ImageNetC datasets.

	m = 1	m = 2	m = 3	m = 4	m = 5	m = 6	m = 7	m = 8
CIFAR10C	14.8	14.5	14.7	14.9	15.3	15.5	16.1	16.8
CIFAR100C	27.9	27.8	28.2	28.3	28.4	28.7	29.5	30.1
ImageNetC	58.1	57.9	58.1	58.3	58.6	59.1	59.7	60.9

Table 9. Performance results (error rate %) with different values of fixed m and dynamic m on CIFAR10C, CIFAR100C, and ImageNetC datasets.

	m = 1	m = 2	m = 3	m = 4	m = 5	m = 6	m = 7	m = 8	m = Dynamic
CIFAR10C	15.3	15.2	14.9	15.5	15.9	16.3	16.7	17.3	14.5
CIFAR100C	28.2	28.1	28.5	28.9	29.3	29.7	30.2	31.1	27.8
ImageNetC	58.7	58.2	58.5	58.9	59.3	59.9	60.3	61.2	57.9

4.6. Analysis of Virtual Decision Boundary

We evaluate the effectiveness of our virtual decision boundary (VDB) method in mitigating feature crossing the decision boundary by analyzing inter-class and intra-class distances. Our VDB method is proposed to alleviate the issue of feature clustering at or slightly crossing the decision boundary. First, we utilized T-SNE to visualize the sample features generated by three different methods (DSS [8], CoTTA [5], RMT [6]) and our method in the Gaussian domain of the CIFAR10C dataset. As shown in Figure 8, the results demonstrate that the features generated by our method exhibit better clustering performance, with smaller intra-class distances and larger inter-class distances. Second, if class feature shifts are well controlled, the inter-class distance should increase, and the intra-class distance should decrease. This will effectively prevent features from crossing the decision boundary. The intra-class distance is expressed as $d_{intra} = \sum_i^C ||z_i - \overline{z_i}||_2$, and the inter-class distance is expressed as $d_{inter} = \sum_i^C \sum_{j \neq i}^C ||z_i - \overline{z_j}||_2$, where $\overline{z}$ represents the mean feature of the class. As shown in Figures 9 and 10, we compare our approach with three other methods include DSS [8], CoTTA [5], and RMT [44]. The results demonstrate that our method excels in managing class shifts during CTTA, consistently achieving more desirable intra-class and inter-class distances across all domains. In particular, to evaluate the effectiveness of our virtual decision boundary in mitigating feature crossing the decision boundary during the early stages of domain adaptation, we counted the number of features crossing the decision boundary for each method at the early stage of domain adaptation. As shown in Figure 11, our method consistently outperforms others across all domains.

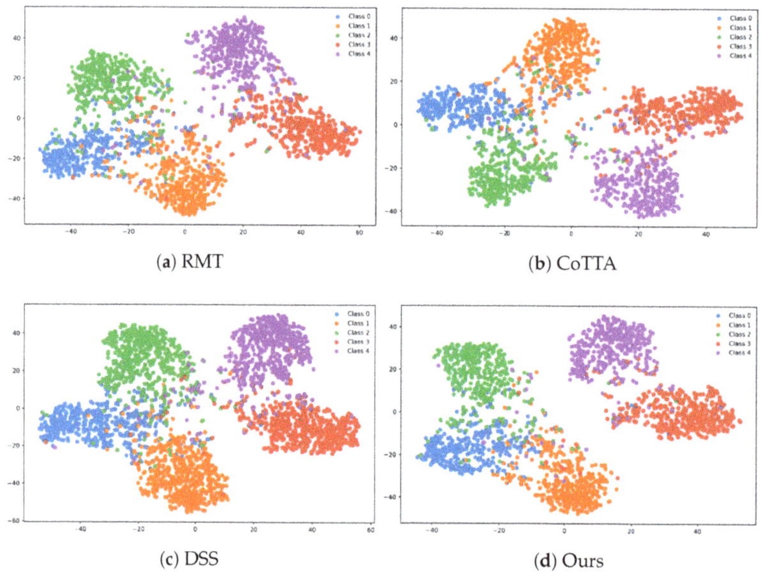

(a) RMT (b) CoTTA (c) DSS (d) Ours

Figure 8. Visualization on the Gaussian domain of the CIFAR10C dataset.

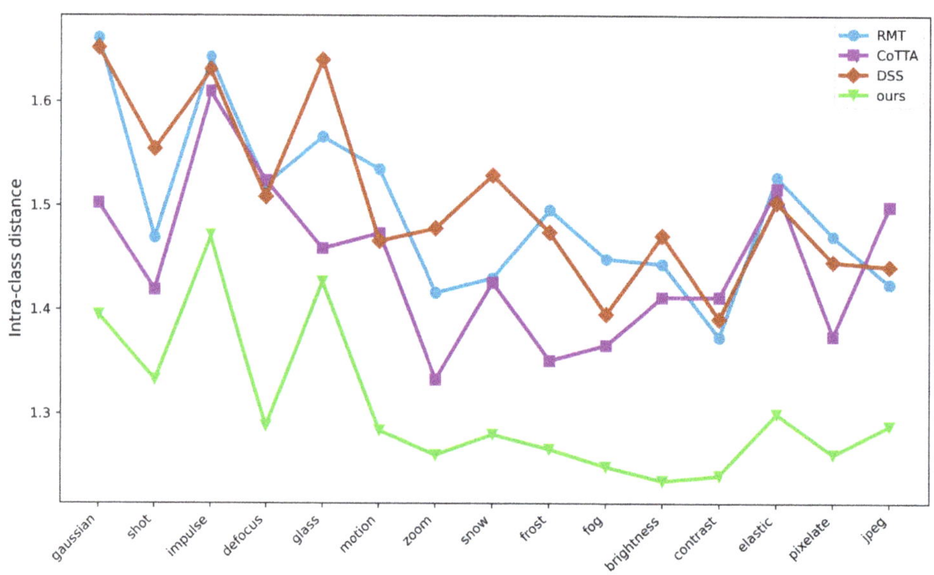

Figure 9. Intra-class distance in all domains.

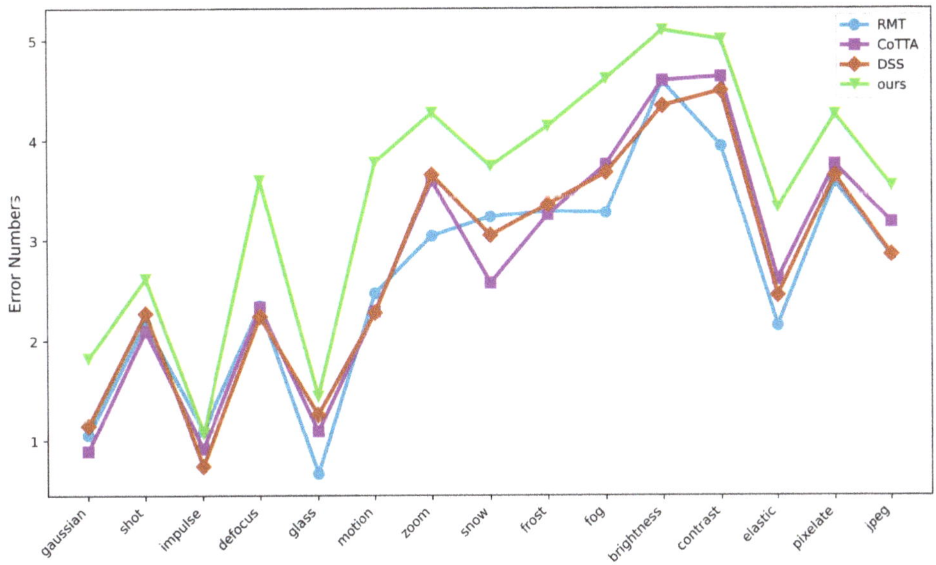

Figure 10. Inter-class distance in all domains.

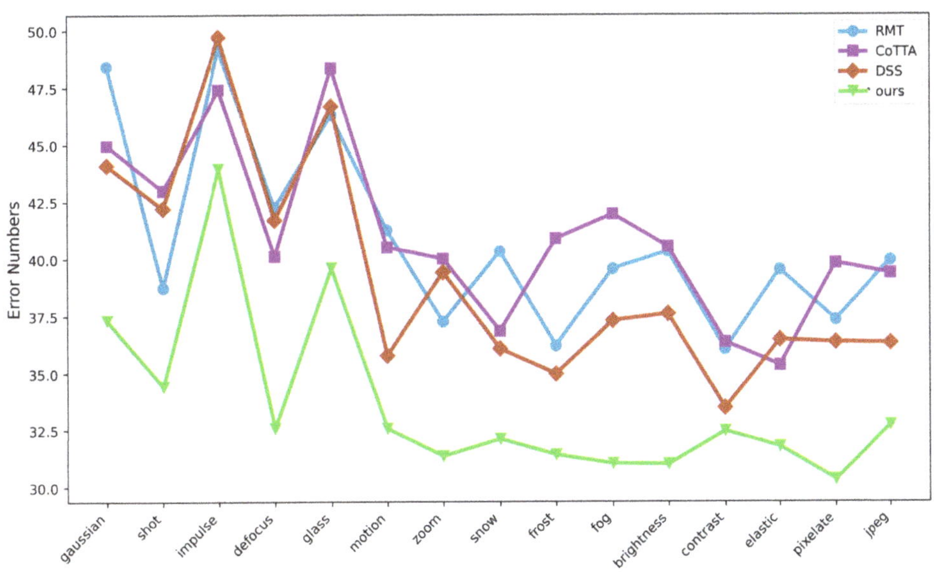

Figure 11. Error numbers at the domain begin.

4.7. Time and Parameter Complexity Evaluation

In this section, we focus on analyzing the time and space complexity of our method compared with other methods. The datasets used include CIFAR10C, CIFAR100C, and ImageNetC, and the methods for comparison include TENT [12], CoTTA [5], DSS [8], and RMT [6]. As shown in Table 10, the time required by our method is only 3 min, 4 min, and 13 min more than the most time-consuming DSS [8] method on the CIFAR10C, CIFAR100C, and ImageNetC datasets, respectively. The reason for the additional time consumption is that our method requires the computation of the channel feature map from the previous model (without computing gradients) and the calculation of the sensitivity loss function. As shown in Figure 12, compared with CoTTA [5], DSS [8], and RMT [6], our method requires more GPU memory at runtime, approximately 0.8GB, due to the additional storage of feature maps computed by the previous model. This additional time and space complexity overhead are small and can be largely ignored, which leads to significant performance improvements.

Table 10. Time required for different methods on CIFAR10C, CIFAR100C, and ImageNetC datasets.

	TENT	CoTTA	DSS	RMT	Ours
CIFAR10C	7 min	15 min	16 min	15 min	19 min
CIFAR100C	9 min	17 min	19 min	18 min	23 min
ImageNetC	33 min	71 min	73 min	72 min	86 min

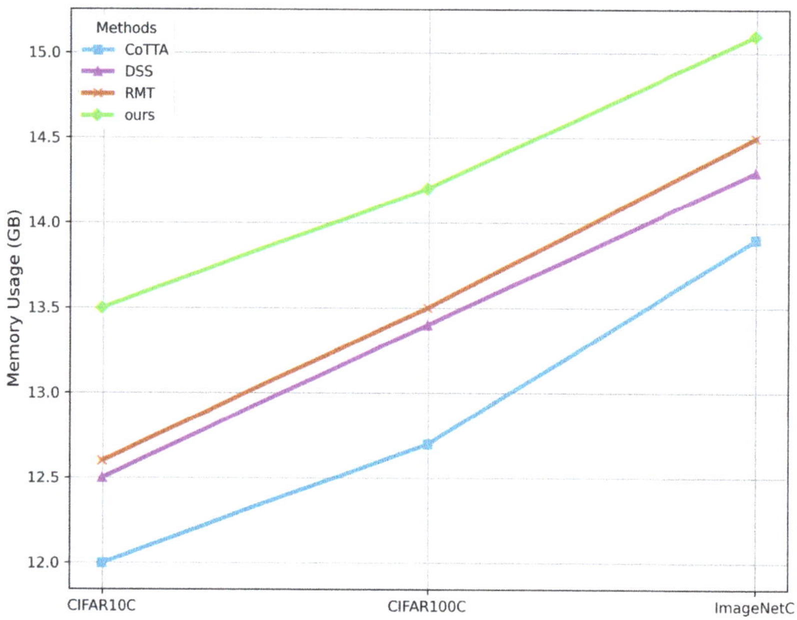

Figure 12. Memory usage for different methods on various datasets.

5. Conclusions

Distinguishing itself from previous research that tends to concentrate on isolated domain adaptation and seeks local optima. Instead, this paper focuses on the often overlooked sustained capability in Continual Test-Time Adaptation. To address this issue, we propose a Dual Constraint method that combines parameter constraints based on channel representations and virtual decision boundary constraints based on features. The parameter constraint penalizes the model's reliance on domain-specific knowledge, forcing it to learn domain-invariant knowledge while alleviating channel instability and mutual interference caused by overfitting to domain-sensitive parameters. This approach also theoretically enhances the model's generalization ability. Additionally, we provide theoretical evidence that our method can effectively enhance the model's generalization ability and promote the learning of domain-invariant knowledge. Meanwhile, the virtual decision boundary constraint pushes features away from the original decision boundary, forming a virtual margin that buffers against domain shifts, effectively reducing the mutual interference between domain-specific knowledge. Our extensive experiments on multiple CTTA benchmark datasets have demonstrated the effectiveness of our proposed methods.

Author Contributions: Conceptualization, P.L. and Y.W.; Methodology, Y.S. and P.L.; Software, P.L.; Investigation, Y.W.; Writing—review & editing, Y.S. and P.L.; Visualization, Y.W.; Supervision, Y.S.; Funding acquisition, Y.S. All authors have read and agreed to the published version of the manuscript.

Funding: This work was supported by the National Natural Science Foundation of China under the Grant No. 62002330.

Data Availability Statement: The original contributions presented in this study are included in the article. Further inquiries can be directed to the corresponding author.

Conflicts of Interest: The authors declare no conflicts of interest.

References

1. Hu, C.; Hudson, S.; Ethier, M.; Al-Sharman, M.; Rayside, D.; Melek, W. Sim-to-real domain adaptation for lane detection and classification in autonomous driving. In Proceedings of the 2022 IEEE Intelligent Vehicles Symposium (IV), Aachen, Germany, 4–9 June 2022; IEEE: New York, NY, USA, 2022; pp. 457–463.
2. Shi, Z.; Su, T.; Liu, P.; Wu, Y.; Zhang, L.; Wang, M. Learning Frequency-Aware Dynamic Transformers for All-In-One Image Restoration. *arXiv* **2024**, arXiv:2407.01636. [CrossRef]
3. Chen, W.; Miao, L.; Gui, J.; Wang, Y.; Li, Y. FLsM: Fuzzy Localization of Image Scenes Based on Large Models. *Electronics* **2024**, *13*, 2106. [CrossRef]
4. Xu, E.; Zhu, J.; Zhang, L.; Wang, Y.; Lin, W. Research on Aspect-Level Sentiment Analysis Based on Adversarial Training and Dependency Parsing. *Electronics* **2024**, *13*, 1993. [CrossRef]
5. Wang, Q.; Fink, O.; Van Gool, L.; Dai, D. Continual test-time domain adaptation. In Proceedings of the IEEE/CVF Conference on Computer Vision and Pattern Recognition, New Orleans, LA, USA, 18–24 June 2022; pp. 7201–7211.
6. Döbler, M.; Marsden, R.A.; Yang, B. Robust mean teacher for continual and gradual test-time adaptation. In Proceedings of the IEEE/CVF Conference on Computer Vision and Pattern Recognition, Vancouver, BC, Canada, 17–24 June 2023; pp. 7704–7714.
7. Brahma, D.; Rai, P. A probabilistic framework for lifelong test-time adaptation. In Proceedings of the IEEE/CVF Conference on Computer Vision and Pattern Recognition, Vancouver, BC, Canada, 17–24 June 2023; pp. 3582–3591.
8. Wang, Y.; Hong, J.; Cheraghian, A.; Rahman, S.; Ahmedt-Aristizabal, D.; Petersson, L.; Harandi, M. Continual test-time domain adaptation via dynamic sample selection. In Proceedings of the IEEE/CVF Winter Conference on Applications of Computer Vision, Waikoloa, HI, USA, 3–8 January 2024; pp. 1701–1710.
9. Liu, J.; Yang, S.; Jia, P.; Zhang, R.; Lu, M.; Guo, Y.; Xue, W.; Zhang, S. Vida: Homeostatic visual domain adapter for continual test time adaptation. *arXiv* **2023**, arXiv:2306.04344.
10. Zhu, Z.; Hong, X.; Ma, Z.; Zhuang, W.; Ma, Y.; Dai, Y.; Wang, Y. Reshaping the Online Data Buffering and Organizing Mechanism for Continual Test-Time Adaptation. In Proceedings of the European Conference on Computer Vision, Dublin, Ireland, 22–23 October 2025; Springer: Berlin/Heidelberg, Germany, 2025; pp. 415–433.
11. Liu, J.; Xu, R.; Yang, S.; Zhang, R.; Zhang, Q.; Chen, Z.; Guo, Y.; Zhang, S. Continual-MAE: Adaptive Distribution Masked Autoencoders for Continual Test-Time Adaptation. In Proceedings of the IEEE/CVF Conference on Computer Vision and Pattern Recognition, Seattle, WA, USA, 16–22 June 2024; pp. 28653–28663.
12. Wang, D.; Shelhamer, E.; Liu, S.; Olshausen, B.; Darrell, T. Tent: Fully test-time adaptation by entropy minimization. *arXiv* **2020**, arXiv:2006.10726.
13. DeVore, R.A.; Temlyakov, V.N. Some remarks on greedy algorithms. *Adv. Comput. Math.* **1996**, *5*, 173–187. [CrossRef]
14. Knowles, J.D.; Watson, R.A.; Corne, D.W. Reducing local optima in single-objective problems by multi-objectivization. In Proceedings of the International Conference on Evolutionary Multi-Criterion Optimization, Zurich, Switzerland, 7–9 March 2001; Springer: Berlin/Heidelberg, Germany, 2001; pp. 269–283.
15. Wang, D.Z.; Lo, H.K. Global optimum of the linearized network design problem with equilibrium flows. *Transp. Res. Part B Methodol.* **2010**, *44*, 482–492. [CrossRef]
16. Zhou, W.; Zhou, Z. Unsupervised Domain Adaption Harnessing Vision-Language Pre-Training. *IEEE Trans. Circuits Syst. Video Technol.* **2024**, *34*, 8201–8214. [CrossRef]
17. Wilson, G.; Cook, D.J. A survey of unsupervised deep domain adaptation. *ACM Trans. Intell. Syst. Technol. (Tist)* **2020**, *11*, 1–46. [CrossRef]

18. Long, M.; Cao, Y.; Wang, J.; Jordan, M. Learning transferable features with deep adaptation networks. In Proceedings of the International Conference on Machine Learning, Lille, France, 6–11 July 2015; PMLR: San Diego, CA, USA, 2015; pp. 97–105.
19. Long, M.; Zhu, H.; Wang, J.; Jordan, M.I. Unsupervised domain adaptation with residual transfer networks. *Adv. Neural Inf. Process. Syst.* **2016**, *29*.
20. Long, M.; Zhu, H.; Wang, J.; Jordan, M.I. Deep transfer learning with joint adaptation networks. In Proceedings of the International Conference on Machine Learning, Sydney, Australia, 6–11 August 2017; PMLR: San Diego, CA, USA, 2017; pp. 2208–2217.
21. Fan, X.; Wang, Q.; Ke, J.; Yang, F.; Gong, B.; Zhou, M. Adversarially adaptive normalization for single domain generalization. In Proceedings of the IEEE/CVF Conference on Computer Vision and Pattern Recognition, Nashville, TN, USA, 20–25 June 2021; pp. 8208–8217.
22. Yang, F.E.; Cheng, Y.C.; Shiau, Z.Y.; Wang, Y.C.F. Adversarial teacher-student representation learning for domain generalization. *Adv. Neural Inf. Process. Syst.* **2021**, *34*, 19448–19460.
23. Zhu, W.; Lu, L.; Xiao, J.; Han, M.; Luo, J.; Harrison, A.P. Localized adversarial domain generalization. In Proceedings of the IEEE/CVF Conference on Computer Vision and Pattern Recognition, New Orleans, LA, USA, 18–24 June 2022; pp. 7108–7118.
24. Liang, J.; Hu, D.; Feng, J. Do we really need to access the source data? source hypothesis transfer for unsupervised domain adaptation. In Proceedings of the International Conference on Machine Learning, Virtual, 13–18 July 2020; PMLR: San Diego, CA, USA, 2020; pp. 6028–6039.
25. Lowd, D.; Meek, C. Adversarial learning. In Proceedings of the Eleventh ACM SIGKDD International Conference on Knowledge Discovery in Data Mining, Chicago, IL, USA, 21–24 August 2005; pp. 641–647.
26. Gopnik, A.; Glymour, C.; Sobel, D.M.; Schulz, L.E.; Kushnir, T.; Danks, D. A theory of causal learning in children: Causal maps and Bayes nets. *Psychol. Rev.* **2004**, *111*, 3. [CrossRef] [PubMed]
27. Hospedales, T.; Antoniou, A.; Micaelli, P.; Storkey, A. Meta-learning in neural networks: A survey. *IEEE Trans. Pattern Anal. Mach. Intell.* **2021**, *44*, 5149–5169. [CrossRef] [PubMed]
28. Cubuk, E.D.; Zoph, B.; Shlens, J.; Le, Q.V. Randaugment: Practical automated data augmentation with a reduced search space. In Proceedings of the IEEE/CVF conference on Computer Vision and Pattern Recognition Workshops, Seattle, WA, USA, 14–19 June 2020; pp. 702–703.
29. Zhou, K.; Yang, Y.; Hospedales, T.; Xiang, T. Deep domain-adversarial image generation for domain generalisation. In Proceedings of the AAAI Conference on Artificial Intelligence, New York, NY, USA, 7–12 February 2020; Volume 34, pp. 13025–13032.
30. Gan, Y.; Bai, Y.; Lou, Y.; Ma, X.; Zhang, R.; Shi, N.; Luo, L. Decorate the newcomers: Visual domain prompt for continual test time adaptation. In Proceedings of the AAAI Conference on Artificial Intelligence, Washington, DC, USA, 7–14 February 2023; Volume 37, pp. 7595–7603.
31. Wang, H.; Ge, S.; Lipton, Z.; Xing, E.P. Learning robust global representations by penalizing local predictive power. *Adv. Neural Inf. Process. Syst.* **2019**, *32*.
32. Guo, J.; Qi, L.; Shi, Y. Domaindrop: Suppressing domain-sensitive channels for domain generalization. In Proceedings of the IEEE/CVF International Conference on Computer Vision, Paris, France, 2–3 October 2023; pp. 19114–19124.
33. Zhao, Y.; Zhang, H.; Hu, X. Penalizing gradient norm for efficiently improving generalization in deep learning. In Proceedings of the International Conference on Machine Learning, Baltimore, MD, USA, 17–23 July 2022; PMLR: San Diego, CA, USA, 2022; pp. 26982–26992.
34. Cong, W.; Cong, Y.; Sun, G.; Liu, Y.; Dong, J. Self-Paced Weight Consolidation for Continual Learning. *IEEE Trans. Circuits Syst. Video Technol.* **2024**, *34*, 2209–2222. [CrossRef]
35. Zhang, W.; Huang, Y.; Zhang, W.; Zhang, T.; Lao, Q.; Yu, Y.; Zheng, W.S.; Wang, R. Continual Learning of Image Classes With Language Guidance From a Vision-Language Model. *IEEE Trans. Circuits Syst. Video Technol.* **2024**, *34*, 13152–13163. [CrossRef]
36. Yu, D.; Zhang, M.; Li, M.; Zha, F.; Zhang, J.; Sun, L.; Huang, K. Contrastive Correlation Preserving Replay for Online Continual Learning. *IEEE Trans. Circuits Syst. Video Technol.* **2024**, *34*, 124–139. [CrossRef]
37. Li, H.; Liao, L.; Chen, C.; Fan, X.; Zuo, W.; Lin, W. Continual Learning of No-Reference Image Quality Assessment With Channel Modulation Kernel. *IEEE Trans. Circuits Syst. Video Technol.* **2024**, *34*, 13029–13043. [CrossRef]
38. Li, K.; Chen, H.; Wan, J.; Yu, S. ESDB: Expand the Shrinking Decision Boundary via One-to-Many Information Matching for Continual Learning With Small Memory. *IEEE Trans. Circuits Syst. Video Technol.* **2024**, *34*, 7328–7343. [CrossRef]
39. Shi, Z.; Liu, P.; Su, T.; Wu, Y.; Liu, K.; Song, Y.; Wang, M. Densely Distilling Cumulative Knowledge for Continual Learning. *arXiv* **2024**, arXiv:2405.09820. [CrossRef]
40. Wu, Y.; Chi, Z.; Wang, Y.; Plataniotis, K.N.; Feng, S. Test-time domain adaptation by learning domain-aware batch normalization. In Proceedings of the AAAI Conference on Artificial Intelligence, Philadelphia, PA, USA, 25 February–4 March; Volume 38, pp. 15961–15969.
41. Zhang, J.; Qi, L.; Shi, Y.; Gao, Y. Domainadaptor: A novel approach to test-time adaptation. In Proceedings of the IEEE/CVF International Conference on Computer Vision, Paris, France, 2–3 October 2023; pp. 18971–18981.

42. Chen, K.; Gong, T.; Zhang, L. Camera-Aware Recurrent Learning and Earth Mover's Test-Time Adaption for Generalizable Person Re-Identification. *IEEE Trans. Circuits Syst. Video Technol.* **2024**, *34*, 357–370. [CrossRef]
43. Wang, Y.; Chen, H.; Heng, Q.; Hou, W.; Fan, Y.; Wu, Z.; Wang, J.; Savvides, M.; Shinozaki, T.; Raj, B.; et al. Freematch: Self-adaptive thresholding for semi-supervised learning. *arXiv* **2022**, arXiv:2205.07246.
44. Yuan, L.; Xie, B.; Li, S. Robust test-time adaptation in dynamic scenarios. In Proceedings of the IEEE/CVF Conference on Computer Vision and Pattern Recognition, Vancouver, BC, Canada, 17–24 June 2023; pp. 15922–15932.
45. Shi, B.; Zhang, D.; Dai, Q.; Zhu, Z.; Mu, Y.; Wang, J. Informative dropout for robust representation learning: A shape-bias perspective. In Proceedings of the International Conference on Machine Learning, Virtual, 13–18 July 2020; PMLR: San Diego, CA, USA, 2020; pp. 8828–8839.
46. Klinker, F. Exponential moving average versus moving exponential average. *Math. Semesterber.* **2011**, *58*, 97–107.
47. Dinh, L.; Pascanu, R.; Bengio, S.; Bengio, Y. Sharp minima can generalize for deep nets. In Proceedings of the International Conference on Machine Learning, Sydney, Australia, 6–11 August 2017; PMLR: San Diego, CA, USA, 2017; pp. 1019–1028.
48. Yang, S.; Wu, J.; Liu, J.; Li, X.; Zhang, Q.; Pan, M.; Zhang, S. Exploring sparse visual prompt for cross-domain semantic segmentation. *arXiv* **2023**, arXiv:2303.09792.
49. Sakaridis, C.; Dai, D.; Van Gool, L. ACDC: The adverse conditions dataset with correspondences for semantic driving scene understanding. In Proceedings of the IEEE/CVF International Conference on Computer Vision, Montreal, QC, Canada, 10–17 October 2021; pp. 10765–10775.
50. Li, Y.; Wang, N.; Shi, J.; Liu, J.; Hou, X. Revisiting batch normalization for practical domain adaptation. *arXiv* **2016**, arXiv:1603.04779. [CrossRef]
51. Chakrabarty, G.; Sreenivas, M.; Biswas, S. Sata: Source anchoring and target alignment network for continual test time adaptation. *arXiv* **2023**, arXiv:2304.10113. [CrossRef]
52. Zagoruyko, S.; Komodakis, N. Wide residual networks. *arXiv* **2016**, arXiv:1605.07146. [CrossRef]
53. Xie, S.; Girshick, R.; Dollár, P.; Tu, Z.; He, K. Aggregated residual transformations for deep neural networks. In Proceedings of the IEEE Conference on Computer Vision and Pattern Recognition, Honolulu, HI, USA, 21–26 July 2017; pp. 1492–1500.
54. He, K.; Zhang, X.; Ren, S.; Sun, J. Deep residual learning for image recognition. In Proceedings of the IEEE Conference on Computer Vision and Pattern Recognition, Las Vegas, NV, USA, 26 June–1 July 2016; pp. 770–778.

electronics

MDPI

Article

Joint Event Detection with Dynamic Adaptation and Semantic Relevance

Xi Zeng [1,2,*], Guangchun Luo [1] and Ke Qin [1]

1 School of Computer Science and Engineering, University of Electronic Science and Technology of China, Chengdu 611731, China; gcluo@uestc.edu.cn (G.L.); qinke@uestc.edu.cn (K.Q.)
2 The 30th Research Institute of China Electronics Technology Group Corporation, Chengdu 610000, China
* Correspondence: 202112090811@std.uestc.edu.cn

Abstract: Event detection is a crucial task in natural language processing, and it plays a significant role in numerous applications, such as information retrieval, question answering, and situational awareness. Real-world tasks typically require robust models that can dynamically adapt to changing data distributions and seamlessly accommodate emerging event types while maintaining high accuracy and efficiency. However, existing methods often face catastrophic forgetting, a significant challenge where models lose previously acquired knowledge when learning new information. This phenomenon hinders models from balancing performance with adaptability, limiting their ability to generalize across dynamic data landscapes. This paper proposes a novel event detection framework, DASR, which aims to enhance the flexibility and diversity of event detection through joint learning and guidance that dynamically adapts to new events and extracts semantic relevance. Firstly, we utilize pre-trained language models (PLMs) trained on general corpora to obtain existing event and type information as global knowledge. Secondly, during prompt fine-tuning for specific tasks, we incorporate an incremental learning module to design incremental prompt templates for newly introduced event types and read out their representations within the PLM. Subsequently, we perform entity recognition and event trigger word detection to extract semantic relevance. In this case, a graph attention mechanism is introduced to enhance the long-distance dependencies within the text (modeled as message passing in the type graph). Additionally, feature fusion integrates entity and event trigger word information into a unified representation. Finally, we validate the effectiveness of the proposed framework through extensive experiments. The experimental results demonstrate that the proposed framework effectively mitigates catastrophic forgetting and significantly improves the accuracy and adaptability of event detection when dealing with evolving data distributions and newly introduced event types.

Keywords: event detection; incremental learning; pre-trained language model; graph neural networks

1. Introduction

In recent years, event detection and extraction [1] have garnered significant attention within the natural language processing (NLP) community [2,3]. These tasks are pivotal for understanding textual information, as they enable the identification and analysis of events from textual data. This, in turn, facilitates numerous downstream applications, such as information retrieval [4], question answering [5], and real-time decision-making systems [6]. Existing works have made considerable progress in this domain, focusing on enhancing model accuracy through advanced techniques such as deep learning [7], attention mechanisms [8], and multi-task learning [9]. Despite these advancements, the dynamic nature of real-world applications presents a common challenge, i.e., the continual emergence and evolution of new events [10–13]. The constantly evolving data distribution and the emergence of novel event types pose significant challenges in maintaining detection accuracy and adaptability [14,15]. These challenges often lead to catastrophic forgetting [16,17], where models fail to retain previously learned knowledge effectively when acquiring new information. Continuous innovation is essential for building resilient event detection systems in dynamic environments.

Incremental learning (IL) [18–22] has been widely employed to address issues related to evolving and expanding datasets. This approach allows models to update their knowledge base without retraining from scratch, preserving old knowledge while learning new information. This characteristic is particularly beneficial for tasks requiring continual adaptation to new data [23]. However, employing IL for event detection [24–26] raises unique and multifaceted challenges. For instance, a key challenge lies in retaining the capability to accurately recognize previously known events with high fidelity, while simultaneously identifying and integrating newly emerging events. Furthermore, this process must skillfully manage the intricate temporal and causal relationships between events, which inevitably increases the complexity of detection methods. Pre-trained language models (PLMs) [27–29] trained on diverse language tasks provide an effective starting point. This extensive training endows them with a profound understanding of language, allowing for fine-tuning on specific tasks such as event detection. The inherent versatility and robust linguistic comprehension of PLMs make them particularly well-suited for IL scenarios, emphasizing the critical need to adapt dynamically without catastrophic forgetting. As such, leveraging PLMs in the realm of IL for event detection holds significant promise, providing a foundation upon which more sophisticated and adaptive models can be developed to tackle the ever-evolving landscape of textual information.

In addition to providing the necessary context information, PLMs need to explore deeper event information to achieve target detection as the diversity of events increases. Existing event detection research focuses on treating multiple event instances as independent data instances and identifying them one by one [11,30–33]. However, these approaches overlook the interactive correlations among event instances within a sentence. This concept, inspired by message aggregation in graph neural networks (GNNs) [34–39], can empirically offer further semantic representation guidance for event detection. GNNs are models designed to process data structured as graphs, enabling the capture of relationships and dependencies between nodes. As shown in Figure 1, compared to traditional event detection models, our model enhances adaptability to new events and extracts useful information from complex correlations through prompt-based incremental fine-tuning and event correlation modeling, rather than being limited to inherent knowledge. In summary, **integrating dynamic event adaptation and diverse semantic relevance aims to address dual challenges: maintaining accuracy within evolving data distributions and capturing the relationships between new event types and existing events.**

To address these challenges, we propose a novel joint event detection method with dynamic adaptation and semantic relevance, named **DASR**, which consists of several key components that are meticulously designed to tackle the unique problems inherent in event detection. First, during the pretraining phase, we utilize PLMs to establish a preliminary and comprehensive understanding of language from general corpora, laying a solid foundation for subsequent tasks. During the fine-tuning phase for specific tasks, we introduce an innovative incremental learning module and design prompt templates to effectively model events. Additionally, to simultaneously handle entity recognition and event trigger word detection, we model the graph structure of event types based on word importance and event-type relevance. We employ an attention mechanism—a model component that assigns varying importance to different parts of the input text—to capture dependencies, facilitating a deeper understanding of the context. We validate this framework through extensive experiments, demonstrating the performance of DASR in mitigating catastrophic forgetting and enhancing adaptability to new event types. DASR's key innovation lies in its combination of prompt-based incremental learning and semantic relevance modeling using graph neural networks. Unlike traditional event detection systems, which often struggle to adapt to evolving event types, DASR's approach allows for continuous adaptation to new events, significantly reducing the risk of catastrophic forgetting. Furthermore, the use of GNNs enables DASR to capture complex event relationships that are typically overlooked by conventional methods, leading to more accurate and robust event detection in dynamic environments.

- We propose a novel framework that seamlessly integrates prompt-based incremental learning and semantic relevance in event detection, thereby enhancing the accuracy and adaptability of event detection.
- Through prompt engineering, type relevance modeling, and disentangled prediction of type interactions, we capture dynamic event relationships, quantify event relevance to provide interpretability for predictions, and mitigate the challenges of entity recognition with long-distance dependencies.
- Extensive experiments show that DASR improves event detection performance, mitigates catastrophic forgetting, and adapts to evolving data distributions.

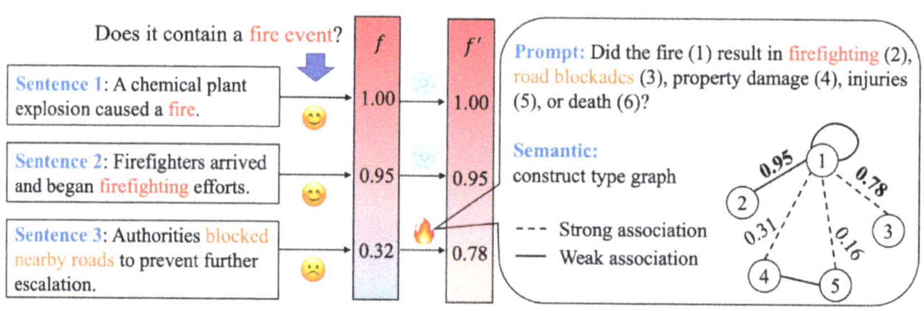

Figure 1. A toy example of event detection. Traditional model f struggles to capture new event correlations and makes erroneous judgments when dealing with temporal context. The improved model f' enhances detection accuracy through incremental prompting and semantic relevance.

The organization of this paper is organized as follows: We describe the overall proposed method in detail and explain the key components and techniques employed in our approach in Section 2. We present and analyze the experimental results, demonstrating the effectiveness of our method through various evaluations and comparisons in Section 3. Finally, we summarize the key findings and contributions of our work and discuss potential future research directions in Section 4.

2. Materials and Methods

2.1. Preliminaries

This section outlines the key concepts and methodologies underpinning DASA, which integrates state-of-the-art techniques in event detection and incremental learning. It incorporates LLMs for their rich contextual representations, memory-augmented networks for knowledge retention, and prompt learning for rapid adaptation to new event types.

Event detection. Event detection is a critical task within NLP that is focused on identifying events and their attributes in textual data. Traditional approaches have relied on static models that struggle to adapt to evolving data distributions and new event types. We aim to capture better syntactic and semantic features for more accurate event detection. Given a sequence of textual input, the task is to identify and classify event triggers and their associated arguments. Formally, let $\mathcal{T} = \{t_1, t_2, \ldots, t_n\}$ represent the event types, and $X = \{\mathbf{x}_1, \mathbf{x}_2, \ldots, \mathbf{x}_m\}$ denote several candidate triggers. The objective is to predict a set of events $E = \{e_1, e_2, \ldots, e_n\}$, where each event e_i consists of a true event trigger $\mathbf{x}_i$ and event type t.

Incremental learning. The incremental learning component seeks to adapt the event detection model to new data without requiring retraining from scratch, all while preserving previously learned knowledge. We aim to develop a robust framework for event detection that leverages incremental learning to maintain performance in dynamic environments. For each new dataset, $D' = (X', Y')$, the model should update its parameters to minimize the loss on D', denoted as $\mathcal{L}(X', Y'; \theta)$, while minimizing the forgetting factor on previous datasets D. In addition, the IL aims to update the model $f_\theta : ([\mathbf{x}; \mathbf{x}'] \rightarrow [\mathbf{y}; \mathbf{y}'] \in Y)$ from the sequence of events.

Prompt engineering. This is also known as contextual prompting, a method that guides LLMs to locate knowledge relevant to specific targets without updating the model's weights or parameters. Formally, the prompt function $x' = g(x)$ is applied to the input x to obtain the corresponding prompt x', which includes x, the intermediate answer $\mathbf{z}$, and discrete or continuous task-specific tokens as task descriptors. Given a prompt, a PLM model can generate the answer $\mathbf{z}$. Prompts with random or true values for $\mathbf{z}$ are referred to as filled prompts and answered prompts, respectively.

2.2. Methodology

This section introduces the event detection framework with joint dynamic adaptation and semantic relevance (DASR), which includes the following: (1) Prompt-based incremental learning: We leverage prompts to fine-tune pre-trained language models, generating incremental representations for newly introduced event types. (2) Trigger relevance estimation: We assess the semantic relevance between trigger words and event types to determine their importance. (3) Instance-guided prediction: We integrate the representations of event types and optimize both the incremental learning loss and classification loss to predict event types accurately. The DASR framework is illustrated in Figure 2.

2.2.1. Prompt-Based Incremental Tuning

Fine-tuning PLMs on new tasks carries the risk of losing information from previously learned data, a problem known as catastrophic forgetting. Class-incremental learning requires the model to retain knowledge of previously learned classes when new classes are added. We aim to use a more convenient prompt-based [40,41] fine-tuning paradigm to jointly assist in the dynamic updating of PLMs. First, the original definition of the prompt template is as follows:

$$f_p^E(x) = (x, E, y), \tag{1}$$

where x denotes the natural language describing the event, E denotes the event-related prompt word, and y denotes the event-type label. When we introduce incremental words IE (i.e., prompt words for new event types) in the prompt design, and aim to obtain prompts for all event types seen so far, the predefined prompt template can be further written as follows:

$$f_p^{IE}(x) = (x, IE, y), \tag{2}$$

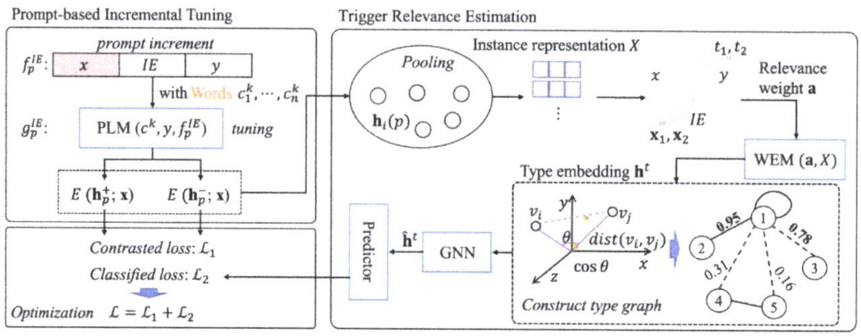

Figure 2. The framework of DASR. **Step 1**: New event types are incorporated into prompts and mapped to event representations jointly with PLM in contrastive training. **Step 2**: DASR captures the correlations of triggers in non-Euclidean space by modeling an event-type graph. **Step 3**: DASR is jointly optimized for the event detection loss.

For example, the initial template is designed for Sports Events, such as "{event_trigger} took place during the {event_type} at {location} on {date}", where the {event_type} refers to Sports Events, such as basketball games or football matches, and the trigger words are game or match. When a new event type, Political Protest is introduced, the template is automatically adjusted to the following: "{event_trigger} occurred during the {event_type} in {location} on {date}". In this case, the event type is Political Protest and the trigger word is protest.

We then adopt the replay strategy from initial incremental learning, utilizing PLMs to generate high-quality pseudo labels for new class events. For each event e, we obtain the words of the top k term frequency sequence from the corresponding sentence to initialize the prompt sentence. The prompts generated for the label words and the corresponding responses are as follows:

$$g_p^{IE}\left(y, c^k\right) = \left(y, c^k, x, IE, y\right), \tag{3}$$

where $c_i^k \in C$ represents the concept sequence of event e_i, which is used to evaluate the importance of words in the sentence. In short, DASR uses a two-step process to generate pseudo-labels for newly introduced event types. First, the model identifies candidate event triggers in unannotated data. Then, it applies the pre-trained PLM to generate high-confidence pseudo-labels for these events based on the identified triggers and context. These pseudo-labels are then used in incremental fine-tuning, allowing the model to adapt to new event categories without the need for manually labeled data. The process involves selecting the top-k most frequent event triggers and using PLMs to predict the event type associated with these triggers. This allows the model to efficiently adapt to new event types as they emerge.

Subsequently, we uniformly adopt BERT as the textual encoder for both prompt templates f and g. Leveraging the transformer architecture, BERT [42] generates contextual text representations conditioned on the given prompt, thereby preserving rich textual information. To enhance the label localization for novel events during the prompt tuning process, we introduce a contrastive learning loss constraint to clarify decision boundaries. We construct positive and negative sample pairs randomly, aiming for incremental prompt learning to pull similar samples of the new events closer and push dissimilar samples further apart in the feature space (using cosine similarity to measure similarity $\phi(,)$), thereby improving the effectiveness of PLMs in event relevance extraction. The loss function is as follows:

$$\mathcal{L}_1 = -\frac{1}{N}\sum_{i=1}^{N}\log\frac{\exp(\phi(\mathbf{h}_i,\mathbf{h}_i^+)/\tau)}{\sum_{j=1}^{N}\exp(\phi(\mathbf{h}_i,\mathbf{h}_j)/\tau)} \quad (4)$$

where N denotes the number of sample pairs, $\mathbf{h}(f,g)$ denotes the sample representation in the latent space, and τ represents the temperature coefficient. BERT is a representative and widely used model in the current NLP field. We use BiLSTM and other models similar to BERT as benchmarks in our comparison experiments to ensure that the experimental results are representative. More complex pre-trained models (e.g., GPT series) require higher computational resources and longer training times, which are not conducive to quickly verifying the effectiveness of DASR.

2.2.2. Trigger Relevance Estimation

It is widely acknowledged that event types benefit the model in learning the semantic correlations between instances. Therefore, we combine event types with prompt embeddings to model a type-related graph structure. Then, leveraging the message-passing mechanism of GNNs, we capture type correlations to represent each event type as prior knowledge for instance type prediction.

For a sentence with the prompt phrase p and the candidate trigger m (including incorrect candidate instances), we input p into BERT to obtain the hidden representation $\mathbf{h}$ as prompt vectors (this is a simplified description of Section 2.2.1). Then, a pooling operation is applied to read out the representation for each candidate instance as a candidate trigger:

$$\mathbf{x}_i = \sigma(W\text{MAXPOOLING}\mathbf{h}_i(p)) \quad (5)$$

where W denotes the learnable parameter and σ denotes the activation function. In particular, a sentence instance is denoted as $X = [\mathbf{x}_1, \mathbf{x}_2, \ldots, \mathbf{x}_m]$.

Relevance representation. Events are semantically composed of words, estimating the importance of words within events helps us quantify the underlying patterns of events. We introduce integrated gradients [43] to evaluate the importance of input features to the model's predictions. For the input vector $x \in \mathbb{R}^n$ and the model $\mathcal{F}: \mathbb{R}^n \to [0,1]$, the attribution value for the i-th dimension is as follows:

$$\mathbf{a}_i = (\mathbf{x}_i - \mathbf{x}_i') \times \int_{\alpha=0}^{1}\frac{\partial\mathcal{F}(\mathbf{x}' + \alpha\times(\mathbf{x}-\mathbf{x}'))}{\partial\mathbf{x}_i}d\alpha, \quad (6)$$

where $\mathbf{x}'$ is an all-zero vector [44] used to align the input range, and the right term represents the gradient of $\mathcal{F}(\mathbf{x}_i)$ for $\mathbf{x}_i$. We then normalize Equation (6) to represent globally reasonable weights:

$$\mathbf{a}_i = \frac{\exp^{\|\mathbf{a}_i\|_2}}{\sum_{n=1}^{N}\exp^{\|\mathbf{a}_n\|_2}} \quad (7)$$

where $\|\cdot\|$ denotes the L_2 norm. Note that our goal always focuses on uncovering the semantic information provided by event types. Therefore, based on word relevance, for a specific event type, $T \in \mathcal{T}$, the type embedding is represented as follows:

$$\mathbf{h}^t = \frac{1}{|T|}\sum_{i=1}^{m}\mathbf{a}_i^T \times WEM(X) \quad (8)$$

where $|T|$ represents the event number of type. The left term of the multiplication denotes the learnable weights with semantic awareness, while the right term represents the event type with textual description encoded by the word-embedding model $WEM(\cdot)$ [45].

Type graph construction. As the initialization of edges in the type graph, we treat the types as nodes in the graph and calculate the possibility of edges existing between nodes using cosine similarity $\phi(,)$:

$$e_{ij} = \begin{cases} \phi\left(\mathbf{h}_i^t, \mathbf{h}_j^t\right), & \text{if} \phi\left(\mathbf{h}_i^t, \mathbf{h}_j^t\right) > \gamma \\ -\text{inf}, & \text{otherwise} \end{cases} \tag{9}$$

where γ is a hyperparameter threshold that satisfies the γ-neighbor strategy to select the most likely edges with the highest similarity scores, thus controlling the scale of the graph during construction.

Type representation. By modeling the graph structure, we can leverage the powerful relational representation capabilities of GNNs to further refine type representations. Specifically, after normalizing $\mathbf{E} = \{e_{ij} | i, j \in \{1, \ldots, n\}\}$ as an adjacency matrix, we learn node representations using a graph attention network [46]:

$$\mathbf{z}_i^{(l,k)} \leftarrow \sum_{j \in N_i \cup i} (\delta_{ij}^k \mathbf{E}) \mathbf{v}_j^{(l-1)}, \tag{10}$$

where δ_{ij}^k denotes normalized attention coefficients computed by the k-th attention mechanism, and $\mathbf{v}$ denotes randomly initialized embeddings. For the transformation, the k intermediate representations corresponding to K attention mechanisms are concatenated after the aggregation through a non-linear transformation to obtain the representation $\hat{\mathbf{h}}$ at layer l as follows:

$$\hat{\mathbf{h}}_i^{(l,t)} = ||_{k=1}^K \sigma\left(\mathbf{z}_i^{(l,k)} \mathbf{W}^{(k,l)}\right), \tag{11}$$

where $||$ denotes the concatenation operator and $\mathbf{W}$ denotes the learnable weight matrix. Thus, after message passing, we acquire informational knowledge from the neighbors associated with specific types, improving the type embeddings. The attention mechanism quantitatively enhances semantic relevance extraction by assigning higher importance to event types that are more semantically relevant to the current task, improving long-range dependency modeling. The final event representation is a weighted combination of these semantic dependencies, leading to more precise event detection.

2.2.3. Instance-Guided Prediction

In the previous design, we injected contextual instances with types into candidate triggers to obtain instance representations X and type representations $\hat{\mathbf{h}}^t$. Therefore, to disentangle the correlations in instances for downstream prediction tasks, we utilize token reconstruction based on contextual instances to predict the target instance types. This naturally covers the exploration of correlations between prompt sentences and semantic representations of instances.

Instance representation. To achieve robust type correlation extraction, we generate a set of handcrafted type representations $\hat{\mathbf{y}}$. Specifically, we evaluate the correlation by comparing the dot product scores between instance embeddings and type embeddings one by one, and after weighted aggregation, the probabilities of the generated predicted types are determined. This process is formally represented as follows:

$$\hat{\mathbf{y}}_i = \sum_{j=1}^{N+1} \left(\frac{\exp\left(\mathbf{x}_i \cdot \hat{\mathbf{h}}_j^t\right)}{\sum_{k=1}^{N+1} \left(\mathbf{x}_i \cdot \hat{\mathbf{h}}_k^t\right)} \hat{\mathbf{h}}_j^t \right), \tag{12}$$

where $N+1$ represents the number of types, indicating that the reconstructed instance type t_k is introduced. Next, we generate sequential handcrafted type representations to predict the target instance type based on the context. For the target instance x_i, we replace its type representation with $\mathbf{e}_{mask}$, and the corresponding sequential instance type representation is as follows:

$$\hat{\mathbf{Y}}^{(i)} = [\hat{\mathbf{y}}_1, \hat{\mathbf{y}}_2, \ldots, \hat{\mathbf{y}}_{i-1}, \mathbf{e}_{\text{mask}}, \hat{\mathbf{y}}_{i+1}, \ldots, \hat{\mathbf{y}}_m]. \tag{13}$$

To generate more correlated target type representations when fusing instance type representations with instance text representations, we employ a neural network for linear fusion, allowing the candidate instances to be re-represented in a learnable manner:

$$\mathbf{Z}^{(i)} = \sigma\left(\hat{\mathbf{W}}\left[\hat{\mathbf{Y}}^{(i)}; X\right]\right), \tag{14}$$

where $\hat{\mathbf{W}}$ denotes the linear fusion parameter and $[;]$ denotes the element-wise vector concatenation.

Type prediction. To identify the handcrafted event types, t_k, and the real event types, t_j, within the candidate instances, we obtain the posterior probabilities in the final predictor as follows:

$$p(\hat{\mathbf{h}}^t|\mathbf{x}_i) = \text{SOFTMAX}(\hat{\mathbf{h}}^t \cdot \hat{\mathbf{z}}_i). \tag{15}$$

This design provides meaningful guidance for event detection. Specifically, when the target type is t_k, the model predicts a higher $p(\hat{\mathbf{h}}_k^t|\mathbf{x}_i)$ to improve generalization ability when encountering new types. Conversely, when the target type is a known event type, the model no longer distinguishes precisely which specific type it is, instead exhibiting a uniformly lower $p(\hat{\mathbf{h}}_k^t|\mathbf{x}_i)$. Overall, the final prediction result is the type with the highest posterior probability. We define the binary cross-entropy loss for type relevance learning as follows:

$$\begin{aligned} \mathcal{L}_2 = -\sum_{\mathbf{x}_i \in X}\sum_{t_j \in \mathcal{T}} &\mathbf{y}_{ij} \log p\left(\hat{\mathbf{h}}_j^t \mid \mathbf{x}_i\right) \\ &+ (1-\mathbf{y}_{ij})\log\left(1 - p\left(\hat{\mathbf{h}}_j^t \mid \mathbf{x}_i\right)\right) \end{aligned} \tag{16}$$

Our overall model optimization objective is as follows:

$$\mathcal{L} = (1-\lambda)\mathcal{L}_1 + \lambda\mathcal{L}_2 \tag{17}$$

where λ is a balancing coefficient used to weigh the impact of incremental learning and type representation on model performance.

3. Results

This section conducts experiments to evaluate the performance of DASR. The experimental results demonstrate that the proposed framework effectively mitigates catastrophic forgetting and significantly improves the accuracy and adaptability of event detection when dealing with evolving data distributions and newly introduced event types.

3.1. Experimental Setup

Datasets. Our experiment was conducted on the publicly available, large-scale dataset, i.e., the Twitter dataset collected by Twitter API for incremental event detection, and the MAVEN and ACE-2005 datasets for offline event detection to evaluate the effectiveness of DASR.

The Twitter dataset, after cleaning out repeated, still contains 68,841 manually labeled tweets related to 503 event classes distributed over a period of approximately 4 weeks (29 days). To simulate evolving data distributions, the Twitter dataset is divided into sequential message blocks by date. Each block introduces new event types, ensuring that the model is exposed to a continuously changing environment. In our experiments, we split the Twitter dataset by date. Specifically, we use the messages of the first week to form an initial message block, M_0, and the messages of the remaining days to form the following message blocks, $M_1, M_2, \ldots, M_{21}$. Furthermore, MAVEN is a newly constructed large-scale fine-grained corpus, which contains 4480 documents with 168 fine-grained event types. We also use the similar ACE-2005 dataset for evaluation, which contains 599 documents with 33 event types.

Baselines. To demonstrate the significant performance of our model, we select the following baselines. **BERT** [42], which is based on the Transformer architecture and deeply represents natural language through pre-training and finetuning; **BiLSTM** [47], which learns long-term bidirectional semantic dependencies; **EventX** [48], a fine-grained event detection method based on community detection suitable for online scenarios; **EE-GCN** [49], which simultaneously exploits syntactic structure and typed dependency label information to perform ED; **KPGNN** [25], a knowledge-preserving incremental heterogeneous graph neural network model for streaming social event detection.

Settings. We set the number of attention heads for the relation aggregation GNN to 8, the embedding dimension to 64, the total number of layers to 2, the learning rate to 0.001, and the optimizer to Adam. The training spans 100 epochs with a patience of 5 for early stopping. The dimension of the $WEM(\cdot)$ embedding is set to 300, the threshold γ is 0.4, λ is 0.6, and the dropout rate is 0.5. We select the best model for inference. BiLSTM and other GNN-based baselines generally use the same configuration, except that the dimension is set to 32. For EventX, we adopt the hyperparameters as suggested in the original paper [48]. We repeat all experiments five times and report the mean and standard variance of the results.

Evaluation metrics. For incremental event detection, to evaluate the performance of all models, we use adjusted mutual information (AMI) [50] and adjusted rand index (ARI) [50] to measure the similarity between ground-truth clusters and their detected message clusters. AMI measures the mutual information between two clusters and adjusts it based on chance. ARI considers all predicted label pairs and chance and calculates pairs allocated in the same or different clusters. Adaptation and mitigation of the catastrophic forgetting problem are assessed by comparing performance on new event types introduced in incremental learning scenarios while ensuring that the performance on previously learned types remains stable. For offline event detection, we employ a comprehensive set of evaluation metrics including precision (P), which measures the accuracy of positive predictions; recall (R), which evaluates the model's ability to identify all relevant instances; and the F1 score, which provides a balanced measure of both precision and recall. The metrics used follow the setup of most of the existing research.

3.2. Performance Comparison

Incremental evaluation. We evaluate DASR in the incremental event detection scenario. As shown in Table 1, we summarize the performances of several state-of-the-art methods in terms of AMI and ARI metrics. The results demonstrate stable performance as new event types emerge, with a peak AMI of 0.80 and ARI of 0.77 on block M6, where the diversity of event types is the highest. The GNN-based methods have achieved advantages across different message blocks, with DASR showing the most significant improvement. Compared to the best-performing model in the baseline, DASR improved the average AMI by 4.3% and the average ARI by 4.0%, with the best results showing a 27% increase in AMI for M1 and a 20% increase in ARI for M15. This indicates that our design effectively retains the most recent knowledge during incremental training and better mitigates the catastrophic forgetting issue as messages accumulate. In contrast, EE-GCN and KPGNN might sometimes be distracted by outdated information. The generally low performances of BERT and BiLSTM are attributed to their neglect of the semantic relevance of events in non-Euclidean space, while EventX's performance in incremental scenarios is limited due to its sole reliance on community detection. In addition, DRSA performs poorly at M19 due to fewer instances of certain event types in the training data, and the model may not be able to adequately learn the features of these types, resulting in poor performance.

Offline evaluation. To further demonstrate the advantages of DASR, we analyze the results of all methods on two offline datasets across three metrics, as shown in Table 2. We follow the standard data split method used in GNNs [34], randomly selecting 70% for training, 10% for validation, and 20% for testing. DASR achieves the top rankings among all methods, with an average improvement of 2.1% across all metrics, and a maximum improvement of 6.2%. This indicates that our approach maintains excellent performance on offline data as well. The additional prompt design does not cause classification bias, and the deep semantic information provides the model with higher-quality information.

3.3. Further Analysis

In this section,we perform further ablation studies, visualization, and hyperparameter analysis of DASR.

Ablation study. As shown in Figure 3, we conducted ablation experiments on the dynamic adjustment (w/o prompt) and semantic relevance (w/o relevance) components of DASR separately to illustrate the effectiveness of each component. Figure 3a,b present the AMI and ARI performances after removing the prompt increment method from DASR. DASR exhibits relatively stable performance across most blocks, with a significant AMI advantage in the 5th, 10th, and 15th blocks. EventX performs significantly lower, with ARI remaining below 0.2 in almost all blocks. EE-GCN and KPGNN show similar performance, generally matching or slightly below that of DASR. Figure 3c,d show the AMI and ARI performances after removing the semantic relevance in DASR. DASR demonstrated high levels and stability, significantly outperforming competing methods in multiple blocks. Although BERT and BiLSTM did not perform as well as DASR, BERT maintained a secondary optimal performance in several blocks, while BiLSTM showed weaker and more fluctuating performance.

Overall, prompt-based incremental learning successfully conveys semantic information across event types by introducing dynamic prompt templates, while avoiding the destruction of existing knowledge by full model retraining. The ablation results show that the module is indispensable for handling dynamic data distributions and new event types. Graph-based semantic relevance modeling effectively captures the complex dependencies between trigger words and event types through the structure of event-type graphs in non-Euclidean space. This module plays a crucial role in improving semantic representation and enhancing overall performance.

Visualization. We visualize the model's predictive outputs, leveraging t-SNE [51] for dimensionality reduction. Figure 4 illustrates the differentiation of data points in the M_3 block using colors to distinguish between different classes. It can be observed that DASR demonstrates the best performance, with its attention mechanism and incremental learning capabilities contributing to exceptional semantic representation and class distinction, resulting in clear clustering boundaries. EE-GCN and KPGNN follow as the next best performers. These models, through graph networks and their enhancement methods, effectively capture semantic relationships, resulting in compact clustering with clear class boundaries. BERT shows moderate performance. It generates relatively compact clusters through the bi-directional representation of the Transformer architecture but still exhibits overlap between categories. BiLSTM and EventX perform poorly. BiLSTM relies on word order information but is weak in semantic clustering, while EventX has the worst clustering effect. In summary, the DASR method exhibits high clarity and separability, further highlighting the method's classification advantages.

Table 1. Summary of incremental evaluation AMI and ARI on the Twitter dataset. The best results are marked in bold. the mean and standard deviation of results are reported with 5 trials.

Method	BERT		BiLSTM		EventX		EE-GCN		KPGNN		DASR	
Metric	AMI	ARI	AMI	ARI	AMI	ARI	AMI	ARI	AMI	ARI	AMI	ARI
M_1	0.34 ± 0.00	0.03 ± 0.00	0.12 ± 0.00	0.03 ± 0.00	0.06 ± 0.00	0.01 ± 0.00	0.35 ± 0.01	0.05 ± 0.00	0.37 ± 0.00	0.07 ± 0.01	**0.64 ± 0.01**	**0.20 ± 0.00**
M_2	0.76 ± 0.00	0.64 ± 0.00	0.41 ± 0.00	0.49 ± 0.00	0.29 ± 0.02	0.45 ± 0.02	0.77 ± 0.00	0.75 ± 0.00	0.78 ± 0.01	0.76 ± 0.02	**0.80 ± 0.00**	**0.77 ± 0.02**
M_3	0.73 ± 0.00	0.43 ± 0.00	0.31 ± 0.00	0.17 ± 0.00	0.18 ± 0.01	0.09 ± 0.01	0.73 ± 0.00	0.48 ± 0.02	0.74 ± 0.00	0.58 ± 0.01	**0.79 ± 0.00**	**0.59 ± 0.00**
M_4	0.55 ± 0.00	0.19 ± 0.00	0.30 ± 0.00	0.11 ± 0.00	0.19 ± 0.01	0.07 ± 0.01	0.60 ± 0.00	0.23 ± 0.00	0.64 ± 0.01	**0.29 ± 0.01**	**0.68 ± 0.00**	0.26 ± 0.03
M_5	0.71 ± 0.00	0.44 ± 0.00	0.33 ± 0.00	0.19 ± 0.00	0.14 ± 0.00	0.04 ± 0.00	0.71 ± 0.01	0.45 ± 0.00	**0.71 ± 0.01**	0.47 ± 0.03	0.71 ± 0.00	**0.49 ± 0.00**
M_6	0.74 ± 0.00	0.44 ± 0.00	0.36 ± 0.00	0.18 ± 0.00	0.27 ± 00	0.14 ± 0.00	0.76 ± 0.00	**0.74 ± 0.02**	0.79 ± 0.01	0.72 ± 0.03	**0.81 ± 0.02**	0.73 ± 0.01
M_7	0.50 ± 0.00	0.07 ± 0.00	0.20 ± 0.00	0.08 ± 0.00	0.13 ± 0.00	0.02 ± 0.00	0.50 ± 0.02	0.10 ± 0.01	0.51 ± 0.01	0.12 ± 0.00	**0.62 ± 0.00**	**0.18 ± 0.00**
M_8	0.75 ± 0.00	0.50 ± 0.00	0.35 ± 0.00	0.08 ± 0.00	0.21 ± 0.00	0.09 ± 0.00	0.75 ± 0.02	0.58 ± 0.01	**0.76 ± 0.01**	0.60 ± 0.01	0.75 ± 0.01	**0.62 ± 0.00**
M_9	0.66 ± 0.00	0.33 ± 0.00	0.32 ± 0.00	0.27 ± 0.00	0.19 ± 0.00	0.07 ± 0.00	0.68 ± 0.00	0.41 ± 0.03	0.71 ± 0.02	**0.46 ± 0.02**	**0.75 ± 0.01**	0.46 ± 0.01
M_{10}	0.70 ± 0.00	0.44 ± 0.00	0.39 ± 0.00	0.22 ± 0.00	0.24 ± 0.00	0.13 ± 0.00	0.76 ± 0.00	**0.73 ± 0.05**	0.78 ± 0.01	0.70 ± 0.06	**0.80 ± 0.02**	0.73 ± 0.03
M_{11}	0.65 ± 0.00	0.27 ± 0.00	0.37 ± 0.00	0.17 ± 0.00	0.24 ± 0.00	0.16 ± 0.00	0.67 ± 0.02	0.45 ± 0.05	0.71 ± 0.01	0.49 ± 0.03	**0.72 ± 0.01**	**0.51 ± 0.06**
M_{12}	0.56 ± 0.00	0.31 ± 0.00	0.32 ± 0.00	0.13 ± 0.00	0.16 ± 0.00	0.07 ± 0.00	0.62 ± 0.01	0.42 ± 0.03	0.66 ± 0.01	**0.48 ± 0.01**	**0.67 ± 0.00**	0.45 ± 0.00
M_{13}	0.59 ± 0.00	0.14 ± 0.00	0.31 ± 0.00	0.13 ± 0.00	0.16 ± 0.00	0.04 ± 0.00	0.61 ± 0.02	0.25 ± 0.01	0.67 ± 0.01	**0.29 ± 0.03**	**0.69 ± 0.01**	0.23 ± 0.00
M_{14}	0.61 ± 0.00	0.30 ± 0.00	0.34 ± 0.00	0.16 ± 0.00	0.14 ± 00	0.10 ± 0.00	0.62 ± 0.02	0.38 ± 0.01	0.65 ± 0.00	0.42 ± 0.02	**0.72 ± 0.02**	**0.62 ± 0.03**
M_{15}	0.50 ± 0.00	0.10 ± 0.00	0.26 ± 0.00	0.14 ± 0.00	0.07 ± 0.00	0.01 ± 0.00	0.53 ± 0.00	0.14 ± 0.03	0.54 ± 0.00	0.17 ± 0.00	**0.63 ± 0.01**	**0.37 ± 0.02**
M_{16}	0.72 ± 0.00	0.41 ± 0.00	0.41 ± 0.00	0.10 ± 0.00	0.19 ± 0.00	0.08 ± 0.00	0.75 ± 0.00	0.59 ± 0.06	0.77 ± 0.01	0.66 ± 0.05	**0.80 ± 0.02**	**0.78 ± 0.01**
M_{17}	0.60 ± 0.00	0.24 ± 0.00	0.35 ± 0.00	0.17 ± 0.00	0.18 ± 0.00	0.12 ± 0.00	0.67 ± 0.00	0.41 ± 0.00	0.68 ± 0.01	0.43 ± 0.05	**0.72 ± 0.01**	**0.51 ± 0.01**
M_{18}	0.53 ± 0.00	0.24 ± 0.00	0.35 ± 0.00	0.19 ± 0.00	0.16 ± 00	0.08 ± 0.00	0.59 ± 0.02	0.44 ± 0.00	0.66 ± 0.02	0.47 ± 0.04	**0.70 ± 0.01**	**0.49 ± 0.01**
M_{19}	0.63 ± 0.00	0.32 ± 0.00	0.35 ± 0.00	0.16 ± 0.00	0.16 ± 0.00	0.07 ± 0.00	0.67 ± 0.01	0.47 ± 0.01	**0.71 ± 0.01**	**0.51 ± 0.03**	0.68 ± 0.00	0.33 ± 0.00
M_{20}	0.62 ± 0.00	0.33 ± 0.00	0.34 ± 0.00	0.20 ± 0.00	0.18 ± 0.00	0.11 ± 0.00	0.66 ± 0.00	0.49 ± 0.01	0.68 ± 0.02	0.51 ± 0.04	**0.70 ± 0.01**	**0.61± 0.02**
M_{21}	0.57 ± 0.00	0.18 ± 0.00	0.31 ± 0.00	0.16 ± 0.00	0.10 ± 0.00	0.01 ± 0.00	0.57 ± 0.00	0.19 ± 0.01	0.57 ± 0.00	0.20 ± 0.01	**0.61 ±0.00**	**0.38 ± 0.00**

Table 2. Results of event detection on the MAVEN and ACE-2005 datasets. The best results are marked in bold. The mean and standard deviation of the results are reported based on 5 trials.

Dataset	MAVEN			ACE-2005			
Metric	P (%)	R (%)	F1 (%)	P (%)	R (%)	F1 (%)	Avg. Rank
BERT	65.0 ± 0.84	**70.9 ± 0.94**	67.8 ± 0.15	71.3 ± 1.77	77.1 ± 1.99	74.1 ± 1.56	3.8
BiLSTM	63.4 ± 0.70	64.8 ± 0.69	64.1 ± 0.13	77.2 ± 2.08	74.9 ± 2.62	75.4 ± 1.64	5.2
EventX	64.8 ± 0.91	68.5 ± 1.11	65.5 ± 0.27	74.6 ± 1.88	76.0 ± 2.00	74.9 ± 1.31	4.8
EE-GCN	64.3 ± 1.67	68.8 ± 1.38	66.9 ± 0.57	76.9 ± 1.72	78.5 ± 1.62	77.2 ± 1.43	3.5
KPGNN	67.0 ± 0.56	70.1 ± 1.08	68.2 ± 0.35	**77.8 ± 1.52**	77.6 ± 2.30	79.1 ± 1.24	2.0
DASR	**68.2 ± 0.51**	69.9 ± 0.85	**68.9 ± 0.66**	77.5 ± 1.30	78.4 ± 2.01	**79.7 ± 0.57**	1.7

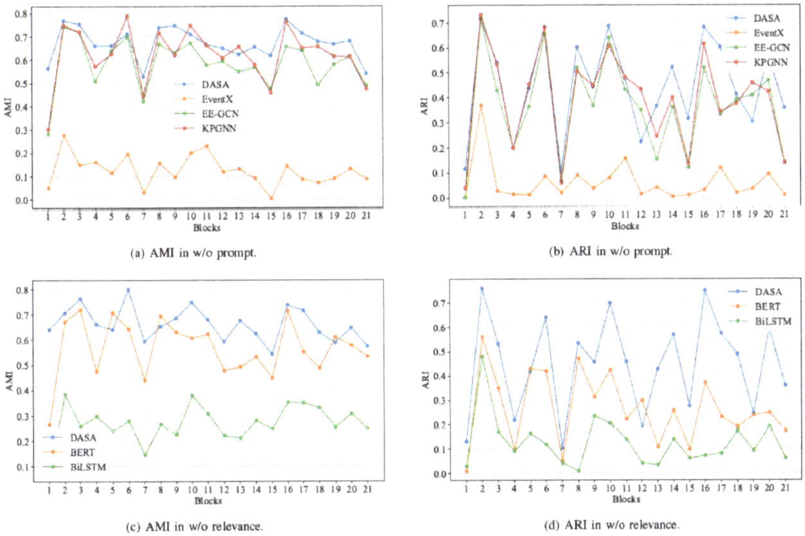

(a) AMI in w/o prompt. (b) ARI in w/o prompt. (c) AMI in w/o relevance. (d) ARI in w/o relevance.

Figure 3. Ablation study on M_3 of the Twitter dataset.

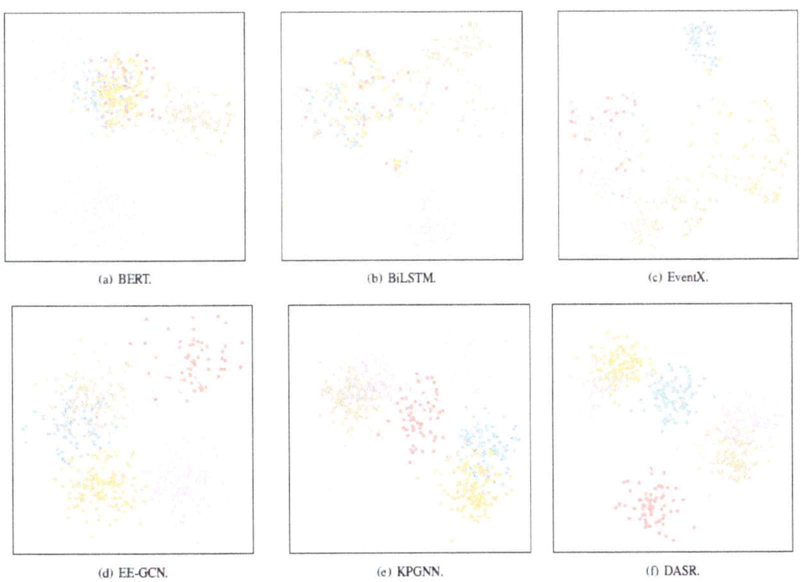

(a) BERT. (b) BiLSTM. (c) EventX. (d) EE-GCN. (e) KPGNN. (f) DASR.

Figure 4. Visualization on M_3 of the Twitter dataset.

Hyperparameter analysis. Figure 5 shows the impact of the hyperparameter γ on the performance of DASR, which represents the threshold for generating edges when constructing the type graph in DASR. As γ increases, the number of edges in the graph grows. The figure includes two curves representing the AMI and ARI scores at different γ values, along with their fitting curves (Fit_AMI and Fit_ARI). From the figure, it can be seen that when γ is around 0.4, both AMI and ARI scores reach their peak, indicating that the graph structure generated at this threshold is the most beneficial for improving the model's performance. As γ continues to increase, the AMI and ARI scores gradually decline, possibly because excessive edges in the network introduce noise, affecting the model's learning efficacy. Figure 6 shows the impact of the hyperparameter γ on the performance of DASR, which represents the trade-off in the loss function between incremental learning and the relevant neural network learning components during DASR training. A larger γ indicates greater emphasis on relevance learning, while a smaller γ emphasizes incremental learning. The figure shows that when γ is around 0.6, the AMI score is the highest, whereas the ARI score reaches a relatively high level when γ is 0.3. Performance is poorer when γ is at an extreme value (0 or 1), suggesting that relying solely on either incremental learning or relevance learning is less ideal. Finding a balance between the two optimizes the model's performance. Overall, by adjusting the hyperparameters γ and γ, the performance of DASR can be effectively enhanced, allowing it to better adapt to different tasks and datasets.

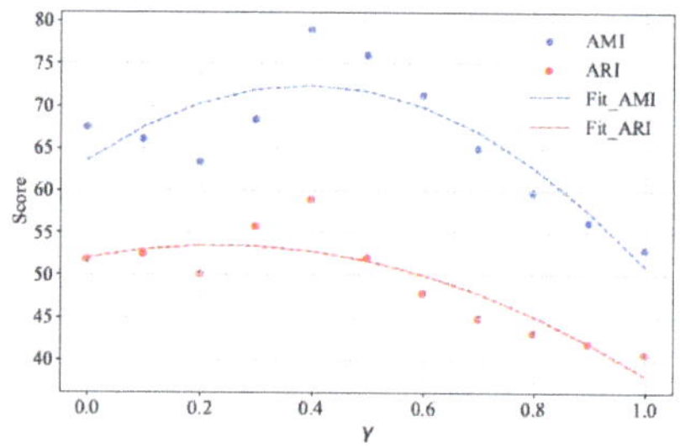

Figure 5. Hyperparameter γ analysis on M_3 of the Twitter dataset.

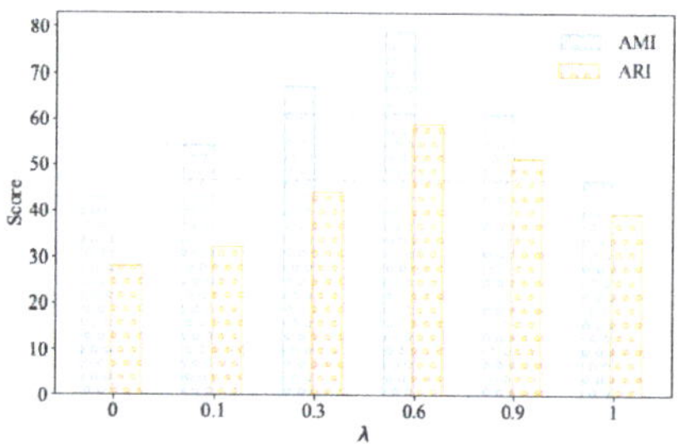

Figure 6. Hyperparameter λ analysis on M_3 of the Twitter dataset.

4. Conclusions

In this paper, we propose an innovative event detection framework (DASR), which cleverly integrates dynamic event adaptation with semantic relevance mining to address the problem of catastrophic forgetting while improving the adaptability and accuracy of the model in event detection. The core components of our approach include leveraging a pre-trained language model for extensive context understanding, integrating an incremental learning module with prompts to model the temporal sequence of events, and utilizing multi-perspective correlations along with a self-attention mechanism to handle entity recognition and event trigger word identification. Extensive experiments validate the efficacy of DASR, demonstrating its outstanding performance in mitigating catastrophic forgetting and adapting to new event types, thereby significantly advancing the state-of-the-art in event detection tasks.

Author Contributions: Conceptualization and design of the study, X.Z., G.L. and K.Q.; Methodology, X.Z. and G.L.; Data curation, X.Z.; formal analysis, X.Z. and K.Q.; writing—original draft preparation, X.Z.; writing—review and editing, X.Z., G.L. and K.Q.; investigation, X.Z.; Project administration, X.Z. All authors have read and agreed to the published version of the manuscript.

Funding: This research received no external funding.

Data Availability Statement: All data presented in this work can be obtained from the corresponding author.

Conflicts of Interest: The authors declare no conflicts of interest with the institutions, organizations, or companies mentioned in the manuscript.

References

1. Nguyen, T.M.; Nguyen, T.H. One for all: Neural joint modeling of entities and events. In Proceedings of the AAAI Conference on Artificial Intelligence, Honolulu, HI, USA, 27 January–1 February 2019; pp. 6851–6858.
2. Khurana, D.; Koli, A.; Khatter, K.; Singh, S. Natural language processing: State of the art, current trends and challenges. *Multimed. Tools Appl.* **2023**, *82*, 3713–3744. [CrossRef] [PubMed]
3. Atefeh, F.; Khreich, W. A survey of techniques for event detection in twitter. *Comput. Intell.* **2015**, *31*, 132–164. [CrossRef]
4. Lewis, D.D.; Jones, K.S. Natural language processing for information retrieval. *Commun. ACM* **1996**, *39*, 92–101. [CrossRef]
5. Hirschman, L.; Gaizauskas, R. Natural language question answering: The view from here. *Nat. Lang. Eng.* **2001**, *7*, 275–300. [CrossRef]
6. Tien, J.M. Internet of things, real-time decision making, and artificial intelligence. *Ann. Data Sci.* **2017**, *4*, 149–178. [CrossRef]
7. Gan, C.; Wang, N.; Yang, Y.; Yeung, D.-Y.; Hauptmann, A.G. Devnet: A deep event network for multimedia event detection and evidence recounting. In Proceedings of the IEEE Conference on Computer Vision and Pattern Recognition, Boston, MA, USA, 7–12 June 2015; pp. 2568–2577.
8. Liu, J.; Chen, Y.; Liu, K.; Zhao, J. Event detection via gated multilingual attention mechanism. In Proceedings of the AAAI Conference on Artificial Intelligence, New Orleans, LA, USA, 2–7 February 2018.
9. Zhao, L.; Sun, Q.; Ye, J.; Chen, F.; Lu, C.-T.; Ramakrishnan, N. Multi-task learning for spatio-temporal event forecasting. In Proceedings of the 21th ACM SIGKDD international Conference on Knowledge Discovery and Data Mining, Sydney, Australia, 10–13 August 2015; pp. 1503–1512.
10. Allan, J.; Papka, R.; Lavrenko, V. On-line new event detection and tracking. In Proceedings of the 21st Annual International ACM SIGIR Conference on Research and Development in Information Retrieval, Melbourne, Australia, 24–28 August 1998; pp. 37–45.
11. Sheng, J.; Sun, R.; Guo, S.; Cui, S.; Cao, J.; Wang, L.; Liu, T.; Xu, H. CorED: Incorporating type-level and instance-level correlations for fine-grained event detection. In Proceedings of the 45th International ACM SIGIR Conference on Research and Development in Information Retrieval, Madrid, Spain, 11–15 July 2022; pp. 1122–1132.
12. Wang, S.; Yu, M.; Huang, L. The art of prompting: Event detection based on type specific prompts. *arXiv*, **2022**, arXiv:2204.07241.
13. Yuan, H.; Sun, Q.; Fu, X.; Ji, C.; Li, J. Dynamic Graph Information Bottleneck. In Proceedings of the ACM on Web Conference 2024, Singapore, 13–17 May 2024; pp. 469–480.
14. Wei, Y.; Yuan, H.; Fu, X.; Sun, Q.; Peng, H.; Li, X.; Hu, C. Poincaré Differential Privacy for Hierarchy-aware Graph Embedding. In Proceedings of the AAAI Conference on Artificial Intelligence, Vancouver, BC, Canada, 20–27 February 2024; pp. 9160–9168.
15. Yang, Z.; Wei, Y.; Li, H.; Li, Q.; Jiang, L.; Sun, L.; Yu, X.; Hu, C.; Peng, H. Adaptive Differentially Private Structural Entropy Minimization for Unsupervised Social Event Detection. In Proceedings of the 33rd ACM International Conference on Information and Knowledge Management, Boise, ID, USA, 21–25 October 2024; pp. 2950–2960.
16. Kirkpatrick, J.; Pascanu, R.; Rabinowitz, N.; Veness, J.; Desjardins, G.; Rusu, A.A.; Milan, K.; Quan, J.; Ramalho, T.; Grabska-Barwinska, A.; et al. Overcoming catastrophic forgetting in neural networks. *Proc. Natl. Acad. Sci. USA* **2017**, *114*, 3521–3526. [CrossRef] [PubMed]
17. Hayes, T.L.; Kafle, K.; Shrestha, R.; Acharya, M.; Kanan, C. Remind your neural network to prevent catastrophic forgetting. In Proceedings of the European Conference on Computer Vision, Glasgow, UK, 23–28 August 2020; pp. 466–483.
18. Van de Ven, G.M.; Tuytelaars, T.; Tolias, A.S. Three types of incremental learning. *Nat. Mach. Intell.* **2022**, *4*, 1185–1197. [CrossRef]
19. Parisi, G.I.; Kemker, R.; Part, J.L.; Kanan, C.; Wermter, S. Continual lifelong learning with neural networks: A review. *Neural Netw.* **2019**, *113*, 54–71. [CrossRef] [PubMed]
20. Wu, T.; Caccia, M.; Li, Z.; Li, Y.F.; Qi, G.; Haffari, G. Pretrained language model in continual learning: A comparative study. In Proceedings of the International Conference on Learning Representations 2022, Virtual Event, 25–29 April 2022.
21. Biesialska, M.; Biesialska, K.; Costa-Jussa, M.R. Continual lifelong learning in natural language processing: A survey. *arXiv* **2020**, arXiv:2012.09823.
22. Yu, P.; Ji, H.; Natarajan, P. Lifelong event detection with knowledge transfer. In Proceedings of the 2021 Conference on Empirical Methods in Natural Language Processing, Punta Cana, Dominican Republic, 7–11 November 2021; pp. 5278–5290.
23. Wang, Z.; Zhang, Z.; Lee, C.-Y.; Zhang, H.; Sun, R.; Ren, X.; Su, G.; Perot, V.; Dy, J.; Pfister, T. Learning to prompt for continual learning. In Proceedings of the IEEE/CVF Conference on Computer Vision and Pattern Recognition, New Orleans, LA, USA, 18–24 June 2022; pp. 139–149.

24. Cao, P.; Chen, Y.; Zhao, J.; Wang, T. Incremental event detection via knowledge consolidation networks. In Proceedings of the 2020 Conference on Empirical Methods in Natural Language Processing (EMNLP), Online, 16–20 November 2020; pp. 707–717.
25. Cao, Y.; Peng, H.; Wu, J.; Dou, Y.; Li, J.; Yu, P.S. Knowledge-preserving incremental social event detection via heterogeneous gnns. In Proceedings of the Web Conference 2021, Ljubljana, Slovenia, 12–23 April 2021; pp. 3383–3395.
26. Peng, H.; Zhang, R.; Li, S.; Cao, Y.; Pan, S.; Philip, S.Y. Reinforced, incremental and cross-lingual event detection from social messages. *IEEE Trans. Pattern Anal. Mach. Intell.* **2022**, *45*, 980–998. [CrossRef] [PubMed]
27. Li, P.; Yu, X.; Peng, H.; Xian, Y.; Wang, L.; Sun, L.; Zhang, J.; Yu, P.S. Relational Prompt-based Pre-trained Language Models for Social Event Detection. *ACM Trans. Inf. Syst.* **2024**, *43*, 1–43. [CrossRef]
28. Yang, S.; Feng, D.; Qiao, L.; Kan, Z.; Li, D. Exploring pre-trained language models for event extraction and generation. In Proceedings of the 57th Annual Meeting of the Association for Computational Linguistics, Florence, Italy, 28 July–2 August 2019; pp. 5284–5294.
29. Wang, Z.; Wang, X.; Han, X.; Lin, Y.; Hou, L.; Liu, Z.; Li, P.; Li, J.; Zhou, J. CLEVE: Contrastive pre-training for event extraction. *arXiv* **2021**, arXiv:2105.14485.
30. Araki, J.; Mitamura, T. Joint event trigger identification and event coreference resolution with structured perceptron. In Proceedings of the 2015 Conference on Empirical Methods in Natural Language Processing, Lisbon, Portugal, 17–21 September 2015; pp. 2074–2080.
31. Li, Q.; Ji, H.; Huang, L. Joint event extraction via structured prediction with global features. In Proceedings of the 51st Annual Meeting of the Association for Computational Linguistics, Sofia, Bulgaria, 4–9 August 2013; Volume 1: Long Papers, pp. 73–82.
32. Nguyen, T.H.; Grishman, R. Event detection and domain adaptation with convolutional neural networks. In Proceedings of the 53rd Annual Meeting of the Association for Computational Linguistics and the 7th International Joint Conference on Natural Language Processing, Beijing, China, 26–31 July 2015; Volume 2: Short Papers, pp. 365–371.
33. Feng, X.; Qin, B.; Liu, T. A language-independent neural network for event detection. *Sci. China Inf. Sci.* **2018**, *61*, 092106. [CrossRef]
34. Kipf, T.N.; Welling, M. Semi-supervised classification with graph convolutional networks. *arXiv* **2016**, arXiv:1609.02907.
35. Wu, Z.; Pan, S.; Chen, F.; Long, G.; Zhang, C.; Philip, S.Y. A comprehensive survey on graph neural networks. *IEEE Trans. Neural Netw. Learn. Syst.* **2020**, *32*, 4–24. [CrossRef]
36. Zhang, Y.; Li, Z.; Liu, Z.; Zheng, H.-T.; Shen, Y.; Zhou, L. Event detection with dynamic word-trigger-argument graph neural networks. *IEEE Trans. Knowl. Data Eng.* **2021**, *35*, 3858–3869. [CrossRef]
37. Nguyen, T.; Grishman, R. Graph convolutional networks with argument-aware pooling for event detection. In Proceedings of the AAAI Conference on Artificial Intelligence, New Orleans, LA, USA, 2–7 February 2018.
38. Fu, X.; Wei, Y.; Sun, Q.; Yuan, H.; Wu, J.; Peng, H.; Li, J. Hyperbolic geometric graph representation learning for hierarchy-imbalance node classification. In Proceedings of the ACM Web Conference 2023, Austin, TX, USA, 30 April–4 May 2023; pp. 460–468.
39. Sun, Q.; Li, J.; Peng, H.; Wu, J.; Fu, X.; Ji, C.; Philip, S.Y. Graph structure learning with variational information bottleneck. In Proceedings of the AAAI Conference on Artificial Intelligence, Vancouver, BC, Canada, 28 February–1 March 2022; pp. 4165–4174.
40. Varshney, V.; Patidar, M.; Kumar, R.; Shroff, G.; Vig, L. Prompt augmented generative replay via supervised contrastive training for lifelong intent detection. In *Findings of the Association for Computational Linguistics: NAACL 2022*; Association for Computational Linguistics: Seattle, WC, USA, 2024.
41. Zhang, S.; Ji, T.; Ji, W.; Wang, X. Zero-shot event detection based on ordered contrastive learning and prompt-based prediction. In *Findings of the Association for Computational Linguistics: NAACL 2022*; Association for Computational Linguistics: Seattle, WC, USA, 2022; pp. 2572–2580.
42. Devlin, J. Bert: Pre-training of deep bidirectional transformers for language understanding. *arXiv* **2018**, arXiv:1810.04805.
43. Sundararajan, M.; Taly, A.; Yan, Q. Axiomatic attribution for deep networks. In Proceedings of the International Conference on Machine Learning, Sydney, Australia, 6–11 August 2017, pp. 3319–3328.
44. Wallace, E.; Tuyls, J.; Wang, J.; Subramanian, S.; Gardner, M.; Singh, S. AllenNLP interpret: A framework for explaining predictions of NLP models. *arXiv* **2019**, arXiv:1909.09251.
45. Pennington, J.; Socher, R.; Manning, C.D. Glove: Global vectors for word representation. In Proceedings of the 2014 Conference on Empirical Methods in Natural Language Processing (EMNLP), Doha, Qatar, 25–29 October 2014; pp. 1532–1543.
46. Veličković, P.; Cucurull, G.; Casanova, A.; Romero, A.; Lio, P.; Bengio, Y. Graph attention networks. *arXiv* **2017**, arXiv:1710.10903.
47. Graves, A.; Schmidhuber, J. Framewise phoneme classification with bidirectional LSTM and other neural network architectures. *Neural Netw.* **2005**, *18*, 602–610. [CrossRef] [PubMed]
48. Liu, B.; Han, F.X.; Niu, D.; Kong, L.; Lai, K.; Xu, Y. Story forest: Extracting events and telling stories from breaking news. *ACM Trans. Knowl. Discov. Data (TKDD)* **2020**, *14*, 31. [CrossRef]
49. Cui, S.; Yu, B.; Liu, T.; Zhang, Z.; Wang, X.; Shi, J. Edge-enhanced graph convolution networks for event detection with syntactic relation. *arXiv* **2020**, arXiv:2002.10757.

50. Vinh, N.; Epps, J.; Bailey, J. Information Theoretic Measures for Clusterings Comparison: Variants, Properties, Normalization and Correction for Chance. *J. Mach. Learn. Res.* **2010**, *11*, 2837–2854.
51. Van der Maaten, L.; Hinton, G. Visualizing data using t-SNE. *J. Mach. Learn. Res.* **2008**, *9*, 2579–2605.

electronics MDPI

Article

Hardness Classification Using Cost-Effective Off-the-Shelf Tactile Sensors Inspired by Mechanoreceptors

Yash Sharma *, Pedro Ferreira and Laura Justham

Wolfson School of Mechanical, Electrical, and Manufacturing Engineering, Loughborough University, Loughborough LE11 3TU, UK; p.ferreira@lboro.ac.uk (P.F.); l.justham@lboro.ac.uk (L.J.)
* Correspondence: y.sharma@lboro.ac.uk

Abstract: Perception is essential for robotic systems, enabling effective interaction with their surroundings through actions such as grasping and touching. Traditionally, this has relied on integrating various sensor systems, including tactile sensors, cameras, and acoustic sensors. This study leverages commercially available tactile sensors for hardness classification, drawing inspiration from the functionality of human mechanoreceptors in recognizing complex object properties during grasping tasks. Unlike previous research using customized sensors, this study focuses on cost-effective, easy-to-install, and readily deployable sensors. The approach employs a qualitative method, using Shore hardness taxonomy to select objects and evaluate the performance of commercial off-the-shelf (COTS) sensors. The analysis includes data from both individual sensors and their combinations analysed using multiple machine learning approaches, and accuracy as the primary evaluation metric was considered. The findings illustrate that increasing the number of classification classes impacts accuracy, achieving 92% in binary classification, 82% in ternary, and 80% in quaternary scenarios. Notably, the performance of commercially available tactile sensors is comparable to those reported in the literature, which range from 50% to 98% accuracy, achieving 92% accuracy with a limited data set. These results highlight the capability of COTS tactile sensors in hardness classification giving accuracy levels of 92%, while being cost-effective and easier to deploy than customized tactile sensors.

Keywords: hardness classification; COTS tactile sensor; Shore hardness scale; mechanoreceptors

1. Introduction

In the robotics domain, various sensor systems, including tactile sensors, cameras, and acoustic sensors, have been utilized to provide perceptual capabilities. These sensors are often integrated into gripper systems, equipped as artificial fingers, layered sensors, or embedded modules. Previous research has explored the use of different sensors for several classifications [1], using sensors individually and in combination. Material classification can be performed to determine the texture of the object or the hardness of the object based on some scale [1,2], whereas object classification studies have involved grasping different objects to analyse their characteristics using sensor values and different resistance values. Techniques such as pressing, sliding sensors across surfaces, and employing squeezing or grasping have been explored in the literature as state-of-the-art methodologies in tactile perception [1,2]. However, the selection of sensors in these studies has typically been driven by application-specific requirements rather than mirroring the architecture or functionality of human mechanoreceptors [3–5]. In humans, tactile information is processed by four distinct mechanoreceptors, each specializing in detecting different tactile signals [6]. Presently, there is a gap in the field concerning the selective integration of sensors which could have the same functionality of human mechanoreceptors to perform hardness classification.

The aim of this study is to explore commercially available tactile sensors that mimic the properties and functionalities of human mechanoreceptors. This research seeks to identify sensors capable of detecting tactile information with a performance comparable to that of human fingers, emphasizing cost-effectiveness. Previous research has primarily focused on utilizing customized tactile sensors that are expensive and application-specific [7–13]. The specialized architecture of these sensors limits their immediate deployment. This work targets the identification of tactile sensors that are readily deployable, requiring minimal installation efforts and offering accessibility without the need for extensive customization. It is important to establish a framework that allows for the description of tactile sensor selection, drawing from the mechanoreceptor concepts described in the existing literature [6,14]. Although prior studies have proposed advanced tactile sensors as artificial mechanoreceptors [5,15–21], their development and implementation in robotics systems are still at an early stage. In contrast, the existing array of tactile sensors in robotics encompasses various modalities, including vibration, thermal, piezoelectric, and piezoresistive sensors [1,22], which collectively provide a rich source of tactile data. However, many of these sensors are customized for specific tasks or applications, and they have not been adequately assessed for classifying materials based on hardness for robotic grasping tasks. Therefore, there is a need to investigate the potential of commercial or cost-effective tactile sensors in hardness classification in robotic manipulation scenarios. Sensors in robotics grippers [8–10,23–25] and receptors in human hands [6,14,26–28] play a crucial role in receiving tactile information, forming the foundation of tactile perception systems in both. In robotics, hardness classification utilizes various tactile sensing methods, where measuring mechanical resistance while squeezing an object is crucial [1,8,29]. Hardness classification judges an object's resistance to deformation, its ability to withstand external forces without distortion, and its tendency to deformation, thereby enabling classification into categories such as hard, soft, or flexible. While most research has primarily focused on binary classifications of hardness and softness [8,9,24,30–32], there remains a significant gap in exploring classifications that involve three or more classes. Developing multi-class systems could offer distinct advantages in real-time applications, particularly in robotic manipulation.

Tactile information obtained through grasping is crucial for understanding an object's tactile properties [8,33], unlike vision systems, which rely on predefined object features and lack predictive abilities where tactile sensors provide numeric data, enabling more feasible and less time-consuming analysis when coupled with machine learning algorithms [34]. This paper proposes the use of cost-effective commercial off-the-shelf tactile sensors, inspired by human mechanoreceptors, to perform hardness classification through grasping methods. Human tactile perception relies on the significant abilities of mechanoreceptors, which are specialized sensors within our bodies that detect pressure, vibration, and texture. These biological sensors enable us to recognize various tactile properties with notable precision. Inspired by this natural mechanism, this paper explores commercial off-the-shelf (COTS) sensors capable of detecting similar tactile information when grasping objects, akin to the capabilities of human hands. The aim is to showcase the effectiveness of available COTS tactile sensor data, as individual sensor data and combinations in classification tasks. The results will demonstrate the potential of individual sensors as mechanoreceptor-inspired tactile receptors performing hardness classification using various machine learning algorithms. Furthermore, this study aims to investigate the potential enhancement in accuracy by combining data from each tactile sensor, demonstrating how collective features from each sensor perform in hardness classification. This is expected to deliver a competitive agile solution for hardness classification that can be incorporated and scaled in robotic systems particularly where vision base solutions are unsuitable.

This paper comprises various sections that identify the required tools, methods, and adaptations to execute machine learning on the collected data, described in the following manner. Section 2 provides the foundational understanding of mechanoreceptors, tactile sensors, and material classification, drawing inspiration for this paper. It serves as the basis for the selection process of commercial off-the-shelf (COTS) sensors, leading to their selectivity based on the functionality of mechanoreceptors. Section 3 presents an approach for COTS sensors identification based on mechanoreceptors inspiration, a further stage to implement sensors with grippers and ML steps to implement an algorithm on data structure obtained from COTS. Section 4 presents the design of the experiment which show case data collection action which connects the approach and experiment setup in collecting data from selected sensors, forming different data configurations from COTS. It explains in depth the grasping method performed using pneumatic grippers utilising pressure to control the gripper system and further also explains how the object was selected and prepared using Shore's qualitative hardness scale (H,S,F,ES) and how data collection was performed. Section 5 showcases three results: 1st for binary (H,S) classification for different ML algorithms with different configurations of sensor data, 2nd for ternary (H,S,F) classification, 3rd for best algorithm outcome with quaternary (H,S,F,ES) classification.

2. Background

Human hands have exceptional tactile capabilities and serve as a significant inspiration for advancing robotic perception. Mechanoreceptors are special sensors/receptors in human skin that detect various tactile sensations like pressure, force, and vibration. They play a crucial role in our ability to perceive and interact with the world around us. Understanding how mechanoreceptors work and their capability to detect hardness is essential for robotics research. Replicating these sensors in robots can provide the capability to perceive and interpret tactile information while grasping. This would enable robots to better understand objects with more precision and perform tasks that require sensitivity to hardness.

2.1. Mechanoreceptors

How a human detects the property of an object using touch, grasping, and picking is based on receptors (biological transducers) which convert any touch to stimulus into tactile information to signal to the brain to judge the object's property. Mechanoreceptors are fundamental sensors that detect different physical/mechanical properties of objects (tactile information) mainly by four receptors [35,36]. Figure 1 illustrates the following: Four of them have the capabilities to detect different physical properties like force, vibration, deformation, indentation, etc. (1) Merkel Cells (Discs): Merkel cells respond to changes in pressure and texture, allowing them to detect variations in surface features such as roughness and indentation, informing of force/pressure. (2) Meissner's Corpuscles: Meissner's corpuscles are sensitive to light touch and low-frequency vibrations, which detect any changes in surface texture and shape. (3) Pacinian Corpuscles: The primary function is to detect sudden changes in vibration and pressure. (4) Ruffini Endings: Ruffini endings are sensitive to sustained pressure and skin stretch, enabling them to detect changes in skin deformation caused by object contact. Based on current understanding, mechanoreceptors comprise tactile receptors that receive multiple pieces of information while grasping or touching the object which processes the tactile information through neurons to the brain to understand the properties and define it or store it in the brain for further reference [37]. Mechanoreceptors are responsible for detecting mechanical stimuli, including pressure, vibration, and deformation, thereby providing humans with valuable tactile information [6,35–37]. In the context of tactile perception, mechanoreceptors play a crucial role in assessing the hardness and softness of objects by responding to various tactile signals. Four important aspects understood from receptors illustrated in Figure 1 are that touch, force, vibration, and sliding are the key physical properties that make humans understand if the object is hard or soft or flexible while grasping or squeezing.

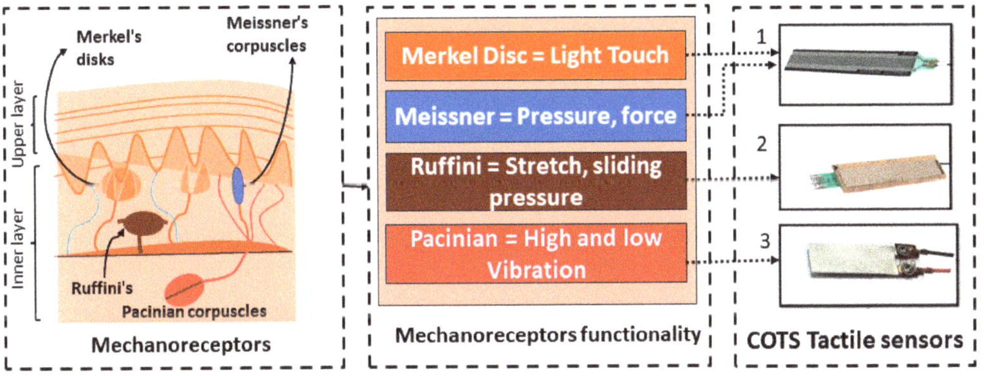

Figure 1. Showcases the mechanoreceptors' architecture and their functionality in detecting different tactile information, where COTS tactile sensors are mapped based on their functionality: (1) FSR force sensor, (2) potentiometer Softpot membrane, (3) piezoelectric thin-film vibration sensor.

2.2. Tactile Sensors

Material classification in robotics is one of the important aspects of development which helps robotic grippers in different environments to handle different objects or perform manipulation. This objective is only possible if sensors embedded within grippers have the capability to detect tactical information to recognize object properties. Material classification can be further subclassified into hardness classification and texture classification. Currently, texture classification has been explored mostly, but hardness classification has not been explored much with different tactile sensors and off-the-shelf sensors. Sensors used in literature are highly customized with shape and dimension which takes time and cost to develop. Parallel research is now more focused in developing sensors that have human capabilities in detecting object characters like texture, hardness, roughness, softness, slippery, etc. [1]. Engineering tools and methods have been developed which have shown some capability to show response type and functionality to human receptors [15–21] but it remains challenging to develop a system or sensors that have the same capability as humans. In the literature, various types of sensors are discussed with respect to their cost, which often varies depending on the technology and complexity involved. Optical sensors and bioinspired sensors are generally considered high-cost options due to their intricate designs and advanced functionalities. Microfluidic sensors and MEMS-based sensors fall into the moderate- to high-cost category, as they involve sophisticated fabrication processes and specialized materials. Similarly, grid capacitive, piezoelectric, and resistive tactile sensors are also moderately priced, reflecting the complexity of their construction and sensing mechanisms. Neuron memristors and ionic tactile sensors are relatively newer technologies, occupying the moderate cost range as they undergo further development and refinement [5,29,38,39]. However, all of them face limitations including high cost, complexity, sensitivity to environmental factors, integration challenges, and training issues with ML models due to the need for extensive datasets and iterative refinement filtration, etc. In comparison, commercially available off-the-shelf (COTS) sensors are easy to use, cost-effective, ready to deploy, flexible to adjust to most grippers, and suitable for exploring with different robotics grippers as tactile sensing techniques. Moreover, these sensors, such as vibration sensors, force-sensitive resistors (FSR), and soft potentiometers, can also play a role as artificial mechanoreceptors based on functionality.

2.3. Material Classification

Material classification involves sorting objects based on various features, much like how humans recognize things by feeling their texture and hardness when they touch or hold them. While texture classification, edge detection, and object classification have received considerable attention, hardness classification has been relatively less explored, despite its significant importance. In the field of material science and engineering, machine learning (ML) techniques have proven invaluable in determining material properties using various methods and approaches, such as the Shore scale taxonomy [7,10,12,36,38,39]. ML enables systems to learn and recognize materials based on their hardness using techniques like grasping, indentation, and resistance methods. By using extensive datasets, ML models discern patterns and correlations, preparing them to accurately classify materials into predefined hardness classes. This iterative refinement process ensures their reliability, aiding industries in material selection and enhancing robotic grasping mechanisms through tactile sensing. Inspired by human grasping mechanisms, ML algorithms serve as analytical tools, processing tactile information to identify object properties. Robotics systems lack the analytical capabilities of the human brain to recognize object properties. However, through the integration of sensor data and machine learning (ML), this approach becomes feasible in advanced tactile sensing for robotics.

These algorithms are built upon various mathematical formulations, which serve as the backend processes. By encapsulating them as functions within ML libraries and implementing them through Python coding, they become readily deployable for data analysis. This facilitates advancements in material classification and robotics. The hardness classification process initiates with a long process of data collection through experiments, gathering information on material properties alongside corresponding hardness measurements. Thorough preprocessing follows, ensuring the quality and integrity of the dataset, while relevant features indicative of hardness properties are selected or extracted to facilitate accurate classification. Subsequently, an appropriate ML model is selected and trained using the pre-processed dataset, allowing it to learn patterns and relationships between input features and hardness classifications. Finally, the trained model's performance is evaluated using established metrics, and iterative refinement and optimization processes are employed to enhance its accuracy and effectiveness in hardness classification tasks. Different types of ML algorithm have been deployed for the purpose of classification which were presented as state-of-art in paper [8,9,24,30–32]. To see how multiple ML algorithms perform in terms of accuracy of hardness classification, grasping data from COTS sensors was adapted against different objects selected and prepared based on the Shore hardness scale.

2.4. Shore Hardness Scale

The Shore hardness scale is a measurement technique used to assess the hardness or softness of material on a wider scale. The Shore hardness scale was adapted in several cases of classification [10] to match the value of an object as a comparison to testify the assumption made with the object as soft or hard while testing. In many classification approaches, hardness and softness were considered because of easiness in testing. It provides a standardized method for quantifying the resistance of a material by imposing force which works similar for humans while considering the object without noticing the values. There are different options of the Shore hardness scale, such as Shore A, Shore D, and Shore OO, each tailored to specific material types and testing conditions. The Shore hardness scale can be explored or used in different ways where one could select the material or object based on the Shore scale that could precisely describe that material shows this property and can be matched to a qualitative scale as extra soft, soft, medium soft, medium hard, hard, and extra hard. In this process, the qualitative scale was adapted to extend the classification beyond binary classification and adapt multiple objects based on the scale beyond hard and soft which could be flexible (deformable but with force). This approach also makes object adaption faster and gives precise results based on ML algorithms' outcomes if the extension of classes improves the outcome or not; also, there is the option to test wider unknown/new test objects' data.

2.5. Summary

Human hands hold exceptional tactile detection capabilities, such as pressure, force, and vibration, due to mechanoreceptors in the skin. These mechanoreceptors are crucial for understanding and interacting with the environment. Inspired by this natural mechanism, which involves converting tactile stimuli into neural signals, this research aims to use tactile capabilities of humans in robotic systems. Unlike existing methods that use highly customized and expensive sensors, this research explores the use of commercial off-the-shelf (COTS) tactile sensors, which are cost-effective and easy to integrate. Hardness classification is particularly important because it enables robots to handle and manipulate objects with precision and can provide capabilities to perform material-type identification and sorting, preventing damage to delicate items and ensuring a secure hold on harder objects, the same as how humans assess material properties through touch or grasping. This capability can improve robot–human interaction such as service robots or robotic assistants; understanding material hardness enhances their ability to perform tasks that involve human-like manipulation skills. By using machine learning (ML) techniques, robots can be trained to recognize material hardness, improving their grasping and manipulation abilities. The Shore hardness scale is employed to provide a standardized measurement of material hardness, aiding in the development of more effective robotic tactile sensing systems.

3. Approach

3.1. Identification of Mechanoreceptor-Inspired Tactile Sensors

Different types of complex and customized tactile sensors were identified in literature which were used in different application and classification tasks. New proposed tactile sensors like Biomimetic tactile sensors, artificial mechanoreceptor tactile sensors, grid-type piezoelectric sensors, and ionics tactile sensors may be under a commercializing state and are not so quickly available [7,13,15,38,40–42]. Based on the current state of the art, it is understood that artificial mechanoreceptors and tactile sensors share a common developmental base, detecting similar physical properties from stimuli. These sensors may vary in their single- or multi-dimensional array configurations. Many studies have focused on emulating spike patterns using artificial mechanoreceptors, with some applied in hardness classification tasks [39,43–45]. Currently, many of these advanced sensors are not available in thin film for multiple tasking within robotics applications. These types of sensors are very sensitive and can quickly degrade by certain actions of grasping and cannot withstand longer periods for data collection with multiple objects. These are advanced, but not much application has been proposed within robotics grasping or in classification. Despite numerous descriptions of artificial mechanoreceptors and multilayer sensors in literature, their use cases remain limited, with commercial availability posing another challenge. Proposed sensors as solutions are customized based on application which does not have further space or scope of development or availability [22,29,38,46]. Thin-film sensors are easy to embed in most robotics grippers, and based on market research, some of the sensors were identified which have some standard dimensions which can be used directly within the robotics application. Therefore, the focus shifts to available tactile sensors in the market, classified based on functionality mirroring natural mechanoreceptors illustrated in Figure 1. These include FSR force-resistive sensors (also known as piezoresistive sensors), vibration sensors (akin to piezoelectric sensors), and soft potentiometers sensors (such as distributed force-resistive sensors). These thin-film sensors can be easily integrated into hard or soft grippers to detect physical changes in force or distributed pressure. For this purpose, the Schunk gripper was adapted due to its capability to accommodate these sensors within its internal structure.

FSR sensors as mechanoreceptor 1 and 2 functionality: FSR sensors, also known as force-sensitive resistors, are tactile sensors designed to detect changes in resistance when force is applied. This change in resistance leads to an increase in voltage, which is typically scaled within the range of zero to five volts. The sensitivity of FSR sensors allows them to detect various levels of force, ranging from light touch to continuous pressure and even high-impact forces, measuring up to 50N. This versatility makes FSR sensors well suited to emulate the functionality of two mechanoreceptors located on the top layer of the skin, namely the Meissner corpuscles and Merkel discs [47]. These receptors are positioned close to the skin's surface, enabling them to accurately perceive touch, pressure, or any form of deformation [47]. When an object deforms due to pressure or manipulation, the mechanoreceptors sense the deformations and send signals in proportion to the degree of deformity. Based on deformation values obtained, soft objects are classified as "soft" and hard objects as "hard" or flexible. This ability of pressure and resistance can be measured by FSR sensors which were also chosen.

Soft Potentiometer sensors as Mechanoreceptor 3 functionality: Potentiometer sensors' capabilities as mechanoreceptors have not yet been fully explored in terms of hardness classification. These sensors are the same as FSR sensors and exhibit a linear change in resistance across their surface. However, they possess a unique capability to detect slip, stretch, or sliding movements based on changes in resistance. This detection capability closely resembles the functionality of mechanoreceptors [6,47] known as Ruffini endings. In this scenario, this off-shelf tactile sensor has the capability to detect changes that are similar in mechanoreceptors to analyse object characteristics or parameters in terms of hardness classification. While squeezing, object deformation will produce stretch across the sensor and will give different output volts as signatures for different objects which will help to classify objects based on hardness.

Vibration sensors as Mechanoreceptor 4 functionality: Vibration sensors, functioning the same as mechanoreceptors, play an essential role in neural pattern generation, classification, and the development of grid sensor architectures. Piezoelectric vibration sensors serve as essential tools for detecting impact and vibrations across a wide spectrum of frequencies. This enables the collection of data crucial for pattern generation and classification tasks. While these sensors may not replicate the intricate architecture of mechanoreceptors, they offer functionality similar to Pacinian receptors, producing data vital for classification purposes. Vibrations generated upon object impact are transmitted to mechanoreceptor junctions. As these mechanoreceptors [6,47] deform in response to detected vibrations, they send electrical impulses to the brain for processing. Higher frequency vibrations typically indicate harder materials, requiring more energy to induce, while lower frequencies correspond to softer materials. This same functionality and process can be performed using thin-film vibration sensors to perform hardness classification.

3.2. Hardness Classification Using Mechanoreceptor-Inspired COTS Sensors

Figure 2 illustrates the proposed approach and steps involved, which are essentially divided into four parts. Firstly, it involves the sensors that are bio inspired by mechanoreceptors. Choosing the right gripper based on dimensions, adapting the grasping method (mechanical resistance) to create an impact on the sensors while grasping objects the same as human pinch grasping, and object selection on the Shore hardness qualitative scale are illustrated in Figure 2. Secondly, it utilizes a machine learning approach to analyse data from sensors, including F-force data, V-vibration data, and P-potentiometer data. Initially, single-sensor (F), (V), and (P) data were considered for analysis, and then different configurations of sensor data were formed from each sensor. This also shows how off-shelf tactile sensors data, selected based on bio-inspired mechanoreceptors, perform a hardness classification. This understanding indicates how an individual and combination of sensors can yield accurate results when attempted. Additionally, it offers insight into how mechanoreceptor-inspired tactile sensors, both individually and in combination, can classify hardness effectively. Accuracy from COTS tactile sensors will set a benchmark for further exploration in layered sensor technology for future scope. These accuracy score data showcase how grasping-based tactile information achieved from sensors in volt as a value can be used to achieve a hardness classification to evaluate their performance. Decisions are made based on the accuracy of each configuration in multi-class scenarios, including dual combinations like (F,V), (F,P), and (P,V), as well as the three-sensor combination (F,V,P). While binary classification has been explored in the literature, this approach also involves different sensors' combination approaches to perform binary (2 classes object), ternary (3 classes object), and quaternary classification (4 classes object) based on object data obtained from testing and Shore hardness scales.

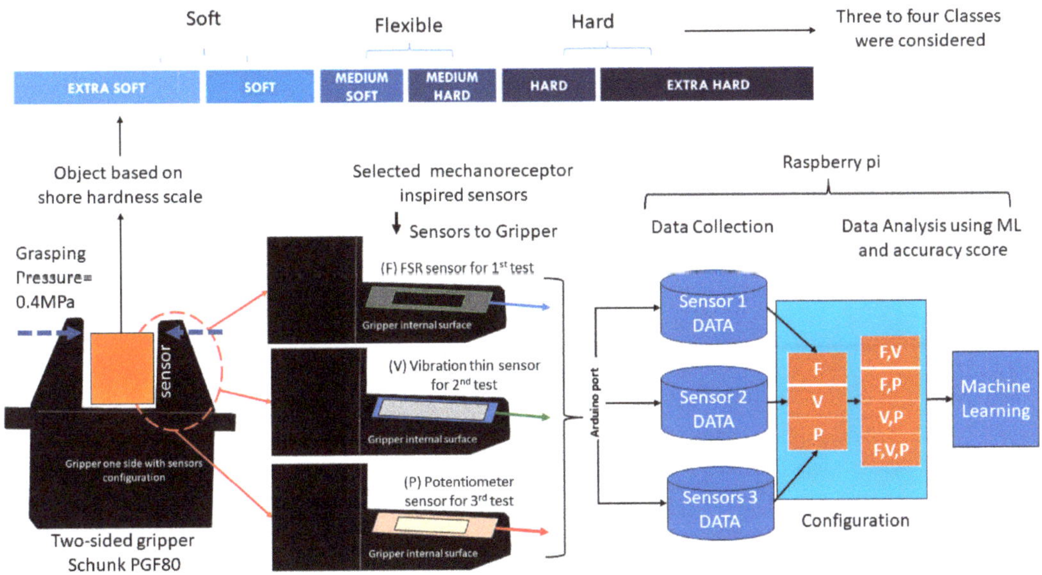

Figure 2. Research approach illustrating firstly object selectivity based on Shore scale considering three to four classes in object; secondly selected sensors embedded on gripper one side; thirdly sensors connected to Arduino to raspberry-pi for data collection; lastly from F,V,P data different set of configurations created to investigate accuracy score using multiple machine learning approaches.

3.3. *Machine Learning Approach for Hardness Classification*

Figure 3a illustrates the process followed in the machine learning approach and Figure 3b shows the data structure of configuration used to train the ML. Machine learning algorithms are essential for hardness classification tasks. These algorithms act like a brain analyser, analysing different types of data such as numeric, text, images, and more. In the context of hardness classification, it helps in understanding the features of datasets obtained from tactile sensors or any other source. The algorithms use various techniques, including supervised learning, where they are trained on labelled data to predict hardness classification. Examples of such techniques are decision tree classification, random forest classification, nearest neighbour classification, including support vector classification, and others. During training, the algorithms learn patterns and relationships within the data, allowing them to accurately classify new inputs. They analyse features extracted from the data to predict whether a material is hard or soft or based on scale. Previously bio signal, FFT, and other digital data were used in relation to objects to train ML [2,38]. In this approach, COTS sensors data were formed in different combinations to train the ML algorithms to understand which sensors' configuration out of F,V,P, (F,V), (F,P), (P,V), (F,P,V) trains ML well, and based on test data, which ML algorithm accuracy comes out to be the best. There are some steps that need to be taken before deploying the machine learning algorithm which are as follows:

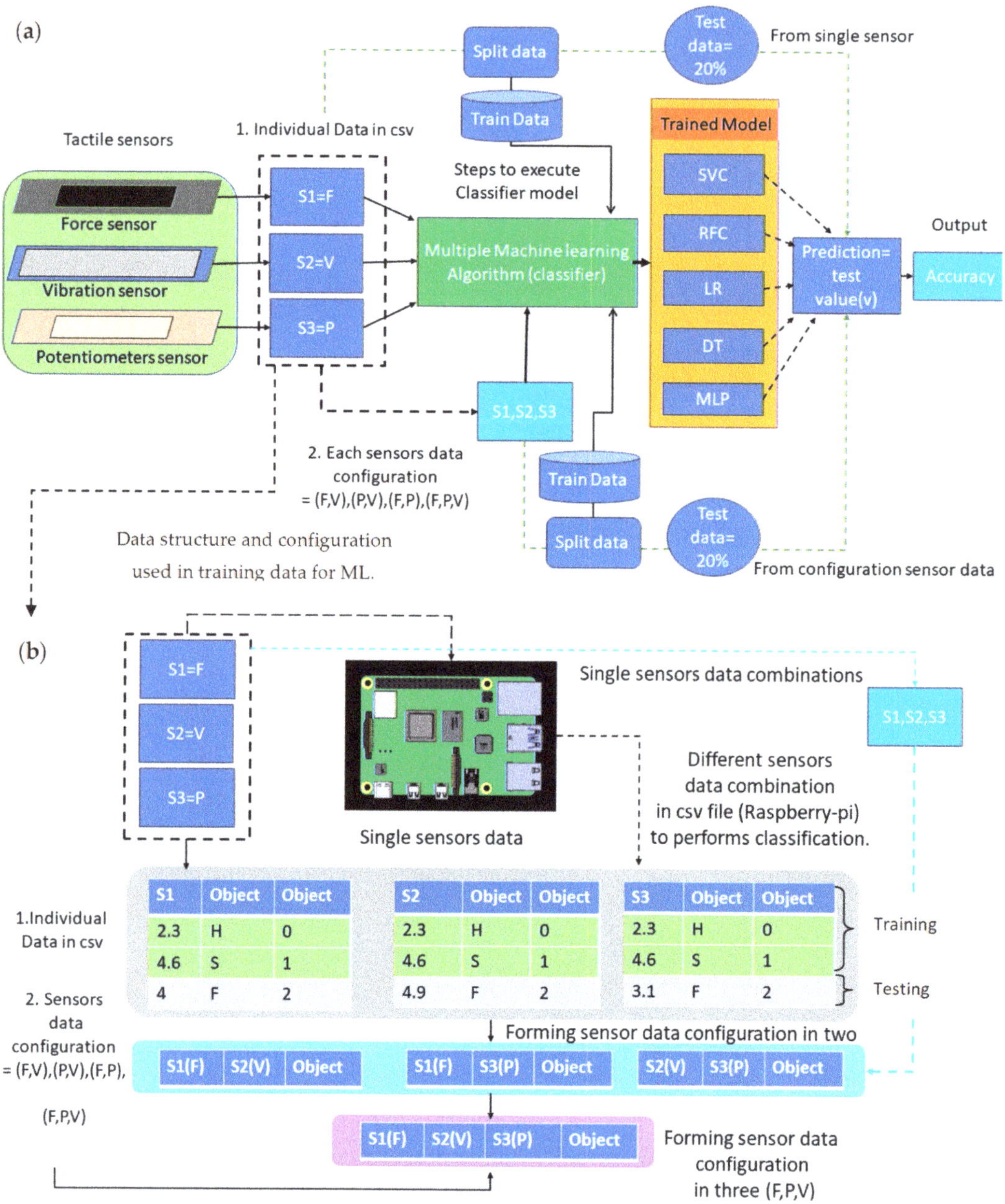

Figure 3. (**a**) Illustrates machine learning framework for COTS sensors data. Diverse machine learning algorithms are employed to train models using these data. The datasets utilized comprise (1) individual sensor data and (2) combinations of sensor data paired with object information denoted as (H,S) and (H,S,F). Subsequently, the test values are inputted into each algorithm, and the accuracy of their outputs is evaluated to test their efficacy. (**b**) Illustrates the data structure of various configurations saved within Raspberry Pi to train ML algorithms to investigate hardness classification.

Python Libraries: Machine learning analysis was conducted using Python using Google Collaboratory. To process all analysis, there were several Python and machine learning libraries which needed to be declared in beginning of the code. In general, libraries like Sklearn, NumPy, Pandas, and Matplotlib were declared in the first step [48–51].

Data importing: The initial step in analysis involves importing the dataset into the Python environment. This dataset serves as the foundation for machine learning tasks, containing the necessary information for training, and evaluating classifier models with a feature variable and a target variable. In this case, Figure 3b showcases different S1, S2, S3 data in single CSV file with the column of sensors' value in volts as X variable and target variable having the information of objects' type as H as 0, S as 1, F as 2 called as Y variable. These data were called and processed and were performed using the Python library called 'Pandas'. In a further case, two combinations (S1,S2), (S2,S3), (S3,S1) of dataset were combined with the help of 'Pandas' to further follow the combination approach.

Preprocessing data: Prior to training machine learning models, it is essential to preprocess the dataset to ensure compatibility with the chosen algorithms and improve model performance. One common preprocessing step involves encoding categorical variables into a numerical format, which is achieved using techniques such as label encoding, also illustrated in Figure 3b. In this case, the target variable needs to be mapped into numerical values using a label encoder, whose function is to transform string values like H as 0, S as 1, and F as 2, in case of ES (extra soft) as 3. Additionally, preprocessing may involve handling missing values, scaling features, and other data transformations required based on dataset preparation using the Imputer library.

Splitting data: To evaluate the performance of machine learning models accurately, it is crucial to divide the dataset into separate training and testing subsets. This process, known as data splitting, helps assess the generalization ability of the trained models by evaluating their performance on unseen data. In this case, data from the configuration F,V,P, (F,V), (F,P), (P,V), (F,P,V) was split accordingly to train the ML algorithm. Typically, a portion of the dataset is reserved for training, while the remaining portion is used for testing. In this case, 80% of the data were used for training and 20% of the data were used for testing, as illustrated in Figure 3a.

Importing ML algorithm: With the dataset prepared and split into appropriate subsets, it further proceeds to import the machine learning algorithms like SVC (support vector classifier), RFC (random forest classifier), DT (decision tree), LR (logistic regression), and MLP (multilayer perceptron) for data analysis. Depending on the nature of the problem and the characteristics of the dataset classifier, the algorithms found to be most suitable are used for training predictive models. As these algorithms take less computational time, they are easy to deploy and analyse in comparison to any form of deep learning or neural network.

Training ML: Once the algorithm(s) are imported, data split into X (sensors value) from different configurations F,V,P, (F,V), (F,P), (P,V), (F,P,V) and Y (target feature = object type (like H as 0, S as 1, and F as 2)) are illustrated in the Figure 3a,b variable used as training data to train the models, where the models learn patterns and relationships within the training data as sensors' values and number of object types. This process involves feeding the training data into the algorithm(s) and iteratively adjusting the model parameters to minimize a predefined loss function. Through this iterative optimization process, the models learn to make predictions or classify new data based on the patterns observed in the training set.

Trained model: After completing the training phase, trained machine learning models were obtained that capture the discerned patterns and correlations between tactile sensors' value and target variables (object types). These relationships form a crucial basis for distinguishing between (1) hard and soft materials within the dataset in case of binary classification and (2) three-object variables, hard, soft, and flexible, in case of ternary classification (H,S,F), as illustrated in Figure 3. With this information, the models are equipped to forecast outcomes for new, unseen instances, particularly in the context of hardness classification tasks.

Prediction: This prediction phase involves feeding unseen data into the trained models and obtaining output predictions based on the learned patterns. The predictions generated by the models represent how machine learning performed predictions based on dataset features and based on different models. In this case, two sets of data will be tested: (1) individual sensor data, and (2) combination of individual sensors from different ML algorithm outcomes in one file, as illustrated in Figure 3.

Validation using accuracy score: To evaluate the performance of the trained machine learning models, validation techniques such as accuracy score are illustrated in Figure 3a,b. In this case, data which was split will use 20% data for testing the model. In this way, models with new data will predict the outcomes as H or S object or for F object. This will be compared to the original test value which will give an accuracy score and performance by comparing it within literature. Also, the accuracy score provides a measure of the overall correctness of the model predictions. This is highlighted in the results section.

4. Design of Experiment

The design of the experiment illustrated in Figure 4 for conducting hardness classification involves both hardware and software components. Firstly, it was crucial to employ COTS tactile sensors to measure the impact of squeezing objects and collect data. This required performing grasping and resistance measurement actions to obtain values from three sensors inspired by mechanoreceptor tactile detection capability. Additionally, the gripper action was facilitated using air pressure, capable of withstanding a force of 0.4 MPa during grasp, controlled through Python coding via Raspberry Pi.

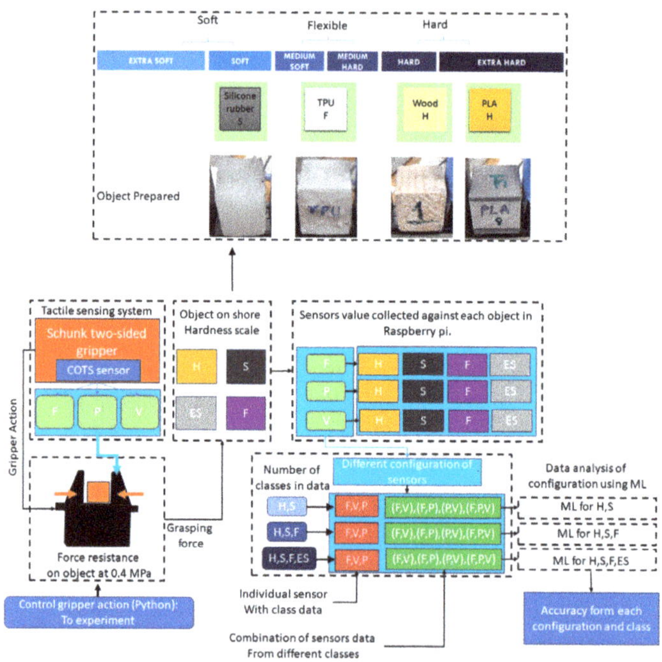

Figure 4. Design of experiment flow process for collecting data from COTS sensors and preparing different set of configuration data to machine learning analysis. Object preparation based on qualitative Shore hardness scale adopted in this study, highlighting the three to four classes considered: H (Hard), S (Soft), and F (Flexible), ES (extra soft). Objects representing each class are showcased based on the scale, with silicone rubber representing Soft, TPU representing Flexible, and Wood and PLA representing Hard materials.

The selection of an appropriate gripper was paramount, requiring compatibility with objects and sensors. A two-finger gripper was chosen for its ability to mimic human grasping and its economic viability for grasping various object types efficiently. The Schunk gripper, available in mechanical and pneumatic types, was selected for compatibility, with tactile sensors shaped to attach easily. Control of pneumatics involved the use of a pressure regulator, solenoid valve, and single electric valve. Integration of Arduino facilitated sensor connection and served as an ADC for data transmission. Additional hardware components, including a 3D printer for object creation and a screen with HDMI port for visualization, completed the experimental setup. Data collected from sensors, denoted as F (force), V (vibration), and P (potentiometer), were saved in Raspberry Pi. These data were further configured into various combinations, such as (F,V), (V,P), (F,P), and (F,P,V), to perform data analysis on individual sensor data and combinations thereof. To examine deeper into the obtained values from COTS sensors, a 'shore taxonomy' process was employed to analyse the deformation resistance to force based on the Shore hardness qualitative scale presented in Figure 4. Additionally, each configuration represented specific conditions for binary, ternary, and quaternary classification. For binary classification, the configurations included H (Hard) and S (Soft) classes. Ternary classification incorporated H (Hard), S (Soft), and F (Flexible) classes, while quaternary classification added an additional class, ES (Extra Soft). This approach was crucial for performing hardness classification against each object, providing insights into the behaviour of future applications in terms of sensor layer operation and the number of classes involved in classification.

4.1. Object Preparation

Illustrated in Figure 4: Objects representing different hardness classes (Soft, Flexible, Hard) are prepared according to the Shore hardness scale for testing. For silicone rubber, Eco flex 30 was used with mould of 3 × 3 cm size and filled with Eco flex liquid and left for around 3 h for best result. After extraction of silicone rubber, it was squeezed/pinch grasped to judge silicone rubber act as soft based on Shore scale. TPU (Thermoplastic polyurethane) is filament used in 3D printing to print objects. In this case, 3D object with 3 × 3 cm dimension was printed; based on property analysed, TPU was considered as flexible with medium hard and medium soft property. In case of wood, a piece was diced to form 3 × 3 cm wooden block. PLA (Polylactic Acid) is also a filament which was used to 3D print 4th object and analysed to be another hard object.

4.2. Hardware

Figure 5a illustrates complete collection of setups which showcase what tools and hardware were used. To conduct hardness classification by squeezing object, also known as grasping, and collecting data, it was necessary to have grippers that effectively have enough space to accommodate object and sensor on gripper. For this purpose, Schunk gripper as mechanical gripper/pneumatic gripper type was selected. Additionally, to collect data from tactile sensors, it was necessary for their shape to be compatible with sticking on the side of the pneumatic gripper. To control pneumatics, further tool requirement was to control air pressure, which was conducted using pressure regulator, solenoid valve, and single electric valve. Main air pressure supply was fed into pressure regulator using supply within IAC lab. To develop a system with both automatic and manual grasping capabilities, a device that could synchronize with Python and collect data simultaneously was required. This synchronization was achieved using Raspberry Pi, which served as the central hub with the resources to combine peripheral devices and control them through digital commands (high and low).

In addition to Raspberry Pi, an Arduino board was utilized to connect sensors and serve as an ADC (analogue to digital converter). This allowed the sensors to be connected to Arduino ports and subsequently linked to Raspberry Pi via serial communication. Other essential components included a pressure regulator to limit air pressure to a certain level, relay modules to control electric components, and solenoid valves to regulate air pressure. An RS power supply was necessary to operate the electric valve and solenoid valve effectively. To perform gripping tasks, a Schunk pneumatic gripper was employed. Furthermore, a 3D printer was utilized to create testing objects from different materials, providing versatility in experimentation. Lastly, a screen with an HDMI port was utilized to visualize the outcome of the experiments, ensuring effective monitoring and analysis of the collected data.

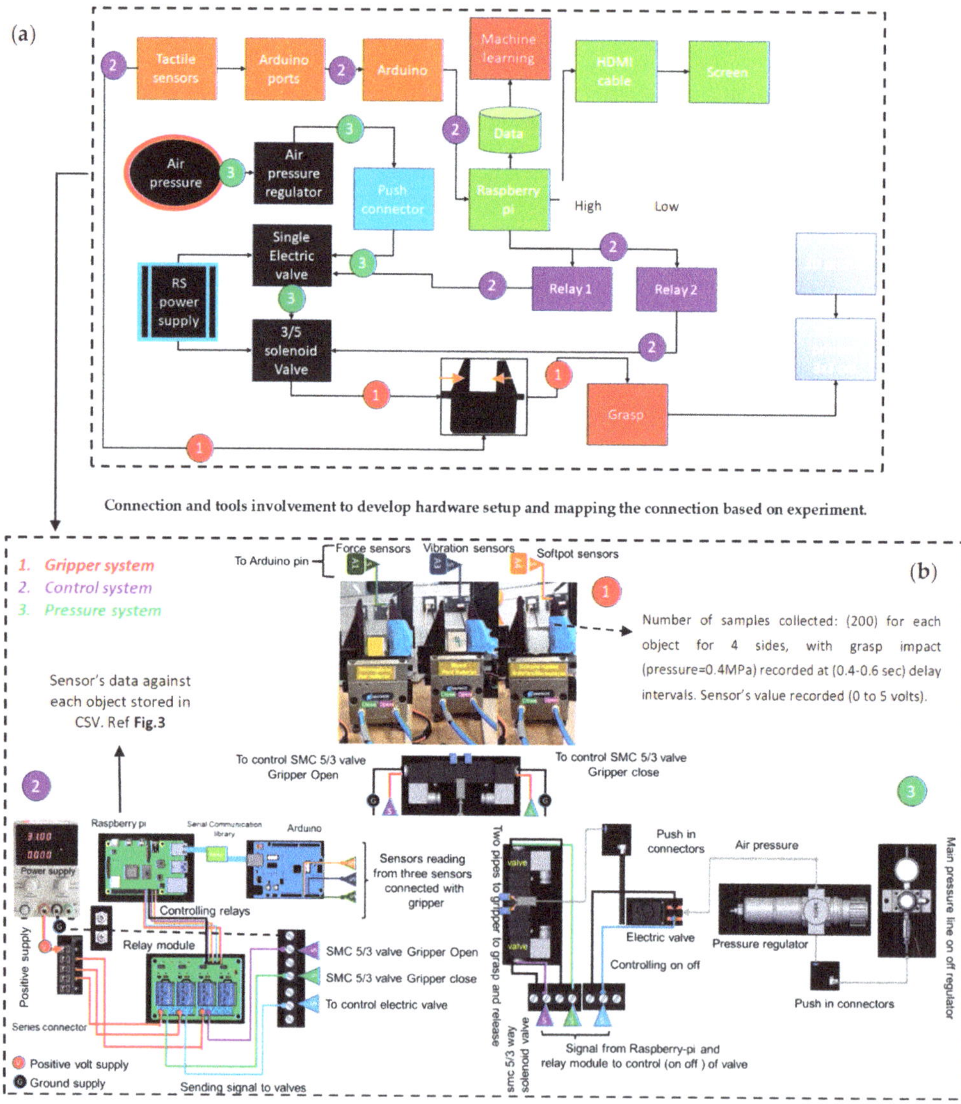

Figure 5. (**a**) Provides an overview or mapping of the experimental setup, demonstrating the connections among the tools involved in developing the prototype; (**b**) presents the complete experimental setup, showcasing three distinct systems: the Gripper system, Control system, and Pressure system. Additionally, three different tactile sensors are positioned on the side of the gripper to collect data alongside various objects. Data collected were saved in Raspberry Pi in CSV format. All data of sensors with objects have around 200 samples.

4.3. Software

To control grasping and collect sensor data for analysis using machine learning algorithms, various Python libraries and machine learning frameworks were employed. To establish communication between the Arduino and Raspberry Pi, the Nanpy library was used, allowing for integration of the two devices in a master–slave configuration to facilitate serial communication [52]. For data collection and processing, the Pandas library has capabilities to manage data and manipulate datasets. Additionally, for the analysis of collected data and the implementation of various machine learning algorithms, the Scikit-learn library was applied, offering a wide range of algorithm selectivity like SVM (support vector classifier), RFC (random forest classifier), and (DT) decision tree, and for measuring performance like accuracy and confusion matrix. Python 3, along with Python idle-Version 3.8 software, Jupyter Notebook Version 5, and Google Collaboratory, were employed for coding purposes. Each platform was selected based on its capabilities and suitability for specific tasks, ensuring efficient analysis and data collection across various scenarios.

4.4. Gripper System

In the exploration of various grippers documented in literature, diverse structures and use cases have been examined. However, the need arose for a general gripper that could mimic pinch grasping same as human hand operation, while also accommodating the dimensions of COTS sensors. Understanding the embedding issues of sensors when using different grippers and sensor variations was crucial. For the experiment depicted in Figure 5b, the Schunk PGF 80 gripper was utilized. One side of the gripper was equipped with sensors, with each sensor being swapped out during the experiment for each object. These sensors were connected to an Arduino to collect data, capturing tactile information from each stimulus object. A pneumatic system controlled by a control system regulated the opening and closing of the gripper by releasing air pressure through a pipe attached to the gripper. The selection of the gripper was based on the space available between its fingers, facilitating easy placement of objects. Additionally, the gripper's adjustable gripping range allowed it to grasp objects larger than its default size. Opting for a two-sided gripper enabled a pinch-type grasping similar to human finger dexterity, occupying minimal space, and swiftly integrating into any robotic arm. The selected sensors were easily installable in these grippers compared to custom-made ones, streamlining the experimental setup, and enhancing efficiency.

4.5. Control System

To control the gripper, an SMC solenoid valve which is controlled through relay module and with python code from Raspberry Pi sends high and low command to GPIO pins. This relay module controls the electric valve in pressure system which controls the main supply of air pressure system at decided delay of 0.4 to 0.6 s. Illustrated in Figure 5b, sensors connected in gripper also connected to Arduino which reads the analogue value from sensors and converts them into voltage on scale of 5 volts. Arduino is connected to Raspberry Pi which works as bridge to sensors and Raspberry Pi. With the help of Python library (Nanpy,) Raspberry Pi and Arduino is connected in master–slave configuration to perform serial communication. Sensors' data were saved in Raspberry Pi from Arduino as CSV file to use in material classification algorithm.

4.6. Pressure System

Air pressure was taken from IAC lab which was connected to central pressure system and pressure regulator with safety knob. This air pressure was passed through regulator which was set around 0.4 MPa approx. This passed through electric valve which was controlled by control system through relay module illustrated in Figure 5b. Main board developed in a way that it can control three multiple air pressure types within this module which can be useful for future scope in terms of pressure variation tests under gripper. In terms of open and close of air flow, described in Figure 5b within gripper system, it is controlled through SMC 5/3 solenoid valve which when one side is open through relay module, gripper opens, and when other side valve is open, gripper closes or grasps. This was achieved by sending high(1) and low(0) command through Raspberry Pi through GPIO pins to 2nd and 3rd relay module.

4.7. Data Collection

The data collection process stated and described in Figure 5b demonstrates the comprehensive connectivity of each module, highlighting how sensors are linked to the gripper and subsequently connected to an Arduino for data transmission to a Raspberry Pi. These data are crucial for further analysis, particularly in material classification research. During the experiment, single sensors were sequentially deployed with the gripper to acquire data, as illustrated in Figure 5, showcasing the formatted data structure illustrated in Figure 4 for subsequent analysis. Each tactile sensor gathered data from three distinct classes of objects grasped by the gripper, with the Raspberry Pi Figure 4 storing these data for each object individually. Gripper grasps every object around 200 samples from each object and sensors with synchronized grasp and releases with duration around 0.4 s to 0.6 s. Utilizing Python's Pandas library, data collection was streamlined through commands that defined the object's state (H, S, or F) based on the Shore hardness quantitative scale. The collected data for each object, captured by three sensors individually connected to Arduino ports, was organized into separate data frames, and saved as CSV files on the Raspberry Pi. Additionally, string labels (H,S,F) were transformed into numerical values using label encoder to perform classifier analysis capabilities. The connection setup detailed in Figure 5a,b illustrates the integration of sensors with the gripper system during testing, providing clarity on how tactile sensor data were stored in CSV files on the Raspberry Pi. To analyse data, Figure 4 explains the data collection process and the resulting data structure and steps for ML algorithms analysis process illustrated in Figure 3.

5. Results Based on Machine Learning

5.1. Result on Hardness Classification Outcome Based on Two Classes (H,S)

In term of hardness-based classification when two classes were considered, (H,S) gives comparable accuracy, the same as in literature which used customized sensors. Figure 6 indicates that binary classification of individual sensors showcases accuracy ranging from 65% to 82% noticed from individual sensors (F), (P), and (V). Accuracy achieved from individual sensors represents that off-shelf tactile sensors can be used in hardness-based classification. In case of two classes (H,S), individual sensors can perform up to 80% of accuracy in predicting 20% of testing data as validation. Literature suggests 80 to 95% [2] of accuracy/prediction while using customized and complex tactile sensors. Ultrasonic sensors [32] used in literature describing the hardness and softness take too much space within the gripping area, which is difficult to adjust in different required environments, which is the same for customized sensors. In terms of the combination of sensors like (F,V) = 85%, (P,F) = 89%, and (V,P) = 87%, accuracy obtained shows that the combination of sensors increases the prediction accuracy which indicates that more tactile information as a feature increases machine learning accuracy. And out of multiple algorithm approaches, it shows that RFC performs best among others. In different configurations, RFC also performs better. Based on three sensors (F,P,V), the configuration achieves around 93% which highest among all configurations of sensors as features. This indicates that an increasing number of features (tactile information or value) from different sensors about objects increases the chance of obtaining optimum accuracy. This also showcases that COTS have the capability to perform hardness classification.

5.2. Hardness Classification Outcome Based on Three Classes (H,S,F)

Hardness classification using three classes showcases results obtained from multiple algorithms presented in Figure 7. In comparison to two classes (H,S), the outcome from three classes (H,S,F) has shown less accuracy among different configurations. In individual sensors, accuracy drops below 70%; in the combination of two (F,V), (P,F), and (V,P), accuracy drops below 80%; only three combinations of individual sensor data (F,P,V) show 82% accuracy by the RFC algorithm which was best among all configurations. This outcome represents that on an increasing number of classes, accuracy decreases among ML algorithms. Accuracy obtained from three classes (H,S,F) was not investigated much in literature; this result shows that multiple sensors' data obtained from COTS as tactile information/ features can perform multiclass hardness classification with 80%+ accuracy with limited data.

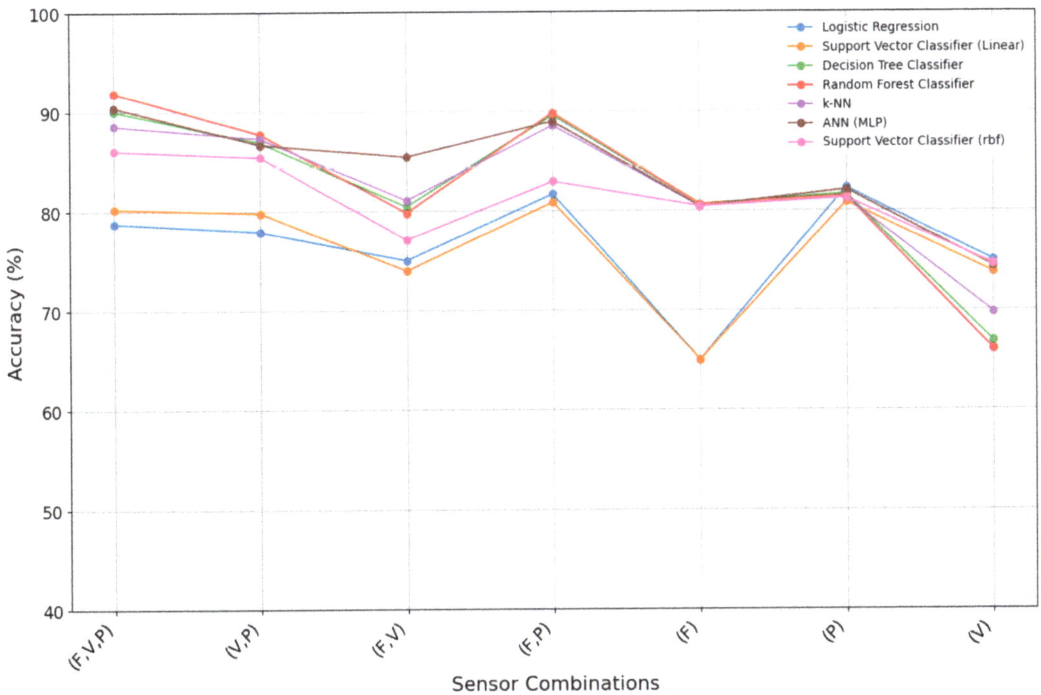

Figure 6. Describes the binary classification (H,S) outcomes among different sensor data configurations or combination using multiple algorithm outputs. Each algorithm is evaluated based on accuracy scores obtained from predictions on test data, which constitutes 20% of the overall dataset remaining unseen during training.

5.3. Result from Best Algorithm and Sensors Configuration—Multiclass Output

From class two (Section 5.1) and class three (Section 5.2) results, it been understood that the output of multiple algorithms indicates in most cases that RFC performs well. Multiple features from sensors with more tactile information (F,P,V) perform overall better. Three classes (H,S,F) with three sensors (F,V,P) configuration results outcome inspires a further look into including a fourth class from the Shore scale which is ES (extra soft). For ES objects, a white sponge was included, and ML models were trained again with all configurations. Figure 8 shows all results obtained from RFC and shows the case that binary classification remains optimum in most configurations where other classes' accuracy falls but shows the capability of performing hardness classification. In four object classes (H,S,ES,F), the (RFC) in (F,V,P) combination performs better among all other configurations achieving accuracy around 79%. Overall, this result also indicates that (F,V,P) combination performs better in all cases.

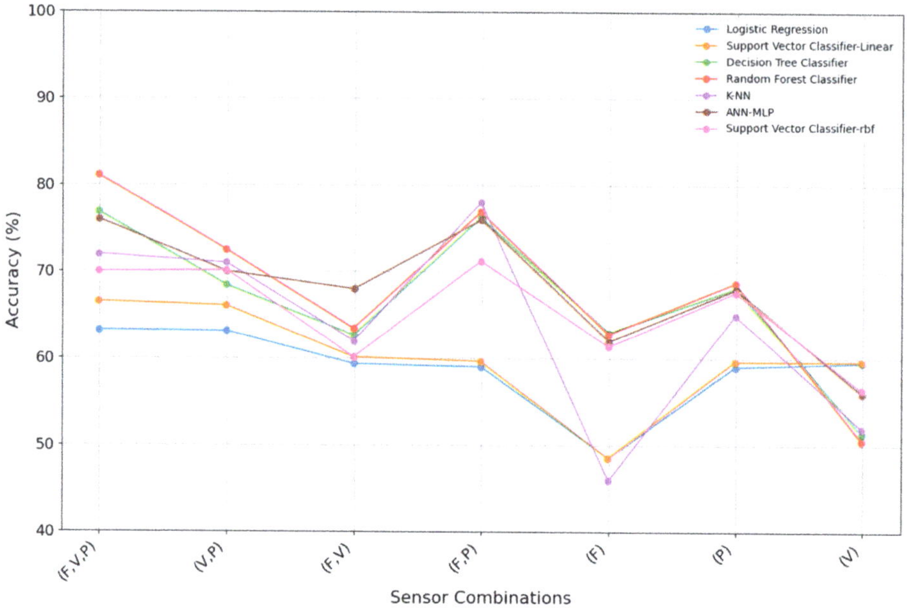

Figure 7. Illustrates the ternary classification (H,S,F) outcomes among different sensor data configurations or combination using multiple algorithm outputs. Each algorithm is evaluated based on accuracy scores obtained from predictions on test data, which constitutes 20% of the overall dataset remaining unseen during training.

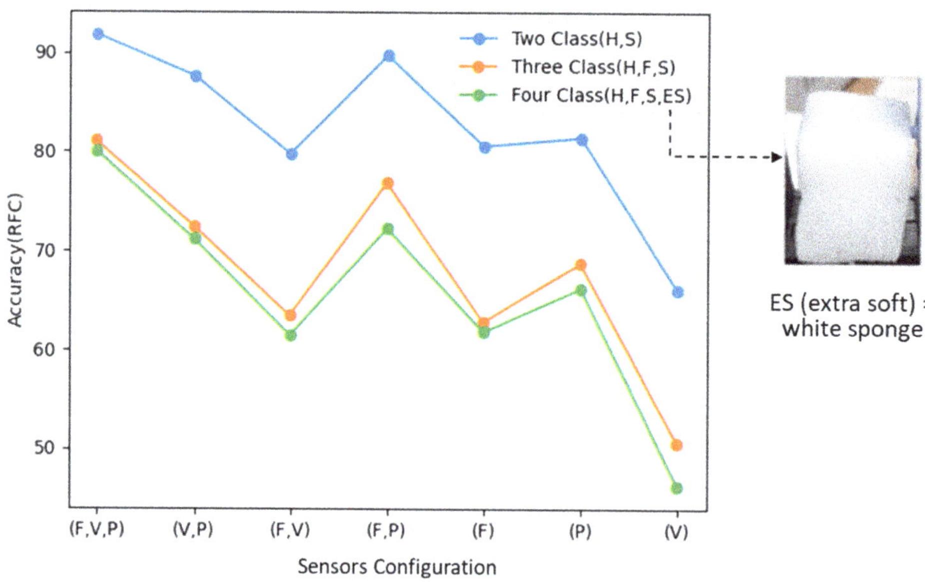

Figure 8. Illustrates the outcomes of multiclass hardness classification accuracy, demonstrating various sensor configurations. The best-performing configuration, utilizing three sensors (F,V,P), achieved optimal results, as highlighted by the outcomes obtained from the random forest classifier (RFC) algorithm.

6. Conclusions and Future Scope

This study demonstrates the feasibility and practicality of using commercial off-the-shelf (COTS) tactile sensors for multiclass hardness classification in robotic applications, specifically showcasing their scalability, cost-effectiveness, and integration ease compared to customized tactile sensors. It particularly underscores the advantages of configurations that incorporate three sensors (F,V,P), which consistently outperform simpler setups with accuracy rates between 80–92% across binary, ternary, and quaternary classifications. These findings highlight not only the feasibility of employing COTS sensors in varied robotic tasks but also their comparability in performance to more complex sensor systems documented in existing literature, where accuracy ranges from 50–97% (binary classification). Notably, the random forest classifier (RFC) was found to be particularly effective, likely due to its robust pattern recognition capabilities within the diverse data sets involving sensor values and target classifications.

In binary classification, individual sensors (F) and (P) have achieved accuracies above 80%, indicating that a single sensor may not be sufficient for comprehensive hardness classification. But individual sensors showcase the possibility of an accuracy score comparable to more complex grid or array sensors documented in the literature, which typically achieve accuracies between 50–90% [41]. This comparison highlights the viability of commercial off-the-shelf (COTS) sensors for rapid deployment in robotic applications, emphasizing their potential for broader use in texture classification and other areas. The selectivity of these sensors, inspired by human mechanoreceptors, is crucial. It enables the capture of diverse force, pressure, and vibration signals from various objects, enhancing hardness classification. This mechanoreceptor-based selectivity is especially beneficial when sensor data from multiple sources (F,V,P) are combined, suggesting future exploration into other bio-inspired sensors such as thermoreceptors for temperature, and optical receptors for light detection.

However, there are notable limitations to consider. The current testing procedures, which primarily involve uniformly shaped square objects, do not fully represent the range of real-world scenarios. The data collected from only four to five objects is limited in quantity, which may restrict the applicability of the tests in real-world and real-time scenarios. This limitation poses a challenge for performing hardness classification with COTS sensors, necessitating tests with a larger number of objects. Including more objects based on the Shore scale could either decrease or increase the accuracy of COTS sensors, which remains uncertain and highlights the need for future research. The resolution and sensitivity of the COTS sensors may not be sufficient for detecting fine differences in hardness, requiring tests with materials of various properties. Furthermore, testing conducted under controlled environments does not account for the variability in real-world conditions, such as temperature, humidity, and platform differences, which could affect sensor performance. Additionally, sensor alignment with various robotic grippers and different objects may yield varying outcomes. This underscores the potential for future work exploring the use of COTS sensors with dexterous robotic grippers to understand their performance across diverse applications. The use of single-sensor data, followed by their combination, can be time-consuming in real-time applications, posing a significant drawback. These factors emphasize the need for more comprehensive testing on objects of different sizes and shapes to accurately assess the capabilities and limitations of COTS sensors in various classification applications. Overall, this study has revealed several significant findings. Firstly, it demonstrated the feasibility and accessibility of hardness classification using commercial off-the-shelf (COTS) sensors, which require minimal processing time and are readily deployable in various robotics environments. Secondly, configurations using three sensors (F,P,V) consistently outperformed others, proving particularly effective in binary classification, although they were less effective in ternary and quaternary scenarios but performed optimally in comparison to others. Notably, among the various machine learning algorithms tested, the random forest classifier (RFC) exhibited optimal performance. This is likely due to RFC's ability to effectively discern patterns within the training data, especially within the subset containing sensor values and target classifications. The achieved accuracy underscores the potential of COTS sensors, yielding results comparable to those documented in existing literature. These findings suggest significant potential for the extensive use of COTS sensors in robotic tactile sensing applications. Additionally, it has shown potential to explore layered or topology of COTS sensors to identify the optimal configurations using a bio-inspired (mechanoreceptor) approach. Future research will focus on analysing tactile information gathered collectively from layered sensors and performed in real-time predictions with unknown or new objects. In future work, we plan to expand the variety of materials testing to include broader parameters like textures, densities, gradient hardness, and composite structures to better evaluate the sensors' capabilities in diverse real-world scenarios. We may also explore advanced machine learning models, deep learning approaches, and ensemble techniques to improve the accuracy and robustness of multiclass classification tasks. In the case of improving ML models, various filtering techniques to enhance the quality of sensor data can also be considered in future work.

Author Contributions: Conceptualization, Y.S. and P.F.; methodology, Y.S. and P.F.; software, Y.S.; validation, Y.S. and P.F.; formal analysis, Y.S. and P.F.; investigation, Y.S.; resources, Y.S.; data curation, Y.S.; writing—original draft preparation, Y.S.; writing—review and editing, Y.S., P.F. and L.J.; visualization, Y.S.; supervision, P.F. and L.J.; project administration, P.F. and Y.S. All authors have read and agreed to the published version of the manuscript.

Funding: This research received no external funding.

Data Availability Statement: Data is contained within the article.

Conflicts of Interest: The authors declare no conflict of interest.

References

1. Jin, J.; Wang, S.; Zhang, Z.; Mei, D.; Wang, Y. Progress on flexible tactile sensors in robotic applications on objects properties recognition, manipulation and human-machine interactions. *Soft Sci.* **2023**, *3*, 8. [CrossRef]
2. Eguíluz, A.G.; Rañó, I.; Coleman, S.A.; McGinnity, T.M. Multimodal Material identification through recursive tactile sensing. *Robot. Auton. Syst.* **2018**, *106*, 130–139. [CrossRef]
3. Yi, Z.; Zhang, Y.; Peters, J. Bioinspired tactile sensor for surface roughness discrimination. *Sens. Actuators A Phys.* **2017**, *255*, 46–53. [CrossRef]
4. Li, G.; Liu, S.; Wang, L.; Zhu, R. Skin-inspired quadruple tactile sensors integrated on a robot hand enable object recognition. *Sci. Robot.* **2020**, *5*, 46–53. [CrossRef]
5. Li, F.; Wang, R.; Song, C.; Zhao, M.; Ren, H.; Wang, S.; Liang, K.; Li, D.; Ma, X.; Zhu, B.; et al. A Skin-Inspired Artificial Mechanoreceptor for Tactile Enhancement and Integration. *ACS Nano* **2021**, *15*, 16422–16431. [CrossRef] [PubMed]
6. Iheanacho, F.; Vellipuram, A.R. *Physiology, Mechanoreceptors*; StatPearls Publishing: Tampa, FL, USA, 2023.
7. Dahiya, R.; Oddo, C.; Mazzoni, A.; Jörntell, H. Biomimetic tactile sensing. In *Biomimetic Technologies*; Elsevier: Amsterdam, The Netherlands, 2015; pp. 69–91. [CrossRef]
8. Amin, Y.; Gianoglio, C.; Valle, M. Embedded real-time objects' hardness classification for robotic grippers. *Future Gener. Comput. Syst.* **2023**, *148*, 211–224. [CrossRef]
9. Song, Y.; Lv, S.; Wang, F.; Li, M. Hardness-and-Type Recognition of Different Objects Based on a Novel Porous Graphene Flexible Tactile Sensor Array. *Micromachines* **2023**, *14*, 217. [CrossRef]
10. Qian, X.; Li, E.; Zhang, J.; Zhao, S.-N.; Wu, Q.-E.; Zhang, H.; Wang, W.; Wu, Y. Hardness Recognition of Robotic Forearm Based on Semi-supervised Generative Adversarial Networks. *Front. Neurorobot.* **2019**, *13*, 73. [CrossRef] [PubMed]
11. Jamali, N.; Sammut, C. Material classification by tactile sensing using surface textures. In Proceedings of the 2010 IEEE International Conference on Robotics and Automation, Anchorage, AK, USA, 3–7 May 2010; IEEE: New York, NY, USA, 2010; pp. 2336–2341.
12. Konstantinova, J.; Cotugno, G.; Stilli, A.; Noh, Y.; Althoefer, K. Object classification using hybrid fiber optical force/proximity sensor. In *2017 IEEE SENSORS*; IEEE: New York, NY, USA, 2017; pp. 1–3. [CrossRef]
13. Dai, K.; Wang, X.; Rojas, A.M.; Harber, E.; Tian, Y.; Paiva, N.; Gnehm, J.; Schindewolf, E.; Choset, H.; Webster-Wood, V.A.; et al. Design of a Biomimetic Tactile Sensor for Material Classification. *arXiv* **2022**, arXiv:2203.15941.
14. Madrigal, D.; Torres, G.; Ramos, F.; Vega, L. Cutaneous mechanoreceptor simulator. In Proceedings of the 2012 IEEE 3rd International Conference on Cognitive Infocommunications (CogInfoCom), Kosice, Slovakia, 2–5 December 2012; IEEE: New York, NY, USA, 2012; pp. 781–786. [CrossRef]
15. Najarian, S.; Dargahi, J.; Mehrizi, A.A. *Artificial Tactile Sensing in Biomedical Engineering*; McGraw-Hill Education: New York, NY, USA, 2009.

16. Chun, S.; Kim, J.-S.; Yoo, Y.; Choi, Y.; Jung, S.J.; Jang, D.; Lee, G.; Song, K.-I.; Nam, K.S.; Youn, I.; et al. An artificial neural tactile sensing system. *Nat. Electron.* **2021**, *4*, 429–438. [CrossRef]
17. Pattnaik, D.; Sharma, Y.; Saveliev, S.; Borisov, P.; Akther, A.; Balanov, A.; Ferreira, P. Stress-induced Artificial neuron spiking in Diffusive memristors. *arXiv* **2023**, arXiv:2306.12853.
18. Kalita, H.; Krishnaprasad, A.; Choudhary, N.; Das, S.; Dev, D.; Ding, Y.; Tetard, L.; Chung, H.-S.; Jung, Y.; Roy, T. Artificial Neuron using Vertical MoS_2/Graphene Threshold Switching Memristors. *Sci. Rep.* **2019**, *9*, 53. [CrossRef] [PubMed]
19. Lucarotti, C.; Oddo, C.; Vitiello, N.; Carrozza, M. Synthetic and Bio-Artificial Tactile Sensing: A Review. *Sensors* **2013**, *13*, 1435–1466. [CrossRef]
20. Spigler, G.; Oddo, C.M.; Carrozza, M.C. Soft-neuromorphic artificial touch for applications in neuro-robotics. In Proceedings of the 2012 4th IEEE RAS & EMBS International Conference on Biomedical Robotics and Biomechatronics (BioRob), Rome, Italy, 24–27 June 2012; IEEE: New York, NY, USA, 2012; pp. 1913–1918. [CrossRef]
21. Bounakoff, C.; Hayward, V.; Genest, J.; Michaud, F.; Beauvais, J. Artificial fast-adapting mechanoreceptor based on carbon nanotube percolating network. *Sci. Rep.* **2022**, *12*, 2818. [CrossRef]
22. Kerr, E.; McGinnity, T.M.; Coleman, S. Material recognition using tactile sensing. *Expert Syst. Appl.* **2018**, *94*, 94–111. [CrossRef]
23. Cirillo, A.; Laudante, G.; Pirozzi, S. Tactile Sensor Data Interpretation for Estimation of Wire Features. *Electronics* **2021**, *10*, 1458. [CrossRef]
24. Drimus, A.; Kootstra, G.; Bilberg, A.; Kragic, D. Design of a flexible tactile sensor for classification of rigid and deformable objects. *Rob. Auton. Syst.* **2014**, *62*, 3–15. [CrossRef]
25. Gao, Z.; Ren, B.; Fang, Z.; Kang, H.; Han, J.; Li, J. Accurate recognition of object contour based on flexible piezoelectric and 9piezoresistive dual mode strain sensors. *Sens. Actuators A Phys.* **2021**, *332*, 113121. [CrossRef]
26. Suslak, T. There and Back again: A Stretch Receptor's Tale. 2015. Available online: https://era.ed.ac.uk/handle/1842/10474 (accessed on 1 April 2024).
27. Hosoda, K.; Tada, Y.; Asada, M. Anthropomorphic robotic soft fingertip with randomly distributed receptors. *Rob. Auton. Syst.* **2006**, *54*, 104–109. [CrossRef]
28. "Somatosensation-collective term for sensory signals from the body; Neuroscience Online. (n.d.). Somatosensory Processes (Section 2, Chapter 5). The University of Texas Medical School at Houston. Available online: https://nba.uth.tmc.edu/neuroscience/m/s2/chapter05.html (accessed on 17 June 2024).
29. Luo, S.; Bimbo, J.; Dahiya, R.; Liu, H. Robotic tactile perception of object properties: A review. *Mechatronics* **2017**, *48*, 54–67. [CrossRef]
30. Shimizu, T.; Shikida, M.; Sato, K.; Itoigawa, K. A New Type of Tactile Sensor Detecting Contact Force and Hardness of an Object. In Proceedings of the Technical Digest. MEMS 2002 IEEE International Conference, Fifteenth IEEE International Conference on Micro Electro Mechanical Systems (Cat. No.02CH37266), Las Vegas, NV, USA, 24–24 January 2002.
31. IEEE Robotics and Automation Society; Institute of Electrical and Electronics Engineers. *ICRA2017: IEEE International Conference on Robotics and Automation: Program, Singapore, 29 May–3 June 2017*; IEEE: New York, NY, USA, 2017.
32. Bouhamed, S.A.; Chakroun, M.; Kallel, I.K.; Derbel, H. Haralick feature selection for material rigidity recognition using ultrasound echo. In Proceedings of the 2018 4th International Conference on Advanced Technologies for Signal and Image Processing (ATSIP), Sousse, Tunisia, 21–24 March 2018; IEEE: New York, NY, USA, 2018.
33. Schmidt, P.A.; Maël, E.; Würtz, R.P. A sensor for dynamic tactile information with applications in human–robot interaction and object exploration. *Rob. Auton. Syst.* **2006**, *54*, 1005–1014. [CrossRef]
34. Würschinger, H.; Mühlbauer, M.; Winter, M.; Engelbrecht, M.; Hanenkamp, N. Implementation and potentials of a machine vision system in a series production using deep learning and low-cost hardware. *Procedia CIRP* **2020**, *90*, 611–616. [CrossRef]
35. Dargahi, J.; Najarian, S. Human tactile perception as a standard for artificial tactile sensing—A review. *Int. J. Med. Robot. Comput. Assist. Surg.* **2004**, *1*, 23–35. [CrossRef] [PubMed]
36. Delmas, P.; Hao, J.; Rodat-Despoix, L. Molecular mechanisms of mechanotransduction in mammalian sensory neurons. *Nat. Rev. Neurosci.* **2011**, *12*, 139–153. [CrossRef] [PubMed]
37. Ding, S.; Bhushan, B. Tactile perception of skin and skin cream by friction induced vibrations. *J. Colloid. Interface Sci.* **2016**, *481*, 131–143. [CrossRef] [PubMed]
38. Yi, Z.; Zhang, Y.; Peters, J. Biomimetic tactile sensors and signal processing with spike trains: A review. *Sens. Actuators A Phys.* **2018**, *269*, 41–52. [CrossRef]
39. Kappassov, Z.; Corrales, J.-A.; Perdereau, V. Tactile sensing in dexterous robot hands—Review. *Rob. Auton. Syst.* **2015**, *74*, 195–220. [CrossRef]
40. Rahiminejad, E.; Parvizi-Fard, A.; Iskarous, M.M.; Thakor, N.V.; Amiri, M. A Biomimetic Circuit for Electronic Skin with Application in Hand Prosthesis. *IEEE Trans. Neural Syst. Rehabil. Eng.* **2021**, *29*, 2333–2344. [CrossRef]
41. Weng, J.; Yu, Y.; Zhang, J.; Wang, D.; Lu, Z.; Wang, Z.; Liang, J.; Zhang, S.; Li, X.; Lu, Y.; et al. A Biomimetic Optical Skin for Multimodal Tactile Perception Based on Optical Microfiber Coupler Neuron. *J. Light. Technol.* **2023**, *41*, 1874–1883. [CrossRef]
42. Kerr, E.; McGinnity, T.M.; Coleman, S. Material classification based on thermal properties—A robot and human evaluation. In Proceedings of the 2013 IEEE International Conference on Robotics and Biomimetics (ROBIO), Shenzhen, China, 12–14 December 2013; IEEE: New York, NY, USA, 2013; pp. 1048–1053. [CrossRef]

43. Pestell, N.; Lepora, N.F. Artificial SA-I, RA-I and RA-II/vibrotactile afferents for tactile sensing of texture. *J. R. Soc. Interface* **2022**, *19*, 20210603. [CrossRef]
44. Liu, F.; Deswal, S.; Christou, A.; Sandamirskaya, Y.; Kaboli, M.; Dahiya, R. Neuro-inspired electronic skin for robots. *Sci. Robot.* **2022**, *7*, eabl7344. [CrossRef]
45. Luque, N.R.; Garrido, J.A.; Ralli, J.; Laredo, J.J.; Ros, E. From Sensors to Spikes: Evolving Receptive Fields to Enhance Sensorimotor Information in a Robot-Arm. *Int. J. Neural. Syst.* **2012**, *22*, 1250013. [CrossRef]
46. Li, P.; Ali, H.P.A.; Cheng, W.; Yang, J.; Tee, B.C.K. Bioinspired Prosthetic Interfaces. *Adv. Mater. Technol.* **2020**, *5*, 1900856. [CrossRef]
47. MacKinnon, C.D. Sensorimotor anatomy of gait, balance, and falls. *Handb. Clin. Neurol.* **2018**, *159*, 3–26. [CrossRef]
48. Hunter, J.D. Matplotlib: A 2D Graphics Environment. *Comput. Sci. Eng.* **2007**, *9*, 90–95. [CrossRef]
49. McKinney, W. Data Structures for Statistical Computing in Python. *SciPy* **2010**, *445*, 56–61. [CrossRef]
50. Harris, C.R.; Millman, K.J.; van der Walt, S.J.; Gommers, R.; Virtanen, P.; Cournapeau, D.; Wieser, E.; Taylor, J.; Berg, S.; Smith, N.J.; et al. Array programming with NumPy. *Nature* **2020**, *585*, 357–362. [CrossRef]
51. Pedregosa, F.; Varoquaux, G.; Gramfort, A.; Michel, V.; Thirion, B.; Grisel, O.; Blondel, M.; Prettenhofer, P.; Weiss, R.; Dubourg, V.; et al. Scikit-learn: Machine learning in Python. *J. Mach. Learn Res.* **2011**, *12*, 2825–2830.
52. Stagi, A. Nanpy Firmware. Available online: https://github.com/nanpy/nanpy-firmware (accessed on 21 May 2024).

electronics

Article

Geometry Optimization of Stratospheric Pseudolite Network for Navigation Applications

Yi Qu [1,2,*], Sheng Wang [1,2], Hui Feng [1] and Qiang Liu [1]

[1] Aerospace Information Research Institute, Chinese Academy of Sciences, Beijing 100094, China; wangsheng@aircas.ac.cn (S.W.); fenghui@aircas.ac.cn (H.F.); liuqiang@aircas.ac.cn (Q.L.)
[2] University of Chinese Academy of Sciences, Beijing 100049, China
* Correspondence: quyi@aircas.ac.cn

Abstract: A stratospheric pseudolite (SP) is a pseudolite installed on a stratospheric airship. A stratospheric pseudolite network (SPN) is composed of multiple SPs, which shows promising potential in navigation applications because of its station-keeping capability, long service duration, and flexible deployment. Most traditional research about SPN geometry optimization has centered on geometric dilution of precision (GDOP). However, previous research rarely dealt with the topic of how SPN geometry configuration not only affects its GDOP, but also affects its energy balance. To obtain an optimal integrated performance, this paper employs the proportion of energy consumption in energy production as an indicator to assess SPN energy status and designs a composite indicator including GDOP and energy status to assess SPN geometry performance. Then, this paper proposes an SPN geometry optimization algorithm based on gray wolf optimization. Furthermore, this paper implements a series of simulations with an SPN composed of six SPs in a specific service area. Simulations show that the proposed algorithm can obtain SPN geometry solutions with good GDOP and energy balance performance. Also, simulations show that in the supposed scenarios and the specific area, a higher SP altitude can improve both GDOP and energy balance, while a lower SP latitude can improve SPN energy status.

Keywords: stratospheric airship; pseudolite network; geometric dilution of precision (GDOP); energy balance; gray wolf optimization

1. Introduction

Stratospheric airships are flight vehicles that can keep their flight altitude by buoyance. They can reside in the lower portion of the stratosphere and perform station-keeping missions for a long time. This capability can provide a very efficient flight for many missions, such as science exploration, communications, earth observations, and navigations.

Researchers have proposed the concept of stratospheric pseudolites (SPs), which means to install transmitters on stratospheric airships to send out GPS-like signals to improve GNSS performance. Furthermore, a stratospheric pseudolite network (SPN) can be constructed by multiple SPs, which can provide independent positioning, GNSS augmentation and other services, especially in case of degraded visibility of GNSS signals. An SPN can provide many advantages such as wide coverage, long service duration, and flexible deployment; therefore, it has attracted abundant attention. Tsujii established a minimum configuration of a GPS/SP system to augment GPS and implemented experiments in both static and kinematic modes [1]. Dovis introduced an SPN system architecture, and discussed some key issues about SPN performance, including SP positioning, pseudo-ephemeris broadcasting, and GDOP improvement [2]. Zheng carried out simulations for SPN in urban areas and proved its effect on improving the horizontal dilution of precision (HDOP) and 3D positioning accuracy [3]. Chandu designed an SPN framework, described its dataflow and mathematical model, and compared the feasibility of various SPN geometry configurations [4]. Dai presented SPN modeling strategies to deal with positioning error sources and proposed geometry optimization solutions for two application scenarios [5].

In previous research about SPN, geometry configuration has been identified as a critical factor affecting its positioning performance significantly. A series of approaches have been discussed to optimize the SPN geometry configuration, which can be divided into empirical methods and meta-heuristic optimization methods.

With empirical methods, Fateev suggested that an ideal pseudolite network should be composed of 5–10 pseudolites, which should be distributed along the edge of the service area and at different altitudes [6]. Sang studied the relationship between network geometry layout and geometric dilution of precision (GDOP) and illustrated three defective layouts that should be avoided in practice [7]. Hu, Gao, and Yang provided different geometry configurations for four SPs, five SPs, and six SPs, respectively, to minimize GDOP based on priori theories [8–10].

As to meta-heuristic optimization methods, Mosavi presented a pseudolite network geometry design approach with multiple evolutionary algorithms, including the genetic algorithm (GA), simulated annealing algorithm (SA), and particle swarm optimization algorithm (PSO) [11]. Shao offered a design strategy for pseudolite network geometry based on PSO and carried out indoor tests [12]. Tang put forward a multi-objective PSO algorithm for pseudolite network geometry design, whose purpose was to maximize the visual area while minimizing GDOP [13]. Yang adopted GA to search for optimal SPN to reduce the ephemeris error, ranging error, and positioning error in GNSS augmentation [14]. Chen utilized improved GA to select the best configuration aiming at enhancing accuracy in mobile positioning [15]. Song proposed an adaptive GA to realize geometry optimization under multiple constraints, especially under orographic terrain constraints and traffic facility constraints [16].

The studies listed above provided rich references for SPN geometry design. However, the energy balance requirement of SPN has rarely been analyzed, with even less consideration of the integrated optimization of SPN energy balance and GDOP.

In fact, the energy system of an SP is quite different from that of other aerial vehicles because it is always in the dynamic variation of energy production and energy consumption in its service duration. The balance between energy production and energy consumption is very vulnerable, since the energy gained from the solar arrays is quite limited, while the energy consumption due to resisting wind is enormous. Once its energy consumption exceeds energy production, an SP will encounter difficulties sustaining normal operation, making its service availability and continuity degrade greatly.

The main problem is that GDOP, SP energy production, and SP energy consumption are all governed by the SPN configuration. If GDOP is regarded as the sole objective during the course of SPN geometry design, an SPN unable to keep energy balance may be obtained. To avoid such an unpractical result, this paper proposes an SPN geometry design algorithm based on gray wolf optimization (GWO), pursuing the integrated optimization of GDOP and SPN energy balance.

This paper assumes the transmitting antennas onboard are directional antennas pointing down to the ground. Furthermore, this paper makes the following assumptions for an SP in its service duration, as Table 1 illustrates:

Table 1. Assumptions for an SP in its service duration.

Symbolic	Physical Meaning	Assumption
λ_j	longitude of the j-th SP	remain constant
Φ_j	latitude of the j-th SP	remain constant
h_j	altitude of the j-th SP	remain constant, 18–20 km as the preferred interval, and 20–28 km as the alternative interval [17,18]

Table 1. *Cont.*

Symbolic	Physical Meaning	Assumption
γ_j	yaw angle of the *j*-th SP	remain constant, do not affect the vector between a user receiver to the SP
α_j	pitch angle of the *j*-th SP	0
β_j	roll angle of the *j*-th SP	0
m_j	mass of the *j*-th SP	remain constant
J_{kaj}	energy spent on keeping flight altitude of the *j*-th SP	0
η_{vj}	photoelectric conversion efficiency	remain constant
J_{loss}	energy loss in the process of transfer, charging, storage, and discharging	0

The rest of this paper is organized as follows. The GDOP of SPN is analyzed in Section 2, an energy balance model of SPN is established in Section 3, an SPN geometry design algorithm based on GWO is proposed in Section 4, simulations and discussions are presented in Section 5, and conclusions and future works finally complete this paper.

2. GDOP of an SPN

GDOP is a widely used indicator in SPN performance assessment. It is defined as the statistics ratio of positioning accuracy, timing accuracy, and ranging accuracy. Given the same pseudo-range error, the smaller the GDOP is, the smaller the positioning error and timing error are.

In this paper, the GDOP of an SPN is defined as the average GDOP of multiple users in the SPN service area, which can be described by Equation (1).

$$GDOP_N = \frac{1}{n_u}\sum_{i=1}^{n_u} GDOP_i \tag{1}$$

In Equation (1), n_u represents the number of observers distributed in the service area. $GDOP_i$ represents the GDOP of the *i*-th user, which can be calculated by Equations (2)–(4) [19–21].

$$GDOP_i = \sqrt{tr(\mathbf{H}_i^T\mathbf{H}_i)}^{-1} \tag{2}$$

$$\mathbf{H}_i = \begin{bmatrix} a_{xi1} & a_{yi1} & a_{zi1} & 1 \\ a_{xi2} & a_{yi2} & a_{zi2} & 1 \\ \vdots & \vdots & \vdots & \vdots \\ a_{xin_p} & a_{yin_p} & a_{zin_p} & 1 \end{bmatrix} \tag{3}$$

$$\begin{aligned} a_{xij} &= \frac{x_j - x_{ui}}{\sqrt{(x_j-x_{ui})^2+(y_j-y_{ui})^2+(z_j-z_{ui})^2}} \\ a_{yij} &= \frac{y_j - y_{ui}}{\sqrt{(x_j-x_{ui})^2+(y_j-y_{ui})^2+(z_j-z_{ui})^2}} \\ a_{zij} &= \frac{z_j - z_{ui}}{\sqrt{(x_j-x_{ui})^2+(y_j-y_{ui})^2+(z_j-z_{ui})^2}} \end{aligned} \tag{4}$$

In Equations (2)–(4), $\mathbf{H}_i$ represents the observation matrix of the *i*-th user, n_p represents the number of SPs in the network, a_{xij}, a_{yij}, a_{zij} represent the vector components between the *i*-th user receiver and the *j*-th SP, (x_{ui}, y_{ui}, z_{ui}) represent the position of the *i*-th user in the ECEF (Earth-Centered, Earth-Fixed) coordinate system, and (x_j, y_j, z_j) represent the position of the *j*-th SP in the ECEF coordinate system, which is gained from (λ_j, Φ_j, h_j) by coordinate transformation.

3. Energy Balance of an SPN

3.1. Energy Consumption of an SP

SP energy is used to support its equipment, such as the propeller, flight controller, TT&C (Telemetry, Tracking, and Command) system, and mission payloads. The propeller consumes a great deal of energy, and it is deeply affected by SP position and attitude. The energy consumption of the other equipment is slightly affected by SP position and attitude, and their requested power is assumed to be constant in this paper.

3.1.1. Energy Consumption of a Propeller

For the j-th SP, the power required by its propeller can be estimated by Equation (5) [22].

$$P_{ej} = \boldsymbol{T}_j \boldsymbol{U}_j / \eta_{pj} / \eta_{ej} \tag{5}$$

In Equation (5):

P_{ej} represents propeller power;

$\boldsymbol{T}_j$ represents thrust generated by the propeller;

$\boldsymbol{U}_j$ represents airspeed;

η_{pj} and η_{ej} represent propeller efficiency and motor efficiency, respectively, which can be assumed as constants according to the analysis in [22].

So, the energy consumption of the propulsion propeller of the j-th SP, represented by J_{ej}, can be expressed by Equation (6).

$$J_{ej} = \int_0^{t_s} \boldsymbol{T}_j \boldsymbol{U}_j / \eta_{pj} / \eta_{ej} dt \tag{6}$$

In Equation (6), t_s represents the station-keeping time of the SP.

To facilitate analysis, SP motion can be decomposed into motion along the axis direction and motion along the normal direction. The propeller is assumed to be able to generate thrusts along the axis direction and along the normal direction independently, as illustrated in Figure 1.

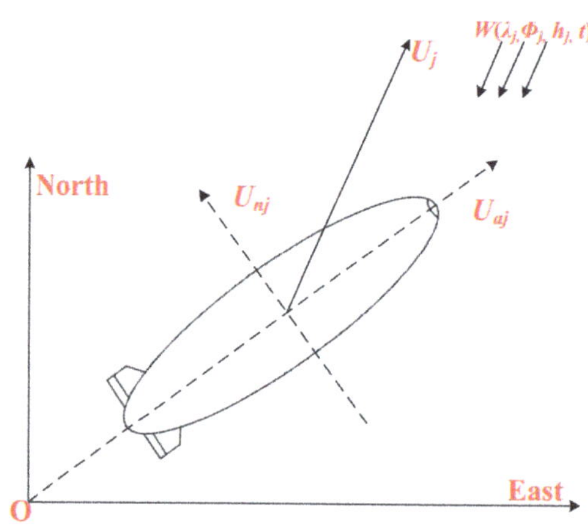

Figure 1. Illustration of an SP motion decomposition in the horizontal plane (since both pitch angle and roll angle of SPs are assumed as 0, this figure only illustrates the motion decomposition in the horizontal plane).

Then, Equation (6) can be rewritten into Equation (7) [23].

$$J_{ej} = \int_0^{t_s} (\boldsymbol{T}_{aj}\boldsymbol{U}_{aj} + \boldsymbol{T}_{nj}\boldsymbol{U}_{nj})/\eta_{pj}/\eta_{ej}dt \tag{7}$$

In Equation (7):

$\boldsymbol{T}_{aj}$ represents the axial component of thrust $\boldsymbol{T}_j$;

$\boldsymbol{U}_{aj}$ represents the axial component of airspeed $\boldsymbol{U}_j$;

$\boldsymbol{T}_{nj}$ represents the normal component of thrust $\boldsymbol{T}_j$;

$\boldsymbol{U}_{nj}$ represents the normal component of airspeed $\boldsymbol{U}_j$;

$\boldsymbol{T}_{aj}$ and $\boldsymbol{T}_{nj}$ can be estimated by aerodynamic resistance, and $\boldsymbol{U}_{aj}$ and $\boldsymbol{U}_{nj}$ can be estimated by local wind speed, as described in Equation (8).

$$\begin{cases} \boldsymbol{T}_{aj} = -\boldsymbol{D}_{aj} = -\frac{1}{2}\rho_{mj}V_j^{2/3}\boldsymbol{U}_{aj}^2 C_{Dj} \\ \boldsymbol{U}_{aj} = -\boldsymbol{W}(\lambda_j, \phi_j, h_j, t)\cos(\gamma_j - \gamma_w) \\ \boldsymbol{T}_{nj} = -\boldsymbol{D}_{nj} = -\frac{1}{2}\rho_{mj}V_j^{2/3}\boldsymbol{U}_{nj}^2 C_{Dj} \\ \boldsymbol{U}_{nj} = -\boldsymbol{W}(\lambda_j, \phi_j, h_j, t)\sin(\gamma_j - \gamma_w) \end{cases} \tag{8}$$

In Equation (8):

$\boldsymbol{D}_{aj}$ represents the aerodynamic resistance along the axis direction;

$\boldsymbol{D}_{nj}$ represents the aerodynamic resistance along the normal direction;

$\boldsymbol{W}(\lambda_j, \Phi_j, h_j, t)$ represents the wind speed at position (λ_j, Φ_j, h_j) and at time t;

ρ_{mj} represents the atmosphere density at altitude h_j;

V_j represents the volume of the j-th SP, and $V_j^{2/3}$ is used to estimate its reference area;

C_{Dj} represents its aerodynamic resistance coefficient;

γ_w represents the local wind direction angle.

Wind speed $\boldsymbol{W}(\lambda_j, \Phi_j, h_j, t)$, wind direction angle γ_w, and atmosphere density ρ_{mj} change enormously with position, as illustrated in Figures 2–4.

Figure 2 illustrates the meridional and zonal wind speeds in a specific area at an altitude of 20 km. It can be seen from Figure 2 that both meridional wind speed and zonal wind speed vary greatly with latitude in the area. The minimum zonal wind speed is about 8 m/s, while the maximum zonal wind speed has doubled to about 16 m/s. The meridional wind speed has undergone a directional reverse.

Figure 3 illustrates the meridional and zonal wind speeds at a certain location within the altitude interval of 18–30 km at four typical times in March, June, September, and December. From Figure 3, it can be seen that the zonal wind speed varies greatly with altitude changes, whose difference can reach tens of meters. Also, the meridional wind speed shows slight changes.

Figure 4 illustrates the atmospheric density variation of the U.S. standard atmosphere model at an altitude interval of 18–30 km. At an altitude of 18 km, the atmosphere density is about 0.12 kg/m^3, while at an altitude of 30 km, the atmosphere density rapidly decays to less than 0.02 kg/m^3. The attenuation of the atmosphere density exceeds 80%, reflecting a significant change.

These differences illustrated in Figures 2–4 can lead to a huge difference in the energy consumption of a propeller.

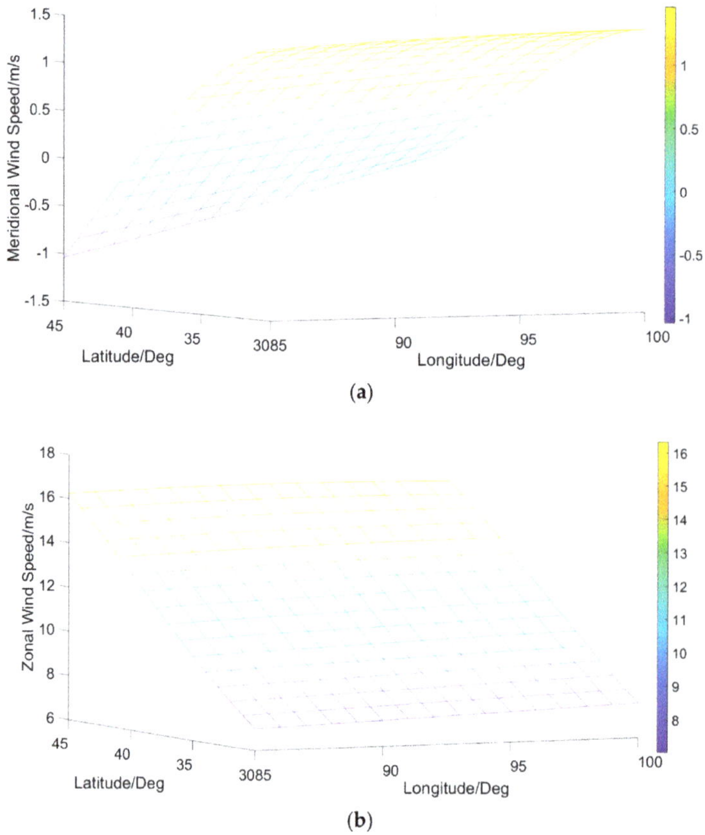

Figure 2. Horizontal wind speeds for a specific region at a certain time: (**a**) meridional wind; (**b**) zonal wind.

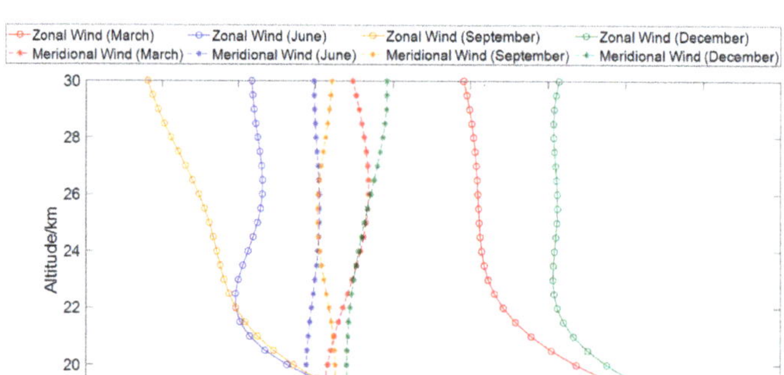

Figure 3. Horizontal winds at different altitudes for a specific region.

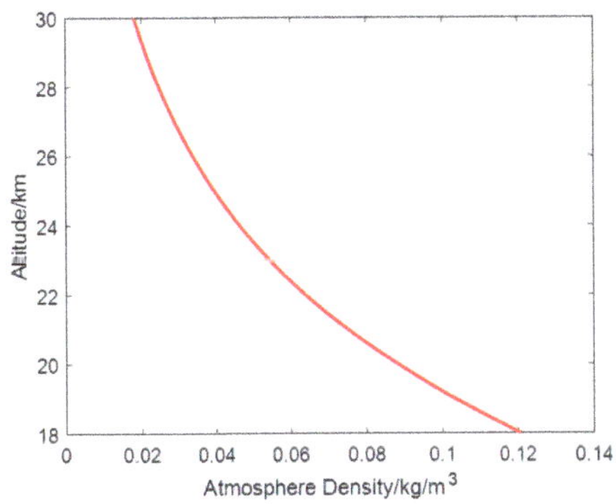

Figure 4. Atmosphere density from 18 km to 30 km according to U.S. Standard Atmosphere, 1976 [24].

3.1.2. Total Energy Consumption of an SP

The energy consumption of other equipment is only slightly affected by the SP position and attitude. Thus, their energy consumption is assumed to be a constant represented by P_{pj}. Their energy consumption, represented by J_{pj}, can be expressed simply by Equation (9).

$$J_{pj} = P_{pj}\, t_s \tag{9}$$

The total energy consumption of the j-th SP, represented by J_j, can be seen as the sum of propeller energy consumption and other equipment energy consumption, which can be expressed by Equation (10).

$$J_j = J_{ej} + J_{pj} \tag{10}$$

3.2. Energy Production of an SP

SP energy production relies on the solar arrays laying on its airship surface, which can convert solar radiation into electricity. Solar radiation harvested by the solar arrays can be divided into direct radiation, scattered radiation, and reflected radiation. Since the effect of scattered radiation and reflected radiation is far less than direct radiation, this paper emphasizes direct radiation and ignores scattered radiation and reflected radiation.

3.2.1. Solar Direct Radiation

The solar direct radiation intensity on the top of the atmosphere in the normal direction, represented by I_{top}, can be expressed by Equation (11) [25].

$$I_{top} = I_{SC}\, E_c \tag{11}$$

In Equation (11), I_{SC} represents the solar constant, and E_c represents a sun–earth distance correction, which can be expressed by Equation (12) [25,26].

$$E_c = 1 + 0.033\cos(2\pi d_n/365) \tag{12}$$

In Equation (12), d_n represents the day number in a year.

The solar direct radiation intensity at different altitudes is affected by the atmosphere transmissivity, which can be expressed by Equation (13) [26].

$$I_{Dj} = I_{top}\tau_j \tag{13}$$

In Equation (13), I_{Dj} and τ_j represent solar direct radiation intensity and atmosphere transmissivity at altitude h_j, respectively. τ_j can be calculated by Equation (14) [27].

$$\tau_j = 0.56(\mathrm{e}^{-0.65\lambda mj} + \mathrm{e}^{-0.95\lambda mj}) \tag{14}$$

In Equation (14), λ_{mj} can be estimated by Equation (15) [27].

$$\lambda_{mj} = \frac{p_j}{p_0}[\sqrt{1229 + (614 \sin \theta_{ej})^2} - 614 \sin \theta_{ej}] \tag{15}$$

In Equation (15), p_j represents the pressure of the atmosphere at altitude h_j; p_0 represents the pressure of the atmosphere at sea level.

θ_{ej} represents the solar elevation angle at latitude Φ_j, which can be expressed by Equation (16) [26].

$$\sin \theta_{ej} = \sin \phi_j \sin \delta + \cos \phi_j \cos \delta \cos \omega_j \tag{16}$$

In Equation (16), δ represents sun declination; ω_j represents sun hour angle.

The sun declination δ is the angle of the sun above or below the equator plane. It changes with the date. It will reach its maximum value (23.45 degree) at the summer solstice in the northern hemisphere, and will reach its minimum value (−23.45 degree) at the winter solstice. δ can be roughly calculated by Equation (17) [25].

$$\delta = \begin{cases} 23.45 \sin(2\pi(d_n - 81)/365) & d_n > 81 \\ 23.45 \sin(2\pi(d_n + 284)/365) & d_n \leq 81 \end{cases} \tag{17}$$

The sun hour angle ω_j is the angle between the sun and the local meridian, which changes 15 degrees per hour and can be calculated by [25].

$$\omega_j = 15(t - 12) \tag{18}$$

3.2.2. Energy Production of a Solar Array

The solar array mounted on the airship surface can be divided into multiple cells to analyze its energy production precisely. If the area of cell k is represented by A_{jk}, the angle of the solar direction vector and the normal vector of cell k is represented by ψ_{jk}, and the output power of cell k produced by solar direct radiation, represented by P_{Djk}, can be calculated by Equation (19) [28,29].

$$P_{Djk} = I_{Dj} A_{jk} cos\psi_{jk} \tag{19}$$

The aggregate output power of all the cells in the solar array, represented by P_{PVj}, can be calculated by Equation (20).

$$P_{PVj} = \sum_{k=1}^{n_c} P_{Djk} \tag{20}$$

In Equation (20), n_c represents the number of solar array cells. The energy production of the j-th SP in its station-keeping time, represented by Q_j, can be expressed as Equation (21).

$$Q_j = \int_0^{t_s} P_{PVj} \eta_{Vj} dt \tag{21}$$

In Equation (21), η_{Vj} represents the photoelectric conversion efficiency of the solar array.

Figure 5 illustrates diurnal energy productions of an SP at different positions, reflecting energy production gaps caused by SP positions.

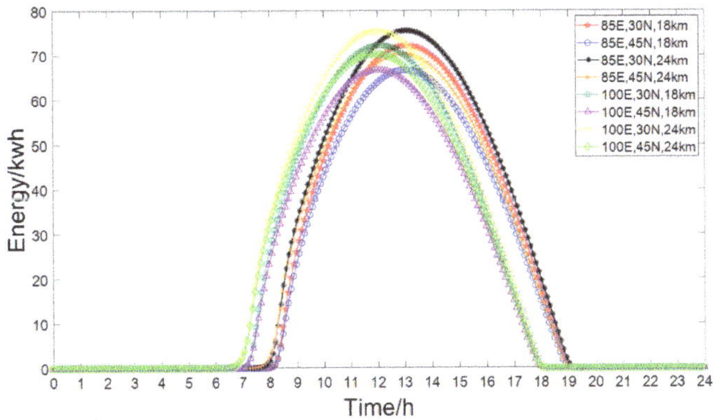

Figure 5. Comparison of energy productions of an SP at different positions on a specific day.

3.3. Energy Balance Indicator of an SPN

According to the analysis above, this paper proposes an energy balance indicator to assess an SP's energy status, which can be formulized as Equation (22).

$$B_j = J_j / Q_j \tag{22}$$

In Equation (22):

J_j represents the energy consumption of the j-th SP, which is defined in Equation (10);

Q_j represents the energy production of the j-th SP, which is defined in Equation (21);

B_j represents the ratio of energy consumption and energy production, reflecting the energy status of the j-th SP.

For an SPN, its energy balance can be assessed by the sum of all the individual energy balance indicators, which can be expressed as Equation (23).

$$B = \sum_{j=1}^{n_p} B_j \tag{23}$$

4. SPN Geometry Optimization Algorithm

4.1. Objective Function and Constraints

SPN design is a multi-objective optimization problem with multiple constraints. This study has two objectives: the first is to minimize SPN GDOP in a given area, which is defined in Equation (1); the second is to minimize the SPN energy balance indicator, which is defined in Equation (23).

Thus, the overall optimization objective function, represented by F, can be expressed as

$$\begin{gathered} F = w_1 GDOP_N + w_2 B \\ w_1 + w_2 = 1 \end{gathered} \tag{24}$$

In Equation (24), w_1 and w_2 represent the weight of GDOP and energy balance in the overall objective function, respectively. Their values can be adjusted within the interval [0, 1] according to requirements, keeping their sum as 1. If w_1 is set as 0, it means that GDOP will be ignored in the process of SPN geometry optimization, and if w_2 is set as 0, it means that energy balance will be ignored.

Two optimization constraints are emphasized in this study. The first is no co-location constraint, meaning that all the SPs in the network should not be deployed in the same position. It can be attributed to the large volume of airships, whose length can reach hundreds of meters. So, a distance is required between SPs to ensure safety, and the distance can be determined according to practical factors such as SP length. The second is the individual energy balance constraint, requiring each SP in the network to maintain its individual energy balance, which means that for any $j \in [1, n_p]$,

$$B_j < 1 \tag{25}$$

4.2. SPN Geometry Optimization Based on GWO

GWO is a widely used optimization algorithm in many fields [30,31]. It is a meta-heuristic optimization algorithm developed by Mirjalili in 2014 that mimics the hunting behavior and leadership hierarchy of gray wolves [30]. Compared with traditional empirical methods, GWO requires neither gradient information nor continuous derivative of objective functions. Compared with other meta-heuristic optimization algorithms, such as GA, PSO, and SA, GWO has fairly competitive performances [30]. Therefore, GWO is employed in this paper to implement SPN geometry optimization.

In GWO, a solution for a problem is regarded as a gray wolf, and all the available solutions are regarded as a wolf population. Gray wolves in the population are divided into four types: alpha, beta, delta, and omega, representing the current best, sub-optimal, the third best, and other solutions, respectively. The search for the optimal solution is performed by three important strategies: approaching, surrounding, and attacking prey. For a detailed discussion about the strategies, please refer to [30].

In this paper, a gray wolf is defined as the positions of all the SPs in an SPN. The fitness of a gray wolf can be calculated by Equation (24), which represents the weighted sum of GDOP and energy balance indicators, reflecting the comprehensive performance of a gray wolf. From the analysis above, it can be seen that the goal of SPN geometry optimization is to find the SP positions that entail minimizing the objective function (24). This is equivalent to finding the gray wolf that obtains the minimum fitness. Specific steps of SPN geometry optimization can be described as follows [30].

Step 1: Initialize GWO parameters, including population size n_w, maximum iteration number n_m, and others. Set the optimal population fitness as infinite and set the iteration counter as 1.

Step 2: Initialize a gray wolf population randomly.

Step 3: For each gray wolf in the current population, check whether it meets the no co-location constraint discussed in Section 4.1. If not, modify the co-located SP positions until the no co-location constraint is met.

Step 4: For each gray wolf in the current population, check whether it meets the individual energy balance constraint discussed in Section 4.1. If not, set the fitness of the gray wolf as infinite.

Step 5: For each gray wolf to meet the constraints in Section 4.1, calculate its fitness defined in Equation (24).

Step 6: Select the minimal fitness as the optimal population fitness. Update the alpha wolf according to the gray wolf with the minimal fitness.

Step 7: Update all the gray wolves in the population with GWO strategies, such as approaching, surrounding, and attacking prey.

Step 8: For each updated gray wolf, check whether it is outside of the search space. If so, put it on the edge of the search space.

Step 9: Increase the iteration counter by 1.

Step 10: Stop the iteration if the iteration counter reaches the maximum iteration number n_m; otherwise, go to step 3 and continue the iteration.

Step 11: Iteration ends. The current alpha wolf is returned as the optimal solution, and an SPN geometry configuration can be achieved based on the optimal solution.

5. Simulation and Discussion

To verify the proposed algorithm, simulations are carried out in the scenario of SPN positioning. The simulation environment is MATLAB 2018b, and the simulation parameters are shown in Table 2. It is assumed that users are distributed in the service area uniformly with an interval of 0.3 degree, which means there are 11 users in both the longitude direction and the latitude direction.

Table 2. Simulation parameter setting.

Symbolic	Physical Meaning	Value
n_p	number of SPs in the network	6
V_{pj}	volume of stratospheric airship/m^3	100,000
C_{Dj}	aerodynamic resistance coefficient	0.027
η_{ej}	electric motor efficiency	0.7
η_{pj}	propeller propulsive efficiency	0.9
η_{vj}	photoelectric conversion efficiency	16%
P_{pj}	other equipment power/W	100
I_{SC}	the solar constant/W/m^2	1367
n_w	wolf population size	100
n_m	maximum iteration number	50
w_1	weight of GDOP in the objective function	0.5
w_2	weight of energy balance in the objective function	0.5
λ_{min}	minimum longitude/degree	90E
λ_{max}	maximum longitude/degree	93E
Φ_{min}	minimum latitude/degree	37N
Φ_{max}	maximum latitude/degree	40N
h_{min}	minimum altitude to deploy SP/km	18
h_{max}	maximum altitude to deploy SP/km	20
d_{st}	start day of the station-keeping (day number in a year)	80
d_{ed}	end day of the station-keeping (day number in a year)	80
t_{st}	start time of the station-keeping (hour in a day)	0
t_{ed}	end time of the station-keeping (hour in a day)	24
n_u	the number of observers distributed in the service area	121

5.1. Comparison of Simulations with/without Consideration of Energy Balance Requirement

To compare the proposed algorithm and traditional algorithms without consideration of the energy balance requirement, two simulations are carried out with almost the same steps and conditions as listed above, except for two differences.

The first difference is the weight assignment in the optimization object function, i.e., w_1 and w_2 in Equation (24). In the simulation without considering the energy balance requirement, w_1 is set as 1 and w_2 is set as 0. In the simulation with consideration of the energy balance requirement, both w_1 and w_2 are set as 0.5.

The second difference is the individual energy balance constraint. In the simulation without consideration of the energy balance requirement, the constraint is ignored, i.e., step 3 of the optimization procedure in Section 4.2 is omitted. In the simulation with consideration of the energy balance requirement, step 3 is executed.

The simulation results are listed in Table 3 and Figures 6 and 7.

In Table 3 and subsequent simulation result tables, columns λ_j, Φ_j, and h_j list the longitudes, latitudes, and altitudes of all the SPs in the SPN, column B_j lists the values of the energy balance indicators for all the SPs in the SPN, column B lists the value of the energy balance indicator for the SPN, column $GDOP_N$ lists the GDOP value for the SPN, and column F lists the fitness of the SPN. For columns B_j, B, $GDOP_N$, and F in these tables, the smaller their value is, the better the SPN geometry performance is.

Table 3. Comparison of simulations with/without consideration of energy balance requirement.

	λ_j/deg, Φ_j/deg, h_j/km	Bj	B	$GDOP_N$	F
with consideration of energy balance requirement	92.7E, 37.0N, 19.5	0.35	1.75	7.52	4.64
	92.4E, 39.4N, 20.0	0.29			
	90.6E, 38.5N, 20.0	0.25			
	92.1E, 37.9N, 20.0	0.23			
	90.0E, 39.7N, 20.0	0.29			
	90.6E, 37.0N, 19.5	0.34			
without consideration of energy balance requirement	91.8E, 38.8N, 18.5	1.11	5.88	7.36	7.36
	90.3E, 39.4N, 18.5	1.12			
	90.3E, 37.6E, 18.5	1.03			
	91.5E, 37.9N, 19.0	0.64			
	93.0E, 37.0N, 18.0	1.67			
	93.0E, 40.0N, 20.0	0.31			

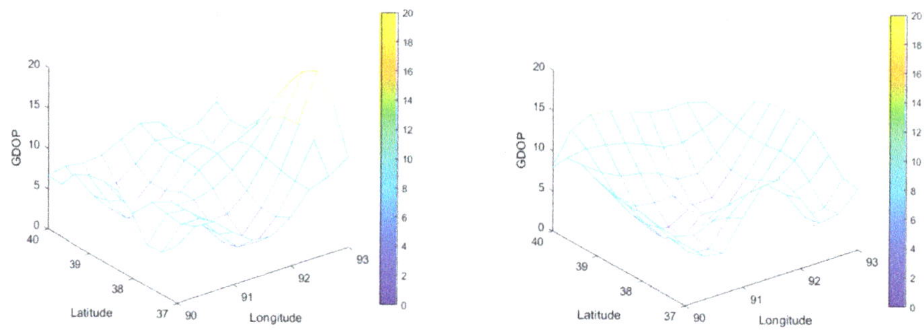

(**a**) (**b**)

Figure 6. GDOP distribution of SPNs: (**a**) SPN with consideration of energy balance requirement; (**b**) SPN without consideration of energy balance requirement.

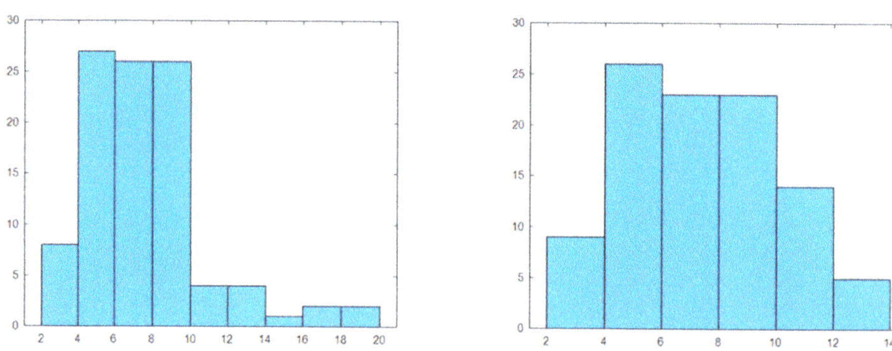

(**a**) (**b**)

Figure 7. GDOP histogram of SPNs: (**a**) SPN with consideration of energy balance requirement; (**b**) SPN without consideration of energy balance requirement.

Column B_j in Table 3 implies that geometry configuration has a significant impact on SPN energy balance. For the same SP, their energy balance indicator can differ by several times at different locations.

Column B in Table 3 also proves the necessity of energy balance analysis in SPN geometry design. If it is ignored, an SPN with unacceptable energy performance might be obtained, as shown in the 7th, 8th, 9th, and 11th row of Table 3 (B_j = 1.11, 1.12, 1.03, and 1.67, respectively). The station-keeping capacity of such an SPN is very poor, which will degrade the SPN availability and continuity.

In contrast, by assigning a weight to the energy balance indicator properly in the objective function and implementing an individual energy balance constraint, the proposed algorithm can avoid unacceptable results effectively. Furthermore, in terms of GDOP, the results of the algorithm considering energy are not much worse than those of algorithms that do not consider energy.

5.2. Comparison among Different Altitude Intervals

In this section, the values of h_{min} and h_{max} in Table 2 are adjusted to compare SPN geometry performances at different altitudes. Simulations in this section and subsequent sections are implemented with consideration of the energy balance requirement.

Simulation results are shown in Table 4 and Figures 8 and 9.

Table 4. Comparison of SPNs in different altitude intervals.

Altitude Interval/km	λ_j/deg, Φ_j/deg, h_j/km	Bj	B	$GDOP_N$	F
18~20	92.7E, 37.0N, 19.5	0.35	1.75	7.52	4.64
	92.4E, 39.4N, 20.0	0.29			
	90.6E, 38.5N, 20.0	0.25			
	92.1E, 37.9N, 20.0	0.23			
	90.0E, 39.7N, 20.0	0.29			
	90.6E, 37.0N, 19.5	0.34			
20~22	90.9E, 38.5N, 21.0	0.10	0.52	7.12	3.82
	90.0E, 40.0N, 21.5	0.10			
	90.6E, 37.0N, 21.5	0.05			
	92.4E, 39.1N, 22.0	0.06			
	92.1E, 37.6N, 20.5	0.13			
	93.0E, 37.6N, 21.0	0.08			
22~24	92.4E, 37.9N, 22.0	0.04	0.23	6.29	3.26
	93.0E, 40.0N, 24.0	0.03			
	91.8E, 39.1N, 22.0	0.06			
	91.2E, 39.7N, 23.0	0.04			
	90.0E, 37.0N, 22.0	0.03			
	90.9E, 37.9N, 23.0	0.03			
24~26	90.9E, 38.2N, 26.0	0.01	0.11	5.58	2.84
	93.0E, 40.0N, 26.0	0.02			
	92.1E, 37.6N, 25.0	0.01			
	90.0E, 39.7N, 25.0	0.03			
	90.0E, 37.0N, 24.0	0.02			
	92.1E, 39.1N, 25.5	0.02			
26~28	93.0E, 40.0N, 28.0	0.01	0.06	5.35	2.71
	90.6E, 39.7N, 27.5	0.01			
	92.4E, 37.6N, 26.5	0.01			
	90.0E, 37.0N, 26.0	0.01			
	92.1E, 39.1N, 28.0	0.01			
	90.6E, 37.9N, 27.5	0.01			

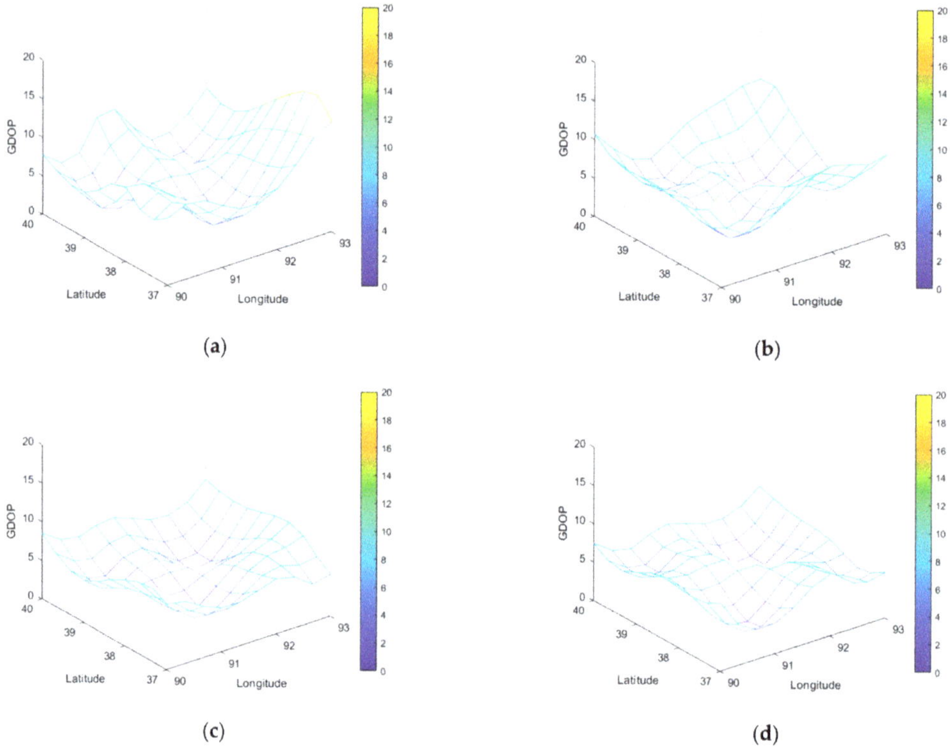

Figure 8. GDOP distribution of SPNs in different altitude intervals: (**a**) SPNs in 20~22 km; (**b**) SPNs in 22~24 km; (**c**) SPNs in 24~26 km; (**d**) SPNs in 26~28 km (for GDOP distribution of SPNs in 18~20 km, please refer to Figure 6a).

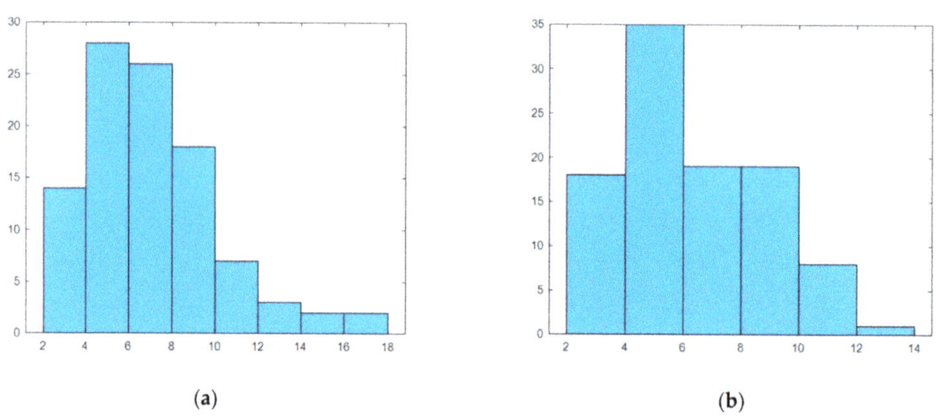

Figure 9. *Cont.*

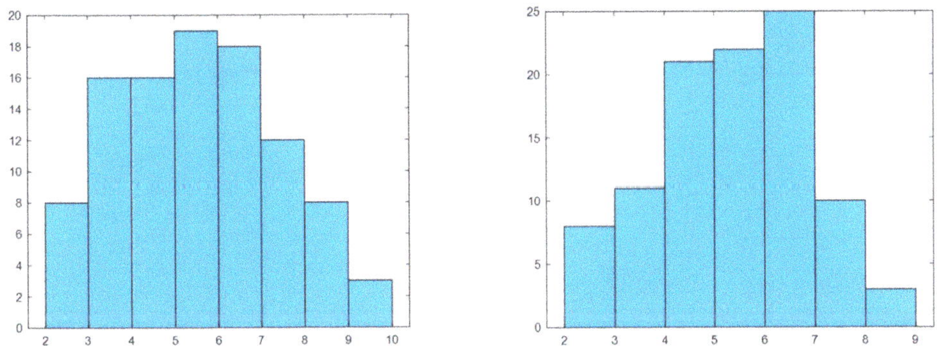

Figure 9. GDOP histogram of SPNs in different altitude intervals: (**a**) SPNs in 20~22 km; (**b**) SPNs in 22~24 km; (**c**) SPNs in 24~26 km; (**d**) SPNs in 26~28 km (for GDOP histogram of SPNs in 18~20 km, please refer to Figure 7a).

The results in Table 4 and Figures 8 and 9 imply that under the given conditions, both GDOP and the energy balance indicator show a decreasing trend with the altitude increasing from 18 km to 28 km, meaning that SPN geometry performance improves as altitude increases in this altitude interval.

The decrease in the energy balance indicator can be attributed to wind, atmosphere, solar radiation, and other factors.

From Figure 3, it can be seen that in March, as the altitude increases from 18 km to 28 km, the zonal wind speed decreases from over 20 m/s to less than 10 m/s, while the increase in the meridional wind speed is less than 5 m/s, resulting in a decrease in energy consumption for the SP. From Figure 4, it can be seen that as the altitude increases, the atmosphere density decreases, which is also beneficial for reducing energy consumption.

From Figure 10, it can be seen that the atmosphere pressure decreases as the altitude increases. From Equations (13)–(15), it can be inferred that solar direct radiation intensity increases as the altitude increases, which can lead to an increase in SP energy production, as illustrated in Figure 5.

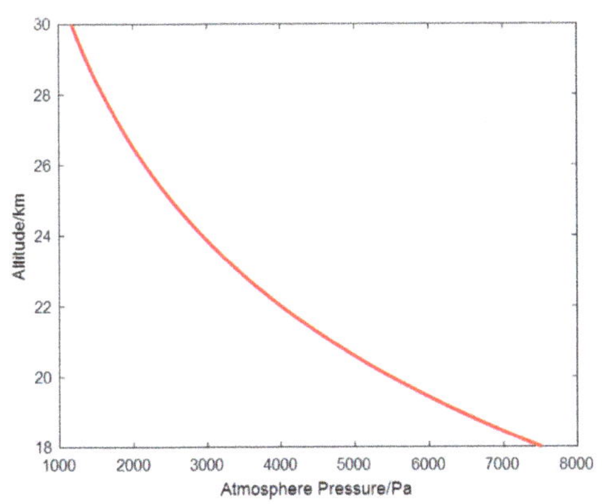

Figure 10. Atmosphere pressure from 18 km to 30 km according to U.S. Standard Atmosphere, 1976 [24].

Therefore, the energy balance indicator, which is the ratio of energy consumption and energy production, decreases as the altitude increases.

In addition, as the altitude increases, the value of the SPN elevation angle increases, which can improve the GDOP of the SPN. This has been deduced and explained in detail in [8,10].

Therefore, both the energy balance indicator and GDOP will improve as the altitude increases, as shown in Table 4, Figure 8, and Figure 9. So, an SPN with a higher station-keeping altitude is expected in order to achieve a better performance under the given conditions.

5.3. Comparison among Different Latitude Intervals

In this section, the values of Φ_{min} and Φ_{max} in Table 2 are adjusted to compare network performances in different latitudes. Simulation results are shown in Table 5.

Table 5. Comparison among SPNs of different latitude intervals.

Latitude Interval/deg	λ_j/deg, Φ_j/deg, h_j/km	Bj	B	$GDOP_N$	F
33~36N	91.5E, 35.1N, 20.0 92.4E, 33.0N, 19.5 93.0E, 36.0N, 20.0 90.0E, 34.2N, 20.0 90.0E, 36.0N, 19.0 91.8E, 33.9N, 20.0	0.15 0.20 0.17 0.12 0.54 0.12	1.30	7.15	4.23
35~38N	90.6E, 35.3N, 19.5 92.4E, 35.3N, 19.0 92.1E, 36.8N, 20.0 93.0E, 38.0N, 20.0 91.2E, 36.2N, 20.0 91.2E, 37.7N, 20.0	0.28 0.52 0.19 0.23 0.17 0.22	1.61	7.32	4.47
37~40N	92.7E, 37.0N, 19.5 92.4E, 39.4N, 20.0 90.6E, 38.5N, 20.0 92.1E, 37.9N, 20.0 90.0E, 39.7N, 20.0 90.6E, 37.0N, 19.5	0.35 0.29 0.25 0.23 0.29 0.34	1.75	7.52	4.64
39~42N	92.1E, 41.4N, 19.5 90.9E, 39.0N, 20.0 93.0E, 42.0N, 20.0 90.9E, 39.9N, 19.5 92.1E, 40.2N, 20.0 90.3E, 41.4N, 19.5	0.55 0.27 0.40 0.47 0.32 0.54	2.55	7.12	4.83
41~44N	90.9E, 41.9N, 20.0 93.0E, 44.0N, 20.0 90.0E, 41.0N, 20.0 90.3E, 42.8N, 20.0 91.8E, 43.1N, 19.5 92.4E, 42.2N, 19.0	0.39 0.49 0.35 0.43 0.63 0.85	3.14	6.89	5.02

The results detailed in Table 5 imply that low latitude intervals tend to be beneficial to the network energy balance while having little impact on SPN GDOP under the given conditions.

From Figure 2, it can be seen that within the service area listed in Table 5, as the latitude decreases, the zonal wind speed decreases significantly, which can reduce SP energy consumption. From Figure 5, it can be seen that lower latitude can help SPs obtain more energy production. Therefore, the SPN energy balance indicator shows a decreasing tendency with decreasing latitude.

6. Conclusions and Future Works

SPN is a novel aerial network with promising potential. Geometry design is a critical problem affecting its service performance significantly. This paper focuses on SPN geometry design to pursue a satisfactory performance for both GDOP and energy balance. In the assumed service area and under the given simulation conditions, the following conclusions can be drawn:

Geometry configuration has a significant impact on both SPN energy balance and GDOP. Consequently, neither of them can be ignored in SPN geometry design.

The energy balance requirement of an SPN can be met by properly assigning weights on the energy balance indicator in the objective function and implementing the energy balance constraint of individual SPs.

Both GDOP and energy balance can be improved by raising the station-keeping altitude towards the altitude interval of 26~28 km.

GDOP shows no substantial improvement when the deployment space is slightly adjusted southward and northward. Energy balance tends to improve gently when deployment space moves southward.

Some issues can be analyzed in the future.

In this paper, the photoelectric conversion efficiency of the solar array is assumed to be constant. However, it changes with the thermal conditions in practice. In the future, the influence of photoelectric conversion efficiency fluctuation on SP energy production and energy balance should be analyzed.

Also, uncertain wind is ignored in this paper since the wind at the altitude interval of SP is relatively stable, but uncertain wind exists according to [32], and it may have an impact on the energy consumption of SPs. Further analysis can be conducted next.

In addition, this paper implements simulations currently just for a small area. More simulations for larger areas can be carried out according to requirements.

Author Contributions: Conceptualization, Y.Q.; methodology, Y.Q. and S.W.; software, Y.Q.; validation, H.F. and Q.L.; writing—original draft preparation, Y.Q.; writing—review and editing, S.W. and H.F.; visualization, Y.Q.; supervision, S.W. All authors have read and agreed to the published version of the manuscript.

Funding: This work was funded by the National Key Research and Development Program of China (Grant No. 2022YFB3901805).

Data Availability Statement: Data are contained within the article.

Conflicts of Interest: The authors declare no conflicts of interest.

References

1. Tsujii, T.; Rizos, C.; Wang, J.; Dai, L.; Roberts, C. A navigation/positioning service based on pseudolites installed on stratospheric airships. *Jpn. Soc. Aeronaut. Space Sci.* **2003**, *50*, 36. [CrossRef]
2. Dovis, F.; Presti, L.; Mulassano, P. Support infrastructures based on high altitude platforms for navigation satellite systems. *IEEE Wirel. Commun.* **2005**, *12*, 106–112. [CrossRef]
3. Zheng, H.; Atia, M.; Yanikomeroglu, H. High Altitude Platform Station (HAPS)-Aided GNSS for Urban Areas. *arXiv* **2023**. [CrossRef]
4. Chandu, B.; Pant, R.S.; Moudgalya, K. Modeling and simulation of a precision navigation system using pseudolites mounted on airships. In Proceedings of the 7th AIAA ATIO Conf, 2nd CEIAT Int'l Conf on Innov and Integr in Aero Sciences, 17th LTA Systems Tech Conf, Belfast, UK, 18–20 September 2007. [CrossRef]
5. Dai, L.W.; Wang, J.L.; Tsujii, T.; Rizos, C. Pseudolite applications in positioning and navigation: Modelling and geometric analysis. *J. Navig.* **2004**. Available online: https://www.researchgate.net/profile/Jinling_Wang2/publication/2865922_Pseudolite_applications_in_positioning_and_navigation_Modelling_and/links/546d98510cf26e95bc3cb846.pdf?__cf_chl_tk=NFi8SdAKc5gBoKKe9i2BXt_BS1bZjkE2X5PooWMY56U-1717999627-0.0.1.1-8446 (accessed on 4 June 2024).
6. Fateev, Y.L.; Ratuschnyak, V.N.; Kartson, I.N.; Tyapkin, V.N.; Dmitriev, D.D.; Goncharov, A.E. Analyzing measurement errors for navigation parameters in onground short-range navigation systems based on pseudolites. *IOP Conf.* **2016**, *155*, 012016. [CrossRef]
7. Sang, W.G.; He, X.F.; Chen, Y.Q. Configuration of pseudolite-alone positioning system based on DOP geometry structure. *Bull. Surv. Mapp.* **2013**, *9*, 1–4.
8. Hu, W.; Yang, J.J.; He, P. Study on pseudolite configuration scheme based on near space airships. *Radio Eng.* **2009**, *39*, 24–27.

9. Gao, S.S.; Zhao, F.; Xie, M.L. Research on the geometric configuration scheme of near space pseudolite-only positioning system. *J. Navig. Position.* **2013**, *1*, 21–25.
10. Yang, Y.; Gao, S.S.; Yan, H.F. Design on geometric configuration schemes of pseudolite in near space. *Syst. Eng. Electron.* **2014**, *36*, 532–538.
11. Mosavi, M.R.; Divband, M. Calculation of geometric dilution of precision using adaptive filtering technique based on evolutionary algorithms. In Proceedings of the 2010 International Conference on Electrical and Control Engineering, Wuhan, China, 25–27 June 2010. [CrossRef]
12. Shao, K.; Li, K.; Wang, J. PSO-Based pseudolite layout strategy. *Commun. Technol.* **2017**, *50*, 2454–2459.
13. Tang, W.J.; Chen, J.P.; Yu, C.; Ding, J.; Wang, R. A new ground-based pseudolite system deployment algorithm based on MOPSO. *Sensors* **2021**, *21*, 5364. [CrossRef]
14. Yang, L.; Zhou, J.H.; Chen, J.P. The study of optimization of formation flying navigation augmentation platforms based on genetic algorithm. *GNSS World China* **2008**, *33*, 9–13.
15. Chen, C.; Chen, K.; Huang, J.; Li, Y. Using genetic algorithms to approximate weighted geometric dilution of precision. In Proceedings of the 2016 International Symposium on Computer, Consumer and Control (IS3C), Xi'an, China, 4–6 July 2016. [CrossRef]
16. Song, J.; Hou, C.; Xue, G.; Ma, M. Study of constellation design of pseudolites based on improved adaptive genetic algorithm. *J. Commun.* **2016**, *11*, 879–885. [CrossRef]
17. Araripe, D.F.; De, M.F.; Campos, D.T. High-altitude platforms—Present situation and technology trends. *J. Aerosp. Technol. Manag.* **2016**, *8*, 249–262. [CrossRef]
18. Nickol, C.L.; Guynn, M.D.; Kohout, L.L.; Ozoroski, T.A. High Altitude Long Endurance UAV Analysis of Alternatives and Technology Requirements Development. In Proceedings of the 45th AIAA Aerospace Sciences Meeting and Exhibit, Reno, NV, USA, 8–11 January 2007. [CrossRef]
19. Yang, X.; Wang, F.; Liu, W.; Xiao, W.; Ye, X. A Layout Method of Space-Based Pseudolite System Based on GDOP Geometry. *Chin. J. Electron.* **2023**, *32*, 1050–1058. [CrossRef]
20. Sultana, Q.; Sunehra, D.; Srinivas, V.S.; Sarma, A.D. Effects of Pseudolite Positioning on DOP in LAAS. *Positioning* **2010**, *1*, 18–26. [CrossRef]
21. Jiang, M.; Li, R.; Liu, W. Research on geometric configuration of pseudolite positioning system. *Comput. Eng. Appl.* **2017**, *53*, 271–276.
22. Zhao, Y.; Garrard, W.; Mueller, J. Benefits of Trajectory Optimization in Airship Flights. In Proceedings of the AIAA 3rd "Unmanned Unlimited" Technical Conference, Workshop and Exhibit, Chicago, IL, USA, 20–23 September 2004. [CrossRef]
23. Wu, L.; Li, Y.; Li, Z. The research of route planning for stratospheric airships based on genetic algorithms. *Spacecr. Recovery Remote Sens.* **2011**, *32*, 1–6.
24. *NOAA-S/T-76-1562*; U.S. Standard Atmosphere, 1976. National Oceanic and Atmospheric Administration; National Aeronautics and Space Administration; United States Air Force: Washington, DC, USA, 1976.
25. Stanciu, C.; Stanciu, D. Optimum tilt angle for flat plate collectors all over the World—A declination dependence formula and comparisons of three solar radiation models. *Energy Convers. Manag.* **2014**, *81*, 133–143. [CrossRef]
26. Zhang, L.; Li, J.; Jiang, Y.; Du, H.; Zhu, W.; Lv, M. Stratospheric airship endurance strategy analysis based on energy optimization. *Aerosp. Sci. Technol.* **2020**, *100*, 105794. [CrossRef]
27. Long, Y.; Deng, X.; Yang, X.; Hou, Z. Trajectory simulation of stratosphere aerostats in polar vortex wind field. *Comput. Simul.* **2021**, *38*, 37–42.
28. Lv, M.; Li, J.; Du, H.; Zhu, W.; Meng, J. Solar array layout optimization for stratospheric airships using numerical method. *Energy Convers. Manag.* **2017**, *135*, 160–169. [CrossRef]
29. Zhang, Y.; Li, J.; Lv, M.; Tan, D.; Zhu, W.; Sun, K. Simplified analytical model for investigating the output power of solar array on stratospheric airship. *Int. J. Aeronaut. Space Sci.* **2016**, *17*, 432–441. [CrossRef]
30. Mirjalili, S.; Mirjalili, S.M.; Lewis, A. Grey wolf optimizer. *Adv. Eng. Softw.* **2014**, *69*, 46–61. [CrossRef]
31. Arora, S.; Singh, H.; Sharma, M.; Sharma, A.; Anand, P. A new hybrid algorithm based on grey wolf optimization and crow search algorithm for unconstrained function optimization and feature selection. *IEEE Access* **2019**, *7*, 26343–26361. [CrossRef]
32. Conway, J.P.; Bodeker, G.E.; Waugh, D.W.; Murphy, D.J.; Cameron, C.; Lewis, J. Using project Loon superpressure balloon observations to investigate the inertial peak in the intrinsic wind spectrum in the midlattitude stratosphere. *J. Geophys. Res. Atmos.* **2019**, *124*, 8594–8604. [CrossRef]

electronics

MDPI

Article

FLsM: Fuzzy Localization of Image Scenes Based on Large Models

Weiyi Chen [1,*], Lingjuan Miao [1], Jinchao Gui [2], Yuhao Wang [1] and Yiran Li [1]

1 School of Automation, Beijing Institute of Technology, Beijing 100081, China; miaolingjuan@bit.edu.cn (L.M.)
2 Aerospace Information Research Institute, Chinese Academy of Sciences, Beijing 100101, China; guijinchao@aircas.ac.cn
* Correspondence: 3220195102@bit.edu.cn

Abstract: This article primarily focuses on the study of image-based localization technology. While traditional methods have made significant advancements in technology and applications, the emerging field of visual image-based localization technology demonstrates tremendous potential for research. Deep learning has exhibited a strong performance in image processing, particularly in developing visual navigation and localization techniques using large-scale visual models. This paper introduces a sophisticated scene image localization technique based on large models in a vast spatial sample environment. The study involved training convolutional neural networks using millions of geographically labeled images, extracting image position information using large model algorithms, and collecting sample data under various conditions in elastic scene space. Through visual computation, the shooting position of photos was inferred to obtain the approximate position information of users. This method utilizes geographic location information to classify images and combines it with landmarks, natural features, and architectural styles to determine their locations. The experimental results show variations in positioning accuracy among different models, with the most optimal model obtained through training on a large-scale dataset. They also indicate that the positioning error in urban street-based images is relatively small, whereas the positioning effect in outdoor and local scenes, especially in large-scale spatial environments, is limited. This suggests that the location information of users can be effectively determined through the utilization of geographic data, to classify images and incorporate landmarks, natural features, and architectural styles. The study's experimentation indicates the variation in positioning accuracy among different models, highlighting the significance of training on a large-scale dataset for optimal results. Furthermore, it highlights the contrasting impact on urban street-based images versus outdoor and local scenes in large-scale spatial environments.

Keywords: localization of image scenes; fuzzy localization; large models; image processing; deep learning

1. Introduction

In recent years, the Beidou Satellite Navigation System (BDS), Galileo Satellite Navigation System (Galileo), modern Global Navigation System (GPS), and Global Navigation System (GLONASS) have been developed. There are over 140 GNSS satellites available [1]. Global Navigation Satellite Systems (GNSSs) are increasingly used for outdoor navigation [2]. The strap-on inertial navigation system (SINS) can automatically measure the user's position, speed, and attitude [3]. However, the inertial navigation system is subject to its own limitations and high costs. Its primary drawback is the increase in error over time, leading to drift. The aforementioned systems represent traditional navigation and positioning technologies. Over years of development, these technical systems have essentially established a relatively comprehensive technical framework, which is extensively utilized in human activities and daily life. As human science and technology advance, the need for navigation and positioning continues to evolve. Robust visual localization over long periods of time is one of the biggest challenges for the long-term navigation of mobile robots [4]. Monocular visual inertial navigation systems (VINSs) are widely used in fields such as robot navigation, autonomous driving, and augmented/virtual reality [5]. The current emergence of various intelligent robots not only significantly facilitates our lives, but also presents higher performance requirements.

Visual navigation is a navigation method using visible and invisible imaging technology, which has the advantages of good concealment, strong autonomy, fast and accurate measurement, low cost, and high reliability [6]. At present, the emergence of diverse intelligent robots not only significantly facilitates human life, but also presents elevated demands for robot performance, thereby establishing a prerequisite for the advancement of visual positioning technology. Visual information is increasingly utilized in navigation applications. With the introduction of numerous new concepts, methods, and theories, image processing technology based on deep learning has progressively matured. Visual navigation technology is expected to be developed and widely used in the fields of aircraft, unmanned aerial vehicles, various cruise missiles, deep space probes, indoor mobile navigation, and so on [7]. Therefore, visual navigation technology has high research and application value in the field of navigation. Usually, visual navigation on robots is achieved by installing a monocular or binocular camera to obtain local images of the environment and to make navigation decisions. The research on intelligent robots began in the late 1960s, marked by Shakey, the first mobile robot developed by Stanford Research Institute (SRI) [8]. Its main objective is to study the real-time control of robot systems in complex environments. Representative examples include urban robots and tactical robots developed by Jet Propulsion Laboratory [9–11]. These robots are equipped with binocular stereo vision systems for obstacle detection. Visual navigation and positioning can also be applied to spacecraft or interplanetary detectors, such as lunar probes. The lunar rover has a high degree of autonomy and is suitable for performing exploration tasks in a complex and unstructured lunar environment [12–15]. The stereo vision system of the lunar rover is the most direct and effective tool for close-range and high-altitude moon detection. It serves as a tool for understanding the lunar environment and provides crucial information for lunar rover survival in complex environments. Using the stereo vision system, we can not only reconstruct the terrain of the environment in real time to avoid obstacles, but also use the obtained stereo sequence images to estimate the movement of the rover itself. Therefore, the application of visual navigation in the field of robotics is extremely extensive and significant [16–18].

The essence of visual navigation is to obtain the two-dimensional image information of the scene through one or more cameras, and then determine the operation information of navigation by using image processing, computer vision, pattern recognition, and other algorithms. The techniques involved include camera calibration, stereo image matching, path identification, and 3D reconstruction [19,20]. Inspired by the process of robot positioning and navigation, we contemplate the extraction of positional data from images and the potential for a single image to facilitate robot navigation and positioning tasks. Addressing this challenge, we have conducted extensive research. Our primary focus is on the feasibility and precision of obtaining location information from images. If it is possible to probabilistically infer geographic location data from images, it would constitute a highly significant area of study. Visual positioning can play an important role in satellite navigation failures and has a broad application value. When satellite navigation signals are obstructed or unavailable, visual positioning systems can provide reliable positioning information, suitable for many fields, such as autonomous driving and unmanned vehicles. With the help of high-precision visual positioning systems, autonomous vehicles can accurately locate and navigate in environments without satellite signals, ensuring safe driving in cities. Indoor navigation—in indoor environments, satellite signals are usually weak or unavailable. By using visual positioning technology, people can accurately navigate and locate within large buildings such as shopping malls, airports, and hospitals. Industrial robots—in the fields of manufacturing and logistics, industrial robots require precise positioning information to perform various tasks. Visual positioning systems can provide real-time location information to help robots accurately perform tasks. Search and rescue—in the event of a disaster, satellite navigation systems may be disrupted or damaged. Visual positioning technology can help search and rescue personnel locate trapped individuals, without satellite signals. Military applications—military departments can use visual positioning systems for precise positioning and navigation, without being disturbed or monitored by hostile forces. Mobile devices—smartphones and tablets can use cameras and sensors for visual positioning, providing users with indoor navigation, augmented reality experiences, and other functions.

This paper focuses on geographical fuzzy positioning using image information mining, consolidates the relevant technical accomplishments in navigation and positioning, and scrutinizes the limitations of current navigation and positioning technology based on their characteristics. Addressing the requirements of visual navigation and positioning, this paper aims to achieve visual positioning capability in a lightweight manner for various scenarios and applications. It introduces a method of image fuzzy positioning based on a large model, enabling scene location determination through images in GPS failure environments. A schematic diagram of FLsM, based on image localization, is shown in Figure 1. The key contributions of this paper are summarized as follows:

- The concept of elastic scale space is introduced, which refers to a coupling space between large-scale scenes and fine scenes, emphasizing the variability and unpredictability of the environment.
- A vision-based fuzzy positioning technology is suggested, emphasizing the semantic information extraction from the visual image itself and providing geographic location.
- By leveraging multiple models for image training, employing advanced deep learning models, and utilizing a large dataset of Internet data for pre-training, we can efficiently match images and texts and accomplish the fuzzy positioning of images.

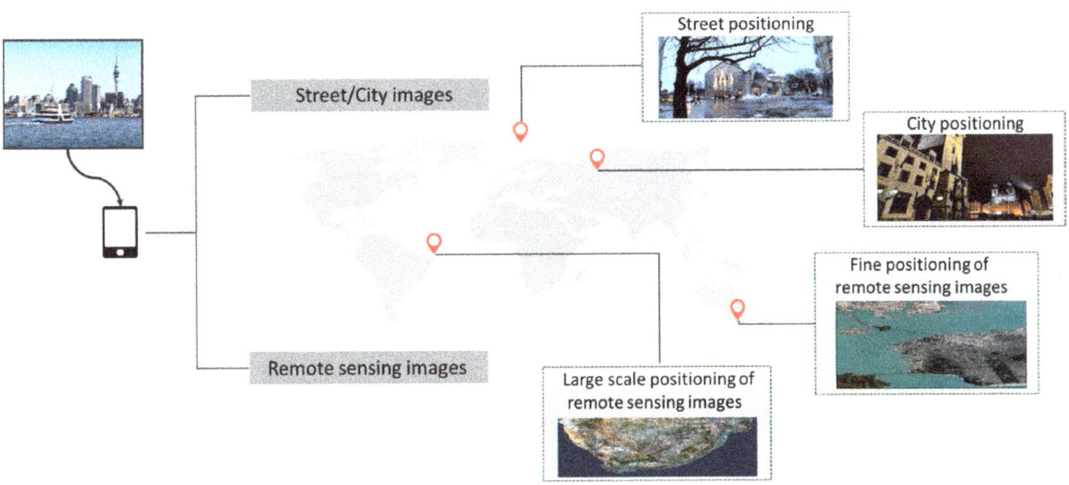

Figure 1. Schematic diagram of image fuzzy positioning.

2. Related Works

In this section, we introduce the related technologies of visual navigation and positioning. Including traditional slam technology, deep learning models, geographic information positioning research, and so on.

2.1. Traditional SLAM Technology

The traditional craft of Simultaneous Localization and Mapping (SLAM) techniques has been a beacon in the odyssey of robotic navigation and mapping [21]. These methods typically harness the power of sensor fusion, utilizing data from cameras, lidars, and other sensors to simultaneously decipher the robot's location and construct a map of its environment. Algorithms such as ORB-SLAM2 and ORB-SLAM3 have been luminaries in this domain, demonstrating significant advancements in accuracy and stability over the past decade [22]. However, as underscored in recent scholarly pursuits, challenges persist in achieving long-term robustness, especially in the face of diverse and dynamic environmental perturbations.

2.2. Deep Learning Models for Visual Navigation

The advent of deep learning has revolutionized the landscape of visual navigation [23,24]. Contemporary approaches leverage the prowess of neural networks to learn complex mappings between visual inputs and navigational outputs. CLIP, as discussed in the previous section, emerges as a model capable of learning transferable visual representations from the vast expanse of natural language supervision [25,26]. It extends the scope of computer vision systems by directly learning from the raw text about images, showcasing promising results across a myriad of tasks, without the need for task-specific training.

Moreover, recent scholarly endeavors delve into the comparison of open-source visual SLAM approaches, evaluating algorithms based on factors such as accuracy, computational performance, and robustness. This reflects the ongoing quest to enhance the performance of visual navigation systems and address the specific challenges posed by different scenarios and datasets.

2.3. Geographic Information Positioning Research

Geographic Information Systems (GISs) and positioning technologies are the compass and sextant of modern navigation systems. Recent works, such as GeoCLIP, integrate CLIP-inspired techniques to chart the course for effective worldwide geo-localization. By encoding GPS information and employing hierarchical learning [27,28], GeoCLIP demonstrates a state-of-the-art performance, navigating the challenges associated with the diversity of global landscapes. This underscores the importance of geographic information in refining the accuracy and reliability of visual navigation systems.

3. Materials and Methods

3.1. Overview

A visual fuzzy positioning method in elastic scale space is proposed for different application requirements of various scenes, based on an in-depth analysis of image information. This method differs from traditional visual positioning methods. High-resolution remote sensing images are used for photogrammetry in large-scale scenes to generate image maps of different scales. Fine-scale scenes are divided into indoor and outdoor areas, and environmental image data are collected from natural target sample data using mobile measuring equipment. The expectation is that these two types of data rely solely on the information carried by the image itself, and the positioning requirements can be fulfilled using deep learning algorithms with large models. The concept of elastic scale space involves leveraging the randomness and lightness of the image, focusing on mining the information value of the image and obtaining rough positioning information. The significance of this work lies in its lightweight design, which does not impose strict requirements on the image itself and emphasizes model training. The image captured by the terminal's camera is used to determine the inclusion of a specific target and then the user's precise position is calculated visually. The entire process is shown in Figure 2.

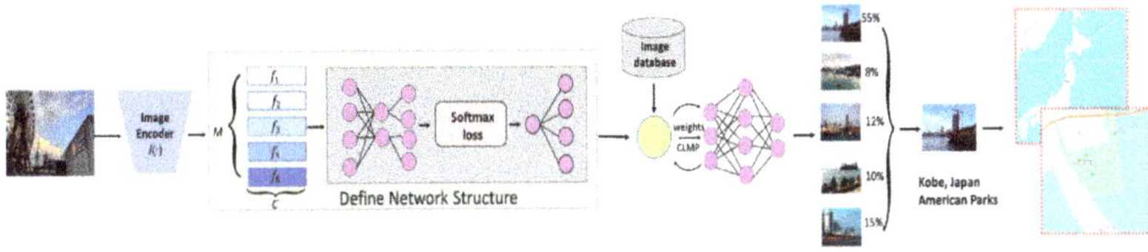

Figure 2. Flow description of image positioning algorithm.

3.2. Constructing an Elastic Scale Spatial Environment

We propose the concept of an elastic scale space, which is a space between the coupling of large-scale and fine-grained scenes, highlighting the arbitrariness and randomness of the environment, as shown in Figure 3.

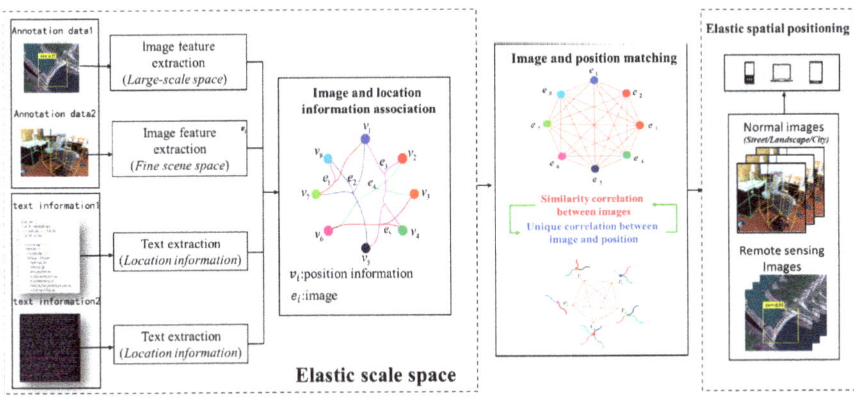

Figure 3. Overall framework of elastic spatial positioning.

Traditionally, in large-scale scenes, it is necessary to make large-scale scene image base maps and develop service engines. The research process is as follows:

- Select an appropriate reference point

The choice of reference points should be clearly defined as a clear landmark, such as road intersections, building corners, etc., that can be considered as optional reference points. After selecting the datum point, high-precision surveying and mapping should be carried out on the selected datum point.

- Image map of construction environment

Utilize aviation professional equipment (drones and aerial photography planes) for photogrammetry work, collect high-resolution remote sensing images of the environmental field, and produce image maps at scales of 1:1000, 1:500, etc.

- Unified spatiotemporal benchmark construction

The determination of the spatiotemporal baseline of the environmental field usually adopts the WEB Mercator projection method, whose core is the transformation of the coordinate system, mainly the transformation of the image into a plane coordinate system, which should be consistent with the current general map projection.

For the construction of fine scenes, the following process is required.

- Target sample collection and data construction

In order to more accurately represent environmental information, it is divided into two parts—indoor environment and outdoor environment. The outdoor environment dataset utilizes satellite positioning technology to collect and store the location information of identifiable targets in the scene. At the same time, the mobile terminal is equipped with a high-definition camera to capture the target from various angles and to collect image information of the target. Then, utilize onboard or airborne measurement systems to collect local RGBD information around the target and construct a data resource lake for the scene through various technical means. Similarly, in indoor environments, it is necessary to establish a unified indoor coordinate system. The camera is used to capture the target from various angles, collect the image information of the target, collect the location information of identifiable targets indoors, and save it.

- Model library construction

According to the requirements, use a deep learning framework to train the collected image information, obtain a proprietary model library, and obtain parameter models that meet the requirements.

3.3. The Concept of Visual Blur Localization

In order to achieve precise positioning, this article suggests a vision-based fuzzy positioning technology that is integrated with satellite navigation and other techniques. In order to categorize or detect input scene photographs, visual blur localization focuses on mining the semantic information included in the visual images themselves, supplying positional range information, and utilizing deep learning network methods [29] in conjunction with large-scale model structures. Semantic segmentation, which effectively pulls information from the scene, is the main focus. The marked position data is extracted from the saved position data, based on the recognition findings. It should be noted that the input for future precise positioning can come from fuzzy positioning data. Transform the fuzzy position information that was previously acquired into fine scene data. Extract the target's local RGBD information from the location data that have been saved and compare it with the scene map based on the recognition findings. Use the 2D–3D visual solving algorithm to obtain the user's precise position. In Figure 4, the technical procedure is displayed.

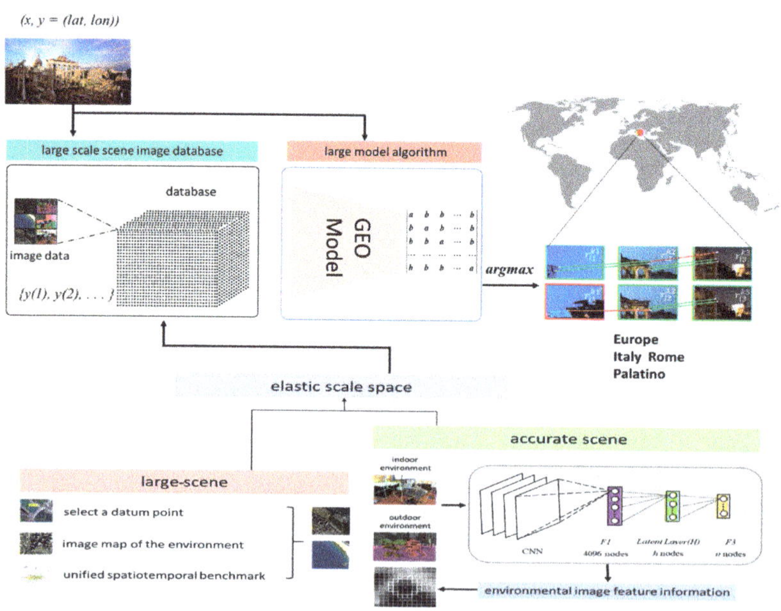

Figure 4. Visual fuzzy positioning process based on large model.

3.4. Multi-Source Image Data Source Matching

Picture retrieval and matching algorithms are essential for the search, matching, and display of location information derived from visual images. These techniques enable the rapid matching of feature data and corresponding picture data. In addition to image encoding and quick image retrieval for large amounts of data, there are numerous important technologies that still need to be resolved. These include fast feature extraction technology for multi-source image data, unified spatiotemporal benchmarks, and image encoding. Aerial surveys, other sensor image data, and satellite remote sensing photos are some of the multi-source image data used to create traditional large-scale scenes. Data integration is based on the unification of spatiotemporal benchmarks and the conversion of multi-source picture data to a common scale. The CGCS2000 coordinate system is the source of the coordinate system [30], terrain feature points, etc., in the image; it projects the image data and feature data in a plane according to a universal map. Uniformly project onto the WEB Magic Card to form a consistent spatiotemporal baseline. Complete feature extraction and spatiotemporal matching processing of multi-source image data.

3.5. Building a Large-Scale Complex Scene Graph Database

This project uses a YFCC100M dataset to obtain metadata containing Geo information, and combines scene image data such as SUN2010, Places2, and Google StreetView to construct a large-scale complex scene graph database [31–33]. Firstly, the YFCC100M metadata is processed to extract data with geo labels from the original dataset. Then, the Geo labeled dataset is transformed using the GEOPY tool to obtain an image dataset with actual location information. Utilize SIFT for feature extraction on scene datasets such as SUN2010, Places2, and Google StreetView to obtain a large database of scene images. YFCC100M obtains a text data document database after data preprocessing, which can be used as a data source for future model training. In particular, this text data contains the Geo information of each image, which is crucial for accurate scene positioning. SUN2010, Places2, and Google StreetView have undergone SIFT algorithm feature processing to obtain a relatively rich set of scene-based feature datasets, providing a reliable data source for training scene recognition models. There is an urgent need for rapid detection, autonomous warning, information confrontation, and on-site disposal of remote targets. The YFCC100M database is an imaging database that has been based on Yahoo Flickr since 2014. The library consists of 100 million pieces of media data generated between 2004 and 2014, including 99.2 million photo data and 800,000 video data [31]. The YFCC100M dataset does not contain photo or video data and each row in the document contains metadata for a photo or video. Among them, Photos/video identifiers, Longitude, and Latitude are used. Geo information refers to geographic location information, which can record the geographic location information at the time of photo shooting, namely longitude and latitude. But not all metadata contains Geo information, so it is necessary to filter out metadata that does not contain Geo information. Then, use the geopy toolkit to convert longitude and latitude to actual addresses. It is easy to obtain the geographic coordinates of a street address, city, country, and land parcel worldwide using geopy, and to parse them through third-party geoencoders and data sources.

3.6. Data Feature Extraction

The CLIP (Contrastive Language–Image Pre-Training) model is used to extract features from the YFCC100M, im2gps3k, and Google BigEarthNet datasets [34–36]. The CLIP model consists of two parts, a visual encoder and a text encoder, in which the visual encoder is used to process image information and the text encoder is used to process the address position information after reverse geocoding. The features extracted from the CLIP model have many advantages. First of all, because the CLIP model can handle both images and texts, it can understand and make use of the association between images and texts, thus extracting richer and more representative features. Secondly, the characteristics of the CLIP model are highly robust, and can remain stable even in the face of various changes (such as illumination changes, visual angle changes, etc.). In addition, the features extracted from the CLIP model have a good generalization ability and can be applied to various tasks and scenes. Finally, the features of the CLIP model have a high degree of discrimination, which can effectively distinguish different objects and scenes. These advantages make the CLIP model perform well in various visual and language tasks. CLIP's visual encoder and text encoder are its core components. The visual encoder is responsible for extracting features from images, while the text encoder is responsible for extracting features from texts. These two encoders can extract the features of text and image, respectively, and then calculate the similarity between the text vector and the image vector to predict whether they match. This design enables the CLIP model to process both text and image at the same time, thus achieving the joint understanding of image and text. This is a major feature of the CLIP model and it is also the key to its outstanding performance in various tasks. This has provided strong support for our work.

3.7. Design of CNN-Based Visual Scene Localization and Recognition

In this technical roadmap, the basic idea is to use deep neural networks to train complex scene data to obtain a deep learning model FLsM, which predicts the approximate position and scene type of the captured photo based on the image. The schematic diagram is shown in Figure 5. When it needs to achieve fast communications between two arbitrary global points, the satellites in the air platform are used for forwarding communications. The solutions provide differentiated services for the ground user according to the quality, content, and priority. In Figure 5, image and text information fusion processing positioning can be seen.

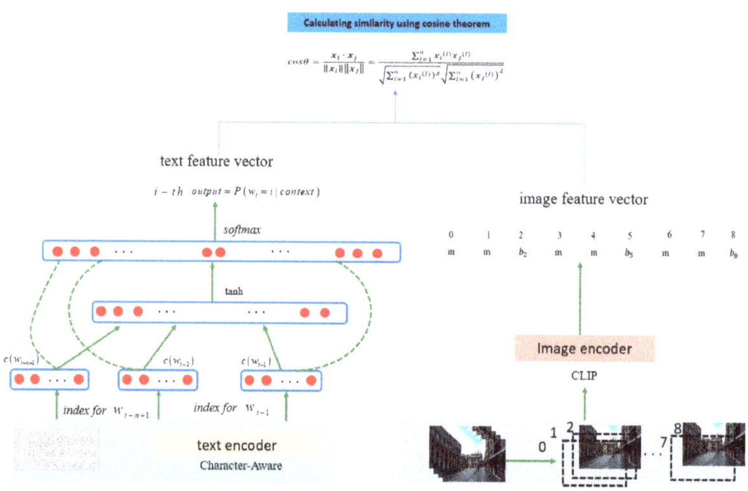

Figure 5. Image and text information fusion processing positioning.

In this technical roadmap, the basic idea is to use deep neural networks to train complex scene data to obtain a deep learning model, FLsM, which predicts the approximate position and scene type of the captured photo, based on the image in Figure 6. According to actual needs, images with GNSS labels are trained on a large amount of data to complete a set of deep learning methods for image localization. The geographic localization problem is transformed into a classification problem; by quantifying all image data with GNSS labels into a fixed number of classes, the GNSS labels are converted into class labels, so that each class represents a physical region in the real world. Then, the classification results are converted into GNSS coordinates of the corresponding region. In this study, in order to obtain a more accurate positioning model, we use multiple models including OpenAI double CLIP model for training [37]. Based on the ability of efficient matching between images and texts, the large model enhances the generalization ability of the visual image positioning model. More than 400 million pairs of image text data are used for pre-training through a large number of Internet data, which cover a wealth of topics and scenes and provide a wide range of samples for model training. A unique method is used in the training process of the visual image positioning model. First, a batch_size image text pair is selected, and then the image is encoded using Image Encoder and the text is encoded using Text Encoder. Next, the cosine similarity between the encoded image and the text vector is calculated to verify the matching between the image and the text. Thanks to its powerful pre-training ability and effective matching verification method, it can be seen through experiments that the positioning accuracy of the model in multiple scenes has reached the current best performance (SOTA). Based on the CLIP model, the visual image positioning model consists of two parts—a visual encoder and a text encoder. The visual encoder is the part used to process the image, which converts the input image into a vector representation of fixed length. The visual encoder can choose to use either the CNN-based ResNet or the Transformer-based ViT. The text encoder is the part used to process text, which converts the input natural language text sequence into a fixed-length vector representation. The text encoder uses the Transformer model. Both encoders are trained to map the input information into the same embedding space and make similar images and texts closer in the embedding space. Model parameters—in different versions of the CLIP model, the number of parameters is different. To ensure transparency and reproducibility in our study, we provide a summary of the datasets used in our experiments in Table 1. This summary includes key information such as the dataset names, their respective sources, and brief descriptions.

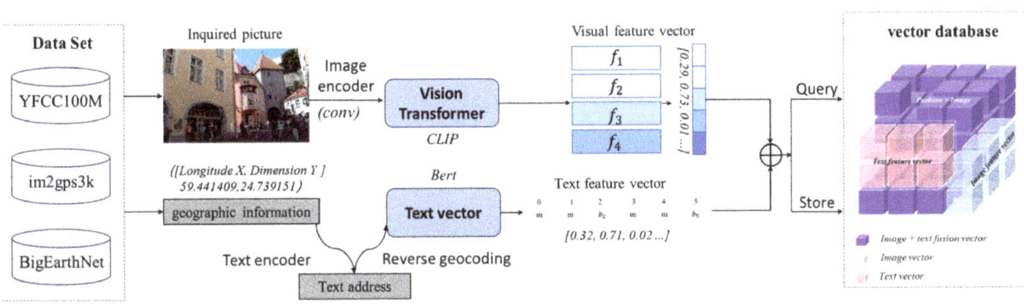

Figure 6. Overall framework of the FLsM structure, integrating image and text large models.

Table 1. Summary of datasets.

Dataset Name	Source	Description
Google Landmarks	cvdfoundation/ google-landmark	5 million landmark-labeled images.
BigEarthNet	Technische Universität Berlin	Contains 590,326 image pairs from Sentinel-1 and Sentinel-2.
Im2GPS3k	TIBHannover/GeoEstimation	Comprises 3000 geotagged images that span a variety of scenes and locations worldwide.
GeoYFCC	abhimanyudubey/GeoYFCC	Comprises a total of 1,147,059 images from 1261 categories across 62 countries.

4. Results

This section delves into investigating the influence of larger models on accuracy through a multi-model and multi-sample approach, using the Google Landmarks Dataset. Specifically, two CLIP models, StreetCLIP (with 420 million parameters) and MetaClip (with 980 million parameters) [38,39], are employed. The experiment adopts a traditional hierarchical search method to facilitate CLIP in deducing the geographical locations of the images. By juxtaposing the performance of these models, particularly highlighting the substantial difference in parameter size between StreetCLIP and MetaClip, valuable insights into the impact of model size on accuracy can be gleaned. The methodology unfolds as follows:

- Step 1: Reverse geocode the latitude and longitude coordinates of the image to obtain textual geographical location information, including country, first-level administrative region, second-level administrative region, address, and detailed address.
- Step 2: Employ the model to predict the country of the image and compare it with the country information obtained from the textual data.
- Step 3: Utilize the model to predict the first-level administrative region of the image and compare it with the corresponding information from the textual data.

  ...

- Continue this iterative process until the detailed address is determined. Then, juxtapose it with the actual region of the image to derive accuracy metrics. The experimental findings are summarized in Table 2 below.

Table 2. Experimental results of the StreetCLIP and MetaClip models.

Model	Country	First-Level Administrative Region	Second-Level Administrative Region	Address	Detailed Address
StreetCLIP	**20.65%**	**6.02%**	1.79%	0.71%	0.59%
MetaCLIP	20.45%	5.99%	**1.93%**	**0.95%**	**0.89%**

Note: Bold green indicates the part with the highest accuracy.

Through experiments, we can see that the positioning accuracy is extremely low and replacing a larger model will not significantly improve the positioning accuracy. Based on this, we change the dataset and select im2gps3k, GeoYFCC, and BigEarthNet for experimental verification [40].

The Im2GPS3k dataset is a subset of the original Im2GPS dataset, which is used for testing in the field of photo geographic positioning estimation. This dataset is an important part of the estimation benchmark of photo geographic location. The purpose of using this dataset is to determine the exact latitude and longitude of the photo shooting place, which is a challenging but widely applicable task in the field of computer vision. There are about 3000 pictures and the dataset distribution, as shown in the figure, of GeoYFCC is a geographical subset of YFCC, which ensures that each country has 20,000–30,000 pictures, so the geographical distribution is more uniform. It contains about 1 million pictures and the dataset is distributed as shown in the figure. BigEarthNet is a large-scale remote sensing dataset based on Sentinel-2 satellite images, which contains 5.9 million image blocks in Europe, each with a size of 120 × 120 pixels, with 13 spectral bands covering 43 land cover/use types. The dataset distribution is shown in Figure 7.

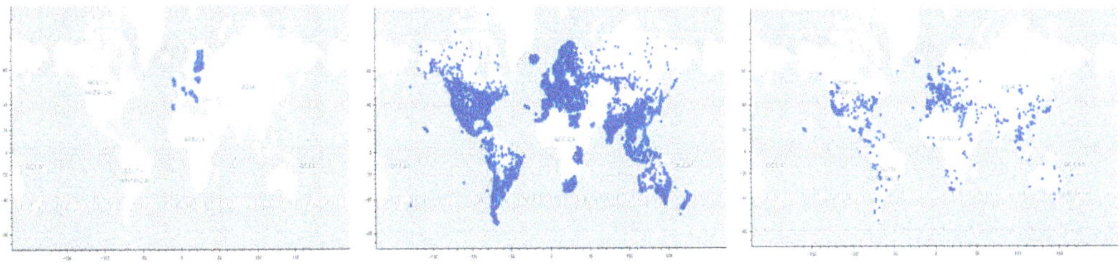

Figure 7. Schematic diagram of global distribution of image dataset.

In the small-scale space under the scene of street and living environment, we studied and performed experiments on two datasets. Firstly, based on the im2gps3k dataset, this study carried out a set of comparative experiments. The traditional hierarchical search is used to traverse the positioning through hierarchical query. First, start the query from a larger area, such as the country, then narrow down the scope one by one, such as the first-level administrative region and the second-level administrative region, and finally obtain the detailed address. The contrast experiment is carried out using a brand-new similarity calculation method. This experiment utilized Milvus as the vector database framework. The specific experimental process is as follows. Initially, each image in the dataset was processed. Every image is associated with a unique image ID and latitude–longitude information. The latitude–longitude coordinates were converted into textual location information using reverse geocoding. Simultaneously, the textual location information was encoded using a model to generate textual feature vectors. These components were abstracted into Milvus entities and stored in the database. These steps were repeated for each image in the dataset. Subsequently, the dataset was traversed again. The model was used to extract image feature vectors from each picture. Then, the database was queried to find the Milvus entity corresponding to the textual feature vector with the highest cosine similarity to the image feature vector. The latitude–longitude coordinates of the image and the corresponding entity were compared to calculate the predicted distance error. Finally, the errors were categorized into different scales (e.g., errors less than 1 km were categorized as within 1 km, errors less than 25 km were categorized as within 25 km, where 25 km includes 1 km). This process aims to establish associations between images and geographical location information. Through the efficient vector search capabilities of the Milvus database, the positioning accuracy of the models was evaluated across different scales. The comparison between the new method using StreetCLIP and the traditional method using StreetCLIP shows that the comparison between the new method and the traditional method is obviously improved on a smaller scale, but there is little difference on a larger scale. Then, new methods are used to test the performance of other models in image positioning, including ViTbigCLIP (GeoDE dataset performs well) and EVAplusCLIP (model parameters are large, reaching 5 billion). The performance can be further improved after replacing other better models (e.g., ViTbigCLIP performs better on GeoDE, and EVAplusCLIP model parameters are larger). The following experimental results are shown in Table 3. Thermal maps of positioning accuracy are shown in Figure 8a. On the GeoYFCC dataset, we use the same method to test StreetCLIP, ViTbigCLIP, EVAplusCLIP, and GeoCLIP (because CeoCLIP has no text encoder, so we use the traditional hierarchical search method for this model). The experimental results are shown in Figure 8b,c. Based on remote sensing images in large-scale space, the positioning accuracy of the model is tested on the BigEarthNet dataset, as shown in Figure 8d. Through experiments, it can be found that the overall accuracy of positioning based on remote sensing images is low, and it only improves slightly at the scale of 750 km, but the partial area of this dataset is mostly around 2500 km, which leads to the soaring accuracy of the last 2500 km, so it is more effective in large-scale space.

Table 3. Experimental results under different models.

Dataset	Model	1 km_Accuracy	25 km_Accuracy	200 km_Accuracy	750 km_Accuracy	2500 km_Accuracy
BigEarthNet	StreetCLIP	**3.38796×10^{-06}**	**0.004997239**	0.067010432	0.177152963	**0.99427435**
BigEarthNet	ViTbigCLIP	2.20217×10^{-05}	0.004595766	**0.071401226**	**0.274355187**	0.950119087
BigEarthNet	EVAplusCLIP	1.18579×10^{-05}	0.004094348	0.042068281	0.18899896	0.956994949
Im2GPS3k	StreetCLIP	0.171838505	0.326659993	0.464798131	0.635969303	0.794127461
Im2GPS3k	ViTbigCLIP	**0.256256256**	**0.450784117**	**0.589255923**	**0.724724725**	**0.851851852**
Im2GPS3k	EVAplusCLIP	0.249249249	0.431097764	0.544210878	0.688688689	0.829496163
mix_feature	StreetCLIP	**0.074074074**	**0.212545879**	**0.297297297**	**0.464130797**	**0.650650651**
mix_feature	ViTbigCLIP	0.016016016	0.082749416	0.122455789	0.23023023	0.448114781
mix_feature	EVAplusCLIP	0.048381715	0.125792459	0.18685352	0.294627961	0.515181849

Note: The accuracy values represent the proportion of correctly identified geographical locations. The spatial scales are in kilometers. Bold green indicates the part with the highest accuracy.

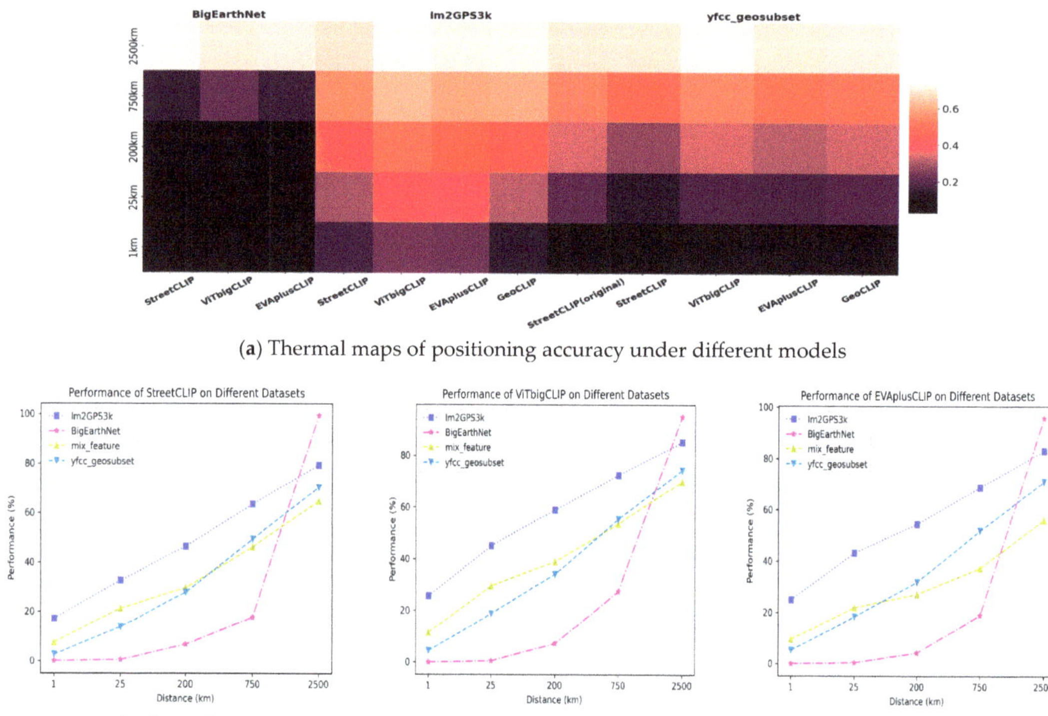

(a) Thermal maps of positioning accuracy under different models

(b) StreetCLIP (c) ViTbigCLIP (d) EVAplusCLIP

Experimental results of positioning progress of different datasets under different models.

Figure 8. Experimental results under different models.

Examine the four charts in Figure 9a–d to see how each model performs under various distance settings and datasets. In most cases, the enhanced version of StreetCLIP (designated as "Our method") has demonstrated superior performance, particularly in large-scale spaces where it operates more flawlessly. On some datasets, the StreetCLIP model outperforms the VitbigCLIP and EVAplusCLIP models; however, the enhanced version of StreetCLIP exhibits a more consistent performance growth in a number of areas. In every scenario, the StreetCLIP original version displayed the slowest performance growth. Figure 9e–h shows the accuracy performance of many models under various distance parameters on four distinct datasets. The variation in the model's performance is represented by the scatter's size. "Our method" (StreetCLIP) generally shows rather large scatter points in all datasets, especially when covering big distances (750 km and 2500 km), which suggests high accuracy. In comparison to other models, "Our method" exhibits a notable improvement in accuracy, particularly on the lm2GPS3k and yfcc_geosubset datasets.

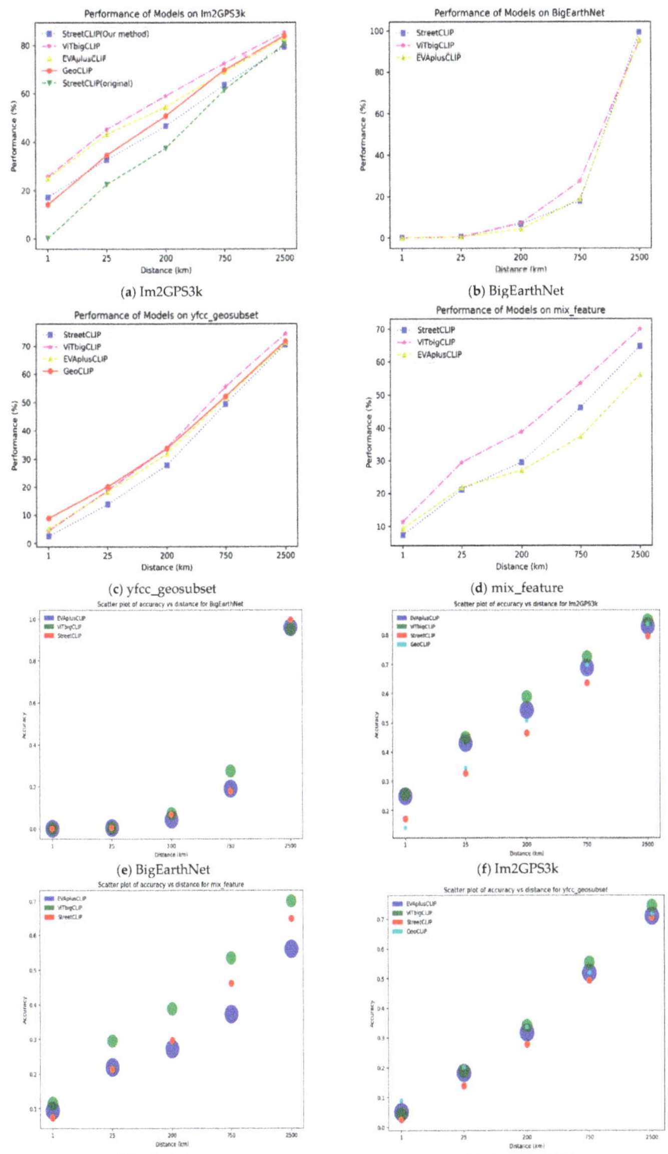

(a) Im2GPS3k (b) BigEarthNet (c) yfcc_geosubset (d) mix_feature (e) BigEarthNet (f) Im2GPS3k (g) mix_feature (h) yfcc_geosubset

Figure 9. Relationship between positioning accuracy of different datasets under different models.

5. Discussion

To investigate visual blur positioning using large models, we utilized conventional street/environmental image and remote sensing image datasets as elastic scale space samples. Experiments were conducted across various large models to analyze their performance. The results indicate that models with larger parameters, such as EVAplusCLIP, exhibit enhanced positioning accuracy on elastic scale space sample datasets, with more stable outcomes aligning with the requirements of visual blur in elastic scale space positioning. The key advantage of EVAplusCLIP lies in its larger model parameters, enabling better data complexity capture and representation, thereby enhancing model generalization. Notably, the output vector dimension of the model significantly impacts its performance, as evidenced by the ViTbigCLIP model having the highest positioning accuracy score on the GeoDE dataset, due to its larger vector dimensions providing more information for improved prediction accuracy. However, optimizing model parameters and vector dimensions alone may not suffice to meet all requirements. To further enhance model accuracy, alternative improvement methods should be considered, such as introducing a vector database to optimize vector retrieval and utilization for improved model accuracy.

6. Conclusions

In this research, we introduce a fuzzy positioning approach for images based on large models in elastic scale space. Our comparative experiments demonstrate that the EVAplusCLIP model achieves a higher positioning accuracy and can effectively serve the image positioning function across various scale spaces. This work represents an exploratory research endeavor with several areas open for future improvement. Potential research directions include optimizing model stability through further experiments with increased model parameters, enhancing model performance on specific datasets by expanding the output vector dimension and training on more relevant data, and exploring additional improvement methods such as vector databases to enhance model accuracy. These paths present critical avenues for our ongoing in-depth investigation and optimization of this research. Through conducting additional experiments and exploring further improvement methods, we can enhance the stability and accuracy of our image positioning approach. Our research has produced promising results, indicating the potential for further advances in model stability through increased parameter experiments. Furthermore, expanding output vector dimensions and training on more relevant data offer exciting opportunities for enhancing model performance across specific datasets. Incorporating vector databases as an improvement method also introduces new possibilities for optimizing positioning accuracy. These areas provide essential directions for our ongoing in-depth investigation and advancement of this research.

Author Contributions: Conceptualization, W.C.; Methodology, W.C.; Formal analysis, Y.W.; Resources, Y.L.; Writing—original draft, W.C.; Supervision, L.M. and J.G. All authors have read and agreed to the published version of the manuscript.

Funding: This research was funded by [China's Ministry of Science and Technology National Key R&D Program Beidou xing Energy] grant number [E33514060C].

Data Availability Statement: The data presented in this study are available in this article.

Conflicts of Interest: The authors declare no conflict of interest.

References

1. Li, X.; Wang, B.; Li, X.; Huang, J.; Lyu, H.; Han, X. Principle and performance of multi-frequency and multi-GNSS PPP-RTK. *Satell. Navig.* **2022**, *3*, 7. [CrossRef]
2. Al Hage, J.; Najjar, M.E.B.E. Improved Outdoor Localization Based on Weighted Kullback-Leibler Divergence for Measurements Diagnosis. *IEEE Intell. Transp. Syst. Mag.* **2018**, *12*, 41–56. [CrossRef]
3. Li, M.-H.; Jiang, P.; Yu, D.-J.; Sun, J.-H. Position and attitude determination by integrated GPS/SINS/TS for feed support system of FAST. *Res. Astron. Astrophys.* **2020**, *20*, 140. [CrossRef]

4. Shi, Q.; Wu, J.; Lin, Z.; Qin, N. Learning a Robust Hybrid Descriptor for Robot Visual Localization. *J. Robot.* **2022**, *2022*, 9354909. [CrossRef]
5. Wang, Z.; Cheng, X. Adaptive optimization online IMU self-calibration method for visual-inertial navigation systems. *Measurement* **2021**, *180*, 109478. [CrossRef]
6. Cao, M.; Tang, F.; Ji, P.; Ma, F. Improved Real-Time Semantic Segmentation Network Model for Crop Vision Navigation Line Detection. *Front. Plant Sci.* **2022**, *13*, 898131. [CrossRef]
7. Critchley-Marrows, J.J.; Wu, X.; Cairns, I.H. An architecture for a visual-based PNT alternative. *Acta Astronaut.* **2023**, *210*, 601–609. [CrossRef]
8. Bellamy, B.R. The Robotic Imaginary: The Human and the Price of Dehumanized Labor by Jennifer Rhee. *Sci. Fict. Stud.* **2019**, *46*, 655–657. [CrossRef]
9. Cass, S. Ayanna Howard: Robot wrangler. *IEEE Spectr.* **2005**, *42*, 21–22. [CrossRef]
10. Carpenter, K.; Wiltsie, N.; Parness, A. Rotary Microspine Rough Surface Mobility. *IEEE/ASME Trans. Mechatron.* **2015**, *21*, 2378–2390. [CrossRef]
11. Tang, C.; Ma, W.; Li, B.; Jin, M.; Chen, H. Cephalopod-Inspired Swimming Robot Using Dielectric Elastomer Synthetic Jet Actuator. *Adv. Eng. Mater.* **2019**, *22*, 1901130. [CrossRef]
12. Ning, X.; Liu, L. A Two-Mode INS/CNS Navigation Method for Lunar Rovers. *IEEE Trans. Instrum. Meas.* **2014**, *63*, 2170–2179. [CrossRef]
13. Ning, X.; Fang, J. A new autonomous celestial navigation method for the lunar rover. *Robot. Auton. Syst.* **2009**, *57*, 48–54. [CrossRef]
14. Wang, W.-R.; Ren, X.; Wang, F.-F.; Liu, J.-J.; Li, C.-L. Terrain reconstruction from Chang'e-3 PCAM images. *Res. Astron. Astrophys.* **2015**, *15*, 1057–1067. [CrossRef]
15. Sutoh, M.; Wakabayashi, S.; Hoshino, T. Influence of atmosphere on lunar rover performance analysis based on soil parameter identification. *J. Terramech.* **2017**, *74*, 13–24. [CrossRef]
16. Choi, I.-S.; Ha, J.-E. Simple method for calibrating omnidirectional stereo with multiple cameras. *Opt. Eng.* **2011**, *50*, 43608. [CrossRef]
17. Gamarra, D.F.T.; Pinpin, L.K.; Laschi, C.; Dario, P. Forward Models Applied in Visual Servoing for a Reaching Task in the iCub Humanoid Robot. *Appl. Bionics Biomech.* **2009**, *6*, 345–354. [CrossRef]
18. Zhang, M.; Cui, J.; Zhang, F.; Yang, N.; Li, Y.; Li, F.; Deng, Z. Research on evaluation method of stereo vision measurement system based on parameter-driven. *Optik* **2021**, *245*, 167737. [CrossRef]
19. Huang, W.; Fajen, B.R.; Fink, J.; Warren, W.H. Visual navigation and obstacle avoidance using a steering potential function. *Robot. Auton. Syst.* **2006**, *54*, 288–299. [CrossRef]
20. Bulanon, D.; Burks, T.; Alchanatis, V. Image fusion of visible and thermal images for fruit detection. *Biosyst. Eng.* **2009**, *103*, 12–22. [CrossRef]
21. Kuo, B.-W.; Chang, H.-H.; Chen, Y.-C.; Huang, S.-Y. A Light-and-Fast SLAM Algorithm for Robots in Indoor Environments Using Line Segment Map. *J. Robot.* **2011**, *2011*, 257852. [CrossRef]
22. Lv, K.; Zhang, Y.; Yu, Y.; Wang, Z.; Min, J. SIIS-SLAM: A Vision SLAM Based on Sequential Image Instance Segmentation. *IEEE Access* **2022**, *11*, 17430–17440. [CrossRef]
23. Zhao, X.; Wang, T.; Li, Y.; Zhang, B.; Liu, K.; Liu, D.; Wang, C.; Snoussi, H. Target-Driven Visual Navigation by Using Causal Intervention. *IEEE Trans. Intell. Veh.* **2023**, *9*, 1294–1304. [CrossRef]
24. Li, J.; Yin, J.; Deng, L. A robot vision navigation method using deep learning in edge computing environment. *EURASIP J. Adv. Signal Process.* **2021**, *2021*, 22. [CrossRef]
25. Zhou, K.; Yang, J.; Loy, C.C.; Liu, Z. Learning to Prompt for Vision-Language Models. *Int. J. Comput. Vis.* **2022**, *130*, 2337–2348. [CrossRef]
26. Xing, Y.; Wu, Q.; Cheng, D.; Zhang, S.; Liang, G.; Wang, P.; Zhang, Y. Dual Modality Prompt Tuning for Vision-Language Pre-Trained Model. *IEEE Trans. Multimed.* **2023**, *26*, 2056–2068. [CrossRef]
27. Sun, B.; Liu, G.; Yuan, Y. F3-Net: Multiview Scene Matching for Drone-Based Geo-Localization. *IEEE Trans. Geosci. Remote Sens.* **2023**, *61*, 3278257. [CrossRef]
28. Vicente Vivanco, C.; Nayak, G.K.; Shah, M. GeoCLIP: Clip-Inspired Alignment between Locations and Images for Effective Worldwide Geo-localization. *arXiv* **2023**, arXiv:2309.16020.
29. Gao, H.; Zhu, M.; Wang, X.; Li, C.; Xu, S. Lightweight Spatial-Spectral Network Based on 3D-2D Multi-Group Feature Extraction Module for Hyperspectral Image Classification. *Int. J. Remote Sens.* **2023**, *44*, 3607–3634. [CrossRef]
30. Cheng, P.; Cheng, Y.; Wang, X.; Wu, S.; Xu, Y. Realization of an Optimal Dynamic Geodetic Reference Frame in China: Methodology and Applications. *Engineering* **2020**, *6*, 879–897. [CrossRef]
31. Thomee, B.; Shamma, D.A.; Friedland, G.; Elizalde, B.; Ni, K.; Poland, D.; Borth, D.; Li, L.-J. Yfcc100m: The new data in multimedia research. *arXiv* **2016**, arXiv:1503.01817. [CrossRef]
32. Alsubai, S.; Dutta, A.K.; Alkhayyat, A.H.; Jaber, M.M.; Abbas, A.H.; Kumar, A. Hybrid deep learning with improved Salp swarm optimization based multi-class grape disease classification model. *Comput. Electr. Eng.* **2023**, *108*, 108733. [CrossRef]
33. Anguelov, D.; Dulong, C.; Filip, D.; Frueh, C.; Lafon, S.; Lyon, R.; Ogale, A.; Vincent, L.; Weaver, J. Google Street View: Capturing the World at Street Level. *Computer* **2010**, *43*, 32–38. [CrossRef]

34. Steven, B.; Ayton, A. Text-to-Image Synthesis with Self-supervision via Contrastive Language-Image Pre-Training (CLIP). Available online: https://www.researchgate.net/publication/369299175_Text-to-Image_Synthesis_with_Self-supervision_via_Contrastive_Language-Image_Pre-training_CLIP (accessed on 13 May 2024).
35. Vo, N.; Jacobs, N.; Hays, J. Revisiting im2gps in the deep learning era. In Proceedings of the IEEE International Conference on Computer Vision, Venice, Italy, 22–29 October 2017.
36. Sumbul, G.; Wall, A.; Kreuziger, T.; Marcelino, F.; Costa, H.; Benevides, P.; Caetano, M.; Demir, B.; Markl, V. BigEarthNet-MM: A large-scale, multimodal, multilabel benchmark archive for remote sensing image classification and retrieval [software and data sets]. *IEEE Geosci. Remote Sens. Mag.* **2021**, *9*, 174–180. [CrossRef]
37. Roumeliotis, K.I.; Tselikas, N.D. ChatGPT and Open-AI Models: A Preliminary Review. *Futur. Internet* **2023**, *15*, 192. [CrossRef]
38. Haas, L.; Silas, A.; Michal, S. Learning generalized zero-shot learners for open-domain image geolocalization. *arXiv* **2023**, arXiv:2302.00275.
39. Parashar, S.; Lin, Z.; Liu, T.; Dong, X.; Li, Y.; Ramanan, D.; Caverlee, J.; Kong, S. The Neglected Tails of Vision-Language Models. *arXiv* **2024**, arXiv:2401.12425.
40. Dubey, A.; Ramanathan, V.; Pentland, A.; Mahajan, D. Adaptive methods for real-world domain generalization. In Proceedings of the IEEE/CVF Conference on Computer Vision and Pattern Recognition, Online, 19–25 June 2021.

electronics MDPI

Article

IPCB: Intelligent Pseudolite Constellation Based on High-Altitude Balloons

Yi Qu [1,2,*], Sheng Wang [1,2], Tianshi Pan [1] and Hui Feng [1]

1 Aerospace Information Research Institute, Chinese Academy of Sciences, Beijing 100094, China
2 University of Chinese Academy of Sciences, Beijing 100049, China
* Correspondence: quyi@aircas.ac.cn

Abstract: IPCBs (Intelligent Pseudolite Constellations based on high-altitude balloons) are a novel type of air-based pseudolite application with many advantages. Compared with ground-based pseudolites and traditional air-based pseudolites, IPCBs have a wider coverage and a lower energy requirement. Compared with LEO satellite constellations, IPCBs have a stronger signal, a lower cost, and a shorter deployment period. These merits give promising potential to IPCBs. In IPCB applications, one of the key factors is geometry configuration, which is deeply influenced by the balloon's unique features. The basic idea of this paper is to pursue a strategy to improve IPCB geometry performance by using diverse winds at different altitudes and balloons' capability of altering flight altitude intelligently. Starting with a brief introduction to IPCBs, this paper defines an indicator to assess IPCB geometry performance, an approach to adjust IPCB geometry configuration and an IPCB geometry configuration planning algorithm. Next, a series of simulations are implemented with an IPCB composed of six pseudolites in winds with/without a quasi-zero wind layer. Some IPCB geometry configurations are analyzed, and their geometry performances are compared. Simulation results show the effectiveness of the proposed algorithm and the influence of the quasi-zero wind layer on IPCB performance.

Keywords: intelligent pseudolite; constellation; high altitude balloon

1. Introduction

Pseudolites are transmitters that can emit navigation signals to improve GNSS performance or to provide navigation service independently [1–7]. An intelligent pseudolite based on high-altitude balloons (IPBs) is a type of novel air-based pseudolite that utilizes a high-altitude balloon as a platform. An intelligent pseudolite constellation based on high-altitude balloons (IPCBs) is composed of multiple IPBs, which can provide emergency positioning service and regional positioning service independently.

As an excellent solution for regional positioning, ICPBs have many advantages. Compared with ground-based pseudolites and traditional air-based pseudolites, IPCBs have a wider coverage and can serve more users because their flight altitude can reach tens of kilometers [8,9]. Furthermore, IPCBs have a low energy requirement since they can accomplish its flight primarily relying on buoyancy and wind rather than oil or electricity [10,11]. Compared with LEO satellite constellations, IPCBs have a stronger signal, a lower cost, and a shorter deployment period [12,13]. In addition, the continuous residence duration of an IPB over a service area is longer than that of an LEO satellite, which can reduce navigation signal lock-lose and cycle slip caused by satellite switching [14–16]. Also, the IPCB operation and maintenance burden is less than that of an LEO satellite constellation because an LEO satellite constellation usually comprises a large number of satellites [17,18]. These merits give promising potential to IPCB applications.

In the application of IPCBs, geometry configuration plays a critical role since it affects IPCB service performance significantly [19–22]. However, the problem becomes very complicated because of the unique dynamic features of IPBs. Most traditional research studies about pseudolite geometry configuration are designed for static pseudowires, which are not suitable for IPCBs [23–27].

This paper centers on the problem of IPCB geometry configuration and proposes a planning algorithm that emphasizes utilizing different winds smartly. The proposed algorithm can fit the dynamic flight of IPCBs adaptively and is easy to implement. Moreover, it can achieve performance improvements by controlling IPB valves and fans only, without extra hardware cost.

The rest of this paper is organized as follows. An overview of IPCBs is described in Section 2, a performance indicator of IPCB geometry configuration is defined in Section 3, an IPCB geometry configuration adjustment approach is designed in Section 4, a series of constraints are discussed in Section 5, a planning algorithm based on whale optimization algorithm (WOA) is proposed in Section 6, simulations and discussions are presented in Section 7, and conclusions are stated in Section 8.

2. Overview of IPCB

An IPB is illustrated in Figure 1. It utilizes a high-altitude balloon as a platform, which is composed of balloon, cable, parachute, gondola, balloon controller, payloads, and other attachments, as shown in Figure 1a. The balloon controller, payloads, and other attachments are installed in the gondola. Furthermore, the balloon consists of a main helium bag and an air ballonet, which are separated by a membrane, as Figure 1b illustrates [28–30]. The main helium bag is filled with helium to provide buoyancy to the IPB, and the air ballonet is filled with air to adjust the IPB mass. Fans and valves are installed on the bottom of the balloon, which can pump air into or release air from the air ballonet. The balloon controller can perceive and adjust IPB flight status intelligently. In particular, the balloon controller can manipulate the fans and valves flexibly, enabling the IPB to adjust its mass and flight altitude in a range.

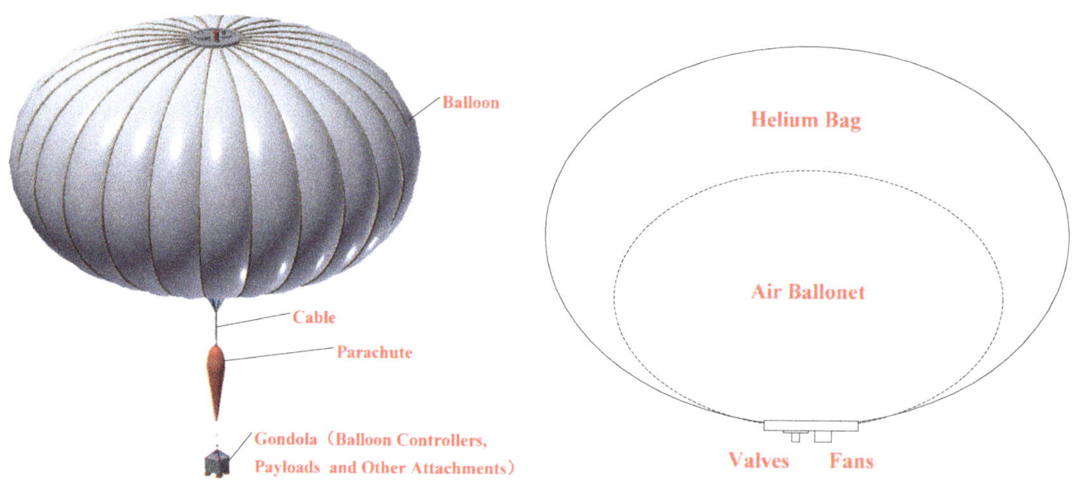

(a) (b)

Figure 1. Structure illustration of IPB studied in this paper: (**a**) structure overview; (**b**) detailed structure of the balloon.

An IPCB is illustrated in Figure 2. It consists of multiple IPBs and can form coverage over a service area. When payloads in the IPBs normally send out navigation signals, the IPCB can provide positioning service for the area independently. During the process of

IPCB service, its geometry configuration is always changing with the wind, and the changes are highly nonlinear, which brings difficulty to IPCB geometry configuration planning.

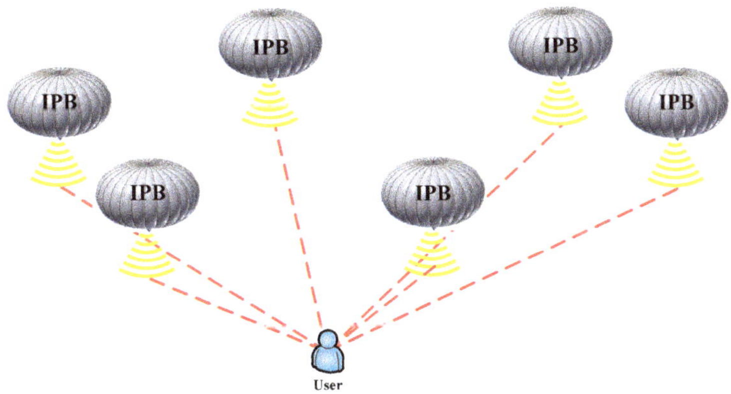

Figure 2. Illustration of IPCB in regional positioning.

3. Performance Indicator of IPCB Geometry Configuration

Assuming that an IPCB composed of n_p IPBs is deployed above the service area initially and that n_u users are selected as samples to assess IPCB geometry performance. Then, the pseudo-range equation from the j-th IPB to the i-th user at time t can be expressed by Equation (1) [31–33].

$$\rho(i,j,t) = \sqrt{(x(j,t)-x_u(i,t))^2 + (y(j,t)-y_u(i,t))^2 + (z(j,t)-z_u(i,t))^2} + ct_u(i,t) \quad (1)$$

In Equation (1), $\rho(i,j,t)$ represents the pseudo-range from the j-th IPB to the i-th user at time t; $(x(j,t), y(j,t), z(j,t))$ represents the position of the j-th IPB at time t; $(x_u(i,t), y_u(i,t), z_u(i,t))$ represents the position of the i-th user at time t; c represents the speed of light; $t_u(i,t)$ represents the clock difference of the i-th user at time t.

Equation (1) can be rewritten as Equation (2) after the first-order Taylor expansion.

$$\Delta\rho(i,j,t) = a_x(i,j,t)\Delta x_u(i,t) + a_y(i,j,t)\Delta y_u(i,t) + a_z(i,j,t)\Delta z_u(i,t) - c\Delta t_u(i,t) \quad (2)$$

In Equation (2), 'Δ' represents the difference between the Taylor expansion point and its neighborhood, $(a_x(i,j,t), a_y(i,j,t), a_z(i,j,t))$ represents the direction cosine from the i-th user to the j-th IPB at time t, which can be calculated by Equation (3).

$$\begin{aligned} a_x(i,j,t) &= \frac{x(j,t)-x_u(i,t)}{\sqrt{(x(j,t)-x_u(i,t))^2+(y(j,t)-y_u(i,t))^2+(z(j,t)-z_u(i,t))^2}} \\ a_y(i,j,t) &= \frac{y(j,t)-y_u(i,t)}{\sqrt{(x(j,t)-x_u(i,t))^2+(y(j,t)-y_u(i,t))^2+(z(j,t)-z_u(i,t))^2}} \\ a_z(i,j,t) &= \frac{z(j,t)-z_u(i,t)}{\sqrt{(x(j,t)-x_u(i,t))^2+(y(j,t)-y_u(i,t))^2+(z(j,t)-z_u(i,t))^2}} \end{aligned} \quad (3)$$

Equation (2) can be expanded from the j-th IPB to all the IPBs in the IPCB, which can be simplified as

$$\boldsymbol{\Delta\rho}(i,t) = \boldsymbol{H}(i,t)\boldsymbol{\Delta x}(i,t) \quad (4)$$

In Equation (4), $\boldsymbol{\Delta\rho}(i,t)$ is a vector representing pseudo-range measurement error, $\boldsymbol{\Delta x}(i,t)$ is a vector representing positioning error, and $\boldsymbol{H}(i,t)$ is an observation matrix.

Equation (4) can be rewritten as Equation (5) by the least square method.

$$\boldsymbol{\Delta x}(i,t) = (\boldsymbol{H}(i,t)^T\boldsymbol{H}(i,t))^{-1}\boldsymbol{H}(i,t)^T\boldsymbol{\Delta\rho}(i,t) \quad (5)$$

If the pseudo-range noises of different IPBs are linearly independent, with an average of 0 and a variance of σ^2, then the covariance of $\boldsymbol{\Delta x}(i,t)$ can be expressed by Equation (6) [33].

$$\mathrm{cov}(\boldsymbol{\Delta x}(i,t)) = \sigma^2\left(\boldsymbol{H}(i,t)^{\mathrm{T}}\boldsymbol{H}(i,t)\right)^{-1} \tag{6}$$

From Equation (6), it can be concluded that $(\boldsymbol{H}(i,t)^{\mathrm{T}}\boldsymbol{H}(i,t))^{-1}$ reveals the magnification from a user pseudo-range measurement error to its positioning error. Given the same user pseudo-range measurement error, the smaller $(\boldsymbol{H}(i,t)^{\mathrm{T}}\boldsymbol{H}(i,t))^{-1}$ is, the smaller the positioning error is. Consequently, the square root of the trace of $(\boldsymbol{H}(i,t)^{\mathrm{T}}\boldsymbol{H}(i,t))^{-1}$ is usually treated as an important indicator to assess the influence of a constellation geometry configuration on its positioning error, named GDOP (geometric dilution of precision), which can be expressed by Equation (7) [33–36].

$$GDOP(i,t) = \sqrt{tr\left(\boldsymbol{H}(i,t)^{T}\boldsymbol{H}(i,t)\right)^{-1}} \tag{7}$$

It is obvious that the value of $GDOP(i,t)$ will fluctuate due to IPCBs' dynamic movements. In particular, some IPBs in the constellation may leave approved airspace as time goes on. In such cases, they cannot continue emitting navigation signals (detailed discussion in Section 5.1), which will decrease the number of available IPBs in the constellation. Once the number of available IPBs in the constellation drops below 4, the GDOP of the IPCB is defined as infinity in this paper.

The geometry performance of an IPCB at time t can be assessed by the average GDOP of multiple users distributed in the service area, as described in Equation (8) [37].

$$GDOP(t) = \frac{1}{n_u}\sum_{i=1}^{n_u} GDOP(i,t) \tag{8}$$

It can be concluded that the objective of IPCB geometry configuration planning is to obtain the minimal $GDOP(t)$ for the whole service duration. However, this approach encounters difficulties when dealing with infinite values. To avoid such difficulties, this paper defines the performance indicator as Equation (9).

$$F = \frac{1}{n_t}\sum_{t=1}^{n_t} \frac{1}{GDOP(t)} \tag{9}$$

In Equation (9), n_t represents the expected service duration of the IPCB. The ultimate objective of IPCB geometry configuration planning is to maximize F defined in Equation (9).

4. Adjustment Approaches of IPCB Geometry Configuration

As discussed in Section 2, IPBs have little actuation capability since they are not equipped with propellers. The primary actuation they can implement is to control their valves and fans, which cannot change the IPCB geometry configuration directly. Therefore, this paper adopts an indirect adjustment approach.

In the vertical direction, IPBs can actively change their masses and flight altitudes by switching their valves and fans. Specific adjustments of each IPB can be managed by its balloon controller. In the horizontal direction, IPBs can change their trajectories with the help of local winds at different altitudes [38–40]. By combining the adjustments in the two directions, IPBs can change their flight trajectories, and the IPCB can modify its geometry configuration [9,28,41,42].

Subsequent detailed analyses will be presented based on the vertical adjustment and the horizontal adjustment of individual IPBs, respectively. The following assumptions are made to simplify the analysis.

(1) The flight altitude of an IPB is fully controllable, and its variation depends on a rise rate and a sink rate, represented by v_{rise} and v_{sink}, respectively.
(2) The horizontal velocity of an IPB is proportional to the local horizontal wind velocity.

(3) Wind velocities vary with altitude but not with horizontal location. The wind velocities and atmosphere environment are steady during IPCB service time.
(4) The influences of balloon volume variation and thermal effect on IPB motions are ignored.

4.1. Vertical Adjustment of an IPB

An IPB usually keeps its flight altitude by maintaining the balance between gravity and buoyancy, as Equation (10) shows:

$$\boldsymbol{B} = \boldsymbol{G} \tag{10}$$

In Equation (10), $\boldsymbol{B}$ represents its buoyancy, and $\boldsymbol{G}$ represents its gravity. Buoyancy $\boldsymbol{B}$ is related to the atmospheric density, volume of the balloon, and gravitation acceleration, as shown in Equation (11).

$$\boldsymbol{B} = \rho_{air}(h)V\boldsymbol{g} \tag{11}$$

In Equation (11), V represents the volume of the balloon, $\boldsymbol{g}$ represents gravitation acceleration, h represents the flight altitude of the IPB, and $\rho_{air}(h)$ represents the air density at altitude h. The air density is not a constant, and it varies in a wide range. According to the standard atmosphere model (U.S. Standard Atmosphere, 1976), its change with altitude is illustrated in Figure 3.

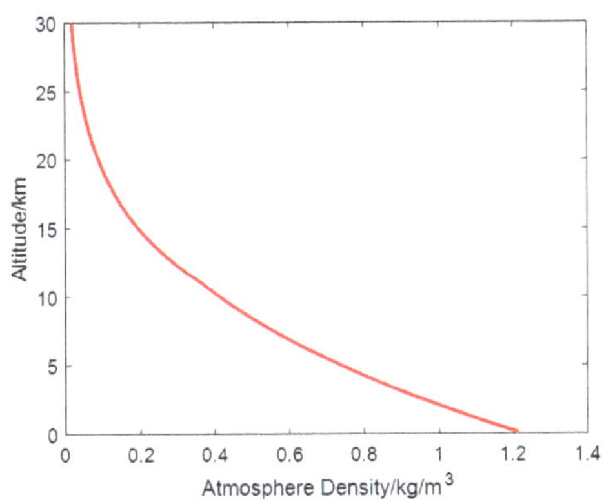

Figure 3. Atmosphere density from 0 km to 30 km.

Gravity $\boldsymbol{G}$ is related to helium mass, air mass, and the masses of others, as shown in Equation (12).

$$\boldsymbol{G} = (m_{He} + m_{air}(h) + m_{other})\boldsymbol{g} \tag{12}$$

In Equation (12), m_{He} represents the helium mass in the main helium bag, $m_{air}(h)$ represents the air mass in the air ballonet at altitude h, m_{other} represents the gross mass of balloon envelop, cable, parachute, gondola, balloon controller, payloads and other attachments.

So if a flight altitude decline is needed, an IPB can pump air into its air ballonet, which will increase its air mass $m_{air}(h)$ and gravity $\boldsymbol{G}$, making $\boldsymbol{G}$ greater than $\boldsymbol{B}$. The change in air mass can be estimated by Equation (13).

$$\Delta m = (\rho_{air}(h_s) - \rho_{air}(h_d))V \tag{13}$$

In Equation (13), h_s and h_d represent the flight altitude before adjustment and after adjustment, respectively. Due to the assumption (1) in this section, the time cost of the flight altitude adjustment Δt can be estimated by Equation (14).

$$h_s - h_d = v_{\sin \mathrm{k}}\Delta t \tag{14}$$

In contrast, if a flight altitude ascent is needed, the IPB should release air from its air ballonet, and the time cost can be estimated by Equation (15).

$$h_s - h_d = v_{\text{rise}} \Delta t \tag{15}$$

4.2. Horizontal Adjustment of an IPB

According to assumption (2) in this section, the horizontal velocity of an IPB is proportional to the local horizontal wind velocity [38–40]. Then, the horizontal kinematics of an IPB can be described by Equation (16).

$$\begin{aligned} \dot{\lambda} &= \gamma \cdot w_z(\lambda, \phi, h, t) \\ \dot{\phi} &= \gamma \cdot w_m(\lambda, \phi, h, t) \end{aligned} \tag{16}$$

In Equation (16), λ represents the longitude of IPB position; $\dot{\lambda}$ represents the variation of λ; ϕ represents the latitude of IPB position; $\dot{\phi}$ represents the variation of ϕ; $w_z(\lambda, \phi, h, t)$ represents zonal wind velocity at position (λ, ϕ, h) and at time t; $w_m(\lambda, \phi, h, t)$ represents meridional wind velocity at position (λ, ϕ, h) and at time t; γ represents a drag coefficient of IPB in the horizontal plane.

Since it is assumed that wind velocities vary with altitude but not with horizontal location, and wind velocities are steady in IPCB service time, the relation between altitude and wind velocity is emphasized in this paper. In general, wind velocity increases as altitude increases in the troposphere, reaching a maximum of about 10~15 km. Then, wind velocity decreases, reaching a minimum in the lower portion of the stratosphere at about 18~25 km [43,44]. For simplicity, a seventh-order polynomial is employed to fit the relation between altitude and wind velocity in this paper, as described below [45–47].

$$\begin{aligned} w_m &= c_{m0} + c_{m1} h_{std} + c_{m2} {h_{std}}^2 + \cdots + c_{m7} {h_{std}}^7 \\ w_z &= c_{z0} + c_{z1} h_{std} + c_{z2} {h_{std}}^2 + \cdots + c_{z7} {h_{std}}^7 \end{aligned} \tag{17}$$

In Equation (17), $\boldsymbol{c_m}$ and $\boldsymbol{c_z}$ represent meridional wind coefficients and zonal wind coefficients, respectively. h_{std} represents normalized altitude, which can be calculated by Equation (18).

$$h_{std} = (h - \mu_d)/\sigma_d \tag{18}$$

In Equation (18), μ_d and σ_d are both normalized parameters. Figure 4 illustrates wind fittings for a specific area in March, June, September, and December.

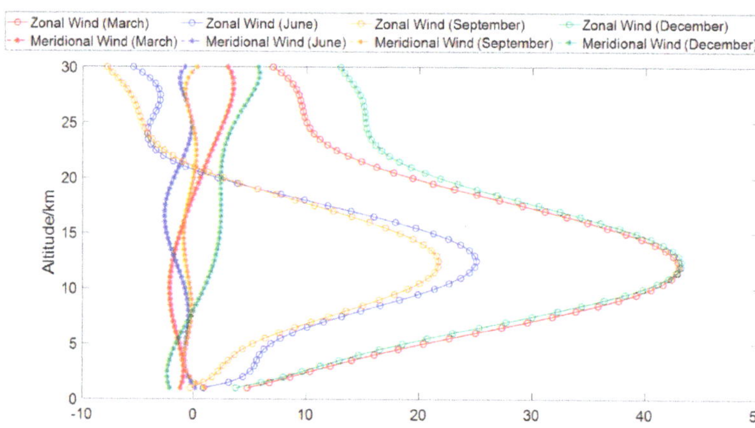

Figure 4. Wind fittings from 1 km to 30 km for a specific region.

From Figure 4, it can be seen that wind velocity changes remarkably with altitude, which provides opportunities for IPBs to adjust their horizontal trajectories utilizing different winds. In particular, a so-called "quasi-zero wind layer" existed at an altitude of about 21 km in June and September, as shown in Figure 4. At altitudes up and down the quasi-zero wind layer, the direction of zonal wind reverses, and the magnitude of meridional wind is small, which is beneficial to IPBs to lengthen their flight time in specific airspace [48–53].

5. Constraints of IPCB Geometry Configuration

IPCB geometry configuration faces many constraints because of its unique features and management strategy, such as airspace constraints, flight altitude constraints, and time interval constraints.

5.1. Airspace Constraint

Airspace is the space in which an IPCB is approved to fly at a certain time. Within the approved airspace, an IPB can fly with its gondola, and its payloads, such as signal generators and transmitters, can run normally. Once an IPB flies out of the approved airspace, its gondola will be cut off from its balloon, and the payloads in the gondola will be switched off, making it impossible for the IPB to emit navigation signals. So, the longitudes and latitudes of each IPB in the IPCB should vary depending on the extent of the approved airspace.

5.2. Flight Altitude Constraint

As discussed above, the flight altitude adjustment of an IPB is achieved by changing its air ballonet volume, which cannot change infinitely. The air ballonet volume of an IPB can only vary in a feasible range, and so does its flight altitude. During IPCB geometry configuration planning, the flight altitudes of all the IPBs should not go beyond the range.

5.3. Time Interval Constraint

Due to the low-density atmosphere at IPCB flight altitudes and the limited capacity of fans, IPCB geometry configuration adjustment requires a long time. So, the time interval between adjacent adjustment actuations should be greater than the maximum trajectory adjustment time required by all the IPBs in the constellation. In this paper, Equations (14) and (15) are used to estimate the time required for an IPB trajectory adjustment.

6. Planning Algorithm of IPCB Geometry Configuration

Based on the analysis above, it can be inferred that if fans and valves can be controlled properly, IPCB geometry configuration can be achieved by utilizing winds at different altitudes effectively. So, the IPCB geometry configuration planning problem can be considered a flight altitude combination problem that can be solved by heuristic algorithms.

WOA is a famous heuristic algorithm that has achieved success in many applications because of its strong robustness, effective searchability, and convenient parameter settings [54–58]. Compared with PSO (particle swarm optimization) or DE (differential evolution algorithm), WOA does not consider subjective parameter settings, such as the inertial coefficient, acceleration coefficient, and other parameters in PSO, scale factor, crossover probability, and other parameters in DE [59,60]. In addition, the performance of WOA, PSO, and DE are compared in reference [54], and WOA displays its excellent capability [54]. Therefore, this paper adopts WOA to realize IPCB geometry configuration planning.

In the planning algorithm, the flight altitudes of all the IPBs in an IPCB can be treated as a whale agent. The general procedure of the algorithm can be described as follows:

Initialize all the whale agents in the current whale population randomly (i.e., initialize the flight altitudes of all the IPBs randomly);

Acquire horizontal winds corresponding to the flight altitudes (i.e., the whale agents just initialized) by Equation (17) or from other data sources;

Calculate horizontal trajectories of all the IPBs in the IPCB by Equation (16);

Adjust the flight altitudes and horizontal trajectories of the IPCB by approaches defined in Section 4;

If the flight altitudes, horizontal trajectories, or adjusted time intervals (calculated by Equation (14) or (15)) do not meet the constraints listed in Section 5, the fitness of the whale agent is defined as 0, meaning that the corresponding IPCB geometry configuration is not feasible in the assumed conditions;

To a whale agent complying with the constraints listed in Section 5, calculate its fitness by Equation (9);

Calculate the fitness of all the whale agents and select the best whale agent in the current whale population;

Update all the whale agents in the current whale population by strategies defined in WOA, such as the "encircling prey" strategy, "bubble-net attacking" strategy, and "search for prey" strategy [54];

Implement updating iteration according to the procedure of WOA, which has been described in detail in reference [54];

When the iteration ends, an IPCB flight altitude can be obtained from the best whale agent. The flight trajectory and constellation GDOP can also be calculated from the best whale agent, forming a complete IPCB geometry configuration.

7. Simulations and Discussions

To verify the effect of the proposed algorithm, simulations are carried out in Matlab 2018b, with the context of IPCB providing independent regional positioning services.

7.1. Simulation Settings

The parameters used in the simulations are listed in Table 1.

Table 1. Parameters used in simulations.

Symbol	Physical Meaning	Value
n_p	number of IPBs in a constellation initially	6
n_t	expected service duration of an IPCB	24 h
c_m	meridional wind coefficients	−0.1338, 1.3189, −1.9669, −2.3772, 4.0187, 1.4032, −1.4290, −0.5504
c_z	zonal wind coefficients	2.1927, −7.6660, −3.0280,28.8161, −2.6879, −41.5979, 2.0248, 21.8084
λ_{min}	minimal longitude of the approved airspace	107° E
λ_{max}	maximal longitude of the approved airspace	109° E
ϕ_{min}	minimal latitude of the approved airspace	39° N
ϕ_{max}	maximal latitude of the approved airspace	41° N
h_{min}	feasible minimal flight altitude of an IPB	20 km
h_{max}	feasible maximal flight altitude of an IPB	24 km

The initial geometry configuration of the IPCB is listed in Table 2, as Figure 5a,b illustrates. Users are distributed uniformly in the service area, as Figure 5c illustrates. The average GDOP of the IPCB with initial geometry configuration is 7.47, whose distribution is illustrated in Figure 5d.

Table 2. Initial geometry configuration of the IPCB.

IPB	Longitude	Latitude	Flight Altitude
IPB 1	107.4° E	39.4° N	21 km
IPB 2	108.6° E	39.4° N	21 km
IPB 3	108.6° E	40.6° N	21 km
IPB 4	107.4° E	40.6° N	21 km
IPB 5	107.8° E	40° N	22 km
IPB 6	108.2° E	40° N	22 km

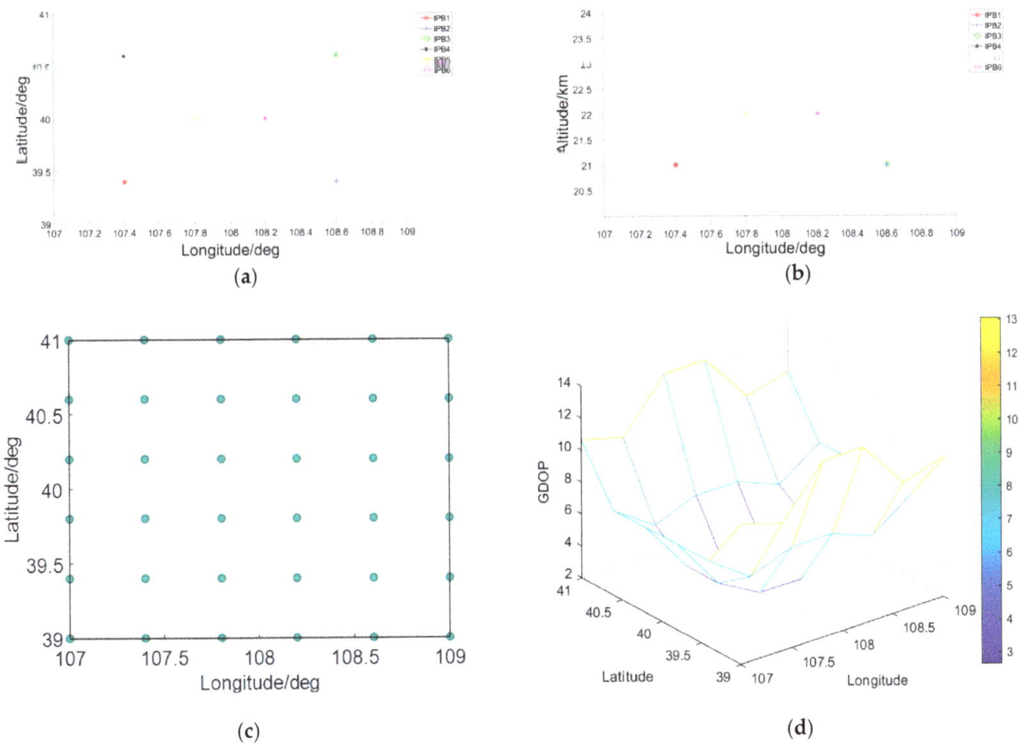

Figure 5. Initial geometry configuration, GDOP distribution of the IPCB, and user distribution: (**a**) initial horizontal layout of the IPCB; (**b**) initial flight altitude of the IPCB; (**c**) user distribution in the service area; (**d**) GDOP distribution of the IPCB with initial geometry configuration.

7.2. Simulation Result

Simulations are carried out with parameters defined in Section 7.1. Figures 6–8 illustrate the planning result. If an IPB flies out of the approved airspace, its subsequent data will not be displayed in figures.

From Figures 6–8, it can be seen that the proposed planning algorithm takes two measures to improve IPCB geometry configuration.

The first measure is to adjust IPB flight altitudes by changing their masses. It makes most IPBs in the constellation fly at an altitude of around 21 km, where wind velocity is small. This measure lengthens IPBs' flight time in the approved airspace, benefiting from keeping the number of available IPBs in the IPCB.

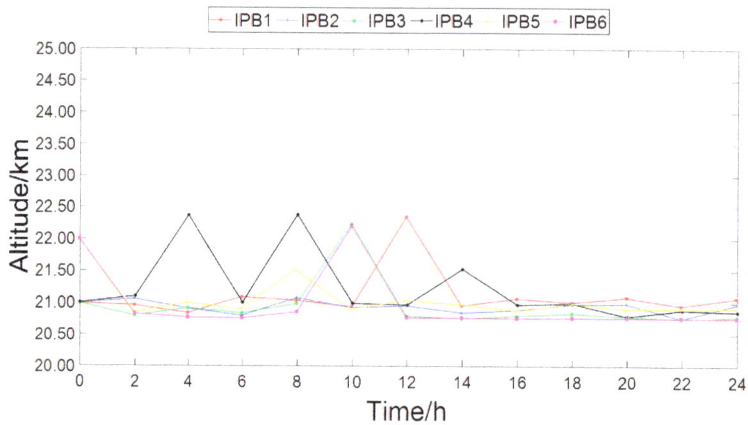

Figure 6. Flight altitude of the IPCB in its service duration.

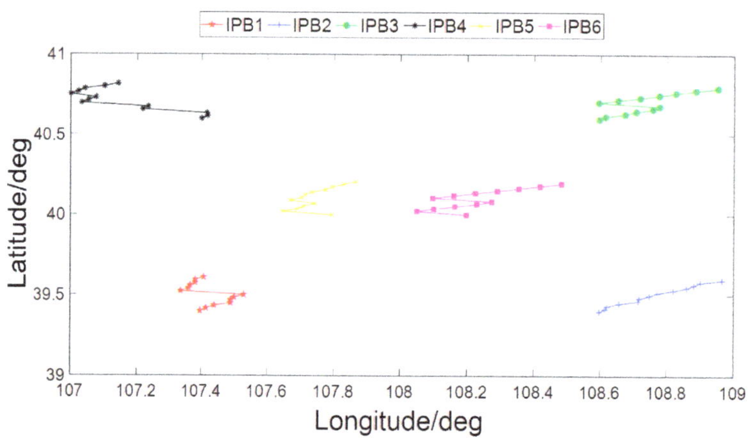

Figure 7. Horizontal trajectory of the IPCB in its service duration.

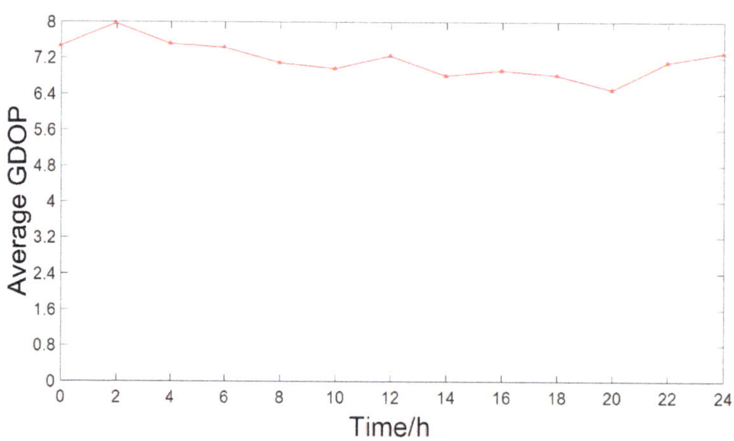

Figure 8. GDOP of the IPCB in its service duration.

The second measure is to make IPBs fly across the quasi-zero wind layer. This measure utilizes the reverse zonal wind direction to change IPBs' movement direction, which can extend IPBs' flight time in the approved airspace and adjust IPBs' horizontal trajectories, thus improving IPCB geometry configuration.

7.3. Discussion about IPCBs with Different Initial Flight Altitudes

IPBs' initial flight altitudes have a significant impact on IPCB geometry performance. In this section, the initial flight altitudes of IPB5 and IPB6 in Table 2 are modified to 20 km, 21 km, 22 km, 23 km, and 24 km, respectively, while other conditions remain untouched. The planning results of different IPCBs are listed in Figures 9–11.

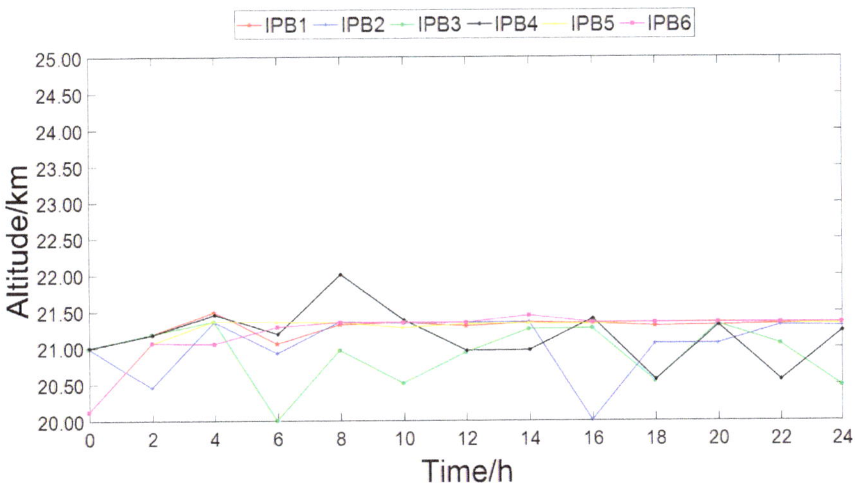

(a)

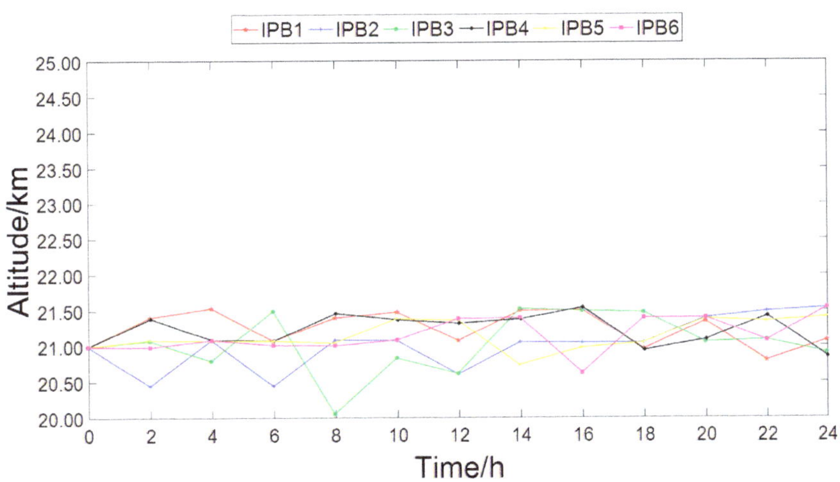

(b)

Figure 9. *Cont.*

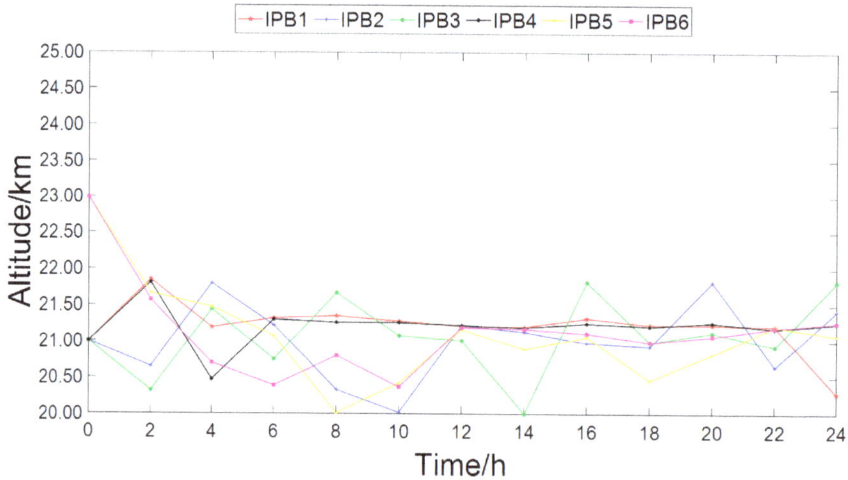

(c)

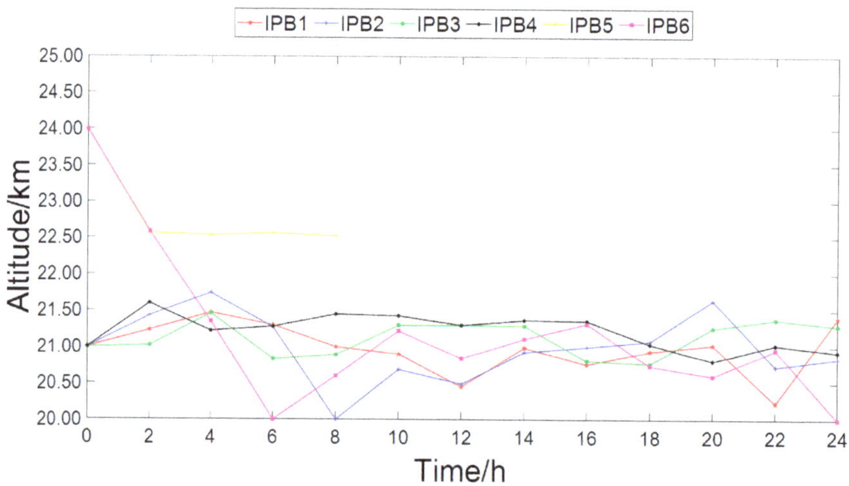

(d)

Figure 9. Flight altitude comparison for IPCBs with different initial flight altitudes: (**a**) IPB5 and IPB6 at 20 km initially; (**b**) IPB5 and IPB6 at 21 km initially; (**c**) IPB5 and IPB6 at 23 km initially; (**d**) IPB5 and IPB6 at 24 km initially. (Flight altitude for IPCB with IPB5 and IPB6 at 22 km initially, please refer to Figure 6).

From the comparison of Table 3, it can be seen that there are no significant performance differences in 20 km, 21 km, 22 km, and 23 km, but there is a declining tendency in 24 km. The result of 24 km can probably be attributed to the short service time of IPB5 caused by the big wind velocity. All of these are reflected in Figures 9–11.

Table 3. Simulation results for IPCBs with different initial flight altitudes.

Initial Flight Altitude of IPB5 and IPB6/km	Average GDOP
20	7.26
21	7.14
22	7.32
23	7.47
24	8.64

In addition, from Figure 11, it can be seen that for the listed initial flight altitudes, the higher the initial flight altitude is, the better the initial GDOP is. However, the IPCB with the highest initial flight altitude has the fastest performance deterioration due to the big wind velocity at the highest altitude. Therefore, if short-term performance is pursued, higher initial flight altitudes may be preferred over lower altitudes. If long-term performance is pursued, initial flight altitudes near the quasi-zero wind layer may be preferred.

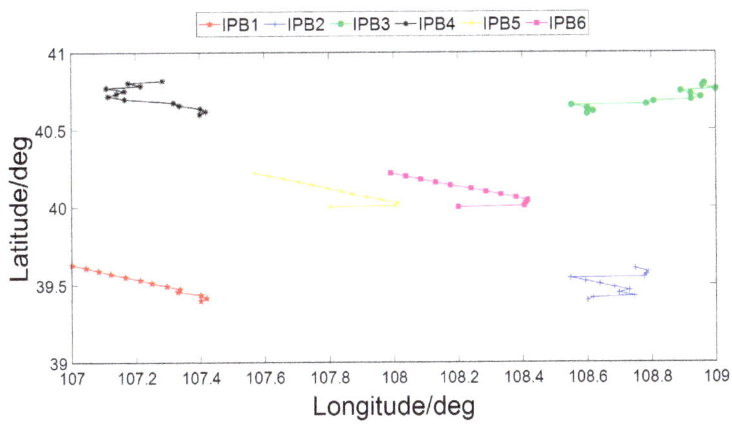

(**a**)

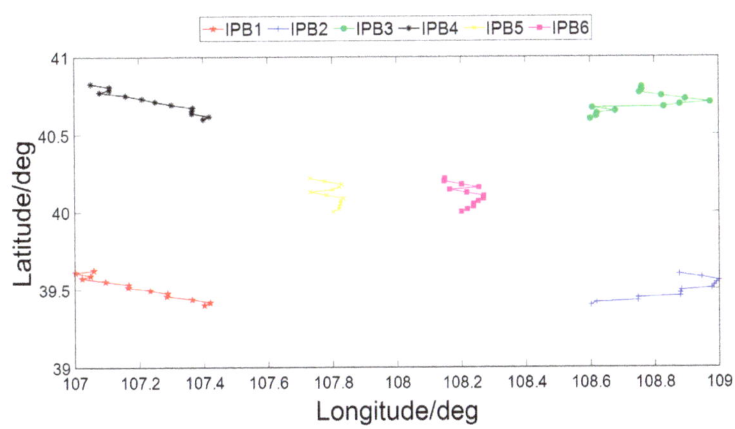

(**b**)

Figure 10. *Cont.*

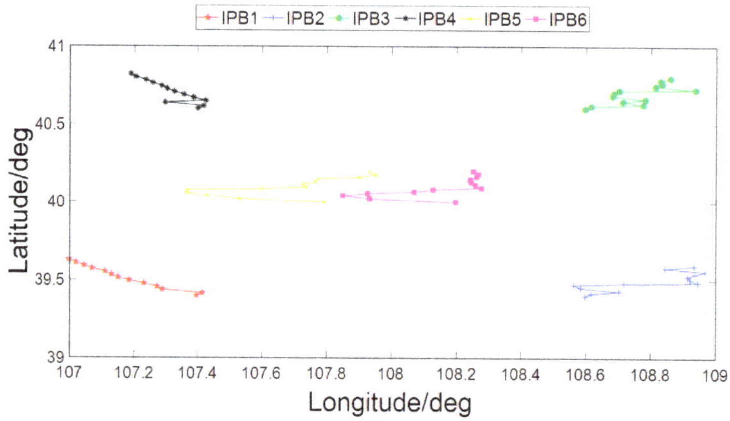

(c)

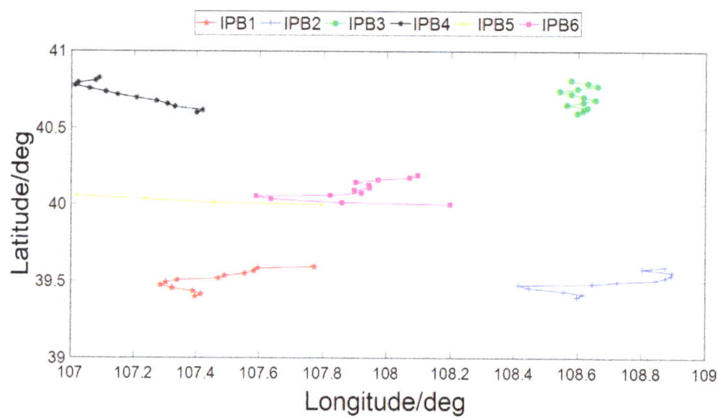

(d)

Figure 10. Horizontal trajectory comparison for IPCBs with different initial flight altitudes: (**a**) IPB5 and IPB6 at 20 km initially; (**b**) IPB5 and IPB6 at 21 km initially; (**c**) IPB5 and IPB6 at 23 km initially; (**d**) IPB5 and IPB6 at 24 km initially. (Horizontal trajectory for IPCB with IPB5 and IPB6 at 22 km initially, please refer to Figure 7).

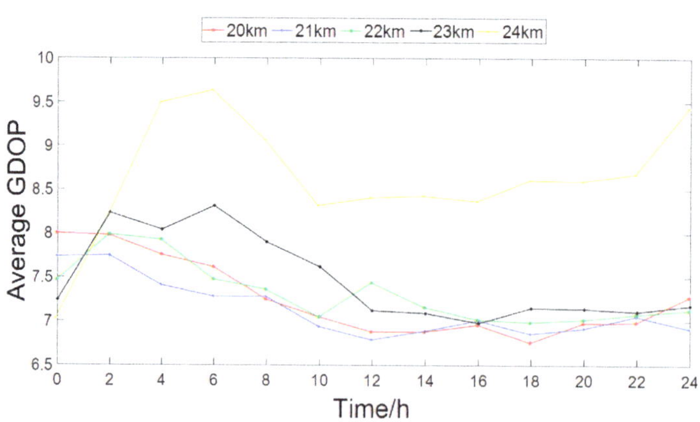

Figure 11. GDOP comparison for IPCBs with different initial flight altitudes.

7.4. Discussion about IPCBs with Different Initial Horizontal Layouts

The initial horizontal layout of an IPCB also has a significant impact on its geometry performance. This section adjusts the initial latitude of IPB5 and IPB6 from south to north, as illustrated in Figure 12, while other conditions remain untouched. The planning results are compared in Figures 13–15.

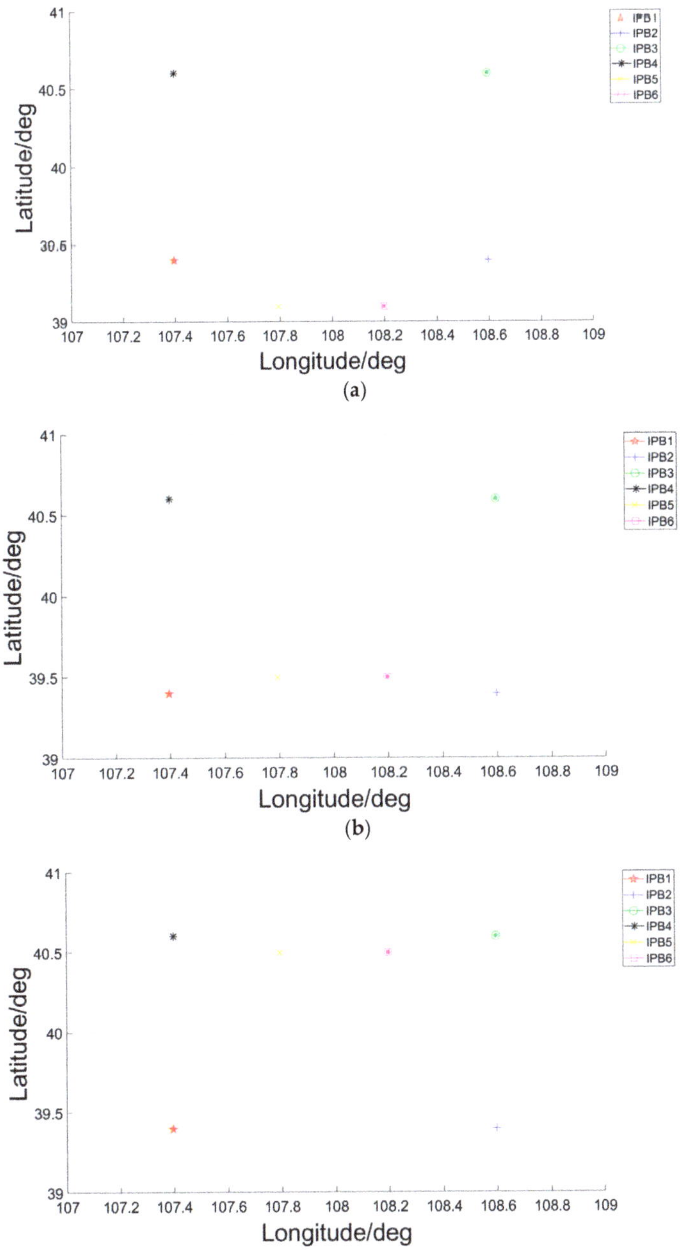

Figure 12. *Cont.*

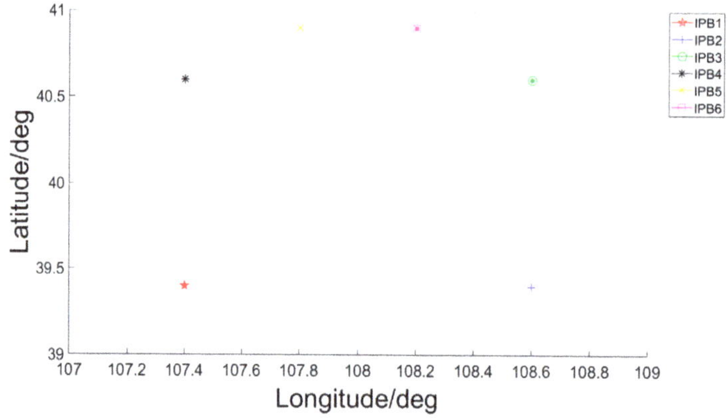

(**d**)

Figure 12. Illustration of different initial horizontal layouts: (**a**) Layout 1 (IPB5 and IPB6 at 39.1° N); (**b**) Layout 2 (IPB5 and IPB6 at 39.5° N); (**c**) Layout 4 (IPB5 and IPB6 at 40.5° N); (**d**) Layout 5 (IPB5 and IPB6 at 40.9° N). (Layout 3 (IPB5 and IPB6 at 40° N), please refered to Figure 5a).

Figures 13–15 and Table 4 show that IPCB can obtain better performance in deploying IPBs with high altitudes to positions near the airspace center (layout 3) than to positions near airspace borders (Layout 1 and Layout 5).

Table 4. Simulation results of IPCBs with different initial horizontal layouts.

Initial Horizontal Layout	Average GDOP
Layout 1	10.88
Layout 2	9.53
Layout 3	7.32
Layout 4	10.83
Layout 5	22.14

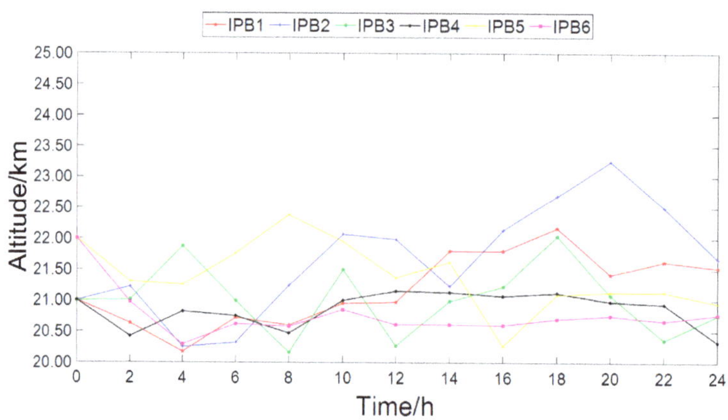

(**a**)

Figure 13. *Cont.*

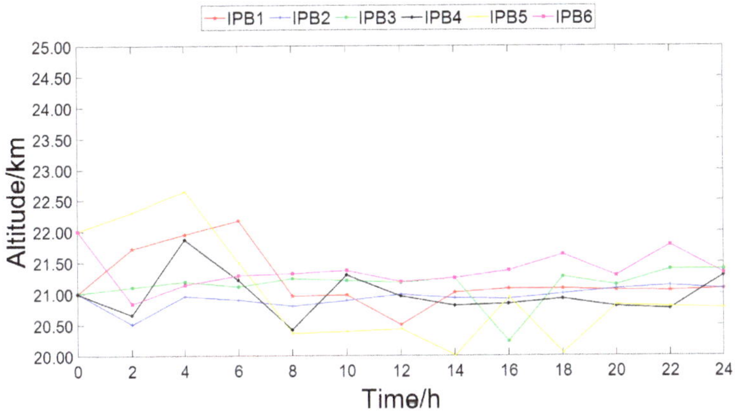

(b)

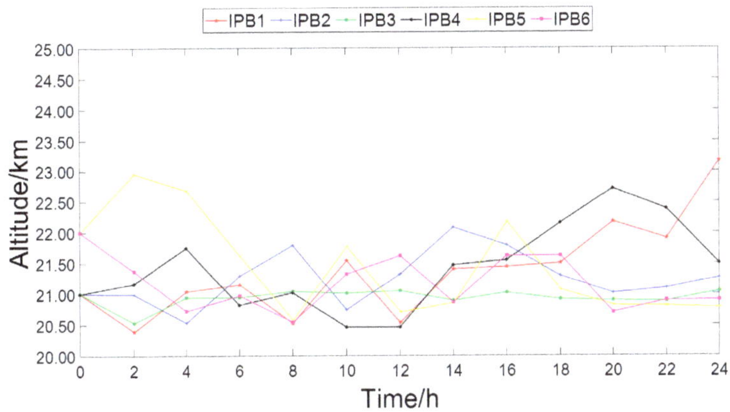

(c)

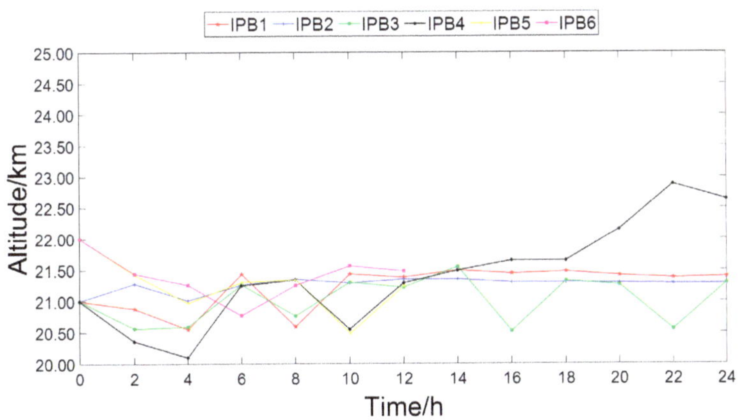

(d)

Figure 13. Flight altitude comparison for IPCBs with different initial horizontal layouts: (**a**) Layout 1; (**b**) Layout 2; (**c**) Layout 4; (**d**) Layout 5. (Flight altitude for IPCB with initial Layout 3, please refer to Figure 6).

Furthermore, Figure 15 shows that IPCBs with Layout 1 and Layout 5 have approximate initial GDOPs. However, their performance displays a different tendency as time goes on. A similar case occurs in IPCBs with Layout 2 and Layout 4. Data in Table 5 also shows that Layout 1 performs better than Layout 5, and Layout 2 performs better than Layout 4. These phenomena may be derived from the winds used in the simulations. In the feasible flight altitude range (21~24 km), the meridional winds are all southerly, leading to all the IPBs moving northward. So, IPCBs with low initial latitudes perform better than IPCBs with high initial latitudes. It is especially obvious in Layout 5, in which IPB5 and IPB6 fly out of the approved airspace quickly due to the short distance between their initial positions and the north border of the airspace.

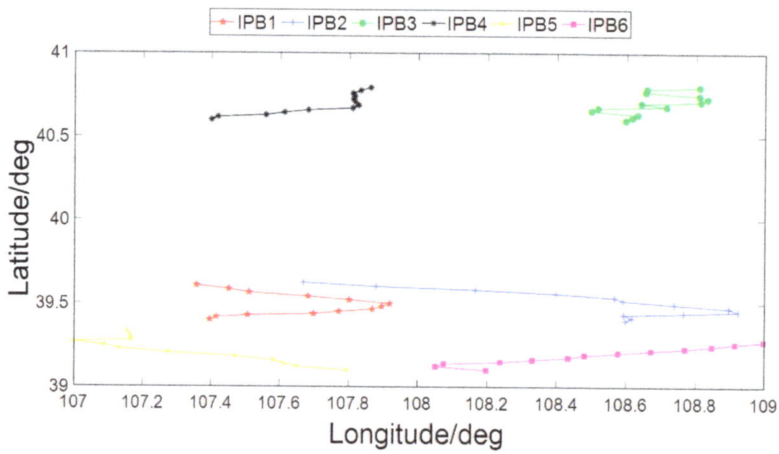

(a)

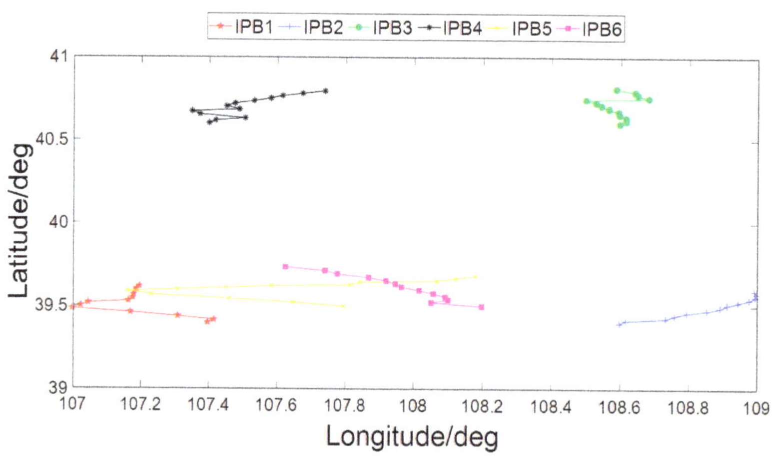

(b)

Figure 14. *Cont.*

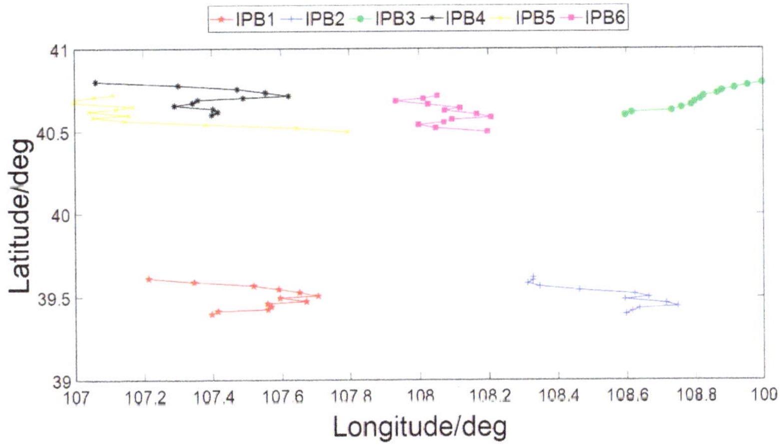

(c)

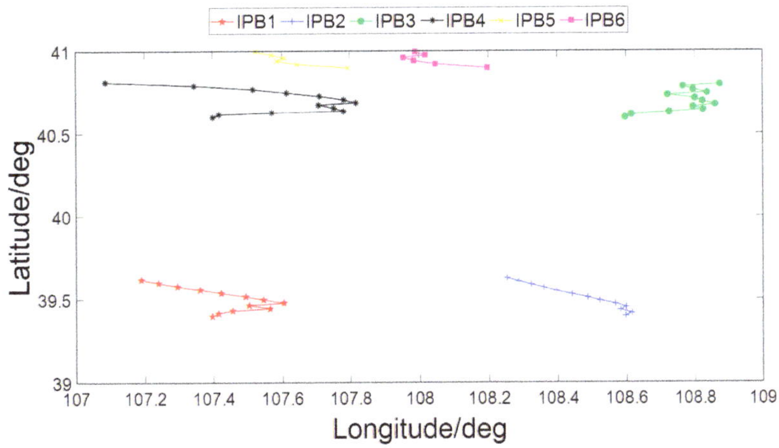

(d)

Figure 14. Horizontal trajectory comparison for IPCBs with different initial horizontal layouts: (**a**) Layout 1; (**b**) Layout 2; (**c**) Layout 4; (**d**) Layout 5. (Horizontal trajectory for IPCB with initial Layout 3, please refer to Figure 7).

In winds with quasi-zero wind layers, the direction reversion rule of zonal wind can be employed to improve IPCB geometry configuration. In contrast, no proper rule of meridional wind can be employed. To achieve good performance throughout the whole service duration, the wind is suggested to be treated as a notable factor in the initial horizontal layout design of IPCB.

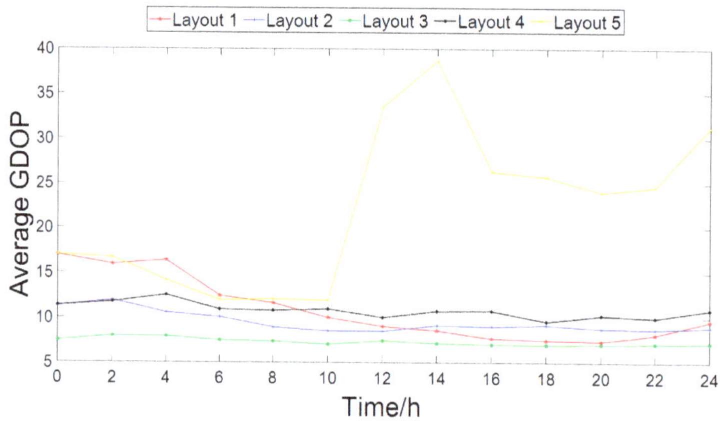

Figure 15. GDOP comparison for IPCBs with different initial horizontal layouts.

Table 5. Coefficients used to fit wind without quasi-zero wind layer.

Symbol	Physical Meaning	Value
c_m	meridional wind coefficients	−0.5066, 4.3171, −14.6477, 27.7923, −36.4969, 32.8921, −13.9287, −0.2023
c_z	zonal wind coefficients	−40.2337, 321.8428, −1018.3275, 1597.1910, −1263.8719, 461.2160, −94.6881, 40.2845

7.5. Discussion about IPCBs in Winds with/without Quasi-Zero Wind Layer

From previous discussions, it can be seen that the quasi-zero wind layer plays an important role in IPCB geometry configuration. However, the quasi-zero wind layer does not always exist. In this section, wind without a quasi-zero wind layer is used to implement the planning. The wind coefficients used in this section are listed in Table 5. The comparison of wind in this section and wind in Section 7.2 is illustrated in Figure 16. The planning result is shown in Figures 17–19.

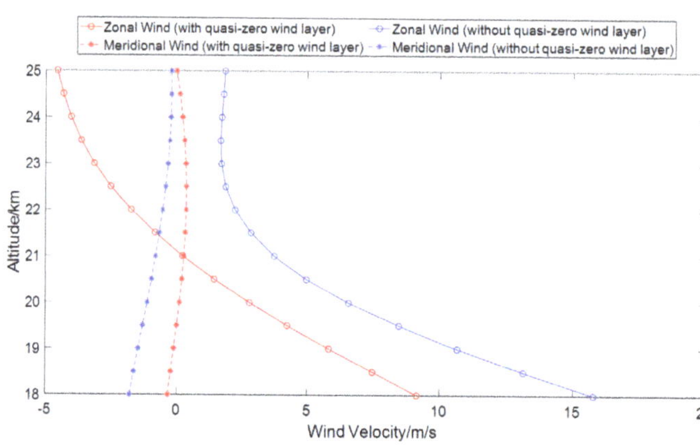

Figure 16. Wind comparison (with/without quasi-zero wind layer).

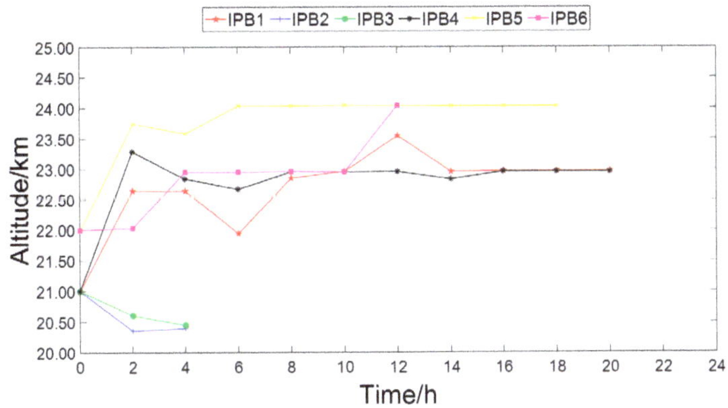

Figure 17. Flight altitude of the IPCB in the wind without a quasi-zero wind layer.

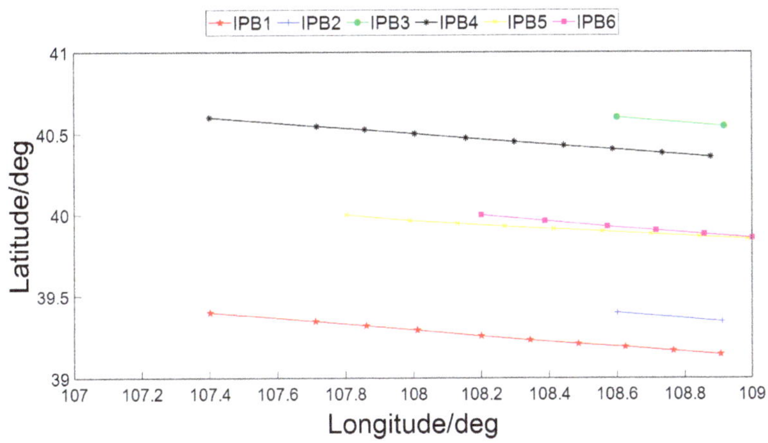

Figure 18. Flight trajectory of the IPCB in the wind without a quasi-zero wind layer.

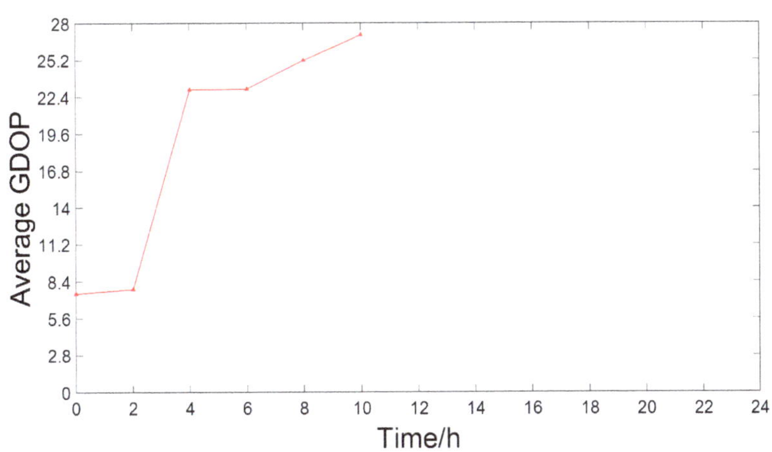

Figure 19. GDOP comparison for IPCBs in winds with/without quasi-zero wind layers.

From Figures 17–19, it can be seen that in winds without a quasi-zero wind layer, zonal trajectory direction reversion does not occur since wind direction reversion does not occur. In addition, the wind velocity is bigger than the wind velocity in Section 7.2, so some IPBs fly out of the approved airspace in a short time, such as IPB2 and IPB3 in Figure 18. This leads to a decrease in the number of available IPBs in the IPCB and deterioration in the IPCB geometry configuration. In order to lengthen IPBs' flight time in the approved airspace, the planning algorithm tends to adjust IPBs' flight altitude to 23~24 km, where wind velocity is small, as IPB1, IPB 4, IPB 5, and IPB6 in Figure 17 illustrate.

By comparing planning results in this section with results in Section 7.2, it can be seen that the results in winds with quasi-zero wind layers are significantly better than those in winds without quasi-zero wind layers. It indicates the significance of the quasi-zero wind layer in improving IPCB geometry performance.

To improve IPCB performance in winds without quasi-zero wind layers, measures such as altering the initial layout, increasing the initial number of IPBs, or supplementing IPBs dynamically can be taken.

Since IPCB achieves its flight mainly by buoyancy and winds, it is sensitive to its running environment. Environment fluctuations or different environment models may bring different results, which can be analyzed in detail in the future.

7.6. Discussion about Uncertain Environment

The simulations above are based on the assumption of a deterministic environment in which wind velocities and atmospheric density are steady. However, uncertain and unknown factors exist in the IPCB running environment.

Conway has investigated horizontal velocity in the midlatitude stratosphere using the observation data of Project Loon and has found the existence of horizontal wind velocity perturbations [61]. Wolf has proposed a model for modeling uncertain winds, which employed a Von Mises distribution to simulate the direction of uncertain wind and a Gaussian distribution to simulate the magnitude of uncertain wind [39]. However, in general, direct observations of winds in the stratosphere are sparse, so a precise model of stratospheric wind is very challenging.

Since this paper focuses on the problem of IPCB geometry configuration, we will not discuss the environment model in detail.

An uncertain environment may degrade the effect of the proposed algorithm because the uncertainty may make IPBs' trajectories deviate from expectation, and as a result, the geometry configuration of IPCB cannot reach the ideal state.

To improve robustness, data filters or environment prediction models can be developed. When perceiving uncertain factors, data filters and prediction models can help to filter out data errors and keep real environment changes. The IPCB can update the environment model and implement the planning algorithm with the new environment model.

8. Conclusions

IPCBs are a novel pseudolite application with many advantages and unique features. Compared with traditional ground-based pseudolites and other air-based pseudolites, IPCB uses high-altitude balloons to achieve higher altitudes and wider coverage. Compared with pseudolites based on powered platforms, IPCBs can save energy costs greatly by utilizing buoyancy and wind. When bringing advantages to applications, these features also bring great challenges to IPCB geometry configuration.

This paper proposes an IPCB geometry configuration planning algorithm that considers the unique features of the IPCB and implements simulations to verify the effectiveness of the proposed algorithm. Furthermore, this paper implements simulations with some typical IPCB geometry configurations and compares their performances.

Simulations show that, in the vertical direction, it can achieve better performance to deploy IPCBs at the altitude of the local quasi-zero wind layer; if the expected service duration is short, IPCBs can be deployed at higher altitudes, and if the expected service duration is long, IPCBs can be deployed at the altitude of local quasi-zero wind layer. In the horizontal direction, the direction of local wind should be treated as an important factor in designing the initial constellation geometry configuration. A quasi-zero wind layer is helpful in improving IPCB geometry performance.

In the future, attention can be paid to approaches to enhance robust algorithms and enhance tolerance to different environment models. Improvements in algorithm performance are also desired to realize real-time control.

Author Contributions: Conceptualization, Y.Q.; methodology, Y.Q. and S.W.; software, Y.Q.; validation, T.P. and H.F.; writing—original draft preparation, Y.Q.; writing—review and editing, S.W. and T.P.; visualization, Y.Q. and T.P.; supervision, S.W. All authors have read and agreed to the published version of the manuscript.

Funding: This work was funded by the National Key Research and Development Program of China (Grant No. 2022YFB3901805).

Data Availability Statement: Data are contained within the article.

Conflicts of Interest: The authors declare no conflicts of interest.

References

1. Klein, D.; Parkinson, B.W. The use of pseudo-satellites for improving GPS performance. *Navig. J. Inst. Navig.* **1984**, *31*, 303–315. [CrossRef]
2. Morley, T.; Lachapelle, G. GPS augmentation with pseudolites for navigation in constricted waterways. *Navig. J. Inst. Navig.* **1997**, *44*, 359–372. [CrossRef]
3. Wang, J. Pseudolite applications in positioning and navigation: Progress and problems. *J. Glob. Position. Syst.* **2002**, *1*, 48–56. [CrossRef]
4. Dixon, C.S.; Morrison, R.G. A pseudolite-based maritime navigation system: Concept through to demonstration. *J. Glob. Position. Syst.* **2008**, *7*, 9–17. [CrossRef]
5. Ma, C.; Yang, J.; Chen, J. Satellite-Ground Joint Positioning System Based on Pseudolite. In Proceedings of the 2018 IEEE CSAA Guidance, Navigation and Control Conference, CGNCC, Xiamen, China, 10–12 August 2018. [CrossRef]
6. Sheng, C.; Gan, X.; Yu, B.; Zhang, J. Precise point positioning algorithm for pseudolite combined with GNSS in a constrained observation environment. *Sensors* **2020**, *20*, 1120. [CrossRef]
7. Liu, T.; Liu, J.; Wang, J.; Zhang, H.; Zhang, B.; Ma, Y.; Sun, M.; Lv, Z.; Xu, G. Pseudolites to support location services in smart cities: Review and prospects. *Smart Cities* **2023**, *6*, 2081–2105. [CrossRef]
8. Kayhan, Ö.; Yücel, Ö.; Hastaoğlu, M.A. Simulation and control of serviceable stratospheric balloons traversing a region via transport phenomena and PID. *Aerosp. Sci. Technol.* **2016**, *53*, 232–240. [CrossRef]
9. Jiang, Y.; Lv, M.; Zhu, W.; Du, H.; Zhang, L.; Li, J. A method of 3-D region controlling for scientific balloon long-endurance flight in the real wind. *Aerosp. Sci. Technol.* **2020**, *97*, 1–11. [CrossRef]
10. Zhu, W.; Xu, Y.; Du, H.; Li, J. Thermal performance of high-altitude solar powered scientific balloon. *Renew. Energy* **2019**, *135*, 1078–1096. [CrossRef]
11. Jiang, Y.; Lv, M.; Qu, Z.; Zhang, L. Performance evaluation for scientific balloon station-keeping strategies considering energy management strategy. *Renew. Energy* **2020**, *156*, 290–302. [CrossRef]
12. Li, C.; Luo, R.; Chen, T. New idea for stratospheric communications—Google Loon. *Commun. Technol.* **2015**, *48*, 125–129. [CrossRef]
13. Deng, X.; Yang, X.; Zhu, B.; Ma, Z.; Hou, Z. Simulation research and key technologies analysis of intelligent stratospheric aerostat Loon. *Acta Aeronaut. Et Astronaut. Sin.* **2023**, *44*, 127412. [CrossRef]
14. Rabinowitz, M.; Parkinson, B.W.; Cohen, C.E.; O'Connor, M.L.; Lawrence, D.G. A system using LEO telecommunication satellites for rapid acquisition of integer cycle ambiguities. In Proceedings of the Position Location and Navigation Symposium, Palm Springs, CA, USA, 20–23 April 1996; pp. 137–145. [CrossRef]
15. Li, X.; Li, X.; Ma, F.; Yuan, Y.; Zhang, K.; Zhou, F.; Zhang, X. Improved PPP ambiguity resolution with the assistance of multiple LEO constellations and signals. *Remote Sens.* **2019**, *11*, 408. [CrossRef]
16. Ge, H.; Li, B.; Jia, S.; Nie, Y.; Wu, T.; Yang, Z.; Shang, J.; Zheng, Y.; Ge, M. LEO enhanced global navigation satellite system (LeGNSS): Progress, opportunities, and challenges. *Geo-Spat. Inf. Sci.* **2022**, *25*, 1–13. [CrossRef]
17. Yuan, H.; Chen, X.; Luo, R.; Wan, H.; Zhang, Y.; Li, R.; Yang, G. Review of the development trend of LEO-based navigation system. *Navig. Position. Timing* **2022**, *9*, 1–11. [CrossRef]
18. Zhang, Y.; Fan, L.; Liu, J.; Li, Z. Feasibility analysis of commercial broadband LEO constellation incorporated into the national comprehensive PNT system. *Navig. Position. Timing* **2022**, *10*, 26–36. [CrossRef]

19. Dai, L.; Wang, J.; Tsujii, T.; Rizos, C. Pseudolite Applications in Positioning and Navigation: Modelling and Geometric Analysis. In Proceedings of the International Symposium on Kinematic Systems in Geodesy, Geomatics & Navigation (KIS2001), Banff, AB, Canada, 5–8 June 2001.
20. Chandu, B.; Pant, R.S.; Moudgalya, K. Modeling and Simulation of a Precision Navigation System Using Pseudolites Mounted on Airships. In Proceedings of the 7th AIAA Aviation Technology, Integration and Operations Conference, Belfast, Northern Ireland, 18–20 September 2007.
21. Oktay, H. Airborne pseudolites in a global positioning system degraded environment. In Proceedings of the 5th International Conference on Recent Advances in Space Technologies–RAST2011, Istanbul, Turkey, 9–11 June 2011. [CrossRef]
22. Tsujii, T.; Harigae, M.; Barnes, J.; Wang, J.; Rizos, C. Experiments of inverted pseudolite positioning for airship-based GPS augmentation system. In Proceedings of the 15th International Technical Meeting of the Satellite Division of the U.S. Institute of Navigation, Portland, OR, USA, 24–27 September 2002; pp. 1689–1695.
23. Sultana, Q.; Sunehra, D.; Ratnam, D.V.; Rao, P.S.; Sarma, A.D. Significance of instrumental biases and dilution of precision in the context of GAGAN. *Indian J. Radio Space Phys.* **2007**, *36*, 405–410.
24. Quddusa, S.; Dhiraj, S.; Vemuri, S.S.; Achanta, D.S. Effects of pseudolite positioning on DOP in LAAS. *Positioning* **2010**, *1*, 18–26. [CrossRef]
25. Wang, J.; Li, H.; Lu, J.; Li, K.; Li, H.; Yang, L.; Li, Y. A PSO-Based Layout Method for GNSS Pseudolite System. In Proceedings of the ICIT 2017, Singapore, 27–29 December 2017. [CrossRef]
26. Jiang, M.; Li, R.; Liu, W. Research on geometric configuration of pseudolite positioning system. *Comput. Eng. Appl.* **2017**, *53*, 271–276.
27. Tang, W.; Chen, J.; Yu, C.; Ding, J.; Wang, R. A new ground-based pseudolite system deployment algorithm based on MOPSO. *Sensors* **2021**, *21*, 5364. [CrossRef]
28. Du, H.; Lv, M.; Li, J.; Zhu, W.; Zhang, L.; Wu, Y. Station-keeping performance analysis for high altitude balloon with altitude control system. *Aerosp. Sci. Technol.* **2019**, *92*, 644–652. [CrossRef]
29. Du, H.; Li, J.; Zhu, W.; Qu, Z.; Zhang, L.; Lv, M. Flight performance simulation and station-keeping endurance analysis for stratospheric super-pressure balloon in real wind field. *Aerosp. Sci. Technol.* **2019**, *86*, 1–10. [CrossRef]
30. Du, H.; Lv, M.; Zhang, L.; Zhu, W.; Wu, Y.; Li, J. Energy management strategy design and station-keeping strategy optimization for high altitude balloon with altitude control system. *Aerosp. Sci. Technol.* **2019**, *93*, 1–9. [CrossRef]
31. Song, J.; Hou, C.; Xue, G.; Ma, M. Study of constellation design of pseudolites based on improved adaptive genetic algorithm. *J. Commun.* **2016**, *11*, 879–885. [CrossRef]
32. Tian, R.; Cui, Z.; Zhang, S.; Wang, D. Overview of navigation augmentation technology based on LEO. *Navig. Position. Timing* **2021**, *8*, 66–81. [CrossRef]
33. Kaplan, E.D.; Hegarty, C.J. *Understanding GPS: Principles and Applications, 2nd ed*; Artech House Inc.: Norword, MA, USA, 2006; p. 02062.
34. Teng, Y.; Wang, J.; Huang, Q. Mathematical minimum of geometric dilution of precision (GDOP) for dual-GNSS constellations. *Adv. Space Res.* **2016**, *57*, 183–188. [CrossRef]
35. Nalineekumari, A.; Sasibhushana, R.G.; Ashok, K.N. GDOP analysis with optimal satellite using GA for southern region of Indian subcontinent. *Procedia Comput. Sci.* **2018**, *143*, 303–308. [CrossRef]
36. Teng, Y.; Wang, J. A closed-form formula to calculate geometric dilution of precision (GDOP) for multi-GNSS constellations. *GPS Solut.* **2016**, *20*, 331–339. [CrossRef]
37. Li, D.; Deng, P.; Liu, B.; Qu, Y.; Zeng, L.; Liu, T. Research on the Dynamic Configuration of Air-Based Pseudolite Network. In *China Satellite Navigation Conference (CSNC) 2015 Proceedings*; Springer: Berlin/Heidelberg, Germany, 2015; Volume II, pp. 357–367.
38. Blackmore, B.; Kuwata, Y.; Wolf, M.T.; Assad, C.; Fathpour, N.; Newman, C.; Elfes, A. Global Reachability and Path Planning for Planetary Exploration with Montgolfirer Balloons. In Proceedings of the 2010 IEEE International Conference on Robotics and Automation, Anchorage Convention District, Anchorage, AK, USA, 3–8 May 2010.
39. Wolf, M.T.; Blackmore, L.; Kuwata, Y.; Fathpour, N.; Newman, C. Probabilistic Motion Planning of Balloons in Strong, Uncertain Wind Fields. In Proceedings of the IEEE International Conference on Robotics & Automation, Anchorage, AK, USA, 3–7 May 2010. [CrossRef]
40. Zhai, J.; Yang, X.; Deng, X.; Long, Y.; Zhang, J.; Bai, F. Global path planning of stratospheric aerostat in uncertain wind field. *J. Beijing Univ. Aeronaut. Astronaut.* **2023**, *49*, 1116–1126. [CrossRef]
41. Bellemare, M.G.; Candido, S.; Castro, P.S.; Gong, J.; Machado, M.C.; Moitra, S.; Ponda, S.S.; Wang, Z. Autonomous navigation of stratospheric balloons using reinforcement learning. *Nature* **2020**, *588*, 77–82. [CrossRef]
42. Furfaro, R.; Lunine, J.I.; Elfes, A.; Reh, K. Wind-based navigation of a hot-air balloon on Titan: A feasible study. In Proceedings of the SPIE Defense and Security Symposium, Orlando, FL, USA, 18–20 March 2008. [CrossRef]
43. Hedin, A.E.; Biondi, M.A.; Burnside, R.G.; Hernandez, G.; Johnson, R.M.; Killeen, T.L.; Mazaudier, C.; Meriwether, J.W.; Salah, J.E.; Sica, R.J.; et al. Revised global model of thermosphere winds using satellite and ground-based observations. *J. Geophys. Res.* **1991**, *96*, 7657–7688. [CrossRef]
44. Mueller, J.B.; Zhao, Y.J.; Garrard, W.L. Optimal ascent trajectories for stratospheric airships using wind energy. *J. Guid. Control. Dyn.* **2009**, *32*, 1232–1245. [CrossRef]

45. Lee, S.; Bang, H. Three-dimensional ascent trajectory optimization for stratospheric airship platforms in the jet stream. *J. Guid. Control. Dyn.* **2007**, *30*, 1341–1352. [CrossRef]
46. Sun, S.; Li, Z.; Tian, K. Modeling and trajectory planning of return process for a class of airship with wind field. *Aerosp. Control. Appl.* **2014**, *40*, 37–41. [CrossRef]
47. Zhang, X.; Chen, L.; Quan, T. Study of ascent trajectory for stratospheric airship in the jet stream. *Electron. Des. Eng.* **2014**, *22*, 11–13.
48. Belmont, A.D.; Dartt, D.G.; Nastrom, G.D. Variations of stratospheric zonal winds, 20–65 km, 1961–1971. *J. Appl. Meteorol.* **1975**, *14*, 585–591. [CrossRef]
49. Tao, M.; He, J.; Liu, Y. Analysis of the characteristics of the stratospheric quasi-zero wind layer and the effects of the quasi-biennial oscillation on it. *Clim. Environ. Res.* **2012**, *17*, 92–102. [CrossRef]
50. Chang, X.; Bai, Y.; Fu, W.; Yan, J. Research on fixed-point aerostat based on its special stratosphere wind field. *J. Northwestern Polytech. Univ.* **2014**, *32*, 12–17.
51. Deng, X.; Li, K.; Yu, C.; Yang, X.; Hou, Z. Station-keeping performance of novel near-space aerostat in quasi-zero wind layer. *J. Natl. Univ. Def. Technol.* **2019**, *41*, 5–12. [CrossRef]
52. Chen, B.; Liu, Y.; Liu, L.; Shen, X.; Zhang, Y. Characteristics of spatial–temporal distribution of the stratospheric quasi-zero wind layer in low-latitude regions. *Clim. Environ. Res.* **2018**, *23*, 657–669. [CrossRef]
53. Roney, J.A. Statistical wind analysis for near-space applications. *J. Atmos. Sol. Terr. Phys.* **2007**, *69*, 1485–1501. [CrossRef]
54. Mirjalili, S.; Lewis, A. The whale optimization algorithm. *Adv. Eng. Softw.* **2016**, *95*, 51–67. [CrossRef]
55. Mafarja, M.M.; Mirjalili, S. Hybrid whale optimization algorithm with simulated annealing for feature selection. *Neurocomputing* **2017**, *260*, 302–312. [CrossRef]
56. Xiong, G.; Zhang, J.; Shi, D.; He, Y. Parameter extraction of solar photovoltaic models using an improved whale optimization algorithm. *Energy Convers. Manag.* **2018**, *174*, 388–405. [CrossRef]
57. Ling, Y.; Zhou, Y.; Luo, Q. Lévy flight trajectory-based whale optimization algorithm for global optimization. *IEEE Access* **2017**, *5*, 6168–6186. [CrossRef]
58. Yan, Z.; Wang, S.; Liu, B.; Li, X. Application of Whale Optimization Algorithm in Optimal Allocation of Water Resources. In Proceedings of the 2018 3rd International Conference on Advances in Energy and Environment Research (ICAEER 2018), Guilin, China, 10–12 August 2018. [CrossRef]
59. Kennedy, J.; Eberhart, R. Particle Swarm Optimization. In Proceedings of the ICNN95–International Conference on Neural Networks, Perth, WA, Australia, 27 November–1 December 1995. [CrossRef]
60. Neri, R.; Tirronen, V. Scale factor local search in differential evolution. *Memetic Comp.* **2009**, *1*, 153–171. [CrossRef]
61. Conway, J.P.; Bodeker, G.E.; Waugh, D.W.; Murphy, D.J.; Cameron, C.; Lewis, J. Using project Loon superpressure balloon observations to investigate the inertial peak in the intrinsic wind spectrum in the midlattitude stratosphere. *J. Geophys. Res. Atmos.* **2019**, *124*, 8594–8604. [CrossRef]

electronics

MDPI

Article

Research on Aspect-Level Sentiment Analysis Based on Adversarial Training and Dependency Parsing

Erfeng Xu [1,2], Junwu Zhu [1,*], Luchen Zhang [3,*], Yi Wang [1,2] and Wei Lin [1,2]

1 School of Information Engineering, Yangzhou University, Yangzhou 225127, China; mz120220937@stu.yzu.edu.cn (E.X.); mz220220329@stu.yzu.edu.cn (Y.W.); mx120220559@stu.yzu.edu.cn (W.L.)
2 Key Laboratory of Intelligent Information Processing, Institute of Computing Technology, Chinese Academy of Sciences, Beijing 100190, China
3 National Computer Network Emergency Response Technical Team/Coordination Center of China, Beijing 100190, China
* Correspondence: jwzhu@yzu.edu.cn (J.Z.); zlc@cert.org.cn (L.Z.)

Abstract: Aspect-level sentiment analysis is used to predict the sentiment polarity of a specific aspect in a sentence. However, most current research cannot fully utilize semantic information, and the models lack robustness. Therefore, this article proposes a model for aspect-level sentiment analysis based on a combination of adversarial training and dependency syntax analysis. First, BERT is used to transform word vectors and construct adjacency matrices with dependency syntactic relationships to better extract semantic dependency relationships and features between sentence components. A multi-head attention mechanism is used to fuse the features of the two parts, simultaneously perform adversarial training on the BERT embedding layer to enhance model robustness, and, finally, to predict emotional polarity. The model was tested on the SemEval 2014 Task 4 dataset. The experimental results showed that, compared with the baseline model, the model achieved significant performance improvement after incorporating adversarial training and dependency syntax relationships.

Keywords: multi head attention mechanism; dependency syntactic relationships; adjacency matrix; adversarial training

1. Introduction

The advent of the Internet and the proliferation of social media platforms have led to an exponential increase in the creation and dissemination of textual content on a daily basis. These data contain rich emotional information, which are crucial for understanding users' attitudes and emotional changes towards products, services, or events. Sentiment analysis, a core component of natural language processing, seeks to automatically discern and extract emotional inclinations from textual data. Its significance has grown notably, finding utility in various sectors including information retrieval, social media analysis, and public opinion monitoring [1,2].

Aspect-level sentiment analysis is a subtask of text sentiment classification. In contrast to general sentiment analysis tasks, which focus on predicting the overall sentiment of a text, aspect-based sentiment analysis tasks necessitate predicting emotional polarity towards specific aspects mentioned within a sentence [3]. Aspect-level sentiment analysis presents a unique challenge wherein different aspect words within the same sentence may exhibit varying emotional polarities. For instance, in the sentence "The food was delicious but the service was bad", "food" and "service" represent distinct aspects. While the evaluation of the food is positive ("delicious"), the evaluation of the service is negative ("bad"). The presence of multi-sentiment scenarios amplifies the complexities inherent in aspect-level sentiment analysis. Models tasked with this challenge must possess the capability to effectively distinguish between different aspects within a sentence and accurately predict the emotional polarity associated with each aspect.

Conventional methodologies for aspect-level sentiment analysis often employed statistical machine learning methods such as naive Bayes or SVM [4], which typically rely on manually designed features for modeling. While these methods have achieved some success to a certain extent, they often rely heavily on the quality and quantity of feature engineering. and struggle with handling complex semantic information and syntactic structures.

The advancement of deep learning, particularly with the emergence of pre-trained language models, has propelled significant strides in aspect-level sentiment analysis. Models such as BERT, RoBERTa, and XLNet [5–7], trained on extensive datasets using self-supervised learning techniques, offer enhanced modeling capabilities for aspect-level sentiment analysis. Mao et al. conducted an empirical study analyzing biases in pre-trained language models (PLMs) for calculating sentiment analysis and emotion detection tasks [8]. It found that RoBERTa outperforms other PLMs in these tasks and proposed methods to mitigate biases.

However, most current aspect-level sentiment analysis methods based on pretrained language models still have limitations. First, these methods often focus on the overall emotional polarity of a sentence while ignoring the relationships between words. Second, the robustness and generalization ability of these models are relatively limited, and they may lead to incorrect classification when exposed to external perturbations.

To address these shortcomings, the presented paper introduces a novel aspect-level sentiment analysis model that combines adversarial training with dependency parsing. The model leverages BERT for word vector conversion and employs an adjacency matrix to capture syntactic dependencies. Multi-head attention combines these features, while adversarial training enhances robustness. This approach enables accurate sentiment polarity predictions at the aspect level.

The primary contributions of this paper can be summarized as follows:

1. The introduction of dependency parsing information in aspect-level sentiment analysis. By constructing an adjacency matrix of syntactic dependency relations, the model can more precisely capture the semantic correlations between different aspects in the text, thereby improving the precision and accuracy of sentiment analysis;
2. To better integrate the features of both BERT and syntactic dependency relations, a multi-head attention mechanism is adopted. This mechanism considers different feature word vectors simultaneously, allowing the model to comprehend the semantic information of the text more comprehensively, thereby enhancing the performance;
3. In order to bolster the robustness and generalizability of the model, an adversarial training mechanism is introduced. By applying small perturbations to the BERT embedding layer, FGM (fast gradient method) can make the model better resist attacks from adversarial samples, thus improving the model's stability and reliability in real-world applications.

2. Related Work

2.1. Aspect-Level Sentiment Analysis

Aspect-level sentiment analysis is a vital task within sentiment analysis, concentrating on the sentiment polarity of particular aspect terms within a sentence. Traditional sentiment analysis methods often target entire documents or single sentences, whereas aspect-based sentiment analysis pays closer attention to more refined sentiment evaluations of specific entities. In past research, the use of traditional machine learning methods for sentiment classification has been a common practice. For instance, Kiritchenko et al. used SVM to detect aspect terms and sentiment in customer reviews [9]; Akhtar et al. employed SVM and CRF for Hindi sentiment classification with good results [10]; Patra et al. used CRF for aspect-level sentiment classification in the domains of Laptop and Restaurant datasets, providing valuable references for consumers and manufacturers [11]. However, these methods require manual feature selection and semantic information extraction, which can reduce the error of opinion word matching but still have limitations. For example, feature extraction from dataset texts requires a significant amount of labor, and the final sentiment analysis results are highly dependent on feature quality, but are incapable of modeling the dependencies between the provided aspect terms and their surrounding contexts.

Comparatively, deep neural networks possess more intricate model architectures and stronger feature extraction capabilities, eliminating the necessity for manual feature extraction, reducing labor costs. With the improvements in computer hardware performance and the widespread use of the Internet, deep neural networks are no longer limited by hardware computing power and data samples. In the realm of sequence models with a focus on attention, researchers have proposed a variety of methodologies. For example, Cheng et al. improved the feature extraction capacity of the Transformer bidirectional encoder through an extended context module and proposed a component focusing module to address the issue of average pooling [12]. Huang et al. proposed the AGSNP model, which combined attention mechanisms and achieved good results [13]. Ayetiran proposed a CNN and BiLSTM variant that combined high-level semantic feature extraction and sentiment polarity prediction [14]. In models focusing on syntactic information, Zeng et al. utilized affective knowledge to enhance word representations, forming a heterogeneous graph based on dependency trees, and designed a multi-level Semantic-HGCN to encode the graph for sentiment prediction [15]. Gu et al. proposed the EK-GCN model, which uses an external sentiment dictionary to assign sentiment scores to individual words within a sentence, constructing an emotional matrix to partially compensate for the shortcomings of the syntactic dependency tree [16]. In models focusing on contextual modeling, Xiao et al. proposed a novel GNN-based deep learning model, leveraging a POS-guided syntactic dependency graph for RGAT to eliminate noise and designing a syntactic distance attention-guided layer for DCGCN to extract semantic dependencies between contextual words [17]. Mewada et al. utilized affective knowledge to enhance word representations, forming a heterogeneous graph based on dependency trees, and then designing a multi-level Semantic-HGCN to encode the graph for sentiment prediction [18]. Xu et al. proposed a sentiment analysis model based on dynamic local context and dependency clusters, which dynamically captured the scope of local context and extracted semantic information, achieving good results [19]. Mao et al. proposed a multi-task learning approach, incorporating a novel gated bridging mechanism (GBM), which achieved superior performance in aspect-based sentiment analysis by effectively filtering irrelevant information and dynamically extracting features for each subtask using a weighted-sum pooling strategy [20].

2.2. Dependency Analysis

Dependency parsing, also known as dependency syntax analysis [21], aims to identify the interdependent relationships between words in a given text and find the corresponding dependent words (tail nodes) for each word (head node), which facilitates a deeper comprehension of the entire sentence's meaning. This is also one of the more critical technologies in the field of NLP. The representation is through directed arrows from the central word to its dependent words, forming directed graphs. Dependency projection trees, and dependency trees are common ways to express dependency structures. Taking the sentence "The iced Americano at this airport tastes good" as an example, the expression of its dependency tree is as follows (Figure 1):

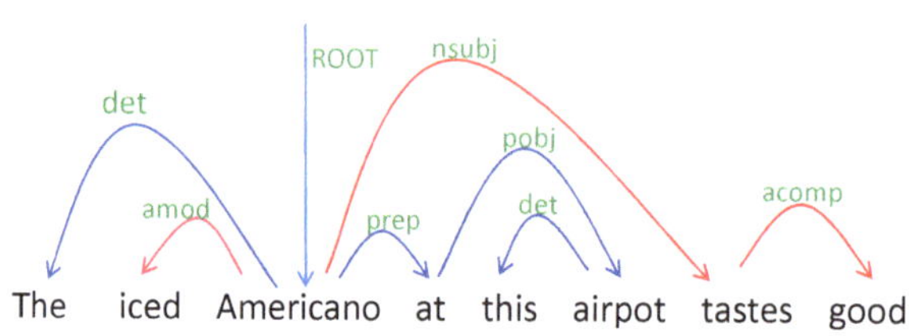

Figure 1. Example diagram of dependency syntax tree.

Dependency syntax analysis is typically represented as a tree structure, where the nodes of the tree represent words, the edges represent the dependency relationships between words, and the parent node of the tree indicates the governor. Some commonly used dependency relation labels, and their meanings in dependency syntax analysis, are presented in Table 1.

Table 1. Partial dependency relationship labels and their meanings.

Labels	Meanings
ROOT	Root node
det	Dependency
amod	Adjectives
nsubj	Noun subjects
prep	Prepositional modifiers
pobj	Object of a preposition
acomp	Complement of an adjective

2.3. Adversarial Training

In the domain of computer vision (CV), it is essential to enhance the robustness of models through adversarial attacks and defenses. For instance, in autonomous driving systems, it is crucial to prevent models from misclassifying red lights as green due to random noise. Similarly, in natural language processing (NLP), adversarial training exists, primarily as a regularization technique aimed at enhancing model generalization.

In 2014, Szegedy et al. introduced the concept of adversarial examples, which is considered a pioneering work in the field [22]. For models processing text input data, the added perturbations can be categorized into two types: discrete, where perturbations are directly applied to the text; and continuous, where tiny perturbations are introduced into the word vector matrix. This paper employs the latter approach for adversarial training. Current popular adversarial training methods include the fast gradient sign method (FGSM) [23], fast gradient method (FGM), projected gradient descent (PGD) [24], free adversarial training (FreeAT) [25], and free large-batch (FreeLB) [26].

The core of adversarial training lies in constructing perturbations that enable the model to recognize diverse adversarial examples. Adversarial training algorithms first generate perturbations using adversarial attacks, then combine these perturbations with original samples to create adversarial examples. Subsequently, the model parameters are adjusted via backpropagation to minimize the loss function. This process can be defined as a max–min optimization problem, where the maximization problem involves finding perturbations that maximize the loss function for generating adversarial examples, while the minimization problem involves minimizing the loss function and updating model parameters, thereby endowing the model with robustness to adapt to such perturbations. Adversarial training can be uniformly represented as a min-max formula, as shown in the following equation:

$$min_{\theta}\mathbb{E}_{(x,y)\sim\mathcal{D}}[max_{\Delta x\in\Omega}L(x+\Delta x,y;\theta)] \tag{1}$$

where $\mathcal{D}$ represents the dataset, x represents the inputs, y represents the labels, and θ is the model parameter that represents the parameter vector of the neural network, $L(x+\Delta x,y;\theta)$ is a single sample of *loss*, Δx is the perturbation and Ω is the perturbation space. Then after the neural network function, the loss obtained by comparing with the label y, $max_{\Delta x\in\Omega}L()$ denotes the optimization objective.

2.4. Attention Mechanisms

In 2014, the Google Mind team's paper brought attention mechanisms into the spotlight [27]. Initially introduced for image processing tasks, attention mechanisms have proven to be effective in other fields as well. Experimental validations have demonstrated the theoretical feasibility of attention mechanisms, and empirical results in the field of

NLP have shown their efficacy in sentiment analysis tasks, highlighting their significant research value. This method is capable of effectively extracting key features, and as such, it is currently widely employed to enhance the performance of sentiment analysis models. The attention mechanism simulates the cognitive process of the human brain, quickly extracting valuable information from extensive text data and assigning higher weights to important information while assigning lower weights to other information.

Bahdanau et al. were the first to introduce attention mechanisms into machine translation based on the encoder–decoder model, successfully translating long sentences [28]. Despite potential issues with the encoding quality, attention mechanisms addressed this by allocating distinct weights to words in the encoding module based on their importance, leading to notable experimental results. The introduction of attention mechanisms has solved the problem of poor coding module quality for machine translation of long sentences, and this technology has been widely applied in the field of NLP, playing an especially important role in sentiment analysis tasks.

The unified computation method of the attention mechanism can be represented as follows:

$$Attention(Q, K, V) = softmax\left(QK^T\right)V \tag{2}$$

In attention mechanisms, Q represents the query vector, K denotes the key vectors within a sentence, typically used for relevance calculations, and V represents the value vectors. Attention weights are obtained through a normalization method, which fundamentally maps the query vector to a series of relationships among key-value pairs. The structure can be visualized as follows (Figure 2):

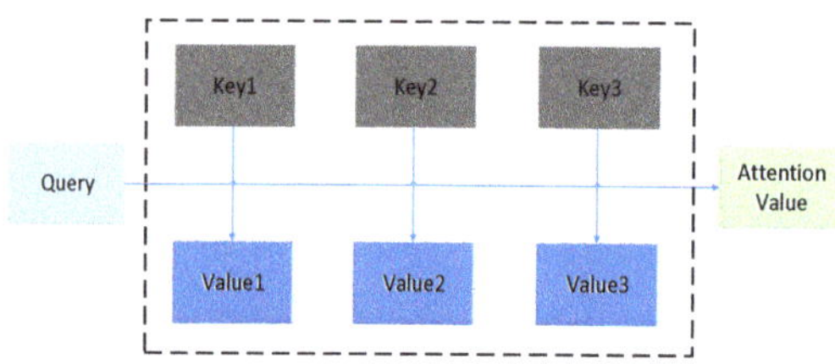

Figure 2. Attention mechanism structure.

3. Overall Model Design

This section begins with a description of the aspect-level sentiment analysis task, followed by the model structure.

3.1. Task Definition

The model input is the given text $W = \{w_1, w_2, \ldots a, \ldots o, \ldots w_n\}$, where a denotes an aspect word and o denotes an opinion word, and the model outputs the sentiment polarity $y \in \{positive, negative, neutral\}$ corresponding to the aspect. Our model leverages the pre-trained language model BERT to generate and train word vectors.

3.2. Model Architecture

The model discussed in the paper comprises the following six main components: a text embedding layer, BERT encoding layer, syntactic dependency relation information layer, adversarial training layer, multi-head attention layer, and an output layer. The model's overall structure is depicted in Figure 3.

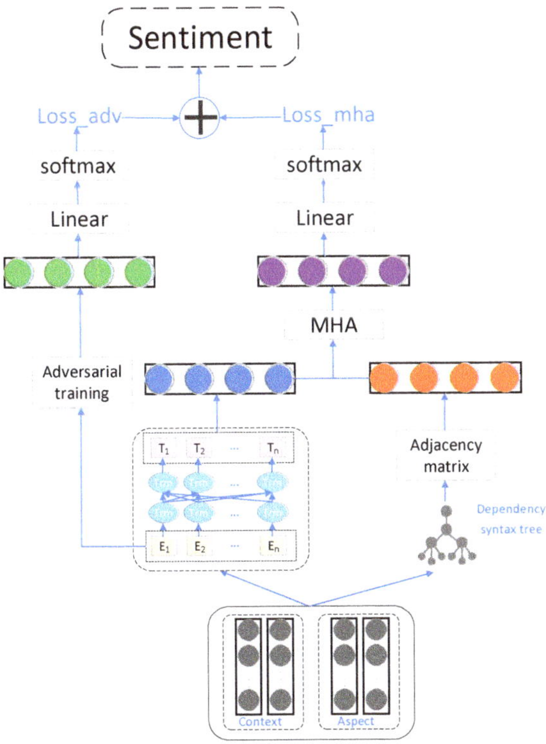

Figure 3. Model structure diagram.

3.2.1. Text Embedding Layer

For a sentence $W = \{w_1, w_2, \ldots a, \ldots o, \ldots w_n\}$, use the pre-training model BERT to map each word onto an embedding vector $e_i \in R^{d\times 1}$, where d represents the dimension of the word vector:

$$W_{Bert} = Bert(W) \tag{3}$$

To fully leverage the power of BERT in model training, the text is formatted into the structure of "[CLS] + context + [SEP] + target + [SEP]". In this format, "[CLS]" and "[SEP]" are special token markers utilized by BERT. "[CLS]" serves as a unique classification token marker that encapsulates classification-related information, while "[SEP]" functions as a separator to demarcate distinct sequences when multiple sequences are input. By adhering to the formatting requirements specified by BERT for text classification tasks, the effectiveness of BERT is maximized.

3.2.2. BERT Encoding Layer

The BERT encoder is constructed using Transformer blocks from the Transformer model [29]. For BERT-BASE, these blocks are employed in 12 layers, each consisting of 12 multi-head attention blocks. After passing through the BERT model, the output is a new sequence with the same length as W_{Bert}, represented as $H_{Bert} = \{h_{CLS}, h_1, \ldots, h_{n-1}, h_{SEP}, h_a, h_{SEP}\}$ as the representation of hidden vectors. Here, "h_{CLS}" is the hidden vector for the classification token, "h_1" to "h_{n-1}" are the hidden vectors for the context tokens, "h_{SEP}" represents the hidden vectors for the separator tokens, and "h_a" represents the hidden vectors for the aspect words.

3.2.3. Dependency Syntax Relation Information Layer

The text is simultaneously processed to establish syntactic dependency relations. In this paper, the StanfordCoreNLP tool is used to obtain the syntactic dependency tree of the text [30]. This is done by capturing the grammatical structure of sentences to extract dependency analysis; the output is a list containing multiple tuples. For example, in the sentence "The iced Americano at this airport tastes good", the output is [('ROOT',0,3), ('det',3,1), ('amod',3,2), ('nsubj',7,3), ('prep',4,3), ('pobj',6,4), ('det',6,5), ('acomp',8,7)]. In this sentence, there are a total of eight elements, so ('amod',3,2) indicates an adjective, where "Americano" depends on "iced". Words in the sentence are encoded starting from 1 to the end of the sentence. The numbers in the tuple represent the positions of the words, and the numbers before and after represent the dependency relationship, where the first number is the head and the second number is the child, indicating that the latter depends on the former. Then, the dependencies are mapped onto a directed graph. The syntactic dependency tree can be conceptualized as graph G with n nodes, where the nodes correspond to the words in the sentence, and the edges represent the syntactic dependencies between words. The dependency parse tree of a sentence is represented as $G = \{V, A\}$, where V stands for all the nodes, which are the words $\{w_1, w_2, \ldots a, \ldots o, \ldots w_n\}$; and $A \in \mathbb{R}^{n \times n}$ is the adjacency matrix, where $A_{ij} = 1$ if there is a syntactic dependency between word w_i and word w_j, and $A_{ij} = 0$ otherwise. Each word in the sentence is adjacent to itself, which implies setting all diagonal elements of the adjacency matrix to 1 [31].

Here is how the syntactic dependency tree and its transformed adjacency matrix are depicted (Figure 4):

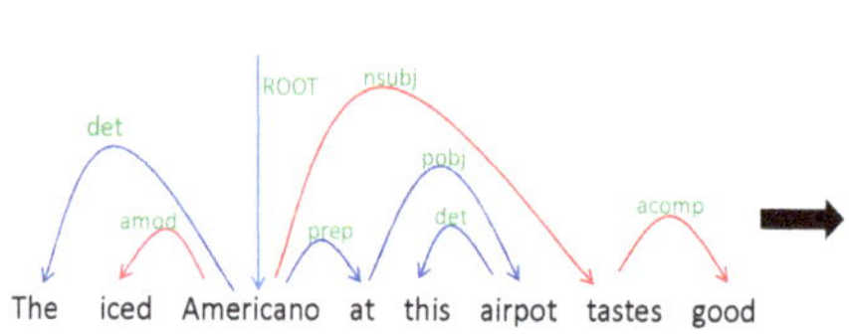

	The	iced	Americano	at	this	airpot	tastes	good
The	1	0	1	0	0	0	0	0
iced	0	1	1	0	0	0	0	0
Americano	1	1	1	1	0	0	1	0
at	0	0	1	1	0	1	0	0
this	0	0	0	0	1	1	0	0
airpot	0	0	0	1	1	1	0	0
tastes	0	0	1	0	0	0	1	1
good	0	0	0	0	0	0	1	1

Figure 4. Syntactic dependency and adjacency matrix.

Next, the adjacency matrix is expanded into a one-dimensional vector and connected to the elements in the matrix row by row or column by column. The unfolded vector is used as an input for the next step of model processing. This converts the information of the adjacency matrix into a vector V_{adj}.

3.2.4. Adversarial Training Layer

The model uses the FGM (fast gradient method) for adversarial training on the BERT embedding layer vectors. FGM stands out from other methods due to its simplicity, ease of use, and computational efficiency. It generates adversarial samples with minimal parameter updates, making it practical for real-world applications with low computational costs, especially with large datasets and complex models. Despite potential variations in performance, FGM typically enhances model robustness against common adversarial attacks. Thus, FGM is a practical choice, particularly in resource-constrained scenarios or where rapid implementation is crucial. By performing gradient ascent based on the specific gradients, it aims to obtain better adversarial samples without significantly altering the distribution of the original samples, thereby allowing the model to adapt to such perturbations. Assuming that the embedding layer vectors $V = \{v_1, v_2, \ldots, v_n\}$ of the input text sequence are x, the perturbation on the embedding layer is as follows:

$$\Delta x = \epsilon \frac{g}{\| g \|_2} \tag{4}$$

$$g = \nabla_x L(x, y; \theta) \tag{5}$$

$$V_{adv} = V + \Delta x \tag{6}$$

After the adversarial training, the obtained feature vectors are denoted as V_{adv}.

3.2.5. Multi-Head Attention Mechanism Layer

After flattening the hidden features H_{Bert} obtained from BERT's output, we obtain H'_{Bert}. Then, we concatenate it with the feature vector obtained after adversarial training to obtain the new hidden feature $Z = \left[H'_{Bert}, V_{adj}\right]$. Z represents the input to the multi-head attention module. By utilizing three different weights W_q, W_k, W_v in the attention layer, we can calculate the resulting vector q, k, v. The steps of the multi-head attention mechanism involve linearly transforming the query (Q), key (K), and value (V) through parameter matrices. Then, scaled dot product operations are performed multiple times before concatenating the results. First, the score for each input feature is calculated: $score = k \times q$. Then, each score is normalized by dividing it by the square root of the dimension of the weight matrix $\sqrt{d_k}$. Next, the softmax function is applied to the normalized scores. Finally, the softmax result is multiplied by the value V. The formula is as follows:

$$a = softmax\left(\frac{QK^T}{\sqrt{d_k}}\right)V \tag{7}$$

The multi-head attention mechanism assigns weighted attention scores to each word in the sentence using multiple attention mechanisms. By increasing the weight coefficients of important information, the model focuses more on words crucial for sentiment analysis, thus further enhancing the accuracy of sentiment analysis. The multi-head attention mechanism consists of multiple heads, each capable of generating different attention distributions, thereby addressing long-range dependencies. Built upon the attention mechanism, the multi-head attention mechanism significantly outperforms standard attention mechanisms, allowing for parallel processing of information in different positional and representational subspaces. With each set of attention projected into different spaces, and considering m as the number of attention heads, the calculation formula is as follows:

$$head_i = a_i \tag{8}$$

$$e = MultiHead(Q, K, V) = Concat(head_1, head_2, \ldots, head_m)W^O \tag{9}$$

Wherein, W^O represents the weight vector, which can be learned through the training process. The Concat function indicates the concatenation of the vectors after the attention computation, and $head_i$ represents the i-th attention mechanism.

Finally, all encoding vectors are weighted and summed to obtain a comprehensive hidden expression e.

3.2.6. Output Layer

Considering that the adversarial perturbations in adversarial training are relatively small values, to prevent the word vectors from becoming too large, which could cause the tiny perturbations to lose their effectiveness, it is necessary to normalize the word vectors. Normalization ensures that the values of the word vectors remain within a reasonable range, allowing the model to be sensitive to the small adversarial perturbations. It is described as follows:

$$V'_{\mathrm{adv}} = \frac{V_{adv} - E(V_{adv})}{\sqrt{Var(V_{adv})}} \tag{10}$$

$$E(V_{adv}) = \sum_{i=1}^{K} f_i V_{advi} \tag{11}$$

$$Var(V_{adv}) = \sum_{i=1}^{K} f_i (V_{advi} - E(V_{adv}))^2 \tag{12}$$

where V_{adv} denotes the original word vector V'_{adv} denotes the normalized word vector and f_i denotes the frequency of the ith word in the training sample.

The fused features and adversarial features of the multi-head attention mechanism are, respectively, used as inputs to the *Softmax* classifier, after which the fused features of the multi-head attention mechanism and the real labels can be calculated as the classification loss $Loss_{mha}$, which is calculated by the following formula:

$$Loss_{mha} = \sum_{i=1}^{N} \left\{ y_i log\hat{y}_i + (1 - y_i) log\left(1 - \hat{y}_i\right) \right\} \tag{13}$$

Wherein y_i represents the true category, $\hat{y}_i$ represents the predicted category, and N is the overall number of samples.

Subsequently, the adversarial features and the true labels are used as inputs to the classifier for calculating the adversarial training loss $Loss_{adv}$, with the following formula:

$$Loss_{adv} = -\frac{1}{N} \sum_{n=1}^{N} logp(y_n \mid x + \Delta x, \theta) \tag{14}$$

In this loss function, the variable is $x + \Delta x$, where Δx represents the adversarial perturbation, N is the overall number of samples, y_n is the corresponding label, and θ is the model's parameters that represents the parameter vector of the neural network. Therefore, the actual loss of the model is as follows:

$$Loss = Loss_{mha} + Loss_{adv} \tag{15}$$

Furthermore, the gradients of $Loss_{mha}$ and $Loss_{adv}$ with respect to the model parameters are computed first. Subsequently, these gradients, along with a predefined learning rate, are utilized to update the model parameters, aiming to progressively decrease the overall loss. Until it satisfies the predetermined maximum number of iterations, this iterative process continues. The generated adversarial training samples are used together with the original samples for model training. This approach can expand the dataset size and effectively enhance the model's generalization performance and classification accuracy.

4. Experimental Analysis

4.1. Experimental Dataset and Experimental Environment

The model in this paper was mainly evaluated on the SemEval2014 Task4 public dataset, which consists of reviews from two domains, Laptops and Restaurants [32]; these datasets are partitioned into a training set and a test set. The aspect words and their corresponding sentiment polarity in the dataset have been labelled, where −1 represents negative, 0 represents neutral and 1 represents positive. The dataset's fundamental statistics are provided in Table 2.

Table 2. Basic statistical information of the dataset.

Datasets	Negative		Neutral		Postive	
	Train	Test	Train	Test	Train	Test
Laptops	851	128	455	167	976	337
Restaurants	807	196	637	196	2164	727

Table 3 illustrates the pertinent configuration of the experimental environment in this paper.

Table 3. Configuration of experimental environment.

Experimental Environment Configuration Table	Configuration Information
Operating System CPU	AMD Ryzen 7 7735H with Radeon Graphics 3.20 GHz
Graphics card	NVIDIA GeForce RTX 4060
Deep Learning Framework	Pytorch
Development Environment	Pycharm

4.2. Experimental Parameter Setting

The experiment used the pretrained language model BERT to generate word vectors. The generated word vectors have a dimension of 768, with a hidden-layer dimension of 300. The dropout rate is set to 0.1, and the learning rate is 2×10^{-5}. The batch size for each input data is 32, and the optimizer used is Adam [33].

4.3. Evaluation Indicators

In the experiment, the evaluation metrics used were *Accuracy* and *Macro-averaged F*1 score [34,35]. Accuracy denotes the proportion of correctly classified positive and negative samples to the total number of samples. The *F*1 score is the harmonic mean of precision and recall, encompassing both precision and recall in the evaluation of the model. The macro-averaged *F*1 score is the average of the *F*1 scores for each category, which helps to avoid the issue of artificially high accuracy due to imbalanced data. The specific formulas are as follows:

$$Accurary = \frac{TP + TN}{TP + TN + FP + FN} \tag{16}$$

$$Precision = \frac{TP}{TP + FP} \tag{17}$$

$$Recall = \frac{TP}{TP + FP} \tag{18}$$

$$F1 = \frac{2 \cdot Precision \cdot Recall}{Precision + Recall} \tag{19}$$

$$MF1 = \frac{1}{C} \sum_{k=1}^{c} F1_k \tag{20}$$

wherein *TP* represents the number of positive samples correctly predicted as positive, *FN* denotes the number of positive samples mistakenly predicted as negative, *FP* indicates the number of negative samples erroneously predicted as positive, and *TN* signifies the number of negative samples accurately predicted as negative. Precision and Recall denote the precision rate and recall rate, respectively, while *C* represents the number of sentiment categories.

To evaluate the significance of the improved results, we also added kappa consistency as a statistical test indicator. The Kappa coefficient, which is a statistical measure of consistency ranging between 0 and 1, is elaborated upon in Table 4. A larger coefficient signifies greater precision in data classification. Its calculation formula is as follows:

$$K = \frac{P_o - P_e}{1 - P_e} \tag{21}$$

Table 4. Kappa coefficient table.

Coefficient	0.8–1.0	0.6–0.8	0.4–0.6	0.2–0.4	0–0.2
Level	Almost perfect	Substantial	Moderate	Fair	Slight

P_o represents the overall classification accuracy. The calculation formula for P_e is as follows:

$$P_e = \frac{a_1 \times b_1 + a_2 \times b_2 + \cdots + a_k \times b_k}{n \times n} \tag{22}$$

a_k represents the actual sample size of class k, b_k represents the predicted sample size of class k, and the total sample size is n.

4.4. Comparative Experiments

The paper selected seven representative aspect-level sentiment analysis models to compare with the model provided in this paper, and their descriptions are as follows:

(1) LSTM [36] is an aspect-level sentiment analysis model based on long short-term memory networks that uses a recurrent neural network structure for modeling and can capture temporal information in text. It performs sentiment classification by integrating the target word and context relationships through two LSTM layers that depend on the target;

(2) TD-LSTM [37] utilizes LSTM to encode the contexts on both sides of the aspect term from different directions, and performs sentiment classification by concatenating the resulting feature representations;

(3) MemNet [38] is a deep memory network model combined with an attention mechanism. By constructing multiple computational layers, each input layer adaptively selects deeper-level information and captures the correlation between each context word and the aspect via attention layers. The output of the final attention layer is utilized for sentiment polarity assessment;

(4) IAN [39] utilizes two LSTM layers to acquire the hidden representations of the context and aspect terms. To precisely capture the semantic relationship between context words and the aspect term, an interactive attention mechanism is incorporated;

(5) RAM [40] is a memory neural network model based on a recurrent attention mechanism that can effectively obtain the sentiment features between words that are farther apart;

(6) AEN [41] utilizes an encoder with an attention mechanism to establish a sentiment analysis model between the context and its corresponding aspect term;

(7) ASGCN [42] constructs a graph convolutional network on the sentence's dependency tree to extract syntactic information. By integrating attention with masked aspect vectors and semantic information, it enhances sentiment classification performance;

(8) GPT3+Prompt [43] is a language model that can be guided to perform aspect-level sentiment analysis tasks and generate relevant text by adding prompts.

Among all comparison models, the accuracy of the ASGCN model reached 75.55% and 80.77% on both datasets, respectively. This is because the ASGCN model constructs a graph convolutional network on the dependency tree of sentences, utilizing syntactic information to extract semantic relationships and improving the accuracy of sentiment classification.

The accuracy rates of LSTM and TD-LSTM on the two datasets reached 66.77% and 74.29%; and 67.71%, 75.36%, respectively. TD-LSTM improved the LSTM model, but because LSTM cannot reflect the interaction information between aspect words and text sentences, and LSTM processes sentences in the order of text sequences, the semantic information learned is not comprehensive enough, and too-long sentences can cause slow the gradient descent. The MemNet model's attention mechanism for selecting deeper-level information may falter in filtering out noisy context words, potentially leading to reduced classification performance. Despite the IAN model's precise capture of semantic relationships, it may face challenges with highly ambiguous context terms. The AEN model's focus on context information might overlook subtle sentiment nuances. Lastly, the RAM model's recurrent attention mechanism may introduce computational complexity and training instability. The above reasons have led to the poor performance of these models.

As indicated in Table 5, the BAMD model surpasses other models in both Accuracy and Macro-F1 scores. On the two datasets, the accuracy rates of the BAMD model reached 76.02% and 83.04%, respectively. Our model offers several advantages over baseline models: first, by integrating dependency parsing information, we accurately capture semantic correlations between different aspects in the text. Second, employing a multi-head attention mechanism enables a comprehensive understanding of semantic information within the text. Lastly, the introduction of adversarial training enhances the model's stability and reliability in real-world applications.

Table 5. Comparing the experimental results of the model on two publicly available datasets.

Comparative Models	Laptops			Restaurants		
	Accurary	Macro-F1	Kappa	Accurary	Macro-F1	Kappa
LSTM	66.77	61.78	-	74.29	62.58	-
TD-LSTM	68.81	64.67	-	76.00	64.51	-
MemNet	70.64	65.17		79.61	69.64	-
IAN	71.20	66.69	-	76.86	66.71	-
RAM	72.32	67.90	0.6745	76.92	68.71	0.7148
AEN	73.69	68.59	0.6886	77.06	69.35	0.7262
ASGCN	75.55	71.05	0.6904	80.77	72.02	0.7377
GPT3 + Prompt	**77.87**	**73.04**	-	**85.45**	**78.96**	-
BAMD(Ours)	76.02	71.54	**0.7171**	83.04	76.61	**0.7853**

Although our model is only 1 to 2.5 percentage points less effective than the closed-source GPT3+Prompt, we acknowledge this difference. Our research suggests that, while our model may be lightweight, with fewer parameters and a smaller memory footprint, this lightweight nature makes it more feasible for deployment and operation in resource-constrained environments, with lower computational costs. While our model may slightly lag behind larger models in performance, its lightweight characteristics provide greater flexibility and feasibility for specific applications in certain scenarios. We will continue to strive for improvement and look forward to achieving better results in future research.

Moreover, it can be clearly seen from the chart that our model's Kappa value is significantly better than the compared models. This indicates that our model can still maintain high classification consistency while considering randomness. The significance of this improvement is not only reflected in the Kappa value, but also in the robustness and generalization ability of the model on different datasets. Therefore, our model performs more reliably and stably in solving this classification task. The optimal performance metrics of each model on two datasets have been bolded in the table.

4.5. Ablation Experiment

To verify the importance of the three major modules designed in this paper, a series of ablation experiments were conducted.

For each ablation experiment, we can infer the importance of each component to model performance by the degree of degradation in the evaluation metrics:

The ablation experiment without Adversarial Training (w/o AT) exhibited a decrease in performance when compared to the original model. This is because adversarial training plays a crucial role in enhancing model robustness and generalization capabilities. Without adversarial training, the model is more susceptible to the influence of biased or noisy samples, leading to a decrease in performance. Therefore, adversarial training is vital for improving the robustness of the model.

Our model significantly outperformed the version without multi-head attention (w/o MHA). The multi-head attention mechanism aids in better integrating features from BERT and syntactic dependency relations, enhancing the model's attention to different aspects of the text and its representational power. If the multi-head attention mechanism is removed, the model may not effectively capture sentiment information across different aspects, resulting in a decline in performance. It is evident that multi-head attention is important for enhancing the model's representational capabilities.

The absence of syntactic dependency relations resulted in varying degrees of decline in both Accuracy and Macro-F1 scores. Syntactic dependency relations provide structural information between words in the text, which helps the model to better understand the semantic and logical relationships within sentences. If syntactic dependency relations are removed, the model may not effectively utilize the structural information of the sentence, leading to a decrease in performance. Therefore, syntactic dependency relations are important for enhancing the model's semantic comprehension.

In summary, adversarial training, multi-head attention mechanism, and syntactic dependency relations each play a significant role in improving model performance. Together, they constitute the key components of the model proposed in this paper.

4.6. Analysis of Model Parameters

To investigate the impact of the constraint radius of the perturbation constraint space S, i.e., the value of ε, on model performance in adversarial training, this paper set ε values to 0.01, 0.1, 0.5, 1, and 2. The accuracy and MF1 scores were tested on both the 14Lap and 14Rest datasets (Figure 5). The experimental results, as shown in the figure below, indicate that introducing adversarial samples during the training stage can enhance the model's resilience to attacks. The model performs optimally when the ε value is 0.1; however, when the ε value is too large, both the model's accuracy and MF1 scores exhibit a downward trend. This phenomenon may be due to the larger perturbation values added, which resulted in significant differences between the generated adversarial samples and the original samples. Although they shared the same label, the model's accuracy in identifying these adversarial samples decreased, subsequently leading to a decline in model performance.

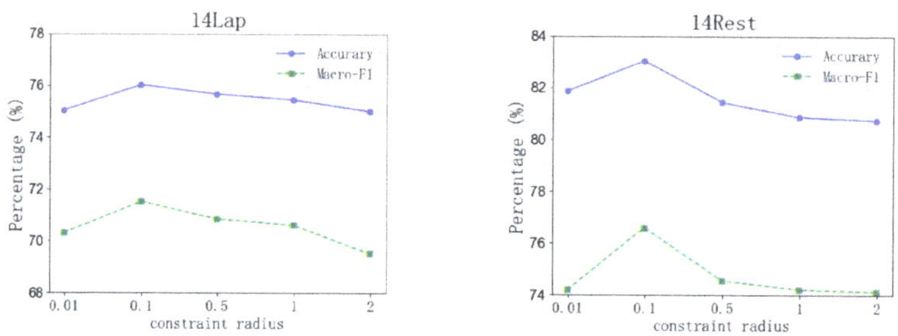

(a) (b)

Figure 5. (**a**) The accuracy of the model under different constraint radii (14Lap); and (**b**) the accuracy of the model under different constraint radii (14Rest).

4.7. Case Study

In order to reflect the effectiveness of the proposed approach, several specific examples were analyzed. Based on Table 6, we extracted the classification results of some typical examples for comparative analysis (Table 7).

Table 6. Results of ablation experiment.

Models	Laptops		Restaurants	
	Accurary	Macro-F1	Accurary	Macro-F1
w/o DS	74.52	70.31	80.57	76.22
w/o AT	73.45	69.63	78.63	73.25
w/o MHA	73.58	69.97	79.78	74.56
BAMD	76.02	71.54	83.04	80.26

Table 7. Typical data experiment examples.

Num	Examples	TD-LSTM	ASGCN	BAMD	Label
1	The *food* is great but the service was dreadful!	Negative (×)	Positive (√)	Positive (√)	Positive
2	I'm delighted to return to the familiar embrace of *Apple's operating system*.	Positive (√)	Negative (×)	Positive (√)	Positive
3	Did not enjoy the new *Windows 8* and touchscreen functions.	Natural (×)	Positive (×)	Negative (√)	Negative

"√" in the table represents the model's correct judgment of emotional polarity, while "×" represents the model's incorrect judgment of emotional polarity.

For the first example sentence, due to the existence of two aspect terms, namely "food" and "service", TD-LSTM focused on the opinion word "dreadful" related to "service", considering it as the opinion word for the aspect term "food", leading to an incorrect matching between aspect terms and opinion words, resulting in a negative sentiment judgment. In the second example sentence, the syntactic distance between "Apple's operating system" and its opinion word "delighted" is too great. The aspect sentiment graph convolutional network (ASGCN) model failed to capture the relationship between them based on syntactic information, which resulted in an incorrect sentiment polarity judgment. The third example also contains two aspect terms, where TD-LSTM failed to accurately match aspect terms with opinion words, and ASGCN failed to capture the feature representation of the negation word "did not". In contrast, BAMD combines both adversarial training and dependency syntax information, and thus can make accurate judgments.

5. Conclusions

This paper introduces an aspect-level sentiment analysis model that leverages adversarial training in conjunction with dependency syntax parsing. By employing BERT for word vector transformation, integrating feature extraction from syntactic dependency relations, and utilizing multi-head attention mechanisms along with adversarial training techniques, the proposed model is capable of predicting the sentiment polarity of specific aspects within sentences. On two public aspect-level sentiment analysis datasets, our model achieves higher accuracy and MF1 scores compared to the baseline models, validating the effectiveness of our approach. However, the model presented in this paper has certain limitations. For instance, the generated dependency syntax relations may contain data noise, and the influence of part-of-speech tags and other syntactic information on the task is not considered. The choice of the adversarial training method can be adjusted to optimize model performance for specific datasets. Future work will focus on further improving and enhancing the model to address these challenges. Specifically, we will explore methods to reduce data noise in generated dependency relations, incorporate part-of-speech tags and other syntactic information, and optimize adversarial training methods for specific datasets. These advancements aim to enhance the model's performance and applicability in aspect-level sentiment analysis, thereby promoting its development and application in various domains.

Author Contributions: Conceptualization, E.X.; methodology, L.Z.; software, E.X.; validation, W.L.; formal analysis, E.X.; investigation, E.X.; resources, Y.W.; data curation, Y.W.; writing—original draft preparation, E.X.; writing—review and editing, E.X. and J.Z. All authors have read and agreed to the published version of the manuscript.

Funding: This research was supported by the National Key Research and Development Program of China (2022YFC3302300), Advanced Research Project (7090201050307) and National 242 Information Security Program (2023A105).

Data Availability Statement: The data presented in this study can be provided upon request.

Acknowledgments: The author thanks my supervisors and colleagues for their help, which enabled me to complete this article.

Conflicts of Interest: The authors declare no conflicts of interest.

References

1. Saberi, B.; Saad, S. Sentiment analysis or opinion mining: A review. *Int. J. Adv. Sci. Eng. Inf. Technol.* **2017**, *7*, 166–1666.
2. Medhat, W.; Hassan, A.; Korashy, H. Sentiment analysis algorithms and applications: A survey. *Ain Shams Eng. J.* **2014**, *5*, 1093–1113. [CrossRef]
3. Thet, T.T.; Na, J.C.; Khoo, C.S.G. Aspect-based sentiment analysis of movie reviews on discussion boards. *J. Inf. Sci.* **2010**, *36*, 823–848. [CrossRef]
4. Cortes, C.; Vapnik, V. Support-vector networks. *Mach. Learn.* **1995**, *20*, 273–297. [CrossRef]
5. Devlin, J.; Chang, M.W.; Lee, K.; Toutanova, K. Bert: Pre-training of deep bidirectional transformers for language understanding. *arXiv* **2018**, arXiv:1810.04805.
6. Zhuang, L.; Wayne, L.; Ya, S.; Jun, Z. A robustly optimized BERT pre-training approach with post-training. In Proceedings of the 20th Chinese National Conference on Computational Linguistics, Huhhot, China, 13–15 August 2021; pp. 1218–1227.
7. Yang, Z.; Dai, Z.; Yang, Y.; Carbonell, J.; Salakhutdinov, R.R.; Le, Q.V. Xlnet: Generalized autoregressive pretraining for language understanding. *Adv. Neural Inf. Process. Syst.* **2019**, *32*, 5753–5763.
8. Mao, R.; Liu, Q.; He, K.; Li, W.; Cambria, E. The biases of pre-trained language models: An empirical study on prompt-based sentiment analysis and emotion detection. *IEEE Trans. Affect. Comput.* **2022**, *14*, 1743–1753. [CrossRef]
9. Kiritchenko, S.; Zhu, X.; Cherry, C.; Mohammad, S. Detecting aspects and sentiment in customer reviews. In Proceedings of the 8th International Workshop on Semantic Evaluation (SemEval), Dublin, Ireland, 23–24 August 2014; pp. 437–442.
10. Akhtar, M.S.; Ekbal, A.; Bhattacharyya, P. Aspect based sentiment analysis in Hindi: Resource creation and evaluation. In Proceedings of the Tenth International Conference on Language Resources and Evaluation (LREC'16), Portorož, Slovenia, 23–28 May 2016; pp. 2703–2709.
11. Patra, B.G.; Mandal, S.; Das, D.; Bandyopadhyay, S. Ju_cse: A conditional random field (crf) based approach to aspect based sentiment analysis. In Proceedings of the 8th International Workshop on Semantic Evaluation (SemEval 2014), Dublin, Ireland, 23–24 August 2014; pp. 370–374.
12. Cheng, L.C.; Chen, Y.L.; Liao, Y.Y. Aspect-based sentiment analysis with component focusing multi-head co-attention networks. *Neurocomputing* **2022**, *489*, 9–17. [CrossRef]
13. Huang, Y.; Peng, H.; Liu, Q.; Yang, Q.; Wang, J.; Orellana-Martín, D.; Pérez-Jiménez, M.J. Attention-enabled gated spiking neural P model for aspect-level sentiment classification. *Neural Netw.* **2023**, *157*, 437–443. [CrossRef]
14. Ayetiran, E.F. Attention-based aspect sentiment classification using enhanced learning through CNN-BiLSTM networks. *Knowl.-Based Syst.* **2022**, *252*, 109409. [CrossRef]
15. Zeng, Y.; Li, Z.; Chen, Z.; Ma, H. Aspect-level sentiment analysis based on semantic heterogeneous graph convolutional network. *Front. Comput. Sci.* **2023**, *17*, 176340. [CrossRef]
16. Gu, T.; Zhao, H.; He, Z.; Li, M.; Ying, D. Integrating external knowledge into aspect-based sentiment analysis using graph neural network. *Knowl. Based Syst.* **2023**, *259*, 110025. [CrossRef]
17. Xiao, L.; Xue, Y.; Wang, H.; Hu, X.; Gu, D.; Zhu, Y. Exploring fine-grained syntactic information for aspect-based sentiment classification with dual graph neural networks. *Neurocomputing* **2022**, *471*, 48–59. [CrossRef]
18. Mewada, A.; Dewang, R.K. SA-ASBA: A hybrid model for aspect-based sentiment analysis using synthetic attention in pre-trained language BERT model with extreme gradient boosting. *J. Supercomput.* **2023**, *79*, 5516–5551. [CrossRef]
19. Xu, M.; Zeng, B.; Yang, H.; Chi, J.; Chen, J.; Liu, H. Combining dynamic local context focus and dependency cluster attention for aspect-level sentiment classification. *Neurocomputing* **2022**, *478*, 49–69. [CrossRef]
20. Mao, R.; Li, X. Bridging towers of multi-task learning with a gating mechanism for aspect-based sentiment analysis and sequential metaphor identification. In Proceedings of the AAAI Conference on Artificial Intelligence, Virtually, 2–9 February 2021; Volume 35, pp. 13534–13542.
21. Nguyen, D.Q.; Verspoor, K. An improved neural network model for joint POS tagging and dependency parsing. *arXiv* **2018**, arXiv:1807.03955.
22. Szegedy, C.; Zaremba, W.; Sutskever, I.; Bruna, J.; Erhan, D.; Goodfellow, I.; Fergus, R. Intriguing properties of neural networks. *arXiv* **2013**, arXiv:1312.6199.
23. Goodfellow, I.J.; Shlens, J.; Szegedy, C. Explaining and harnessing adversarial examples. *arXiv* **2014**, arXiv:1412.6572.
24. Madry, A.; Makelov, A.; Schmidt, L.; Tsipras, D.; Vladu, A. Towards deep learning models resistant to adversarial attacks. *arXiv* **2017**, arXiv:1706.06083.
25. Shafahi, A.; Najibi, M.; Ghiasi, M.A.; Xu, Z.; Dickerson, J.; Studer, C.; Davis, L.S.; Taylor, G.; Goldstein, T. Adversarial training for free! *Adv. Neural Inf. Process. Syst.* **2019**, *32*, 3358–3369.
26. Zhu, C.; Cheng, Y.; Gan, Z.; Sun, S.; Goldstein, T.; Liu, J. Freelb: Enhanced adversarial training for natural language understanding. *arXiv* **2019**, arXiv:arxiv:1909.11764.
27. Mnih, V.; Heess, N.; Graves, A. Recurrent models of visual attention. *Adv. Neural Inf. Process. Syst.* **2014**, *27*, 2204–2212.

28. Bahdanau, D.; Cho, K.; Bengio, Y. Neural machine translation by jointly learning to align and translate. *arXiv* **2014**, arXiv:1409.0473.
29. Vaswani, A.; Shazeer, N.; Parmar, N.; Uszkoreit, J.; Jones, L.; Gomez, A.N.; Kaiser, Ł.; Polosukhin, I. Attention is all you need. *Adv. Neural Inf. Process. Syst.* **2017**, *30*, 5998–6008.
30. Manning, C.D.; Surdeanu, M.; Bauer, J.; Finkel, J.R.; Bethard, S.; McClosky, D. The Stanford CoreNLP natural language processing toolkit. In Proceedings of the 52nd Annual Meeting of the Association for Computational Linguistics: System Demonstrations, Baltimore, MD, USA, 23–24 June 2014; pp. 55–60.
31. Kipf, T.N.; Welling, M. Semi-supervised classification with graph convolutional networks. *arXiv* **2016**, arXiv:1609.02907.
32. Kirange, D.K.; Deshmukh, R.R. Emotion classification of restaurant and laptop review dataset: Semeval 2014 task 4. *Int. J. Comput. Appl.* **2015**, *113*, 17–20.
33. Kingma, D.P.; Ba, J. Adam: A method for stochastic optimization. *arXiv* **2014**, arXiv:1412.6980.
34. Chinchor, N.; Sundheim, B.M. MUC-5 evaluation metrics. In Proceedings of the Fifth Message Understanding Conference (MUC-5), Baltimore, MD, USA, 25–27 August 1993.
35. Yang, Y.; Liu, X. A re-examination of text categorization methods. In Proceedings of the 22nd Annual International ACM SIGIR Conference on Research and Development in Information Retrieval, Berkeley, CA, USA, 15–19 August 1999; pp. 42–49.
36. Greff, K.; Srivastava, R.K.; Koutník, J.; Steunebrink, B.R.; Schmidhuber, J. LSTM: A search space odyssey. *IEEE Trans. Neural Netw. Learn. Syst.* **2016**, *28*, 2222–2232. [CrossRef]
37. Tang, D.; Qin, B.; Feng, X.; Liu, T. Effective LSTMs for target-dependent sentiment classification. *arXiv* **2015**, arXiv:1512.01100.
38. Tang, D.; Qin, B.; Liu, T. Aspect level sentiment classification with deep memory network. *arXiv* **2016**, arXiv:1605.08900.
39. Ma, D.; Li, S.; Zhang, X.; Wang, H. Interactive attention networks for aspect-level sentiment classification. *arXiv* **2017**, arXiv:1709.00893.
40. Chen, P.; Sun, Z.; Bing, L.; Yang, W. Recurrent attention network on memory for aspect sentiment analysis. In Proceedings of the 2017 Conference on Empirical Methods in Natural Language Processing, Copenhagen, Denmark, 7–11 September 2017; pp. 452–461.
41. Song, Y.; Wang, J.; Jiang, T.; Liu, Z.; Rao, Y. Attentional encoder network for targeted sentiment classification. *arXiv* **2019**, arXiv:1902.09314.
42. Zhang, C.; Li, Q.; Song, D. Aspect-based sentiment classification with aspect-specific graph convolutional networks. *arXiv* **2019**, arXiv:1909.03477.
43. Fei, H.; Li, B.; Liu, Q.; Bing, L.; Li, F.; Chua, T.S. Reasoning implicit sentiment with chain-of-thought prompting. *arXiv* **2023**, arXiv:2305.11255.

MDPI AG
Grosspeteranlage 5
4052 Basel
Switzerland
Tel.: +41 61 683 77 34

Electronics Editorial Office
E-mail: electronics@mdpi.com
www.mdpi.com/journal/electronics

www.ingramcontent.com/pod-product-compliance
Lightning Source LLC
Chambersburg PA
CBHW061630210926
51850CB00009B/2498
9783725864188

Advances in Multi-GNSS Technology and Applications

Guest Editors

Bobin Cui

Jungang Wang

MDPI

Basel • Beijing • Wuhan • Barcelona • Belgrade • Novi Sad • Cluj • Manchester

Guest Editors

Bobin Cui
College of Geological
Engineering and Geomatics
Chang'an University
Xi'an
China

Jungang Wang
Faculty VI - Planning
Building Environment
Technische Universität Berlin
Berlin
Germany

Editorial Office
MDPI AG
Grosspeteranlage 5
4052 Basel, Switzerland

This is a reprint of the Special Issue, published open access by the journal *Remote Sensing* (ISSN 2072-4292), freely accessible at: https://www.mdpi.com/journal/remotesensing/special_issues/62C1N264PK.

For citation purposes, cite each article independently as indicated on the article page online and as indicated below:

Lastname, A.A.; Lastname, B.B. Article Title. *Journal Name* **Year**, *Volume Number*, Page Range.

ISBN 978-3-7258-7569-6 (Hbk)
ISBN 978-3-7258-7570-2 (PDF)
https://doi.org/10.3390/books978-3-7258-7570-2

Contents

About the Editors vii

Preface ix

Wei Xie, Kan Wang, Wen Lai, Mengjun Wu, Mengyuan Li and Xuhai Yang
LEO-Enhanced Multi-GNSS Real-Time PPP Time Transfer
Reprinted from: *Remote Sens.* **2025**, *17*, 3549, https://doi.org/10.3390/rs17213549 . . . . **1**

Le Wang, Jinbo She, Bobin Cui, Ziwei Wang, Weicong Yang and Yimin Wang
Mitigating Integrity Risk in SBAS Positioning Using Enhanced IGG III Robust Estimation
Reprinted from: *Remote Sens.* **2025**, *17*, 3067, https://doi.org/10.3390/rs17173067 . . . . **19**

Jin Wang, Xiaohang Chang, Rui Tu, Shiwei Yan, Shengli Wang and Pengfei Zhang
Retrieval of Wave Parameters from GNSS Buoy Measurements Using Spectrum Analysis: A Case Study in the Huanghai Sea
Reprinted from: *Remote Sens.* **2025**, *17*, 2869, https://doi.org/10.3390/rs17162869 . . . . **47**

Damian Kiliszek
GPS and Galileo Precise Point Positioning Performance with Tropospheric Estimation Using Different Products: BRDM, RTS, HAS, and MGEX
Reprinted from: *Remote Sens.* **2025**, *17*, 2080, https://doi.org/10.3390/rs17122080 . . . . **73**

Wei Xu, Pengfei Zhang, Lei Wang, Chao Yan and Jian Chen
Performance Assessment of Undifferenced GPS/Galileo Precise Time Transfer with a Refined Clock Model
Reprinted from: *Remote Sens.* **2025**, *17*, 1910, https://doi.org/10.3390/rs17111910 . . . . **91**

Yu Cao, Le Wang, Zhiwei Qin, Wen Lai, Shi Du and Yuanyuan Wang
High-Precision Satellite Clock Offset Estimated by SRIF Based on Epoch-Wise Updated Orbit
Reprinted from: *Remote Sens.* **2025**, *17*, 1391, https://doi.org/10.3390/rs17081391 . . . . **110**

Zhaochen Li, Yueling Cao, Shanshi Zhou, Xiaogong Hu and Ran Liu
BeiDou Satellite-Based Augmentation System Algorithm Optimization and Performance Validation of Ionospheric Degradation Parameters with RTCA Protocol
Reprinted from: *Remote Sens.* **2025**, *17*, 1110, https://doi.org/10.3390/rs17071110 . . . . **127**

Daniele Borio
Global Navigation Satellite System Meta-Signals with an Arbitrary Number of Components
Reprinted from: *Remote Sens.* **2025**, *17*, 571, https://doi.org/10.3390/rs17040571 . . . . **146**

Xiuxun Liu, Zuping Tang and Jiaolong Wei
Multi-Layer Perceptron Model Integrating Multi-Head Attention and Gating Mechanism for Global Navigation Satellite System Positioning Error Estimation
Reprinted from: *Remote Sens.* **2025**, *17*, 301, https://doi.org/10.3390/rs17020301 . . . . **170**

Wenhao Xiong, Yumiao Tian, Xiaolei Dai, Qichao Zhang, Yibing Liang and Xiongwei Ruan
Analysis of Multi-GNSS Multipath for Parameter-Unified Autocorrelation-Based Mitigation and the Impact of Constellation Shifts
Reprinted from: *Remote Sens.* **2024**, *16*, 4009, https://doi.org/10.3390/rs16214009 . . . . **191**

Wenquan Zhuang, Yuhang Li, Ming Hao, Shangwu Song, Baiyun Liu and Lihong Fan
Quantifying Creep on the Laohushan Fault Using Dense Continuous GNSS
Reprinted from: *Remote Sens.* **2024**, *16*, 3746, https://doi.org/10.3390/rs16193746 . . . . **215**

Yubo Li, Zhuojun Jiang, Chuang Qian, Wenjing Huang and Zeen Yang
A Deep-Learning Based GNSS Scene Recognition Method for Detailed Urban Static Positioning Task via Low-Cost Receivers
Reprinted from: *Remote Sens.* **2024**, *16*, 3077, https://doi.org/10.3390/rs16163077 . . . . **233**

Zi Chen, Guanwen Huang and Yongzhi Zhang
Prediction of Deformation in Expansive Soil Landslides Utilizing AMPSO-SVR
Reprinted from: *Remote Sens.* **2024**, *16*, 2483, https://doi.org/10.3390/rs16132483 . . . . **251**

About the Editors

Bobin Cui

Bobin Cui is an Associate Professor at Chang'an University. He earned his doctoral degree from the Technical University of Berlin and the GFZ German Research Centre for Geosciences. His core research interests lie in the field of Global Navigation Satellite System (GNSS) real-time kinematic Precise Point Positioning (PPP-RTK), with a specific focus on atmospheric modeling, precise positioning, and clock offset estimation techniques. He has presided over multiple national and provincial/ministerial research projects, including the National Natural Science Foundation and the Key Research and Development Program. He acts as a peer reviewer for more than 20 SCI-indexed journals and has published over 50 academic papers. He contributes actively to the scientific community and has been recognized as a "Outstanding Reviewer" by the journal *Measurement Science and Technology*.

Jungang Wang

Jungang Wang is a research scientist in space geodesy at Technische Universität Berlin. His research focuses on the integrated analysis of space geodetic techniques, including Global Navigation Satellite Systems (GNSS), Very Long Baseline Interferometry (VLBI), and Satellite Laser Ranging (SLR). His work contributes to the realization and improvement of terrestrial and celestial reference frames, the determination of Earth orientation parameters, high-precision orbit determination, and geoscientific applications of GNSSs. His scientific interests include advanced analysis strategies for space geodetic observations, innovative approaches for multi-technique combination, and improved modeling of tropospheric delays. He has extensive expertise in both post-processing and real-time GNSS data analysis and has been actively involved in international collaborations related to multi-technique integration and space–time reference systems. He contributes actively to the scientific community and has been recognized as a "Top Reviewer" by the journal *Advances in Space Research*.

Preface

This Reprint presents a curated collection of 13 original research papers from the Special Issue entitled "Advances in Multi-GNSS Technology and Applications", published in the open access journal *Remote Sensing*. As the Guest Editors of this Special Issue, we are pleased to introduce this Reprint to the global research community, detailing the core focus, research boundaries, foundational goals, driving motivations, and intended readership of the work in this collection.

At its core, this Reprint centers on theoretical breakthroughs, technological advancements, and cross-domain application expansion of multi-constellation Global Navigation Satellite System (multi-GNSS) technology. Against the backdrop of integrated development across four core global navigation systems (GPS, GLONASS, Galileo, and BDS), this body of work addresses the key technical bottlenecks and real-world application challenges that multi-GNSS faces when delivering high-precision, high-reliability, and high-integrity Positioning, Navigation, and Timing (PNT) services in complex and challenging environments.

Spanning the full technical pipeline of multi-GNSS research from fundamental theoretical modeling to real-world engineering validation, the work compiled in this Reprint covers six interconnected research pillars aligned with the original Special Issue's vision. These pillars include: high-precision multi-GNSS positioning and time transfer algorithms, such as LEO constellation-enhanced PPP time transfer, refined clock model-optimized time transfer, and high-precision satellite clock offset estimation; navigation signal processing and integrity monitoring, including arbitrary-component GNSS meta-signal construction, SBAS positioning integrity risk mitigation, BDSBAS ionospheric degradation parameter optimization, and multi-GNSS multipath effect suppression; atmospheric and space environment modeling tailored for GNSS applications, covering tropospheric delay estimation in PPP and ionospheric anomaly response for satellite-based augmentation systems; intelligent algorithm empowerment for GNSS performance enhancement, including deep learning-based positioning error estimation and urban navigation scene recognition for low-cost GNSS receivers; marine and geoscience applications of GNSS technology, such as wave parameter retrieval from GNSS buoy measurements and fault creep quantification using dense continuous GNSS arrays; and geological hazard monitoring and early warning applications, including GNSS-driven deformation prediction for expansive soil landslides.

Our primary aim in curating this Reprint is to systematically compile and disseminate the latest peer-reviewed advances in multi-GNSS research, closing the gap between theoretical innovation and real-world engineering implementation across the satellite navigation sector. We pursue this goal with three core purposes in mind: first, to showcase cutting-edge research outcomes that solve the core technical challenges of multi-GNSS in harsh and complex scenarios, from urban canyons and open ocean surfaces to seismically active zones; second, to foster academic exchange and interdisciplinary collaboration across satellite navigation, geodesy, geoscience, artificial intelligence, and transportation engineering, while advancing the establishment of unified technical standards and international cooperation in the multi-GNSS field; third, to deliver a systematic, authoritative technical reference for the industrialization and large-scale deployment of high-precision multi-GNSS technology in safety-critical fields including civil aviation, autonomous driving, and geological disaster prevention and mitigation.

This Special Issue and its accompanying Reprint were born from a critical inflection point in the global multi-GNSS field, where unprecedented opportunities for innovation coincide with pressing technical and application challenges. The integration of multi-constellation navigation systems has dramatically improved the geometric configuration, observation redundancy, and service coverage of satellite navigation, while also placing more rigorous demands on algorithm design, signal processing, error modeling, and scenario adaptation. Meanwhile, the rapid growth of emerging industries, including intelligent connected vehicles, the low-altitude economy, and smart ocean management, has created urgent demand for PNT services with higher precision, stronger reliability, and better environmental robustness. Even so, notable technical bottlenecks persist in current multi-GNSS research: slow convergence of precise positioning in complex environments, insufficient integrity assurance for safety-critical applications, limited adaptability of low-cost receivers in harsh scenarios, and a need for deeper integration between GNSS technology and cross-disciplinary fields. It was against this backdrop that we launched this Special Issue to gather global research efforts to tackle these challenges, and compiled this Reprint to preserve and widely disseminate these valuable peer-reviewed research outcomes, with the long-term goal of driving continued innovative development of multi-GNSS technology and its cross-domain applications.

This Reprint is crafted to serve a diverse, global audience spanning academia, industry, and policy within the navigation and geospatial sectors. It is primarily intended for researchers, scholars, and graduate students specializing in global navigation satellite systems, geodesy and geomatics, remote sensing, and space geodesy. It also serves as a key resource for engineering and technical personnel engaged in the research, development, and application of PNT technology across civil aviation, intelligent transportation, marine engineering, geological hazard monitoring, and geographic information services. In addition, this work provides valuable insights for policy makers and standard setters working in the satellite navigation and geospatial information industry.

We would like to express our sincere gratitude to all authors who contributed their high-quality research to this Special Issue, to the peer reviewers for their rigorous and professional comments that enhanced the quality of the included papers, and to the editorial team of *Remote Sensing* for their strong support and meticulous work throughout the publication process of the Special Issue and the compilation of this Reprint. We hope this Reprint will serve as a valuable resource and a solid foundation for future research and application innovation in the multi-GNSS field.

Bobin Cui and Jungang Wang

Guest Editors

remote sensing

MDPI

Article

LEO-Enhanced Multi-GNSS Real-Time PPP Time Transfer

Wei Xie [1,2], Kan Wang [1,2,3,*], Wen Lai [4], Mengjun Wu [5], Mengyuan Li [1,2] and Xuhai Yang [1,2,3]

[1] National Time Service Center, Chinese Academy of Sciences, Xi'an 710600, China
[2] Key Laboratory of Time Reference and Applications, Chinese Academy of Sciences, Xi'an 710600, China
[3] University of Chinese Academy of Sciences, Beijing 100049, China
[4] College of Geology Engineering and Geomatics, Chang'an University, Xi'an 710054, China
[5] School of Physics and Astronomy, Beijing Normal University, Beijing 100875, China
* Correspondence: wangkan@ntsc.ac.cn

Highlights

What are the main findings?

- Integrating GNSS and LEO observations can significantly reduce the convergence time and improve the precision of time links.
- Using 1 s observations, the time link errors can converge to 1 ns within 1 min.

What are the implications of the main findings?

- With LEO augmentation, time links can achieve high precision with less time.
- Not only fast, but also high-performance time links can be achieved with LEO augmentation.

Abstract

GNSS Precise Point Positioning (PPP) technology has been applied to the time transfer for a long time, enabling time synchronization between two arbitrary stations on a global scale. Over the past decade, Low Earth Orbit (LEO) satellite constellations have been developed to enhance GNSS, offering rapid geometry configuration variations that can accelerate PPP convergence and enhance the time link performance. In this contribution, LEO observations are integrated into GNSS to enhance the real-time PPP time transfer. Simulated LEO constellations with varying numbers of satellites are used to assess their impact on real-time PPP time transfer performance. One week of observation data from 11 globally distributed stations is used to generate 10 time links, and five experimental schemes are designed: (1) GPS/BDS-3/Galileo solution (GCE), (2) GCE with 120 LEO satellites (GCE+120L), (3) GCE with 180 LEO satellites (GCE+180L), (4) GCE with 240 LEO satellites (GCE+240L), and (5) GCE with 300 LEO satellites (GCE+300L). Results showed that compared to the GCE solution, integrating 120, 180, 240, and 300 LEO satellites increases the average number of observed satellites from 23.4 to 30.6, 34.1, 37.7, and 41.3, respectively, while reducing Time Dilution of Precision (TDOP) values from 0.547 to 0.424, 0.391, 0.363, and 0.342, respectively. Using 30 s observations, the average convergence time to STD of time link errors better than 0.1 ns is reduced from 7.95 to 5.94, 4.83, 4.46, and 4.45 min in static mode, with improvements of 25.3%, 39.2%, 43.9%, and 44.0%, respectively, and from 8.75 to 6.18, 5.17, 4.89, and 4.72 min in kinematic mode, with improvements of 29.3%, 40.8%, 44.1%, and 46.0%, respectively. Using 1 s observations, Scenarios GCE+120L, GCE+180L, GCE+240L, and GCE+300L can achieve 1 ns convergence within 1 min. The time link precision was also found to be significantly improved, i.e., from 0.337 to 0.243 ns in static mode with improvements of 27.9%, and from 0.377 to 0.253 ns in kinematic mode with improvements of 32.9%. The time link stability is significantly enhanced for averaging times between 60 and 20,000 s in both static and kinematic modes, with a maximum improvement of nearly 50%. These results have demonstrated that integrating LEO satellites can significantly enhance real-time PPP time transfer performance.

Keywords: LEO; PPP; time transfer; convergence time; time link stability

1. Introduction

Time transfer is the transmission of high-precision time information between two or more locations, enabling clocks in different sites to achieve time synchronization [1]. Precise time transfer is essential for various fields, such as communication networks and power transmission [2].

GPS Precise Point Positioning (PPP) technology has been utilized by the Bureau International des Poids et Mesures (BIPM) for International Atomic Time (TAI) comparison since 2008 [3] due to its high precision, low cost, and global scale coverage. Multi-GNSS observations were afterward incorporated into time transfer to enhance redundancy and robustness [4,5]. Despite the convenience brought by PPP time transfer, the large altitude difference (about 20,000 km or more) between the GNSS Medium Earth Orbit (MEO) satellites and the ground stations results in slow satellite geometry changes and strong temporal correlations in estimated parameters. Consequently, PPP time transfer requires nearly 10 min (or more) to converge [6], during which the time link stability is degraded. This is especially crucial for kinematic PPP time transfer.

In recent years, Low Earth Orbit (LEO) navigation satellite constellations have been widely developed [7], such as Xona Space [8], Centispace [9], and Geely [10]. Some LEO satellites will broadcast GNSS-like navigation signals to ground users, enabling LEO-enhanced GNSS ground Positioning, Navigation, and Timing (PNT) services. The LEO satellite orbit altitude is about 400–1500 km [11]. The lower altitudes compared to those of GNSS satellites imply faster motion and geometry change, which also helps to decrease the correlation among the estimated parameters between adjacent epochs and shorten the convergence time.

Simulated LEO constellations have been investigated for augmenting Real-Time Kinematic (RTK) [12], GNSS PPP [13], PPP-Ambiguity Resolution (AR) [14], and PPP-RTK [15]. Compared to the multi-GNSS solutions, the convergence time, the positioning accuracy, and the Time To First Fix (TTFF) in AR are remarkably improved after using LEO-enhanced GNSS. Real LEO observations from CENTISPACE™ LEO mission [16,17] have been used for LEO-enhanced multi-GNSS PPP, confirming both the improved convergence time and positioning performance.

In the PPP time transfer, the receiver clocks are simultaneously estimated with the station coordinates. The LEO-enhanced GPS PPP time transfer was investigated by Liu et al. [6], in which one-day observation data with 160 LEO satellites in fixed station mode was used, showing that the short-term stability and the convergence time can be improved. Nowadays, the fixed station coordinates mode is not sufficient considering applications in kinematic scenarios, such as autonomous driving. The LEO enhancement in time transfer is worthwhile to investigate in multi-GNSS cases under both static and kinematic scenarios across multiple days, with varying LEO constellation sizes, considering the different phases of the constellation development. The performance of LEO-enhanced PPP time transfer under different observation sampling intervals, e.g., 30 s and 1 s, can also be compared.

This contribution focuses on LEO-enhanced multi-GNSS real-time PPP time transfer. One week of observation data from 11 Multi-GNSS Experiment (MGEX) stations [18] were selected to generate 10 time links, and the LEO constellations with 120, 180, 240, and

300 satellites were simulated and used to conduct the experiments. After this introduction, the LEO-enhanced multi-GNSS real-time PPP time transfer methods are introduced. Next, the LEO constellation simulation, data and processing strategies are presented. Thereafter, the performance of LEO-enhanced multi-GNSS PPP time transfer is evaluated in terms of the number of satellites and Time Dilution of Precision (TDOP), convergence time, time link precision and stability in both the static and kinematic scenarios. Finally, the conclusion is provided.

2. Methods

This section first introduces the time transfer model based on multi-GNSS PPP. The LEO-enhanced multi-GNSS PPP model is then presented. The time transfer method is given at the end.

2.1. Multi-GNSS-Based PPP Model

The dual-frequency Ionospheric-Free (IF) observation equations from a GNSS satellite s to receiver r for system g (GPS, BDS-3, Galileo) can be expressed as:

$$\delta_{r,IF}^{s,g} = \rho_r^{s,g} + c\left(t_{r,IF}^{g} - t_{IF}^{s,g}\right) + T_r^{s,g} + \varepsilon_{r,pc}^{s,g} \tag{1}$$

$$\varphi_{r,IF}^{s,g} = \rho_r^{s,g} + c\left(t_{r,IF}^{g} - t_{IF}^{s,g}\right) + \lambda_{IF}^{g} N_{r,IF}^{s,g} + T_r^{s,g} + e_{r,lc}^{s,g} \tag{2}$$

where δ and φ represents the code and carrier phase IF observations, respectively. $\rho_r^{s,g}$ is the geometric distance between the GNSS satellite and the receiver. c denotes the speed of light. $t_{r,IF}^{g}$ and $t_{IF}^{s,g}$ are the receiver and satellite clocks containing the corresponding IF code biases [19], respectively. $T_r^{s,g}$ is the slant troposphere delay, and λ_{IF}^{g} is the IF wavelength. $N_{r,IF}^{s,g}$ is the IF combined ambiguity in cycles, the carrier phase and code delay on the satellite and receiver end were absorbed by ambiguity [20]. $\varepsilon_{r,pc}^{s,g}$ and $e_{r,lc}^{s,g}$ represent the measurement noise for code and carrier phase observations, respectively. The antenna Phase Center Offsets (PCOs) and Variations (PCVs) on the satellite and receiver end, the relativistic effects, the tidal loadings, and the phase windup were corrected using the existing models in Equations (1) and (2) [21]. Note that in addition to the traditional IF form, the PPP can also be formed in the Undifferenced and Uncombined (UDUC) form [22].

The observation equations can be transformed into error equations for PPP, expressed as:

$$v_{r,pc}^{s,g} = \mu \cdot \hat{X} + c \cdot \hat{t}_{r,IF}^{g} + m_r^{s,g} \hat{Z}_w - l_{r,pc}^{s,g} \tag{3}$$

$$v_{r,lc}^{s,g} = \mu \cdot \hat{X} + c \cdot \hat{t}_{r,IF}^{g} + \lambda_{IF}^{g} \hat{N}_{r,IF}^{s,g} + m_r^{s,g} \hat{Z}_w - l_{r,lc}^{s,g} \tag{4}$$

where pc and lc indicate the code and phase IF combination, respectively. v and l represent the posterior and the Observed-Minus-Computed (OMC) terms, respectively. μ is the unit direction vector from the satellite to the receiver. $\hat{X}$ is the vector of the coordinate increments relative to their a priori positions. $\hat{t}_{r,IF}^{g}$ is the estimated receiver clock, $m_r^{s,g}$ and $\hat{Z}_w$ are the tropospheric mapping function, and the estimated Zenith Wet Delay (ZWD) of the troposphere.

By integrating GPS, BDS-3, and Galileo observations, the multi-GNSS-based PPP models can be formed and expressed as:

$$\begin{cases} v_{r,pc}^{s,G} = \mu \cdot \hat{X} + c \cdot \hat{t}_{r,IF}^{G} + m_r^{s,G} \hat{Z}_w - l_{r,pc}^{s,G} \\ v_{r,pc}^{s,C} = \mu \cdot \hat{X} + c \cdot \hat{t}_{r,IF}^{G} + I\hat{S}B_r^{CG} + m_r^{s,C} \hat{Z}_w - l_{r,pc}^{s,C} \\ v_{r,pc}^{s,E} = \mu \cdot \hat{X} + c \cdot \hat{t}_{r,IF}^{G} + I\hat{S}B_r^{EG} + m_r^{s,E} \hat{Z}_w - l_{r,pc}^{s,E} \end{cases} \tag{5}$$

$$\begin{cases} v_{r,lc}^{s,G} = \mu \cdot \hat{X} + c \cdot \hat{t}_{r,IF}^{G} + \lambda_{IF}^{G} \hat{N}_{r,IF}^{s,G} + m_r^{s,G} \hat{Z}_w - l_{r,lc}^{s,G} \\ v_{r,lc}^{s,C} = \mu \cdot \hat{X} + c \cdot \hat{t}_{r,IF}^{G} + \lambda_{IF}^{C} \hat{N}_{r,IF}^{s,C} + I\hat{S}B_r^{CG} + m_r^{s,C} \hat{Z}_w - l_{r,lc}^{s,C} \\ v_{r,lc}^{s,E} = \mu \cdot \hat{X} + c \cdot \hat{t}_{r,IF}^{G} + \lambda_{IF}^{E} \hat{N}_{r,IF}^{s,E} + I\hat{S}B_r^{EG} + m_r^{s,E} \hat{Z}_w - l_{r,lc}^{s,E} \end{cases} \quad (6)$$

where $I\hat{S}B_r^{CG}$ and $I\hat{S}B_r^{EG}$ stand for the Inter-System Biases (ISB) between GPS and BDS, and between GPS and Galileo, respectively.

2.2. LEO-Enhanced Multi-GNSS PPP Model

The LEO observations are integrated into the multi-GNSS observations to enhance the PPP time transfer, which can be expressed as:

$$\begin{cases} v_{r,pc}^{s,G} = \mu \cdot \hat{X} + c \cdot \hat{t}_{r,IF}^{G} + m_r^{s,G} \hat{Z}_w - l_{r,pc}^{s,G} \\ v_{r,pc}^{s,C} = \mu \cdot \hat{X} + c \cdot \hat{t}_{r,IF}^{G} + I\hat{S}B_r^{CG} + m_r^{s,C} \hat{Z}_w - l_{r,pc}^{s,C} \\ v_{r,pc}^{s,E} = \mu \cdot \hat{X} + c \cdot \hat{t}_{r,IF}^{G} + I\hat{S}B_r^{EG} + m_r^{s,E} \hat{Z}_w - l_{r,pc}^{s,E} \\ v_{r,pc}^{s,L} = \mu \cdot \hat{X} + c \cdot \hat{t}_{r,IF}^{G} + I\hat{S}B_r^{LG} + m_r^{s,L} \hat{Z}_w - l_{r,pc}^{s,L} \end{cases} \quad (7)$$

$$\begin{cases} v_{r,lc}^{s,G} = \mu \cdot \hat{X} + c \cdot \hat{t}_{r,IF}^{G} + \lambda_{IF}^{G} \hat{N}_{r,IF}^{s,G} + m_r^{s,G} \hat{Z}_w - l_{r,lc}^{s,G} \\ v_{r,lc}^{s,C} = \mu \cdot \hat{X} + c \cdot \hat{t}_{r,IF}^{G} + \lambda_{IF}^{C} \hat{N}_{r,IF}^{s,C} + I\hat{S}B_r^{CG} + m_r^{s,C} \hat{Z}_w - l_{r,lc}^{s,C} \\ v_{r,lc}^{s,E} = \mu \cdot \hat{X} + c \cdot \hat{t}_{r,IF}^{G} + \lambda_{IF}^{E} \hat{N}_{r,IF}^{s,E} + I\hat{S}B_r^{EG} + m_r^{s,E} \hat{Z}_w - l_{r,lc}^{s,E} \\ v_{r,lc}^{s,L} = \mu \cdot \hat{X} + c \cdot \hat{t}_{r,IF}^{G} + \lambda_{IF}^{L} \hat{N}_{r,IF}^{s,L} + I\hat{S}B_r^{LG} + m_r^{s,L} \hat{Z}_w - l_{r,lc}^{s,L} \end{cases} \quad (8)$$

where L indicates the observations from LEO satellites. $I\hat{S}B_r^{LG}$ is the estimated receiver ISB between GPS and LEO signals. The estimated parameters can be expressed:

$$Y = \left[\hat{X}, \hat{t}_{r,IF}^{G}, I\hat{S}B_r^{CG}, I\hat{S}B_r^{EG}, I\hat{S}B_r^{LG}, \hat{Z}_w, \hat{N}_{r,IF}^{s,G}, \hat{N}_{r,IF}^{s,C}, \hat{N}_{r,IF}^{s,E}, \hat{N}_{r,IF}^{s,L}\right] \quad (9)$$

2.3. Time Transfer Method

The estimated receiver clock $\hat{t}_{r,IF}^{G}$ is used to conduct precise time transfer. The time difference between the receiver clocks of two stations can be expressed as [23]:

$$\Delta dt = \left(\hat{t}_A - t_{ref}\right) - \left(\hat{t}_B - t_{ref}\right) = \hat{t}_A - \hat{t}_B \quad (10)$$

where $\hat{t}_A$ and $\hat{t}_B$ are the estimated receiver clocks at stations A and B, and t_{ref} represents the time reference, which depends on the GNSS satellite clocks used in the PPP. The time reference of $\hat{t}_A$ and $\hat{t}_B$ is the same because the same satellite clocks were applied when estimating the receiver clocks.

3. Data and Processing Strategies

This section first introduces the simulation of the LEO satellite constellations to be used in the processing. The experimental settings are then briefly explained.

3.1. LEO Satellite Constellations

Due to the limited availability of public LEO satellite navigation signals, navigation signals from LEO satellites are simulated for the PPP processing. The Walker constellation is applied in this contribution due to the uniform coverage of its geometric distribution to the Earth [24,25]. As shown in Figure 1, the Walker constellations 120/12/1, 180/12/1, 240/12/1, and 300/12/1 are to be used, with the number of LEO satellites amounting to 120, 180, 240, and 300, respectively. The satellite orbits have an altitude of 1175 km and an inclination of 85° [24,25]. In this study, only the near-polar orbit configuration is employed to simulate a relatively conservative situation at the beginning phase of the LEO constellation development, while adding more LEO satellites with inclined orbits should further improve the overall geometry for the LEO PNT services.

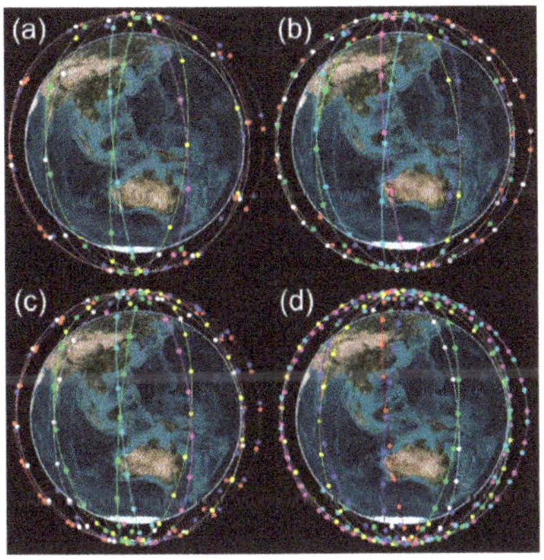

Figure 1. Simulated LEO satellite constellations with 120 (**a**), 180 (**b**), 240 (**c**), and 300 satellites (**d**).

The initial orbits of the LEO satellites are calculated using the six Keplerian parameters, and the satellite orbits can then be numerically integrated to the desired time points. Because the variations in GNSS satellite clocks present nearly linear characteristics [26], the LEO satellite clocks are simulated using a linear model and random noise [27], which can be expressed as:

$$clk_i^s = a_0^s + a_1^s(t_i - t_0) + \varepsilon_i^s \tag{11}$$

where clk_i^s is the LEO satellite clock at the epoch i for satellite s. a_0^s, and a_1^s represent the clock bias and clock drift for satellite s, respectively. t_i and t_0 indicate the time at epoch i and reference epoch, respectively. ε_i^s is the random noise term with the Standard Deviation (STD) of 0.001 ns [28]. The real LEO satellite clocks contain other systematic effects as described in [29–32]. For LEO-enhanced PPP, however, only the differences between the simulated and the used ones (in the PPP), i.e., the real-time LEO satellite orbital/clock estimation errors, are of concern. Their influences on the PPP are investigated in detail in our other work [33], and are not further studied in this contribution.

For the observation simulations, the downlink signal frequencies are assumed to be the same as the BDS B1I (1575.42 MHz) and B3I (1268.52 MHz) [24,25], with a sampling of 30 s. The observation noise is set to 3 mm for the carrier phase and 0.3 m for the code observations. The relativistic effects [34,35], the phase windups [36], and the Earth rotation corrections are simulated using the existing models and are assumed to be corrected in the PPP processing. The PCOs and PCVs of the LEO satellite are set to 0 considering they are known and corrected. The ionospheric delay is simulated using the Global Ionosphere Map (GIM) provided by the Center for Orbit Determination in Europe (CODE) [37]. The slant tropospheric delay consists of hydrostatic and wet components [38], both calculated using the Saastamoinen model [39] and the Global Mapping Function (GMF) [40]. The station coordinates are obtained from International GNSS Services (IGS) weekly solutions, and the receiver PCO and PCV for LEO signals are also set to 0. The receiver clocks are applied using values from European Space Agency Multi-GNSS (ESM) final products [41], with the ISBs between LEO and GPS signals set to zeros.

3.2. Experimental Settings

A total of 11 MGEX stations [18] were selected to perform the LEO-enhanced multi-GNSS real-time PPP time transfer experiments, as shown in Figure 2. The details of these stations are listed in Table 1, in which all stations are equipped with high-performance atomic clocks. The station CEBR, equipped with an external H-Maser, is selected as the central node (red dot in Figure 2), while the other stations (blue triangles in Figure 2) are used to conduct the time transfer experiments. The time link distance ranges from 1741.1 to 12,484.1 km, i.e., long to very long baselines.

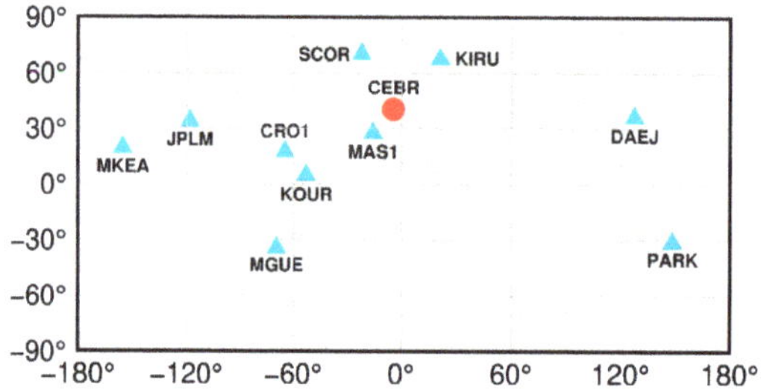

Figure 2. Station distribution of the 11 MGEX stations used for the time transfer experiments.

Table 1. Station information used for LEO-enhanced multi-GNSS PPP real-time transfer.

Station	Receiver Type	Clock	Location	Distance (km)
CEBR	SEPT POLARX5TR	External H-Maser	Spain	/
CRO1	SEPT POLARX5TR	External H-Maser	USA	5998.0
DAEJ	TRIMBLE NETR9	External Cesium	South Korea	9134.5
JPLM	SEPT POLARX5	External Rubidium	USA	8517.1
KIRU	SEPT POLARX5TR	External Cesium	Sweden	3373.9
KOUR	SEPT POLARX5TR	External H-Maser	France	5959.4
MAS1	SEPT POLARX5	External Cesium	Spain	1741.1
MGUE	SEPT POLARX5TR	External H-Maser	Argentina	9511.0
MKEA	SEPT POLARX5	External H-Maser	USA	10,709.9
PARK	SEPT POLARX5TR	External H-Maser	Australia	12,484.1
SCOR	SEPT POLARX5	External Rubidium	Denmark	3449.0

Real GNSS and simulated LEO observations from DOY 119 to 125, 2024, were used. The dual-frequency signals from GPS L1/L2, BDS-3 B1I/B3I, Galileo E1/E5a, and LEO B1I/B3I, sampled at 30 s, were employed to generate the IF combination. Real-time GNSS satellite orbits and clocks released from the Centre national d'études spatiales (CNES) [42], and simulated LEO satellite orbits and clocks were applied and fixed. The GNSS satellite PCOs and PCVs were corrected using values from IGS antenna files, while the LEO satellite PCOs and PCVs were not corrected, as they are set to zeros in LEO signal simulations. Relativistic effects [34], phase-windup [36], and Earth rotation are calculated by applying the existing models as mentioned before. The tropospheric slant hydrostatic delay is corrected using the Saastamoinen model [39] and the GMF [40]. Receiver PCOs and PCVs were only corrected for GNSS observations, as those for LEO signals are simulated to be zeros. The Kalman filter [43,44] was used for estimating the parameters epoch-by-epoch, and the station coordinates were estimated as constants in the static mode [45] and white noise in the kinematic mode [45]. Both the receiver clocks and ISBs are estimated as white noise [46]. The variance setting is $60 \times 60\ \mathrm{m}^2$ for both kinematic positioning and ISBs parameters. The ZWD is estimated using a random walk model with the process noise standard deviation of $\left(1 \times 10^{-4} \times \sqrt{30}\right)$ m [45]. Ambiguities were treated as constants unless a cycle slip or loss of lock occurred. The simulated LEO satellite orbits and clocks are directly used in the PPP, with their errors not considered in this study.

Five schemes were designed with different numbers of LEO satellites: (1) GPS+BDS-3+Galileo solution (GCE), and (2) GPS+BDS-3+Galileo solution with 120 (GCE+120L), (3) 180 (GCE+180L), (4) 240 (GCE+240L), and (5) 300 LEO satellites (GCE+300L). The program was initialized every 3 h to collect more samples to investigate the performance of the convergence time and the time link precision. Additionally, daily solutions were applied to evaluate the time link stability.

4. Experimental Results

The experimental results were evaluated in terms of the number of satellites and TDOP, the convergence time, the time link precision, and the time link stability in this section.

4.1. Number of Satellites and TDOP

The performance of the PPP time transfer is directly influenced by the number of observed satellites and the TDOP values. Figure 3 presents the time series of satellite visibility and TDOP variations using different schemes over 7 days for three representative stations at different latitudes: KOUR (low latitude), CEBR (mid-latitude), and KIRU (high latitude). Under the GCE solution (blue lines), the number of visible satellites ranges from 20 to 32 for KOUR, 17 to 29 for CEBR, and 21 to 34 for KIRU. The integration of LEO constellations (120, 180, 240, and 300 satellites) significantly enhances satellite visibility. When incorporating 300 LEO satellites (magenta lines), the number of observed satellites can be remarkably increased to 28–46 for KOUR, 30–45 for CEBR, and 57–76 for KIRU. Notably, high-latitude stations exhibit significantly more satellite visibility compared to those of low- and mid-latitude counterparts. This spatial disparity is caused by two factors: First, the near-polar orbital configuration of the used Walker constellations results in satellites frequently passing over high-latitude regions. Second, high-latitude areas serve as intersection zones for multiple orbital planes, thereby increasing satellite visibility.

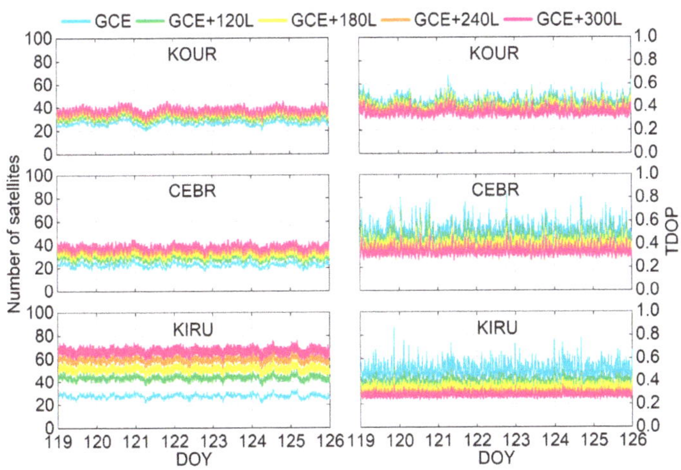

Figure 3. Number of observed satellites and TDOP time series for stations KOUG, CEBR, and KIRU from DOY 119 to 125, 2024.

For the TDOP time series (see the right panel of Figure 3), the values range from 0.354 to 0.677 for KOUR, 0.341 to 0.830 for CEBR, and 0.355 to 0.875 for KIRU under the GCE solution. These values exhibit a significant reduction after integrating LEO observations. Notably, distinct spikes in the TDOP time series are observed in the GCE solution, particularly at the mid- and high-latitude stations (CEBR and KIRU). However, these phenomena are mitigated when integrating LEO observations. A clear inverse relationship is observed between satellite visibility and TDOP values, i.e., the larger the number of observed satellites, the smaller the TDOPs. A higher number of observed satellites can improve the geometric configuration and enhance the observation model. These factors collectively reduce uncertainties in receiver clock estimation, thereby lowering the TDOP values. The integration of LEO satellites into GNSS observations is expected to significantly enhance time transfer performance based on the model strength improvement.

The average number of observed satellites and the TDOP values across all stations are presented in Figure 4. For the GCE solution (blue bars), the number of observed satellites ranges from 13.7 to 27.8 for different stations, while the TDOP values vary between 0.459 and 0.865. After integrating 120, 180, 240, and 360 LEO satellites, the number of observed satellites increases to 19.5–43.5, 22.4–51.4, 25.3–59.3, and 28.2–67.2, respectively. Correspondingly, TDOP values can significantly decrease to 0.360–0.541, 0.325–0.481, 0.296–0.433, and 0.275–0.400, owing to the increased number of observed satellites and the enhanced observation model strength. The two high-latitude stations, KIRU and SCOR, exhibit significantly higher satellite visibility compared to stations located at mid- and low latitudes. Notably, the number of observed satellites at station DAEJ is lower than that of other stations, with only 13.7 satellites being observed for GCE solutions, resulting in a higher TDOP value of 0.865. By checking observation files, the BDS-3 satellite Pseudorange Random Noise (PRN) larger than C32 cannot be received by this station, which may be related to hardware or software updates to be performed. However, its TDOP is reduced to 0.400 when integrating 300 LEO satellites, demonstrating the effectiveness of LEO-enhanced GNSS observations in improving time transfer performance, especially for stations not tracking enough GNSS observations, or in single- or dual-system situations.

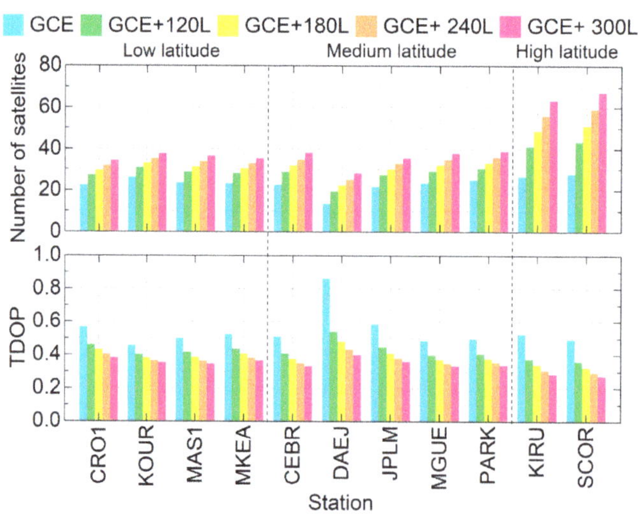

Figure 4. Average numbers of satellites (**top**) and TDOPs (**bottom**) for ground stations.

The average number of observed satellites and the TDOPs of all stations are listed in Table 2. Under the GCE solutions, the average number of satellites amounts to 23.4, with a corresponding TDOP of 0.547. After integrating 120, 180, 240, and 300 LEO satellites, the number of observed satellites is increased to 30.6, 34.1, 37.7, and 41.3, with improvements of 30.8%, 45.7%, 61.1%, and 76.5%, respectively, Meanwhile, the TDOP values decrease to 0.424, 0.391, 0.363, and 0.342, with a reduction rate of 22.5%, 28.5%, 33.6%, and 37.5%, respectively.

Table 2. Average numbers of satellites and TDOPs for all stations from DOY 119 to 125, 2024.

	GCE	GCE+120L	GCE+180L	GCE+240L	GCE+300L
Number of satellites	23.4	30.6	34.1	37.7	41.3
TDOP	0.547	0.424	0.391	0.363	0.342

4.2. Convergence Time

The time link error is defined as the time link difference between the estimated values and those based on the European Space Agency Multi-GNSS (ESM) final products. Figure 5 presents the time series of time link errors for three time links using the five schemes without or with different LEO constellations over the first three hours on DOY 122, 2024. The top and bottom panels represent the static and kinematic modes, respectively. The time series of the time link errors for the GCE solutions (blue lines) exhibits larger fluctuations than the other schemes in both the static and kinematic modes, particularly for the CEBR-CRO1 and CEBR-JPLM time links during the convergence period. After incorporating LEO observations, these fluctuations are significantly reduced or sometimes even eliminated. The improvements are especially significant for the kinematic mode (bottom panels) with weaker observation models.

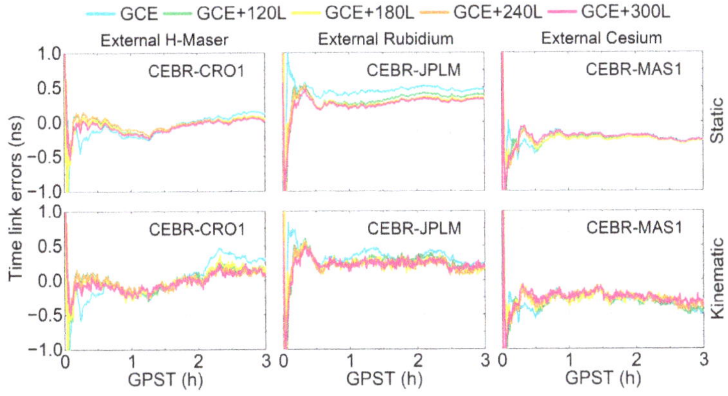

Figure 5. Time link errors of three time links in static (**top**) and kinematic modes (**bottom**) for the first 3 h on DOY 122, 2024.

Also observed in Figure 5 is the accelerated convergence process after integrating LEO observations. When the time link errors STD between the current and the next 10 epochs can be kept within 0.1 ns (approximately 3 cm), the time elapsed from the start of data processing to the current epoch is regarded as the convergence time [47]. To evaluate the convergence performances, the program is restarted every 3 h, resulting in a total of 560 solution pieces. The percentage distribution of the convergence times at 2 min intervals for all pieces is presented in Figure 6 for both the static and kinematic modes. For the GCE solutions, the percentage of the convergence times within 6 min amounts to 31.1% for the static mode, and 23.4% for the kinematic mode. After integrating 120, 180, 240, and 300 LEO satellites, it can be improved to 57.9%, 78.2%, 82.9%, and 82.9% in the static mode, and 53.9%, 72.3%, 77.5%, and 79.5% in the kinematic mode, respectively. For convergence times within 8 min, the percentage increases from 59.3% (static) and 49.6% (kinematic) in the GCE solutions to over 90% after integrating 180, 240, and 300 LEO satellites in both modes. Furthermore, the percentage of solutions with convergence times exceeding 10 min is 18.6% for the static mode and 27.0% for the kinematic mode in the GCE solutions. They are reduced to below 4% after integrating LEO observations. The convergence performance in the static mode is overall better than that of the kinematic mode, adding or not adding the LEO observations.

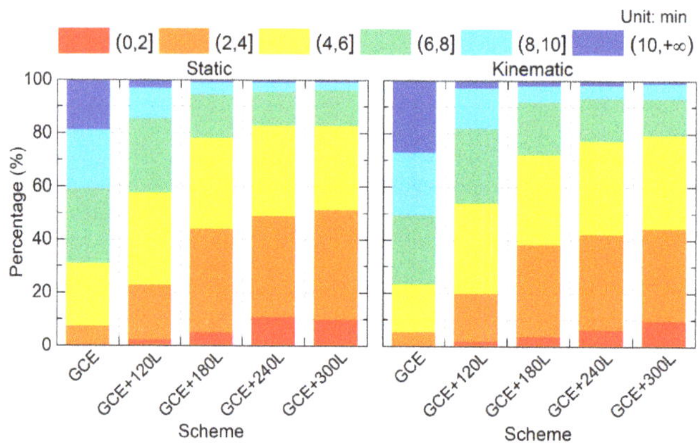

Figure 6. Percentage of the convergence times across each 2 min interval for different schemes.

Figure 7 presents the average convergence times for each time link in both the static and kinematic modes across different schemes. It can be observed that the GCE solutions require at least 7 min to achieve convergence in both modes, with the longest convergence time observed for the CEBR-DAEJ time link, which may be attributed to the lower number of observed satellites and their poor TDOPs (see Figure 4). After integrating LEO observations, the convergence time is significantly reduced for all time links in both the static and kinematic modes. Specifically, the average convergence time decreases to approximately 5 min when utilizing a LEO constellation with at least 180 satellites. As the number of LEO satellites further increases, the improvement becomes marginal. The convergence time differences among the GCE+180L, GCE+240L, and GCE+300L schemes are negligible in both static and kinematic modes.

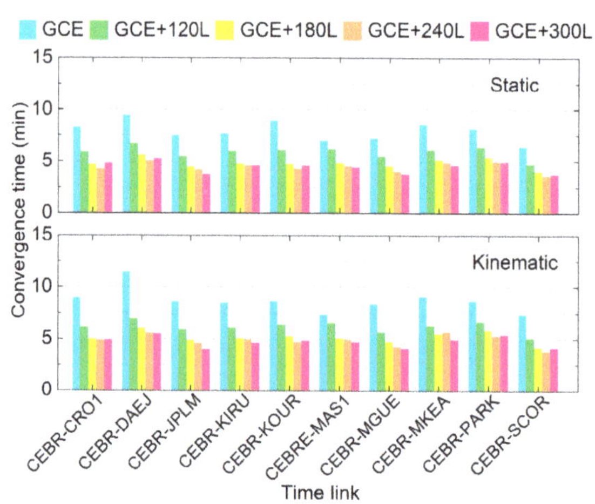

Figure 7. Average convergence times of different time links applying the five schemes.

The average convergence times of all time links are listed in Table 3. When using the GCE solutions, the average convergence time amounts to 7.95 min in static and 8.74 min in kinematic mode. By integrating 120, 180, 240, and 300 LEO satellites, the convergence time is reduced to 5.94, 4.83, 4.46, and 4.45 min in the static mode, and to 6.18, 5.17, 4.89, and 4.72 min in the kinematic mode, with improvements of 25.3%, 39.2%, 43.9%, and 44.0% in static mode, and 29.3%, 40.8%, 44.1%, and 46.0% in kinematic mode, respectively.

Table 3. Average convergence times of different schemes for static and kinematic mode (unit: min).

	GCE	GCE+120L	GCE+180L	GCE+240L	GCE+300L
Static	7.95	5.94	4.83	4.46	4.45
Kinematic	8.74	6.18	5.17	4.89	4.72

The convergence times presented in this section were derived from observations with a 30 s sampling interval. Further reducing the sampling interval, i.e., to 1 s, should shorten the convergence time for all schemes. This will be discussed in Section 5.

4.3. Time Link Precision

The time link precision is evaluated using the Standard Deviation (STD) of the time link errors (including the convergence period). Consistent with the convergence time analysis, a total of 560 solution instances were examined. Figure 8 presents the percentage distribution of the STD values in 0.1 ns increments from 0 to 0.5 ns, and those exceeding 0.5 ns. The percentage of solutions within the 0–0.4 ns range is 73.2%, 82.9%, 89.8%, 92.1%, and 92.3% for the static GCE, GCE+120L, GCE+180L, GCE+240L, and GCE+300L solutions, respectively. In the kinematic mode, the corresponding percentages are 64.8%, 81.4%, 88.64%, 90.5%, and 91.1%. The improvement in STD percentages confirms that integrating LEO observations can significantly enhance the time link precision.

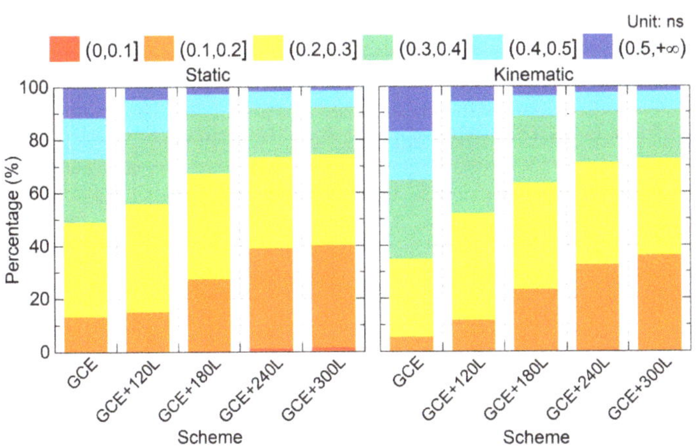

Figure 8. STD percentage across each 0.1 ns interval for the static (**left**) and the kinematic mode (**right**).

Figure 9 presents the STD of each time link under different schemes for both the static and kinematic modes. Compared to the GCE solutions, the integration of LEO observations can significantly improve the time link STD, particularly for links with sub-optimal precision, e.g., the CEBR-DAEJ time link. The difference in time link STD between the GCE+240L and GCE+300L solutions is marginal.

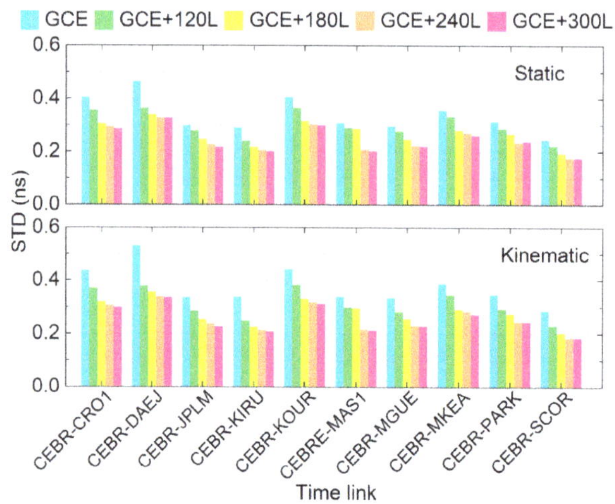

Figure 9. STD of each time link across different schemes for the static and kinematic modes.

The average time link precision of each scheme for the static and kinematic modes is listed in Table 4, including and excluding the first hour. Having the convergence period included, the STDs of the static and kinematic GCE solutions are 0.337 and 0.377 ns. After integrating 120, 180, 240, and 300 LEO satellites, the STD is reduced to 0.302, 0.266, 0.248, and 0.243 ns in the static mode, and 0.313, 0.277, 0.259, and 0.253 ns in the kinematic mode. This implies improvements of 10.4%, 21.1%, 26.4%, and 27.9% in the static mode, and 17.0%, 26.5%, 31.3%, and 32.9% in the kinematic mode. After removing the first hour of the time link errors, i.e., excluding the convergence period, the time link STDs amount to 0.064 ns for the GCE solution in the static mode, and are improved to 0.058, 0.055, 0.055, and 0.054 ns after integrating 120, 180, 240, and 300 LEO satellites, with improvements of 9.4% to 15.6%. For the kinematic mode, the STDs amount to 0.092, 0.084, 0.081, 0.080, and 0.079 ns for the five schemes, with the improvements amounting to 8.7% to 14.1%. This demonstrated that even after convergence, the improvements are still significant after integrating LEO satellites in the multi-GNSSS time link performances.

Table 4. Average time link precision across each scheme for the static and kinematic modes, including and excluding the first processing hour (unit: ns).

	GCE	GCE+120L	GCE+180L	GCE+240L	GCE+300L
Static	0.337	0.302	0.266	0.248	0.243
Static (excl. 0–1 h)	0.064	0.058	0.055	0.055	0.054
Kinematic	0.377	0.313	0.277	0.259	0.253
Kinematic (excl. 0–1 h)	0.092	0.084	0.081	0.080	0.079

4.4. *Time Link Stability*

To further evaluate the performance of the estimated time links, the Modified Allan Deviation (MDEV) [48] is applied to evaluate the time link stability. The average time link stability (considering also the convergence period) of three time links equipped with different types of clocks is presented in Figure 10 for both the static and kinematic modes. The LEO-enhanced solutions exhibit poorer short-term stabilities, i.e., at an averaging time of 30 s in the static mode and 30–60 s in the kinematic mode. This can be attributed to the increased number of estimated parameters after integrating LEO observations, leading to higher short-term noise. However, a notable improvement was observed for averaging times between 60 and approximately 1000 s in the static mode. For the kinematic mode, improvements are observed for averaging times ranging from 90 to 10,000 s, with the most significant enhancements occurring between 1000 and 10,000 s. Since the kinematic PPP model strength is weaker than its static counterpart, a longer convergence time is required. As a result, the improvements in the 1000–10,000 s range are more pronounced.

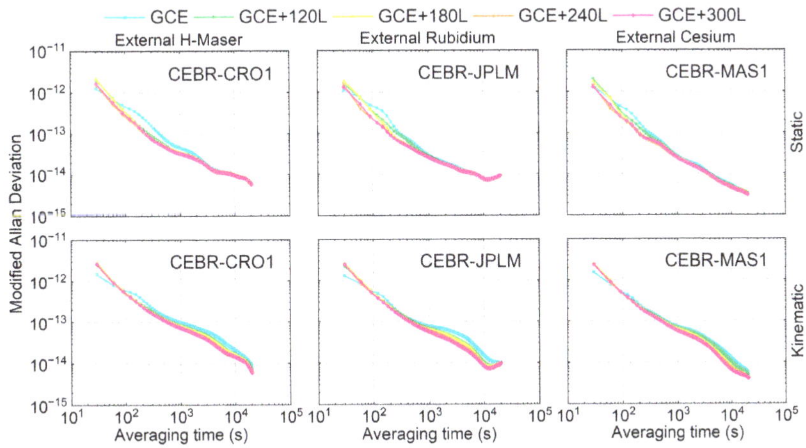

Figure 10. Average MDEVs of three time links in the static (**top**) and kinematic modes (**bottom**).

The average improvements in the MDEV for all time links at different averaging times are presented in Figure 11 for both the static and kinematic modes. Similar to Figure 10, the time link stability improvement is not observed for averaging times of 30 s in static mode and 30–60 s in the kinematic mode after integrating LEO-enhanced multi-GNSS observations. However, positive improvements are evident for averaging times longer than 90 s. The maximum improvement occurs at an averaging time of approximately 210 s, with improvements of 35.3%, 45.0%, 49.8%, and 48.6% for the GCE+120L, GCE+180L, GCE+240L, and GCE+300L solutions, respectively. For the kinematic mode, the maximum improvement is observed at an averaging time of approximately 5600 s, with improvements of 27.0%, 36.7%, 42.8%, and 46.8% for GCE+120L, GCE+180L, GCE+240L, and GCE+300L schemes, respectively. The time link stability improvements achieved with GCE+240L and GCE+300L solutions are superior to those of GCE+120L and GCE+180L. Similar to the time link STD results, the difference between the GCE+240L and GCE+300L solutions is marginal in both the static and kinematic modes.

The time link stabilities were also evaluated using the time link errors after removing the first convergence hour. The time link stability improvements in the LEO-enhanced multi-GNSS solutions compared to those of GNSS-only solutions are presented in Figure 12. It can be observed that the improvements are less than 5% and can be ignored for the static mode. For the kinematic mode, the improvements can be observed when the averaging time becomes longer than 480 s. The maximum improvement is about 45% for the GCE+300L solutions.

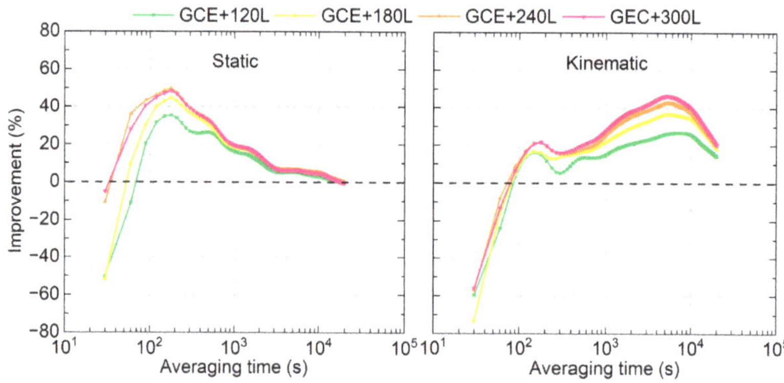

Figure 11. Average improvements in the time link stabilities after integrating LEO observations in multi-GNSS PPP time transfer.

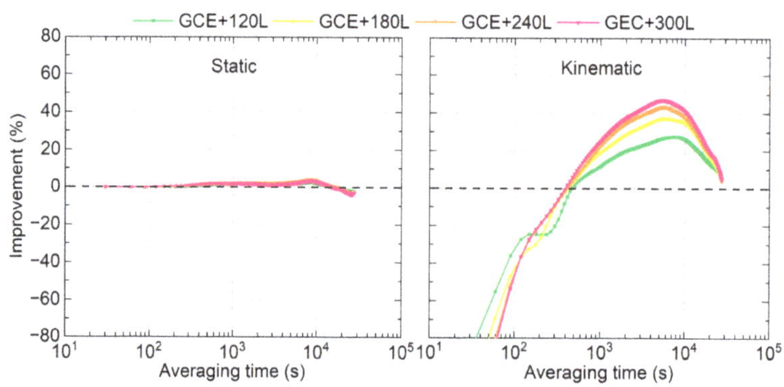

Figure 12. Average time link stability improvements in the LEO-enhanced multi-GNSS real-time PPP time transfer compared to those of multi-GNSS solutions.

5. Discussions

In this study, the sampling interval was set to 30 s, as often used in GNSS-based PPP [1,2]. Compared to using 30 s observations, it was reported that the PPP convergence can be shortened from 7.9, 5.3, and 6.9 min to 4.4, 3.1 and 3.7 min in the east, north and up coordinates, respectively, after using 1 s observations in LEO-enhanced GPS/BDS PPP [49], which concluded that a higher sampling rate is beneficial for further shorten the convergence time. Therefore, in this contribution, we have additionally conducted LEO-enhanced multi-GNSS PPP time transfer using 1 s observations to evaluate the impacts. One-day observations on DOY 124, 2024, were processed with the program restarted every 30 min, resulting in 480 PPP pieces for the assessment. The continuous network solutions from DOY 118 to 125, 2024, were processed using the ESM final GPS satellite orbits, in which the receiver clocks on DOY 124 were used as references. The detailed clock estimation strategy can be found in [50–52].

With the mean offsets removed, the 50%, 68%, and 95% percentile lines of the time link errors are shown in Figure 13. The GCE solutions require 68, 108 and 159 s to converge to 1 ns in the static mode for the three percentile lines, and 74, 119 and 162 s for the kinematic mode. After integrating 120, 180, 240 and 300 LEO satellites, the required times are shortened to 13–26, 18–46, 19–47 s for the static mode, and 13–26, 18–47, and 19–46 s in the kinematic mode. The results show that the 1-sigma (68%) percentile line of the time link errors can converge to 1 ns within 1 min in using LEO-enhanced multi-GNSS real-time PPP time transfer.

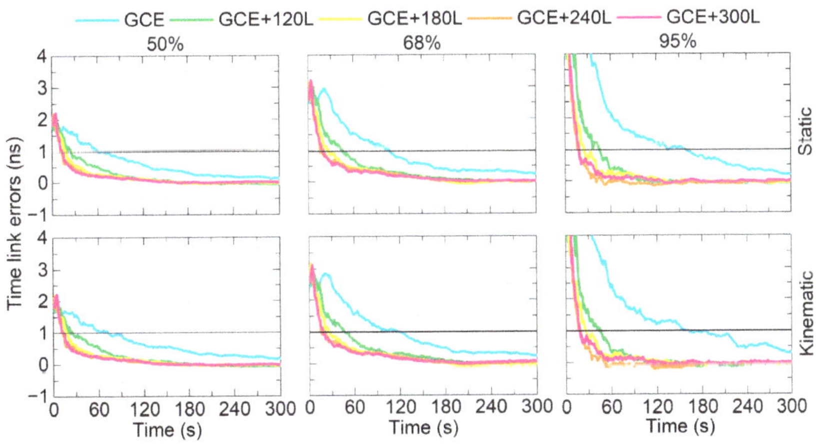

Figure 13. Percentile lines of time link errors of LEO-enhanced multi-GNSS PPP time transfer with different numbers of LEO satellites for the static (**top**) and kinematic mode (**bottom**).

6. Conclusions

GNSS PPP time transfer suffers from a long initialization time due to the slow geometry change in GNSS satellites. The fast motion of LEO satellites helps accelerate changes in satellite geometry, thereby improving the convergence speed. This contribution investigates the performance of LEO-enhanced multi-GNSS real-time PPP time transfer. Simulated LEO constellations consisting of different numbers of LEO satellites (120, 180, 240, and 300) are integrated into the multi-GNSS scenario to perform time transfer in both the static and kinematic modes. The analysis is based on one week of observations from 11 globally distributed stations. 3 h solutions are used to assess the convergence times and the time link precision, while the MDEVs are evaluated using a daily based solution.

Compared to the GCE solutions, the integration of 120, 180, 240, and 300 LEO satellites can increase the average number of observed satellites from 23.4 to 30.6, 34.1, 37.7, and 41.3, respectively, while reducing the TDOP values from 0.547 to 0.424, 0.391, 0.363, and 0.342. Using 30 s observations, the average convergence times for all time links are shortened from 7.95 to 5.94, 4.83, 4.46, and 4.45 min for the static mode, and from 8.75 to 6.18, 5.17, 4.89, and 4.72 min in the kinematic mode. Additionally, the time link precision in the static mode is improved from 0.337 to 0.302, 0.266, 0.248, and 0.243 ns, while it is from 0.377 to 0.313, 0.277, 0.259, and 0.253 for the kinematic mode. The time link stabilities (MDEVs) are enhanced in both the static and kinematic modes compared to the GCE solutions. Using 1 s observations, the 1-sigma percentile line of the time link errors can converge to 1 ns within 1 min after using LEO-enhanced multi-GNSS real-time PPP time transfer.

Author Contributions: Conceptualization, W.X. and K.W.; methodology, W.X.; software, W.X., W.L. and M.W.; validation, W.X., W.L. and M.W. writing—original draft preparation, W.X.; writing—review and editing, K.W., W.L., M.W., M.L. and X.Y.; supervision, K.W.; funding acquisition, W.X. and K.W. All authors have read and agreed to the published version of the manuscript.

Funding: This work is supported by the National Natural Science Foundation of China (Nos. 42404033, 12473078), the International Partnership Program of the Chinese Academy of Sciences (CAS) (Grant No. 021GJHZ2023010FN), and the Special Research Assistant Funding Project, CAS (No. 110400T0XW).

Data Availability Statement: The observations can be found at ftp://igs.gnsswhu.cn/ (accessed on 25 September 2025). The DCB files can be found at https://data.bdsmart.cn/pub/product/bias/ (accessed on 25 September 2025). The GNSS satellite orbits and clocks can be found at http://www.ppp-wizard.net/products/REAL_TIME/ (accessed on 25 September 2025). The receive clocks can be found at the http://navigation-office.esa.int/products/gnss-products/ (accessed on 25 September 2025).

Acknowledgments: We acknowledge the support of the international GNSS monitoring and assessment system (iGMAS) at the National Time Service Center, and the National Space Science Data Center, National Science & Technology Infrastructure of China (http://www.nssdc.ac.cn (accessed on 25 September 2025)). Wuhan University is acknowledged for providing the Multi-GNSS observation data. The CAS is acknowledged for providing the Multi-GNSS DCB products. The CNES is acknowledged for providing the Multi-GNSS real-time orbits and clocks. The ESM is acknowledged for providing the receiver clocks.

Conflicts of Interest: The authors declare no conflicts of interest.

References

1. Ge, Y.; Qin, W.; Su, K.; Yang, X.; Ouyang, M.; Zhou, F.; Zhao, X. A new approach to real-time precise point positioning timing with International GNSS Service real-time service products. *Meas. Sci. Technol.* **2019**, *30*, 125104. [CrossRef]
2. Tu, R.; Zhang, P.; Zhang, R.; Liu, J.; Lu, X. Modeling and assessment of precise time transfer by using BeiDou navigation satellite system triple-frequency signals. *Sensors.* **2018**, *18*, 1017. [CrossRef] [PubMed]
3. Petit, G.; Jiang, Z. Precise Point Positioning for TAI Computation. *Int. J. Navig. Obs.* **2008**, *2008*, 562878. [CrossRef]
4. Defraigne, P.; Baire, Q. Combining GPS and GLONASS for time and frequency transfer. *Adv. Space Res.* **2011**, *47*, 265–275. [CrossRef]
5. Zhang, P.; Tu, R.; Lu, X.; Fan, L.; Zhang, R. Performance of Multi-GNSS Real-Time UTC(NTSC) Time and Frequency Transfer Service Using Carrier Phase Observations. *Remote Sens.* **2021**, *13*, 4184. [CrossRef]
6. Liu, M.; Tu, R.; Chen, Q.; Li, Q.; Chen, J.; Zhang, P.; Lu, X. Study of fast and reliable time transfer methods using low Earth orbit enhancement. *Remote Sens.* **2024**, *16*, 2044. [CrossRef]
7. Reid, T.G.R.; Neish, A.M.; Walter, T.; Enge, P.K. Broadband LEO constellations for navigation. *Navig. J. Inst. Navig.* **2018**, *65*, 205–220. [CrossRef]
8. Çelikbilek, K.; Saleem, Z.; Ferre, R.M.; Praks, J.; Lohan, S. Survey on Optimization Methods for LEO-Satellite-Based Networks with Applications in Future Autonomous Transportation. *Sensors* **2022**, *22*, 1421. [CrossRef] [PubMed]
9. Yang, L. The CentiSpace-1: A leo satellite-based augmentation system. In Proceedings of the 14th meeting of the international committee on global navigation satellite systems, Beijing, China, 10 December 2019; p. 12.
10. Yang, Z.; Liu, H.; Qian, C.; Shu, B.; Zhang, L.; Xu, X.; Zhang, Y.; Lou, Y. Real-Time Estimation of Low Earth Orbit (LEO) Satellite Clock Based on Ground Tracking Stations. *Remote Sens.* **2020**, *12*, 2050. [CrossRef]
11. Li, X.; Ma, F.; Li, X.; Lv, H.; Bian, L.; Jiang, Z.; Zhang, X. LEO constellation-augmented multi-GNSS for rapid PPP convergence. *J. Geod.* **2019**, *93*, 749–764. [CrossRef]
12. Li, X.; Lv, H.; Ma, F.; Li, X.; Liu, J.; Jiang, Z. GNSS RTK Positioning Augmented with Large LEO Constellation. *Remote Sens.* **2019**, *11*, 228. [CrossRef]
13. Ke, M.; Lv, J.; Chang, J.; Dai, W.; Tong, K.; Zhu, M. Integrating GPS and LEO to accelerate convergence time of precise point positioning. In Proceedings of the 7th international conference on wireless communications and signal, Nanjing, China, 15–17 October 2015; pp. 1–5.
14. Li, X.; Li, X.; Ma, F.; Yuan, Y.; Zhang, K.; Zhou, F.; Zhang, X. Improved PPP Ambiguity Resolution with the Assistance of Multiple LEO Constellations and Signals. *Remote Sens.* **2019**, *11*, 408. [CrossRef]
15. Li, X.; Yuan, Y.; Han, X.; Li, X.; Fu, Y. Toward wide-area and high-precision positioning with LEO constellation augmented PPP-RTK. *IEEE Trans. Instrum. Meas.* **2024**, *73*, 5500213. [CrossRef]
16. Li, W.; Yang, Q.; Du, X.; Li, M.; Zhao, Q.; Yang, L.; Qin, Y.; Chang, C.; Wang, Y.; Qin, G. LEO augmented precise point positioning using real observations from two CENTISPACE™ experimental satellites. *GPS Solut.* **2024**, *28*, 44. [CrossRef]
17. Xu, S.; Yang, Q.; Du, X.; Xu, X.; Zhao, Q.; Yang, L.; Qin, Y.; Guo, J. Multi-GNSS Precise Point Positioning enhanced by the real navigation signals from CENTISPACE™ LEO mission. *Adv. Space Res.* **2024**, *73*, 4175–4186. [CrossRef]
18. Montenbruck, O.; Steigenberger, P.; Prange, L.; Deng, Z.; Zhao, Q.; Perosanz, F.; Romero, I.; Noll, C.; Stürze, A.; Weber, G.; et al. The multi-GNSS experiment (MGEX) of the international GNSS service (IGS)-achievements, prospects and challenges. *Adv. Space Res.* **2017**, *59*, 1671–1697. [CrossRef]

19. Xie, W.; Huang, G.; Fu, W.; Li, P.; Cui, B. An efficient clock offset datum switching compensation method for BDS real-time satellite clock offset estimation. *Adv. Space Res.* **2021**, *68*, 1802–1813. [CrossRef]
20. Yue, C.; Wang, H.; Xu, C.; Dang, Y. M_FCB: An open-source software for multi-GNSS fractional cycle bias estimation. *GPS Solut.* **2025**, *29*, 51. [CrossRef]
21. Kouba, J.; Héroux, P. Precise point positioning using IGS orbit and clock products. *GPS Solut.* **2001**, *5*, 12–28. [CrossRef]
22. Tu, R.; Zhang, P.; Zhang, R.; Liu, J.; Lu, X. Modeling performance analysis of precise time transfer based on BDStriple-frequency un-combined observations. *J. Geod.* **2019**, *93*, 837–847. [CrossRef]
23. Ge, Y.; Wang, Q.; Wang, Y.; Lyu, D.; Cao, X.; Shen, F.; Meng, X. A new receiver clock model to enhance BDS-3 real-time PPP time transfer with the PPP-B2b service. *Satell. Navig.* **2023**, *4*, 8. [CrossRef]
24. Lai, W.; Huang, G.; Wang, L.; Qin, Z.; Li, R.; Xie, S.; She, H. Precise orbit determination of integrated BDS-3 and LEO satellites with ambiguity fixing under regional ground stations. *Meas. Sci. Technol.* **2024**, *35*, 116305. [CrossRef]
25. Lai, W.; Huang, G.; Wang, L.; Xie, S.; She, H. Beidou UPD estimation and assessment based on LEO-assisted regional stations observations. *GPS Solut.* **2025**, *29*, 110. [CrossRef]
26. Li, H.; Luojie, D.; Ding, H. Real-time service performances of BDS-3 and Galileo constellations with a linear satellite clock correction models. *Satell. Navig.* **2023**, *4*, 23. [CrossRef]
27. Huang, G.; Zhang, Q.; Xu, G. Real-time clock offset prediction with an improved model. *GPS Solut.* **2014**, *18*, 95–104. [CrossRef]
28. Cao, Y.; Huang, G.; Xie, W.; Xie, S.; Wang, H. Assessment and comparison of satellite clock offset between BeiDou-3 and other GNSSs. *Acta Geod. Geophys.* **2021**, *56*, 303–319. [CrossRef]
29. Wang, J.; Wu, M.; Wang, K.; Zou, M.; Yang, X. Influencing factors on real-time determination of LEO satellite clocks. *Meas. Sci. Technol.* **2025**, *36*, 066315. [CrossRef]
30. Wu, M.; Wang, K.; Wang, J.; Xie, W.; Liu, J.; Chen, B.; Ge, Y.; El-Mowafy, A.; Yang, X. Impact of latency and continuity of GNSS products on filter-based real-tTime LEO satellite clock determination. *Remote Sens.* **2024**, *16*, 4315. [CrossRef]
31. Xie, W.; Su, H.; Wang, K.; Liu, J.; Wu, M.; Zou, M.; El-Mowafy, A.; Yang, X. Real-time LEO satellite clock estimation with predicted LEO satellite orbits constrained. *GPS Solut.* **2024**, *28*, 172. [CrossRef]
32. Xie, W.; Wang, K.; Wang, J.; Chen, B.; Wu, M.; El-Mowafy, A.; Yang, X. Impact of ISB stochastic models on LEO satellite clock estimation with onboard multi-GNSS observations. *Meas. Sci. Technol.* **2025**, *36*, 046313. [CrossRef]
33. Ye, Z.; Wang, K.; Wei, C.; Chen, B.; Wang, J.; Wu, M. The impact of LEO satellite hardware delays on LEO-augmented precise point positioning. *GPS Solut.* **2025**. *under review*.
34. Kouba, J. Relativity effects of galileo passive hydrogen maser satellite clocks. *GPS Solut.* **2019**, *23*, 117. [CrossRef]
35. Wu, M.; Wang, K.; Liu, J.; Zhu, Y. Relativistic effects of LEO satellite and its impact on clock prediction. *Meas. Sci. Technol.* **2023**, *34*, 095005. [CrossRef]
36. Wu, J.; Wu, S.; Hajj, G.A.; Bertiger, W.I.; Lichten, S.M. Effects of antenna orientation on GPS carrier phase. *Manuscr. Geod.* **1993**, *18*, 91–98. [CrossRef]
37. Dach, R.; Brockmann, E.; Schaer, S.; Beutler, G.; Meindl, M.; Prange, L.; Bock, H.; Jäggi, A.; Ostini, L. GNSS processing at CODE: Status report. *J. Geod.* **2009**, *83*, 353–365. [CrossRef]
38. Davis, J.L.; Herring, T.A.; Shapiro, I.I.; Rogers, A.E.E.; Elgered, G. Geodesy by radio interferometry: Effects of atmospheric modeling errors on estimates of baseline length. *Radio Sci.* **1985**, *20*, 1593–1607. [CrossRef]
39. Saastamoinen, J. Contributions to the theory of atmospheric refraction—Part II Refraction corrections in satellite geodesy. *Bull. Géod.* **1973**, *47*, 13–34. [CrossRef]
40. Boehm, J.; Niell, A.; Tregoning, P.; Schuh, H. Global Mapping Function (GMF): A new empirical mapping function based on numerical weather model data. *Geophys. Res. Lett.* **2006**, *33*, L07304. [CrossRef]
41. Schoenemann, E.; Gini, F.; Mayer, V.; Springer, T. ESOC MGNSS Processing Strategy. Available online: http://navigation-office.esa.int/products/gnss-products/esm.pdf (accessed on 11 October 2025).
42. Laurichesse, D.; Privat, A. An open-source PPP client implementation for the CNES PPP-WIZARD demonstrator. In Proceedings of the 28th International Technical Meeting of the Satellite Division of The Institute of Navigation (ION GNSS+ 2015), Tampa, FL, USA, 14–18 September 2015.
43. Kalman, R.E. A new approach to linear filtering and prediction problems. *J. Basic Eng.* **1960**, *82*, 35–45. [CrossRef]
44. Jing, C.; Bertolesi, E.; Huang, G.; Li, X.; Zhang, Q.; Zhai, W.; Liu, G.; Li, H. Predicting Accelerometer Baseline Correction and Nondivergent Deformation Velocity Based on Convolutional Neural Network (CNN) During GNSS Downgrade. *IEEE Sens. J.* **2025**, *25*, 11982–11994. [CrossRef]
45. Li, P.; Zhang, X. Integrating GPS and GLONASS to accelerate convergence and initialization times of precise point positioning. *GPS Solut.* **2014**, *18*, 461–471. [CrossRef]
46. Xie, W.; Wang, K.; Cui, B.; Yang, X. Regional multi-station real-time time transfer using an undifferenced multi-GNSS network solution. *GPS Solut.* **2024**, *28*, 198. [CrossRef]

47. Fu, W.; Wang, L.; Chen, R.; Han, Y.; Zhou, H.; Li, T. Combined BDS-2/BDS-3 real-time satellite clock estimation with the overlapping B1I/B3I signals. *Adv. Space Res.* **2021**, *68*, 4470–4483. [CrossRef]
48. Allan, D.W.; Barnes, J.A. A modified 'Allan variance' with increased oscillator characterization ability. In Proceedings of the 35th Annual Frequency Control Symposium, Philadelphia, PA, USA, 27–29 May 1981; pp. 470–475.
49. Ge, H.; Li, B.; Ge, M.; Zang, N.; Nie, L.; Shen, Y.; Schuh, H. Initial assessment of precise point positioning with LEO enhanced Global Navigation Satellite Systems (LeGNSS). *Remote Sens.* **2018**, *10*, 984. [CrossRef]
50. Fu, W.; Huang, G.; Zhang, Q.; Gu, S.; Ge, M.; Schuh, H. Multi-GNSS real-time clock estimation using sequential least square adjustment with online quality control. *J. Geod.* **2019**, *93*, 963–976. [CrossRef]
51. Xie, W.; Huang, G.; Fu, W.; Du, S.; Cui, B.; Li, M.; Tan, Y. Rapid estimation of undifferenced multi-GNSS real-time satellite clock offset using partial observations. *Remote Sens.* **2023**, *15*, 1776. [CrossRef]
52. Xie, W.; Wang, K.; Fu, W.; Xie, S.; Cui, B.; Li, M. Real-time estimation of BDS-3 satellite clock offset with ambiguity resolution using B1C/B2a signals. *Remote Sens.* **2024**, *16*, 1666. [CrossRef]

remote sensing

MDPI

Article

Mitigating Integrity Risk in SBAS Positioning Using Enhanced IGG III Robust Estimation

Le Wang [1,2,3], Jinbo She [1,*], Bobin Cui [1,2], Ziwei Wang [1,4], Weicong Yang [1] and Yimin Wang [1]

1 School of Geological Engineering and Geomatics, Chang'an University, 126 Yanta Road, Xi'an 710054, China; wangle18@chd.edu.cn (L.W.)
2 Engineering Research Center of Intelligent BDS, Universities of Shaanxi Province, 126 Yanta Road, Xi'an 710054, China
3 Shaanxi Yellow River Science Research Institute, 126 Yanta Road, Xi'an 710054, China
4 Research Group of Astronomy and Geomatics (gAGE), Universitat Politècnica de Catalunya (UPC), 08034 Barcelona, Spain
* Correspondence: 2023226009@chd.edu.cn

Highlights

What are the main findings?

- The improved IGG III robust estimation algorithm significantly enhances the user-end positioning performance of SBAS. In various scenarios, it greatly improves the positioning accuracy in both the horizontal and vertical directions, reduces the integrity risk, and enhances the availability.
- This method effectively suppresses the influence of outliers in the observed data, avoids overly conservative protection level (PL) estimation, and obtains more reasonable and reliable protection level values without affecting real-time performance.

What is the implication of the main finding?

- This research offers a powerful and computationally efficient solution that can enhance the service reliability of SBAS in challenging environments such as urban dynamics and UAV maneuvering flight, making it highly suitable for safety-critical applications in aviation and intelligent transportation.
- This algorithm has strong adaptability and stability in multiple practical scenarios, providing an important reference for the performance optimization and integrity guarantee of future SBAS user-end under poor data quality conditions.

Abstract

To address the limitations in positioning accuracy and the risk of integrity degradation in Satellite-Based Augmentation Systems (SBAS) user-end after applying augmentation information, this study proposes a positioning algorithm integrating an improved IGG III robust estimation method. By using integrity information from SBAS, this method improves protection level calculations and better adjusts observed weights by adding new factors to the weight function model. This improvement allows for better discrimination between reliable and anomalous measurements, thereby enhancing positioning accuracy, reducing integrity risks, and improving availability. Experimental results show that, compared to conventional SBAS user positioning, the proposed method achieves notable performance improvements across various scenarios. In static environments, it reduces horizontal integrity risk by up to 6.7%, increases availability by up to 6.6%, and improves positioning accuracy by up to 71.3%. In urban vehicular environments, horizontal integrity risk is reduced by 0.5%, availability is increased by 0.5%, and accuracy improves by up to 58.7%. In Unmanned Aerial Vehicle flight scenarios, horizontal integrity risk is reduced by 2.8%, availability increases by 2.8%, and accuracy improves by up to 50.38%. In all scenarios, vertical integrity risk is completely eliminated and availability improves slightly. Additionally, compared to the conventional IGG III estimator, the improved method offers more effective control over weight adjustment during solution estimation, thereby avoiding excessive down-weighting and mitigating overbounding of protection levels. These results demonstrate the potential of the proposed method to improve the performance and reliability of SBAS user-end under both static and dynamic conditions.

Keywords: robust estimation; SBAS; positioning; protection level; integrity

1. Introduction

Satellite-Based Augmentation Systems (SBAS) have been widely implemented to enhance the accuracy, integrity, and availability of Global Navigation Satellite Systems (GNSS) positioning, particularly for safety-critical applications such as aviation and intelligent transportation [1]. Despite the provision of correction and integrity information, SBAS user-end may still experience degraded positioning performance in challenging environments [2], where observation quality is uneven and outliers are more frequent. In such cases, ensuring robust positioning and reliable Protection Level (PL) estimation becomes critical to maintain user trust and service availability. This motivates the development of improved positioning algorithms that can effectively handle observation anomalies while preserving integrity assurance.

At present, with the continuous development of SBAS technology, its user-end positioning method has become more mature and can show better positioning results and integrity performance when the data quality is better. However, in further expanding the application field of SBAS, the complexity of the environment has become more prominent, and the unstable quality of observation data has become more frequent, which leads to poor positioning results and integrity risks despite the use of enhancement information in messages to improve positioning accuracy. Therefore, the SBAS user-end positioning method still needs to be further improved in scenarios with more complex environments and poor data quality.

In recent years, there have been studies from the aspect of multi-system fusion and weighted model optimisation in SBAS user-end positioning. To improve the accuracy of Unmanned Aerial Vehicle (UAV) positioning, Krasuski et al. integrated the augmented data from multiple European Geostationary Navigation Overlay Service (EGNOS) satellites through an average weighted model to improve the accuracy of positioning [3]. Meanwhile, a proposed weighting algorithm based on dual GNSS receivers optimises the weighting model by taking the number of satellite tracking as a weighting factor, which effectively improves the positioning accuracy of System for Differential Correction and Monitoring (SDCM) and EGNOS in air navigation [4]. Thereafter, in order to better apply SBAS to UAV, a positioning accuracy improvement algorithm based on the joint solution of multiple augmentation systems is proposed, which integrates the ionospheric correction data provided by multiple SBAS through the average weighting model to enhance the positioning accuracy of UAV [5]. SBAS in urban environments where Geostationary Earth Orbit (GEO) visibility is poor and SBAS signals are obscured; Yoon et al. proposed online transmission of SBAS correction information through the communication link of the ground control station, which achieves the improvement of positioning accuracy in complex environments [6].

Liang Li et al. used SBAS enhancement parameters for real-time single-frequency Precise Point Positioning (PPP), which significantly improved the positioning accuracy compared to the traditional SBAS method [7]. Park et al. proposed an efficient multi-GNSS and multi-SBAS positioning method by independently calculating the positioning results of each GNSS and each SBAS combination, and then determining the final position by weighted summing, which ultimately improves the positioning accuracy [8]. To address the problem that single-frequency SBAS is affected by ionospheric storms and scintillation, Li et al. proposed a new dual-frequency user positioning algorithm, which improves the positioning accuracy and integrity [9]. Based on the concept of flexible Positioning Navigation Timing (PNT) [10], Zhang et al. proposed an integration framework without ground network, which achieves multi-system error correction by flexibly integrating B1C, B2a and broadcast ephemeris messages, and improves the positioning stability of BeiDou Satellite-Based Augmentation System (BDSBAS) in complex environments [11]. To address the problem that the overlapping areas between different SBAS systems cannot be effectively fused, Wang et al. proposed a fusion positioning algorithm for BDSBAS and Multi-functional Satellite Augmentation System (MSAS) based on fully connected neural networks, which can provide more stable and high-precision augmented navigation services [12]. Wang et al. proposed a novel service weighting model for road transport SBAS Dual-Frequency Multi-Constellation (DFMC), which integrates the Carrier-to-Noise Ratio (CNR) and the smoothing time factor to dynamically adjust the weights of the observed data, which effectively reduces the protection level and improves the availability [13]. Zheng et al. proposed two improved integrity monitoring approaches: one based on covariance analysis and another based on pseudorange residuals combined with satellite-station geometry [14,15]. These methods significantly enhanced system integrity and usability. Wang et al. employed a Kalman filter to amplify the covariance matrix of observation noise, thereby reducing the observation weight when the number of stations decreased. This adjustment improved both the accuracy and continuity of the system [16]. Shao et al. introduced a Dual-Frequency Range Error (*DFRE*) estimation technique based on the projection method, which establishes an envelope for the maximum correction error with a specified probability [17]. In addition, a dual-frequency SBAS integrity optimization algorithm and an improved over-limit pairing method have been developed [18,19]. These approaches enhance system availability by optimizing weight assignments and modeling error characteristics more precisely. More recently, with the rapid increase in Low Earth Orbit (LEO) satellites and their global coverage, new opportunities have emerged. Xin et al. improved monitoring performance and service coverage by optimizing the observation link, effectively mitigating the impact of satellite entry and exit, and substantially enhancing integrity performance [20]. In summary, there are fewer current improved SBAS user-end positioning methods, and most of them still use traditional user-end positioning methods. In addition, the existing weighted model optimisation has unspecified positioning performance in more complex environments with poorer data quality, or the weighted model is complex and dependent on smoothing time. And although multi-system fusion can improve the positioning performance of SBAS users, it cannot reduce the effect of coarseness in the observations when the data quality is poor, and it is not applicable to single-system users, which has certain limitations.

To address these limitations, this study introduces an improved robust estimation algorithm into the SBAS user positioning process, based on the IGG III framework. The proposed method is designed with consideration of both the weighting behavior of the user algorithm and the PL computation. Through extensive experiments in static, urban vehicular, and UAV flight scenarios, the method is shown to significantly reduce integrity risk, improve availability, and enhance positioning accuracy compared to traditional approaches. The results also demonstrate that the improved estimator avoids overbounding by maintaining controlled weight variation compared to the traditional IGG III robust estimation, thereby providing a more balanced and effective solution for SBAS user-end positioning.

The remainder of this paper is organized as follows. Method presents the improved IGG III-based robust estimation algorithm, including the design of the weight function and its integration with PL computation. The results describe the experimental setup and test scenarios, covering static, urban, and UAV environments and further discusses the experimental results, focusing on positioning accuracy, integrity risk, and availability performance. Finally, we conclude the study and outline potential directions for future work.

2. Method

2.1. IGG III Variance Expansion Function

The IGG III variance expansion function divides the observed data into three categories according to quality standards: effective information, available information and harmful information. Based on the principle of robust estimation, differentiates information of different quality, that is, effective information retains the original weight value, the available information is estimated with a down-weighting, and the harmful information is given a zero-weighting [21].

The IGG III weight function can be expressed as follows [22]:

$$\overline{P}_i = \begin{cases} P_i & |\tilde{v}_i| \le k_0 \\ P_i \frac{k_0}{|\tilde{v}_i|}\left(\frac{k_1-|\tilde{v}_i|}{k_1-k_0}\right)^2 & k_0 < |\tilde{v}_i| \le k_1 \\ 0 & |\tilde{v}_i| > k_1 \end{cases} \tag{1}$$

where P_i is the weight of the i-th satellite, k_0 and k_1 are harmonic coefficients that partition the standardized residual $|\tilde{v}_i|$ into three intervals, corresponding, respectively, to effective information, available information, and harmful information. In practical applications, k_0 is typically set in the range of 1.0 to 2.0, k_1 is chosen between 2.5 and 3.0 and $\tilde{v}_i$ is the standardised residuals.

It can be seen from the weight function that the selection of harmonic coefficients directly influences whether observation weights are appropriately reassigned, thereby enabling the effective utilization of reliable observations, the cautious handling of suspicious data, and the exclusion of outliers. Assuming that the observation residuals v_i follow a Gaussian distribution, the corresponding test statistic can be expressed as follows:

$$\tilde{v}_i = \frac{v_i - \mu}{\sigma_0\sqrt{q_{v_i}}} \sim N(0,1) \tag{2}$$

where q_{v_i} denotes the weight of the i-th observation, σ_0 is the standard deviation of the residuals, and μ is the mean of the residuals from all observations. Based on the statistical principle of three-sigma σ outlier detection, the IGG III weight function is applied with harmonic coefficients set as $k_1 = 3\sigma$ and $k_0 = 2\sigma$, where σ is the standard deviation of the residuals. Under this configuration, the standardized residuals are divided into three intervals: outliers $(-\infty, -3\sigma) \cup (3\sigma, +\infty)$, accounting for approximately 0.3% of the data; suspicious values $(-3\sigma, -2\sigma) \cup (2\sigma, 3\sigma)$, accounting for about 4.3%; and valid values $(-2\sigma, 2\sigma)$, covering roughly 95.4% of the total. This classification effectively distinguishes between outliers, questionable data, and reliable observations, enhancing the robustness of the estimation process.

2.2. SBAS User-End Positioning and Protection Level Calculation

The SBAS user-end consists primarily of two components: the augmented positioning solution and the PL. The augmented positioning solution utilizes GNSS observation data, broadcast ephemerides, and SBAS correction messages to estimate the user position and receiver clock offset using a least squares estimation approach. The PL computation relies on the integrity information provided by SBAS. If PL exceeds a predefined Alarm Limit (AL), an alert is triggered to prevent potential hazards during navigation. The relationship among PE, PL, and AL, along with the corresponding system service status, is summarized in Table 1. The integrity risk rate refers to the probability that the location information error provided by the navigation system exceeds the specified tolerance range within a given service time, but the system fails to issue an alarm to the user. Availability rate is the ability of a navigation system to provide the required functionality and performance within the expected service time, usually expressed in percentage.

Table 1. Relationship among PE, PL, and AL.

Relationship	Service State	Affect
$PE \leq PL \leq AL$	Available	None
$PL < PE \leq AL$	Misleading Information (MI)	Integrity risk
$AL < PL < PE$	MI	Integrity risk
$PL < AL < PE$	Hazardous MI (HMI)	Integrity risk (missing detections)
$PE < AL < PL$	Unavailable	Alarm

(1) Enhanced positioning solution

SBAS user can improve satellite orbit and satellite clock accuracy using satellite-based augmentation corrections. The pseudorange single-point positioning observation equation (dual-frequency ionosphere-free combination) using the satellite-based enhancement corrections can be expressed as follows [23]:

$$P_j^i(IF) = \rho_j^i + c * \left(\Delta t_j - \Delta t^i\right) + T_j^i + \varepsilon_j^i + cor^i \tag{3}$$

where ρ_j^i is the geometric distance from the station to the satellite, c is speed of light, Δt_j and Δt^i are the receiver and satellite clock offsets, respectively, T_j^i is tropospheric slant delay, ε_j^i is the pseudorange noise, and cor^i is the correction using SBAS corrections.

The specific process is as follows: first, the unit direction vector between the receiver and each satellite is computed; then, the satellite orbit corrections are projected onto the line-of-sight vector from the receiver to the satellite; finally, these are combined with the satellite clock corrections to form the total range correction. Unlike conventional single-point positioning, SBAS user positioning incorporates integrity information to determine the variance of the spatial signal error, which is used in the weighting of observations.

$$P = \begin{bmatrix} \frac{1}{\sigma_1^2} & \cdots & 0 \\ \vdots & \ddots & \vdots \\ 0 & \cdots & \frac{1}{\sigma_n^2} \end{bmatrix} \tag{4}$$

where $\sigma_i^2 = \sigma_{i,DFC}^2 + \sigma_{i,trop}^2 + \sigma_{i,air\_DF}^2 + \sigma_{i,iono}^2$, where $\sigma_{i,DFC}^2$ is the variance of the spatial signal error (specifically, the variance of the orbital clock corrections mapped in the line-of-sight direction), $\sigma_{i,trop}^2$ is the variance of tropospheric residual error, σ_{i,air_DF}^2 is the variance of the dual-frequency measurement noise, multipath, and antenna group delay errors, and

$\sigma^2_{i,iono}$ is the variance of the residuals in the ionosphere from the dual-frequency ionosphere combination, and they can be expressed as

$$\begin{cases} \sigma^2_{i,DFC} = (\sigma_{DFRE} * \delta_{DFRE})^2 + \varepsilon^2_{corr} + \varepsilon^2_{er} \\ \sigma^2_{i,trop} = 0.12 * \frac{1.001}{\sqrt{0.002001 + \sin(Ele)^2_{i,deg}}} \\ \sigma^2_{i,air\_DF} = 0.36 + 0.4 * e^{-\frac{Ele_i}{14^\circ}} \\ \sigma^2_{i,iono} = \frac{40.0}{261.0 + (ele_{i,deg})^2} + 0.018 \end{cases} \tag{5}$$

where σ_{DFRE} is the sigma value of Dual Frequency Range Error ($DFRE$), which can be retrieved from the correspondence table of $DFRE$ and Dual Frequency Range Error Indicator (DFREI) [24], ε^2_{corr} and ε^2_{er} are both degradation parameters, and δ_{DFRE} can be computed from the user positional information and the covariance array information of the satellite.

(2) Protection level solution

The calculation of the PL must balance both integrity and availability requirements. As such, the PL should neither be overly conservative (loose) nor overly optimistic (tight). The PL computation formula adopted in this study is given as follows [25]:

$$\begin{cases} HPL = K_H * d_{major} \\ VPL = K_V * d_U \end{cases} \tag{6}$$

where K_H and K_V are constants. For non-precision approach, K_H = 6.18 and no PL is solved in vertical direction. For precision approach, K_H = 6.0 and K_V = 5.33 [26]; d_{major} and d_U are obtained from the elements in the covariance matrix $invG$, and they can be expressed as

$$\begin{cases} d_{major} = \sqrt{\frac{d^2_{east} + d^2_{north}}{2} + \sqrt{\left(\frac{d^2_{east} - d^2_{north}}{2}\right)^2 + d^2_{EN}}} \\ d^2_{east} = invG(1,1) \\ d^2_{north} = invG(2,2) \\ d_U = invG(3,3) \end{cases} \tag{7}$$

The covariance matrix can be expressed as follows:

$$invG = \left(G^{\mathrm{T}}WG\right)^{-1} \tag{8}$$

where G is the spatial geometric matrix and W is the weight matrix for position solving as follows [27]:

$$G = \begin{bmatrix} -\cos\left(Ele^1\right) * \sin\left(Azl^1\right) & -\cos\left(Ele^1\right) * \cos\left(Azl^1\right) & -\sin\left(Ele^1\right) & 1 \\ \vdots & \vdots & \vdots & \vdots \\ -\cos\left(Ele^n\right) * \sin(Azl^n) & -\cos\left(Ele^n\right) * \cos(Azl^n) & -\sin(Ele^n) & 1 \end{bmatrix} \tag{9}$$

$$W = \begin{bmatrix} \frac{1}{\sigma^2_{1,DFC} + \sigma^2_{1,trop} + \sigma^2_{1,air\_DF} + \sigma^2_{1,iono}} & \cdots & 0 \\ \vdots & \ddots & \vdots \\ 0 & \cdots & \frac{1}{\sigma^2_{n,DFC} + \sigma^2_{n,trop} + \sigma^2_{n,air\_DF} + \sigma^2_{n,iono}} \end{bmatrix} \tag{10}$$

where Ele and Azl are the elevation angle and azimuth angle of the satellite relative to the user, n is the number of satellites.

2.3. Improved SBAS User Positioning Algorithm Combined with IGG III Robust Estimation

In SBAS user positioning, the observation weight is generally composed of three parts: spatial signal influence, signal propagation process and receiver-related effects. In addition to position estimation, the PL must also be calculated using the same observation weight applied during positioning. When the standardized residual exceeds the threshold, although the traditional IGG III method can reduce the weight of unreliable data and improve positioning accuracy by increasing the observation variance, it adjusts the above three parts at the same time, which has the risk of excessive weight changes, which may lead to the calculated PL overbounding issue. Therefore, directly applying the traditional IGG III robust estimation method is not suitable for SBAS user positioning. To solve this problem, this study uses the spatial signal error variance accurately calculated from integrity information to divide the total variance into two parts: (1) the spatial signal influence, expressed as $\sigma^2_{i,DFC}$; (2) signal propagation, and the comprehensive influence of the receiver-related factors is expressed as $\sigma^2_{i,trop} + \sigma^2_{i,air\_DF} + \sigma^2_{i,iono}$. The former has higher accuracy, while the latter is mainly related to the height angle. In robust estimation, the two components are adjusted according to the variance of spatial signal error.

To facilitate the selection of appropriate harmonic coefficient values for robust estimation, it is necessary to further analyze the distribution characteristics of the spatial signal error variance, $\sigma^2_{i,DFC}$. In this study, statistical analysis was performed on the values of $\sigma^2_{i,DFC}$ for all GPS satellites on Day of Year (DOY) 159 in 2023. The Cumulative Distribution Function (CDF) of these values is presented in Figure 1. It shows the probability that a random variable takes a value less than or equal to a certain value. The horizontal axis represents the value range of the random variable, and the vertical axis represents the cumulative probability, that is, the probability that the value of the random variable is less than or equal to the corresponding value on the horizontal axis. It can be observed that in 90% of the cases, $\sigma^2_{i,DFC}$ remains below 0.25, with only a small proportion of values exceeding this threshold. The rapid rise in the curve indicates that $\sigma^2_{i,DFC}$ is highly concentrated near a specific value, which further highlights the characteristics of $\sigma^2_{i,DFC}$ distribution: most of the values are closely clustered in a concentrated area, while only a few values are scattered far away from the concentrated area.

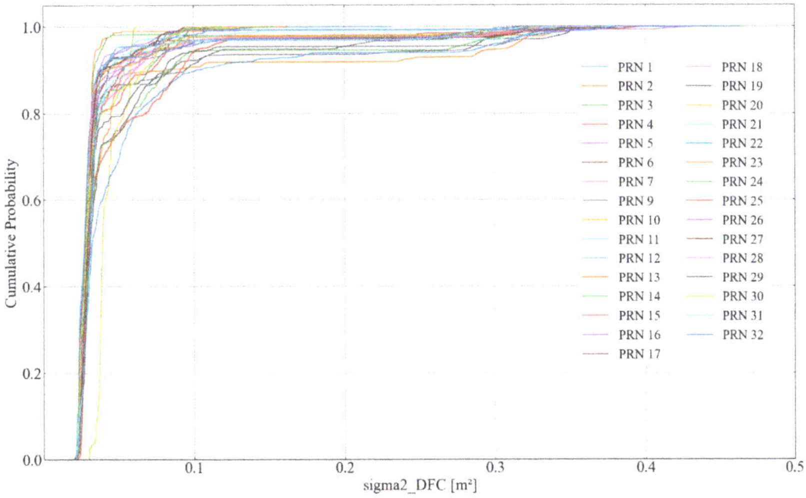

Figure 1. Cumulative distribution function of $\sigma^2_{i,DFC}$ for different GPS.

Meanwhile, $\sigma_{i,DFC}^2$ reflects the correction accuracy of the satellite orbit and clock information. When the correction accuracy is low, the value of $\sigma_{i,DFC}^2$ becomes larger, indicating that the residual ephemeris errors have a more significant impact on user positioning accuracy. Therefore, when the pseudorange residual satisfies the robust estimation condition, the magnitude of $\sigma_{i,DFC}^2$ can be used to selectively adjust the observation variance, thereby controlling the extent of deweighting. This approach helps to avoid excessive suppression of valid measurements. Based on this idea, the improved weight function for SBAS user positioning is formulated as follows:

$$\overline{P}_i = \begin{cases} P_i & |\tilde{v}_i| \leq k_0 \\ \dfrac{1}{\sigma_{i,DFC}^2 \Big/ \frac{k_0}{|\tilde{v}_i|}\left(\frac{k_1-|\tilde{v}_i|}{k_1-k_0}\right)^2 + \sigma_{i,trop}^2 + \sigma_{i,air\_DF}^2 + \sigma_{i,iono}^2} & k_0 < |\tilde{v}_i| \leq k_1, \tilde{\sigma}_{i,DFC}^2 \geq k_2 \\ \dfrac{1}{\sigma_{i,DFC}^2 + (\sigma_{i,trop}^2 + \sigma_{i,air\_DF}^2 + \sigma_{i,iono}^2) \Big/ \frac{k_0}{|\tilde{v}_i|}\left(\frac{k_1-|\tilde{v}_i|}{k_1-k_0}\right)^2} & k_0 < |\tilde{v}_i| \leq k_1, \tilde{\sigma}_{i,DFC}^2 < k_2 \\ 0 & k_1 < |\tilde{v}_i| \end{cases} \tag{11}$$

For ease of representation, the spatial signal error variance $\sigma_{i,DFC}^2$ is standardized in this paper and denoted as $\tilde{\sigma}_{i,DFC}^2$. The threshold for normalization is defined as twice the median value of the error. The harmonic coefficients k_0 and k_1 are retained as in the traditional IGG III formulation, where k_0 is typically chosen between 1.0 and 2.0, and k_1 is set between 2.5 and 3.0. Here, $\tilde{v}_i$ denotes the standardized residual, P_i is the weight of the i-th satellite. $\sigma_{i,trop}^2$, σ_{i,air_DF}^2 and $\sigma_{i,iono}^2$ have explained in Equation (4). In addition, it can be seen from the calculation formula of the protection level in Equations (6)–(10) that the covariance matrix of the positioning robust estimation output will be used to calculate the protection level. Therefore, the improved IGG III robust estimation also has corresponding optimization effects on the protection level calculation while finely adjusting the weight.

The flowchart of the improved robust algorithm for SBAS user positioning is shown in Figure 2. First, an initial positioning solution is obtained, and convergence is checked. If the solution has not yet converged, the pseudorange residuals are computed and standardized, while the spatial signal error variances are also statistically analyzed and standardized. Next, a decision is made on whether to apply down-weighting based on whether the standardized residual exceeds a predefined threshold. If down-weighting is required, the observation variance is selectively adjusted according to the magnitude of the standardized spatial signal error variance. The updated weights are then used in the augmented positioning solution to improve positioning accuracy. If no weight adjustment is needed, the original weights are retained. After convergence is reached through iterative updates, the PL is computed accordingly.

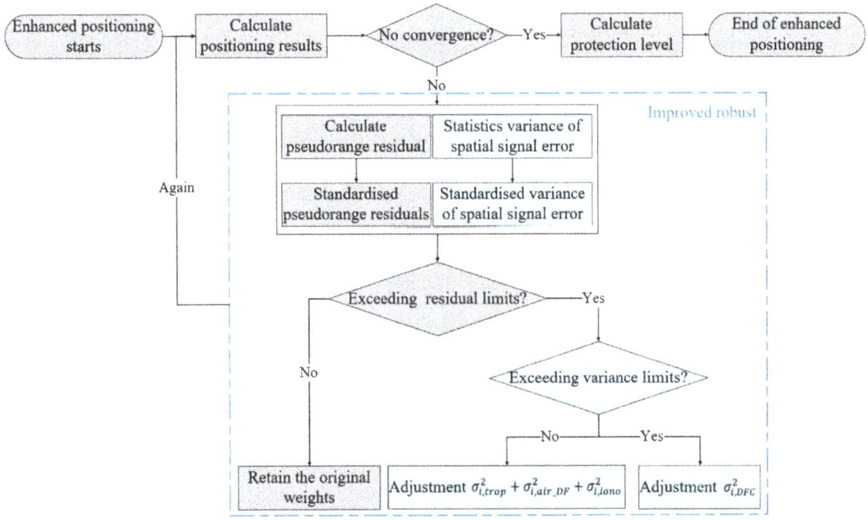

Figure 2. The flowchart of improved SBAS user positioning robust algorithm.

3. Experiment and Result Analysis

3.1. Processing Strategy and Data Quality Analysis

This study conducts experimental analysis through three scenarios: static, urban environment, and UAV maneuvering flight to verify the effectiveness of the algorithm in terms of accuracy, availability, integrity, and PL. Among them, the static experimental data selects the DOY 159–165 in 2023 actual measured data of three monitoring stations in China (green) CSH2, FUZ2, and KAS2. The sampling rate is 1 s. There are 21 enhanced stations (red). They are used to calculate the correction information of SBAS and serve the user-end positioning. The station distribution is shown in Figure 3.

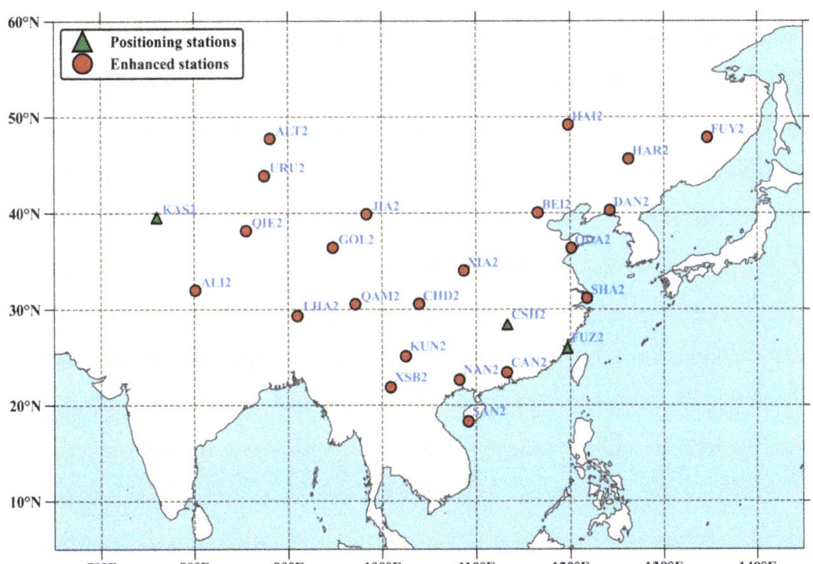

Figure 3. Distribution of stations.

To verify that the corrective values employed in this study are neither anomalous nor in violation of SBAS end-user requirements, data from the enhanced stations illustrated in Figure 3 were used to compute the corrections. The methodology and procedural flow are outlined in Figure 4, and the key steps are summarized as follows: First, dual-frequency ionosphere-free combinations of pseudorange observations were generated to eliminate the majority of ionospheric delays. The combined observations were subsequently smoothed, after which the effects of satellite clock errors and tropospheric delays were removed, yielding pseudorange residuals that primarily contained errors from satellite orbit, satellite clock, and receiver clock. Following clock synchronization, the receiver clock error was eliminated. Finally, the post-synchronization residuals were utilized to estimate the satellite orbit and clock corrections. It should be emphasized that the corrections derived in this study are directly integrated into the user positioning algorithm rather than being broadcast. In most practical applications, such corrections are typically transmitted to users through data messages; however, this delivery mechanism does not compromise the accuracy of the estimated parameters.

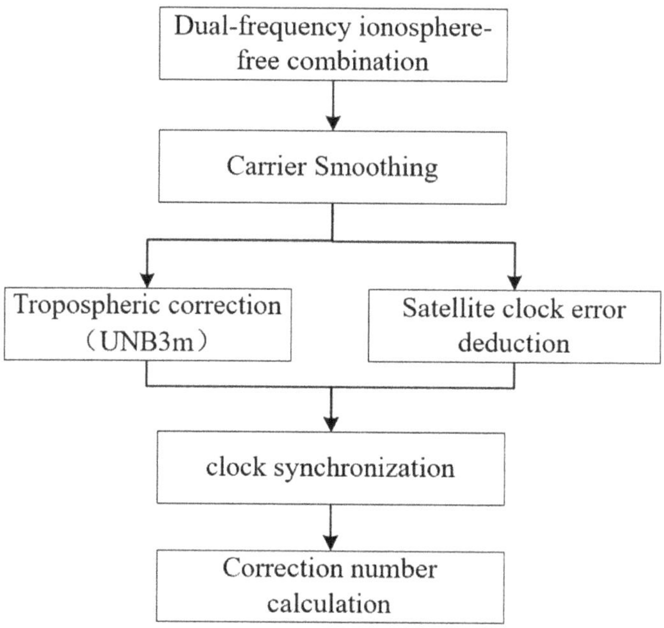

Figure 4. The process of correcting the number calculation.

Figure 5 counts the pseudorange residuals before and after enhancement of all GPS satellites over a total of 159–165 days, and uses the pseudorange residuals before and after enhancement as the evaluation criteria. The specific calculation is as follows: first calculate the average value of the pseudorange residuals before enhancement from all enhancement stations for a given satellite, and then calculate the average value of the pseudorange residuals after enhancement using the correction number, repeating the process until all satellites are evaluated. Compared with the values before enhancement, the pseudorange residuals after enhancement for all satellites are reduced to varying degrees, and the values of the pseudorange residuals after enhancement fall between 0.1 m and 0.3 m, indicating that the corrections used in this study are not abnormal and the accuracy is in line with the current level.

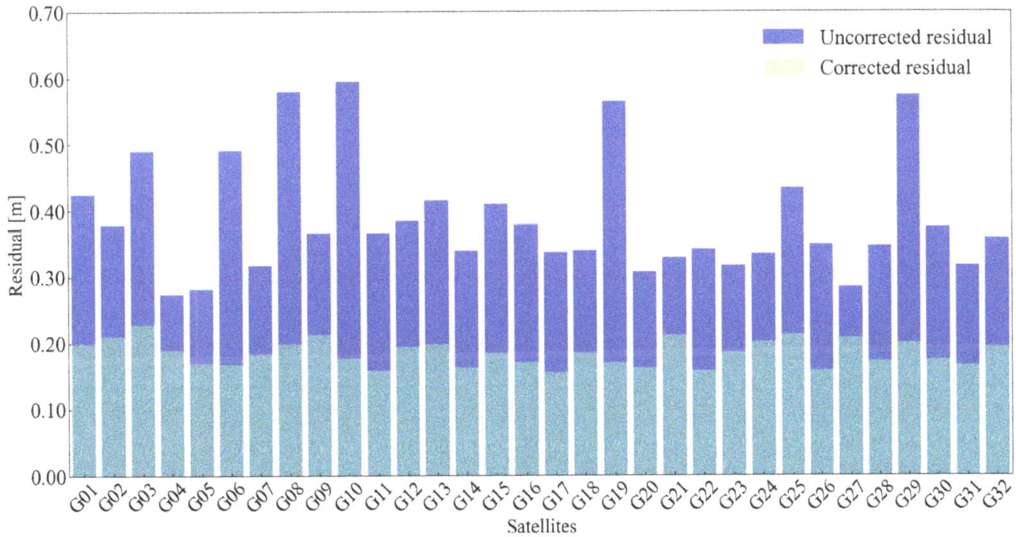

Figure 5. DOY 159-165 pseudorange residuals before and after enhancement of GPS satellites.

To demonstrate that the improved IGG III robust estimation method proposed in this study does not introduce substantial computational overhead for SBAS users, Figure 6 presents a comparison of the computation delays of SBAS-enhanced positioning with and without the improved algorithm. The results indicate that the average delays are 28.3 ms and 32.69 ms, respectively. The overall difference is minimal, corresponding to an average increase of 4.4 ms. Such an increment is negligible relative to current hardware capabilities and therefore does not compromise real-time applicability. Moreover, when deployed on high-performance servers, the delay difference is expected to be further reduced.

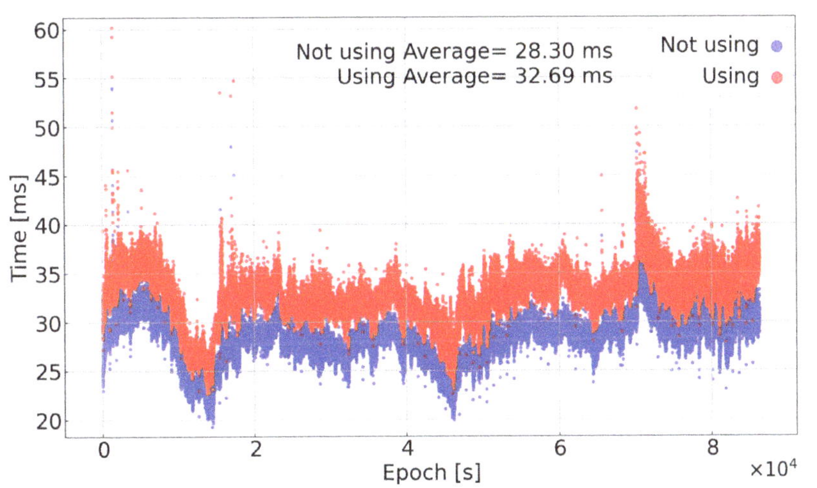

Figure 6. The computational delay of not using the improved algorithm and using the improved algorithm.

The vehicle dataset in the urban environment is collected using a passenger car driving through city streets. The data acquisition lasts for approximately one hour, with a sampling interval of 1 s. The collection environment and trajectory are illustrated in Figure 7. The

UAV maneuvering flight dataset is obtained from a flight conducted in a relatively open environment, lasting around 20 min, also with a sampling rate of 1 s. It is worth noting that only GPS data are used in this experiment, with no additional onboard sensors. Therefore, the data collection environment is relatively unobstructed and does not include areas with severe signal occlusion. The experimental equipment in the two scenarios is shown in Figure 8.

Figure 7. Scenario and real tracks.

Figure 8. Vehicle-mounted and UAV collection equipment.

A comparison of the positioning strategies and solutions for dual-frequency SBAS user is summarized in Table 2. Solution 1 represents enhanced positioning without the application of robust estimation, while Solution 2 incorporates the improved robust algorithm into the positioning process. In both cases, the PL is calculated using the observation weights derived from their respective enhanced positioning strategies.

Table 2. Solution and enhanced positioning strategy.

Item	Solution 1	Solution 2
Signal frequency	GPS L1&L2	GPS L1&L2
Orbital correction	Broadcast ephemeris + DFMC SBAS	Broadcast ephemeris + DFMC SBAS
Clock correction	Broadcast ephemeris + DFMC SBAS	Broadcast ephemeris + DFMC SBAS
Ionospheric correction	Dual-frequency combination	Dual-frequency combination
Tropospheric correction	UNB3m	UNB3m
Elevation cut-off	10°	10°
Robust	No	Yes

In order to show that the experimental data quality in the three scenarios meets the premise of poor data quality, the GPS (L1 and L2) observation data in the three scenarios are analyzed in terms of Signal-to-Noise Ratio (SNR), multipath error, and Position Dilution of Precision (PDOP) value. This study uses a box plot to count different values. The upper and lower boundaries of the box represent the first quartile (Q1) and the third quartile (Q3), respectively. A line in the box represents the median (Q2) of the data. The values beyond the whiskers are considered outliers, which are values other than 2.7 times the median value of the error. The data quality results are illustrated in Figure 9. Experimental data are affected to varying degrees by environmental factors, which cause outliers to appear in different indicators. In the static scenario, CSH2, KAS2, and FUZ2 stations show large outliers in PDOP values and multipath errors. Although there are few outliers in the SNR (L2), the dispersion is relatively high. In the on-board scenario of urban environments, there are large outliers in the SNR (L1), PDOP value, and multipath error, among which the outliers in the SNR are the most significant. In the drone maneuvering flight scenario, outliers appear in the SNR, PDOP value, and multipath error, and the range is very large in the SNR (L1). In addition, this study also counts the average values of each indicator under different scenarios in Table 3. It can be seen that except for some outliers, the overall quality of observation data in each scenario can meet the positioning requirements, and the overall quality of observation data in the drone maneuvering flight scenario is the worst. In summary, the data quality in the three scenarios is poor, and many outliers appear, which eventually have a great impact on the positioning results and integrity. This can be used as experimental data in this paper to effectively verify the effectiveness of the algorithm.

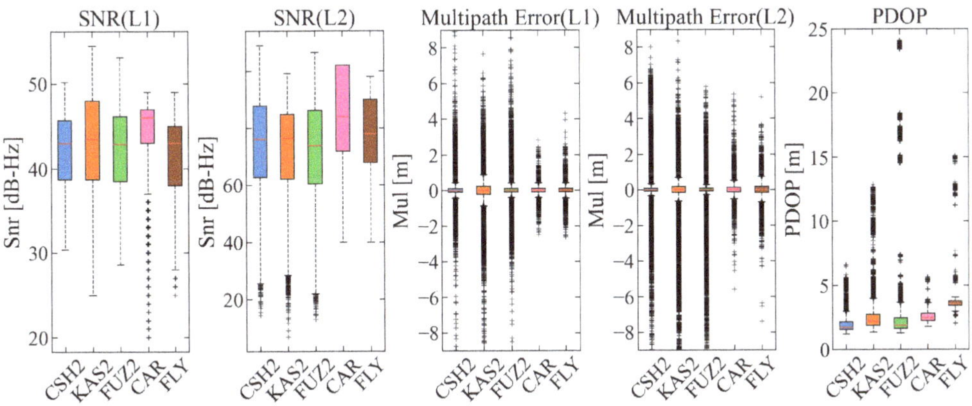

Figure 9. GPS (L1, L2) data quality analysis chart in different scenarios.

Table 3. Average of data quality analysis indicators in different scenarios.

Station ID	SNR (L1)/dB-Hz	SNR (L2)/dB-Hz	Multipath Error (L1)/m	Multipath Error (L2)/m	PDOP/m
CSH2	42.16	37.25	0.17	0.14	1.98
KAS2	43.29	36.34	0.35	0.25	2.44
FUZ2	42.42	36.47	0.16	0.14	2.2
CAR	44.15	41.00	0.13	0.17	2.51
FLY	41.59	38.83	0.17	0.31	4.13

3.2. Static Scenario

The three stations used two methods of solution 1 and solution 2 for positioning and protection level calculation, and compared with the APV-I standard (HAL is 40 m, VAL is 50 m), the percentage of availability improvement and the percentage of integrity risk reduction in solution 2 is DOY 159-165 compared with solution 1, as shown in Figure 10. The CSH2, KAS2 and FUZ2 stations used the solution 2 method to reduce the percentage of horizontal integrity risk by 6.5, 11.15 and 0.66, respectively, and the percentage of horizontal availability by 6.5, 11.12 and 0.66, respectively, and the percentage of vertical integrity risk by 5.18, 2.45 and 0.28, respectively, and the percentage of vertical availability by 5.08, 2.09 and 0.18, respectively. It shows that in static scenarios, adding an improved robust algorithm can effectively reduce the risk of integrity and improve availability.

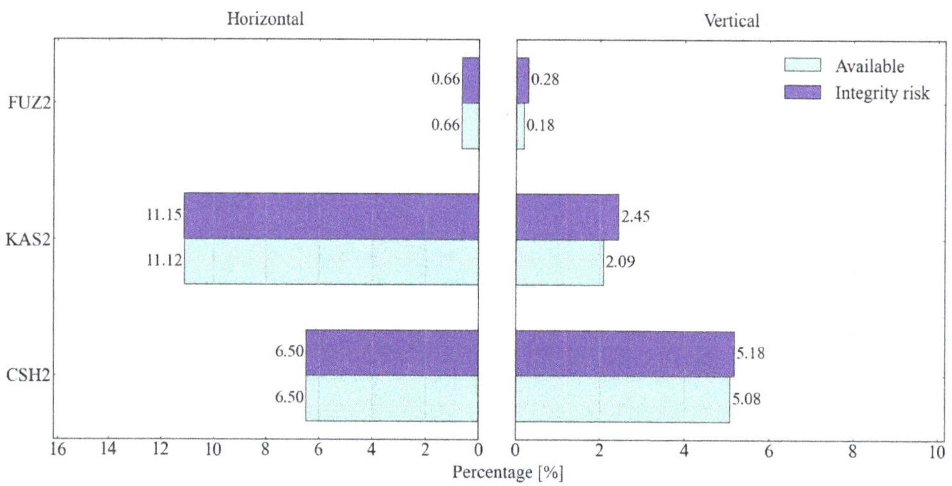

Figure 10. Percentage of availability improvement and integrity risk reduction for three stations using improved robust algorithm.

Taking the annual accumulation date of 159 as an example, we will analyze the performance of solution 2 in three aspects: integrity, positioning accuracy and protection level. In this study, Stanford diagrams were used to calculate its integrity performance according to the APV-I standard, as shown in Figures 11–13. The horizontal axis is the positioning error, the vertical axis is the protection level, and the red line is the alarm limit. The figure is divided into 5 sub-graphs according to the size relationship between them, representing different navigation states (Table 1). The area in the upper left corner of the figure indicates that the system is unavailable when the position error exceeds the protection level. The area in the lower left corner indicates that the positioning information provided by the system is available within the protection level. The orange area in the lower right corner indicates that the position error is below the protection level, but the positioning information provided by the system may mislead the user. The red area in the lower right corner indicates that the position error exceeds the protection level. The positioning information provided by the system will not only mislead the user, but may also pose danger.

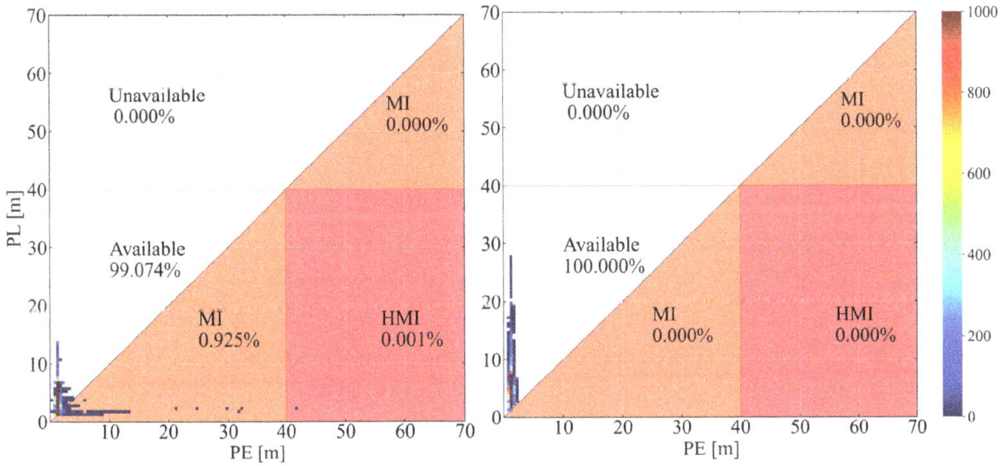

(**a**) Horizontal

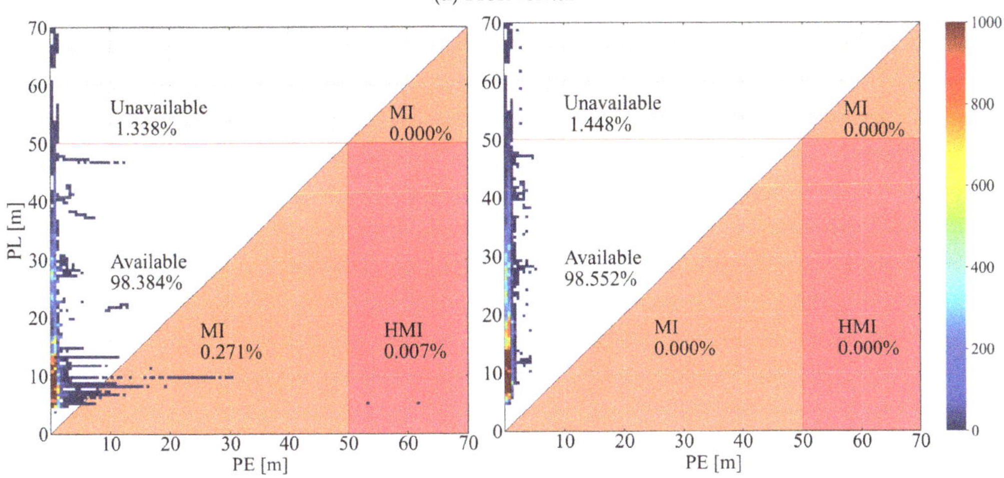

(**b**) Vertical

Figure 11. Stanford diagram of CSH2 station pre-robust (**left**) and post-robust (**right**).

Solution 1: Although the three stations use SBAS correction information, the positioning error (PE) still large, and there are still occurring MI and HMI. Solution 2: The PE of the three stations is significantly improved, reducing the percentage of MI and HMI, and the availability rate is also improved. Details are outlined in Table 4. Specifically, in the horizontal direction, MI rates dropped from 0.931%, 0.863%, and 10.441% to 0%, 0.002%, and 3.72%—decreasing by approximately 0.93%, 0.85%, and 6.72%, respectively—while HMI was eliminated entirely. Availability correspondingly rose from 99.07%, 98.82%, and 89.49% to 100%, 99.66%, and 96.12%, an improvement of 0.93%, 0.84%, and 6.62%. Vertically, MI and HMI were completely eliminated, and availability increased modestly from 98.38% to 98.55%, from 98.13% to 98.27%, and from 95.91% to 96.06%, gains of 0.17%, 0.14%, and

0.15%. The results show that the improved robust algorithm can further reduce the missing detections rate and integrity risk rate while weakening the impact of the coarse errors, so as to achieve effective alarms, and to improve the availability rate.

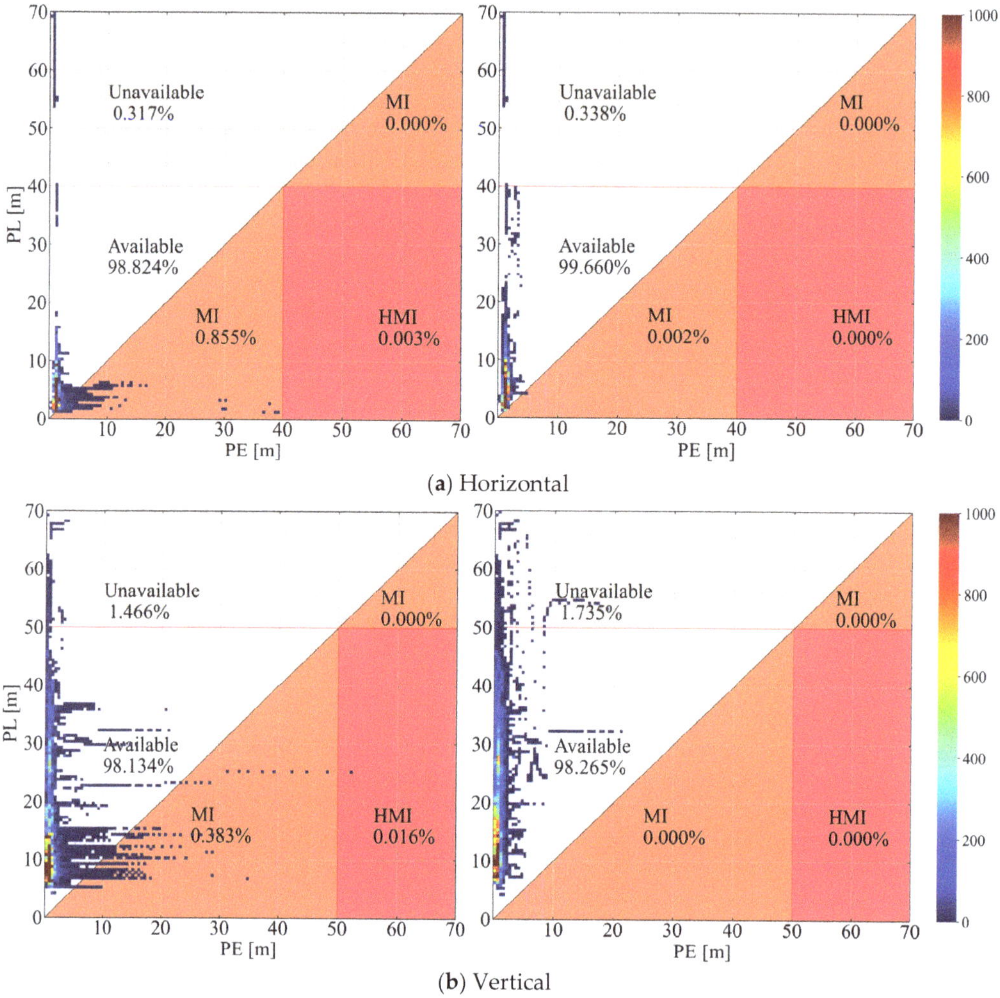

Figure 12. Stanford diagram of FUZ2 station pre-robust (**left**) and post-robust (**right**).

Table 4. Statistics of integrity monitoring performance of three stations.

ID	Solution	Horizontal			Vertical		
		Availability	**HMI**	**MI**	**Availability**	**HMI**	**MI**
CSH2	Solution 1	99.07%	0.001%	0.93%	98.38%	0.007%	0.27%
	Solution 2	100%	0%	0%	98.55%	0%	0%
FUZ2	Solution 1	98.82%	0.003%	0.86%	98.13%	0.02%	0.38%
	Solution 2	99.66%	0%	0.002%	98.27%	0%	0%
KAS2	Solution 1	89.49%	0.001%	10.44%	95.91%	0.002%	0.48%
	Solution 2	96.12%	0%	3.72%	96.06%	0%	0%

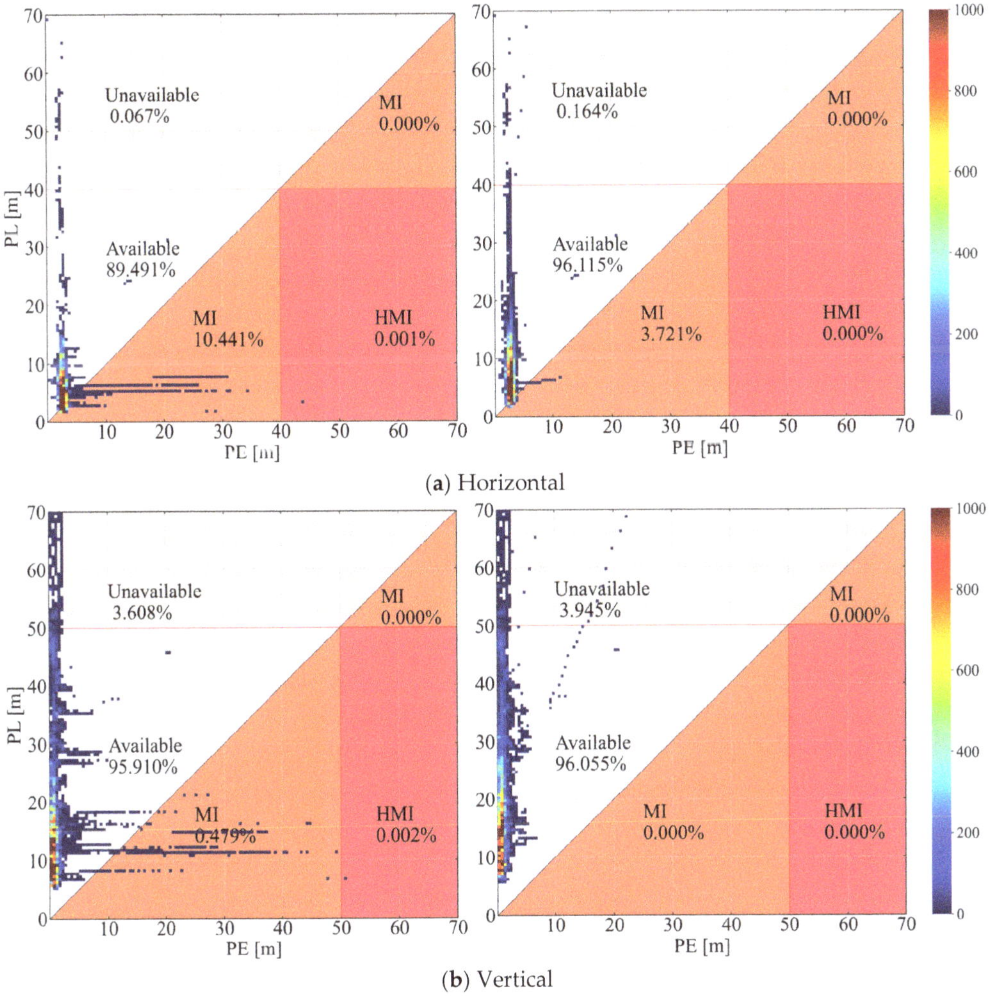

Figure 13. Stanford diagram of KAS2 station pre-robust (**left**) and post-robust (**right**).

Through the presence of outliers in the data quality indicators, it can be observed that under the influence of the surrounding environment, the quality of certain observations is significantly degraded, leading to substantial errors. These inaccuracies are likely attributable to factors such as receiver quality, short-term adverse atmospheric conditions, signal reflections from buildings in urban settings, and temporary unfavorable satellite geometric configurations. In order to more clearly analyze the improvement effect of the improved robust algorithm on the positioning results, we also compared and analyzed with the M-estimators method, and the PE sequence diagrams of the horizontal and vertical directions of the three stations are drawn and their internal and external consistency accuracy are counted, as shown in Figures 14–16 and Table 5. The positioning results of Solution 2 are smoother than those of Solution 1, which effectively weakens the influence of coarse errors. As shown in Table 5, the Standard Deviation (STD) and Root Mean Square Error (RMSE) of horizontal and vertical directions of the three stations in Solution 2 are reduced compared with that of Solution 1. Among them, the horizontal and vertical internal consistency accuracy of the KAS2 station increased from 1.22 to 0.35 and from 1.87 to 0.62, respectively, which increased by 71.31% and 66.84%, respectively; the external consistency accuracy increased from 2.73 to 2.4, and from 1.99 to 0.82, respectively, which increased by 12.09% and 58.79%, respectively. The horizontal and vertical internal consistency accuracy of the FUZ2 station increased from 0.52 to 0.2 and from 1.28 to 0.62, respectively, which increased by 61.54% and 51.56%, respectively; the external consistency accuracy increased from 1.13 to 1.01 and from 1.48 to 0.82, which increased by 10.62% and 44.59%, respectively. The horizontal and vertical internal consistency accuracy of the CSH2 stations increased from 0.42 to 0.14 and from 0.95 to 0.3, respectively, which increased by 66.67% and 68.42%, respectively; the external consistency accuracy increased from 1.08 to 0.99 and from 1.04 to 0.43, which increased by 8.33% and 58.65%, respectively. The results show that when the data quality is poor and the PE is still large even with the addition of SBAS correction information, adding the improved robust algorithm to the SBAS user positioning can effectively weaken the effect of coarse errors and improve the positioning accuracy. The M-estimators method can also effectively reduce the impact of coarse errors, but compared with the improved IGG III robust estimation algorithm, the improvement effect is weaker. After the robust estimation, a very small number of elements still have large errors, and this study considers that there are two reasons: firstly, the number of satellites participating in the positioning of this element is relatively small, and no robust treatment; secondly, this element has too many satellites involved in robust estimation, which makes the robust estimation ineffective.

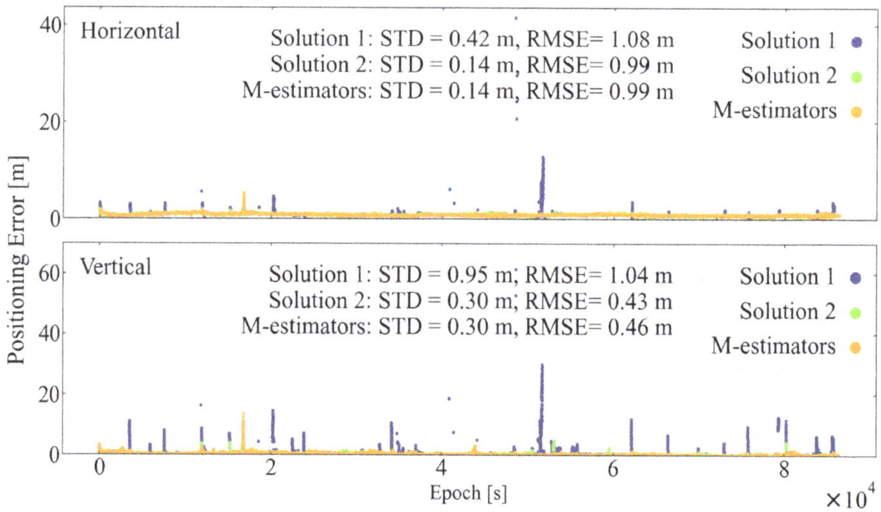

Figure 14. Comparison of positioning errors of CSH2.

Table 5. Statistics of positioning errors of three stations.

Station ID	Solution 1	Horizontal		Vertical	
		STD/m	RMSE/m	STD/m	RMSE/m
KAS2	Solution 1	1.22	2.73	1.87	1.99
	Solution 2	0.35	2.4	0.62	0.82
	M-estimators	0.85	2.64	1.43	1.61
FUZ2	Solution 1	0.52	1.13	1.28	1.48
	Solution 2	0.2	1.01	0.62	0.82
	M-estimators	0.29	1.04	0.89	1.07

Table 5. *Cont.*

Station ID	Solution 1	Horizontal		Vertical	
		STD/m	RMSE/m	STD/m	RMSE/m
CSH2	Solution 1	0.42	1.08	0.95	1.04
	Solution 2	0.14	0.99	0.3	0.43
	M-estimators	0.14	0.99	0.3	0.46

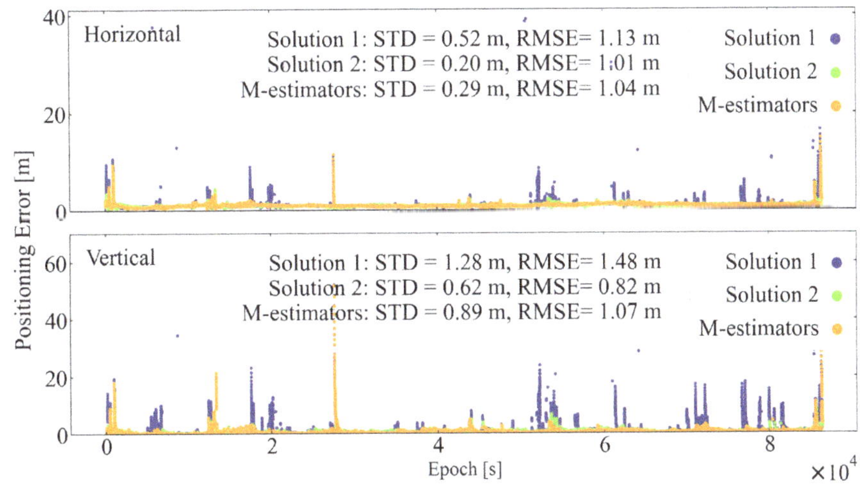

Figure 15. Comparison of positioning errors of FUZ2.

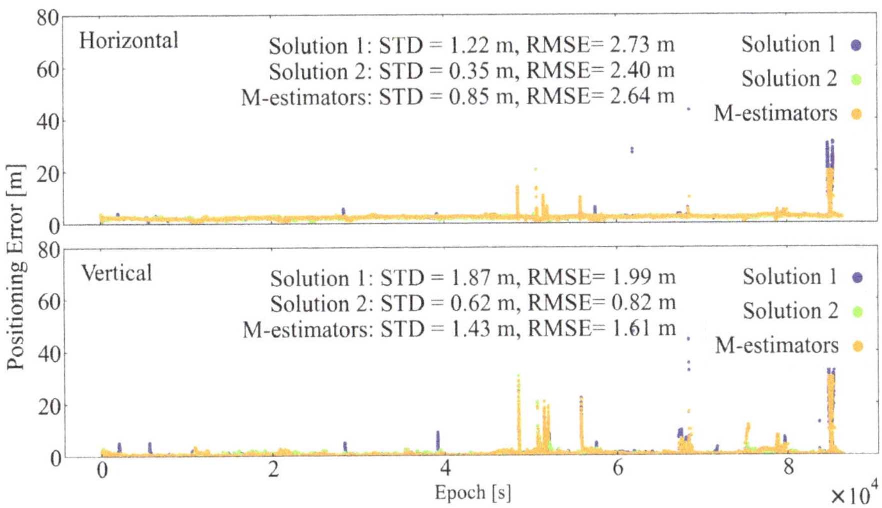

Figure 16. Comparison of positioning errors of KAS2.

To verify that the proposed improved robust algorithm offers better control over weight adjustment in SBAS user positioning compared to the traditional IGG III method, and to prevent excessive conservatism in protection level (PL) calculation caused by robustness strategies, a comparative analysis was also conducted with the PL results derived from the M-estimators method. The protection levels calculated using the traditional IGG III method, the improved robust algorithm, and the M-estimators approach are presented in Figures 17–19. The three stations use the Horizontal Protection Level (HPL) and Vertical Protection Level (VPL) of the improved robust algorithm in part of the epoch is smaller than the traditional IGG III robust algorithm, which indicates that the improved robust algorithm will not be too 'loose' on the PL solving under the premise of meeting the integrity and positioning accuracy. The PL calculation under the M-estimators method is the most conservative, further explaining the superiority of the improved algorithm. In other epochs, the PL difference between the traditional IGG III robust algorithm and the improved IGG III robust algorithm is not obvious, because when the enhancement information is better, the variance of the corrections is smaller compared with the variance of other errors, and at this time, the variance of the corrections has less influence in the deweighting, which makes the PL calculated by the two algorithms comparable.

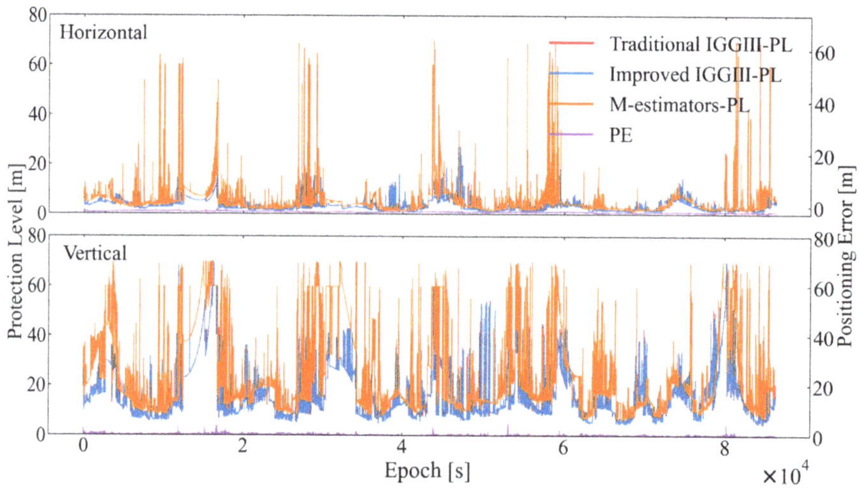

Figure 17. Comparison of protection level of CSH2 (Red is the protection level calculated by traditional IGG III robust algorithm, blue is the protection level calculated by the improved robust algorithm in such pictures, orange is the protection level calculated by the M-estimators method in such pictures, and purple is the positioning error).

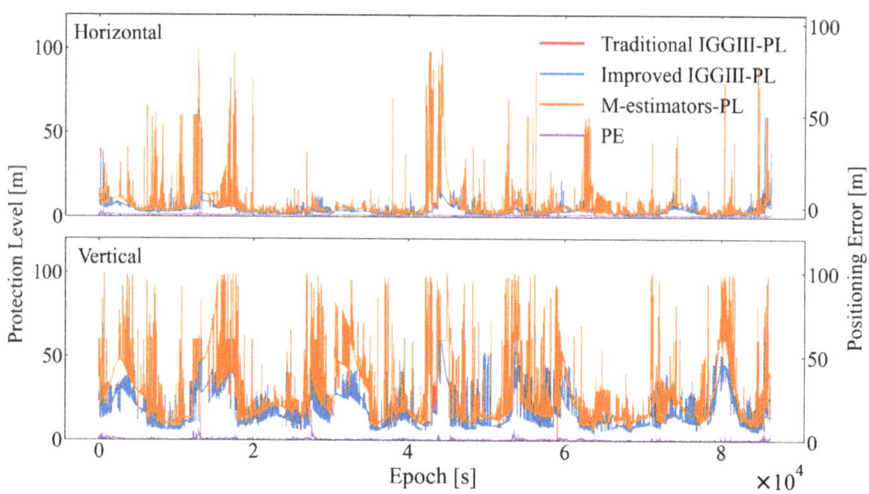

Figure 18. Comparison of protection level of FUZ2.

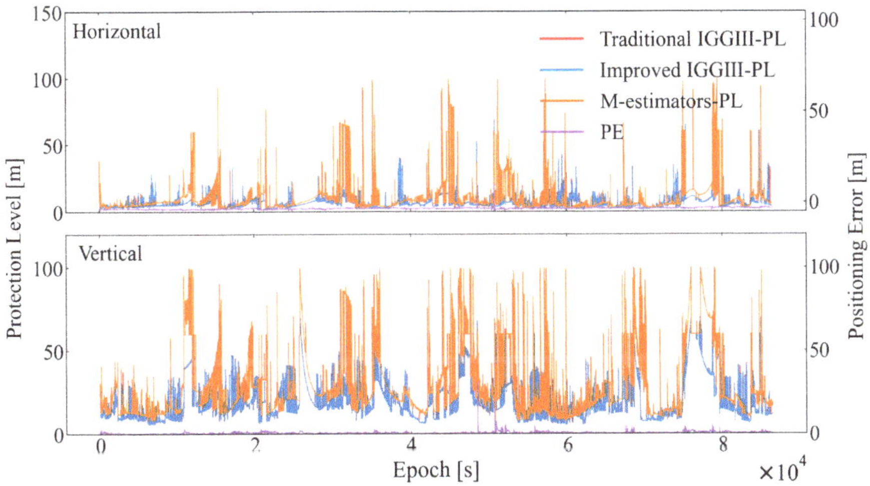

Figure 19. Comparison of protection level of KAS2.

3.3. Urban Environment Vehicle Scenario

The data in the urban environment are used to perform positioning and protection-level solutions using two methods: Solution 1 and Solution 2, respectively. The RTK result is used as the true value, and the APV-I standard is selected to count its integrity performance. The results are shown in Figure 20. In Solution 1, in the urban environment, which is susceptible to environmental interference, using SBAS correction information, the PE still exists in a large situation, and the MI and HMI occurs; in Solution 2, the horizontal and vertical PE is improved, the percentage of MI and HMI are reduced, and the availability rate is improved. The percentage of MI in the horizontal direction decreased from 0.6% to 0.08%, a decrease of 0.52%, and the availability increased from 99.39% to 99.91%, an increase of 0.52%. The vertical percentages of MI and HMI both decreased to 0%, and the availability increased from 98.73% to 98.79%, an increase of 0.06%. The results show that the improved robust algorithm can reduce the missing detections rate and integrity risk by weakening the effect of the coarse errors in the SBAS-enhanced urban environment vehicle scenario, achieve effective alarms, and improve the availability rate.

In order to more clearly analyze the improvement effect of the improved robust algorithm on the positioning results, the PE sequence and three-dimensional motion trajectory in the horizontal and vertical directions, which are illustrated in Figure 21. Compared with Solution 1, the positioning results of Solution 2 are more stable, eliminating larger error jumps. In addition, the STD and RMSE in Solution 2 are both lower than Solution 1. Significant improvements were observed across all accuracy metrics. The horizontal internal consistency accuracy improved by 23.28%, increasing from 0.73 to 0.56. The horizontal external consistency accuracy showed a 5.4% enhancement, rising from 1.48 to 1.4. In the vertical direction, the internal consistency accuracy increased by 58.77%, from 2.28 to 0.94, while the external consistency accuracy improved by 49.8%, from 2.63 to 1.32. The M-estimators method effectively improves positioning accuracy, but the improvement effect is not as good as Solution 2. Figure 22 further demonstrates the effect of the improved robust algorithm in the motion trajectory, with a smoother elevation direction. The results show that the use of improved robust algorithms in the positioning of the SBAS-enhanced urban environment vehicle scenario can effectively weaken the influence of coarse errors and improve positioning accuracy.

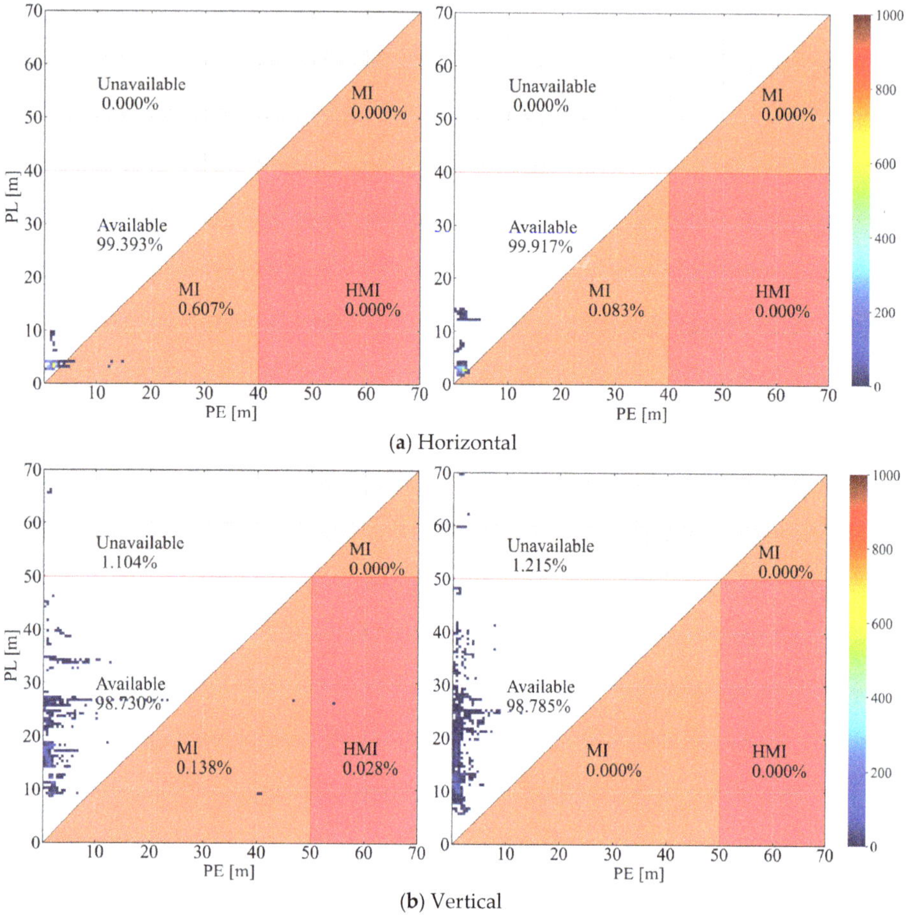

Figure 20. Stanford diagram of CAR pre-robust (**left**) and post-robust (**right**).

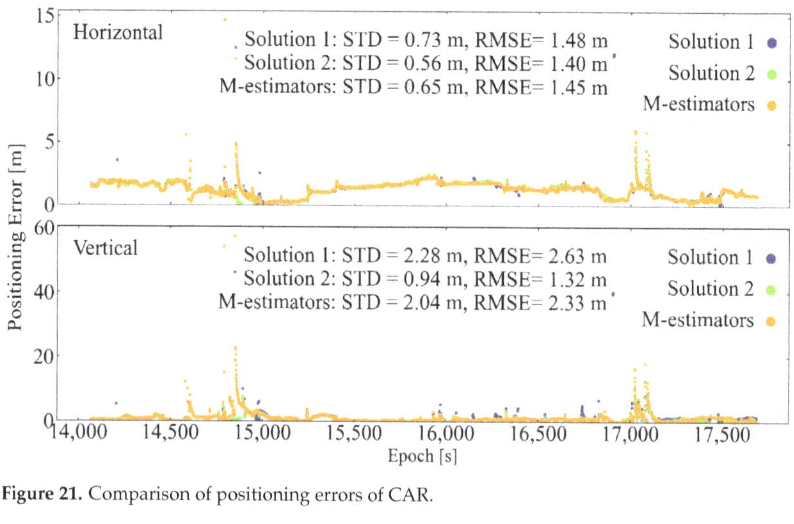

Figure 21. Comparison of positioning errors of CAR.

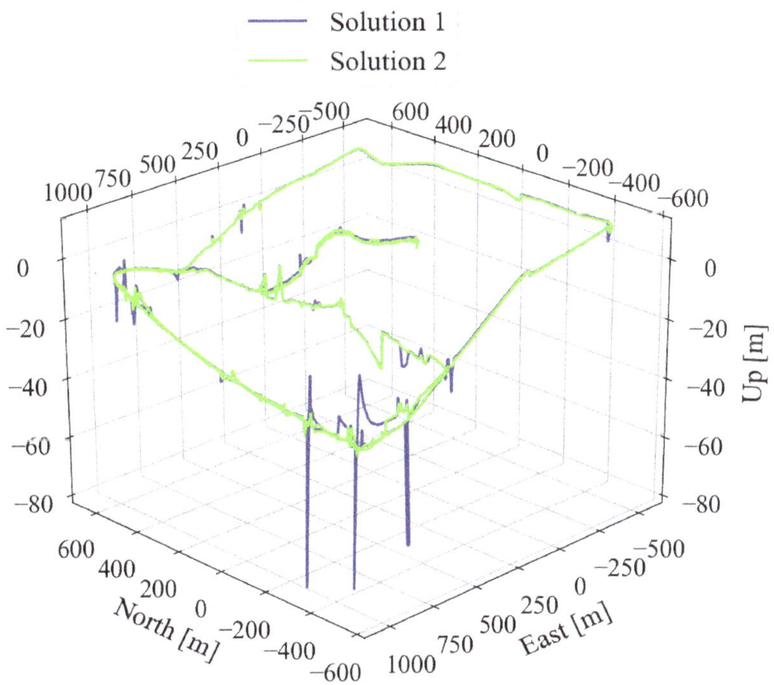

Figure 22. Three-dimensional trajectory of CAR.

The PL results calculated by the positioning weights after robust estimation using the traditional IGG III robust algorithm, the improved robust algorithm and the M-estimators method are shown in Figure 23. In the SBAS-enhanced Urban environment vehicle scenario, compared with traditional IGG III robust algorithm, HPL and VPL using improved robust algorithms have been reduced in most epochs. The result show that the improved robust algorithm is better in the calculation of PL under the premise of integrity and positioning accuracy. The M-estimators method can envelop positioning errors, but the results are too conservative and will reduce usability.

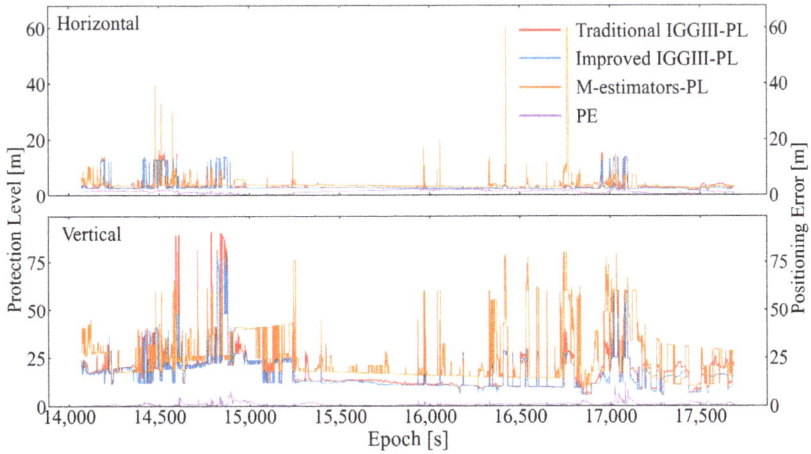

Figure 23. Comparison of protection level of CAR.

3.4. UAV Maneuvering Flight Scenario

Using the data in the UAV maneuvering flight scenario to perform positioning and PL solutions using the two methods of Solution 1 and Solution 2, respectively, the RTK results are taken as the true values and the APV-I standard are chosen to count the integrity, and the result are shown in Figure 24. In Solution 1, using SBAS correction information, the PE is still large and there are MI and HMI. In Solution 2, the horizontal and vertical PE are improved, the percentage of MI and HMI are reduced, and the availability rate is improved. The horizontal percentage of MI is reduced by 2.76%, and the availability rate is improved by 2.76%; the vertical percentages of MI are reduced by 1.71%, and the availability rate is improved by 1.65%. The results show that the improved robust algorithm can reduce the missing detections rate and integrity risk by weakening the effect of the coarse errors in the SBAS-enhanced UAV maneuvering flight scenario, achieve effective alarms, and improve the availability rate.

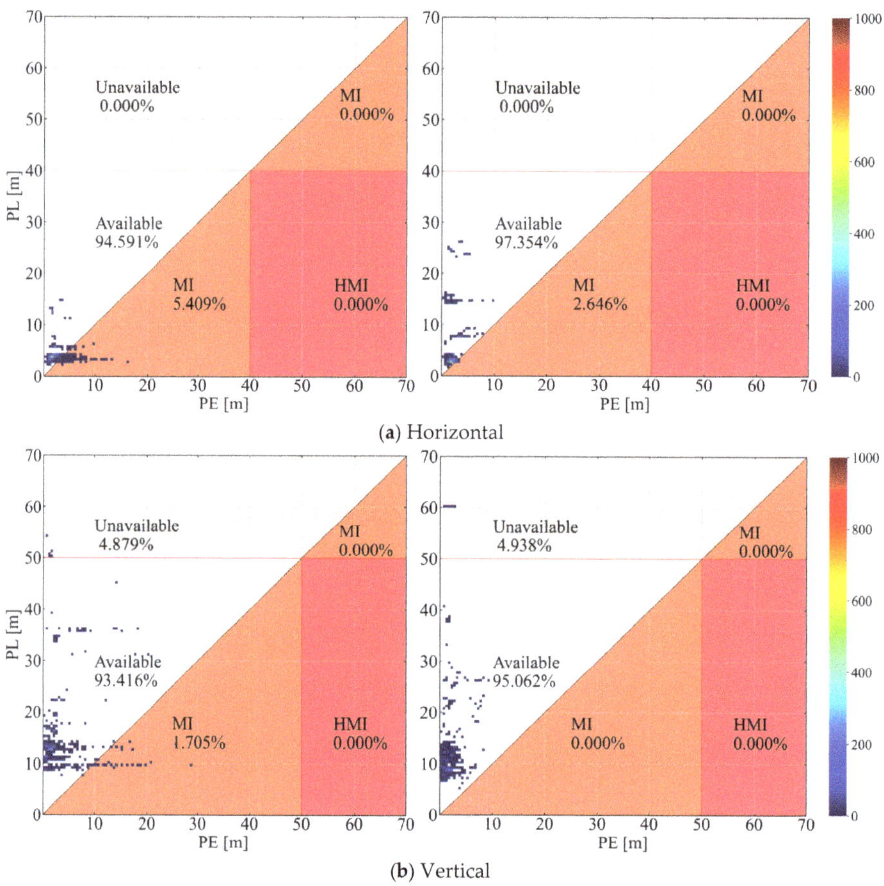

Figure 24. Stanford diagram of FLY pre-robust (**left**) and post-robust (**right**).

The positioning error sequences in the horizontal and vertical directions are illustrated in Figure 25, respectively, showing in detail the improved effect of the improved robust algorithm and M-estimators methods on the positioning results. Compared with Solution 1, the positioning results of Solution 2 are smoother and the larger error jump points are eliminated; the positioning accuracy of the M-estimators method has been improved, but the improvement effect is not as good as that of Solution 2. In addition, the STD and RMSE of Solution 2 in the horizontal and vertical directions are reduced compared with that of Solution 1, of which the internal and external consistency accuracies in the horizontal direction can be improved by 47.1% and 16.45%, the internal and external consistency accuracies in the vertical direction can be improved by 50.38% and 38.15%, respectively. The results show that in the SBAS-enhanced UAV maneuvering flight scenario, the use of the improved robust algorithm in the user-end positioning can effectively weaken the influence of the coarse errors and improve the positioning accuracy.

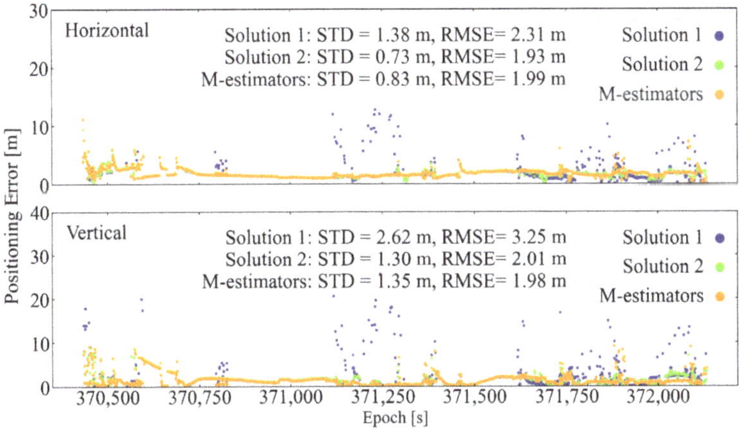

Figure 25. Comparison of positioning errors of FLY.

The results of the PL calculations using the traditional IGG III robust algorithm, the improved robust algorithm and M-estimators method in the SBAS-enhanced UAV maneuvering flight scenario are shown in Figure 26. In SBAS-enhanced UAV maneuvering flight scenario, compared with traditional IGG III robust algorithm, HPL and VPL using improved robust algorithms are reduced most of epochs. The result show that, while ensuring integrity and positioning accuracy, the improved robust algorithm is better than the traditional IGG III robust algorithm when calculating the PL. In addition, the M-estimators method is too conservative in PL calculations, which will affect the availability of positioning users, further explaining the advantages of improved robust algorithms in PL calculations.

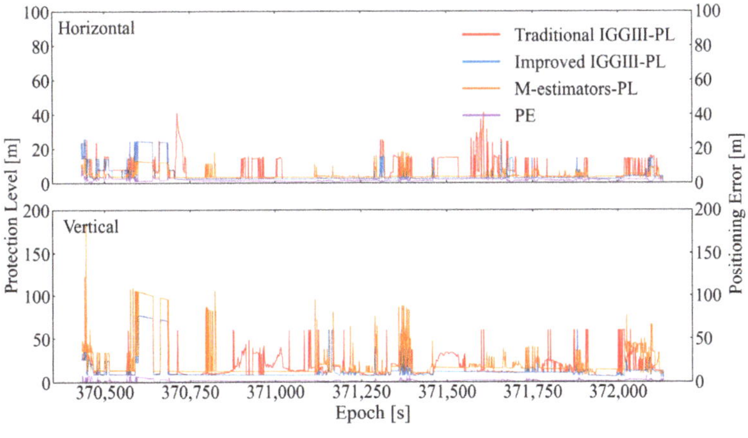

Figure 26. Comparison of protection level of FLY.

3.5. Statistical Test

This study conducted a statistical analysis (t-test) on the experimental results of the above three scenarios. The differences between the positioning errors and protection level results of Solution 1 and Solution 2 are shown in Table 6. The t-test results show that, except for the horizontal positioning error results in the urban scenario, there are significant differences in the experimental results of all other scenarios, and the degree of effect is above moderate. The differences in the results of horizontal positioning errors in urban scenarios did not reach statistical significance, but a moderate effect was observed, indicating that there might be differences of practical significance.

Table 6. t-test of the experimental results in three scenarios.

Item	T	p	D
CAR-vertical-errors	10.509	<0.001	2.151
CAR-vertical-PL	18.525	<0.001	5.916
CAR-horizontal-errors	0.477	0.633	0.542
CAR-horizontal-PL	21.942	<0.001	0.551
FLY-vertical-errors	6.705	<0.001	2.466
FLY-vertical-PL	26.699	<0.001	0.931
FLY-horizontal-errors	2.084	<0.037	1.244
FLY-horizontal-PL	16.346	<0.001	2.083
FUZ2-vertical-errors	51.247	<0.001	1.145
FUZ2-vertical-PL	22.317	<0.001	1.920
FUZ2-horizontal-errors	5.282	<0.001	0.498
FUZ2-horizontal-PL	−1.860	0.063	0.689
KAS2-vertical-errors	22.743	<0.001	1.811
KAS2-vertical-PL	5.642	<0.001	1.552
KAS2-horizontal-errors	16.181	<0.001	1.18
KAS2-horizontal-PL	11.656	<0.001	1.087
CSH2-vertical-errors	35.629	<0.001	0.933
CSH2-vertical-PL	17.022	<0.001	1.56
CSH2-horizontal-errors	11.257	<0.001	0.408
CSH2-horizontal-PL	4.91	<0.001	0.454

($p > 0.05$: Not Significant; $p \leq 0.05$: Significant; $p \leq 0.01$: Highly Significant; $p \leq 0.001$: Extremely Significant).

4. Discussion

This study addresses the issues of degraded positioning performance, increased integrity risk, and higher rates of missed detection events under poor data quality conditions in SBAS user-end positioning by proposing an improved IGG III robust estimation positioning algorithm. Through validation using real-world data from three scenarios—static, urban vehicular, and UAV maneuvering flight—several noteworthy observations were made.

In terms of system performance, the improved robust algorithm significantly enhances the availability of SBAS and effectively reduces integrity risks and missed detection rates in both horizontal and vertical directions. This improvement markedly strengthens system stability in complex environments. The influence of coarse errors in observation data on positioning accuracy is also effectively suppressed by the improved robust estimation, leading to substantial improvements in both horizontal and vertical positioning accuracy. Particularly noteworthy is the outstanding performance of the improved algorithm in protection level computation, avoiding the common issue of overbounding seen in traditional methods and resulting in more reasonable and reliable protection level estimates.

Comparative analysis with the M-estimators method further demonstrates the unique advantages of the proposed improved IGG III robust estimation for SBAS user-end applications. It not only effectively addresses the limitations of traditional methods in suppressing coarse errors but also avoids potential overconservatism in protection level computation.

From the perspective of current research trends, previous improvements in SBAS performance have predominantly focused on the system end, such as data preprocessing, time synchronization, and model optimization. While these efforts have enhanced overall system performance, there has been relatively little attention paid to user-end performance under challenging conditions, especially when performance remains unsatisfactory even after enhancements. Although various methods exist to suppress coarse errors and improve positioning accuracy, few are well-suited for SBAS user-end applications. This study focuses on this practical issue and offers a new approach to addressing positioning problems at the user end in degraded environments.

It should be noted that this study has certain limitations regarding datasets. Due to constraints in data acquisition, the current research is validated only based on data from China. This geographical limitation may affect the assessment of the algorithm's applicability in other regions, and future work will expand the study area to further verify the method's universality.

5. Conclusions

This study demonstrates that the enhanced IGG III robust estimation-based SBAS user-end positioning algorithm effectively improves positioning performance under challenging environmental conditions. The proposed method not only preserves strong robustness but also leads to notable enhancements in system availability, positioning accuracy, and the reliability of protection level calculations. It offers a practical and efficient solution for mitigating positioning challenges in degraded operational environments. Future work will focus on extending the validation to broader geographical regions and more diverse scenarios to further assess the algorithm's adaptability and robustness.

Author Contributions: Conceptualization, L.W. and J.S.; methodology, L.W. and J.S.; software, J.S.; validation, L.W., J.S. and B.C.; investigation, W.Y. and Z.W.; resources, L.W. and B.C.; writing—original draft preparation, J.S. and Y.W.; writing—review and editing, L.W., J.S. and B.C.; funding acquisition, L.W. All authors have read and agreed to the published version of the manuscript.

Funding: This work was supported by the Programs of the Fundamental Research Funds for the Central Universities (CHD 300102265503, 300102261301, 300102264105), National Natural Science Foundation of China (42374025), Key Research and Development Program of Shaanxi (2021LLRH-06),and Shaanxi Post-doctoral Scientific Research Funding (2024BSHSDZZ217).

Data Availability Statement: As this study involves privacy, participants in this study did not agree to publicly share their data, and support data was not available.

Conflicts of Interest: The authors declare no conflict of interest.

References

1. Wang, L.; She, J.; Zhang, X.; Wang, Z.; Shao, B. A review on the development of satellite-based augmentation technology. *Nav. Posit. Time* **2025**, *12*, 35–49.
2. Jing, C.; Bertolesi, E.; Huang, G.; Li, X.; Zhang, Q.; Zhai, W.; Liu, G.; Li, H. Predicting accelerometer baseline correction and non-divergent deformation velocity based on convolutional neural network (CNN) during GNSS Downgrade. *IEEE Sens. J.* **2025**, *25*, 11982–11994. [CrossRef]
3. Krasuski, K.; Wierzbicki, D. Application the SBAS/EGNOS Corrections in UAV Positioning. *Energies* **2021**, *14*, 739. [CrossRef]
4. Krasuski, K.; Ciećko, A.; Bakuła, M.; Grunwald, G. Accuracy Examination of the SDCM Augmentation System in Aerial Navigation. *Energies* **2022**, *15*, 7776. [CrossRef]
5. Krasuski, K.; Bakuła, M.; Gołda, P.; Grunwald, G.; Gołda, P.; Mrozik, M.; Kozuba, J. Algorithms to improve unmanned aerial vehicle positioning accuracy using European Geostationary Navigation Overlay Service and System for Differential Corrections and Monitoring ionospheric corrections. *Adv. Sci. Technol. Res. J.* **2025**, *19*, 284–300. [CrossRef] [PubMed]

6. Yoon, H.; Seok, H.; Lim, C.; Park, B. An Online SBAS Service to Improve Drone Navigation Performance in High-Elevation Masked Areas. *Sensors* **2020**, *20*, 3047. [CrossRef] [PubMed]
7. Li, L.; Jia, C.; Zhao, L.; Cheng, J.; Liu, J.; Ding, J. Real-Time Single Frequency Precise Point Positioning Using SBAS Corrections. *Sensors* **2016**, *16*, 1261. [CrossRef]
8. Park, K.W.; Cho, M.G.; Park, C. An efficient positioning method for multi-GNSS with multi-SBAS. *J. Position. Navig. Timing* **2018**, *7*, 245–253.
9. Li, K.; Wang, X.; Zheng, S.; Chen, S.; Gao, M.; Yang, P.; Jin, X.; Yang, J. IFree filter-based user positioning for satellite-based augmentation system. *Int. J. Aeronaut. Space Sci.* **2024**, *25*, 687–697. [CrossRef]
10. Yang, Y.; Yao, Z.; Mao, Y.; Xu, T.; Wang, D. Resilient satellite-based PNT system design and key technologies. *Chinese Sci. Bull.* **2025**, *55*, 686–698. [CrossRef]
11. Zhang, Z.; Wang, H. Integrated BeiDou satellite-based augmentation system framework combining B2a and B1C services. *GPS Sol.* **2025**, *29*, 74. [CrossRef]
12. Wang, E.; Wang, Y.; Yu, T.; Zhang, S.; Liu, L.; Jia, S.; Lei, H.; Chen, Y. Research on the Fusion Positioning Algorithm of BDSBAS and MSAS Based on Fully Connected Neural Network. *J. Aeronaut. Astronaut. Aviat.* **2025**, *57*, 207–222.
13. Wang, K.; El-Mowafy, A.; Rizos, C.; Wang, J. SBAS DFMC service for road transport: Positioning and integrity monitoring with a new weighting model. *J. Geod.* **2021**, *95*, 29. [CrossRef]
14. Zheng, S.; Gao, M.; Huang, Z.; Jin, X.; Li, K. Satellite integrity monitoring for satellite-based augmentation system: An improved covariance-based method. *Satell. Navig.* **2022**, *3*, 9. [CrossRef]
15. Zheng, S.; Gao, M.; Li, K.; Jin, X.; Qin, J.; Xia, D. Potential performance of satellite-based augmentation system based on CMONOC in China. In Proceedings of the China Satellite Navigation Conference (CSNC2022), Beijing, China, 22–25 May 2022; pp. 231–243.
16. Wang, Q.; Li, R.; Li, M. Research on the impact of the decrease in observation stations on SBAS. In Proceedings of the China Satellite Navigation Conference (CSNC 2022), Beijing, China, 22–25 May 2022; pp. 96–105.
17. Shao, B.; Ding, Q.; Wu, X.B. Estimation method of SBAS dual-frequency range error integrity parameter. *Satell. Navig.* **2020**, *1*, 663–673. [CrossRef]
18. Li, X.; Qiao, Y.; Wang, T. Analysis on the integrity simulation of dual-frequency WAAS. In Proceedings of the 2017 IEEE 2nd Advanced Information Technology, Electronic and Automation Control Conference (IAEAC), Chongqing, China, 25–26 March 2017; pp. 2078–2082.
19. Zhang, Y.; Xiao, Z.; Li, P.; Tang, X.; Ou, G. Conservative Sensor Error Modeling Using a Modified Paired Overbound Method and its Application in Satellite-Based Augmentation Systems. *Sensors* **2019**, *19*, 2826. [CrossRef]
20. Xin, J.; Wang, D.; Li, K. An Integrity Monitoring Method for Navigation Satellites Based on Multi-Source Observation Links. *Remote Sens.* **2024**, *16*, 4574. [CrossRef]
21. Liu, C.; Liu, C.J.; Wang, S.; Xu, L.; Zhang, X. Two New Improvements of IGGIII Scheme. *Bull. Surv. Mapp.* **2016**, *4*, 54–57.
22. Yang, Y. *The Theory of Tolerance Estimation and Its Application*; Bayi Publishing House: Beijing, China, 1993; pp. 107–110.
23. She, H.; Huang, G.; Wang, L.; Lai, W.; Xie, S. Enhancing LEO orbit determination and prediction through constraints on atmospheric drag scale factor. *GPS Sol.* **2025**, *29*, 94. [CrossRef]
24. Chen, J.; Huang, Z.; Li, R. Computation of satellite clock-ephemeris augmentation parameters for dual-frequency multi-constellation satellite-based augmentation system. *J. Syst. Eng. Electron.* **2018**, *29*, 1111–1123.
25. Wang, Z.; Wang, L.; Xie, W.; Huang, G.; Yang, W.; Tian, Y. DFMC SBAS service performance analysis of multi-GNSS based on BDS-3 in different regions. *Meas. Sci. Technol.* **2024**, *35*, 116310. [CrossRef]
26. Krasuski, K.; Mrozik, M.; Wierzbicki, D.; Ćwiklak, J.; Kozuba, J.; Ciećko, A. Designation of the Quality of EGNOS+SDCM Satellite Positioning in the Approach to Landing Procedure. *Appl. Sci.* **2022**, *12*, 1335. [CrossRef]
27. Liu, Y.; Cao, Y.; Shao, B.; Tang, C.; Zhou, S.; Hu, X.; Yang, J.; Liu, J.; Li, P. Research on performance improvement method of BDSBAS multi-GNSS service with DFMC protocol. *Adv. Space Res.* **2023**, *72*, 2283–2296. [CrossRef]

remote sensing

MDPI

Article

Retrieval of Wave Parameters from GNSS Buoy Measurements Using Spectrum Analysis: A Case Study in the Huanghai Sea

Jin Wang [1,2], Xiaohang Chang [1], Rui Tu [1,*], Shiwei Yan [1], Shengli Wang [3] and Pengfei Zhang [1]

1 College of Geodesy and Geomatics, Shandong University of Science and Technology, Qingdao 266590, China; wangjin@sdust.edu.cn (J.W.); 202383020041@sdust.edu.cn (X.C.); 202283020009@sdust.edu.cn (S.Y.); zhangpengfei@ntsc.ac.cn (P.Z.)
2 Chinese Academy of Surveying and Mapping, Beijing 100830, China
3 College of Ocean Science and Engineering, Shandong University of Science and Technology, Qingdao 266590, China; shlwang@sdust.edu.cn
* Correspondence: turui@sdust.edu.cn

Abstract

Global Navigation Satellite System (GNSS) buoys are widely used to retrieve wave parameters such as significant wave heights (SWHs) and dominant wave periods. In addition to the statistical methods employed to estimate wave parameters, spectral-analysis-based approaches are also frequently utilized to analyze them. This study presents statistical and spectral methods for retrieving wave parameters at GNSS buoy positioning resolution in the Huanghai Sea area. To verify the method's effectiveness, the zero-crossing method and three spectral analysis techniques (periodogram, autocorrelation function, and autoregressive model methods) were used to estimate wave height and period for comparison. The vertical positioning resolution was decomposed into low-frequency ocean-tide level information and high-frequency wave height and period information with the Complete Ensemble Empirical Mode Decomposition (CEEMD) method and moving average filtering. The horizontal positioning results and velocity parameters were used to determine the wave direction using directional spectrum analysis. The results show that the three spectral methods yield consistent effective wave heights, with a maximum difference of 0.02 s in the wave period. Compared with the zero-crossing method results, the wave height and period obtained through spectral analysis differ by 0.05 m and 0.79 s, respectively, while the average wave height and period differ by 0.09 m and 0.08 s, respectively. The GNSS-derived wave heights also closely match tidal gauge observations, confirming the method's validity. Directional spectrum analysis indicates that wave energy is concentrated in the 0.2–0.25 Hz frequency band and within a directional range of $0^\circ \pm 30^\circ$, with a dominant northward propagation trend. These findings demonstrate that the proposed approach can provide high accuracy and physical consistency for GNSS-based wave monitoring under complex sea conditions.

Keywords: GNSS buoy; RTK technology; wave inversion; spectral analysis; directional spectrum analysis

1. Introduction

Ocean waves, primarily driven by wind and tidal forces, are dynamic surface phenomena with significant variability across time and space scales. They play a crucial role in shaping marine ecosystems and designing coastal infrastructure, and they hold vast potential for renewable energy [1–3]. Accurate wave measurement and analysis are essential for ensuring maritime safety and supporting ocean energy exploitation [4].

Various techniques are used for wave measurement, each with distinct advantages and limitations. A wave staff, a straightforward tool fixed vertically to the seabed or platform, measures wave height via calibrated markings but is limited to stationary point measurements [5]. Pressure sensors, which determine wave height by measuring seabed pressure variations, are suitable for deep-water zones but exhibit reduced accuracy in shallow waters [6]. Acoustic wave gauges utilize sound wave reflections, making them suitable for various water depths, but they are costly and sensitive to changes in water temperature and salinity [7]. Optical wave measurement systems employ lasers or cameras to capture spatial wave profiles and require appropriate lighting conditions [8]. Shipborne wave measurement systems estimate wave parameters by analyzing vessel motion, allowing for real-time observations during navigation, though they are less accurate [9]. Radar systems provide all-weather, large-area wave monitoring capabilities but involve expensive equipment and complex data processing [10]. Satellite remote sensing provides global wave data coverage, but its temporal resolution is limited [11]. Currently, no single method achieves an optimal balance between high accuracy, low cost, and real-time measurement capabilities for comprehensive wave monitoring.

Wave buoys enable real-time measurement of key wave parameters, such as wave height, period, and direction, and are widely used for ocean wave monitoring. The first generation of laboratory wave buoys was developed in 1953 by the UK National Oceanography Centre to measure wave height and period [12]. In the 1970s, Datawell Corporation introduced the Waverider series of wave buoys, which provided high precision, real-time data transmission, and multi-parameter measurements, fulfilling diverse scientific research needs in ocean wave studies [13,14]. With continuous technological advancements, wave buoys have evolved from single-parameter measurement devices to multifunctional instruments capable of simultaneously determining wave direction, spectral characteristics, and marine environmental parameters [15]. In addition to high measurement accuracy, these buoys offer operational flexibility, allowing for their deployment in both open ocean and coastal areas and thus providing continuous wave data records [16].

In recent years, GNSS-equipped wave buoys have gained significant attention as cost-effective and accurate tools for ocean wave monitoring [17]. By integrating GNSS receivers into buoy systems, three-dimensional position and velocity data can be acquired through Precise Point Positioning (PPP) or Real-Time Kinematic (RTK) techniques, enabling effective wave inversion [18,19]. GNSS buoys offer advantages such as low cost, high precision, all-weather capability, and ease of deployment, making them particularly suitable for large-scale ocean surface wave monitoring [20]. Lutsenko et al. employed GNSS signals for wave monitoring and analyzed the feasibility of monitoring surface disturbances [21]. Yang et al. evaluated the feasibility and accuracy of using the single-frequency time-differential carrier phase (TDCP) to determine wave height and period, achieving high-precision and low-cost wave monitoring [22]. Using GNSS technology for wave measurement and directional wave spectrum analysis, Gu et al. demonstrated that GNSS buoys, through carrier phase and Doppler frequency shift data, can accurately extract parameters such as wave height, period, and direction, generating high-resolution directional spectra [23]. Liang et al. deployed and tested GNSS buoys in the open ocean; the results verified the ability of GNSS buoys to continuously measure sea level and waves in the open ocean [24].

In addition to their standalone capabilities, GNSS buoys have been increasingly integrated with complementary technologies to enhance measurement accuracy, spatial resolution, and functionality in ocean wave monitoring. GNSS Reflectometry (GNSS-R), which analyzes signals reflected from the sea surface, has proven effective in retrieving wave height and wavelength [25,26]; Ruffini et al. further demonstrated its centimeter-level accuracy in low-altitude aerial experiments [27]. GNSS–acoustic positioning combines GNSS signals with underwater acoustics for high-precision monitoring of seabed displacement and wave dynamics [28], with Tomita et al. improving vertical accuracy by addressing uncertainties in sound speed profiles [29]. Shore-based GNSS reference networks, which deploy high-precision GNSS receivers along coastlines, are widely applied in marine observation and deformation monitoring [30,31]. Virtual Base Station Real-Time Kinematic (VBS-RTK) technology, which uses virtual reference stations derived from surrounding continuously operating GNSS stations to achieve precise real-time positioning, was validated by Lin et al. in the Su'ao waters of Taiwan—demonstrating agreement with both tide gauge data and ATC (accelerometer–inclinometer–compass) sensor results [32]. Shipborne GNSS platforms provide flexible real-time observation solutions for complex sea states; for example, Wang et al. successfully applied shipborne GNSS-R during Antarctic expeditions to invert wave height and wind speed from reflected signals [33,34]. Through the integration of these technologies, GNSS buoys have evolved into more comprehensive and versatile observation platforms for high-resolution, multi-parameter ocean wave monitoring under diverse conditions.

Accurate inversion of ocean wave parameters is essential for understanding sea state dynamics, supporting marine engineering, and ensuring navigational safety. With the growing availability of high-resolution measurement platforms such as GNSS buoys, robust and precise wave inversion algorithms have become a research focus. These algorithms typically extract and interpret wave signals using two primary analytical frameworks: statistical techniques and spectral analysis [35]. Statistical methods describe wave motion characteristics under simplified conditions but are limited to time and spatial domains [36,37]. Waves can be modeled as the superposition of cosine components with different amplitudes, frequencies, orientations, and phases, with the wave energy distribution depending deterministically on wave frequencies [38,39]. Wave spectrum analysis and directional spectrum analysis are widely applied to overcome the limitations of statistical methods [40,41]. Spectral analysis decomposes waves into frequency components and provides wave energy distribution versus frequency, enabling derivation of statistical parameters such as wave height and period via spectral moments [42]. Besides the commonly used Fast Fourier Transform (FFT) and Welch method, other techniques, such as the periodogram (a direct Fourier-based method), autocorrelation function (based on signal correlation), and autoregressive (AR) model (a parametric method using linear prediction), have also been applied to wave analysis [43–45]. While spectral shapes may vary depending on the method, the derived statistical features are generally consistent [46]. Directional spectrum analysis extends this method by estimating wave energy distribution in terms of both frequency and direction, offering insights into wave propagation [47]. Key approaches include the direct Fourier transform, parametric modeling, maximum likelihood and entropy methods, wavelet-based techniques, Bayesian estimation, and extended eigenvector methods [48,49].

This study proposes a multi-parameter wave inversion framework based on GNSS buoys that integrates time-domain statistical and frequency-domain spectral techniques to extract key wave parameters—including height, period, and direction—and characterize the two-dimensional wave energy distribution. GNSS buoys effectively combine accuracy and cost efficiency: by acquiring the absolute position and velocity of the buoy in real time, they avoid introducing systematic bias during wave measurement. As GNSS receiver technology becomes more affordable, such buoys enable low-cost, high-density deployments over wide marine areas, significantly expanding observational coverage and improving monitoring efficiency. Field tests in the Yellow Sea demonstrate the method's stability and reliability for large-scale, real-time monitoring. The results show the enhanced accuracy of wave parameter inversion, offering reliable data support for applications such as marine disaster mitigation and resource development.

In this study, a 10 m GNSS buoy—originally developed for long-term marine meteorological observation and satellite altimetry calibration—was used as the experimental platform. Given its proven operational stability and deployment history, the platform provides a practical basis for validating the proposed GNSS-based wave monitoring approach. Rather than conducting simulation-based assessments of the buoy's wave-following characteristics, we focused on evaluating its real-world measurement performance under actual sea conditions. This design reflects the study's emphasis on feasibility verification under actual field constraints.

2. Methods and Principles

2.1. Statistical Inversion of Wave Period and Wave Height

Statistical analysis is used to characterize and examine the statistical properties of ocean waves. These methods provide quantitative representations of wave data, facilitating a deeper understanding of wave variability and temporal patterns. The zero-crossing method is a statistical approach commonly employed in wave analysis, primarily to derive wave periods and heights from time series data. This technique relies on zero-crossing points within wave signals to capture wave periodicity and height characteristics [35].

First, the wave signal is analyzed to detect zero-crossing points, which occur when the signal transitions from negative to positive, indicating the start or end of a wave cycle. By measuring the time interval between consecutive zero-crossing points, the wave period can be determined. The duration between two successive zero-crossings represents one complete wave period, and averaging multiple periods offers insights into the wave's periodic characteristics. Subsequently, wave height is estimated using the zero-crossing method. After identifying the wave's crests (highest points) and troughs (lowest points), the vertical distance between them is calculated to represent the wave height. This method effectively determines wave height by analyzing the amplitude variations between successive crests and troughs.

To obtain the wave height sequence $H_i(i = 1, 2, \cdots\cdots, N)$, the wave heights are arranged in descending order, which represents the ranking of wave height magnitudes. Correspondingly, the wave period sequence $H_i(i = 1, 2, \cdots\cdots, N)$ can be derived, though it should be noted that the wave period sequence is not ordered by period magnitude but rather follows the ranking obtained for the wave height sequence. Table 1 provides the commonly used characteristic values in wave statistical analysis [36].

Table 1. Commonly used characteristic values in wave statistical analysis.

Characteristic Value	Calculation Formula	Characteristic Value	Calculation Formula
Maximum wave height	$H_{\max} = H_1$	Maximum wave period	$T_{\max} = T_1$
Highest one-tenth wave height	$H_{1/10} = \frac{10}{N} \sum_{i=1}^{\frac{N}{10}} H_i$	Mean period of the highest one-tenth waves	$T_{1/10} = \frac{10}{N} \sum_{i=1}^{\frac{N}{10}} T_i$
Significant wave height	$H_{1/3} = \frac{3}{N} \sum_{i=1}^{\frac{N}{3}} H_i$	Significant wave period	$T_{1/3} = \frac{3}{N} \sum_{i=1}^{\frac{N}{3}} T_i$
Mean wave height	$\overline{H} = \frac{1}{N} \sum_{i=1}^{N} H_i$	Mean wave period	$\overline{T} = \frac{1}{N} \sum_{i=1}^{N} T_i$

The highest one-third wave height ($H_{1/3}$) and highest one-third wave period ($T_{1/3}$) are commonly referred to as the significant wave height (H_s) and significant wave period (T_s), respectively, serving as generalized representations of sea state characteristics.

2.2. Wave Period and Wave Height Inversion Based on Spectral Analysis

Wave spectrum analysis is used to investigate the frequency composition of ocean waves, specifically how wave energy is distributed across different frequencies. Through spectral analysis, it is possible to identify the dominant frequency components within a wave signal and characterize the periodic nature of the waves. The wave spectrum not only provides the energy distribution as a function of frequency but also enables the derivation of various first-order statistical wave parameters, such as wave height and wave period, from the spectral moments [38].

2.2.1. Theoretical Basis of Wave Spectrum Analysis

Wave spectrum analysis is based on random wave theory, the core of which is the treatment of sea surface fluctuations as a random process composed of simple harmonic waves of different frequencies, amplitudes, and phases. According to Longuet-Higgins' linear wave theory, the vertical displacement of the sea surface can be expressed as follows:

$$\zeta(t) = \sum_{n=1}^{\infty} a_n \cos(\omega_n t + \varepsilon_n) \tag{1}$$

where a_n represents the wave amplitude, ω_n denotes the angular frequency, and ε_n is defined as a uniformly distributed random phase. According to random wave theory, the total energy of a wave in the frequency interval $(\omega, \omega + \Delta\omega)$ is as follows:

$$\sum_{\omega}^{\omega+\Delta\omega} \frac{1}{2} a_n^2 = S(\omega)\Delta\omega \tag{2}$$

where $S(\omega)$ is the wave power spectral density, which characterizes the distribution of wave energy in different frequency components [42].

This theoretical framework lays a solid foundation for wave spectrum analysis and provides a theoretical basis for subsequent spectral estimation methods.

2.2.2. Wave Spectrum Estimation Method

The process of deriving wave spectra from measured wave data using various estimation techniques is referred to as wave spectrum estimation. The wave spectrum primarily characterizes the relationship between wave energy and frequency. The corresponding spectral estimation methods can be categorized into classical spectral estimation and modern spectral estimation.

Classical spectral estimation methods are based on the Fourier transform and are suitable for analyzing stationary random processes. These methods obtain the power spectral density by calculating the autocorrelation function and subsequently performing a Fourier transform. Commonly used classical spectral estimation methods include the autocorrelation function, periodogram, Bartlett, and Welch methods [43,50].

Modern spectral estimation methods, on the other hand, introduce parametric models to describe the signal generation process and obtain the power spectral density by estimating the model parameters. These methods are typically applied to process short data records and non-stationary signals. In modern spectral estimation, the most commonly used models include the AR (Autoregressive) model, MA (Moving Average) model, and ARMA (Autoregressive Moving Average) model. Among these, the AR model is particularly effective in providing high-resolution spectrum estimates with relatively simple computational implementation, making it suitable for rapid application and analysis, especially for wave data with short time series [45].

(1) Classical Spectral Estimation Methods

Classical spectral estimation methods are based on the Fourier transform. These methods feature straightforward computation and clear physical interpretation. Commonly used classical spectral estimation techniques include the autocorrelation function and periodogram methods.

1. Autocorrelation Function Method

The autocorrelation function method uses indirect processing. First, the wavefront time series is processed to obtain its autocorrelation function. Then, the autocorrelation function is processed using the Fourier transform to obtain the estimated value of the coarse spectrum, and finally, the obtained coarse spectrum is smoothed. The formula of the autocorrelation function is as follows:

$$R(\tau) = \frac{1}{N-v}\sum_{n=1}^{N-v} \zeta(t_n)\zeta(t_n+\tau), v = 1,2,\cdots,m \tag{3}$$

where N represents the number of samples, τ denotes the time delay at which the autocorrelation is computed, ν is an auxiliary variable that adjusts the summation range to ensure that $t_n + \tau$ does not exceed the signal length N, and m is the maximum lag order, which defines the upper limit of v and τ.

$$S(\omega) = \frac{2}{\pi}\int_0^\infty R(\tau)\cos(\omega\tau)d\tau \tag{4}$$

The spectral value can be obtained by substituting Equation (3) into the above equation and integrating the corresponding values, with L_n used to represent the rough spectral estimate at frequency f_n. The trapezoidal formula is used in the numerical integration, and the spectral value is as follows:

$$\begin{gathered} L_n = \frac{2}{\pi}\left[\frac{1}{2}R(0) + \sum_{v=1}^{m-1} R(v\Delta t)\cos(2\pi f_n v\Delta t) + \frac{1}{2}R(m\Delta t)\cos(2\pi f_n m\Delta t)\right] \\ n = 0,1,2,\cdots,m \end{gathered} \tag{5}$$

The frequency interval used in the above expression is as follows:

$$\begin{cases} \Delta f = f_N/m = \frac{1}{2m\Delta t} \\ f_n = n\Delta f = \frac{n}{m}\frac{1}{2\Delta t} \end{cases} \tag{6}$$

Substituting the above equation into Equation (5) yields the following:

$$\begin{gathered} L_n = \frac{2\Delta t}{\pi}\left[\frac{1}{2}R(0) + \sum_{v=1}^{m-1} R(v\Delta t)\cos\frac{\pi v n}{m} + \frac{1}{2}R(m\Delta t)\cos \pi n\right] \\ n = 0,1,2,\cdots,m \end{gathered} \tag{7}$$

The spectral values obtained from the above estimation may exhibit local fluctuations due to the finite sample length and numerical effects. To suppress these variations and enhance spectral smoothness, a three-point weighted moving average filter is applied. The weights used are 0.23, 0.54, and 0.23, assigned to the current point and its two immediate neighbors.

This specific set of weights is not arbitrary; it is chosen to satisfy the following desirable properties: (1) symmetry, which ensures zero phase distortion; (2) normalization, preserving the amplitude scale of the spectrum; (3) a larger central weight that emphasizes the main frequency component while still smoothing out local fluctuations.

Among possible choices that meet these criteria, the values {0.23, 0.54, 0.23} represent a practical and balanced solution and are widely adopted in engineering practice for smoothing tasks [51,52]. The smoothing operation is expressed as follows:

$$S(f_n) = 0.23L_{N-1} + 0.54L_N + 0.23L_{N+1} \tag{8}$$

2. Periodogram Method

The periodogram method, also known as the direct method, is a classical spectral estimation technique. The basic idea is to directly apply the Fourier transform to the wave surface displacement sequence to obtain the spectral density. Specifically, this method first calculates the autocorrelation function of the wave surface displacement and then applies the Fourier transform to obtain the spectrum. During this process, the periodogram method effectively estimates the spectral distribution of the signal [43].

Assuming that the waves are a stationary process with statistical stationarity, observing the waves allows for the acquisition of a wave surface height sequence $\zeta(t)$. The spectral energy density function of this sequence can be expressed as follows:

$$\hat{S}(\omega) = \lim_{T\to\infty} \frac{1}{2\pi T}|X(\omega)|^2 \tag{9}$$

where ω is the angular frequency, $\hat{S}(\omega)$ is the spectral density, T is the total observation period, and $X(\omega)$ is the Fourier transform of $\zeta(t)$. For the discrete value of $\zeta(t)$, the following formula is obtained:

$$X(\omega) = \sum_{n=1}^{N} \zeta_n e^{-jn\omega\Delta t}\Delta t \tag{10}$$

where ζ_n represents the sample values of $\zeta(t)$ for $n = 1, 2, \cdots, N$; Δt is the sampling interval; j is the imaginary unit; and $T = N\Delta t$. Substituting Equation (10) into Equation (9) yields the following formula:

$$\hat{S}(\omega) = \frac{\Delta t}{2\pi N}\left|\sum_{n=1}^{N} \zeta_n e^{jn\omega}\right|^2, |\omega| < \pi/\Delta t \tag{11}$$

When $\Delta t = 1$, the above formula can be changed to the following formula:

$$\hat{S}_1(\omega) = \frac{1}{2\pi N}\left|\sum_{n=1}^{N} \zeta_n e^{jn\omega}\right|^2, |\omega| < \pi \tag{12}$$

where ζ_n is the wavefront height obtained by sampling per unit time. The above two formulas can be used to obtain the following formula:

$$\hat{S}(\omega) = \Delta t \hat{S}_1(\omega\Delta t) \tag{13}$$

Under the condition of $\Delta t \neq 1$, the initial computation may be performed using Equation (12), with subsequent conversion to $\hat{S}_1(\omega)$ via Equation (13). To enable discrete-value computation, the following formula is introduced:

$$\omega'_r = 2\pi r/N \tag{14}$$

where r is an integer, and $0 \leq r \leq \frac{N}{2}$. Thus, the equation becomes the following:

$$\hat{S}_1(\frac{2\pi r}{N}) == \frac{1}{2\pi N}\left|\sum_{n=1}^{N}\zeta_n e^{j\frac{2\pi r}{N}(n-1)} e^{j\frac{2\pi}{N}}\right|^2 = \frac{1}{2\pi N}\left|\sum_{n=1}^{N}\zeta_n e^{j\frac{2\pi r}{N}(n-1)}\right|^2 \tag{15}$$

To align with the functional formulation of the Fourier transform, we adjust the index of ξ_n by defining $\xi_n = \xi_{n-1}$, yielding the following:

$$\hat{S}_1(\frac{2\pi r}{N}) = \frac{1}{2\pi N}\left|\sum_{n=1}^{N}\zeta_{n-1} e^{j\frac{2\pi r}{N}(n-1)}\right|^2 = \frac{1}{2\pi N}|A_r|^2, r = 0, 1, 2, \cdots, N/2 \tag{16}$$

The expanded form of A_r is $A_r = \sum_{k=1}^{N}\zeta_k e^{j\frac{2\pi r}{N}k}$; the spectrum of the above formula is the value of $\omega \in (0, \pi/\Delta t)$, and the value of $\omega \in (-\pi/\Delta t, 0)$ can be obtained from the symmetry.

Consequently, the problem of power spectral estimation reduces to the computation of A_r. As with the autocorrelation function method, the raw spectrum obtained using this approach may exhibit local fluctuations due to numerical artifacts. To mitigate these effects and improve spectral continuity, a three-point weighted moving average filter is applied to smooth the spectrum. This post-processing step enhances the stability and accuracy of the spectral estimates.

(2) Modern Spectral Estimation Methods

Modern spectral estimation methods are based on stochastic process modeling, where sampled data are used to construct parametric models that provide improved spectral resolution. Common models include the Autoregressive (AR), Moving Average (MA), and Autoregressive Moving Average (ARMA) models. For power spectral estimation, the MA model generally provides lower resolution, while the ARMA model involves more complex and cumbersome parameter estimation. Therefore, the AR model is typically preferred in modern spectral analysis due to its computational efficiency and relatively simple implementation [45].

In the AR parametric method, the input sequence $u(n)$ is assumed to be a white noise series with zero mean and variance σ^2. This sequence drives a linear system with the transfer function $H(z)$, producing the output sequence $x(n)$. The system parameters can be estimated from either $x(n)$ or its autocorrelation function $r_x(m)$, enabling power spectral estimation. The input–output relationship is as follows:

$$\zeta_n = -\sum_{k=1}^{p} a_k\zeta(n-k) + \sum_{k=0}^{q} b_k u(n-k), b_0 = 1 \tag{17}$$

The transfer function and power spectrum of the AR model are as follows:

$$\zeta_n = -\sum_{k=1}^{p} a_k\zeta(n-k) + u(n) \tag{18}$$

In the above equation, the white noise sequence $u(n)$ has a zero mean and a variance of σ^2, where p denotes the model order, and a_k represents the model parameters. The transfer function and power spectrum of this model are, respectively, given by

$$H(z) = \frac{1}{\left|1 + \sum_{k=1}^{p} a_k e^{-j2\pi f k\Delta t}\right|^2} \tag{19}$$

$$P_x(f) = \frac{\sigma^2 \Delta t}{\left|1 + \sum_{k=1}^{p} a_k e^{-j2\pi f k \Delta t}\right|^2} \tag{20}$$

The autocorrelation function $r_x(m)$ and AR coefficients a_k satisfy the following:

$$r_x(m) = \begin{cases} -\sum_{k=1}^{p} a_k r_x(m-k) & m \geq 1 \\ -\sum_{k=1}^{p} a_k r_x(k) + \sigma^2 & m = 0 \end{cases} \tag{21}$$

which can be expressed in matrix form (using the even symmetry of $r_x(m)$ as follows:

$$\begin{bmatrix} r_x(0) & r_x(1) & r_x(2) & \cdots & r_x(0) \\ r_x(1) & r_x(0) & r_x(1) & \cdots & r_x(p-1) \\ r_x(2) & r_x(0) & r_x(0) & \cdots & r_x(p-2) \\ \vdots & \vdots & \vdots & \vdots & \vdots \\ r_x(p) & r_x(p-1) & r_x(p-2) & \cdots & r_x(0) \end{bmatrix} \begin{bmatrix} 1 \\ a_1 \\ a_2 \\ \vdots \\ 0 \end{bmatrix} = \begin{bmatrix} \sigma^2 \\ 0 \\ 0 \\ \vdots \\ 0 \end{bmatrix} \tag{22}$$

This is the Yule–Walker equation, the cornerstone of AR parameter estimation.

In practical AR spectral estimation, the model order p is a critical parameter influencing the accuracy and stability of the spectral results. In this study, the AR model order was empirically determined using the Burg algorithm, with a relatively high order (order = 100) adopted. Sensitivity analyses were conducted by comparing spectral estimations with AR orders ranging from 10 to 100. These analyses indicated that the spectral shapes and derived wave parameters stabilized for AR orders above approximately 50, while orders below 20 produced noticeable spectral distortions. Therefore, an order of 100 was selected as a practical trade-off between spectral resolution and computational complexity.

2.2.3. Calculation of Wave Characteristic Elements

The wave spectrum not only describes the distribution of wave energy in different frequencies but also enables the derivation of first-order statistical wave parameters through its spectral moments [22,23]. The relationships between these statistical parameters and the spectral moments m_n are as follows:

$$\begin{cases} \overline{H} = \sqrt{2\pi m_0} \\ H_s = H_{1/3} = 4\sqrt{m_0} \\ \overline{T} = 1.2T_{0,2} = 2.4\pi\sqrt{m_0/m_2} \\ T_s = T_{H_{1/3}} = 0.937T_p \end{cases} \tag{23}$$

where $\overline{H}$ is the mean wave height, H_s is the significant wave height, $\overline{T}$ is the mean wave period, T_s is the significant wave period, T_p is the peak period (corresponding to maximum spectral density), and $T_{0,2}$ is the mean period calculated from the zeroth and second spectral moments [37] as follows:

$$T_{0,2} = 2\pi\sqrt{m_0/m_2} \tag{24}$$

where $m_n(n = 0, 1, 2 \cdots)$ is the n-th-order spectral moment of the wave spectrum $S(\omega)$, and its expression is as follows:

$$m_n = \int_0^\infty \omega^n S(\omega) d\omega \tag{25}$$

where ω^n represents the n-th power of angular frequency ω, and other symbols maintain their previously defined meanings.

2.3. Wave Direction Analysis Based on Directional Spectrum

Wave spectrum analysis provides the one-dimensional distribution of wave energy with respect to frequency. However, to obtain the two-dimensional distribution of wave energy in terms of both frequency and direction, directional spectrum analysis must be employed. Observing and analyzing directional spectra typically require synchronous measurements of wave characteristics from multiple observation points, making the process significantly more complex than frequency spectrum analysis, with considerably higher observation costs. When the wave frequency spectrum is already known, the directional spectrum can be estimated by utilizing the relationship between the one-dimensional frequency spectrum and the two-dimensional directional spectrum [16].

The directional spectrum can be expressed in terms of the wave frequency spectrum $S(f)$ and the directional spreading function $G(f,\theta)$ as follows:

$$S(f,\theta) = S(f)\cdot G(f,\theta) \tag{26}$$

$$\int_{-\pi}^{\pi} G(f,\theta)d\theta = 1 \tag{27}$$

where $S(f)$ is the spectrum; $G(f,\theta)$ is the directional distribution function, which cannot be negative, and the integral from $-\pi$ to π is 1; and the variables f and θ correspond to wave frequency and direction, respectively. It is evident that once $S(f)$ is determined, $G(f,\theta)$ becomes the key to solving for the directional spectrum.

The directional spreading function $G(f,\theta)$ can be obtained through cross-spectral analysis of the wave displacement time series measured in two horizontal directions. The function may be expressed as follows:

$$G(f,\theta) = \frac{1}{\pi}\left[\frac{1}{2} + a_1\cos(\theta) + b_1\sin(\theta) + a_2\cos(2\theta) + b_2\sin(2\theta) + \cdots\right] \tag{28}$$

This formulation can be derived from wave cross-spectra.

Wave cross-spectra represent frequency-domain characteristics of wave signals. In wave analysis, cross-spectra not only reveal frequency coherence between two time series but also indicate the directional distribution of wave energy, making them essential tools for directional spectral estimation.

The cross-spectrum between any two wave signals is given by the following [23]:

$$S_{p,q}(f) = \frac{1}{T}\overline{F}_p(f)F_q(f) = C_{pq}(f) - iQ_{pq}(f)p, q = 1,2,3,\cdots \tag{29}$$

where $S_{p,q}(f)$ denotes the cross-spectrum between wave measurements p and q. Here, $F_q(f)$ is the Fourier transform of signal q, while $\overline{F}_p(f)$ represents the complex conjugate of the Fourier transform of signal p. The terms $C_{pq}(f)$ and $Q_{pq}(f)$ correspond to the co-spectrum and quadrature spectrum, respectively.

The wave surface elevation sequence and the displacement time series (u, w, n) in the east–west and north–south directions are Fourier transformed to obtain the real and imaginary parts of their frequency-domain Fourier coefficients as follows:

$$F_{uf} = \alpha_{uf} + i\beta_{uf}, F_{wf} = \alpha_{wf} + i\beta_{wf}, F_{nf} = \alpha_{nf} + i\beta_{nf} \tag{30}$$

The isotropic spectrum $C_{pq}(f)$ and anisotropic spectrum $Q_{pq}(f)$ can be derived as follows:

$$\begin{cases} C_{nu} = \overline{A}_{nf} \times \overline{A}_{uf} = \alpha_{nf}\alpha_{uf} + \beta_{nf}\beta_{uf} \\ Q_{nu} = \overline{A}_{nf} \times \overline{A}_{uf} = \alpha_{nf}\alpha_{uf} - \beta_{nf}\beta_{uf} \end{cases} \tag{31}$$

From the definition of the anisotropic spectrum, we know that $Q_{uu} = Q_{ww} = Q_{nn} = 0$, and from the physical properties of waves, we know that $Q_{nw} = Q_{wn} = 0$; thus, we can obtain the 3×3 matrices of isotropic and anisotropic spectra as follows:

$$\begin{bmatrix} C_{ww} & C_{wn} & C_{wu} \\ C_{nw} & C_{nn} & C_{nu} \\ C_{uw} & C_{un} & C_{uu} \end{bmatrix} \text{ and } \begin{bmatrix} 0 & 0 & Q_{wu} \\ 0 & 0 & Q_{nu} \\ Q_{uw} & Q_{un} & 0 \end{bmatrix} \quad (32)$$

Based on the joint analysis of isotropic and anisotropic spectra, the directional distribution characteristics of wave energy are determined by substituting the calculated Fourier coefficients a_1, a_2, b_1, and b_2 into the directional distribution function (28). The peak direction of the directional distribution function $G(f,\theta)$ is the dominant propagation direction of the frequency component.

3. Wave Data Acquisition and Signal Extraction

3.1. GNSS-Buoy-Based Data Collection and Processing

The experimental data used in this study were acquired from a GNSS buoy deployed in the coastal waters of Qingdao, China, on 27 April 2023. The location of the field experiment is illustrated in Figure 1.

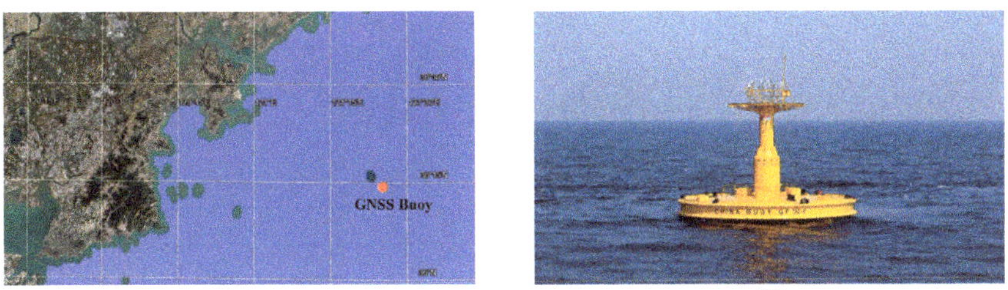

(a) (b)

Figure 1. Experimental site and setup. (**a**) GNSS buoy test area; (**b**) GNSS buoy.

The experiment employed a large GNSS buoy (10 m buoy body) deployed 3 km from the reference station. The buoy was equipped with a Hi-Target Vnet8 receiver (Hi-Target, Shenzhen, China) as the GNSS module and a high-precision HZZACF-S808 GNSS antenna (Hi-Target, Shenzhen, China), with data sampled at 1 Hz. To validate the effectiveness of the proposed wave measurement method, wave monitoring results obtained from other sensors installed on the buoy were used as reference data.

The GNSS buoy data were processed using Real-Time Kinematic (RTK) technology to derive coordinate time series. Data were processed using RTKLIB (version 2.4.3), an open-source GNSS positioning software developed by the Tokyo University of Marine Science and Technology. Key parameter settings used for data processing are summarized in Table 2 (unlisted parameters retained their default values).

The raw GNSS buoy observations were processed using RTK positioning to obtain displacement time series in the North–East–Up coordinate system, as shown in Figure 2. Through position differencing and coordinate transformation, velocity information in the NEU directions was derived. Subsequent integration of these velocity components, followed by Complete Ensemble Empirical Mode Decomposition (CEEMD) filtering—a data-driven signal decomposition technique effective for extracting non-stationary and nonlinear wave components—and moving average filtering, yielded the final wave surface displacement measurements from the GNSS buoy.

Table 2. Parameter configuration.

Parameter	Configuration
Position Mode	Kinematic
Frequencies	L1
Solution Type	Forward
Elevation Mask	15.0 deg
Ephemeris Data	Broadcast
Navigation Systems	GPS/GLONASS/Galileo/BDS
Ambiguity Resolution	Instantaneous
Integer Validation Threshold	3.0

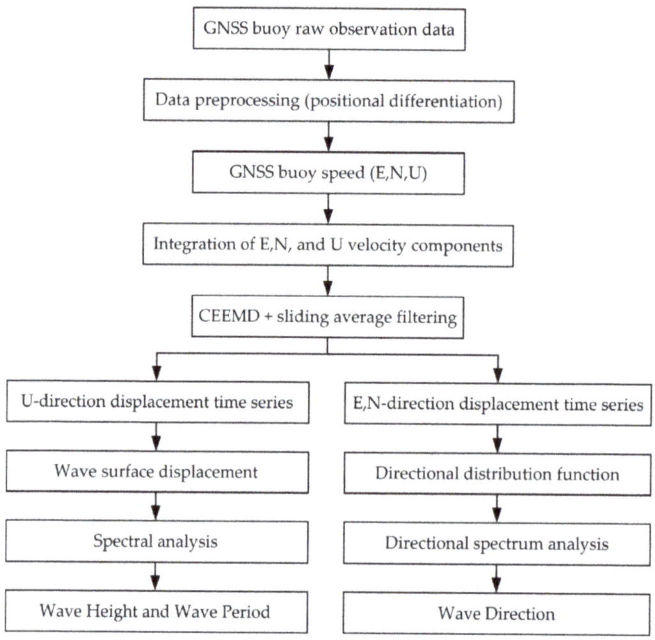

Figure 2. Diagram of ocean wave measurement process using GNSS buoy.

The three-dimensional wave surface displacement time series of the GNSS buoy were obtained by integrating the velocity components in the NEU (North–East–Up) directions, as shown in Figure 3.

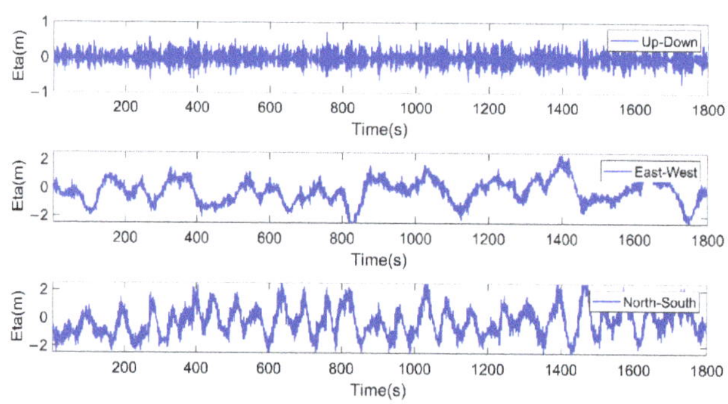

Figure 3. Three-dimensional wavefront displacement time series from GNSS buoy.

Spectral analysis of the vertical (U-direction) displacement time series yields key wave parameters, including wave height and period. By applying Fourier transforms to both vertical and horizontal displacement components, cross-spectral analysis can be performed to derive the directional spreading function. Combined with the frequency spectrum results, this function enables the determination of directional wave spectrum characteristics.

3.2. GNSS-Based Ocean Wave Signal Data Processing

3.2.1. Ocean Wave Signal Separation and Extraction

By isolating the vertical (U-direction) displacement time series, sea surface fluctuations can be obtained. Figure 4 presents a 30 min U-direction wave surface displacement sequence from 11:30 to 12:00 UTC on 27 April 2023. The time series exhibits distinct low-frequency components, primarily attributable to tidal signals. Wave and tidal signals differ in their periodicities, so appropriate filtering methods are used to effectively separate these components. Commonly employed approaches include time-domain filtering and spatial filtering techniques.

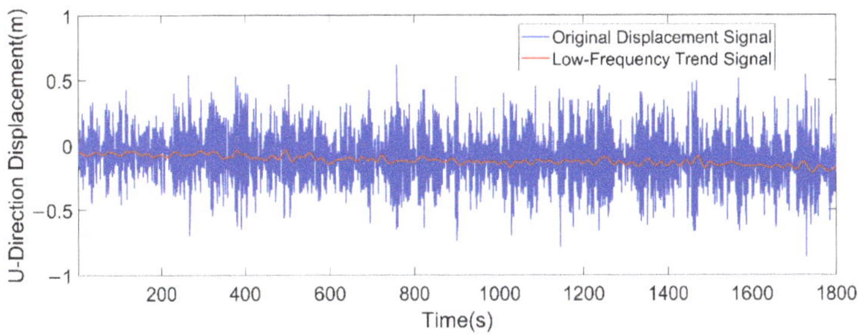

Figure 4. U-direction wavefront displacement sequence (sampling rate 1 Hz).

Figure 5 demonstrates the filtering results of the original wave surface displacement using two methods: CEEMD filtering combined with moving average filtering (time domain) and Butterworth high-pass filtering (frequency domain). Both approaches effectively extract the wave-induced fluctuations from the original signal.

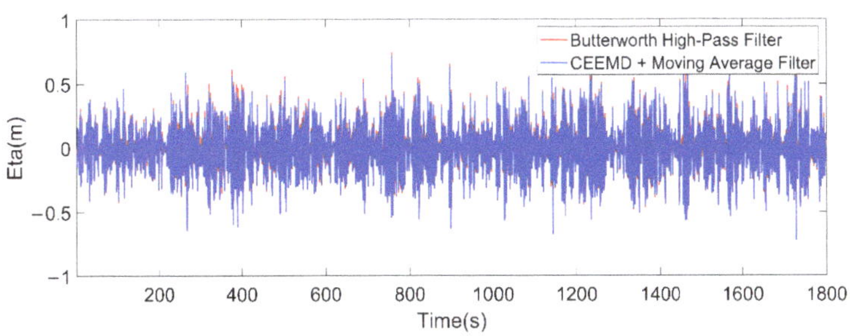

Figure 5. Wavefront displacement sequence filtering (time domain and frequency domain).

The combination of CEEMD and the moving average filtering method leverages the adaptive decomposition capability of CEEMD and the smoothing effect of moving average filtering, enabling precise analysis of trend and periodic characteristics in the signal. This approach achieves effective separation of wave and tidal signals, and subsequent wave analysis will be based on the wave–tide separation results obtained through this combined filtering method.

3.2.2. Correction of Directional Spreading Function

The cross-spectra obtained through direct Fourier transform yield preliminary estimates, resulting in inaccuracies in the calculated parameters (a_1, a_2, b_1, and b_2) and, consequently, an imprecise directional spreading function. Therefore, necessary corrections must be applied to the directional spreading function.

As shown in Figure 6, small negative values appear in the estimated directional spreading function due to numerical artifacts introduced during spectral estimation. These negative values are non-physical, as the directional spreading function is theoretically non-negative and must integrate to unity. To correct this while minimizing the impact on directional accuracy, we apply the following procedure [23]:

(1) Set all negative values in the directional spreading function to zero.
(2) Redistribute the integrated area of the negative-valued portions to the positive-valued regions proportionally based on their respective weights.
(3) Verify that the corrected directional spreading function satisfies the normalization condition (integral equals unity).

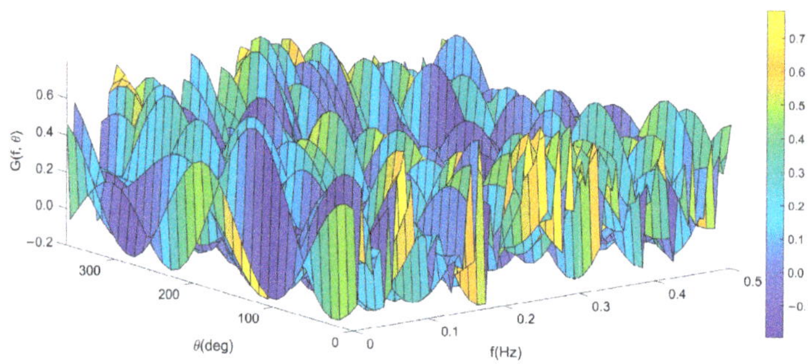

Figure 6. Direction distribution function before correction.

The corrected directional spreading function, as shown in Figure 7, now ensures both non-negativity and normalization properties, making it suitable as input for ocean wave directional spectrum calculations.

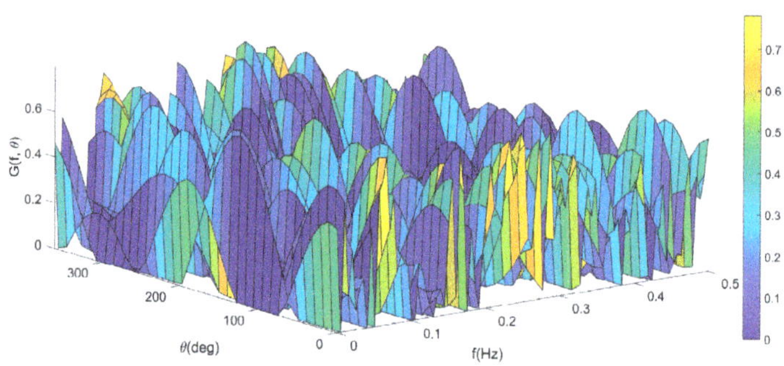

Figure 7. Revised directional distribution function.

4. Analysis of Measured Wave Results

4.1. Wave Parameter Retrieval Based on Statistical Analysis

Ocean wave statistical analysis primarily employs statistical and spectral methods. Statistical approaches characterize wave properties through quantitative descriptors, elucidating wave variability and trends. The zero-upcrossing method, a statistical technique for wave analysis, estimates wave periods and heights by detecting zero-crossing points in time series data [50].

Wave Period and Height Estimation via Zero-Upcrossing Method

To evaluate the feasibility of retrieving wave parameters from GNSS buoy data, measurements were conducted from 08:00 to 12:00 on 27 April 2023, divided into eight 30 min intervals. Figure 8 displays the first 8 min of wave surface elevation data for each interval, demonstrating clear wave height variations recorded by the GNSS buoy. Throughout the observation period, the buoy data exhibited high stability.

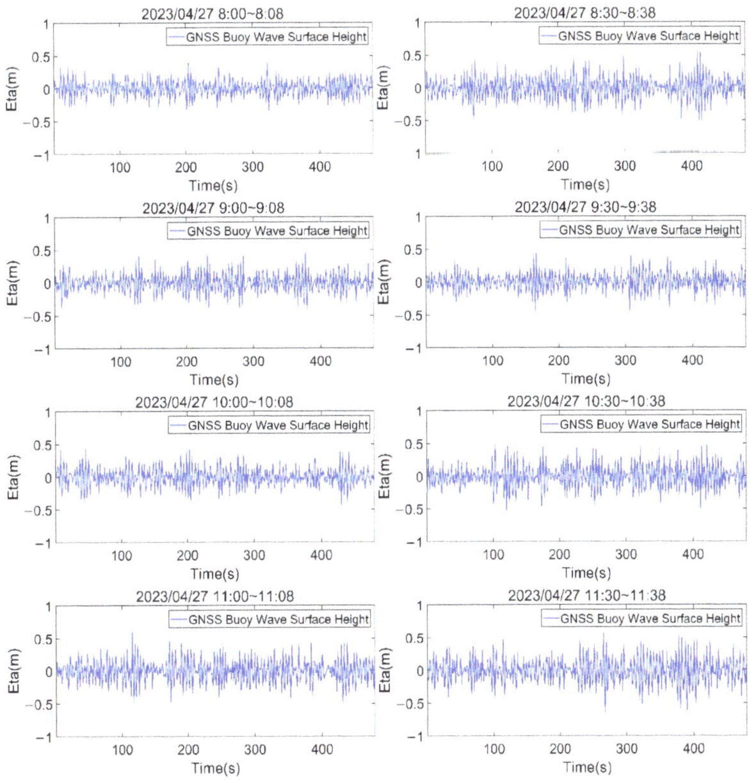

Figure 8. GNSS buoy wavefront height variation (data duration: 8 min).

For each time interval, wave statistical analysis was conducted using approximately 100 consecutive waves as a standard sample. The following figure (Figure 9), which consists of eight subplots labeled a to h, presents the wave statistical results obtained from the GNSS buoy in the eight different time periods. Each subplot for the time interval consists of three parts: the first part shows the crest and trough heights of each wave arranged chronologically; the second part displays the wave height distribution, with all waves sorted in descending order, highlighting the top one-third of the wave sequence using a pink dashed line to represent the significant waves; and the third part, based on the sorting in the second part, presents the corresponding wave periods using asterisks to denote all individual periods, while the significant wave period is explicitly identified in the legend.

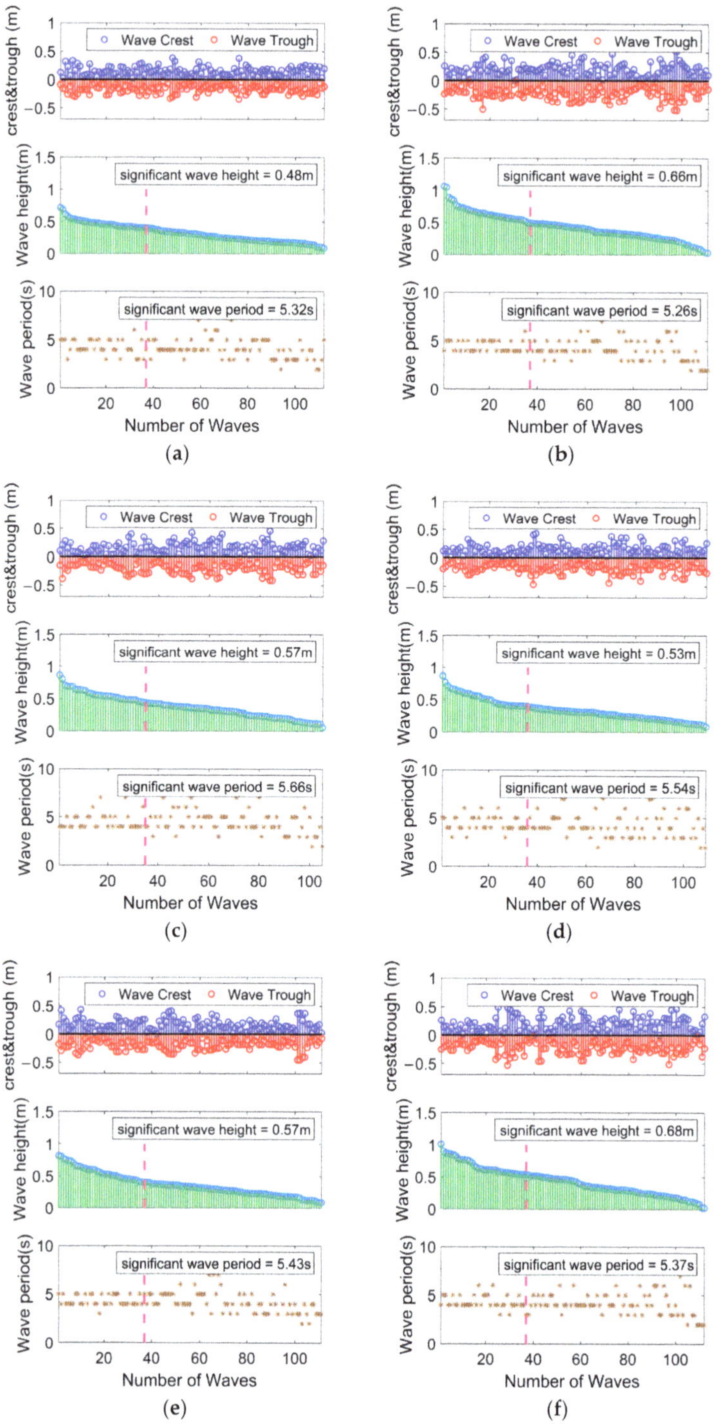

Figure 9. *Cont.*

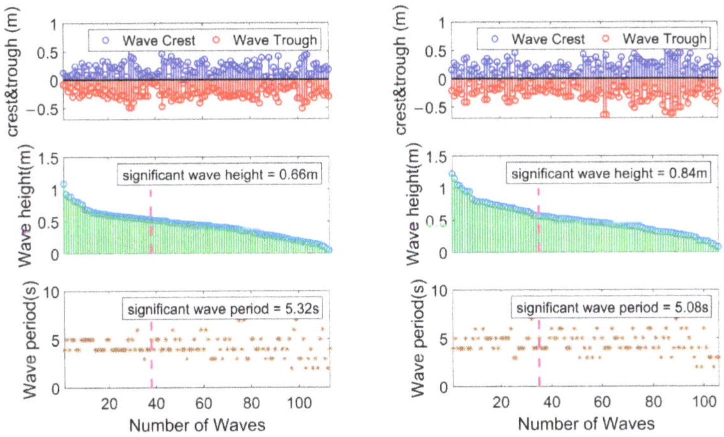

(g) (h)

Figure 9. GNSS buoy wave crest, trough, height, and period statistics. (**a**) 27 April 2023 8:00~8:08; (**b**) 27 April 2023 8:30~8:38; (**c**) 27 April 2023 9:00~9:08; (**d**) 27 April 2023 9:30~9:38; (**e**) 27 April 2023 10:00~10:08; (**f**) 27 April 2023 10:30~10:38; (**g**) 27 April 2023 11:00~11:08; (**h**) 27 April 2023 11:30~11:38. The asterisks (*) in figure represent the sorted wave periods (in seconds) corresponding to wave heights arranged in descending order.

Based on the statistical wave characteristics presented in the figure above, the characteristic wave heights and corresponding wave periods were calculated for each dataset.

For each data segment, wave sequences were statistically analyzed using the zero-upcrossing method. The maximum wave height represents the absolute crest value within the sequence, while the highest one-tenth wave height denotes the average of the top 10% largest waves—both exhibiting relatively strong randomness. In contrast, the significant wave height (highest one-third of the waves' average) and mean wave height (average of all waves) better reflect the statistical characteristics of the wave field [23]. The mean wave parameters particularly demonstrate a more stable representation of the prevailing sea state.

Theoretical analysis indicates that shorter wave samples tend to yield larger mean wave heights and periods. For instance, the 09:00–09:08 segment containing 105 waves showed greater mean values compared to the 111-wave (08:30–09:08) and 110-wave (09:30–09:38) segments. These results confirm that GNSS buoys can effectively capture wave height and period parameters through zero-upcrossing analysis and reliably record oceanic wave details.

It is worth noting that the 10 m GNSS buoy used in this study, while offering excellent attitude stability and robustness for GNSS positioning, is not specifically optimized for wave-following performance, especially in low-sea-state conditions. This inherent design characteristic may reduce its sensitivity to small-amplitude wave fluctuations, leading to limited correlation between significant wave height and significant wave period, as observed in Tables 3 and 4. This phenomenon is consistent with the expected behavior of large-diameter buoys and reflects a trade-off between positional accuracy and wave-tracking responsiveness. Therefore, the results discussed here should be interpreted within the context of the buoy's physical limitations. This highlights the need for further comparative studies involving different buoy configurations and sea conditions.

Table 3. Statistical values of characteristic wave height and number of waves.

Data Period	Statistical Wave Height Parameters (m)				Number of Waves
	H_{max}	$H_{1/10}$	$H_{1/3}$	H_{mean}	
08:00–08:08	0.72	0.57	0.48	0.32	112
08:30–08:38	1.06	0.82	0.66	0.44	111
09:00–09:08	0.87	0.70	0.57	0.38	105
09:30–09:38	0.87	0.68	0.53	0.35	110
10:00–10:08	0.82	0.72	0.57	0.37	111
10:30–10:38	1.01	0.84	0.68	0.44	113
11:00–11:08	1.09	0.82	0.66	0.45	113
11:30–11:38	1.23	0.98	0.84	0.54	106

Table 4. Statistical value of characteristic wave period.

Data Period	Statistical Wave Period Parameters (s)			
	T_{max}	$T_{1/10}$	$T_{1/3}$	T_{mean}
08:00–08:08	8.00	6.00	5.32	4.19
08:30–08:38	7.00	5.83	5.26	4.22
09:00–09:08	8.00	6.73	5.66	4.52
09:30–09:38	8.00	6.36	5.54	4.34
10:00–10:08	7.00	6.33	5.43	4.30
10:30–10:38	8.00	6.25	5.37	4.22
11:00–11:08	8.00	6.00	5.32	4.22
11:30–11:38	7.00	6.18	5.08	4.33

4.2. Wave Period and Height Retrieval Based on Spectral Analysis

Section 4.1 analyzed GNSS buoy measurements from 08:00 to 12:00 on 27 April 2023, using the zero-upcrossing method. This section focuses on the 11:30–12:00 dataset, employing spectral analysis to calculate and examine wave parameters. The following presents spectral analysis results of U-direction displacement time series obtained through three methods: the periodogram, autocorrelation function, and AR model approaches, as shown in Figure 10.

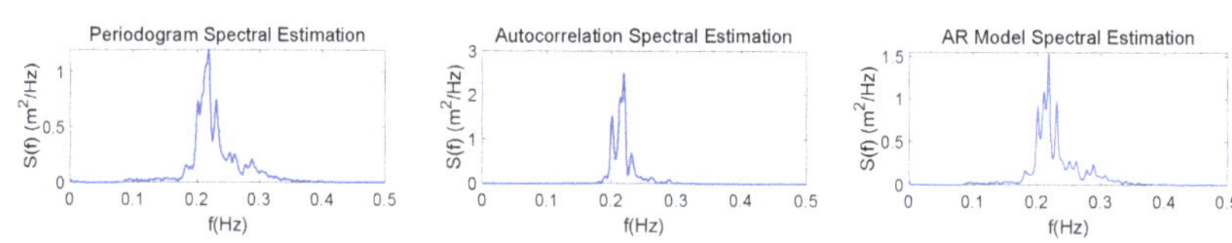

Figure 10. Spectral estimation diagram of measured data.

To better visualize the differences and similarities between the three spectral estimation methods, we normalized the power spectral densities to eliminate amplitude variations, enabling direct comparison on a unified scale. The normalized spectra clearly show the frequency-domain characteristics of each method, facilitating identification of their respective strengths and limitations. By plotting the spectral results from all three methods together, we can intuitively compare their spectral shapes and distribution patterns, thoroughly analyzing their performance in different frequency bands and potential biases, as shown in Figure 11.

Comparison of the power spectral density obtained using the periodogram, autocorrelation function, and AR model methods reveals nearly identical spectral shapes. While minor differences exist in certain areas due to variations in wave data processing methods, they do not affect the calculation of first-order wave characteristics. Table 5 presents wave parameters obtained from the three spectral analysis methods, showing identical values for significant and mean wave heights across methods, with only minor differences in wave periods: 0.02 s for the significant period and 0.01 s for the mean period. Compared to parameters obtained using the zero-upcrossing method, the spectral analysis results show differences of 0.05 m in significant wave height, 0.79 s in significant wave period, 0.09 m in mean wave height, and 0.08 s in mean wave period.

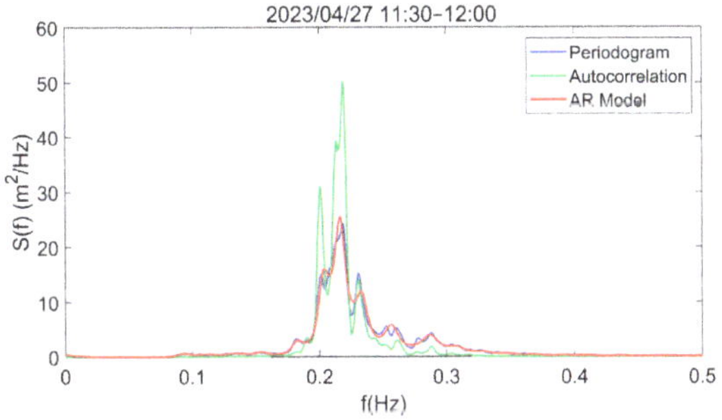

Figure 11. Comparative analysis of power spectral density using periodogram, autocorrelation function, and AR model methods.

Table 5. Comparison of wave characteristic element results using periodogram, autocorrelation function, and AR model methods.

Wave Parameters	Periodogram	Autocorrelation	AR Model	Zero-Crossing
Significant Wave Height (m)	0.89	0.89	0.89	0.84
Significant Wave Period (s)	4.27	4.27	4.29	5.08
Mean Wave Height (m)	0.56	0.56	0.56	0.54
Mean Wave Period (s)	4.25	4.24	4.25	4.33

The comparison between the reference significant wave height and GNSS-buoy-derived wave heights is shown in Figures 12 and 13.

To quantitatively assess agreement between GNSS-derived results and the reference measurements, we calculated the Root Mean Square Error (RMSE) and Pearson correlation coefficient. For significant wave height, the RMSE was 0.0788 m with a correlation of 0.9298. For significant wave period, the RMSE was 0.1926 s with a correlation of 0.9381. These metrics confirm that the GNSS-based inversion method yields reliable and accurate wave parameters [23].

Comprehensive statistical and spectral analyses of the same GNSS buoy data further demonstrate consistency across methods, fully validating the effectiveness and applicability of the proposed inversion framework.

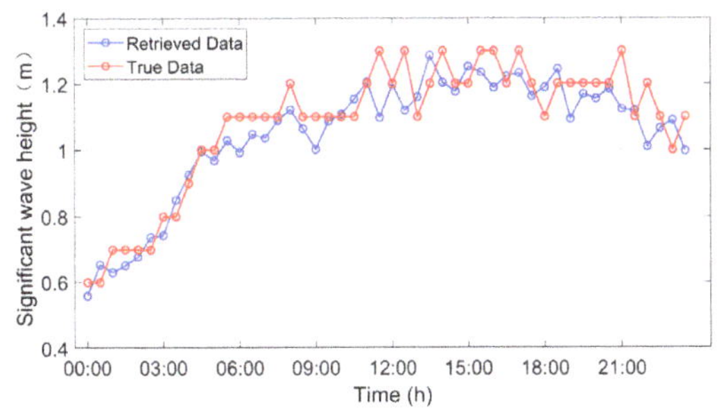

Figure 12. Comparison chart of inverted sea surface wave height and true sea surface wave height.

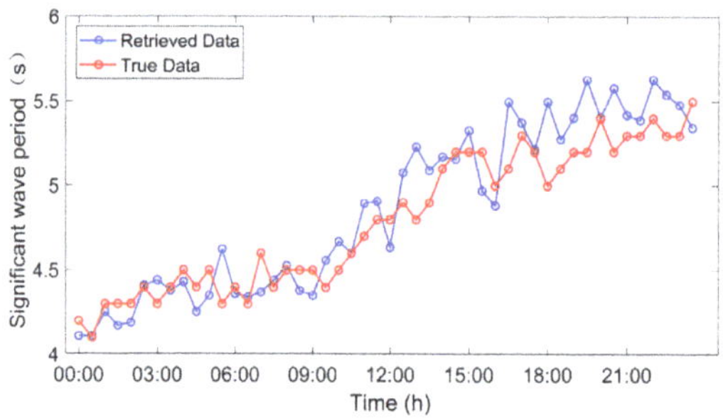

Figure 13. Comparison chart of inverted period and real period.

4.3. Wave Direction Retrieval Based on Directional Spectrum Analysis

Directional spectrum analysis provides the two-dimensional distribution of wave energy with respect to both frequency and direction, offering deeper insight into wave propagation characteristics. In this study, we applied cross-spectral analysis using the horizontal displacement time series of the GNSS buoy to estimate the directional spreading function. This enabled the construction of the directional wave spectrum and the identification of dominant propagation trends under actual sea conditions.

4.3.1. Three-Dimensional Wave Directional Spectrum Analysis

Based on the wave frequency spectrum and directional spreading function, the directional wave spectrum for the corresponding time period is presented below, as shown in Figure 14.

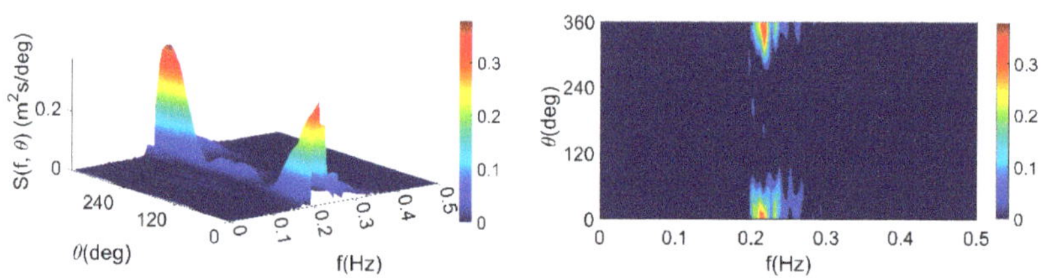

Figure 14. Directional spectrum: 3D and plan views (27 April 2023 11:30–12:00).

During the test, the wind speed and direction were stable, and the overall wave changes were not significant in the short term. The wave direction in the figure is given in the form of wave direction azimuth (0° to the north and clockwise to the positive direction).

The following is a three-dimensional diagram of the directional spectrum from 8:00 to 12:00 on 27 April, with the directional spectrum given every half hour, as shown in Figure 15.

The horizontal axis in the figure is frequency, ranging from 0 Hz to 0.5 Hz, indicating the frequency distribution of waves; the vertical axis is direction, ranging from 0° to 360°, indicating the propagation direction of waves; and the vertical axis is spectral density, indicating the wave energy density.

In the figure, the peak of the spectral density is mainly concentrated in the frequency range of 0.2 Hz to 0.25 Hz, and the direction ranges from 0° to 60° and 300° to 360°, which indicates the frequency band where the wave energy is mainly concentrated; the main propagation direction of the wave is northward. In addition to the dominant spectral density peak, a secondary peak with lower amplitude is also consistently present. The main peak is generally caused by the main wind direction in the sea area, while the secondary peak is generally the result of the combined effect of the main wind direction of the sea area and the long-distance swell [53].

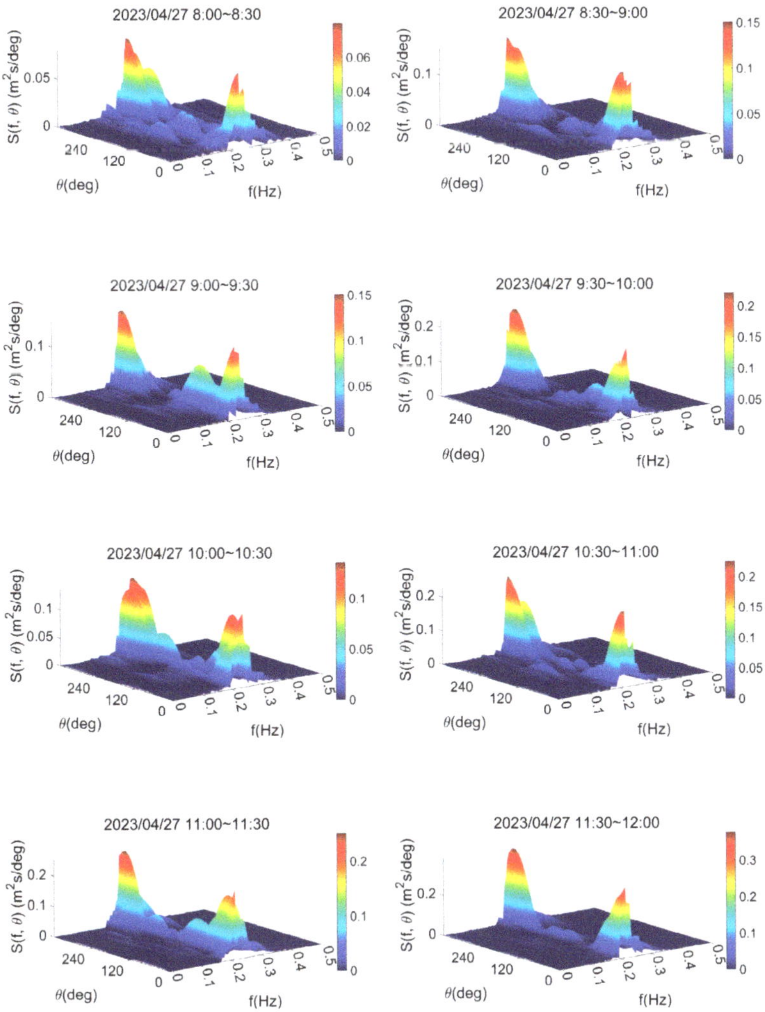

Figure 15. Three-dimensional wave directional spectrum.

4.3.2. Analysis of Wave Direction–Height Relationship

The three-dimensional graph intuitively illustrates the spatial distribution of wave energy; however, limitations in resolution and viewing angle may affect the clarity and interpretability of the visualization. The polar coordinate graph can simply and clearly display multi-dimensional information, making it easier to extract and analyze key information. The following is the corresponding polar coordinate graph of the directional spectrum, as shown in Figure 16.

The figure illustrates the directional distribution of waves over a 24 h period, showing predominant wave directions clustered near 0° (true north). This indicates that wave energy was primarily concentrated in northerly directions throughout the day. Notably, wave heights at 0° were both significantly larger and more concentrated, further confirming the strong directional focusing of wave energy toward the north.

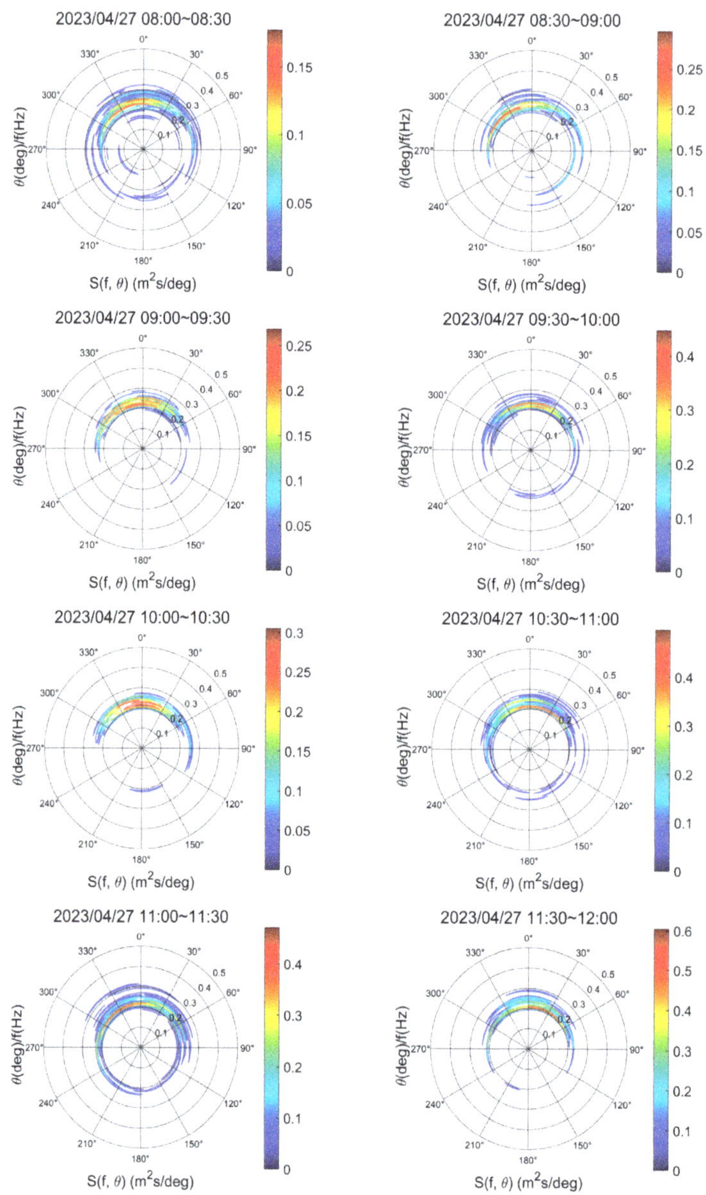

Figure 16. Polar coordinate diagram of wave direction spectrum.

5. Conclusions

GNSS buoy technology, leveraging high-precision satellite positioning and advanced spectral analysis, provides an effective and reliable solution for real-time ocean wave monitoring. This study proposes a comprehensive GNSS-based wave inversion framework and demonstrates its effectiveness using field data collected in the Huanghai Sea. The approach integrates multiple spectral estimation methods and directional spectrum analysis to extract key wave parameters—height, period, and propagation direction—with high accuracy and consistency. Notably, the directional spectrum analysis successfully captured the concentration of energy in specific frequency bands and revealed a predominantly northward wave propagation trend, validating the method's applicability under real sea conditions.

As an observational platform, the GNSS buoy continuously records three-dimensional motion trajectories. Compared to traditional wave gauges and radar-based instruments, GNSS buoys offer advantages such as low cost, easy deployment, and long-term stability, making them highly suitable for long-duration and wide-area marine monitoring.

To overcome limitations in traditional wave signal processing and directional spectral estimation, this study employed a combination of statistical and spectral approaches. The zero-crossing method and three spectral techniques—the periodogram, autocorrelation function, and autoregressive model—were applied to estimate wave height and period, providing a robust cross-validation framework. Horizontal displacement and velocity data were used to retrieve wave propagation characteristics in two dimensions for directional spectral estimation.

Analysis of data collected on 27 April 2023 confirmed the stability and performance of the proposed method. The three spectral approaches produced wave period estimates with a maximum discrepancy of only 0.02 s. Compared with the zero-crossing method, the spectral analysis results show differences of 0.05 m and 0.79 s in significant wave height and period, respectively, with average differences of 0.09 m and 0.08 s. These results demonstrate strong internal consistency and validate the accuracy of the GNSS-based approach. Directional spectrum analysis showed that energy was concentrated in the 0.2–0.25 Hz frequency band and within a directional range of $0° \pm 30°$, reflecting a clear northward propagation pattern, consistent across all observation periods.

Taken together, the results demonstrate that the proposed GNSS-buoy-based wave inversion framework is accurate, consistent, and robust under complex sea conditions. This study highlights the practical potential of GNSS buoys as cost-effective tools for real-time ocean wave monitoring, with broad applications in marine engineering, disaster prevention, and oceanographic research.

Nevertheless, several limitations should be acknowledged. The large size (10 m) of the buoy reduces its wave-following sensitivity, especially for low sea states, potentially affecting accuracy. In rough sea conditions, multipath interference, buoy instability, and antenna tilt may further degrade GNSS positioning. Additional stabilizing structures or more rigid antenna mountings may be required to ensure reliable operation. Moreover, the reliance on uninterrupted GNSS signals and post-processing may limit real-time performance in extreme environments.

Future research will focus on enhancing signal robustness under variable sea conditions, improving hardware stability, and extending validation across diverse platforms and environmental scenarios. These efforts aim to support the development of a scalable, all-weather GNSS-based wave monitoring solution.

Author Contributions: Methodology, J.W.; software, X.C.; validation, J.W., X.C., S.Y. and S.W.; formal analysis, R.T.; investigation, J.W. and S.W.; resources, J.W. and R.T.; data curation, J.W.; writing—original draft preparation, J.W. and R.T.; writing—review and editing, R.T. and P.Z.; supervision, J.W.; and funding acquisition, J.W. All authors have read and agreed to the published version of the manuscript.

Funding: This work was supported by the National Natural Science Foundation of China (No. 42204032, 42274019, 41931076, and 42474014), Laoshan Laboratory (LSKJ202205100 and LSKJ202205105), the Shandong Province Science Foundation for Youths (No. ZR2022QD015), the Hainan Provincial Key Research and Development Project (No. ZDYF2024GXJS289), the National Key Research and Development Program of China (No. 2025YFE0104000), and the Excellent Youth Foundation of Shandong Scientific Committee (No. ZR2024JQ024).

Data Availability Statement: The data presented in this study are available on request from the corresponding author.

Acknowledgments: The authors would like to express their gratitude to the editors and the reviewers for their substantive and valuable comments for the improvement of this paper. We would like to thank the Tokyo University of Marine Science and Technology for providing the open-source GNSS software RTKLIB.

Conflicts of Interest: The authors declare no conflicts of interest.

References

1. Holthuijsen, L.H. *Waves in Oceanic and Coastal Waters*; Cambridge University Press: Cambridge, UK, 2010; ISBN 978-1-139-46252-5.
2. Kinsman, B. *Wind Waves: Their Generation and Propagation on the Ocean Surface*; Courier Corporation: North Chelmsford, MA, USA, 1984; ISBN 0-486-64652-1.
3. Massel, S.R. *Ocean Surface Waves: Their Physics and Prediction*; World Scientific: Singapore, 2017; Volume 45, ISBN 981-322-839-3.
4. He, Q.; Silliman, B.R. Climate Change, Human Impacts, and Coastal Ecosystems in the Anthropocene. *Curr. Biol.* **2019**, *29*, R1021–R1035. [CrossRef]
5. Hasselmann, K.; Barnett, T.P.; Bouws, E.; Carlson, H.; Cartwright, D.E.; Enke, K.; Ewing, J.A.; Gienapp, A.; Hasselmann, D.E.; Kruseman, P. Measurements of Wind-Wave Growth and Swell Decay during the Joint North Sea Wave Project (JONSWAP). *Ergaenzungsheft Dtsch. Hydrogr. Z. Reihe A* **1973**, *8*, 1–15.
6. Marques, O.B.; Feddersen, F.; MacMahan, J. An Effective Water Depth Correction for Pressure-Based Wave Statistics on Rough Bathymetry. *J. Atmos. Ocean. Technol.* **2024**, *41*, 1047–1062. [CrossRef]
7. Mercier, P.; Thiébaut, M.; Guillou, S.; Maisondieu, C.; Poizot, E.; Pieterse, A.; Thiébot, J.; Filipot, J.-F.; Grondeau, M. Turbulence Measurements: An Assessment of Acoustic Doppler Current Profiler Accuracy in Rough Environment. *Ocean Eng.* **2021**, *226*, 108819. [CrossRef]
8. Bruno, F.; Bianco, G.; Muzzupappa, M.; Barone, S.; Razionale, A.V. Experimentation of Structured Light and Stereo Vision for Underwater 3D Reconstruction. *ISPRS J. Photogramm. Remote Sens.* **2011**, *66*, 508–518. [CrossRef]
9. Tannuri, E.A.; Simos, A.N.; Sparano, J.V.; Matos, V.L. Motion-Based Wave Estimation: Small-Scale Tests with a Crane-Barge Model. *Mar. Struct.* **2012**, *28*, 67–85. [CrossRef]
10. Wyatt, L.R.; Thompson, S.P.; Burton, R.R. Evaluation of High Frequency Radar Wave Measurement. *Coast. Eng.* **1999**, *37*, 259–282. [CrossRef]
11. Ma, C.; Pan, L.; Qiu, Z.; Liang, D.; Chen, G.; Yu, F.; Sun, H.; Sun, D.; Wu, W. Ocean Wave Inversion Based on a Ku/Ka Dual-Band Airborne Interferometric Imaging Radar Altimeter. *Remote Sens.* **2022**, *14*, 3578. [CrossRef]
12. Tucker, M.J. *Waves in Ocean Engineering: Measurement, Analysis, Interpretation*; E. Horwood: New York, NY, USA, 1991; pp. 109–111.
13. De Vries, J.J.; Waldron, J.; Cunningham, V. Field Tests of the New Datawell DWR-G GPA Wave Buoy. *Sea Technol.* **2003**, *44*, 50–55.
14. Behrens, J.; Terrill, E.; Thomas, J.O. CDIP Wave Observations during Hurricanes Irma, Jose, and Maria, and a nor'easter. *Shore Beach* **2018**, *86*, 14–20.
15. Bidlot, J.-R.; Holmes, D.J.; Wittmann, P.A.; Lalbeharry, R.; Chen, H.S. Intercomparison of the Performance of Operational Ocean Wave Forecasting Systems with Buoy Data. *Weather Forecast.* **2002**, *17*, 287–310. [CrossRef]
16. Gorman, R.M. Estimation of Directional Spectra from Wave Buoys for Model Validation. *Procedia Iutam* **2018**, *26*, 81–91. [CrossRef]
17. Fund, F.; Perosanz, F.; Testut, L.; Loyer, S. An Integer Precise Point Positioning Technique for Sea Surface Observations Using a GPS Buoy. *Adv. Space Res.* **2013**, *51*, 1311–1322. [CrossRef]
18. Dawidowicz, K. Sea Level Changes Monitoring Using GNSS Technology—A Review of Recent Efforts. *Acta Adriat.* **2014**, *55*, 145–162.
19. Zhu, L.; Yang, L.; Xu, Y.; Yang, F.; Zhou, X. Retrieving Wave Parameters from GNSS Buoy Measurements Using the PPP Mode. *IEEE Geosci. Remote Sens. Lett.* **2020**, *19*, 1–5. [CrossRef]
20. Kato, T.; Terada, Y.; Tadokoro, K.; Kinugasa, N.; Futamura, A.; Toyoshima, M.; Yamamoto, S.; Ishii, M.; Tsugawa, T.; Nishioka, M. Development of GNSS Buoy for a Synthetic Geohazard Monitoring System. *J. Disaster Res.* **2018**, *13*, 460–471. [CrossRef]
21. Lutsenko, V.I.; Lutsenko, I.V.; Popov, I.V.; Luo, Y.; Anh, N.X.; Guo, Q.; Zheng, Y. GNSS Signal Use for Sea Waves Monitoring. In Proceedings of the 2020 IEEE Ukrainian Microwave Week (UkrMW), Kharkiv, Ukraine, 21–25 September 2020; pp. 768–771.
22. Yang, L.; Xu, Y.; Jiang, Y.; Mertikas, S.P.; Wang, Z.; Zhu, L.; Liu, N.; Lin, L. Retrieval of Ocean Wave Characteristics via Single-Frequency Time-Differenced Carrier Phases From GNSS Buoys. *IEEE Trans. Geosci. Remote Sens.* **2024**, *62*, 4204512. [CrossRef]
23. Gu, H.; Zhu, X.; Shan, R.; Zang, J.; Qian, L.; Lin, P. Evaluation of a GNSS for Wave Measurement and Directional Wave Spectrum Analysis. *Ocean Eng.* **2023**, *270*, 113683. [CrossRef]

24. Liang, G.; Li, S.; Bao, K.; Wang, G.; Teng, F.; Zhang, F.; Wang, Y.; Guan, S.; Wei, Z. Development of GNSS Buoy for Sea Surface Elevation Observation of Offshore Wind Farm. *Remote Sens.* **2023**, *15*, 5323. [CrossRef]
25. Shum, C.K.; Lee, H.; Abusali, P.A.M.; Braun, A.; de Carufel, G.; Fotopoulos, G.; Komjathy, A.; Kuo, C. Prospects of Global Navigation Satellite System (GNSS) Reflectometry for Geodynamic Studies. *Adv. Space Res.* **2011**, *47*, 1814–1822. [CrossRef]
26. Jing, C.; Niu, X.; Duan, C.; Lu, F.; Di, G.; Yang, X. Sea Surface Wind Speed Retrieval from the First Chinese GNSS-R Mission: Technique and Preliminary Results. *Remote Sens.* **2019**, *11*, 3013. [CrossRef]
27. Ruffini, G.; Soulat, F.; Caparrini, M.; Germain, O.; Martín-Neira, M. The Eddy Experiment: Accurate GNSS-R Ocean Altimetry from Low Altitude Aircraft. *Geophys. Res. Lett.* **2004**, *31*, L12306. [CrossRef]
28. Guerra, M.; Hay, A.E.; Karsten, R.; Trowse, G.; Cheel, R.A. Turbulent Flow Mapping in a High-Flow Tidal Channel Using Mobile Acoustic Doppler Current Profilers. *Renew. Energy* **2021**, *177*, 759–772. [CrossRef]
29. Tomita, F.; Kido, M.; Iinuma, T.; Ohta, Y. GNSS-Acoustic Positioning Error in the Vertical Component Considering the Uncertainty of a Reference Sound Speed Profile. *Mar. Geophys. Res.* **2025**, *46*, 3. [CrossRef]
30. Pérez Gómez, B.; Vilibić, I.; Šepić, J.; Međugorac, I.; Ličer, M.; Testut, L.; Fraboul, C.; Marcos, M.; Abdellaoui, H.; Álvarez Fanjul, E. Coastal Sea Level Monitoring in the Mediterranean and Black Seas. *Ocean Sci.* **2022**, *18*, 997–1053. [CrossRef]
31. Lin, Y.-P.; Huang, C.-J.; Chen, S.-H. Variations in Directional Wave Parameters Obtained from Data Measured Using a GNSS Buoy. *Ocean Eng.* **2020**, *209*, 107513. [CrossRef]
32. Lin, Y.-P.; Huang, C.-J.; Chen, S.-H.; Doong, D.-J.; Kao, C.C. Development of a GNSS Buoy for Monitoring Water Surface Elevations in Estuaries and Coastal Areas. *Sensors* **2017**, *17*, 172. [CrossRef]
33. Lei, J.; Li, F.; Zhang, S.; Xiao, C.; Xie, S.; Ke, H.; Zhang, Q.; Li, W. Ocean Tides Observed from A GPS Receiver on Floating Sea Ice near Chinese Zhongshan Station, Antarctica. *Mar. Geod.* **2018**, *41*, 353–367. [CrossRef]
34. Wang, F.; Yang, D.; Wang, J.; Xing, J.; Yu, Y. Shipborne GNSS Reflectometry for Monitoring Along-Track Significant Wave Height and Wind Speed. *Ocean Eng.* **2023**, *281*, 114935. [CrossRef]
35. Rychlik, I.; Johannesson, P.; Leadbetter, M.R. Modelling and Statistical Analysis of Ocean-Wave Data Using Transformed Gaussian Processes. *Mar. Struct.* **1997**, *10*, 13–47. [CrossRef]
36. Zhang, J.; Ma, Y.; Benoit, M. Statistical Distributions of Free Surface Elevation and Wave Height for Out-of-Equilibrium Sea-States Provoked by Strong Depth Variations. *Ocean Eng.* **2024**, *293*, 116645. [CrossRef]
37. Gryazin, D.G.; Belova, O.O.; Gleb, K.A.; Kovchin, M.I. Study of the Characteristics of Wave Buoys by Comparison Measurements. *Oceanology* **2018**, *58*, 742–748. [CrossRef]
38. Dehghan, Y.; Sadrinasab, M.; Chegini, V.; Akhyani, M. Wave Spectrum Analysis for Southern Caspian Sea. *Ocean Eng.* **2023**, *272*, 113490. [CrossRef]
39. Krogstad, H.E. Maximum Likelihood Estimation of Ocean Wave Spectra from General Arrays of Wave Gauges. *Model. Identif. Control.* **1988**, *9*, 81–97. [CrossRef]
40. Liu, P.C. Wavelet Spectrum Analysis and Ocean Wind Waves. In *Wavelet Analysis and Its Applications*; Elsevier: Amsterdam, The Netherlands, 1994; Volume 4, pp. 151–166. ISBN 1874-608X.
41. Joodaki, G.; Nahavandchi, H.; Cheng, K. Ocean Wave Measurement Using GPS Buoys. *J. Geod. Sci.* **2013**, *3*, 163–172. [CrossRef]
42. Hashimoto, N. Analysis of the Directional Wave Spectrum from Field Data. *Adv. Coast. Ocean Eng.* **1997**, *3*, 103–144.
43. Torabi, M.; Shahidi, R.; Gill, E.W. Ocean Surface Wave Parameter Estimation Using Periodogram Method. In Proceedings of the XXXVth URSI General Assembly and Scientific Symposium (URSI GASS), Sapporo, Hokkaido, Japan, 19–26 August 2023.
44. Rasool, S.; Muttaqi, K.M.M.K.M. Modeling Ocean Waves and Investigation of Oceanic Wave Spectra for Wave-to-Wire System. *J. Eng. Res.* **2022**, *10*, 1–17. [CrossRef]
45. Valipour, M.; Banihabib, M.E.; Behbahani, S.M.R. Parameters Estimate of Autoregressive Moving Average and Autoregressive Integrated Moving Average Models and Compare Their Ability for Inflow Forecasting. *J. Math. Stat.* **2012**, *8*, 330–338. [CrossRef]
46. Gryazin, D.; Gleb, K. A New Method to Determine Directional Spectrum of Sea Waves and Its Application to Wave Buoys. *J. Ocean Eng. Mar. Energy* **2022**, *8*, 269–283. [CrossRef]
47. Krogstad, H.E.; Magnusson, A.K.; Donelan, M.A. Wavelet and Local Directional Analysis of Ocean Waves. In Proceedings of the ISOPE International Ocean and Polar Engineering Conference, San Francisco, CA, USA, 28 May–2 June 2006; ISOPE: Mountain View, CA, USA, 2006; p. ISOPE-I-06-204.
48. Donelan, M.; Babanin, A.; Sanina, E.; Chalikov, D. A Comparison of Methods for Estimating Directional Spectra of Surface Waves. *J. Geophys. Res. Ocean.* **2015**, *120*, 5040–5053. [CrossRef]
49. Chang, Z.; Han, F.; Sun, Z.; Gao, Z.; Wang, L. Three-Dimensional Dynamic Sea Surface Modeling Based on Ocean Wave Spectrum. *Acta Oceanol. Sin.* **2021**, *40*, 38–48. [CrossRef]
50. Yu, Y.; Liu, S. *Random Wave and Its Applications to Engineering*, 4th ed.; Dalian University of Technology Press: Dalian, China, 2011; ISBN 978-7-5611-0466-8.
51. Smith, S.W. *The Scientist and Engineer's Guide to Digital Signal Processing*; California Technical Pub: Brisbane, CA, USA, 1997.

52. Lyons, R.G. *Understanding Digital Signal Processing, 3/E*; Pearson Education India: Chennai, India, 1997; ISBN 81-317-6436-2.
53. Panahi, R.; Shafieefar, M. Development of a Bi-Modal Directional Wave Spectrum. *Ocean Eng.* **2015**, *105*, 104–111. [CrossRef]

remote sensing

MDPI

Article

GPS and Galileo Precise Point Positioning Performance with Tropospheric Estimation Using Different Products: BRDM, RTS, HAS, and MGEX

Damian Kiliszek

Faculty of Civil Engineering and Geodesy, Military University of Technology, Gen. S. Kaliskiego 2, 00-908 Warsaw, Poland; damian.kiliszek@wat.edu.pl

Abstract: The performance of Precise Point Positioning (PPP) using different Global Navigation Satellite System (GNSS) product sets, including broadcast ephemerides, International GNSS Service Real-Time Service (IGS-RTS) corrections, Galileo High Accuracy Service (HAS) corrections, and precise products from the Center for Orbit Determination in Europe (CODE) Multi-GNSS Experiment (MGEX), has been evaluated. The availability of solutions, convergence time, position accuracy and Zenith Tropospheric Delay (ZTD) estimation across these products were analyzed using simulated real-time and postprocessing static modes, using data from globally distributed stations with a 1 s observation interval. The results indicate that precise products from the MGEX provide the highest accuracy, achieving centimeter-level precision in post-processed mode. Real-time simulated solutions, such as HAS and IGS-RTS, deliver promising results, with Galileo HAS meeting its target accuracy of 20 cm horizontally and 40 cm vertically and a convergence time under 5 min. However, Global Positioning System (GPS) performance within HAS is limited by a significantly lower correction availability—around 67% on average compared to over 95% for Galileo—which negatively impacts PPP performance. ZTD estimation results show that real-time services (HAS, IGS-RTS) achieved errors within 1–3 cm, sufficient for meteorological applications. This study highlights the growing importance of HAS in real-time positioning applications and suggests further improvements in GPS for enhanced performance.

Keywords: HAS; RTS; PPP; GPS; Galileo; ZTD; convergence time

1. Introduction

Over the past three decades, the concept of precise absolute positioning has evolved significantly since its initial formulation [1–5]. Today, the Precise Point Positioning (PPP) method is widely applied across a broad spectrum of scientific and engineering domains [6], including seismology [7], reference frame determination [8], atmospheric sounding [9,10], natural hazard monitoring [11], and the estimation of drone trajectories [12]. PPP can achieve a millimeter- to centimeter-level accuracy in static, post-processed applications and a decimeter-level accuracy in kinematic and real-time scenarios. However, the primary limitation of PPP remains its relatively long convergence time [13].

In response to this limitation, substantial research efforts in recent years have focused on enhancing the efficiency and reliability of absolute positioning. Notable advancements include the development of PPP with Ambiguity Resolution (PPP-AR) [14,15], PPP-Real-Time Kinematic (PPP-RTK) [16,17], different/combined products that are available [18–20], and the use of multi-GNSS and multi-frequency measurements [21–24]. In parallel, interest has grown in single-frequency solutions and low-cost GNSS receivers, particularly for emerging technologies like smartwatches, drones, and smartphones [25,26].

In the coming years, the adoption of PPP is expected to expand further, particularly in real-time applications. This trend is driven by ongoing technological developments, including the increasing use of unmanned aerial systems, the advancement in smart city infrastructures, and the rising demand for reliable, high-precision positioning in autonomous systems. These use cases require independent, globally available, and highly accurate localization solutions.

Recent advancements in GNSS infrastructure have further accelerated the evolution of precise positioning techniques. Next-generation satellites—such as the GPS Block III, which enables additional observations on the L5 frequency [27]—are gradually being deployed. Concurrently, the number of operational Galileo satellites continues to increase [28]. As the Galileo constellation expands, further improvements in positioning accuracy are anticipated, particularly when leveraging onboard ephemeris products [29–31]. Notably, while GPS ephemeris data are updated every two hours, Galileo satellites provide more frequent updates, ranging from every 10 to 60 min.

To facilitate high-accuracy positioning in real time, the International GNSS Service (IGS) initiated the Real-Time Pilot Project in 2007 and officially launched the Real-Time Service (RTS) in 2013 [32,33]. The RTS supports real-time applications by providing access to precise GNSS products, including orbit and clock corrections, as well as code and phase biases [34]. These corrections serve as a viable alternative to ultra-rapid products, thereby enabling global real-time PPP, time synchronization, and natural hazard monitoring.

Currently, IGS RTS offers multi-GNSS combination products—SSRA02IGS1/SSRC02IGS1, which support GPS, GLONASS, and Galileo, and SSRA03IGS1/SSRC03IGS1, which also include BeiDou corrections. These corrections are applied to broadcast data but require an internet connection and compatible software such as the BKG NTRIP Client (BNC) [35]. In addition to the official IGS RTS solutions, several individual Analysis Centers (ACs) provide their own real-time products [36]. A study by Li et al. [37] comparing official combined RTS corrections and individual AC products showed that the highest accuracy was achieved using corrections from CNES and WHU, with CNES offering superior availability.

Despite their utility, the dependence on internet connectivity limits the use of these products in many regions. Recently, new correction services broadcast directly via navigation systems have emerged, offering enhanced accessibility. These include the BeiDou B2b service, Japan's Centimeter-Level Augmentation Service (CLAS) for QZSS [38], and Europe's Galileo High Accuracy Service (HAS) [39]. While B2b and CLAS are broadcast via geostationary satellites and thus limited to the Asia–Pacific region, Galileo HAS data are transmitted by a subset of satellites—currently up to 20—from the overall constellation, enabling global coverage with varying levels of availability [40].

The Galileo HAS will operate at two levels [41]. Level 1 provides global coverage with a target horizontal accuracy below 20 cm and vertical accuracy below 40 cm, with convergence times under 300 s. Level 2, limited to Europe, aims to achieve the same accuracy with a faster convergence time (under 100 s) by incorporating regional atmospheric corrections. HAS's development comprises three phases: Phase 1 (testing and experimentation, 2020–2022); Phase 2 (Initial Service), launched on 24 January 2023; and Phase 3 (Full Service), which will deliver Level 2 capabilities [40].

As a central element of the European GNSS ecosystem, Galileo HAS offers free access to PPP corrections globally, broadcast via the E6-B signal or through the internet. Although the service is relatively new, its potential is already evident. Open-source tools such as HASlib [42], HAS PPP [43], and Python-based decoding toolboxes have been developed to enable the real-time reception and processing of HAS corrections [44]. Preliminary studies using test signals have demonstrated a promising PPP performance [45–47]. Furthermore, integration with tools like RTKLIB [48] highlights the potential of open-source platforms to broaden access to high-precision GNSS positioning. Nevertheless, further research is required to enhance the service's reliability and broaden its applicability.

One of the key advantages of PPP is its capability to estimate tropospheric delay parameters, particularly the Zenith Tropospheric Delay (ZTD), in both post-processed and real-time modes. These estimates are essential for atmospheric studies, as they can be further converted into Precipitable Water Vapor (PWV) and assimilated into Numerical Weather Prediction (NWP) models to improve short-term weather forecasts. Several studies have highlighted the sensitivity of ZTD estimations to the quality of input products and modeling strategies [49]. For real-time applications, particularly those supporting NWP data assimilation, accuracy thresholds of 0.6 cm (target) and 3 cm (maximum) are typically required [10]. Recent advances in real-time GNSS processing, including PPP based on broadcast ephemerides [46,49] and the integration of multiple GNSS constellations [50,51], have shown the potential to meet these requirements. Additionally, current research continues to investigate how spatial resolution and data assimilation methods affect the accuracy of humidity fields derived from GNSS-based PWV products [52], reinforcing the importance of reliable real-time ZTD estimations in operational meteorology.

Currently, several studies on PPP positioning using HAS corrections have been carried out. However, they differ in terms of the time periods, processing software, and models used, and they often do not compare the obtained results with other products and corrections.

The aim of this study is to analyze PPP performance using different types of products and corrections: broadcast ephemerides, CNES RTS corrections, Galileo HAS corrections, and final precise products from the Center for Orbit Determination in Europe (CODE) within the Multi-GNSS Experiment (MGEX). The analysis is carried out using a uniform PPP processing scheme: all test cases use identical 1 s interval observations from globally distributed stations, the same processing software, and consistent modeling assumptions. This setup enables an objective and direct comparison of PPP performance across the different products.

The key performance indicators evaluated include solution availability, convergence time, positioning accuracy, and ZTD estimation for GPS, Galileo, and combined GPS+Galileo solutions. Importantly, HAS data were collected in real time, enabling a realistic simulation of near-real-time PPP performance. BRDM and CNES RTS corrections were also used in post-processing, representing additional simulated real-time scenarios. The final CODE MGEX products serve as the most precise reference solution and are used as a benchmark for the highest achievable PPP accuracy.

This study also examines the availability of HAS products and PPP solutions, highlighting practical challenges in real-time applications. In addition, 1 s observation intervals were employed with final CODE products—an uncommon approach in post-processing that enables a more detailed analysis of short-term convergence behavior.

Following the introduction, Section 2 describes the data, product sets, and processing methodology. Section 3 presents the results of the conducted analyses. Section 4 provides a discussion, and Section 5 offers a summary of the key findings.

2. Methodology and Data

This analysis utilized data from 12 globally distributed MGEX stations (Figure 1 and Table 1), which collected multi-GNSS observations at 1 s intervals over a six-day period, from Day of Year (DoY) 223 to DoY 228 in 2024. These days were selected because real-time HAS product data were successfully collected for them. For each day, 8 three-hour sessions were analyzed, resulting in a total of 48 sessions per station across the six-day span. The computations were performed using the raPPPid software (version 3.1), which employs an Extended Kalman Filter (EKF), in combination with the conventional PPP model [53,54].

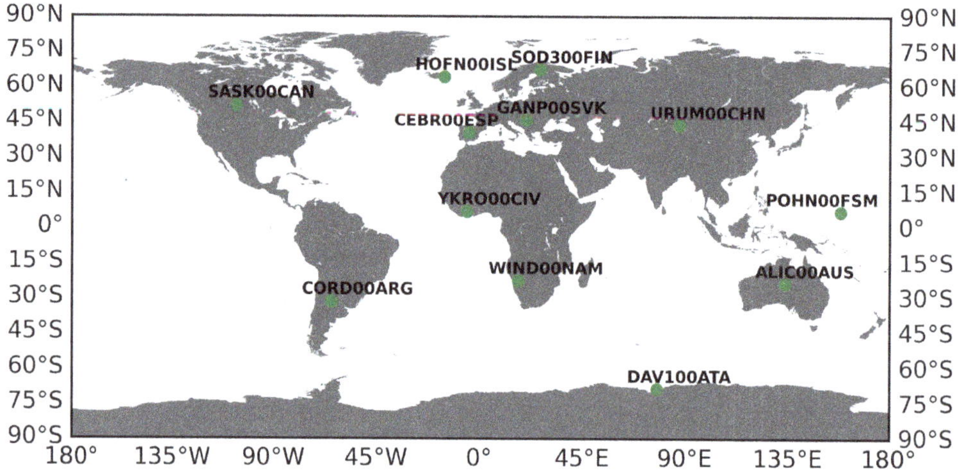

Figure 1. Location of test stations.

Table 1. Equipment of the used IGS stations.

Station	Receiver	Antenna
ALIC00AUS	SEPT POLARX5	TWIVC6050 NONE
CEBR00ESP	SEPT POLARX5TR	SEPCHOKE_B3E6 NONE
CORD00ARG	SEPT POLARX5	TPSCR.G3 NONE
DAV100ATA	SEPT POLARX5	LEIAR25.R3 LEIT
GANP00SVK	TRIMBLE ALLOY	TRM59800.00 SCIS
HOFN00ISL	LEICA GR50	LEIAR25.R4 LEIT
POHN00FSM	SEPT POLARX5	JAVRINGANT_DM NONE
SASK00CAN	JAVAD TRE_G3TH DELTA	NOV750.R4 NOVS
SOD300FIN	JAVAD TRE_3 DELTA	JAVRINGANT_DM SCIS
URUM00CHN	JAVAD TRE_3	JAVRINGANT_G5T NONE
WIND00NAM	JAVAD TRE_3	JAVRINGANT_G5T NONE
YKRO00CIV	JAVAD TRE_3 DELTA	ASH701945C_M NONE

The undifferenced ionosphere-free linear combinations for GPS and Galileo code and carrier-phase observations (in units of length) are expressed as

$$\begin{aligned} PR_{IF,G} &= \rho_r^G + c\left(\delta_{r,clock,G} - \delta_{clock}^G\right) + T_r^G + \varepsilon_{IF} \\ \Phi_{IF,G} &= \rho_r^G + c\left(\delta_{r,clock,G} - \delta_{clock}^G\right) + T_r^G + \lambda_{IF}^G N_{IF}^G + \epsilon_{IF} \\ PR_{IF,E} &= \rho_r^E + c\left(\delta_{r,clock,E} - \delta_{clock}^E\right) + T_r^E + \varepsilon_{IF} \\ \Phi_{IF,E} &= \rho_r^E + c\left(\delta_{r,clock,E} - \delta_{clock}^E\right) + T_r^E + \lambda_{IF}^E N_{IF}^E + \epsilon_{IF} \end{aligned} \tag{1}$$

where

- G, E denote the GPS and Galileo systems, respectively;
- PR_{IF} and Φ_{IF} are ionosphere-free linear combination for code and carrier-phase observations, respectively [m];

- ρ_r^S is the true geometric range between the satellite in the emission time and the receiver in the reception time [m];
- c is the speed of light [m/s];
- $\delta_{r,clock,G}$, $\delta_{r,clock,E}$ represent the receiver clock offsets in the GPS and Galileo system time, respectively [s]
- δ^G_{clock}, δ^E_{clock} denote the satellite clock offsets for GPS and Galileo satellites, respectively [s];
- T_r^S is the slant tropospheric delay [m];
- λ_{IF} is the wavelength for the ionosphere-free linear combination [m];
- N_{IF} is the real value of the ambiguity ionosphere-free linear combination [cycles];
- ε_{IF}, ϵ_{IF} are other errors, for example noise and multipath [m].

This model also requires the inclusion of errors such as receiver and satellite phase center corrections, relativistic correction for code and carrier-phase observations of the ionosphere-free linear combinations and carrier-phase wind-up for phase observations, and the use of site displacement effects, which are defined in the IERS conventions [55].

In this way, the station coordinates, clock correction of the receiver for each system, wet component of the tropospheric delay, and real ambiguity values were estimated. A list of the used products, models, and methods are shown in Tables 2 and 3.

Table 2. Solutions names and input data sources.

Solution Name	Products
BRDM	broadcast products BRD400DLR; GPS LNAV and Galileo INAV
HAS	broadcast products BCEP00BKG0 with HAS corrections SSRA00EUH0; GPS LNAV and Galileo INAV; collected in real-time with BNC software (https://igs.bkg.bund.de/ntrip/bnc (accessed on 14 June 2025))
CNES	broadcast products with IGS RTS CNES corrections converted to daily CNES archive products as SP3, CLK, and BIA (http://www.ppp-wizard.net/products/REAL_TIME/ (accessed on 14 June 2025)
CODE	precise CODE MGEX FINAL products: SP3 with 5 min interval, CLK with 30 s interval

Table 3. Methods and models used. I think the table should be on one page.

Items	Models/Methods
Positioning mode	static mode
PPP model	conventional PPP model using undifferenced dual-frequency code and phase ionosphere-free linear combination
Sessions	eight three-hour sessions per day (48 sessions for each station from six days)
Signals	BRDM, CNES, and CODE solutions: L1/L2 for GPS and E1/E5a for Galileo HAS solution: L1/L2 for GPS and E1/E5b for Galileo
Stochastic modeling	$\sin^2 ele$
Constellation	GPS (G), Galileo (E), GPS+Galileo (GE)
Standard deviation of raw observations	code = 0.300 m, phase = 0.002 m
Cut-off elevation angle	10°
Interval estimation	1-s
Periods	six days: from 223 DoY to 228 DoY of 2024
Reference frame	IGS20

Table 3. *Cont.*

Items	Models/Methods
Reference coordinates	daily IGS SINEX files
PCO and PCV for satellite antenna	igs20.atx for CNES and CODE solutions; for BRDM and HAS the products were referred to APC
PCO and PCV for receiver antenna	igs20.atx
Ionospheric delay	ionosphere-free linear combination
Tropospheric delay	a priori value: GPT3 (VMF3 for CODE solution) estimated: wet component mapping function: GPT3 (VMF3 for CODE solution) gradients: estimated
Solid tides, ocean loading, Shapiro effects, phase wind-up	IERS convention 2010
Ambiguities	float
Receiver clock correction	estimated separately for each system

Four different solutions were computed for GPS, Galileo, and GPS+Galileo constellations:

- **'BRDM' solution**: computed using only broadcast products, with the BRD400DLR broadcast ephemerides downloaded;
- **'HAS' solution**: computed using broadcast ephemerides combined with HAS corrections. The products were collected in real-time using BNC software;
- **'CNES' solution**: computed using CNES archive products. The broadcast ephemerides and real-time CNES corrections were converted to SP3 and CLK files;
- **'CODE' solution**: computed using the final CODE MGEX products. These products were downloaded as SP3 and CLK with 5 min and 30 s intervals, respectively.

The analysis was conducted using a 10° elevation angle cut-off, with the $\sin^2 ele$, function applied for stochastic modeling. The reference coordinates were based on the daily IGS20 SINEX reference frame. The estimated tropospheric delay was compared with the official IGS value.

To estimate the ZTD for the BRDM, HAS, and CNES solutions, the GPT3 model was applied, while the VMF3 model was used for the CODE solution [56]. Additionally, for the CODE solution, the CODE MGEX satellite biases were included as observable-specific signal bias (OSB) files [14]. For the HAS and CNES solutions, bias corrections were obtained from the respective correction streams. In the case of the CNES solution, phase bias corrections were also applied, while for the HAS solution, only code bias corrections were used. For the BRDM solution, the Time Group Delay (TGD) data, as transmitted in the navigation messages, were utilized [57].

3. Results

The analysis focused on four key evaluation criteria: solution availability, convergence time, accuracy, and the precision of ZTD estimation. The following subsections present the results for each of these aspects.

3.1. Solution Availability

Solution availability was defined as the ratio of epochs in which position solutions were successfully derived to the total number of epochs for which position solutions were obtained using the CODE products. The analysis included only epochs where both GNSS observations and the corresponding orbit and clock products (or corrections) were available, which allowed the computation of positions and enabled an objective comparison of the performance of different solutions.

Table 4 presents the results of this analysis, including the average number of satellites used for each solution.

Table 4. Availability of obtained solutions and used satellites. Averaged across all analyzed stations.

Solution	GNSS	Number of Used Satellites	Availability [%]
BRDM	G	8.9	100.0
	E	6.8	99.8
	GE	15.6	100.0
HAS	G	5.0	74.0
	E	6.9	99.9
	GE	10.6	99.1
CNES	G	9.0	100.0
	E	6.5	99.7
	GE	15.5	100.0
CODE	G	9.3	-
	E	8.1	-
	GE	17.4	-

For both the BRDM and CNES solutions, a 100% availability was achieved for the GPS and combined GPS+Galileo constellations, with an average of approximately 9 GPS satellites and approximately 15.5 satellites, respectively, for the combined systems. For Galileo-only solutions, the availability was slightly lower—99.8% for BRDM and 99.7% for CNES—with a slightly reduced number of satellites: 6.8 and 6.5 on average, respectively. This is due to the fact that fewer Galileo satellites were available in BRDM and CNES products; for CNES products, about 20–21 Galileo satellites were available.

Different results were observed for the HAS products. For Galileo, solution availability reached 99.9% with about 6.9 satellites used, while for GPS it was significantly lower—only 74%, with an average of 5 satellites used. The HAS service is not yet fully operational and is primarily focused on the Galileo system, which may explain the lower availability and accuracy of the GPS-based solutions [44]. This is further illustrated in Figure 2, which shows the average availability of HAS orbit and clock corrections per satellite for both the Galileo and GPS constellations, averaged over the entire observation period. The dashed line marks the 90% availability threshold, highlighting that Galileo consistently exceeds this level with average availabilities of 95.49% for orbit corrections and 95.95% for clock corrections. In contrast, GPS average availabilities are much lower, at 67.11% for orbit and 67.64% for clock corrections. Notably, 10 GPS satellites show an availability below 50%, with 4 of them even below 20%—G07 (IIR-M), G13 (IIR), G15 (IIR-M), and G30 (IIF)—which negatively impacts the overall GPS performance in HAS products. While these satellites come from different blocks—IIR, IIR-M, and IIF—there is no clear dependency between satellite block type and correction availability. This indicates that other factors, such as satellite-specific conditions, operational status, or current HAS configuration, may be more influential in determining the performance of individual satellites.

Combining GPS and Galileo for HAS products slightly reduced the overall solution availability to 99.1%, with an average of 10.6 satellites used. This is likely due to the lower accuracy of the HAS products for GPS satellites, negatively affecting the overall quality of position solutions.

The highest number of satellites used was recorded for the CODE solution: 9.3 GPS satellites, 8.1 Galileo satellites, and 17.4 satellites for the combined GPS+Galileo constellation.

Figure 2. Average availability of HAS ORBIT and CLOCK corrections per satellite for the Galileo and GPS constellations. The dashed line marks the 90% availability threshold.

3.2. Convergence Time

This subsection analyzes the convergence times relative to the required accuracy levels. Additionally, the number of observational sessions meeting the specified thresholds was determined. Convergence time was defined as the moment when the errors reached the specified threshold and remained within it for at least the next 30 min. The thresholds were defined as 10 cm for both 2D and 3D for the CNES and CODE solutions. For the BRDM and HAS solutions, the thresholds were set at 20 cm horizontally (2D) and 40 cm vertically (U), which is the main objective of the HAS service; additionally, a 40 cm threshold was applied for spatial (3D). Identical accuracy thresholds were used for both BRDM and HAS to ensure a fair comparison, as the main goal is to highlight the impact of HAS corrections relative to broadcast-only data.

Figure 3 presents the results, and detailed data are provided in Appendix A. Table 5 summarizes the percentage of sessions meeting the required accuracy levels.

Table 5. Percentage of sessions in which the convergence time met the defined accuracy thresholds for all solutions and constellations.

Solution	GNSS	U	2D	3D
BRDM	G	80.7	27.4	35.9
	E	92.0	64.6	70.8
	GE	93.2	62.8	74.0
HAS	G	45.1	22.2	26.0
	E	75.0	70.7	73.3
	GE	75.0	71.0	74.0
CNES	G	99.1	99.5	97.6
	E	96.2	96.7	91.7
	GE	99.6	99.6	99.0
CODE	G	100.0	99.5	99.5
	E	99.8	99.6	98.8
	GE	100.0	99.8	99.6

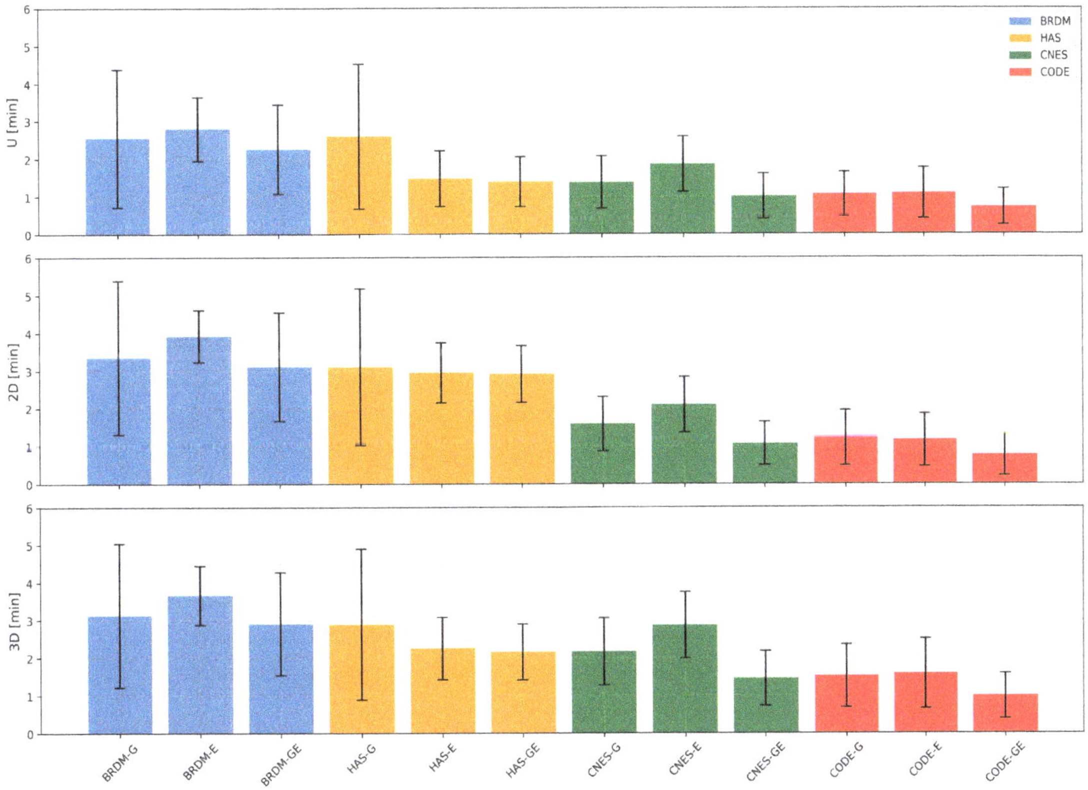

Figure 3. Mean convergence time with standard deviation for all solutions and constellations for U, 2D, and 3D.

Solutions based on CNES and CODE products demonstrated a consistently outstanding convergence performance, with success rates approaching or exceeding 99% in the most favorable configurations. For the CODE solution, convergence times were remarkably short, averaging approximately 1.2 min in 2D and 1.5 min in 3D when using GPS-only signals, with a slightly better performance observed for Galileo signals in 2D. The combined GPS+Galileo (GE) configuration yielded the best results, achieving convergence in less than 1 min in 2D and approximately 1 min in 3D, with success rates reaching up to 99.6%.

The CNES solution, simulating real-time operational conditions, also demonstrated a high percentage of sessions that successfully met the defined convergence criteria, indicating a reliable convergence performance, although convergence times were slightly longer compared to CODE. For GPS-only observations, convergence thresholds were achieved in 99.5% of sessions for 2D and 97.6% for 3D positioning. However, when using Galileo-only signals, the percentage of sessions that successfully reached convergence dropped to 96.7% for 2D and 91.7% for 3D, with average convergence times increasing to 2.09 min in 2D and 2.87 min in 3D. Nevertheless, when combining GPS and Galileo signals, the CNES solution achieved convergence within approximately 1.1 min in 2D and 1.5 min in 3D, with the percentage of sessions successfully reaching convergence closely matching that of the CODE solution—up to 99.6% for 2D and 99.0% for 3D.

In contrast, products based on BRDM and HAS corrections exhibited significantly longer convergence times and a lower percentage of sessions successfully reaching convergence. It should be noted that different convergence thresholds were applied for BRDM and HAS compared to CODE and CNES solutions. For the BRDM solution using GPS-only observations, the average convergence time was approximately 3.17 min for the vertical component (U), 3.35 min in 2D, and 3.13 min in 3D, with a relatively high variability between sessions (standard deviations of ±2.04 min for 2D, ±1.91 min for 3D, and ±2.04 min for U). The percentage of sessions achieving the required thresholds was noticeably lower, amounting to only 80.7% for U, 27.4% for 2D, and 35.9% for 3D positioning. Using Galileo-only signals resulted in slightly worse average times—3.51 min for U, 3.92 min in 2D, and 3.66 min in 3D—but the variability was considerably smaller (±0.69 min for 2D, ±0.78 min for 3D, and ±0.69 min for U), and the percentage of sessions successfully reaching convergence improved to 92.0% for U, 64.6% for 2D, and 70.8% for 3D. A moderate improvement was observed when combining GPS and Galileo signals, with average convergence times of 2.86 min for U, 3.10 min in 2D, and 2.91 min in 3D, although the percentage of sessions successfully converged remained relatively modest.

The HAS solution, benefiting from additional correction data, generally outperformed BRDM, although the differences were not uniform across all configurations. When using GPS-only observations, the convergence performance was noticeably worse, with average times of 2.59 min for U, 3.10 min in 2D, and 2.89 min in 3D, accompanied by a large variability between sessions of ±2.07 min for U, ±2.07 min for 2D, and ±2.00 min for 3D. In this case, the percentage of sessions achieving the required convergence thresholds was very low, amounting to only 45.1% for the vertical component (U), 22.2% for 2D positioning, and 26.05% for 3D. A clear improvement was observed when using Galileo-only signals, where the convergence times decreased to 1.47 min for U, 2.94 min for 2D, and 2.25 min for 3D, with a relatively low variability (±0.80 and ±0.83 min for 2D and 3D and ±0.80 min for U), and success rates exceeded 70%. The best performance for the HAS solution was again obtained when combining GPS and Galileo signals, achieving average convergence times of 1.39 min for U, 2.91 min in 2D, and 2.16 min in 3D, with a relatively small variability (±0.76 and ±0.75 min for 2D and 3D and ±0.76 min for U) and with over 70% of sessions meeting the required accuracy thresholds.

These results indicate that although HAS corrections significantly enhance the positioning performance compared to relying on broadcast data alone (BRDM), the convergence is still slower and less stable than what can be achieved with post-processing (CODE) or with real-time precise products (CNES).

Detailed convergence results for each coordinate component are available in Appendix A.

3.3. Accuracy

The positioning accuracy was evaluated by comparing the coordinates obtained in this study with the daily reference coordinates provided in the IGS20 SINEX files. To mitigate the influence of the initial convergence period, the first 10 min of data were excluded from the analysis. Mean errors and standard deviations are summarized in Tables 6 and 7.

The results obtained for the BRDM, HAS, CNES, and CODE products show significant differences in positioning accuracy, particularly depending on the GNSS signals used. For the BRDM solution, the results indicate a relatively low positioning accuracy, particularly for GPS signals. The mean errors in the N, E, and U components were −1.92 cm, 1.43 cm, and −6.14 cm, respectively, resulting in 2D and 3D accuracies of 64.83 cm and 93.18 cm. Despite relatively small mean errors, the significant standard deviations (±11.38 cm in N, ±24.98 cm in E, and ±31.78 cm in U) indicate considerable variability. For Galileo signals, although the mean errors in N, E, and U slightly increased, the 2D and 3D accuracies improved significantly, reaching 50.33 cm and 71.34 cm, respectively, which was notably better than for GPS. The use of combined GPS and Galileo (GE) signals slightly improved the results, with mean errors of 0.99 cm (N), −1.50 cm (E), and −0.87 cm (U), resulting in 2D and 3D accuracies of 49.57 cm and 69.40 cm, with the standard deviations remained high (±15.66 cm in N, ±26.63 cm in E, ±24.07 cm in U).

Table 6. Mean coordinate difference to reference coordinates of all stations for all stations for all analyzed solutions. Values in cm.

Solution	GNSS	N	E	U	2D	3D
BRDM	G	−1.9	1.4	−6.1	64.8	93.2
	E	−3.4	3.56	−0.7	50.3	71.3
	GE	1.0	−1.5	−0.9	49.6	69.4
HAS	G	0.4	−7.9	−2.5	86.2	128.5
	E	0.2	0.1	1.5	18.9	28.8
	GE	−0.2	0.1	0.4	19.9	29.5
CNES	G	−0.5	−1.2	0.0	5.7	8.0
	E	−1.0	−1.4	−0.2	7.2	11.2
	GE	−0.6	−1.0	0.0	4.0	5.7
CODE	G	0.3	−0.7	−0.2	3.8	5.4
	E	0.1	0.0	−0.4	3.2	4.9
	GE	0.2	−0.2	−0.2	2.2	3.3

Table 7. Standard deviation of coordinate differences to reference coordinates of all stations for all analyzed solutions. Values in cm.

Solution	GNSS	N	E	U	2D	3D
BRDM	G	11.4	25.0	31.8	17.6	21.7
	E	6.8	19.6	14.9	13.1	16.8
	GE	15.7	26.6	24.1	28.5	35.8
HAS	G	21.8	41.8	47.0	38.9	47.4
	E	3.2	6.1	9.1	6.1	10.2
	GE	3.4	7.4	8.6	6.7	10.0
CNES	G	1.7	3.1	2.2	3.0	4.2
	E	1.2	2.6	3.2	2.8	4.7
	GE	1.5	2.2	2.1	2.0	2.8
CODE	G	1.5	2.1	1.4	2.9	4.1
	E	0.6	1.1	1.3	1.9	3.0
	GE	1.2	1.1	0.8	1.8	2.5

For the HAS solution, the results show an intense variability for GPS. The mean errors for GPS were 0.40 cm in N, −7.87 cm in E, and −2.52 cm in U, corresponding to 2D and 3D accuracies of 86.15 cm and 128.50 cm, which are even worse than those for BRDM. GPS results were characterized by particularly large standard deviations (±21.81 cm in N, ±41.82 cm in E, and ±46.98 cm in U). A substantial improvement was observed with Galileo signals, where the mean errors decreased to 0.15 cm (N), 0.06 cm (E), and 1.47 cm (U), achieving 2D and 3D accuracies of 18.91 cm and 28.81 cm. For the GE combination, the mean errors were −0.23 cm in N, 0.09 cm in E, and 0.42 cm in U, with accuracies of 19.93 cm (2D) and 29.50 cm (3D). These results meet the requirements of the HAS service, except for GPS-only. This variability is consistent with the significantly lower availability of HAS corrections for GPS satellites, averaging around 67%, compared to over 95% for Galileo (Section 3.1). The HAS service primarily focuses on Galileo, and several GPS satellites have an availability below 50%, which negatively impacts the GPS positioning accuracy.

The CNES solution, based on real-time precise corrections, achieved a markedly better performance. For GPS, the mean errors were −0.48 cm (N), −1.25 cm (E), and −0.02 cm (U), resulting in 2D and 3D accuracies of 5.66 cm and 7.98 cm. For Galileo, there were mean errors of −0.95 cm (N), −1.39 cm (E), and −0.17 cm (U), leading to accuracies of 7.21 cm (2D) and 11.17 cm (3D), which were not as good as for GPS. The GE combination provided the best performance, with mean errors of −0.64 cm (N), −1.05 cm (E), and −0.04 cm (U) and accuracies of 3.95 cm (2D) and 5.73 cm (3D). Standard deviations were exceptionally low (e.g., ±1.74 cm in N, ±3.11 cm in E, and ±2.21 cm in U for GPS), confirming the high precision and stability of the CNES solution.

The CODE solution, based on precise products, demonstrated the best performance. For GPS, the mean errors were 0.26 cm (N), −0.72 cm (E), and −0.15 cm (U), corresponding to 2D and 3D accuracies of 3.80 cm and 5.39 cm. For Galileo, the mean errors were minimal (0.06 cm in N, 0.00 cm in E, and −0.36 cm in U), with 2D and 3D accuracies of 3.24 cm and 4.88 cm. The GE combination further improved the results, with mean errors of 0.16 cm (N), −0.23 cm (E), and −0.24 cm (U), and 2D and 3D accuracies of 2.21 cm and 3.26 cm, respectively. Standard deviations remained very low across all components, confirming the superior precision and consistency of the CODE solution.

It should be noted that the 2D and 3D values represent the mean absolute differences from the reference coordinates. Therefore, these values may appear large, which is consistent with the large standard deviations.

In summary, worse results were obtained for GPS in both BRDM and HAS solutions. The inclusion of Galileo significantly improved the achieved accuracy. The best results for each of the products were obtained by combining GPS and Galileo signals, which provided the best convergence times and the highest measurement accuracy.

3.4. ZTD (Zenith Tropospheric Delay)

The ZTD analysis was performed by comparing the estimated zenith delays with the reference values provided by the IGS. The results are illustrated in Figure 4. The figure also highlights the 6 mm (target) and 3 cm (maximum acceptable) thresholds commonly required for real-time applications [10].

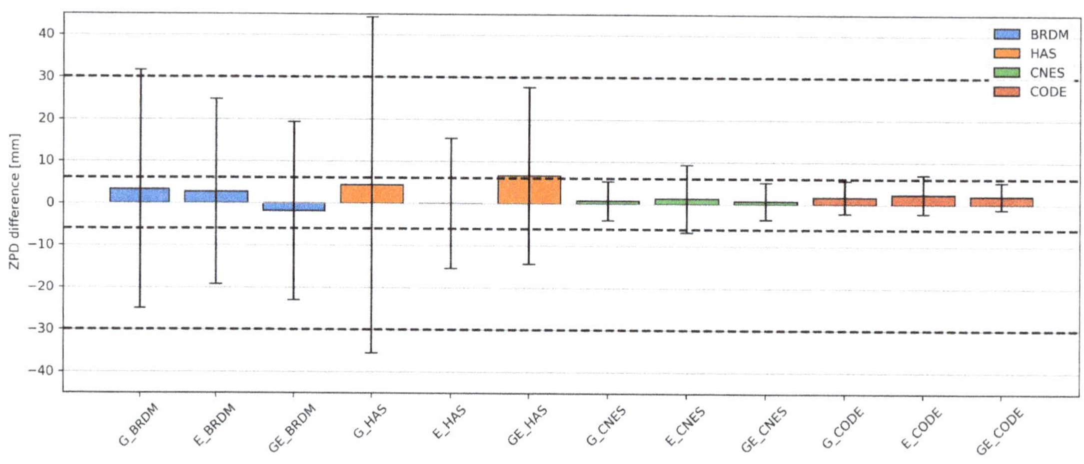

Figure 4. Mean difference and STD of the ZTD estimates at all stations with respect to the IGS reference for all analyzed solutions.

The BRDM solution, based solely on broadcast ephemerides, demonstrated a somewhat lower performance, particularly for GPS signals. The mean ZTD errors reached

3.2 mm for GPS, 2.7 mm for Galileo, and 1.5 mm for the combined GPS+Galileo solution. Although these mean values are within the 6 mm target, the associated standard deviations were relatively high: ±30.6 mm for GPS, ±21.9 mm for Galileo, and ±27.6 mm for GPS+Galileo. Notably, the standard deviation for GPS slightly exceeded the 30 mm threshold, while for Galileo and GPS+Galileo it remained within acceptable limits. This indicates that while the average ZTD estimation was satisfactory, the variability of the results, particularly for GPS, could impact reliability in real-time applications.

In the case of the HAS solution, a larger variability was observed, especially for GPS signals. The mean ZTD errors were 5.0 mm for GPS, −0.2 mm for Galileo, and 7.5 mm for the combined GPS+Galileo solution. The corresponding standard deviations were ±43.1 mm for GPS, ±15.4 mm for Galileo, and ±19.3 mm for GPS+Galileo. These results indicate that the use of Galileo signals, either alone or in combination with GPS, provides a noticeably better stability and precision compared to using GPS signals alone when applying HAS corrections.

The CNES solution, based on real-time precise corrections, demonstrated outstanding performance in ZTD estimation. The mean ZTD errors were 0.7 mm for GPS, 1.3 mm for Galileo, and 0.7 mm for GPS+Galileo. The associated standard deviations were also exceptionally low: ±4.6 mm for GPS, ±8.0 mm for Galileo, and ±4.4 mm for GPS+Galileo.

Similarly, the CODE solution, based on post-processed precise products, yielded very high-quality results. The mean ZTD errors were 1.7 mm for GPS, 2.4 mm for Galileo, and 2.0 mm for GPS+Galileo. The standard deviations were also the lowest among all the solutions analyzed: ±3.9 mm for GPS, ±4.6 mm for Galileo, and ±3.2 mm for GPS+Galileo.

Overall, while BRDM and HAS demonstrated acceptable mean errors, their standard deviations—particularly for GPS signals—slightly exceeded the thresholds, which could pose challenges for high-reliability real-time applications. However, despite these slight exceedances, both solutions remain highly useful and can still be considered reliable for many applications. In contrast, the CNES and CODE solutions clearly outperformed the others, achieving both target mean errors and maintaining a low variability, ensuring robust and accurate ZTD estimations suitable for operational NWP systems. Additionally, as the Galileo HAS service continues to evolve, improvements in its performance are expected, which could further enhance its reliability and precision in the near future.

4. Discussion

In the conducted research, various GNSS solutions were compared under simulated real-time conditions using 1 s interval observations for static positioning. The analysis revealed significant differences in performance, particularly regarding solution availability, convergence time, accuracy, and ZTD estimation. Notable differences were observed between solutions based on broadcast ephemerides (BRDM), real-time precise corrections (CNES RTS), post-processed precise products (CODE MGEX), and the HAS service.

Convergence time was one of the key factors differentiating GNSS solutions based on PPP positioning. For the BRDM and HAS services, more lenient accuracy thresholds were applied (20 cm horizontally (2D) and 40 cm vertically (U) and spatially (3D)), whereas stricter thresholds (10 cm both horizontally and spatially) were used for CNES and CODE. BRDM and HAS exhibited longer and more variable convergence times compared to CNES and CODE, especially when using GPS-only data. However, integrating signals from Galileo (or combined GPS+Galileo) improved performance, although convergence times remained longer. These differences stem from the limitations associated with broadcast ephemerides in BRDM solutions and the focus of the HAS service on the Galileo system.

In contrast, for HAS solutions, the use of Galileo signals significantly improved availability and accuracy, whereas relying solely on GPS limited these parameters. The reduced availability of HAS orbit and clock corrections for GPS satellites indicates the need for the further development of this service. Similarly, Zhao et al. (2024) [44] demonstrated that HAS products for GPS are of a lower accuracy and that the positioning results for GPS were worse compared to those for Galileo. In the case of CNES, slightly worse results were obtained for Galileo signals compared to GPS. However, the combined use of both signals (GPS+Galileo) provided better results, both in terms of positioning accuracy and ZTD estimation. This clearly demonstrates that employing both systems together substantially enhances the overall solution performance.

The ZTD analysis showed that BRDM solutions had a poorer ZTD estimation quality, especially for GPS, where large standard deviations were observed. In comparison, integrating GPS with Galileo improved the results by reducing ZTD variability. A similar dependency was reported by Yang et al. (2024) [49], confirming that the addition of Galileo signals enhances the stability and precision of ZTD retrieval. Moreover, this study also demonstrated that the use of broadcast ephemeris products allows for a ZTD estimation with sufficient accuracy, especially when using the Galileo signal. HAS further showed improvements when using Galileo signals, with standard deviations of ZTD being significantly lower than for GPS. However, the CNES and CODE solutions, which utilize precise corrections and products, proved to be the most accurate and stable in ZTD estimation, making them suitable for high-reliability applications such as weather forecasting.

5. Summary

The research shows that GNSS solutions based on broadcast ephemerides (BRDM) and correction services like HAS exhibit a lower performance in terms of convergence time, accuracy, and ZTD estimation, especially when using GPS-only data. The HAS service demonstrated significant improvement when integrated with Galileo, enhancing solution availability and accuracy. In real-time applications, Galileo consistently provides better results than GPS alone, partly because of more frequent updates of broadcast ephemerides. Nevertheless, solutions such as CNES and CODE, offering real-time precise corrections and post-processed precise products, continue to deliver a higher performance for high-precision applications.

The shortened convergence times observed in this study (1–2 min for CODE and CNES, below 5 min for HAS) were achieved using 1 s interval observations, demonstrating the potential for rapid solution stabilization, which is essential for real-time GNSS applications. While HAS meets its accuracy and convergence time requirements, further development of the service, particularly regarding GPS support, could further enhance its performance.

The presented results are based on real-time/near real-time simulation conditions under ideal circumstances and compared with the best possible results obtained using postprocessed precise CODE MGEX products. It should also be emphasized that the conditions assumed in the study were ideal, and deviations from them could occur in real-world real-time applications, for example, interruptions in receiving corrections or signal outages.

Funding: This research was funded by the Military University of Technology in Warsaw, Faculty of Civil Engineering and Geodesy, Institute of Geospatial Engineering and Geodesy, research grant for 2025.

Data Availability Statement: The original contributions presented in the study, including the downloaded real-time data (HAS and BRDM), are included in the article and openly available on Zenodo at https://zenodo.org/records/15675402 (accessed on 14 June 2025); further inquiries can be directed to the corresponding author.

Acknowledgments: The author thanks the International GNSS Service (IGS) for ensuring observation data and products and the developer of the raPPPid open-source software package.

Conflicts of Interest: The author declares no conflicts of interest.

Abbreviations

The following abbreviations are used in this manuscript:

AC	Analysis Center
APC	Antenna Phase Center
AR	Ambiguity Resolution
BNC	BKG NTRIP Client
CLAS	Centimeter-Level Augmentation Service
CNES	Centre National d'Études Spatiales
CODE	Center for Orbit Determination in Europe
DoY	Day of Year
EKF	Extended Kalman filter
GNSS	Global Navigation Satellite System
GPT	Global Pressure and Temperature
GPS	Global Positioning System
HAS	High Accuracy Service
IERS	International Earth Rotation and Reference Systems Service
IGS	International GNSS Service
NWP	Numerical Weather Prediction
OSB	Observable-specific Signal Bias
PCO	Phase Canter Offset
PCV	Phase Center Variations
PPP	Precise Point Positioning
PWV	Precipitable Water Vapor
QZSS	Quasi-Zenith Satellite System
RTK	Real-Time Kinematic
RTS	Real-Time Service
VMF	Vienna Mapping Functions
TGD	Time Group Delay
ZTD	Zenith Tropospheric Delay

Appendix A

Results of the convergence time analysis presented in tabular form.

Table A1. Mean convergence time of all stations for all analyzed solutions. Values in minutes.

Solution	GNSS	N	E	U	2D	D3
BRDM	G	2.24	3.17	2.55	3.35	3.13
	E	2.34	3.51	2.80	3.92	3.66
	GE	1.82	2.86	2.25	3.10	2.91
HAS	G	2.07	2.82	2.59	3.10	2.89
	E	1.14	2.66	1.47	2.94	2.25
	GE	1.20	2.61	1.39	2.91	2.16
CNES	G	0.61	1.50	1.37	1.58	2.17
	E	0.88	1.83	1.85	2.09	2.87
	GE	0.52	0.98	1.00	1.06	1.45
CODE	G	0.54	1.14	1.06	1.21	1.52
	E	0.50	1.02	1.08	1.15	1.58
	GE	0.40	0.69	0.72	0.76	0.98

Table A2. Standard deviation in convergence time of all stations for all analyzed solutions. Values in minutes.

Solution	GNSS	N	E	U	2D	D3
BRDM	G	1.62	1.97	1.83	2.04	1.91
	E	0.72	0.69	0.85	0.69	0.78
	GE	1.03	1.45	1.18	1.44	1.37
HAS	G	1.64	1.97	1.92	2.07	2.00
	E	0.62	0.82	0.73	0.80	0.83
	GE	0.62	0.74	0.66	0.76	0.75
CNES	G	0.39	0.70	0.70	0.72	0.89
	E	0.41	0.72	0.73	0.74	0.88
	GE	0.41	0.59	0.60	0.57	0.73
CODE	G	0.51	0.69	0.59	0.73	0.83
	E	0.43	0.67	0.67	0.70	0.93
	GE	0.42	0.53	0.48	0.55	0.59

References

1. Héroux, P.; Kouba, J. *GPS Precise Point Positioning with a Difference*; Natural Resources Canada, Geomatics Canada, Geodetic Survey Division: Ottawa, ON, Canada, 1995.
2. Malys, S.; Jensen, P.A. Geodetic Point Positioning with GPS Carrier Beat Phase Data from the CASA UNO Experiment. *Geophys. Res. Lett.* **1990**, *17*, 651–654. [CrossRef]
3. Zumberge, J.F.; Heflin, M.B.; Jefferson, D.C.; Watkins, M.M.; Webb, F.H. Precise Point Positioning for the Efficient and Robust Analysis of GPS Data from Large Networks. *J. Geophys. Res. Solid Earth* **1997**, *102*, 5005–5017. [CrossRef]
4. Kouba, J.; Springer, T. New IGS Station and Satellite Clock Combination. *GPS Solut.* **2001**, *4*, 31–36. [CrossRef]
5. Héroux, P.; Kouba, J. GPS Precise Point Positioning Using IGS Orbit Products. *Phys. Chem. Earth Part A Solid Earth Geod.* **2001**, *26*, 573–578. [CrossRef]
6. Kouba, J.; Lahaye, F.; Tétreault, P. Precise Point Positioning. In *Springer Handbook of Global Navigation Satellite Systems*; Teunissen, P.J.G., Montenbruck, O., Eds.; Springer International Publishing: Cham, Switzerland, 2017; p. 723.
7. Geng, J. *GNSS Seismogeodesy*; Elsevier: Amsterdam, The Netherlands, 2022; pp. 1–331. [CrossRef]
8. Strasser, S.; Mayer-Gürr, T. The IGS Repro3 Contribution by TU Graz: Benefits and Challenges of an Uncombined and Undifferenced Processing Approach 2020. Available online: https://ui.adsabs.harvard.edu/abs/2020AGUFMG005.0003S/abstract (accessed on 12 June 2025).
9. Liu, Z.; Li, Y.; Li, F.; Guo, J. Near Real-Time PPP-Based Monitoring of the Ionosphere Using Dual-Frequency GPS/BDS/Galileo Data. *Adv. Space Res.* **2018**, *61*, 1435–1443. [CrossRef]
10. Ahmed, F.; Václavovic, P.; Teferle, F.N.; Douša, J.; Bingley, R.; Laurichesse, D. Comparative Analysis of Real-Time Precise Point Positioning Zenith Total Delay Estimates. *GPS Solut.* **2016**, *20*, 187–199. [CrossRef]
11. Hohensinn, R.; Aichinger-Rosenberger, M.; Wareyka-Glaner, M.F.; Ravanelli, M. Natural-Hazard Monitoring with Global Navigation Satellite Systems (GNSS). *Adv. Geophys.* **2024**, *65*, 1–123. [CrossRef]
12. Marut, G.; Hadas, T.; Kazmierski, K.; Bosy, J. Affordable Real-Time PPP—Combining Low-Cost GNSS Receivers with Galileo HAS Corrections in Static, Pseudo-Kinematic, and UAV Experiments. *Remote Sens.* **2024**, *16*, 4008. [CrossRef]
13. Li, X.; Barriot, J.P.; Lou, Y.; Zhang, W.; Li, P.; Shi, C. Towards Millimeter-Level Accuracy in GNSS-Based Space Geodesy: A Review of Error Budget for GNSS Precise Point Positioning. *Surv. Geophys.* **2023**, *44*, 1691–1780. [CrossRef]
14. Banville, S.; Geng, J.; Loyer, S.; Schaer, S.; Springer, T.; Strasser, S. On the Interoperability of IGS Products for Precise Point Positioning with Ambiguity Resolution. *J. Geod.* **2020**, *94*, 10. [CrossRef]
15. Lin, J.; Geng, J.; Yan, Z.; Masoumi, S.; Zhang, Q. Correcting Antenna Phase Center Effects to Reconcile the Code/Phase Bias Products from the Third IGS Reprocessing Campaign. *GPS Solut.* **2023**, *27*, 70. [CrossRef]
16. Gao, R.; Liu, Z.; Odolinski, R.; Zhang, B. Improving GNSS PPP-RTK through Global Forecast System Zenith Wet Delay Augmentation. *GPS Solut.* **2024**, *28*, 1–14. [CrossRef]
17. An, X.; Ziebold, R.; Lass, C. PPP-RTK with Rapid Convergence Based on SSR Corrections and Its Application in Transportation. *Remote Sens.* **2023**, *15*, 4770. [CrossRef]
18. Kiliszek, D.; Szołucha, M.; Kroszczyński, K. Accuracy of Precise Point Positioning (PPP) with the Use of Different International GNSS Service (IGS) Products and Stochastic Modelling. *Geod. Cartogr.* **2018**, *67*, 207–238. [CrossRef]

19. Zajdel, R.; Masoumi, S.; Sośnica, K.; Gałdyn, F.; Strugarek, D.; Bury, G. Combination and SLR Validation of IGS Repro3 Orbits for ITRF2020. *J. Geod.* **2023**, *97*, 1–29. [CrossRef]
20. Mansur, G.; Brack, A.; Sakic, P.; Männel, B.; Schuh, H. Utilizing Least Squares Variance Component Estimation to Combine Multi-GNSS Clock Offsets. *GPS Solut.* **2024**, *28*, 1–14. [CrossRef]
21. Su, K.; Jin, S.; Jiao, G. Assessment of Multi-Frequency Global Navigation Satellite System Precise Point Positioning Models Using GPS, BeiDou, GLONASS, Galileo and QZSS. *Meas. Sci. Technol.* **2020**, *31*, 064008. [CrossRef]
22. Hou, Z.; Zhou, F. Assessing the Performance of Precise Point Positioning (PPP) with the Fully Serviceable Multi-GNSS Constellations: GPS, BDS-3, and Galileo. *Remote Sens.* **2023**, *15*, 807. [CrossRef]
23. Kiliszek, D.; Kroszczyński, K. Performance of the Precise Point Positioning Method along with the Development of GPS, GLONASS and Galileo Systems. *Measurement* **2020**, *164*, 108009. [CrossRef]
24. Kiliszek, D.; Kroszczyński, K.; Araszkiewicz, A. Analysis of Different Weighting Functions of Observations for GPS and Galileo Precise Point Positioning Performance. *Remote Sens.* **2022**, *14*, 2223. [CrossRef]
25. Bahadur, B. Real-Time Single-Frequency Precise Positioning with Galileo Satellites. *J. Navig.* **2022**, *75*, 1–124. [CrossRef]
26. Mikoś, M.; Kazmierski, K.; Wachulec, N.; Sośnica, K. Accuracy of Satellite Positioning Using GNSS Receivers in Sports Wat14 June 2025ches. *Measurement* **2024**, *229*, 114426. [CrossRef]
27. GPS Space Segment. Available online: https://www.gps.gov/systems/gps/space/#generations (accessed on 14 June 2025).
28. Galileo Constellation Status. Available online: https://www.gsc-europa.eu/system-service-status/constellation-information (accessed on 14 June 2025).
29. Hadas, T.; Kazmierski, K.; Sośnica, K. Performance of Galileo-Only Dual-Frequency Absolute Positioning Using the Fully Serviceable Galileo Constellation. *GPS Solut.* **2019**, *23*, 108. [CrossRef]
30. Carlin, L.; Hauschild, A.; Montenbruck, O. Precise Point Positioning with GPS and Galileo Broadcast Ephemerides. *GPS Solut.* **2021**, *25*, 1–13. [CrossRef]
31. Alonso, M.T.; Sanz, J.; Juan, J.M.; García, A.R.; Casado, G.G. Galileo Broadcast Ephemeris and Clock Errors Analysis: 1 January 2017 to 31 July 2020. *Sensors* **2020**, *20*, 6832. [CrossRef]
32. Hadas, T.; Bosy, J. IGS RTS Precise Orbits and Clocks Verification and Quality Degradation over Time. *GPS Solut.* **2015**, *19*, 93–105. [CrossRef]
33. Elsobeiey, M.; Al-Harbi, S. Performance of Real-Time Precise Point Positioning Using IGS Real-Time Service. *GPS Solut.* **2016**, *20*, 565–571. [CrossRef]
34. Real-Time Service (RTS)—International GNSS Service. Available online: https://igs.org/rts/ (accessed on 14 June 2025).
35. BKG Ntrip Client—BNC. Available online: https://igs.bkg.bund.de/ntrip/download (accessed on 14 June 2025).
36. RTS Contributors—International GNSS Service. Available online: https://igs.org/rts/contributors/#real-time-analysis-centers (accessed on 14 June 2025).
37. Li, B.; Ge, H.; Bu, Y.; Zheng, Y.; Yuan, L. Comprehensive Assessment of Real-Time Precise Products from IGS Analysis Centers. *Satell. Navig.* **2022**, *3*, 1–17. [CrossRef]
38. Cabinet Office, Government of Japan. Quasi-Zenith Satellite System Interface Specification: Centimeter Level Augmentation Service (IS-QZSS-L6-004). Available online: https://qzss.go.jp/en/technical/download/pdf/ps-is-qzss/is-qzss-l6-004.pdf (accessed on 14 June 2025).
39. Galileo High Accuracy Service (HAS) | European GNSS Service Centre (GSC). Available online: https://www.gsc-europa.eu/galileo/services/galileo-high-accuracy-service-has (accessed on 14 June 2025).
40. EUSPA Galileo High Accuracy Service Service Definition Document (HAS SDD), Issue 1.0. Available online: https://www.gsc-europa.eu/sites/default/files/sites/all/files/Galileo-HAS-SDD_v1.0.pdf (accessed on 14 June 2025).
41. EUSPA Galileo High Accuracy Service Info Note. Available online: https://www.gsc-europa.eu/sites/default/files/sites/all/files/Galileo_HAS_Info_Note.pdf (accessed on 14 June 2025).
42. Horst, O.; Kirkko-Jaakkola, M.; Malkamäki, T.; Kaasalainen, S.; Fernández-Hernández, I.; Moreno, A.C.; Díaz, S.C. HASlib: An Open-Source Decoder for the Galileo High Accuracy Service. In Proceedings of the 35th International Technical Meeting of the Satellite Division of the Institute of Navigation, ION GNSS+ 2022, Denver, CO, USA, 19–23 September 2022; Volume 4, pp. 2625–2633. [CrossRef]
43. Zhang, R.; Tu, R.; Lu, X. HASPPP: An Open-Source Galileo HAS Embeddable RTKLIB Decoding Package. *GPS Solut.* **2024**, *28*, 1–7. [CrossRef]
44. Zhao, L. Python Toolbox for BDS PPP-B2b and Galileo HAS Decoding and Its Products Performance Validation. *GPS Solut.* **2025**, *29*, 1–10. [CrossRef]
45. Naciri, N.; Yi, D.; Bisnath, S.; de Blas, F.J.; Capua, R. Assessment of Galileo High Accuracy Service (HAS) Test Signals and Preliminary Positioning Performance. *GPS Solut.* **2023**, *27*, 1–14. [CrossRef] [PubMed]
46. Zhao, Q.; Wang, W.; Yin, J.; Wu, K.; Zhou, L.; Yao, Y.; Wang, P.; Su, J.; Wang, X.; Wang, H.; et al. Real-Time Retrieval of High-Precision ZTD Maps Using GNSS Observation. *Geod. Geodyn.* **2024**. [CrossRef]

47. Susi, M.; Borio, D.; Azeez, A.; Park, J.; Wezka, K.; Fernandez-Hernandez, I. Assessing the High-Accuracy Service at High Latitudes. *Eng. Proc.* **2023**, *54*, 8. [CrossRef]
48. Prol, F.S.; Kirkko-Jaakkola, M.; Horst, O.; Malkamäki, T.; Bhuiyan, M.Z.H.; Kaasalainen, S.; Fernández-Hernández, I. Enabling the Galileo High Accuracy Service with Open-Source Software: Integration of HASlib and RTKLIB. *GPS Solut.* **2024**, *28*, 1–13. [CrossRef]
49. Xu, Z.; Liu, S.; Xu, Y.; Tang, L.; Yang, N.; Zhang, G. GNSS Real-Time ZTD/PWV Retrieval Based on PPP with Broadcast Ephemerides. *Atmosphere* **2024**, *15*, 1030. [CrossRef]
50. Lu, C.; Li, X.; Nilsson, T.; Ning, T.; Heinkelmann, R.; Ge, M.; Glaser, S.; Schuh, H. Real-Time Retrieval of Precipitable Water Vapor from GPS and BeiDou Observations. *J. Geod.* **2015**, *89*, 843–856. [CrossRef]
51. Yuan, Y.; Zhang, K.; Rohm, W.; Choy, S.; Norman, R.; Wang, C.S. Real-Time Retrieval of Precipitable Water Vapor from GPS Precise Point Positioning. *J. Geophys. Res. Atmos.* **2014**, *119*, 10044–10057. [CrossRef]
52. Li, H.; Choy, S.; Wang, X.; Liang, H.; Purwar, S.; Zhang, K. Investigating the Optimal Spatial Resolution for Assimilating GNSS PWV Into an NWP System to Improve the Accuracy of Humidity Field. *IEEE J. Sel. Top. Appl. Earth Obs. Remote Sens.* **2023**, *16*, 6876–6888. [CrossRef]
53. Glaner, M.F. Towards Instantaneous PPP Convergence Using Multiple GNSS Signals. Ph.D. Thesis, Technische Universität Wien, Wien, Austria, 2022.
54. Glaner, M.F.; Weber, R. An Open-Source Software Package for Precise Point Positioning: RaPPPid. *GPS Solut.* **2023**, *27*, 1–11. [CrossRef]
55. Petit, G.; Brian, L. IERS Conventions 2010. Available online: https://apps.dtic.mil/sti/tr/pdf/ADA535671.pdf (accessed on 14 June 2025).
56. Landskron, D.; Böhm, J. VMF3/GPT3: Refined Discrete and Empirical Troposphere Mapping Functions. *J. Geod.* **2018**, *92*, 349–360. [CrossRef]
57. Wang, N.; Li, Z.; Montenbruck, O.; Tang, C. Quality Assessment of GPS, Galileo and BeiDou-2/3 Satellite Broadcast Group Delays. *Adv. Space Res.* **2019**, *64*, 1764–1779. [CrossRef]

remote sensing

MDPI

Article

Performance Assessment of Undifferenced GPS/Galileo Precise Time Transfer with a Refined Clock Model

Wei Xu [1,2,*], Pengfei Zhang [2], Lei Wang [2], Chao Yan [3] and Jian Chen [4]

[1] School of Geographic Information and Tourism, Chuzhou University, Chuzhou 239000, China
[2] School of Geodesy and Geomatics, Wuhan University, Wuhan 430079, China; pfzhang@sgg.whu.edu.cn (P.Z.); lwangsgg@whu.edu.cn (L.W.)
[3] School of Instrument Science and Engineering, Southeast University, Nanjing 210069, China; chaoyan@seu.edu.cn
[4] School of Geomatics, Anhui University of Science and Technology, Huainan 232001, China; cj_19930815@163.com
* Correspondence: whuxwei@whu.edu.cn

Abstract: Conventional Global Navigation Satellite System (GNSS) time transfer algorithms typically model receiver clock offsets as white noise for estimation, neglecting the physical characteristics of atomic clocks, which consequently limits the performance of GNSS time transfer. To overcome this limitation, this study proposes an undifferenced GPS/Galileo combined Precise Point Positioning (PPP) time transfer model, incorporating both one-state (only clock offset parameter) and two-state (both clock offset and frequency offset parameters) refined clock models with clock instantaneous re-initialization (CIR) strategy at the day boundary epoch. Using observations from International GNSS Service (IGS) Multi-GNSS Experiment (MGEX) stations equipped with external hydrogen masers, precise time transfer performance under refined clock models was evaluated based on undifferenced GPS/Galileo combined PPP float solutions and PPP ambiguity resolutions. Experimental results demonstrate that, compared to traditional models, the refined clock models improve time transfer accuracy and frequency stability by an average of 6.7% and 25.8%, respectively. The improvement is most significant for short term frequency stability, with a maximum enhancement exceeding 85%. As the averaging time increases, the improvement in long term frequency stability gradually diminishes. Notably, the two-state refined clock model slightly outperforms the one-state model in time transfer performance, with the two-state refined clock model improving short-, medium-, and long term frequency stability by 11.5%, 8.0%, and 0.2%, respectively, compared to the one-state refined clock model. These findings strongly advocate adopting the two-state refined clock model to optimize both time transfer precision and short term stability in high-accuracy applications.

Keywords: GNSS; PPP; time transfer; refined clock model; ambiguity resolution

1. Introduction

Time is one of the seven fundamental physical quantities in the International System of Units (SI). Its precise transfer is increasingly critical in communication, satellite navigation and positioning, timing, and other fields. Since Allan and Weiss [1] proposed the Global Navigation Satellite System (GNSS) common-view time transfer method in 1980, GNSS has gradually become a vital tool for high-precision remote time transfer [2]. Notably, the advent of GNSS carrier-phase Precise Point Positioning (PPP) technology over the past two decades has achieved a quantum leap in time transfer accuracy [3], advancing from the initial microsecond level to the sub-nanosecond regime [4]. This technological breakthrough led to the formal adoption of PPP by the International Bureau of Weights and Measures (BIPM) in 2009 as a standard methodology for International Atomic Time (TAI) computation and Coordinated Universal Time (UTC) maintenance [5].

The scientific community has extensively investigated multi-GNSS PPP time transfer capabilities. Defraigne and Baire [6] conducted GPS/GLONASS combined PPP time transfer experiments, demonstrating that the combination enhances robustness. Zhang et al. [7,8], Tu et al. [9] and Lyu et al. [10] studied multi-GNSS combined PPP time transfer models, showing that multi-GNSS integration improves accuracy and frequency stability. Ge et al. [11], Lyu et al. [12], Wang et al. [13], and Xu et al. [14] further validated these advantages in real-time processing scenarios. Researchers have proposed various undifferenced ambiguity resolution methods to recover the integer nature of undifferenced ambiguities in PPP [15,16]. Subsequently, PPP ambiguity resolution time transfers were implemented, proving their superiority over traditional float solutions [17,18]. Petit et al. [19–22] demonstrate that ambiguity resolution time transfer can achieve sub-nanosecond, with frequency stability reaching 1×10^{-16} @5 days and sub 10^{-17}@10 days. Studies by Xu et al. [23], Ren et al. [24], Qin et al. [25], and Zhang et al. [26] indicated that undifferenced ambiguity resolutions marginally enhance long term stability but show limited improvement in short term stability. In addition, with the development of Galileo and BDS multi-frequency signals in recent years, Su et al. [27], Zhang et al. [28], Ge et al. [29], and Xu et al. [30] have tested the models and accuracy of PPP time transfer using Galileo and BDS-3 triple-frequency, quad-frequency, and penta-frequency signals, demonstrating the advantages of multi-frequency signals in improving short term frequency stability.

In the aforementioned GNSS PPP time transfer studies, the receiver clock offset is commonly estimated as high-frequency white noise without considering the correlation between adjacent epochs or the physical noise characteristics of high-precision atomic clocks. Currently, with the development trend of atomic clocks toward low cost, lightweight, and high precision, over 8% of the stations in the International GNSS Service (IGS) Multi-GNSS Experiment (MGEX) network are equipped with cesium atomic masers, and approximately 15% are equipped with hydrogen masers, creating new opportunities for refined clock offset modeling. In response to these developments, Ge et al. [31–33] investigated the performance of post-processing and real-time BDS PPP time transfer under a one-state refined clock model. Lyu et al. [34] proposed a refined two-state clock model incorporating clock offset and frequency offset parameters, demonstrating its significant improvement in short term frequency stability. Qin et al. [35,36] used the refined clock model to predict missing satellite clock offsets in real-time PPP time transfer and used it for mitigating day boundary jumps. Mikoś et al. [37] tested the Galileo and multi-GNSS PPP positioning and time transfer performance with a priori clock offset parameters as constraints, verifying the advantages of the proposed method. Han et al. [38] studied a PPP time transfer model with an adaptive refined clock method.

While previous studies have developed and evaluated refined clock models for PPP time transfer, a critical research gap persists: the comparative analysis of one-state refined clock models (accounting solely for clock offset parameter) versus two-state refined clock models (incorporating both clock offset and drift parameters) and their respective impacts on time transfer accuracy remains unexplored. Moreover, conventional refined clock models necessitate daily initialization of receiver clock parameters by treating them as white noise, inevitably introducing discontinuities in time transfer results. To address this limitation, this study introduces an instantaneous re-initialization strategy based on a refined clock model, which mitigates day-boundary jumps by resetting ambiguity parameters instead of clock parameters, enabling seamless time transfer. Furthermore, with the emergence of undifferenced PPP ambiguity resolution and its demonstrated advantages in high-precision time transfer, there is an urgent need to assess the applicability of these one-state and two-state refined clock models in PPP ambiguity resolution time transfer.

The structure of this research includes the following: Firstly, the GPS/Galileo combined PPP time transfer model and refined receiver clock method are studied. Secondly, three time transfer links with atomic clocks are selected to compare and evaluate the performance of the PPP time transfer with refined clock models. Finally, the research work is discussed, and the conclusion is given.

2. Methodology

2.1. GPS/Galileo Combined PPP Time Transfer Model

The original GPS/Galileo combined code and carrier phase observations can be expressed as:

$$\begin{cases} P_{r,j}^{G,i} = \rho_r^{G,i} + c\left(dt_r^G - dt^{G,i}\right) + M_r^{G,i} \cdot Z_r + \gamma_j^G \cdot I_{r,1}^{G,i} + d_{r,j}^G - d_j^{G,i} + \zeta_{r,j}^{G,i} \\ P_{r,j}^{E,i} = \rho_r^{E,i} + c\left(dt_r^E - dt^{E,i}\right) + M_r^{E,i} \cdot Z_r + \gamma_j^E \cdot I_{r,1}^{E,i} + d_{r,j}^E - d_j^{E,i} + \zeta_{r,j}^{E,i} \\ L_{r,j}^{G,i} = \rho_r^{G,i} + c\left(dt_r^G - dt^{G,i}\right) + M_r^{G,i} \cdot Z_r - \gamma_j^G \cdot I_{r,1}^{G,i} + \lambda_j^{G,i}\left(N_{r,j}^{G,i} + b_{r,j}^G - b_j^{G,i}\right) + \xi_{r,j}^{G,i} \\ L_{r,j}^{E,i} = \rho_r^{E,i} + c\left(dt_r^E - dt^{E,i}\right) + M_r^{E,i} \cdot Z_r - \gamma_j^E \cdot I_{r,1}^{E,i} + \lambda_j^{E,i}\left(N_{r,j}^{G/E,i} + b_{r,j}^E - b_j^{E,i}\right) + \xi_{r,j}^{E,i} \end{cases} \tag{1}$$

where P and L represent the code and carrier phase observations in meters, respectively; superscript i denotes the i-th satellite; superscripts G and E denote the GPS and Galileo system; subscripts r and j denote the GNSS receivers and frequency identifiers, respectively; λ is the wavelength corresponding to the frequency f_j; ρ represents the geometric distance between the satellite and the receiver at the time of satellite signal emission instant; c represents the speed of light in a vacuum; dt_r and dt^s denote the offsets of the receiver clock offset and satellite clock referred to the system time in s, respectively; M denotes the wet mapping function; Z denotes the zenith wet delay; I represents the slant ionosphere delay; γ denotes the frequency-dependent ionosphere delay amplification factor; N is carrier phase integer ambiguity; d denotes the uncalibrated code delay (UCD) in meters; b denotes the uncalibrated phase delay (UPD) in cycles; and ζ and ξ represent the code and carrier phase observation noises, respectively.

After using precise products and eliminating the first-order ionosphere delay by using a dual-frequency IF combination, the linearized GPS/Galileo combined PPP observation equations can be expressed as follows:

$$\begin{cases} p_{r,IF}^{G,i} = \mathbf{u}_r^{G,i} \cdot \mathbf{x} + c\widetilde{dt}_r^G + M_r^{G,i} \cdot Z_r + \zeta_{r,IF}^{G,i} \\ p_{r,IF}^{E,i} = \mathbf{u}_r^{E,i} \cdot \mathbf{x} + c\widetilde{dt}_r^G + M_r^{E,i} \cdot Z_r + ISB^E + \zeta_{r,IF}^{E,i} \\ l_{r,IF}^{G,i} = \mathbf{u}_r^{G,i} \cdot \mathbf{x} + c\widetilde{dt}_r^G + M_r^{G,i} \cdot Z_r + \lambda_{IF}^{G,i} \cdot \widetilde{N}_{r,IF}^{G,i} + \xi_{r,IF}^{G,i} \\ l_{r,IF}^{E,i} = \mathbf{u}_r^{E,i} \cdot \mathbf{x} + c\widetilde{dt}_r^G + M_r^{E,i} \cdot Z_r + \lambda_{IF}^{E,i} \cdot \widetilde{N}_{r,IF}^{E,i} + \xi_{r,IF}^{E,i} \end{cases} \tag{2}$$

with

$$\begin{cases} c\widetilde{dt}_r^G = cdt_r^G + d_{r,IF}^G \\ ISB^E = -d_{r,IF}^G + d_{r,IF}^E \\ \widetilde{N}_{r,IF}^{G/E,i} = \left(N_{r,IF}^{G/E,i} + b_{r,IF}^{G/E} - b_{IF}^{G/E,i}\right) - d_{r,IF}^G / \lambda_{IF}^{G/E,i} \end{cases} \tag{3}$$

where p and l are code and carrier phase observed minus computed values in meters, respectively; subscript "IF" denotes ionospheric-free combination; $\mathbf{u}$ denotes the unit vector of the component from the receiver to the satellite; $\mathbf{x}$ denotes the vector of the receiver position increments; and the symbol "~" denotes the re-parameterized estimate. ISB denotes inter-system bias between the GPS and Galileo systems, which only exists in Galileo's code observation equation.

The satellite UPD in Formula (3) can be corrected using the integer recovery clock and wide-lane satellite bias products provided by Centre National d'Etudes Spatiales (CNES). Meanwhile, the single difference between satellites eliminates the UPD at the receiver end. Subsequently, the integer characteristic of the undifferenced ambiguity is restored, thereby achieving PPP ambiguity resolution.

The CNES integer recovery clock $dt_I^{G/E,i}$ and wide lane satellite bias $\overline{\mu}_{WL}^{G/E,i}$ products can be expressed as:

$$\begin{cases} \overline{\mu}_{WL}^{G/E,i} = \lambda_{WL}^{G/E}\left(b_1^{G/E,i} - b_2^{G/E,i}\right) - \lambda_{NL}^{G/E}\left(d_1^{G/E,i}/\lambda_1 + d_2^{G/E,i}/\lambda_2\right) \\ dt_I^{G/E,i} = dt^{G/E,i} + \lambda_{IF}^{G/E} b_{IF}^{G/E,i} \end{cases} \tag{4}$$

When the CNES integer recovery clock is employed to substitute the satellite clock in the code equation, the linearized GPS/Galileo combined PPP time transfer model can be expressed as:

$$\begin{cases} p_{r,IF}^{E,i} = \mathbf{u}_r^{G,i}\mathbf{x} + c\widetilde{dt}_r^G + M_r^{G,i}\cdot Z_r + e_{L-P}^{G,i} + \zeta_{P,IF}^{G,i} \\ p_{r,IF}^{E,i} = \mathbf{u}_r^{E,i}\mathbf{x} + c\widetilde{dt}_r^G + M_r^{G,i}\cdot Z_r + ISB^E + e_{L-P}^{E,i} + \zeta_{P,IF}^{E,i} \\ l_{r,IF}^{G,i} = \mathbf{u}_r^{G,i}\mathbf{x} + c\widetilde{dt}_r^G + M_r^{G,i}\cdot Z_r + \lambda_{IF}^{G}\overset{\frown}{N}{}_{r,IF}^{G,i} + \zeta_{r,IF}^{G,i} \\ l_{r,IF}^{E,i} = \mathbf{u}_r^{E,i}\mathbf{x} + c\widetilde{dt}_r^G + M_r^{E,i}\cdot Z_r + \lambda_{IF}^{E}\overset{\frown}{N}{}_{r,IF}^{E,i} + \zeta_{r,IF}^{E,i} \end{cases} \tag{5}$$

with

$$\begin{cases} e_{L-P}^{G/E,i} = d_{r,IF}^{i,G/E,i} - \lambda_{IF}^{G/E,i} b_{r,IF}^{G/E} - d_{IF}^{G/E} + \lambda_{IF}^{G/E,i} b_{IF}^{G/E,i} \\ \overset{\frown}{N}{}_{r,IF}^{G/E,i} = N_{r,IF}^{G/E,i} - d_{IF}^{G/E,i}/\lambda_{IF}^{G/E} \end{cases} \tag{6}$$

where e is the difference between the satellite code clock and integer recovery clock products, which code equation residuals will absorb.

2.2. Refined Clock Model with the Characteristics of an Atomic Clock

The noise of a precision atomic clock is typically composed of five types of superimposed noise: Random Walk Frequency Modulation (RWFM), Flicker Frequency Modulation (FFM), White Frequency Modulation (WFM), Flicker Phase Modulation (FPM), and White Phase Modulation (WPM) [39]. The specific expression is as follows [40]:

$$\begin{cases} z(t) = z_{-2}(t) + z_{-1}(t) + z_0(t) + z_1(t) + z_2(t) = \sum\limits_{\alpha=-2}^{2} z_\alpha(t) \\ s_y(f) = h_{-2}f^{-2} + h_{-1}f^{-1} + h_0 f^0 + h_1 f^1 + h_2 f^2 = \sum\limits_{\alpha=-2}^{2} h_\alpha f^\alpha \end{cases} \tag{7}$$

where $z(t)$ represents the total clock noise; $z_\alpha(t)(\alpha = -2, -1, 0, 1, 2)$ corresponds to RWFM, FFM, WFM, FPM, and WPM, respectively; $s_y(f)$ is the power spectral density of the clock frequency (unit: 1/Hz); f is the Fourier frequency, h_α ($\alpha = -2, -1, \ldots, 2$) are the intensity coefficients of the five types of clock noise. WFM and RWFM noise dominate precision atomic clocks (such as hydrogen masers), and the influence of linear frequency drift can be neglected [41].

Introducing clock offset and frequency offset into the Kalman filter means estimating the clock frequency offset parameter simultaneously with the receiver clock offset parameter. This paper refers to it as a two-state refined clock model (RCM2) for ease of expression. The Kalman filter observation equation considering only the clock offset and frequency offset parameters can be simplified as:

$$\boldsymbol{L}_k = \boldsymbol{A}_k \boldsymbol{X}_k + \boldsymbol{V}_k \tag{8}$$

where L_k denotes the observations, X_k denotes the two-state vector of clock offset and frequency offset, A_k is the observation matrix, and V_k represents the observation noise. Assuming the time interval between epoch t_k and t_{k-1} is Δt, the state transition matrix $F_{k,k-1}$ is given by [42]:

$$F_{k,k-1} = \begin{bmatrix} 1 & \Delta t \\ 0 & 1 \end{bmatrix} \tag{9}$$

The corresponding state equation can be expressed as:

$$X_k = F_{k,k-1} X_{k-1} + W_k \tag{10}$$

Among them, W_k represents the two-state process noise vector.

Since phase measurement noise is relatively small and does not accumulate over time, the noise covariance matrix primarily considers the phase noise generated by the accumulation of frequency modulation noise. In this case, the prior noise covariance matrix Q_W corresponding to W_k is given by [43]:

$$Q_W = \begin{bmatrix} q_1 \Delta t + \frac{q_2 \Delta t^3}{3} & \frac{q_2 \Delta t^2}{2} \\ \frac{q_2 \Delta t^2}{2} & q_2 \Delta t \end{bmatrix} \tag{11}$$

where q_1 and q_2 represent the coefficients of WFM noise and RWFM noise, respectively.

Research by Yang et al. [44] demonstrates that the relationship between Allan variance and the noise coefficients is given by:

$$\begin{aligned} \delta_y^2(\tau) = & \left(\sqrt{3q_0}/\tau\right)^2 + \left(\sqrt{q_1/\tau}\right)^2 + \left(\sqrt{q_2\tau/3}\right)^2 \\ & = \left(\delta_y^{WPM}(\tau)\right)^2 + \left(\delta_y^{WFM}(\tau)\right)^2 + \left(\delta_y^{RWFM}(\tau)\right)^2 \end{aligned} \tag{12}$$

where $\delta_y^2(\tau)$ represents the Allan variance at averaging time τ, q_0 denotes the measurement noise coefficient for WPM, $\delta_y^{WPM}(\tau)$, $\delta_y^{WFM}(\tau)$ and $\delta_y^{RWFM}(\tau)$ represent the Allan Deviation (ADEV) for WPM, WFM, and RWFM at averaging time τ, respectively.

In practical applications, the clock offset can first be obtained from IGS final clock products or PPP solutions. Then, the Allan variance of the station clock offset at different averaging times τ is computed. Finally, a least-squares piecewise fitting method is applied to the Allan variance to derive the noise coefficients q_0, q_1, and q_2.

When considering only the one-state clock offset parameter (RCM1), the corresponding prior noise covariance Q_W can be expressed as [45]:

$$Q_W = \left(\delta_y^2(\Delta t) \cdot \Delta t^2\right) = 3q_0 + q_1 \Delta t + \frac{1}{3} q_2 \Delta t^3 \tag{13}$$

Given that the magnitudes of q_0 and q_1 are comparable, while q_2 is substantially smaller than both, the third term on the right-hand side of Equation (13) becomes negligible relative to the first two terms for a small sampling interval Δt. After ignoring the influence of the third term, Equation (13) can be approximately regarded as the one-state representation of Equation (10). Compared to Equation (10), this approximation amplifies the effect of WPM. It should be noted that for high-precision clocks without manual intervention or major maintenance/resets, the noise coefficients of the refined clock model and their corresponding variance-covariance matrices do not need to be recalculated across different periods. Their stability allows for consistent application over extended durations unless significant hardware changes occur.

2.3. Clock Instantaneous Re-Initialization Strategy

Due to discontinuities in precise satellite products at day boundaries, receiver clock offsets estimated by traditional models tend to absorb part of these jumps. However, since observation conditions vary across stations, the amount of jump absorbed by each receiver clock differs, making it impossible to eliminate day-boundary jumps through between-station differencing fully.

Conventional studies on GNSS PPP time transfer with RCM, required initializing clock offset parameters at day boundaries and re-estimating receiver clock offsets as white noise. This approach caused part satellite clock offset jumps to be absorbed into time transfer results, leading to discontinuities at day boundaries and degrading time transfer accuracy. This study employs a Clock Instantaneous Re-initialization (CIR) strategy based on an RCM to address this issue. The strategy leverages the strong correlation between receiver clock offset and ambiguity parameters to absorb day-boundary jumps in clock offset estimations via ambiguity parameters, thereby restoring the continuity of time transfer.

Implementing the CIR strategy involves the following steps: First, reset all ambiguity parameters at the day boundary epoch (k) and assign them relatively large variances (10^4 m^2/s^2). Subsequently, reinitialize the RCM using the method researched by Stein et al. [46]. The prior clock offset dt_k, the initial frequency offset df_k, and the covariance Q_W for initializing the RCM are as follows [46]:

$$dt_k = dt_{k-1} \tag{14}$$

$$df_k = \frac{\tau_1 + \tau_2}{\tau_1}\left[\frac{dt_{k-1} - dt_{k-2}}{\tau_2}\right] - \frac{\tau_2}{\tau_1}\left[\frac{dt_{k-2} - dt_{k-3}}{\tau_1 + \tau_2}\right] \tag{15}$$

$$Q_W = \begin{bmatrix} q_0 & \frac{\tau_1+2\tau_2}{\tau_2(\tau_1+\tau_2)}q_0 \\ \frac{\tau_1+2\tau_2}{\tau_2(\tau_1+\tau_2)}q_0 & \frac{2\tau_1^4+8\tau_1\tau_2+10\tau_1^2\tau_2^2+4\tau_1\tau_2^3+2\tau_2^2}{\tau_1^2\tau_2^2(\tau_1/\tau_2)^2}q_0 + \frac{\tau_1^3+4\tau_1^2\tau_2+4\tau_1\tau_2^2+\tau_2^3}{\tau_1\tau_2(\tau_1+\tau_2)}q_1 + \frac{\tau_2}{3}q_2 \end{bmatrix} \tag{16}$$

where dt_{k-3}, dt_{k-2}, and dt_{k-1} are the estimated receiver clock offsets at the day boundary of the previous three consecutive epochs, where τ_1 represents the time interval between dt_{k-3} and dt_{k-2}, and τ_2 represents the time interval between dt_{k-2} and dt_{k-1}. Apart from the state parameters of the RCM and the ambiguity parameters, the initial values and prior variances of all other parameter estimates can be inherited from the epoch immediately preceding the day boundary.

The receiver clock offset parameter can achieve rapid and accurate convergence using the CIR strategy, thanks to the high-precision prior information. Meanwhile, the reset ambiguity parameters, characterized by relatively large variances, can absorb the discontinuities in satellite clock products at day boundaries. This approach ultimately realizes the objective of continuous receiver clock offset estimation.

3. Data Selection and Processing Strategies

Four stations (BRUX, PTBB, SPT0, and USN7) are selected from the IGS MGEX network to conduct the PPP time transfer experiment with the refined clock model. These stations were chosen primarily for their high-precision hydrogen atomic clocks, critical roles as key nodes in international time transfer networks, and consistently excellent observation data quality and time transfer accuracy. Specifically, BRUX, PTBB, SPT0, and USN7 are stations of the UTC time-keeping laboratories ORB (Observatoire Royal de Belgique), PTB (Physikalisch-Technische Bundesanstalt), RISE (Research Institutes of Sweden AB), and USNO (United States Naval Observatory), respectively. Taking USN7 as the centre, three time and frequency links, namely USN7-BRUX, USN7-PTBB, and USN7-SPT0, are formed to verify the time transfer performance of the refined clock model. The observation period is 8 days (DOYs 279–286 in 2021), and the observation sampling interval is 30 s. The CNES GRG precise clock and orbit products are used for data processing. The geographical distribution of these stations and the basic information are shown in Figure 1 and Table 1.

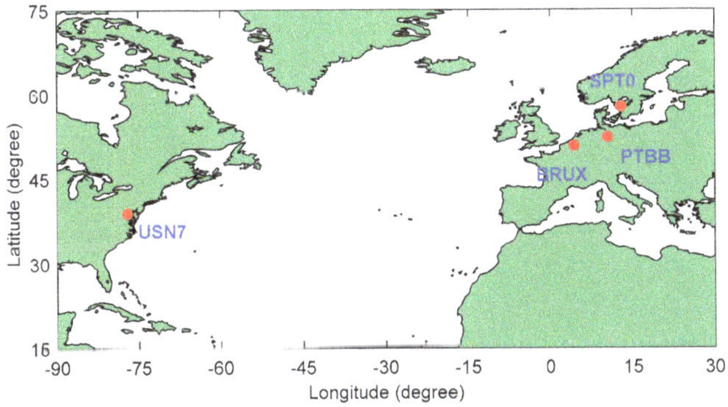

Figure 1. Geographical distribution of experimental stations.

Table 1. Basic information on the experimental stations.

Station	Receiver	Antenna	Country
BRUX	SEPT POLARX5TR	JAVRINGANT_DM	Belgium
PTBB	SEPT POLARX5TR	LEIAR25.R4	Germany
SPT0	SEPT POLARX5TR	JNSCR_C145-22-1	Sweden
USN7	SEPT POLARX5TR	TPSCR.G5	America

To verify the time transfer performance of the refined clock model, this research conducts experiments of PPP using the model without the refined clock model (receiver clock offset estimated as white noise), within the one-state refined clock model and the two-state refined clock model, respectively, which are denoted as RCM0, RCM1, and RCM2. By performing piecewise least squares fitting on the Allan variance of the station clock offsets provided by the IGS, the noise coefficients q_0, q_1, and q_2 of the atomic clocks at the selected stations are calculated. The Allan deviations (ADEV) corresponding to the observation sampling interval of 30 s are obtained from the fitting coefficients, and the specific values are shown in Table 2.

Table 2. Atomic clock noise coefficients and Allan deviations of the selected stations.

Station	q_0	q_1	q_2	ADEV
BRUX	1.67×10^{-24}	2.37×10^{-25}	6.96×10^{-35}	1.16×10^{-13}
PTBB	1.59×10^{-24}	2.97×10^{-25}	1.16×10^{-35}	1.23×10^{-13}
SPT0	1.01×10^{-23}	1.77×10^{-25}	9.32×10^{-35}	2.06×10^{-13}
USN7	9.50×10^{-23}	6.33×10^{-25}	1.03×10^{-33}	2.30×10^{-13}

GPS and Galileo raw code and carrier phase observation noises are set to 0.3 m and 3.0 mm, respectively; elevation-dependent weighting for the observations was applied. Since the IGS antenna correction product does not provide receiver-side antenna phase center offset (PCO) and phase center variation (PCV) correction parameters for Galileo, and considering the similar or identical signal frequencies between GPS and Galileo, this study herein employs GPS receiver-side antenna PCO and PCV correction parameters to approximate those for Galileo's receiver-side antenna. The detailed processing strategies are summarized in Table 3.

Table 3. Details of the PPP time transfer processing strategy.

Item	Strategy
Observations	GPS/Galileo combination; GPS: L1/L2; Galileo: E1/E5a
Elevation cutoff	7°
Satellite orbit & clock	CNES GRG final precise products
Satellite DCB	Corrected by Chinese Academy of Sciences product [47]
Earth's rotation	International Earth Rotation Service (IERS) convention [48]
Relativistic effect	Corrected: elliptical orbit and Shapiro effect [49,50]
Phase windup effect	Corrected by model [51]
Tide effect	Corrected [48,52]
Antenna PCO and PCV	Corrected igs14.atx file
Inter-system bias	Estimated as white noise (10^4 m^2/s^2)
Inter-frequency bias	Estimated as constants
Station coordinate	Static, estimated as constant
Receiver clock	RCM0 (estimated as white noise), RCM1 (refined one-state clock model), and RCM2 (refined two-state clock model)
Ionospheric delay	First-order eliminated by IF combination
Tropospheric delay	Saastamoinen model [53] and estimated as a random walk (10^{-8} m^2/s^2), global mapping function (GMF) [54]
Ambiguity	Estimated as constant, float, and fixed solutions

4. Results

4.1. Quality Assessment of Observation

As shown in Figure 2, the number of satellites at the experimental stations is relatively similar. The number of satellites in the GPS/Galileo combination can reach about 16, and the number of GPS satellites is slightly higher than that of Galileo satellites. The average number of satellites in the GPS/Galileo combination at the stations of BRUX, PTBB, SPT0, and USN7 is 16.3, 16.2, 16.9, and 15.9, respectively, with SPT0 having marginally the highest number and USN7 slightly fewer. The Time Dilution of Precision (TDOP) values across all stations are comparable, predominantly clustered around 0.6. The mean TDOP values for the GPS/Galileo combination at BRUX, PTBB, SPT0, and USN7 stations are 0.61, 0.63, 0.61, and 0.65, respectively. USN7 exhibits a slightly higher TDOP, corresponding to its marginally lower average satellite number than the other three stations.

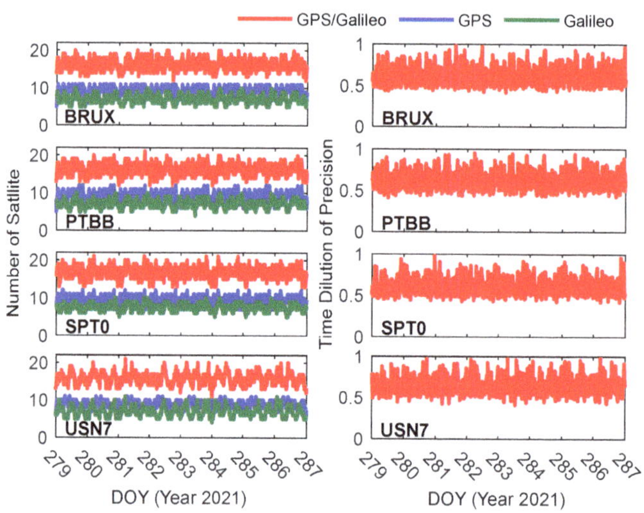

Figure 2. The number of satellites and the TDOP of the experimental station.

As shown in Figure 3, the magnitudes of the MPC (multi-path combination) noise at each station are relatively close, all within 0.5 m. The mean RMS values for GPS L1, L2, and Galileo E1 and E5a are 0.381 m, 0.327 m, 0.291 m, and 0.214 m, respectively. The MPC noise at the L1 frequency is the largest, while that at the E5a frequency is the smallest. The mean MPC noises for BRUX, PTBB, SPT0, and USN7 are 0.253 m, 0.325 m, 0.270 m, and 0.365 m, respectively. Among them, the MPC noise at BRUX is slightly smaller, and the noise at USN7 is slightly larger.

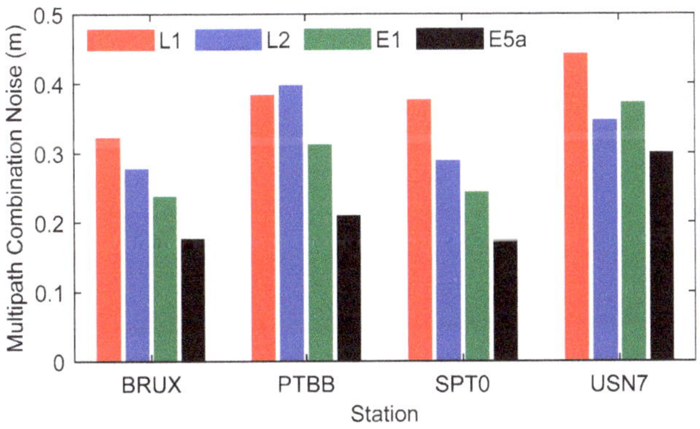

Figure 3. The root mean square (RMS) value of multi-path combination (MPC) noise at the BRUX, PTBB, SPT0, and USN7 stations.

4.2. PPP Time Transfer with Refined Clock Model

Due to space limitations, Figure 4 only shows the receiver clock offset sequences of the BRUX and PTBB stations without and within the CIR strategy. It can be found that the traditional algorithm (without CIR) demonstrates a jump ranging from −0.2 ns to 0.4 ns at the daily boundary. In contrast, the CIR strategy successfully eradicates this jump at the daily boundary, rendering the clock offsets more continuous. This phenomenon validates the efficacy of the CIR strategy in eliminating the daily boundary jump. When the CIR strategy is implemented, the standard deviations (STDs) following epoch-to-epoch differencing at the BRUX and PTBB stations are 0.6 ps and 1.3 ps, respectively. Compared to the STDs of the traditional algorithm, these values are decreased by 88.6% and 77.2%, respectively, which indicates that the CIR strategy reduces the noise in time transfer and enhances the accuracy. Considering the advantage of the CIR strategy in attenuating the jump at the daily boundary, the CIR strategy will be adopted in the refined model in the subsequent experiment.

Figure 5 depicts the clock offset sequences of the three links after eliminating the corresponding trend terms from the initial clock offsets. The clock offset trends in RCM0, RCM1, and RCM2 are generally consistent. Nevertheless, the outcomes of RCM1 and RCM2 exhibit greater smoothness and stability. For instance, around DOY 284.5, there are some abrupt jumps in the data of RCM0, whereas RCM1 and RMC2 have effectively suppressed such sudden changes. Moreover, when juxtaposed with RCM0, the clock offset sequence of RCM1 demonstrates enhanced continuity. This distinctive feature gives rise to an overall divergence between RCM0 on the one hand and RCM1 and RCM2 on the other hand during specific time intervals.

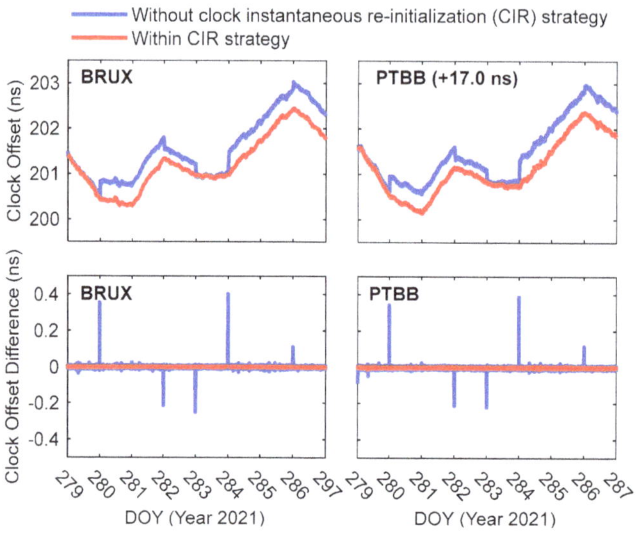

Figure 4. The receiver clock offset sequences of BRUX and PTBB stations.

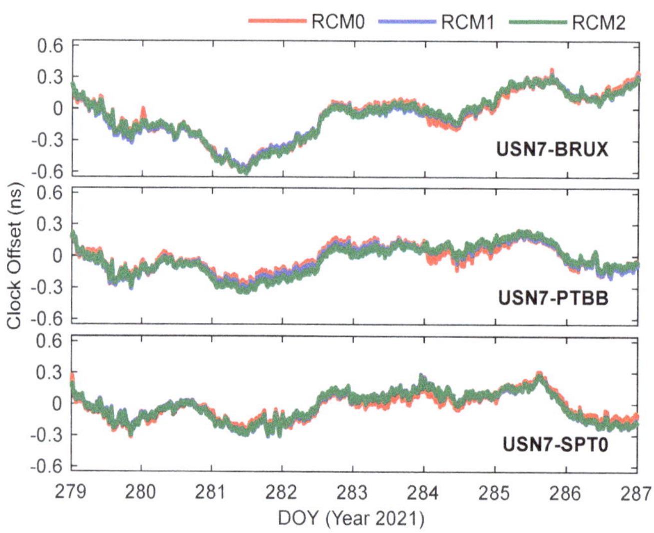

Figure 5. The PPP time transfer clock offset within the refined clock model.

Table 4 presents the STDs of the discrepancies between each scheme's time transfer outcomes and the IGS clock offsets. Evidently, relative to RCM0, the STDs of RCM1 and RCM2 exhibit a downward trend. Specifically, the reduction rates are 7.4% for RCM1 and 10.3% for RCM2 on average. Notably, the STD of RCM2 is marginally higher than that of RCM1. This deviation may be attributed to the accuracy of the fitting coefficients for clock offset noise in RCM2. After implementing the refined clock model, significant decreases are observed in the average STD values. Specifically, the reduction rates are 5.6%, 15.1%, and 5.4%, respectively, in USN7-BRUX, USN7-PTBB, and USN7-SPT0 links.

Table 4. Standard deviations (STD) of the PPP time transfer with the refined clock model (ps).

Link	RCM0	RCM1	RCM2
USN7-BRUX	32.04	30.72	29.80
USN7-PTBB	36.27	31.21	30.36
USN7-SPT0	34.13	32.92	31.69
Mean	34.14	31.62	30.62

As shown in Figure 6, the frequency stability of the time transfer after applying the refined clock model has been significantly improved, especially in terms of short term frequency stability. Taking the USN7-BRUX link as an example, the frequency stabilities of RCM0 at 30 s, 60 s, and 120 s are 2.26×10^{-13}, 1.30×10^{-13}, and 7.67×10^{-14}, respectively. In contrast, the frequency stabilities of RCM1 at the corresponding moments are 4.82×10^{-14}, 4.04×10^{-14}, and 3.43×10^{-14}, respectively, and those of RCM2 at the corresponding moments are 2.78×10^{-14}, 2.44×10^{-14}, and 2.34×10^{-14}, respectively. Compared with RCM0, the frequency stabilities of RCM1 and RCM2 have been improved by more than half an order of magnitude in the short term. The improvement effect of RCM2 on short term frequency stability is better than that of RCM1, which may be because RCM2 adds constraints to the state transition matrix, and its constraints are stronger than those of RCM1. As time increases, the differences in frequency stability among RCM0, RCM1, and RCM2 gradually decrease, and their frequency stabilities at 120,000 s are the same.

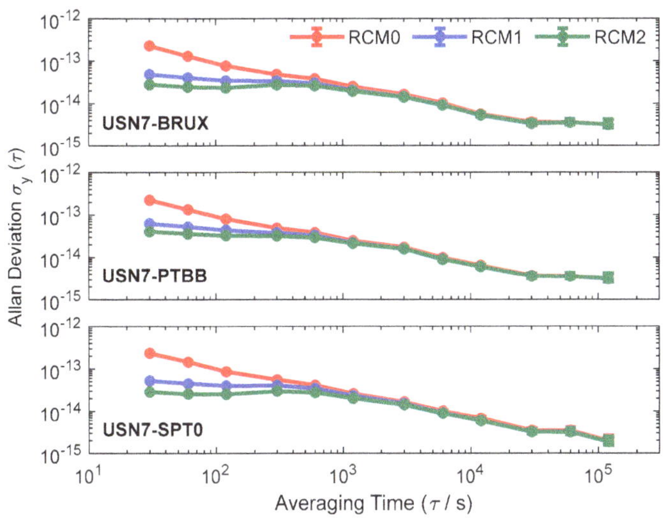

Figure 6. Frequency stability of the PPP time transfer with the refined clock model.

Figure 7 illustrates the mean values of the stability gain at different moments. The average magnitudes of the stability improvement for RCM1 and RCM2 are 23.7% and 30.2%, respectively. Both RCM1 and RCM2 exhibit the most significant improvement in short term stability. For RCM1, the improvement magnitudes at 30 s, 60 s, 120 s, and 240 s are 76.2%, 66.3%, 51.3%, and 26.2%, respectively. Correspondingly, for RCM2, the magnitudes at these moments are 85.8%, 79.0%, 66.5%, and 41.6%, respectively. However, the refined clock model has a relatively minor improvement in long term stability. The improvement magnitudes of RCM1 and RCM2 at 60,000 s and 120,000 s are around 2.0%. Although the improvement magnitude of RCM2 is higher than that of RCM1 before 3000 s, the two become quite close in terms of the improvement magnitude after 12,000 s. Additionally, the improvement magnitudes of RCM2 at 60,000 s are slightly smaller than those of RCM1.

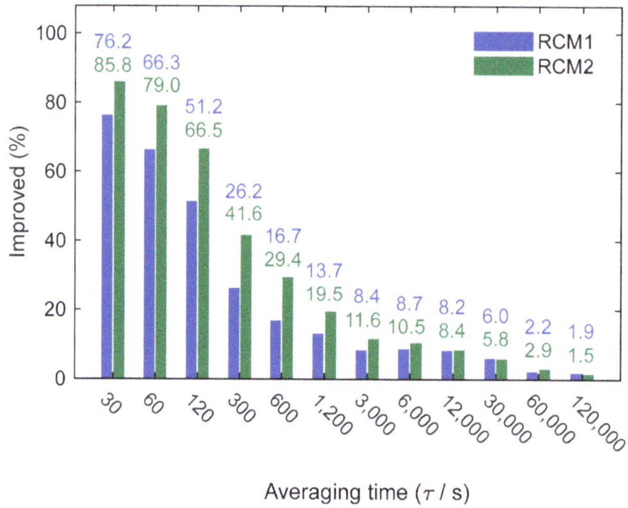

Figure 7. Gain of PPP time transfer stability for refined clock model.

4.3. PPP Time Transfer with Refined Clock Model and Ambiguity Resolution

This study utilizes the wide-lane Satellite Bias and Integer-Recovered Clock products provided by CNES to conduct time transfer experiments using undifferenced ambiguity resolution within and without refined clock models in a GPS/Galileo combined system.

As shown in Figure 8, after adopting the refined clock model, the RMS value of the narrow-lane ambiguity residuals shows a slight downward trend compared with the unrefined scheme, with a decline of approximately 0.001 cycles. Meanwhile, the proportion of residuals distributed within the range of ± 0.15 cycles has increased by about 0.1%. This result indicates that the refined clock model plays a positive role in accurately separating ambiguity parameters, significantly improving the calculation accuracy of ambiguity parameters. In addition, since the CIR strategy resets the ambiguity parameters at the daily boundary, the ambiguity parameters absorb the jumps of the satellite clock products, which may lead to the inability of the ambiguity parameters to be quickly fixed at the epoch of the daily boundary. Furthermore, this situation results in the ambiguity fixing rate when using the refined clock model being slightly lower than that of the conventional scheme.

Figure 9 and Table 5 present the comparative results between the time transfer solutions and the IGS products. Since the IGS only provided the clock offset of the PTBB station on DOY 294, Figure 9 shows the clock offsets of the USN7-PTBB link for a single day. To display the results of each scheme, Figure 9 performs a global shift on the results of the RCM1 and RCM2 schemes. The results reveal that each scheme exhibits highly consistent trends in the variations of clock offsets. Notably, the ambiguity resolution integrated with the refined clock model demonstrates enhanced robustness and reliability compared to the conventional solution. This superiority is particularly evident around DOY 284 for the USN7-BRUX link and around DOY 285 for the USN7-SPT0 link.

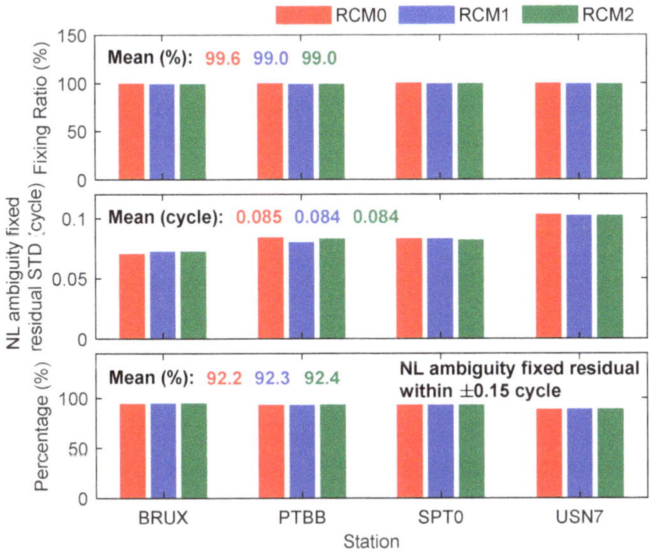

Figure 8. RMS of the narrow-lane ambiguity fractional-cycle residual and fixing rate.

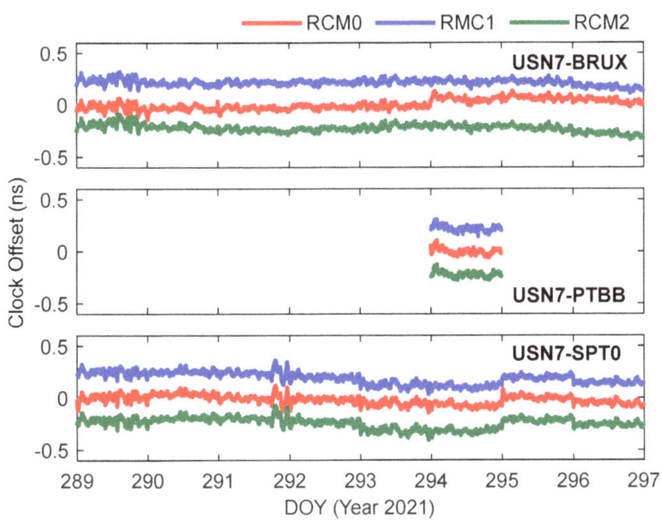

Figure 9. Clock offset sequences of PPP time transfer stability for refined clock model with ambiguity resolution.

Table 5. STDs of PPP time transfer with the refined clock model and ambiguity resolution (ps).

Link	RCM0	RCM1	RCM2
USN7-BRUX	22.32	21.59	20.67
USN7-PTBB	30.52	28.17	29.29
USN7-SPT0	24.47	23.87	23.53
Mean	25.77	24.54	24.49

The statistical results show that the STD of the RCM1 and RCM2 schemes is slightly lower than that of the traditional scheme. Specifically, compared with the RCM0, the STD of RCM1 and RCM2 decreases by approximately 4.7% and 4.9% on average, respectively. Among them, in the BRUX-PTBB link, the improvement effect of RCM1 on the accuracy is the most remarkable. These findings comprehensively confirm the technical advantages of the refined clock model in the PPP ambiguity resolution time transfer. Moreover, through a comparison of Tables 4 and 5, the accuracy of the ambiguity resolution is 22.4% higher than that of the float solution on average, which attests to the superiority of the ambiguity resolution in time transfer applications.

As shown in Figure 10, the frequency stability of the RCM1 and RCM2 has been significantly improved compared with that of the RCM0, especially the improvement effect in the short term, which is the most obvious. For instance, at the USN7-BRUX link, during the averaging time of 30 s to 300 s, the ADEV corresponding to the refined clock model scheme is distinctly lower than that of the traditional scheme. This phenomenon could be attributed to the refined clock model suppressing the observation noise, significantly boosting the short term stability. With the increase of the averaging time, the improvement range of the frequency stability of the refined clock scheme compared with the unrefined clock scheme shows a gradually decreasing trend, and the long term frequency stabilities of the two schemes are the same. Additionally, the frequency stability of RCM2 is superior to that of RCM1, which is generally consistent with the results from PPP float time transfer. This consistency can be attributed to the fact that RCM2 employs a more rigorous model than RCM1.

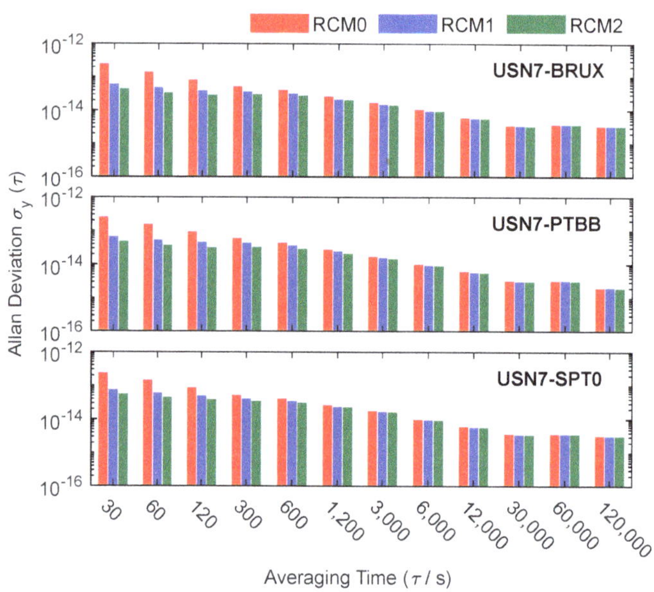

Figure 10. Frequency stability of PPP time transfer with the refined clock model and ambiguity resolution.

Table 6 shows the mean frequency stability characteristics for the three links USN7-BRUX, USN7-PTBB, and USN7-SPT0 in time transfer. It can be observed that when the averaging time is less than 3000 s, the frequency stability improves most rapidly (e.g., the ADEV of RCM0 decreases from 2.2×10^{-13} to 1.6×10^{-14}). When the averaging time exceeds 30,000 s, the frequency stabilities of the RCM0, RCM1, and RCM2 schemes approach each other, with their variation ranges narrowing to extremely small values and gradually tending to a stable state. Statistically, compared with the RCM0 scheme, the mean frequency stabilities of RCM1 and

RCM2 improved by 21.7% and 27.5%, respectively. The improvement is most significant in the short term (averaging time < 1000 s), reaching 60.7% and 70.2%, respectively; in the medium term (1000 s $\leq$ averaging time $\leq$ 10,000 s), the improvements are 13.6% and 21.7%; while in the long term (averaging time > 10,000 s), the improvements are only 2.6% and 2.7%. This result indicates that the RCM strategy mainly optimizes the frequency stability in the short term (with the maximum improvement exceeding 70%). In contrast, its effect on long term stability is relatively limited. Compared to the PPP ambiguity float time transfer solutions, the ambiguity fixed solutions of RCM1 and RCM2 show slightly inferior performance in improving frequency stability, with the stability improvement being approximately 2.5% lower than that of the PPP float solutions. This phenomenon may be related to the ambiguity fixing rate not reaching 100% and, second, to the fact that the receiver clock offsets parameters absorbing the receiver phase biases after ambiguity resolution. In addition, the refined clock model is highly sensitive to prior variance, which may lead to over-constraint problems and affect time transfer performance. Therefore, a reasonable method for determining prior variance still needs further exploration.

Table 6. Mean PPP time transfer frequency stability values with the refined clock model and ambiguity resolution ($\times 10^{-15}$).

Tau/s	RCM0	RCM1	RMC2
30	229.513	64.346	47.524
60	137.617	51.783	37.202
120	82.093	42.716	31.649
300	51.495	38.429	31.080
600	39.307	32.808	27.977
1200	25.264	22.068	20.465
3000	16.210	15.003	14.503
6000	9.554	8.980	8.801
12,000	5.695	5.370	5.359
30,000	3.250	3.133	3.118
60,000	3.339	3.337	3.323
120,000	2.697	2.672	2.682

5. Discussion

This study establishes a time transfer model based on undifferenced GPS/Galileo combined PPP ambiguity float and fixed solutions, comparing the time transfer accuracy and frequency stability between traditional white-noise receiver clock offset estimation and refined first-order (one-state) versus second-order (two-state) clock error models. The findings reveal that refined clock models exhibit significant sensitivity to prior noise parameters covariance configuration. Suboptimal selection of excessively small prior noises may induce over-constraint issues, leading to systematic biases in time transfer. Furthermore, the refined clock model demonstrates applicability only under stable clock performance conditions. Any clock reset or adjustment necessitates a recalibration of prior noise coefficients, constraining the model's operational versatility. Consequently, developing adaptive prior noise coefficient determination methodologies remains a critical research frontier.

Additionally, while this investigation focuses on two primary clock noise components (WFM and RWFM), comprehensive modeling incorporating other clock noise types warrants further exploration. Analysis of time transfer outcomes indicates that although the two-state model offers theoretical rigor—yielding marginally superior time transfer accuracy and frequency stability compared to the one-state model, particularly in short term stability enhancement—it introduces increased computational complexity. In multi-GNSS integration scenarios, the two-state model necessitates additional ISB parameters. Consequently, developing refined ISB parameterization strategies compatible with advanced clock models is a pivotal challenge for future research.

6. Conclusions

Leveraging the physical noise characteristics of atomic clocks, this study established an undifferenced GPS/Galileo PPP time transfer model incorporating one-state and two-state refined clock methods with the CIR method. By selecting three time transfer links composed of four MGEX stations (BRUX, PTBB, SPT0, and USN7), the performance of precise time transfer under refined clock models was evaluated using PPP ambiguity float solutions and ambiguity resolutions. Experimental results reveal that the refined clock strategies effectively mitigate receiver clock offset discontinuities at daily boundaries induced by satellite products, achieving average improvements of 6.7% in time transfer accuracy and 25.8% in frequency stability compared to conventional models. The enhancement is most significant for short term frequency stability, with stability improvements reaching 78.3%, 70.2%, 56.8%, and 33.2% at 30 s, 60 s, 120 s, and 300 s averaging times, respectively. The improvement gradually diminishes as the averaging time increases, showing less than 2% enhancement at 3000 s and 120,000 s averaging times. Additionally, the two-state refined clock model slightly outperforms the one-state model, demonstrating 11.5%, 8.0%, and 0.2% better frequency stability in short term, medium-term, and long term periods, respectively. Consequently, the two-state refined clock model is recommended for applications prioritizing high-precision time transfer and enhanced short term stability performance.

With the continuous advancement of atomic clocks towards miniaturization and cost reduction, equipping many IGS stations with high-precision hydrogen clocks is becoming increasingly feasible. Against this backdrop, the RCM based on clock characteristics is set to see broader applications in high-precision time transfer and positioning. Compared with traditional algorithms, RCM can enhance time transfer accuracy and frequency stability, thereby directly improving time-frequency signals' synchronization and timing capabilities. This holds critical significance for 5G communication synchronization and UTC maintenance fields.

Additionally, in recent years, with the growing research interest in using GNSS time-frequency signals to determine geopotential and orthometric height, the importance of high-precision time-frequency transfer has become increasingly prominent. Theoretical research [55] indicates that the enhancement in time-frequency transfer stability brought by RCM will equivalently improve the accuracy of determining geopotential and orthometric height using time-frequency signals. The undifferenced time transfer technology incorporating RCM is expected to provide critical technical support for achieving centimeter-level cross-sea height transfer and promoting the future unification of global elevation benchmarks.

Author Contributions: Conceptualization, W.X.; methodology, C.Y.; software, W.X.; validation, W.X., C.Y., P.Z. and L.W.; formal analysis, W.X.; investigation, W.X.; resources, C.Y.; data curation, L.W.; writing—original draft preparation, W.X. and J.C.; writing—review and editing, C.Y. and L.W.; visualization, W.X., P.Z. and J.C. All authors have read and agreed to the published version of the manuscript.

Funding: This work is supported by the National Natural Science Foundation of China (No. 42304095; No. 42030105), the Key Project of Natural Science Research in Universities of Anhui Province (No. 2023AH051634), the Chuzhou University Research Initiation Fund Project (No. 2023qd07, No. 2023qd08), and the China Postdoctoral Science Foundation (No. 2024M752480).

Data Availability Statement: The GPS and Galileo observations are available at ftp://igs.gnsswhu.cn/pub/gps/data/daily/ (accessed on 1 April 2025).

Acknowledgments: We thank the IGS and CNES for providing precise satellite orbit and satellite clock offset products and GNSS observations used in this research.

Conflicts of Interest: The authors declare no conflict of interest.

References

1. Allan, D.W.; Weiss, M.A. Accurate time and frequency transfer during common-view of a GPS satellite. In Proceedings of the 34th Annual Symposium on Frequency Control, Philadelphia, PA, USA, 28–30 May 1980; pp. 334–346.
2. Zhang, M.; Lü, J.; Bai, Z.; Jiang, Z.; Chen, B. An overview on GNSS carrier-phase time transfer research. *Sci. China-Tech. Sci.* **2020**, *63*, 589–596. [CrossRef]
3. Zumberge, J.F.; Heflin, M.B.; Jefferson, D.C.; Watkins, M.M.; Webb, F.H. Precise point positioning for the efficient and robust analysis of GPS data from large networks. *J. Geophys. Res.-Solid Earth* **1997**, *102*, 5005–5017. [CrossRef]
4. Ray, J.; Senior, K. IGS/BIPM pilot project: GPS carrier phase for time/frequency transfer and timescale formation. *Metrologia* **2003**, *40*, S270. [CrossRef]
5. Dow, J.M.; Neilan, R.E.; Rizos, C. The international GNSS service in a changing landscape of global navigation satellite systems. *J. Geod.* **2009**, *83*, 191–198. [CrossRef]
6. Defraigne, P.; Baire, Q. Combining GPS and GLONASS for time and frequency transfer. *Adv. Space Res.* **2011**, *47*, 265–275. [CrossRef]
7. Zhang, P.; Tu, R.; Zhang, R.; Gao, Y.; Cai, H. Combining GPS, BeiDou, and Galileo satellite systems for time and frequency transfer based on carrier phase observations. *Remote Sens.* **2018**, *10*, 324. [CrossRef]
8. Zhang, P.; Tu, R.; Han, J.; Gao, Y.; Zhang, R.; Lu, X. Characterization of biases between BDS-3 and BDS-2, GPS, Galileo and GLONASS observations and their effect on precise time and frequency transfer. *Meas. Sci. Technol.* **2020**, *32*, 035006. [CrossRef]
9. Tu, R.; Hong, J.; Zhang, P.; Zhang, R.; Fan, L.; Liu, J.; Lu, X. Multiple GNSS inter-system biases in precise time transfer. *Meas. Sci. Technol.* **2019**, *30*, 115003. [CrossRef]
10. Lyu, D.; Zeng, F.; Ouyang, X. Time transfer algorithm using multi-GNSS PPP with ambiguity resolution. *Chin. Astron. Astrophys.* **2020**, *44*, 371–382.
11. Ge, Y.; Ding, S.; Dai, P.; Qin, W.; Yang, X. Modeling and assessment of real-time precise point positioning timing with multi-GNSS observations. *Meas. Sci. Technol.* **2020**, *31*, 065016. [CrossRef]
12. Lyu, D.; Zeng, F.; Ouyang, X.; Zhang, H. Real-time clock comparison and monitoring with multi-GNSS precise point positioning: GPS, GLONASS and Galileo. *Adv. Space Res.* **2020**, *65*, 560–571. [CrossRef]
13. Wang, S.; Zhao, X.; Ge, Y.; Yang, X. Investigation of real-time carrier phase time transfer using current multi-constellations. *Measurement* **2020**, *166*, 108237. [CrossRef]
14. Xu, W.; Yan, C.; Zhang, P.; Chen, J.; Qiu, D.; Liang, L.; Tao, J. Real-ime undifferenced precision time transfer of multi-GNSS and multi-frequency. *Meas. Sci. Technol.* **2025**, *36*, 025005. [CrossRef]
15. Shi, J.; Gao, Y. A comparison of three PPP integer ambiguity resolution methods. *GPS Solut.* **2014**, *18*, 519–528. [CrossRef]
16. Banville, S.; Geng, J.; Loyer, S.; Schaer, S.; Springer, T.; Strasser, S. On the interoperability of IGS products for precise point positioning with ambiguity resolution. *J. Geod.* **2020**, *94*, 10. [CrossRef]
17. Mi, X.; Zhang, B.; El-Mowafy, A.; Wang, K.; Yuan, Y. Undifferenced and uncombined GNSS time and frequency transfer with integer ambiguity resolution. *J. Geod.* **2023**, *97*, 13. [CrossRef]
18. Baudiquez, A.; Defraigne, P.; Gertsvolf, M.; Guo, J.; Jian, B.; Meynadier, F.; Tagliaferro, G. Comparison between four integer ambiguity resolved PPP GNSS time transfer software solutions. *Metrologia* **2025**, *62*, 025009. [CrossRef]
19. Petit, G.; Kanj, A.; Loyer, S.; Delporte, J.; Mercier, F.; Perosanz, F. 1×10^{-16} frequency transfer by GPS PPP with integer ambiguity resolution. *Metrologia* **2015**, *52*, 301. [CrossRef]
20. Petit, G.; Kanj, A.; Harmegnies, A.; Loyer, S.; Delporte, J.; Mercier, F.; Perosanz, F. GPS frequency transfer with IPPP. In Proceedings of the 2014 European Frequency and Time Forum (EFTF), Budapest, Hungary, 12–16 April 2014; pp. 451–454.
21. Petit, G. Sub-10^{-16} accuracy GNSS frequency transfer with IPPP. *GPS Solut.* **2021**, *25*, 22. [CrossRef]
22. Petit, G.; Defraigne, P. Calibration of GNSS stations for UTC. *Metrologia* **2023**, *60*, 025009. [CrossRef]
23. Xu, W.; Shen, W.; Cai, C.; Li, L.; Wang, L.; Ning, A.; Shen, Z. Comparison and evaluation of carrier phase PPP and single difference time transfer with multi-GNSS ambiguity resolution. *GPS Solut.* **2022**, *26*, 58. [CrossRef]
24. Ren, Z.; Lyu, D.; Gong, H.; Peng, J.; Huang, X.; Sun, G. Continuous time and frequency transfer using robust GPS PPP integer ambiguity resolution method. *GPS Solut.* **2023**, *27*, 82. [CrossRef]
25. Qin, W.; Yang, H.; Zhang, Z.; Wei, P.; Yang, X. The analysis on time transfer of GPS/Galileo/BDS PPP with integer ambiguity resolution. *Phys. Scr.* **2024**, *99*, 025011. [CrossRef]
26. Zhang, R.; Tu, R.; Ge, Y.; Wang, S.; Lu, X. An investigation into multi-frequency ambiguity resolution for real-time PPP time and frequency transfer. *Meas. Sci. Technol.* **2025**, *36*, 045003. [CrossRef]

27. Su, K.; Jin, S. Triple-frequency carrier phase precise time and frequency transfer models for BDS-3. *GPS Solut.* **2019**, *23*, 86. [CrossRef]
28. Zhang, P.; Tu, R.; Gao, Y.; Zhang, R.; Han, J. Performance of Galileo precise time and frequency transfer models using quad-frequency carrier phase observations. *GPS Solut.* **2020**, *24*, 40. [CrossRef]
29. Ge, Y.; Cao, X.; Shen, F.; Yang, X.; Wang, S. BDS-3/Galileo time and frequency transfer with quad-frequency precise point positioning. *Remote Sens.* **2021**, *13*, 2704. [CrossRef]
30. Xu, W.; Yan, C.; Chen, J. Performance evaluation of BDS-2/BDS-3 combined precise time transfer with B1I/B2I/B3I/B1C/B2a five-frequency observations. *GPS Solut.* **2022**, *26*, 80. [CrossRef]
31. Ge, Y.; Zhou, F.; Liu, T.; Qin, W.; Wang, S.; Yang, X. Enhancing real-time precise point positioning time and frequency transfer with receiver clock modeling. *GPS Solut.* **2019**, *23*, 20. [CrossRef]
32. Ge, Y.; Dai, P.; Qin, W.; Yang, X.; Zhou, F.; Wang, S.; Zhao, X. Performance of multi-GNSS precise point positioning time and frequency transfer with clock modeling. *Remote Sens.* **2019**, *11*, 347. [CrossRef]
33. Ge, Y.; Wang, Q.; Wang, Y.; Lyu, D.; Cao, X.; Shen, F.; Meng, X. A new receiver clock model to enhance BDS-3 real-time PPP time transfer with the PPP-B2b service. *Satell. Navig.* **2023**, *4*, 8. [CrossRef]
34. Lyu, D.; Zeng, F.; Ouyang, X.; Yu, H. Enhancing multi-GNSS time and frequency transfer using a refined stochastic model of a receiver clock. *Meas. Sci. Technol.* **2019**, *30*, 125016. [CrossRef]
35. Qin, W.; Ge, Y.; Wei, P.; Yang, X. An approach to a clock offsets model for real-time PPP time and frequency transfer during data discontinuity. *Appl. Sci.* **2019**, *9*, 1405. [CrossRef]
36. Qin, W.; Ge, Y.; Wang, W.; Yang, X. An approach for mitigating PPP day boundary with clock stochastic model. In Proceedings of the 2020 Joint Conference of the IEEE International Frequency Control Symposium and International Symposium on Applications of Ferroelectrics (IFCS-ISAF), New Orleans, LA, USA, 19–23 July 2020; pp. 1–5.
37. Mikoś, M.; Kazmierski, K.; Hadas, T.; Sośnica, K. Stochastic modeling of the receiver clock parameter in Galileo-only and multi-GNSS PPP solutions. *GPS Solut.* **2024**, *28*, 14. [CrossRef]
38. Han, J.; Zhang, J.; Zhong, S.; Lu, R.; Peng, B. PPP time transfer using an adaptive clock constraint model. *Geo-Spatial Inf. Sci.* **2024**, 1–13. [CrossRef]
39. Ai, Q.; Yuan, Y.; Xu, T.; Zhang, B. Time and frequency characterization of GLONASS and Galileo on-board clocks. *Meas. Sci. Technol.* **2020**, *31*, 065003. [CrossRef]
40. Zucca, C.; Tavella, P. The clock model and its relationship with the Allan and related variances. *IEEE Trans. Ultrason. Ferroelectr. Freq. Control.* **2005**, *52*, 289–296. [CrossRef]
41. Zucca, C.; Tavella, P. A mathematical model for the atomic clock error in case of jumps. *Metrologia* **2015**, *52*, 514. [CrossRef]
42. Huang, G.; Zhang, Q. Real-time estimation of satellite clock offset using adaptively robust Kalman filter with classified adaptive factors. *GPS Solut.* **2012**, *16*, 531–539. [CrossRef]
43. Hutsell, S.T. Relating the Hadamard variance to MCS Kalman filter clock estimation. In Proceedings of the 27th Annual Precise Time and Time Interval Systems and Applications Meeting, Bethesda, MD, USA, 29 November–1 December 1995; pp. 291–302.
44. Yang, Y.; Yue, X.; Yuan, J.; Rizos, C. Enhancing the kinematic precise orbit determination of low earth orbiters using GPS receiver clock modelling. *Adv. Space Res.* **2014**, *54*, 1901–1912. [CrossRef]
45. Wang, K.; Rothacher, M. Stochastic modeling of high-stability ground clocks in GPS analysis. *J. Geod.* **2013**, *87*, 427–437. [CrossRef]
46. Stein, S.R.; Filler, R.L. Kalman filter analysis for real time applications of clocks and oscillators. In Proceedings of the 42nd Annual Frequency Control Symposium, Philadelphia, PA, USA, 1–3 June 1988; IEEE: Piscataway, NJ, USA, 1988; pp. 447–452.
47. Wang, N.; Yuan, Y.; Li, Z.; Montenbruck, O.; Tan, B. Determination of differential code biases with multi-GNSS observations. *J. Geod.* **2016**, *90*, 209–228. [CrossRef]
48. Petit, G.; Luzum, B. IERS Conventions (2010). In *IERS Technical Note 36*; Verlag des Bundesamts für Kartographie und Geodäsie: Frankfurt am Main/Berlin, Germany, 2010.
49. Kouba, J. *A Guide to Using International GNSS Service (IGS) Products IGS Central Bureau*; Jet Propulsion Laboratory: Pasadena, CA, USA, 2009; p. 34.
50. Mudrak, A.; De Simone, P.; Lisi, M. Relativistic corrections in the European GNSS Galileo. *Aerotec. Missili Spaz.* **2015**, *94*, 139–144. [CrossRef]
51. Wu, J.-T.; Wu, S.C.; Hajj, G.A.; Bertiger, W.I.; Lichten, S.M. Effects of antenna orientation on GPS carrier phase. *Astron. Astrophys. Trans.* **1991**, *18*, 1647–1660. [CrossRef]
52. Lyard, F.; Lefevre, F.; Letellier, T.; Francis, O. Modelling the global ocean tides: Modern insights from FES2004. *Ocean Dyn.* **2006**, *56*, 394–415. [CrossRef]
53. Saastamoinen, J. Contributions to the theory of atmospheric refraction. *Bull. Géod.* **1972**, *105*, 279–298. [CrossRef]

54. Böhm, J.; Niell, A.; Tregoning, P.; Schuh, H. Global mapping function (GMF): A new empirical mapping function based on numerical weather model data. *Geophys. Res. Lett.* **2006**, *33*, L03304. [CrossRef]
55. Müller, J.; Dirkx, D.; Kopeikin, S.M.; Lion, G.; Panet, I.; Petit, G.; Visser, P.N. High performance clocks and gravity field determination. *Space Sci. Rev.* **2018**, *214*, 5. [CrossRef]

remote sensing

MDPI

Article

High-Precision Satellite Clock Offset Estimated by SRIF Based on Epoch-Wise Updated Orbit

Yu Cao [1,2], Le Wang [1,3,*], Zhiwei Qin [4], Wen Lai [1], Shi Du [5] and Yuanyuan Wang [1]

1 School of Geological Engineering and Geomatics, Chang'an University, Xi'an 710054, China; caoyu0827@chd.edu.cn (Y.C.); laiwnn@chd.edu.cn (W.L.); yuan@chd.edu.cn (Y.W.)
2 Department of Civil and Environmental Engineering, Brunel University London, Uxbridge UB8 3PH, Middlesex, UK
3 Engineering Research Center of Intelligent BDS, Universities of Shaanxi Province, Xi'an 710054, China
4 Shanghai Astronomical Observatory, Shanghai 200030, China; qinzhiwei@shao.ac.cn
5 German Research Centre for Geosciences (GFZ), Telegrafenberg, 14473 Potsdam, Germany; dshicdgnsser@chd.edu.cn
* Correspondence: wangle18@chd.edu.cn

Abstract: High-precision clock offset products directly affect the performance and reliability of precise point positioning (PPP) applications. Currently, real-time clock offset products offered by institutions such as the Centre national d'études spatiales (CNES) rely on ultra-rapid predicted orbits. However, these orbits have limited accuracy and exhibit jumps during updates, constraining the accuracy of real-time clock estimation. To address this issue, we propose an undifferenced ambiguity resolution (UD AR) technique for clock offset estimation based on epoch-wise updated orbits. Clock estimation experiments were performed using both predicted and epoch-wise updated orbits, with square root information filtering (SRIF) applied in three schemes: double-differenced (DD), UD, and float solutions. Compared with predicted orbits, epoch-wise updated orbits provided smoother sequences with higher accuracy, significantly improving clock offset estimation accuracy in all schemes. Moreover, the UD AR solution significantly enhanced clock offset estimation accuracy, and the high-precision epoch-wise updated orbit products increased the narrow-lane fixing rate of the UD solutions. The clock accuracies of BDS-3, Galileo, and GPS reached 0.032 ns, 0.023 ns, and 0.026 ns, respectively, representing improvements of 36%, 34%, and 41% compared with the float solutions and 41%, 30%, 26% compared with the UD solution based on 1 h predicted orbits. Finally, the positioning performance of the proposed method was validated via PPP using 25 stations, showing improvements of 50%, 48%, and 41% in the north, east, and up directions compared with CNES products. Therefore, by combining epoch-wise updated orbit products with the UD AR to improve clock accuracy, this method provides a new approach to generating high-precision clock products, significantly contributing to enhancing PPP services.

Keywords: clock estimation; SRIF; epoch-wise updated orbit; integer ambiguity resolution

1. Introduction

Global Navigation Satellite System (GNSS) technology plays a vital role in high-precision positioning, navigation, and timing (PNT) applications [1]. As core elements of GNSS technology, real-time satellite clock offset products directly determine the quality of real-time applications such as precise point positioning (PPP) and PPP-RTK [2,3]. With the development of GNSS technology, the demand for real-time and high-precision applications has steadily increased, particularly in fields such as autonomous driving, precise agriculture, and geological hazard monitoring [4–8]. Consequently, real-time precise clock offset products have become a significant research focus in the GNSS field.

Currently, real-time clock offset products are primarily generated through two methods: prediction and estimation. Due to the relatively low precision of the prediction method, high-precision clock products predominantly rely on real-time clock estimation [9]. Over the past few years, numerous scholars have extensively researched satellite clock offset estimation. In terms of quality control, Fu et al. proposed an improved quality control algorithm based on the Sequential Least Squares (SLSQ) method, significantly enhancing clock offset accuracy [10]. Meanwhile, Bock et al. effectively eliminated outliers in observation data through post-residual analysis [11]. Regarding computational efficiency, studies have primarily focused on multi-threaded parallel processing, Kalman filtering, and square root information filtering (SRIF), all of which have demonstrated excellent performance in multi-system satellite clock offset estimation [12–14]. Furthermore, numerous studies have shown that ambiguity resolution (AR) techniques can substantially enhance the precision of satellite clock offset estimation [15–18], although fixing rates and computational efficiency require further enhancement.

In real-time clock estimation processing, satellite orbit products are typically considered known information, providing precise satellite positions at specific epochs [19]. As a result, the precision of real-time satellite orbit products directly determines the accuracy and reliability of real-time clock offset estimation [20]. Some institutes, such as the International GNSS Service (IGS), provide ultra-rapid predicted orbit products (e.g., IGU products), but these face challenges, such as discontinuities at update epochs and limited accuracy. These issues constrain further improvement in real-time clock estimation precision and, as a result, impact the performance of real-time PPP [21].

To address these challenges, this study proposes a clock estimation method based on epoch-wise estimated orbit products. Compared with ultra-rapid predicted orbit products, real-time epoch-wise estimated orbit products can effectively avoid discontinuities at update epochs and offer higher orbit accuracy, providing a foundation for high-precision real-time clock offset estimation. Notably, the Shanghai Astronomical Observatory (SHAO) has been releasing epoch-wise, updated, real-time estimated orbit products since November 2024, presenting a significant opportunity for advancing clock offset estimation research. Furthermore, to overcome the limitations of existing clock offset estimation methods, this study incorporates square root information filtering (SRIF) for parameter estimation and ambiguity resolution techniques to enhance the precision of real-time satellite clock offset estimation. The ultimate goal is to obtain high-precision products that support improvements in the service quality of real-time PPP.

This paper is organized as follows: Section 2 details the clock offset estimation model, ambiguity resolution technique, SRIF parameter estimation method, and data processing procedures. Then, the predicted and epoch-wise estimated orbits are analyzed in Section 3.1. Sections 3.2 and 3.3 present an analysis of ambiguity resolution and a comparison of clock estimation based on different orbits. The results are then validated in Section 3.4 using kinematic PPP. Finally, Section 4 summarizes the conclusions.

2. Methods

This section introduces the mathematical models of clock estimation, undifferenced (UD) AR, double-differenced (DD), and real-time satellite clock estimation by SRIF. The sources of orbit data and our clock processing strategy are also illustrated.

2.1. Real-Time Clock Estimation Model

The ionospheric-free (IF) combination is an effective method for mitigating the impact of first-order ionospheric delays in observation data, commonly used in satellite clock offset estimation. For the dual-frequency IF combination observation from satellite (s) to receiver (r), the mathematical expression can be formulated as follows:

$$P^s_{r,IF} = \rho^s_r + c\left(\bar{t}_{r,IF} - \bar{t}^s_{IF}\right) + m^s_r T_r + \varepsilon^s_{r,pc} \tag{1}$$

$$L^s_{r,IF} = \rho^s_r + c\left(\bar{t}_{r,IF} - \bar{t}^s_{IF}\right) + \lambda_{IF}\overline{N}^s_{r,IF} + m^s_r T_r + \varepsilon^s_{r,lc} \tag{2}$$

where the $P^s_{r,IF}$ and $L^s_{r,IF}$ are the IF observations of the code and phase, respectively. ρ^s_r is the distance between the satellite (s) to receiver (r). $\bar{t}_{r,IF}$ and $\bar{t}^s_{IF}$ represent clock offsets of the receiver and satellite after the IF combination, respectively, which include the uncalibrated hardware delays from the receiver $d_{r,IF}$ and satellite d^s_{IF} [22]. These can be expressed as follows: $\bar{t}_{r,IF} = t_{r,IF} + d_{r,IF}$ and $\bar{t}^s_{IF} = t^s_{IF} + d^s_{IF}$. m^s_r and T_r are the mapping function and tropospheric zenith wet delay (ZWD). λ_{IF} is the wavelength. $\overline{N}^s_{r,IF}$ is the re-parameterized phase ambiguity of the IF combination. $\varepsilon^s_{r,pc}$ and $\varepsilon^s_{r,lc}$ are the code and phase measurement noise, and c denotes the speed of light.

2.2. Ambiguity Resolution

GNSS ambiguity resolution methods are primarily divided into DD and UD. Ambiguities lose integer characteristics after IF combinations and can be split into wide-lane (WL) and narrow-lane (NL). Since ambiguity absorbs the hardware delay at both the receiver and the satellite [22], it can be expressed as follows [16]:

$$\lambda_{IF}.\,\overline{N}^s_{r,IF} = \frac{cf_2}{f_1^2 - f_2^2}\overline{N}^s_{r,WL} + \frac{c}{f_1 + f_2}\overline{N}^s_{r,NL} \tag{3}$$

After the IF combination, the WL and NL ambiguities can be expressed as follows:

$$\overline{N}^s_{r,WL} = N^s_{r,WL} + b_{r,WL} - b^s_{WL} \tag{4}$$

$$\overline{N}^s_{r,NL} = N^s_{r,NL} + b_{r,NL} - b^s_{NL} \tag{5}$$

where $N^s_{r,WL}$ and $N^s_{r,NL}$ denote the integer WL and NL ambiguities. $b_{r,WL}$ and b^s_{WL} are the WL uncalibrated phase delay (UPD) on the receiver and satellite sides, respectively. $b_{r,NL}$ and b^s_{NL} are the NL UPD on the receiver and satellite sides, respectively.

The DD AR primarily eliminates the UPD of the satellite and receiver sides by differencing between two satellites, s_1 and s_2, and two receivers, r_1 and r_2. The specific principle is as follows:

$$\Delta\overline{N}^{s_1s_2}_{r_1r_2,WL} = N^{s_1}_{r_1,WL} - N^{s_2}_{r_1,WL} - N^{s_1}_{r_2,WL} + N^{s_2}_{r_2,WL} \tag{6}$$

$$\Delta\overline{N}^{s_1s_2}_{r_1r_2,NL} = N^{s_1}_{r_1,NL} - N^{s_2}_{r_1,NL} - N^{s_1}_{r_2,NL} + N^{s_2}_{r_2,NL} \tag{7}$$

Substituting the above expression into Equation (3), the DD ambiguity can be expressed as follows:

$$\lambda_{IF}.\Delta\overline{N}^{s_1s_2}_{r_1r_2,IF} = \frac{cf_2}{f_1^2 - f_2^2}\Delta\overline{N}^{s_1s_2}_{r_1r_2,WL} + \frac{c}{f_1 + f_2}\Delta\overline{N}^{s_1s_2}_{r_1r_2,NL} \tag{8}$$

The MW (Melbourne–Wübbena) combination is commonly employed to compute float WL ambiguity. The UPD of the receiver and satellite sides can be eliminated by differencing between the satellites and the stations, enabling the determination of the DD WL ambiguity. If the fractional part of the rounded value of $\Delta\overline{N}^{s_1s_2}_{r_1r_2,WL}$ is smaller than the threshold, the ambiguity is successfully fixed. Then, DD NL ambiguity can be calculated using Equation (8), and the fractional component of the rounded value of the float DD NL ambiguity can be compared with the threshold. If the fractional part is smaller than the threshold, the ambiguity is successfully fixed; otherwise, it is not. Finally, the fixed and float DD ambiguities are combined to form the constraint equations. These are incorporated into the original equations for updating, improving the accuracy of the unknown parameters.

In terms of UD AR, the integer values of the UD WL ambiguities can be determined based on the MW combined observations [23,24], leaving the fractional parts, thus representing the WL UPD observation $UPD^{s}_{r,WL}$ as follows:

$$UPD^{s}_{r,WL} = b_{r,WL} - b^{s}_{WL} + \varepsilon_{WL} \tag{9}$$

where ε_{WL} is the noise of the observation. Assuming there are rn stations capable of observing sm satellites, the observation equation is given by the following:

$$\begin{bmatrix} UPD^{s1}_{r1,WL} \\ \vdots \\ UPD^{sm}_{rn,WL} \end{bmatrix} = \begin{bmatrix} -1 & \cdots & 0\ 1 & \cdots & 0 \\ \vdots & \ddots & \vdots\ \vdots & \ddots & \vdots \\ 0 & \cdots & -1 0 & \cdots & 1 \end{bmatrix} \begin{bmatrix} b^{s1}_{WL} \\ \vdots \\ b_{rn,WL} \end{bmatrix} \tag{10}$$

There is a linear relationship between the receiver clock and the satellite clocks, leading to a rank deficiency of 1 in Equation (10). Consequently, one satellite clock is chosen as the reference and set to 0. The least-square adjustment is then applied to estimate the UPD for the stations and satellites. Based on the WL UPD and IF ambiguities, the same processing procedure then can be employed to estimate the NL UPD.

If the WL and NL UPD products are obtained, the UD AR solution can be applied. Initially, the WL ambiguities are fixed. The criterion for determining if the WL ambiguities are fixed is based on whether the fractional part falls below the predefined threshold after accounting for the receiver and satellite UPD estimates. If this value is lower than the threshold, the ambiguities are fixed. In this study, the threshold for fixing UD WL ambiguity is 0.25 cycles. Subsequently, based on Equation (9), UD NL ambiguities can be fixed using the floating ambiguity variance and NL UPD. Once both the wide-lane and narrow-lane ambiguities are fixed, the IF ambiguity with integer characteristics can be resolved. Finally, the fixed ambiguities are combined with the float ambiguities to form a strong constraint added to the original equation for parameter updating, improving the precision of clock estimation.

2.3. Real-Time Clock Estimation by SRIF

SRIF is a state estimation algorithm that enhances numerical stability and computational efficiency by decomposing the square root of the information matrix [25]. Assuming there are m observations and n unknown parameters x at the epoch i, then the observation equation is given by the following:

$$\underset{m\times 1}{l_i} = \underset{m\times n}{H_i}\ \underset{n\times 1}{x_i} + \underset{m\times 1}{\varepsilon_i} \tag{11}$$

where l_i is the observation series, H_i represents the design matrix, and ε_i indicates the measurement noise. The state parameters can be divided into time-varying parameters ΔX_i and deterministic parameters y. The state equation for the time-varying parameters is given by the following:

$$\Delta X_i = \Phi_{i,\, i-1} \Delta X_{i-1} + w_i \tag{12}$$

where $\Phi_{i,\,i-1}$ represents the state transition matrix and w_i denotes the processing noise with $E(w_i) = 0,\ D(w_i) = D_{w_i}$. Given the initial values of the state parameters, the corresponding initial constraint equation is given by the following:

$$x = x_0 + w_0 \tag{13}$$

Since the SRIF requires that the noises must follow a normal distribution N (0,1), the aforementioned equations need to be standardized. The standardized equations are as follows:

$$\begin{cases} z_i = A_i x_i + v_i \\ R_{w_i}\Delta X_i = R_{w_i}\Phi_{i,\,i-1}\Delta X_{i-1} + R_{w_i} w_i \\ z_0 = R_{w_0} x + v_0 \end{cases} \tag{14}$$

where $z_i = R_{v_i} l_i$,$A_i = R_{v_i} H_i$, $v_i = R_{v_i}{}^{-1}\varepsilon_i$, $z_0 = R_{w_0} x_0$, and $v_0 = R_{w_0} w_0$.

In SRIF, the observation equations of the estimated parameters are as follows:

$$\begin{bmatrix} R_0 \\ A \end{bmatrix} x = \begin{bmatrix} z_0 \\ z \end{bmatrix} - \begin{bmatrix} v_0 \\ v \end{bmatrix} \tag{15}$$

The SRIF estimates the parameters epoch by epoch, and its time update relies on the information and parameter states from the previous epoch. The above equation can be expressed after the measurement update:

$$\begin{bmatrix} \hat{R}_{\Delta X_{i-1}} & \hat{R}_{\Delta X_{i-1} y} \\ 0 & \hat{R}_y \end{bmatrix} \begin{bmatrix} \Delta X_{i-1} \\ y \end{bmatrix} = \begin{bmatrix} \hat{z}_{\Delta X_{i-1}} \\ \hat{z}_y \end{bmatrix} \tag{16}$$

Combining the state equation, the matrix after the time update is given by the following expression:

$$\begin{bmatrix} -R_{w_i}\Phi_{i,\,i-1} & R_{w_i} & 0 \\ \hat{R}_{\Delta X_{i-1}} & 0 & \hat{R}_{\Delta X_{i-1} y} \\ 0 & 0 & \hat{R}_y \end{bmatrix} \begin{bmatrix} \Delta X_{i-1} \\ \Delta X_i \\ y \end{bmatrix} = \begin{bmatrix} \hat{z}_{w_i} \\ \hat{z}_{\Delta X_{i-1}} \\ \hat{z}_y \end{bmatrix} \tag{17}$$

After the Householder orthogonal transformation, the above equation can be converted into an upper triangular matrix of the following form [13]:

$$\begin{bmatrix} \widetilde{R}_{\Delta X_{i-1}} & \widetilde{R}_{\Delta X_{i-1}\Delta X_i} & 0 \\ 0 & \widetilde{R}_{\Delta X_i} & \widetilde{R}_{\Delta X_i y} \\ 0 & 0 & \widetilde{R}_y \end{bmatrix} \begin{bmatrix} \Delta X_{i-1} \\ \Delta X_i \\ y \end{bmatrix} = \begin{bmatrix} \widetilde{z}_{w_i} \\ \widetilde{z}_{\Delta X_i} \\ \widetilde{z}_y \end{bmatrix} \tag{18}$$

From the above equation, it is evident that the parameter ΔX_{i-1} can be eliminated. Consequently, the state parameters can be predicted by this equation:

$$\begin{bmatrix} \widetilde{R}_{\Delta X_i} & \widetilde{R}_{\Delta X_i y} \\ 0 & \widetilde{R}_y \end{bmatrix} \begin{bmatrix} \Delta X_i \\ y \end{bmatrix} = \begin{bmatrix} \widetilde{z}_{\Delta X_i} \\ \hat{z}_y \end{bmatrix} \tag{19}$$

Following the time update, the state parameters for the current epoch are integrated into the equation. This updated equation is then used as the initial input for the measurement update, which can be represented as follows:

$$\begin{bmatrix} \widetilde{R}_i \\ A_i \end{bmatrix} x_i = \begin{bmatrix} \widetilde{z}_i \\ z_i \end{bmatrix} \tag{20}$$

By reapplying the Householder orthogonal transformation and updating the observations with the residuals, the equation is transformed into the following:

$$\begin{bmatrix} \hat{R}_i \\ 0 \end{bmatrix} x_i = \begin{bmatrix} \hat{z}_i \\ e_i \end{bmatrix} \tag{21}$$

During the real-time satellite clock estimation, the observation residuals and the variance of unit weight are obtained from the covariance matrix. The residuals are stored in vector $\boldsymbol{e_i}$ and $\boldsymbol{e_i} \sim N(0, I)$. The variance of unit weight is given by the following:

$$\sigma_0 = \sqrt{\frac{V^T V}{m}} = \sqrt{\frac{\boldsymbol{e}_i^T \boldsymbol{e}_i}{m}} \tag{22}$$

By employing hypothesis testing to detect and identify outliers [26–28], and through steps such as the Householder orthogonal transformation, the following equation can be derived [13]:

$$\begin{bmatrix} \hat{R}_i & \hat{R}_{\Delta,i} \\ 0 & \hat{R}_\Delta \\ 0 & 0 \end{bmatrix} \begin{bmatrix} x \\ \Delta \end{bmatrix} = \begin{bmatrix} \tilde{z}_i \\ z_\Delta \\ \hat{e}_i \end{bmatrix} \tag{23}$$

The final solutions for the estimated parameters and covariance matrix are as follows:

$$x_i = \hat{R}_i^{-1} \hat{z}_i \tag{24}$$

$$D_{x_i} = \hat{R}_i^{-1} \hat{R}_i^{-T} \tag{25}$$

The overall process of real-time clock estimation based on SRIF is shown in Figure 1, which provides the main steps and corresponding formulas.

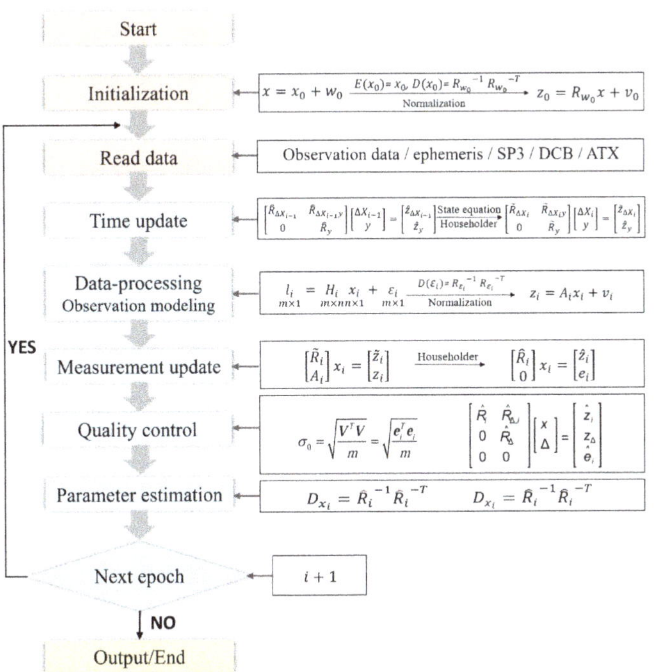

Figure 1. Data processing flowchart.

2.4. Data Collection and Processing Strategy

To evaluate the precision of clock estimation and the effectiveness of ambiguity resolution based on different orbit products, epoch-wise updated and predicted orbit products were collected for the experiment. Epoch-wise updated orbits for 1–31 October 2024 (DOY 275–305), were estimated using the same processing strategy as the real-time orbits recorded by the SHAO. Since the ultra-rapid orbit from IGS is updated every 6 h and lacks BDS data, the predicted orbits were obtained from the Wuhan University IGS analysis center, which generates ultra-rapid predicted orbits that are updated every hour based on observations (https://igs.org/products/#orbits_clocks-2024.09.17, accessed on 21 February 2025). Three sets of predicted orbits were then integrated into daily files with update intervals of 1 h, 3 h, and 6 h. Specifically, these updated orbit products are constructed by extracting segments from each ultra-rapid file starting from its update time. The orbits and clocks are compared to the precise products of the CODE (Center for Orbit Determination in Europe), ensuring the reliability and accuracy of the experimental results.

The station distribution of the clock estimation experiments is illustrated in Figure 2. The orange dots represent 100 stations estimating clocks, and the green stars represent 25 stations validating positioning results.

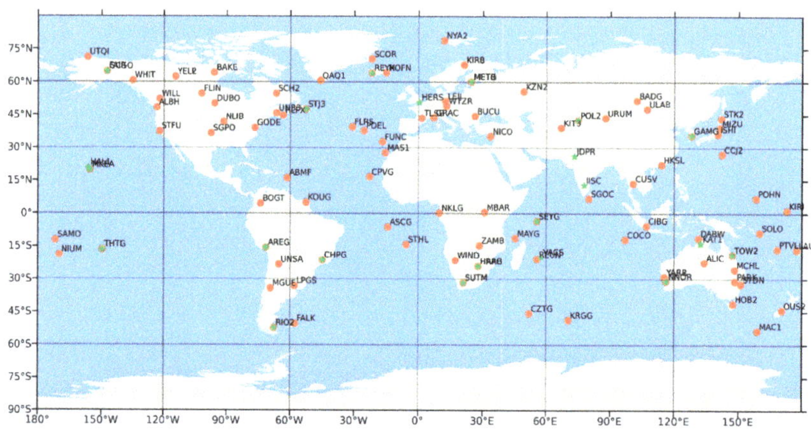

Figure 2. Stations distribution of the experiment.

Table 1 outlines the specific strategy for real-time clock estimation. All experiments were conducted using the PANDA software. The clock estimation experiments for GPS, Galileo, and BDS-3 were conducted in three modes: UD, DD, and float. Due to the characteristics of the GLONASS, only the float solution was implemented.

Table 1. Clock Offset Estimation Strategy.

Item	Strategy
Software	Modified PANDA
Stations	100 IGS/MGEX tracking stations [29,30]
Time interval	DOY 275–DOY 305, 2024
Observation	Undifferenced IF combination observations GPS: L1/L2, BDS-3: B1C/B2a, E: E1/E5a, R: G1/G2
Sampling interval	30 s
Cutoff angle	7°
Satellite antenna error	PCO and PCV corrections from the igs20.atx [31]
Receiver antenna error	PCO corrections from the igs20.atx [31]
Phase wind-up	Model correction [32]

Table 1. *Cont.*

Item	Strategy
Station coordinate	Fix to IGS weekly solution
Station displacement	Solid earth tide, pole tide and ocean tide: IERS Convention 2010 [33,34]
Satellite orbit and ERP	Real-time and predicted orbit products
Satellite and receiver clock	Estimated as white noise
ISB	Estimated as white noise
Estimator	SRIF
Troposphere delay	Saastamoinen model, GMF mapping function, piecewise constants [35,36]
Ambiguity	Estimated to be constant in the absence of cycle slips
Round-off criterion	0.25 cycles for WL, 0.15 cycles for NL
Differential code bias	CODE P1C1 bias product

3. Results

In this study, the accuracies of the predicted and epoch-wise updated orbit products were analyzed. Then, clock offset estimation experiments were conducted using SRIF based on these two products, and a discussion on the effects of UD and DD ambiguities fixing in clock offset estimation is presented. The clock offset accuracies were compared to precise clock products of the CODE. Finally, kinematic PPP was used for validation.

3.1. Analysis of Predicted and Estimated Satellite Orbit Products

To compare the differences between ultra-rapid predicted and real-time estimated satellite orbit products, precise orbits from the CODE were used as a reference to derive orbit residuals and accuracy. Taking BDS-3 as an example, the ultra-rapid orbit residuals at update intervals of 1 h, 3 h, and 6 h in the along (A), cross (C), and radial (R) directions are illustrated in Figure 3. Evidently, regardless of the update interval, significant jumps occurred in orbit residuals at each update epoch, indicating discontinuity in the residual sequences at these epochs. Additionally, the orbit accuracy of the 1 h update interval is superior to that of the 3 h and 6 h intervals, with the 6 h update exhibiting larger residual fluctuations and a wider distribution range. Out of all directions, the radial residuals are the smallest, while the along and cross residuals show larger variability.

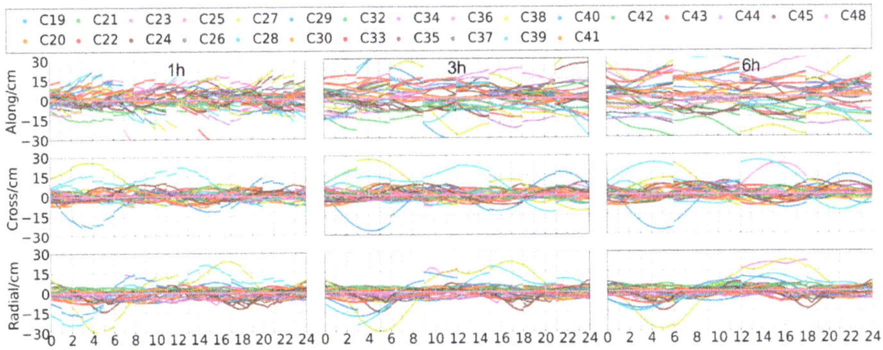

Figure 3. The ultra-rapid orbit residual series for BDS-3 (DOY: 283).

Figure 4 shows the same day residuals of the real-time orbits estimated epoch by epoch for BDS-3. Compared with the predicted orbit residuals, these orbits demonstrate higher accuracy and better continuity. The satellites with larger residuals in the figure are C38, C39, and C40, which are IGSO satellites and exhibit lower orbit accuracy compared with MEO satellites. Excluding these three, the residuals of the remaining satellites in the A, C, and R directions fluctuate within 10 cm, indicating excellent overall performance.

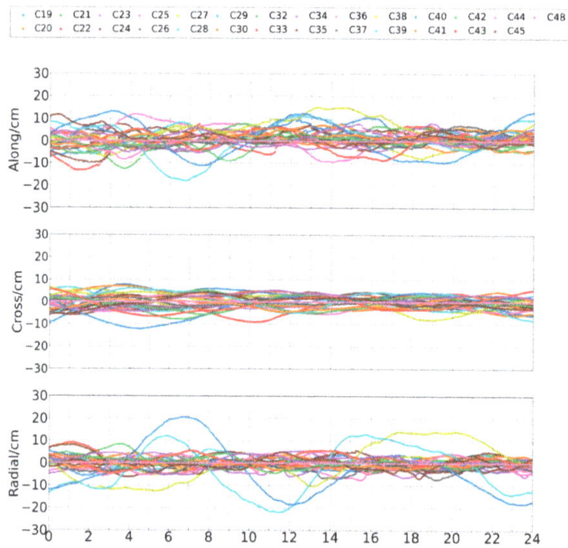

Figure 4. The estimated orbit residual series for BDS-3 (DOY: 283).

Figure 5 shows the average accuracy of the ultra-rapid predicted and epoch-wise updated orbits, where C, E, G, and R represent BDS-3, Galileo, GPS, and GLONASS, respectively. The labels 1 h, 3 h, 6 h, and SRIF represent predicted orbits with update frequencies of 1 h, 3 h, and 6 h, as well as epoch-wise updated orbits derived using SRIF. The accuracy of the along component is the worst, while the radial direction achieves the best accuracy, likely because a portion of radial errors are absorbed into the clock offset. For predicted orbits, the accuracies in the along and cross directions for a 1 h update frequency are significantly better than those with 3 h and 6 h update intervals, while the difference in the radial direction is minimal. The epoch-wise estimated orbit accuracy is clearly superior to that of the predicted orbit, with a particularly notable improvement in the along component. Additionally, the orbit accuracies of GPS and Galileo demonstrate superior performance, while BDS-3 and GLONASS exhibit comparable results. Overall, epoch-wise estimated orbits demonstrate a significant accuracy advantage, with the most noticeable improvement in the along direction.

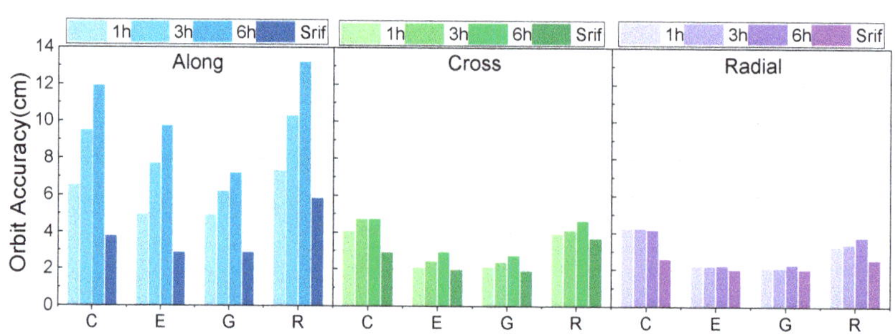

Figure 5. Average accuracy of predicted and epoch-wise updated orbits for BDS-3, Galileo, GPS, and GLONASS.

The average 1D-RMS errors of the GNSS orbits are summarized in Table 2. Clearly, the orbit precision of Galileo and GPS is generally superior to those of BDS-3 and GLONASS. As the update frequency decreases, the accuracy of the predicted orbit becomes worse. Compared with predicted orbits, epoch-wise updated orbits demonstrate significant advantages in both accuracy and continuity.

Table 2. The average 1D-RMS of different orbit products (Unit: cm).

Scheme	BDS-3	Galileo	GPS	GLONASS
1 h	5.23	3.41	3.43	5.28
3 h	6.75	4.88	4.15	6.12
6 h	7.97	6.07	4.78	8.53
Epoch-wise	3.17	2.46	2.45	4.33

3.2. Analysis of Ambiguity Resolution

In real-time clock estimation, orbit accuracy significantly impacts AR. In this section, UD, DD AR, and float solutions in clock estimation are applied to BDS-3, Galileo, and GPS based on predicted and epoch-wise updated orbit products. Then, the UD and DD ambiguity fixing rates of BDS-3, Galileo, and GPS are discussed.

The WL and NL of the UD and DD ambiguity fixing percentages are displayed in Figure 6 based on epoch-wise updated orbits recorded over 10 days. The WL fixing rate of DD is 100% for all systems, while the NL fixing rate significantly decreases. This is because the observation data were preprocessed according to the elevation angle and quality control methods during the DD AR process. For UD, the WL fixing rate is nearly 99%, and the NL fixing rate slightly decreases, with fluctuations of around 95%. Compared with DD, the WL fixing rate is almost similar, while the NL fixing rate is notably high.

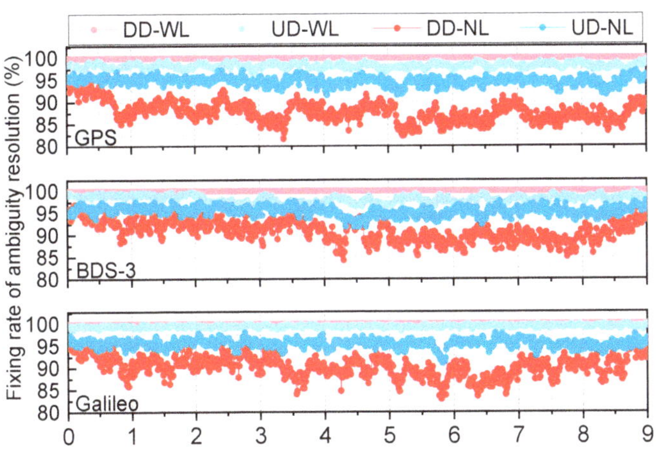

Figure 6. UD and DD ambiguity fixing percentages in clock estimation based on epoch-wise updated orbit.

To explore the impact of different satellite orbits on the DD AR of clock offset estimation, Figure 7 presents DD ambiguity fixing rates in clock estimation based on 1 h, 3 h, and 6 h predicted orbit, as well as epoch-wise updated orbits (marked as SRIF). It can be seen that all the WL fixing rates for BDS-3, Galileo, and GPS based on different orbits reach 100%. Conversely, the average NL fixing rate for DD ambiguities ranges from approximately 84.7% to 90.1%. Epoch-wise updated orbits obviously outperform predicted orbits in NL fixing, and the NL fixing rate slightly decreases as the orbit update frequency decreases, indicating that orbit precision affects NL ambiguity fixing.

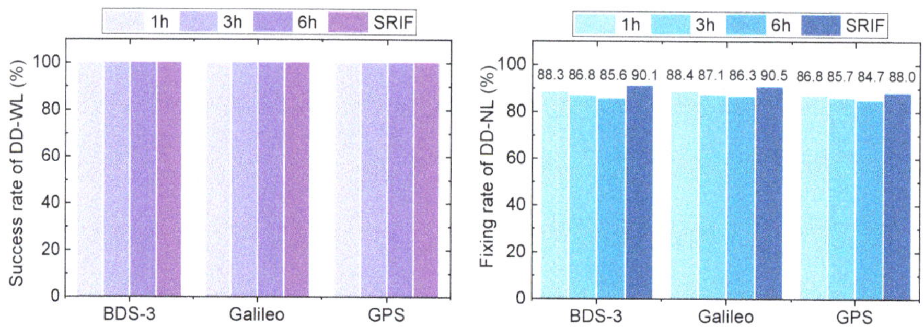

Figure 7. DD ambiguity fixing rate in clock estimation based on different orbits.

The UD ambiguity fixing rates of the predicted orbit are shown in Figure 8. The WL ambiguity fixing rates demonstrated excellent performance, with the rates for BDS-3, Galileo, and GPS all exceeding 98%. The results for the 1 h, 3 h, and 6 h predicted orbits, as well as for the epoch-wise updated orbits, show minimal variation. In terms of NL ambiguity fixing, predicted orbits with a 1 h update frequency exhibited significantly higher fixing rates than those of the 3 h and 6 h orbits. Furthermore, epoch-wise estimated orbits demonstrated a notable improvement over predicted orbits, particularly for BDS-3, where the NL fixing rate improved by 5%, 10%, and 19% compared with the 1 h, 3 h, and 6 h predicted orbits, respectively. Therefore, the quality and precision of orbit products have a remarkable impact on UD NL fixing, with higher-precision orbits leading to better fixing rates.

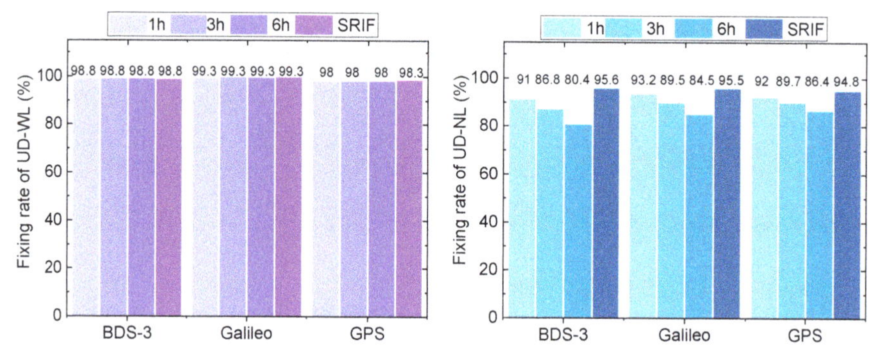

Figure 8. UD ambiguity fixing percentages in clock estimation based on different orbits.

3.3. Comparison of Clock Estimation Results

Real-time clock offset estimation depends on accurate real-time orbit products as known information. Orbit products with high update frequency can promptly capture dynamic variations, thus reducing error accumulation. Therefore, the results for clock offset estimation based on predicted and epoch-wise estimated orbit products are compared below.

Figure 9 displays the average accuracy of satellite clocks estimated on basis of 1 h, 3 h, and 6 h predicted orbit products. The STDs of the clock offsets are calculated by first taking the mean of the single differences between all satellite clocks within the same GNSS system as a virtual reference and then computing the second differences relative to this mean. The standard deviation of these second differences is used as the clock STD.

All results are better than 0.1 ns, except for the float solution with 6 h predicted orbits for BDS-3. The precision of satellite clock offsets significantly decreases as the predicted orbit update frequency reduces. Furthermore, the results for the DD and float solutions are slightly worse than those of the UD solution, indicating that the accuracy of satellite clock estimation can be improved by UD AR to some extent. It is notable that the clock offset accuracies of Galileo and GPS are slightly higher than those of BDS-3, which can be attributed to the more precise and stable ultra-rapid orbit products provided for Galileo and GPS. Therefore, both the accuracy and update frequency of satellite orbit products significantly impact the performance of real-time clock offset estimation.

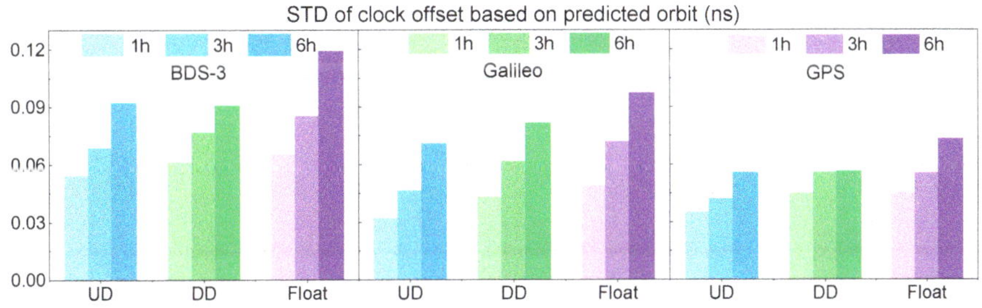

Figure 9. Averaged accuracy of clock offsets based on predicted orbit.

Figure 10 presents the clock estimation results for BDS-3, Galileo, and GPS with three schemes (UD, DD, and float), based on the epoch-wise estimated orbit, to further investigate the impact of orbit accuracy on clock offset estimation. The DD AR and float solutions are quite similar, and the UD solutions exhibit the best accuracy. Of the three systems, Galileo demonstrates the best performance, with a UD solution accuracy for all satellites within 0.03 ns. The GPS results are slightly better than those of BDS-3, with the most significant improvement observed after applying the UD AR. For BDS-3, the clock accuracy of the DD AR solution is worse than that of the float solution. Overall, the UD AR solution demonstrates a significant advantage in improving clock offset estimation accuracy.

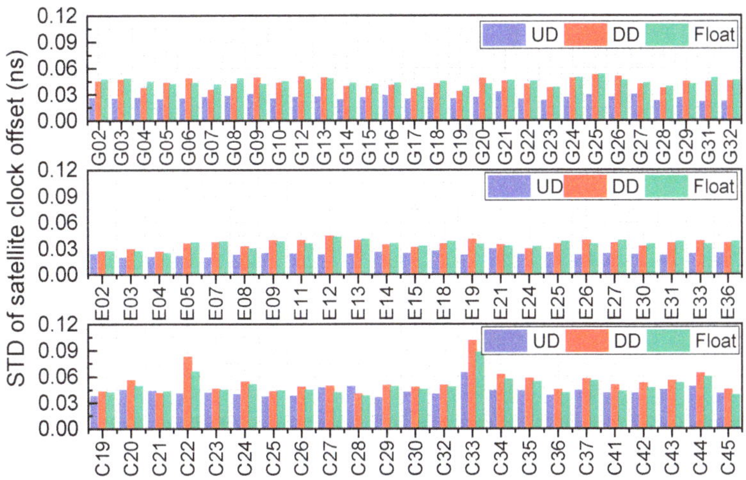

Figure 10. Clock accuracy of each satellite based on the epoch-wise updated orbit.

From Figures 9 and 10, it can be observed that both ambiguity resolution and high-quality orbit products significantly contribute to improving satellite clock offset estimation.

Specifically, with float solutions, the epoch-wise updated orbits consistently outperform ultra-rapid orbit products in clock estimation accuracy across all solution types—float, DD, and UD. This highlights the advantage of the proposed orbits in real-time clock estimation. Moreover, regardless of the orbit used, the UD solution yields better clock estimation performance than the float and DD solutions. When using epoch-wise updated orbits, the NL fixing rate of UD is significantly higher, resulting in more accurate and stable clock estimates. This further demonstrates the effectiveness of the proposed method.

Table 3 shows the mean accuracy of clock offset estimation using SRIF based on estimated orbits. The DD solution clearly does not enhance the precision of real-time satellite clock offset estimation, as its results are nearly identical to those of the float solution. By contrast, the UD solutions provide the best performance. The BDS-3, Galileo, and GPS clock offset accuracies reach 0.032 ns, 0.023 ns, and 0.026 ns, respectively, representing improvements of 36%, 34%, and 41% compared with the float solution. In contrast to the results based on predicted orbits, all the UD, DD, and float solutions exhibit notable improvements. In conclusion, high-precision orbit products and UD AR significantly enhance real-time clock offset estimation.

Table 3. The average accuracy of clock offset based on epoch-wise updated orbit (Unit: ns).

Scheme	BDS-3	Galileo	GPS
Float	0.050	0.035	0.044
DD	0.053	0.035	0.043
UD	0.032	0.023	0.026

3.4. PPP Validation

To further validate the performance of the optimal clock estimation method, kinematic PPP experiments with BDS-3, Galileo, and GPS were conducted using epoch-wise estimated orbit and the optimal clock products over 30 days. The PPP experiment involved 25 stations, as indicated in Figure 2. The IF combination with the same frequency as the clock offset estimation was used in the experiment. Furthermore, CNES real-time satellite orbit and clock products were used for comparison.

Positioning errors in the north (N), east (E), and up (U) directions for station TOW2 from DOY 287 to 293 are shown in Figure 11. The orange dots represent the positioning results using CNES real-time orbit and clock products, and the blue dots represent those of the estimated orbit and clock products combined with UD AR using SRIF. Both sets of results exhibit rapid convergence in all three directions, with errors in the N and E directions stabilizing after convergence. In comparison, the SRIF product provides more stable positioning results with higher accuracy in all three NEU directions, outperforming the CNES results.

Figure 12 presents the PPP positioning results for all stations using SRIF and CNES products. The red bars on the left represent the average convergence time in the N, E, and U directions, and the green bars on the right represent the accuracy in these directions. The convergence time of SRIF is slightly shorter than that of the CNES products for most stations in the N and U components. However, in the E component, some stations show a slower convergence when using SRIF products. Regarding positioning accuracy, the SRIF products outperform the CNES product in the N, E, and U components, demonstrating higher precision overall.

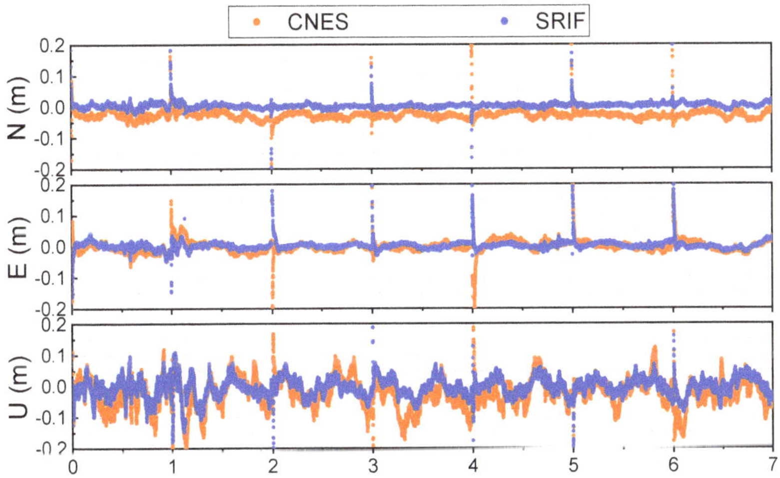

Figure 11. Positioning errors of station TOW2 using CNES and SRIF solution products.

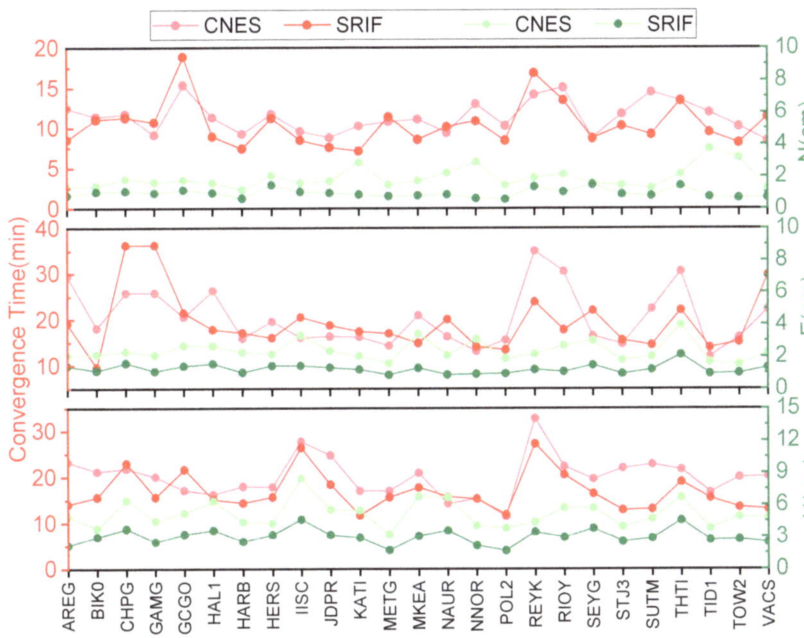

Figure 12. Positioning results and convergence times of SRIF and CNES products for each station.

Table 4 summarizes the average positioning precision and convergence time for the 25 stations. The PPP results demonstrate that the average convergence time using SRIF products is 15.57 min, comparable to the 17.28 min for the CNES products. Regarding positioning accuracy, the SRIF products achieve accuracies of 0.94 cm, 1.22 cm, and 3.01 cm in the N, E, and U directions, respectively, representing improvements of 50%, 48%, and 41% compared with CNES product. Consequently, the epoch-wise estimated orbit and UD AR satellite clock products using SRIF offer faster convergence and significantly enhance positioning accuracy for PPP.

Table 4. Average positioning errors and convergence time for SRIF and CNES.

	N (cm)	E (cm)	U (cm)	Convergence Time (min)
SRIF	0.94	1.22	3.01	15.57
CNES	1.88	2.36	5.13	17.28

4. Conclusions

Real-time clock offset products are among the core components of PPP services. Currently, real-time clock offset products provided by institutions such as the CNES rely on ultra-rapid predicted orbits. However, these predicted orbits have limited accuracy and discontinuous updates, which constrain the precision of real-time clock offset estimation. Therefore, this study proposes a UD AR clock offset estimation method implemented by SRIF based on real-time estimated orbits.

First, a comparison between the predicted orbits and the epoch-wise updated orbits highlighted the significant advantages of the latter in terms of both accuracy and continuity. Following this, three clock offset estimation schemes (UD, DD, and float solutions) were designed. The results show that the clock estimation accuracy and AR fixing rate can be significantly improved with high-precision epoch-wise updated orbits. In addition, the UD AR technique enhances clock offset estimation accuracy, while the DD AR does not significantly improve clock offset estimation. High-precision epoch-wise updated orbit products notably enhance the NL fixing rate of UD, substantially improving clock offset estimation accuracy. Specifically, the clock offset accuracies for BDS-3, Galileo, and GPS reach 0.032 ns, 0.023 ns, and 0.026 ns, respectively. Finally, the effectiveness of the proposed method was validated through PPP, and it was determined that this method outperforms real-time clock offset products based on predicted orbits of CNES. In conclusion, combining epoch-wise updated orbit products with the UD AR method can effectively improve the accuracy of real-time clock offset products, thereby enhancing the service quality of PPP.

Author Contributions: Conceptualization, Y.C. and L.W.; methodology, Y.C. and Z.Q.; software, Z.Q.; validation, Y.C. and L.W.; formal analysis, S.D. and Y.W.; writing—review and editing, Y.C. and Y.W.; visualization, W.L.; supervision, L.W.; All authors have read and agreed to the published version of the manuscript.

Funding: This work was supported by the Programs of the Fundamental Research Funds for the Central Universities (CHD 300102261301), the Special Fund for Basic Scientific Research of Central Colleges (CHD 300102262715), Natural Science Basic Research Program of Shaanxi (2019JC-20), Key Research and Development Program of Shaanxi (2022KW-09, 2021LLRH-06), Innovation Capability Support Program of Shaanxi (2021TD-03).

Data Availability Statement: No new data were created or analyzed in this study. Data sharing is not applicable to this article.

Acknowledgments: The authors would like to thank WHU for data support.

Conflicts of Interest: The authors declare that they have no known competing financial interests or personal relationships that could have appeared to influence the work reported in this paper.

References

1. Kazmierski, K.; Zajdel, R.; Sośnica, K. Evolution of orbit and clock quality for real-time multi-GNSS solutions. *GPS Solut.* **2020**, *24*, 111. [CrossRef]
2. Huang, G.W.; Zhang, Q.; Xu, G.C. Real-time clock offset prediction with an improved model. *GPS Solut.* **2014**, *18*, 95–104. [CrossRef]
3. Kouba, J.; Héroux, P. Precise Point Positioning Using IGS Orbit and Clock Products. *GPS Solut.* **2001**, *5*, 12–28. [CrossRef]

4. Guo, J.; Li, X.; Li, Z.; Hu, L.; Yang, G.; Zhao, C.; Fairbairn, D.; Watson, D.; Ge, M. Multi-GNSS precise point positioning for precision agriculture. *Precis. Agric.* **2018**, *19*, 895–911. [CrossRef]
5. Liu, X.; Du, Y.; Huang, G.; Wang, D.; Zhang, Q. Mitigating GNSS multipath in landslide areas: A novel approach considering mutation points at different stages. *Landslides* **2023**, *20*, 2497–2510. [CrossRef]
6. Zumberge, J.; Heflin, M.; Jefferson, D.; Watkins, M.; Webb, F.H. Precise point positioning for the efficient and robust analysis of GPS data from large networks. *J. Geophys. Res. Solid Earth* **1997**, *102*, 5005–5017. [CrossRef]
7. Liu, J.; Liang, Y.; Xu, D.; Gong, X.; Hyypp, J. A ubiquitous positioning solution of integrating GNSS with LiDAR odometry and 3D map for autonomous driving in urban environments. *J. Geod.* **2023**, *97*, 39. [CrossRef]
8. Wang, D.; Huang, G.; Du, Y.; Zhang, Q.; Bai, Z.; Tian, J. Stability analysis of reference station and compensation for monitoring stations in GNSS landslide monitoring. *Satell. Navig.* **2023**, *4*, 29. [CrossRef]
9. Huang, G.; Cui, B.; Zhang, Q.; Fu, W.; Li, P. An improved predicted model for BDS ultra-rapid satellite clock offsets. *Remote Sens.* **2018**, *10*, 60. [CrossRef]
10. Fu, W.; Huang, G.; Zhang, Q.; Gu, S.; Ge, M.; Schuh, H. Multi-GNSS real-time clock estimation using sequential least square adjustment with online quality control. *J. Geod.* **2019**, *93*, 963–976. [CrossRef]
11. Bock, H.; Dach, R.; Jaeggi, A.; Beutler, G. High-rate GPS clock corrections from CODE: Support of 1 Hz applications. *J. Geod.* **2009**, *83*, 1083–1094. [CrossRef]
12. Dai, Z.; Dai, X.; Zhao, Q.; Bao, Z.; Li, C. Multi-GNSS real-time clock estimation using the dual-thread parallel method. *Adv. Space Res.* **2018**, *62*, 2518–2528. [CrossRef]
13. Zuo, X.; Jiang, X.; Li, P.; Wang, J.; Ge, M.; Schuh, H. A square root information filter for multi-GNSS real-time precise clock estimation. *Satell. Navig.* **2021**, *2*, 28. [CrossRef]
14. Hauschild, A.; Montenbruck, O. Kalman-filter-based GPS clock estimation for near real-time positioning. *Gps Solut.* **2009**, *13*, 173–182. [CrossRef]
15. Dai, X.; Gong, X.; Li, C.; Qing, Y.; Gu, S.; Lou, Y. Real-time precise orbit and clock estimation of multi-GNSS satellites with undifferenced ambiguity resolution. *J. Geod.* **2022**, *96*, 73. [CrossRef]
16. Li, X.; Li, X.; Yuan, Y.; Zhang, K.; Zhang, X.; Wickert, J. Multi-GNSS phase delay estimation and PPP ambiguity resolution: GPS, BDS, GLONASS, Galileo. *J. Geod.* **2018**, *92*, 579–608. [CrossRef]
17. Dai, Z.; Dai, X.; Zhao, Q.; Liu, J. Improving real-time clock estimation with undifferenced ambiguity fixing. *GPS Solut.* **2019**, *23*, 44. [CrossRef]
18. Du, S.; Shu, B.; Xie, W.; Huang, G.; Ge, Y.; Li, P. Evaluation of Real-time Precise Point Positioning with Ambiguity Resolution Based on Multi-GNSS OSB Products from CNES. *Remote. Sens.* **2022**, *14*, 4970. [CrossRef]
19. Zuo, X.; Li, P.; Cui, B.; Ge, M.; Schuh, H. A computational efficient approach for multi-GNSS real-time precise clock estimation with undifferenced ambiguity resolution. *J. Geod.* **2024**, *98*, 73. [CrossRef]
20. Cao, Y.; Huang, G.; Xie, S.; Xie, W.; Liu, Z.; Tan, Y. An evaluation method of GPS satellite clock in-orbit with periodic terms deducted. *Measurement* **2023**, *214*, 112765. [CrossRef]
21. Qin, Z.; Zhang, Q.; Huang, G.; Tang, L.; Wang, J.; Wang, X. BDS Orbit Maneuver Detection Based on Epoch-Updated Orbits Estimated by SRIF. *Remote Sens.* **2023**, *15*, 2558. [CrossRef]
22. Li, X.; Li, X.; Liu, G.; Feng, G.; Guo, F.; Yuan, Y.; Zhang, K. Spatial–temporal characteristic of BDS phase delays and PPP ambiguity resolution with GEO/IGSO/MEO satellites. *GPS Solut.* **2018**, *22*, 123. [CrossRef]
23. Melbourne, W.G. The case for ranging in GPS-based geodetic systems. In Proceedings of the First International Symposium on Precise Positioning with the Global Positioning System, Rockville, MD, USA, 15–19 April 1985; pp. 373–386.
24. Wübbena, G. Software developments for geodetic positioning with GPS using TI-4100 code and carrier measurements. In Proceedings of the First International Symposium on Precise Positioning with the Global Positioning System, Rockville, MD, USA, 15–19 April 1985; pp. 403–412.
25. Bierman, G.J. The treatment of bias in the square-root information filter/smoother. *J. Optim. Theory Appl.* **1975**, *16*, 165–178. [CrossRef]
26. Teunissen, P.J.G. Quality control in integrated navigation systems. In Proceedings of the IEEE Position Location & Navigation Symposium, Las Vegas, NV, USA, 20 March 1990.
27. Teunissen, P.J.G. Distributional theory for the DIA method. *J. Geod.* **2017**, *92*, 59–80. [CrossRef]
28. Yang, Y.; Gao, W.; Zhang, X. Robust Kalman filtering with constraints: A case study for integrated navigation. *J. Geod.* **2010**, *84*, 373–381. [CrossRef]
29. Teunissen, P.J.; Montenbruck, O. *Springer Handbook of Global Navigation Satellite Systems*; Springer: Berlin/Heidelberg, Germany, 2017; Volume 10.
30. Montenbruck, O.; Steigenberger, P.; Prange, L.; Deng, Z.; Zhao, Q.; Perosanz, F.; Romero, I.; Noll, C.; Stürze, A.; Weber, G. The Multi-GNSS Experiment (MGEX) of the International GNSS Service (IGS)–achievements, prospects and challenges. *Adv. Space Res.* **2017**, *59*, 1671–1697. [CrossRef]

31. Yuan, Y.; Li, X.; Yao, Y.; Huang, S.; Wang, Q.; Zhang, K. Estimation of phase center corrections for BDS satellites aligned to the IGS20 frame. *GPS Solut.* **2024**, *28*, 63. [CrossRef]
32. Wu, J.-T.; Wu, S.C.; Hajj, G.A.; Bertiger, W.I.; Lichten, S.M. Effects of antenna orientation on GPS carrier phase. *Astrodynamics* **1992**, *18*, 1647–1660. [CrossRef]
33. Luzum, B.; Petit, G. The IERS Conventions (2010): Reference systems and new models. *Proc. Int. Astron. Union* **2012**, *10*, 227–228. [CrossRef]
34. Lyard, F.; Lefevre, T.; Letellier, F.; Francis, O. Modelling the globalocean tides: Modern insights from FES2004. *J. Geophys. Res. Atmos.* **2004**, *56*, 394–415.
35. Saastamoinen, J. Contributions to the theory of atmospheric refraction. *Bull. Géodésique* **1972**, *105*, 279–298. [CrossRef]
36. Böhm, J.; Niell, A.; Tregoning, P.; Schuh, H. Global Mapping Function (GMF): A new empirical mapping function based on numerical weather model data. *Geophys. Res. Lett.* **2006**, *33*, L07304. [CrossRef]

remote sensing

MDPI

Article

BeiDou Satellite-Based Augmentation System Algorithm Optimization and Performance Validation of Ionospheric Degradation Parameters with RTCA Protocol

Zhaochen Li [1,2], Yueling Cao [2,*], Shanshi Zhou [2], Xiaogong Hu [2] and Ran Liu [3]

1 School of Physics, Northwest University, Xi'an 710127, China; lizhaochen@shao.ac.cn
2 Shanghai Astronomical Observatory, Chinese Academy of Sciences, Shanghai 200030, China; sszhou@shao.ac.cn (S.Z.); hxg@shao.ac.cn (X.H.)
3 Beijing Satellite Navigation Center, Beijing 100094, China; liuran0902@foxmail.com
* Correspondence: caoyueling@shao.ac.cn

Abstract: The BeiDou Satellite-Based Augmentation System (BDSBAS), based on the Radio Technical Commission for Aeronautics (RTCA) protocol, aims to provide high-precision, single-frequency positioning with integrity assurance for civil aviation users in China and surrounding regions. Given the anticipated high solar activity between 2023 and 2025, ionospheric anomalies may degrade positioning accuracy and significantly impact BDSBAS integrity performance. To enhance BDSBAS integrity, this study evaluates and analyzes the system's ionospheric degradation parameters for 2023. The results indicate that during the active ionospheric period in 2023, the rate of ionospheric grid delay changes exceeding the limits of the currently broadcasted parameters increased by 0.86%, posing potential integrity risks compared to 2022. To address this issue, we propose a novel algorithm for ionospheric degradation parameters and assess its applicability, stability, and effectiveness using BDSBAS single-frequency service message data from IGS monitoring stations in China. Statistical analysis in the localization domain demonstrates that the new method reduces the rate of ionospheric degradation parameters exceeding the threshold by 1.10% in 2023–2024. This approach significantly enhances BDSBAS integrity service capabilities, supporting its performance improvement and official deployment.

Keywords: BeiDou satellite-based augmentation system (BDSBAS); integrity performance; degradation parameters; service availability

1. Introduction

With the widespread adoption of Global Navigation Satellite Systems (GNSS), many countries and regions have established satellite-based augmentation systems (SBAS) to meet higher demands for positioning accuracy and integrity [1,2]. Currently, ten SBAS are either operational or under construction worldwide, including, as follows: the Wide Area Augmentation System (WAAS) in the United States; the European Geostationary Navigation Overlay System (EGNOS) in Europe; the Multi-functional Transport Satellite-based Augmentation System (MSAS) in Japan; the GPS Aided Geo Augmented Navigation (GAGAN) in India; the System for Differential Correction and Monitoring (SDCM) in Russia; the BeiDou Satellite-Based Augmentation System (BDSBAS) in China; the SBAS for Africa and the Indian Ocean (A-SBAS); the Algerian Satellite-Based Augmentation System (AL-SBAS); the Southern Positioning Augmentation Network (SPAN) in Australia and New Zealand; and the Korea Augmentation Satellite System (KASS) in Korea [3]. Among these,

WAAS, EGNOS, MSAS, and GAGAN have officially launched RTCA single-frequency services, while BDSBAS, SDCM, A-SBAS, AL-SBAS, SPAN, and KASS are currently in testing or trial operations.

BDSBAS broadcasts comprehensive correction information, including differential corrections for satellite orbits and clocks, ionospheric grid corrections, and integrity data such as User Differential Range Error (UDRE), Grid Ionospheric Vertical Error (GIVE), and degradation parameters [4,5]. These parameters ensure the timeliness and accuracy of differential corrections in both spatial and temporal domains, particularly benefiting users experiencing correction delays or who are operating at the service area's edge [6]. This information is transmitted via three geosynchronous pseudo-random noise satellites (PRN 130, PRN 143, and PRN 144) to augment GPS signals and provide APV-I level services in China and the surrounding regions, with service indicators detailed in Table 1 [7,8].

Table 1. Signal-in-space performance requirements for SBAS on APV-I.

Flight Operation	Horizontal/Vertical Accuracy (95%)	Integrity Level	HPL/VPL	Alarm Time	Continuity	Availability
APV-I	16 m/20 m	2×10^{-7}/approach	40 m/50 m	10 s	8×10^{-6}/15 s	0.99~0.99999

BDSBAS message type 10 retains the spatiotemporal characteristics of the most recent effective differential information, minimizing service degradation even if differential correction data are lost. Table 2 provides a detailed description of the efficacy degradation parameters [9], which are categorized by function into the following four groups: (1) fast-correction degradation parameters, encompassing category B_{rrc}; (2) long-term correction degradation parameters, including variation limit C_{ltc_v0} and update period I_{ltc_v0}, etc.; (3) geosynchronous orbit (GEO) message degradation parameters, consisting of categories update period I_{ltc_v0}, variation limit C_{geo_lsb} and variable speed limit C_{geo_v}; and (4) ionospheric degradation parameters are divided into variation limit C_{iono_step}, update period I_{iono}, and variable speed limit C_{iono_ramp}. In addition, RSS_{iono} and RSS_{UDRE} are, respectively, the root sum square flag and user differential range error flag, and $C_{covariance}$ is equivalent to the residual variance.

Table 2. Message type 10 detailed content.

Parameter	Effective Range	Units	Parameter	Effective Range	Units
B_{rrc}	0–2.046	m	I_{geo}	0–511	s
C_{ltc_lsb}	0–2.046	m	C_{er}	0–31.5	m
C_{ltc_v1}	0–0.05115	m/s	C_{iono_step}	0–1.023	m
I_{ltc_v1}	0–511	s	I_{iono}	0–511	s
C_{ltc_v0}	0–2.046	m	C_{iono_ramp}	0–0.005155	m/s
I_{ltc_v0}	0–511	s	RSS_{UDRE}	0–1	unitless
C_{geo_lsb}	0–0.5115	m	RSS_{iono}	0–1	unitless
C_{geo_v}	0–0.05115	m/s	$C_{covariance}$	0–12.7	unitless

Ionospheric degradation parameters define the change limits of vertical delays at Ionospheric Grid Points (IGPs) broadcast via message type 26 and are disseminated through message type 10. The parameters are as follows: C_{iono_step} denotes the limit on the variation between successive ionospheric grid delay values; I_{iono} indicates the interval between updates to ionospheric delay corrections; and C_{iono_ramp} defines the rate limit for changes in successive ionospheric grid delay values. RSS_{iono} serves as a root sum square flag for the ionospheric degradation parameters. Currently, different SBAS broadcast ionospheric degradation parameters are constant values, as shown in Table 3.

Table 3. Global SBAS ionospheric degradation parameters.

SYSTEM	C_{iono_step}	I_{iono}	C_{iono_ramp}	RSS_{iono}
BDSBAS	0.952	240	0.000000	0
WAAS	0.836	300	0.000000	0
EGNOS	0.657	300	0.000000	0
SDCM	0	300	0.000740	1
MSAS	0.836	300	0.000000	0
GAGAN	0.952	300	0.000000	0
KASS	0.657	300	0.000000	0
SPAN	0.657	300	0.000000	0
ASBAS	0.657	300	0.000000	0

SBAS services aim to enhance the accuracy, integrity, availability, and continuity of GNSS positioning [10]. Integrity performance is a key reliability indicator, evaluated using Horizontal Protection Levels (HPL) and Vertical Protection Levels (VPL) derived from SBAS integrity messages. These protection levels quantify error bounds in positioning, ensuring navigation system integrity and safety.

Degradation parameters contribute to protection level calculations. If the computed HPL and VPL remain below the required Horizontal and Vertical Alert Limits (HAL and VAL) for a given service level, the service is deemed available; otherwise, it is considered unavailable. Additionally, if the user's protection level falls below actual Horizontal and vertical Positioning Errors (HPE and VPE), an integrity risk arises. Thus, ionospheric degradation parameters are crucial for ensuring SBAS service reliability and availability [11,12].

In BDSBAS studies, service performance and integrity have received significant attention [13–15]; however, degradation parameters have been less explored. Ionospheric correction plays a vital role in achieving high-accuracy, single-frequency positioning [6,16,17]. Therefore, this study focuses on ionospheric degradation parameters from message type 10. Section 2.1 analyzes cases where BDSBAS ionospheric degradation parameters failed to provide effective envelopment during peak ionospheric activity, evaluating their impact on service performance. Section 2.2 identifies limitations in BDSBAS ionospheric degradation representation during high solar activity and proposes an improved algorithm. Section 3 details the experimental results validation of this new method and Section 4 discusses the impact of the new degradation parameters on positioning services.

2. Materials and Methods

2.1. Performance Analysis of Ionospheric Degradation Parameters

Ionospheric degradation parameters, defined as an integrity threshold, are designed to encapsulate and mitigate ionospheric delay errors. This capability ensures compliance with APV-I's performance criteria for integrity services. The research presented in this paper indicates that with BDSBAS degradation parameters C_{iono_step} set at 0.952 and C_{iono_ramp} at 0, the probability of exceeding the threshold for BDSBAS remained relatively low in 2021. However, as ionospheric activity has increased in recent years, the probability of integrity parameter leakage has risen annually. The frequency of ionospheric variations $\Delta IGPdelay$ exceeding the threshold $I_{\mathrm{lim}it}$ has also increased. The threshold of ionospheric delay change is determined by the ionospheric degradation parameter in Equation (1), as follows:

$$I_{\text{limit}} = C_{iono\_step} + C_{iono\_ramp} * I_{iono} \quad (1)$$

The parameters C_{iono_step}, C_{iono_ramp} and I_{iono}, in Equation (1) have different fixed values in different systems. The specific values are shown in Table 1. In the previous paragraph, $\Delta IGPdelay$ represents the difference in the same grid points delays between the two adjacent epochs, as follows:

$$\Delta IGPdelay = IGPdelay_t - IGPdelay_{t-1} \quad (2)$$

The probability of ionospheric grid delay variations exceeding the threshold can be calculated by Equation (3), as follows:

$$P(\Delta IGPdelay > I_{\text{limit}}) = \frac{\sum_{t=1}^{T} \sum_{j=1}^{J} (\Delta IGPdelay_{t,j} > I_{\text{limit}})}{T * J} \quad (3)$$

In Equation (3), T is the total number of the epoch and J is the total number of ionospheric grid points. $\Delta IGPdelay_{t,j}$ is the variation in ionospheric delay at the grid point j on the epoch t.

This study selects a grid point at 15°N latitude and 100°E longitude, a region of high ionospheric activity, for detailed analysis. The ionospheric delay variation envelope for the first day of each year from 2021 to 2024 in BDSBAS is shown in Figure 1. The horizontal axis is in hours and the vertical axis is in meters, the blue dots represent the ionospheric delay variation between adjacent epochs at this grid point from 2021 to 2024, and the red line indicates the variation threshold I_{limit}. It is equal to C_{iono_step} because C_{iono_ramp} is the fixed value 0.

In 2021, the degradation parameter threshold nearly encompassed all ionospheric delay variations, as shown in Figure 1a. However, as ionospheric activity intensified in 2022, Figure 1b shows that the degradation parameter could no longer fully envelop the actual variation in ionospheric delay; the probability of exceeding the threshold increased further in 2023 and 2024, as shown in Figure 1c,d.

Table 4 provides more detailed data on the annual probability of exceeding the limits calculated by Equation (3) from 2021 to 2024 for BDSBAS, EGNOS, MSAS, and WAAS. The probability of exceeding the threshold for BDSBAS and MSAS has increased greatly, whereas EGNOS and WAAS have shown no significant rise. This discrepancy is likely due to the higher proportion of low-latitude regions in China, where ionospheric anomalies have a greater impact [18–20]. In contrast, EGNOS and WAAS operate in regions with fewer ionospheric anomalies [21].

Table 4. The annual probability of exceeding the threshold probability in 2021–2024.

	2021	2022	2023	2024
BDSBAS	0.12%	0.78%	1.64%	1.96%
MSAS	0.54%	1.44%	2.42%	3.24%
EGNOS	0.15%	0.36%	0.38%	0.40%
WASS	0.02%	0.07%	0.16%	0.23%

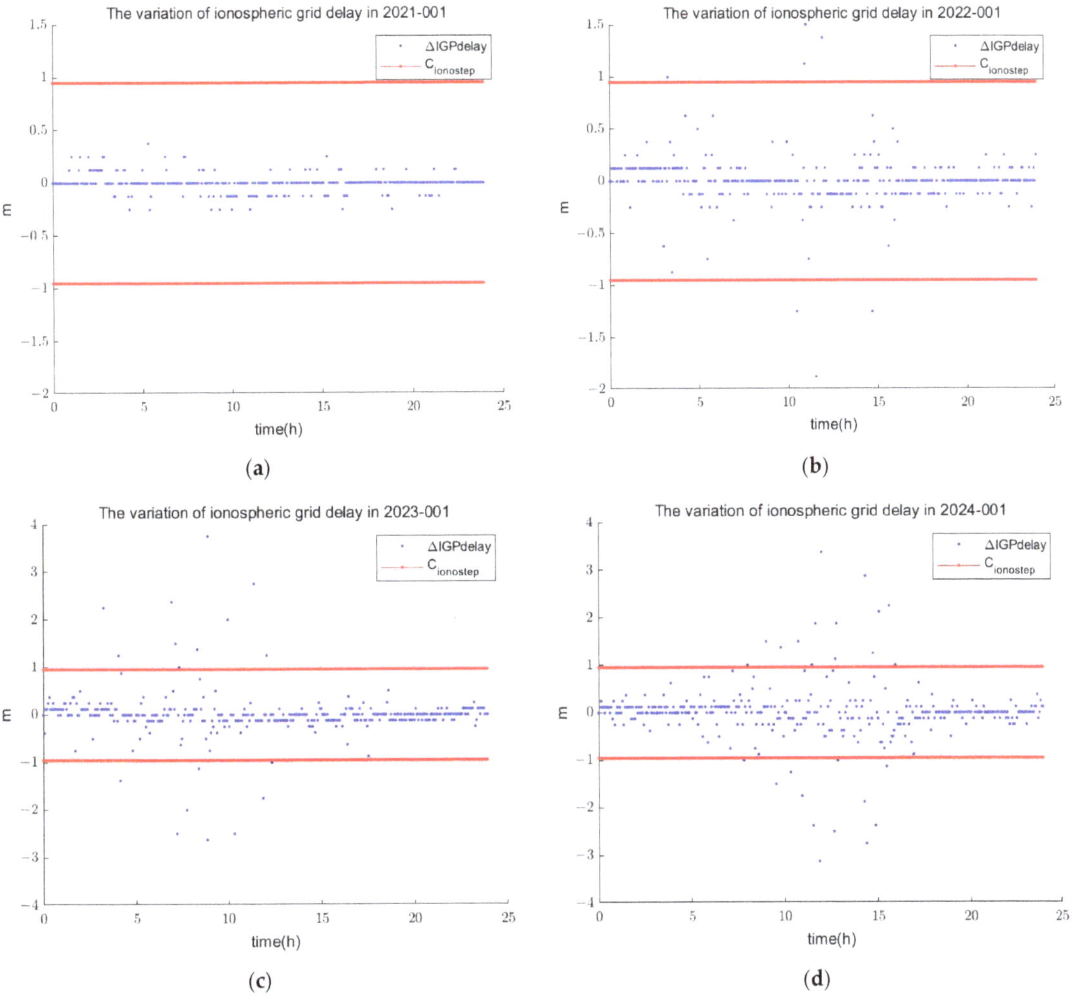

Figure 1. (**a**) Variation in ionospheric delay in the 17th grid point in the 7th band on the first day of 2021; (**b**) variation in ionospheric delay in the 17th grid point in the 7th band on the first day of 2022; (**c**) variation in ionospheric delay in the 17th grid point in the 7th band on the first day of 2023; and (**d**) variation in ionospheric delay in the 17th grid point in the 7th band on the first day of 2024.

The daily probability of exceeding the threshold for BDSBAS and MSAS from 2021 to 2024 is shown Figure 2. The vertical axis represents the probability of exceeding the threshold, while the horizontal axis indicates the number of days. The exceeding threshold rate of MSAS is generally higher than that of BDSBAS, with both showing synchronized variations. The probability of ionospheric changes in both BDSBAS and MSAS exceeding the degradation parameter threshold increases over time. This trend reflects the growing impact of increased solar activity on China and the surrounding region.

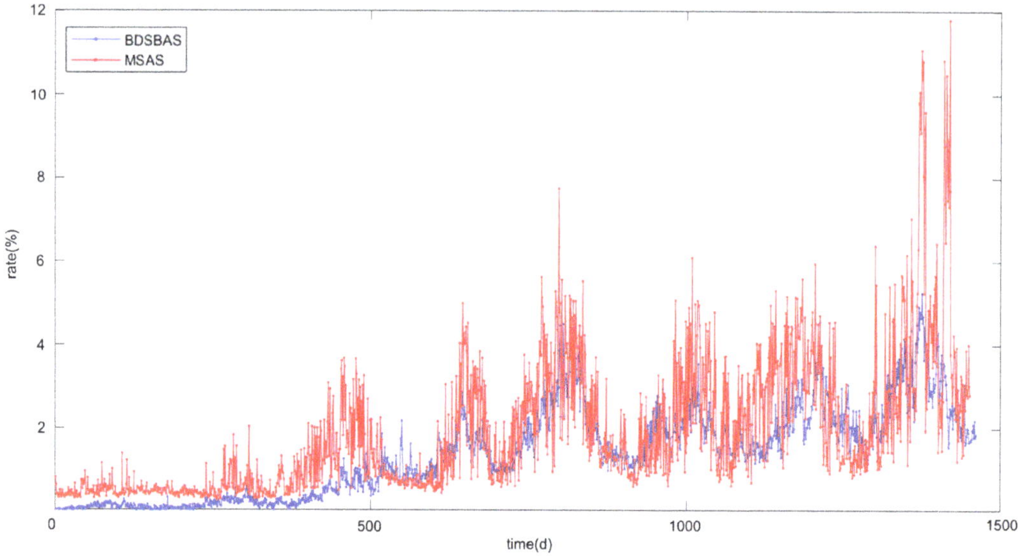

Figure 2. Probability of exceeding the threshold of BDSBAS and MSAS from the years 2021 to 2024.

To further investigate the relationship between ionospheric anomalies and the exceeding threshold rate, as well as their connection to solar activity, a time-series analysis of the BDSBAS exceeding threshold rate and sunspot number (SSN) from 2021 to 2024 was conducted. In Figure 3, the daily exceeding threshold rate is plotted on the left vertical axis, and SSN is plotted on the right vertical axis, with the horizontal axis representing the number of days. SSN, derived from sunspot counts, is a widely used index for quantifying solar activity [22]. It is strongly correlated with most solar activity indicators and serves as a key driver of ionospheric variations [23].

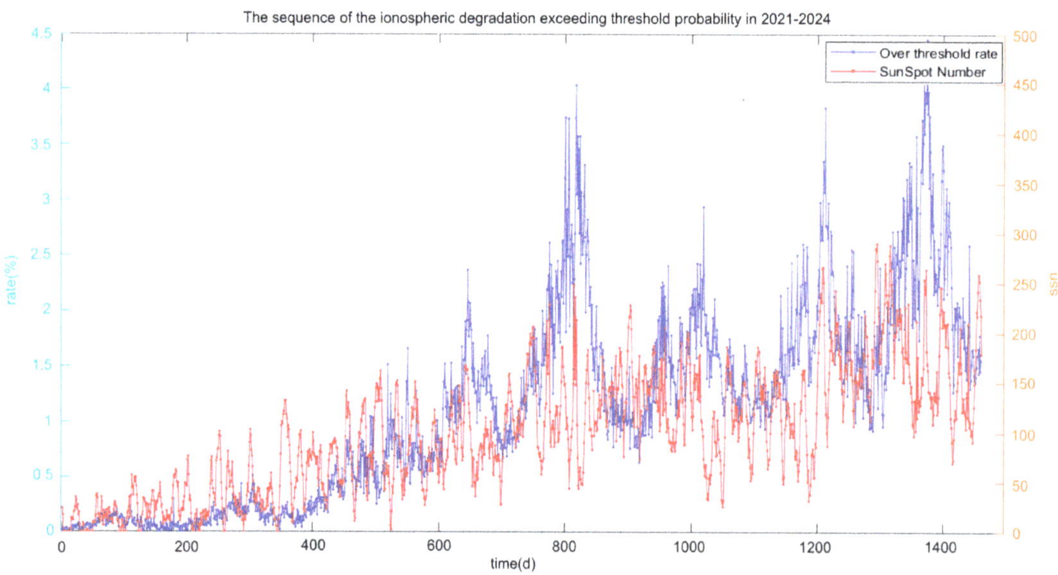

Figure 3. Probability of exceeding the threshold from the years 2021 to 2024 (horizontal axis: days).

The proportion of ionospheric delay variations exceeding the threshold has risen annually, as shown in Figure 3. The exceeding threshold rate of ionospheric degradation parameters correlates with ionospheric activity, which in turn is influenced by solar activity. While SSN and total electron content (TEC) exhibit a strong long-term correlation, their short-term correlation is weaker. Both SSN and the over-limit rate exhibit oscillatory increases over four years; however, local correlations vary due to factors such as geomagnetic variations [21,24]. Additionally, the solar activity index reveals hysteresis and saturation effects in TEC, where TEC values during the ascending and descending phases of the solar cycle differ despite similar levels of solar activity. Saturation implies that beyond a certain threshold, increasing solar activity does not proportionally increase TEC [25,26]. Overall, solar activity and ionospheric changes exhibit a strong correlation.

A statistical analysis of ionospheric grid points in the BDSBAS service area was conducted to examine the exceeding threshold probability across different latitude bands. Figure 4 illustrates the changes in exceeding the threshold daily probability from 2021 to 2024 for latitudes between 5° and 55°. The horizontal axis is in days, and the vertical axis shows the probability of exceeding the threshold. The changes in exceeding the threshold probability vary across latitude bands, with an overall cyclical upward trend displaying seasonal periodicity; it is especially conspicuous in low-latitude areas.

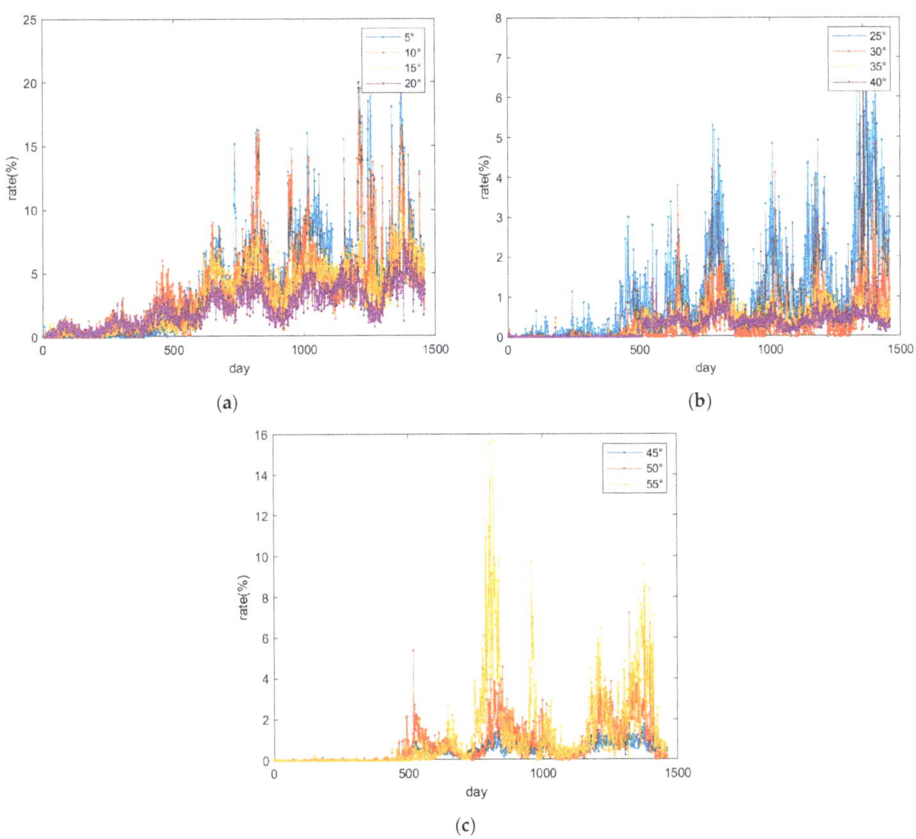

Figure 4. (**a**) Probability of exceeding the threshold in the 5° to 20° latitude band; (**b**) probability of exceeding the threshold in the 25° to 40° latitude band; and (**c**) probability of exceeding the threshold in the 45° to 55° latitude band.

Figure 5 summarizes the relationship between latitude and the probability of exceeding the threshold. The horizontal axis is in latitudes and the vertical axis shows the probability of exceeding the threshold. The lowest probability occurs at the 40° latitude and increases toward the edges, with more significant issues in low-latitude areas. The ionosphere is particularly active below the 35° latitude, directly affecting ionospheric degradation parameters and, consequently, service availability and integrity. Therefore, during periods of ionospheric activity, its impact on the integrity in southern China is significantly heightened.

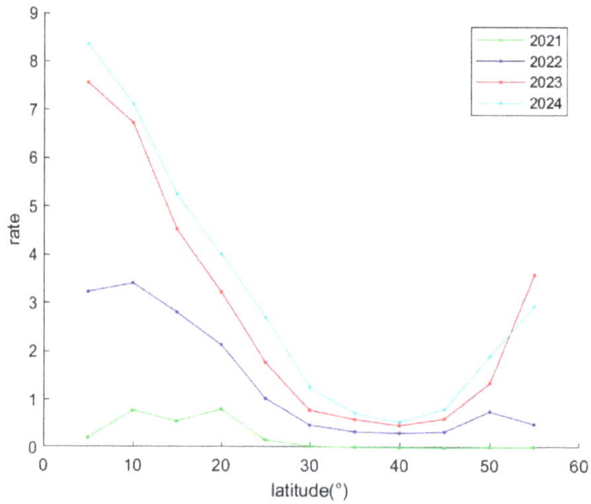

Figure 5. Annual probability of exceeding the threshold in the different latitude from the years 2021 to 2024.

To assess the impact of ionospheric activity on service performance, BDSBAS service data from 2022 and 2023 were analyzed, focusing on enhanced positioning, protection levels, and service availability. The positioning performance and protection levels of selected IGS stations were evaluated using degradation parameters. The algorithm of the protection level refers to the RTCA [27]. Present service availability statistics for the WUH2 station from DOY (day of the year) 001–003 in 2022 are shown in Figure 6, showing the service availability of WUH2 station from DOY 001–003 in 2023, where blue represents the Horizontal Protection Level, yellow denotes the Horizontal Position Error, and red marks the Protection Level threshold, with the horizontal axis in days and the vertical axis in meters. The figures indicate that while positioning errors remained stable, the probability of integrity and misleading information events increases. In addition, the increased protection level can lead to reduced availability in areas with larger protection levels.

The service availability maps in the China region for DOY 001–015 in 2022 and 2023 are shown in Figure 7. The color axis denotes service availability; the horizontal axis is the latitude, and the vertical axis is precision. This shows a decline in single-frequency enhancement service performance in 2023. This decline is particularly pronounced in lower-latitude regions, where ionospheric activity is higher.

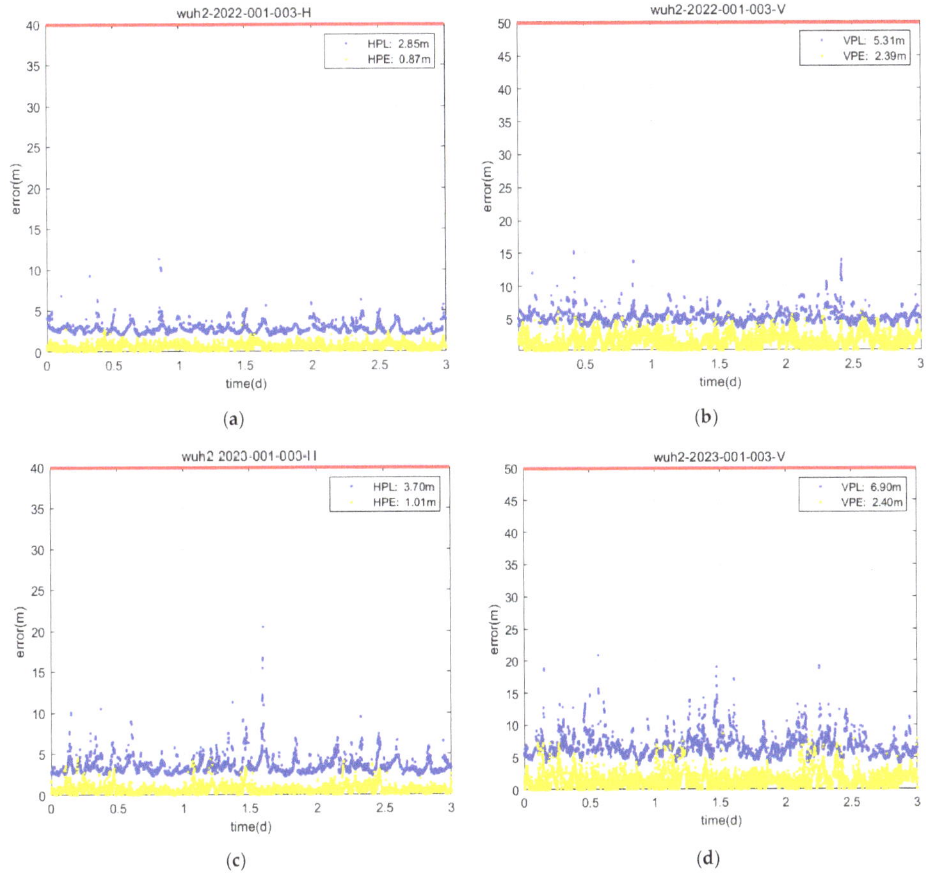

Figure 6. (**a**) HPL and HPE of WUH2 in 2022; (**b**) VPL and VPE of WUH2 in 2022; (**c**) HPL and HPE of WUH2 in 2023; and (**d**) VPL and VPE of WUH2 in 2023.

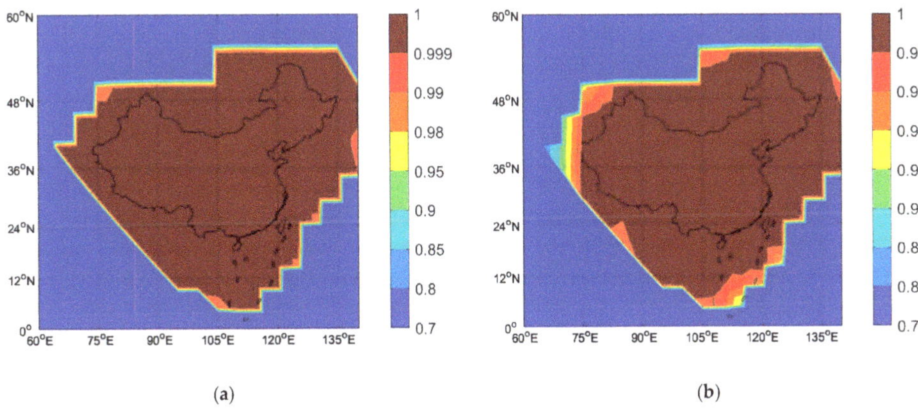

Figure 7. (**a**) Service availability in the China region in 2022; and (**b**) service availability in the China region in 2023.

The data in Table 5 compare BDSBAS service metrics between 2022 and 2023 when broadcasting a fixed degradation parameter value. Performance metrics include enhanced positioning, protection levels, service availability, and integrity risk, using select International GNSS Service observation stations as examples. The table presents differences between the 2023 and 2022 results, with positive values indicating increases and negative values indicating decreases. Each statistical period starts on January 1st, May 1st, and September 1st, covering 15 days.

Table 5. Service performance change table for fixed value degradation parameters in 2023 (unit: meters).

STATION	DATE	HPE	VPE	HPL	VPL	Availability	PMI	PHMI
JFNG	1-1	0.08	−0.23	0.54	1.04	0.00%	0.28%	0.000%
	5-1	−0.10	−0.11	0.06	0.19	0.00%	0.57%	0.000%
	9-1	−0.23	−0.66	−0.02	0.12	0.00%	0.17%	0.000%
URUM	1-1	0.03	−0.15	0.39	0.99	0.00%	0.20%	0.000%
	5-1	−0.04	−0.07	−0.48	−0.24	0.00%	0.12%	0.000%
	9-1	−0.12	−0.36	−0.59	−0.28	0.00%	0.25%	0.000%
WUH2	1-1	0.14	0.01	0.57	1.11	0.00%	0.20%	0.000%
	5-1	−0.03	0.10	0.04	0.20	0.00%	0.52%	0.000%
	9-1	−0.16	−0.38	−0.03	0.11	0.00%	0.39%	0.000%
HKSL	1-1	0.20	0.47	0.66	1.38	−0.01%	0.87%	0.000%
	5-1	0.14	0.34	0.03	0.24	0.00%	1.57%	0.000%
	9-1	0.04	0.18	−0.11	0.06	0.00%	0.64%	0.000%
HKWS	1-1	0.20	0.44	0.65	1.37	−0.01%	0.91%	0.000%
	5-1	0.14	0.35	0.02	0.23	0.00%	1.57%	0.000%
	9-1	0.04	0.19	−0.10	0.07	0.00%	0.98%	0.000%

Table 5 shows the decline in service availability located in edge-service areas in 2023 compared to 2022. Additionally, some misleading information (MI) events are present and the probability of hazardous misleading information (HMI) events remains consistently zero in 2023, particularly in the vertical direction, due to ionospheric influences.

2.2. New Calculation Method of Ionospheric Degradation Parameters

The current method for BDSBAS broadcasting ionospheric degradation parameters is as follows: Set C_{iono_step} as a fixed value of 0.952; set I_{iono} as a fixed value of 240; set C_{iono_ramp} as a fixed value of 0; and set RSS_{iono} as 0. Due to the current ionospheric activity, the predefined threshold for ionospheric variations in the degradation parameter fails to fully encompass actual ionospheric delay changes. This paper proposes a new algorithm for determining the broadcast value of the ionospheric degradation parameter.

Real-time calculation of ionospheric grid point vertical delay values is possible through observations from evenly distributed monitoring stations utilizing a dual-frequency approach to correct ionospheric errors. Consequently, this method enables fitting vertical delay parameters at grid points, relying on ionospheric delay measurements at piercing points.

Initially, dual-frequency analysis is used to extract high-precision ionospheric delay values at piercing points, as follows:

$$IGPdelay = \frac{f_2^2}{f_1^2 - f_2^2}(P_2 - P_1) \tag{4}$$

In Equation (4), f denotes frequency and P represents corrected pseudo-range observations.

Ionospheric delays are determined for piercing points within four quadrants around grid points, based on their distances from the grid points. If a minimum of three piercing points exist within a 5° radius $(\varphi \pm 5^\circ, \lambda \pm 5^\circ)$ of a grid point (φ, λ), an inverse-distance-weighted fitting for zenith ionospheric delay is performed, considering the spherical geometry, as follows:

$$IGPdelay_j = \frac{\sum_{i=1}^{n} \frac{I_j}{I_i} \cdot IPPdelay_i \cdot W_{ij}}{\sum_{i=1}^{n} W_{ij}} \quad (5)$$

In Equation (5), $IGPdelay_j$ is the vertical delay of the ionosphere grid point j, $IPPdelay_i$ is the vertical delay of the piercing point i. I_j and I_i are the ionospheric delay values for the grid point j and ionospheric piercing point i. W_{ij} and W_{kj} are the weights assigned to piercing points i and k in relation to grid point j.

W_{ij} is defined as the reciprocal of the distance from the piercing point i to the grid point j, as follows:

$$W_{ij} = \frac{1}{R \cdot \arccos(\sin(\varphi_i)\sin(\varphi_j)) + \cos(\varphi_i)\cos(\varphi_j)\cos(\lambda_i - \lambda_j)} \quad (6)$$

In Equation (6), R denotes the Earth's radius, (φ_i, λ_i) refers to the latitude and longitude of the piercing point, and (φ_j, λ_j) indicates the latitude and longitude of the grid point.

According to the change in ionospheric grid delay between the adjacent epoch $\Delta IGPdelay$ by Equation (2), the new degradation parameters are calculated according to Equation (7), as follows:

$$\begin{aligned} &\begin{cases} C_{iono\_step} = \Delta IGPdelay_{MAX} \\ C_{iono\_ramp} = 0 \end{cases} && \text{if } \Delta IGPdelay_{MAX} \le (C_{iono\_step})_{LIMIT} \\ &\begin{cases} C_{iono\_step} = (C_{iono\_step})_{LIMIT} \\ C_{iono\_ramp} = (\Delta IGPdelay_{MAX} - C_{iono\_step}) / I_{iono} \end{cases} && \text{if } \Delta IGPdelay_{MAX} > (C_{iono\_step})_{LIMIT} \\ &C_{iono\_ramp} = (C_{iono\_ramp})_{LIMIT} && \text{when } C_{iono\_ramp} > (C_{iono\_ramp})_{LIMIT} \end{aligned} \quad (7)$$

In Equation (7), the $\Delta IGPdelay_{MAX}$ is the maximum ionospheric grid delay increments at all valid ionospheric grid points determined for each epoch. The $(C_{iono\_step})_{LIMIT}$ and $(C_{iono\_ramp})_{LIMIT}$ are specified by the RTCA [27], as follows:

$$\begin{aligned} \Delta IGPdelay_{MAX} &= MAX(|\Delta IGPdelay_{1,2,\cdots,j}|) \\ (C_{iono\_step})_{LIMIT} &= 1.023 \\ (C_{iono\_ramp})_{LIMIT} &= 0.005155 \end{aligned} \quad (8)$$

The C_{iono_step} is determined by the maximum ionospheric grid delay increment $\Delta IGPdelay_{MAX}$ of the delay values from all real-time valid ionospheric grid points. If the $\Delta IGPdelay_{MAX}$ is less than the maximum $(C_{iono\_step})_{LIMIT}$ specified by the RTCA protocol, then $\Delta IGPdelay_{MAX}$ is broadcast and the C_{iono_ramp} is set to 0. If ΔI_{max} exceeds the maximum $(C_{iono\_step})_{LIMIT}$ specified by the RTCA protocol, the $(C_{iono\_step})_{LIMIT}$ is broadcast directly, and C_{iono_ramp} is broadcast within the maximum range $(C_{iono\_ramp})_{LIMIT}$ specified by the RTCA protocol to envelop $\Delta IGPdelay_{MAX}$. Additionally, I_{iono} is set to the actual update cycle of the ionospheric grid, which is 240 s, and the ionospheric square root and factor parameter RSS_{iono} is set to 1. The degradation parameters are synchronized with the ionospheric delay update time, which is also 240 s.

In recent years, ionospheric activity has led to frequent large variations in the ionospheric grid delay between consecutive epochs. Directly setting the maximum ionospheric delay change to meet the envelope requirement is therefore unreasonable. In this study, large ionospheric grid delay jumps were excluded to ensure service integrity. However, frequent meter-level ionospheric delay jumps may reduce availability while maintaining integrity. Despite this, integrity remains ensured with the application of degradation parameters.

3. Results

The comparative analysis of the features between the newly calculated parameter and the existing parameter in the first week of 2024, as shown in Figure 8, is derived from Equation (7). The green line represents the current fixed parameter, while the red dot denotes the new parameter computed in this study. The horizontal axis is in days and the vertical axis is in meters. Since the values of the majority of the new C_{iono_ramp} are zero for the most part, the current C_{iono_ramp} is almost entirely covered in the figure at present. This approach updates degradation parameters in real time rather than using fixed values, allowing for greater flexibility to accommodate significant and occasional ionospheric changes to reduce the probability of exceeding the threshold.

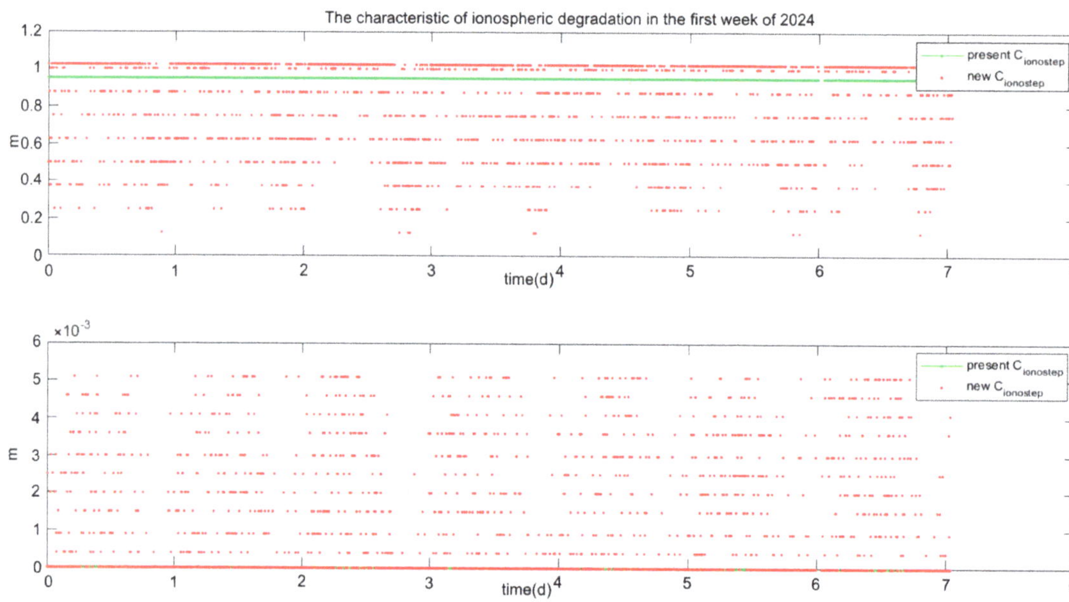

Figure 8. The feature of exceeding threshold probability under different ionospheric degradation parameters in 2024-001.

Real-time calculation of ionospheric degradation parameters using the proposed method significantly improves the envelope rate. As shown in Figure 9 below, the probability of exceeding the threshold at grid points is illustrated for the first day of 2023 and 2024. The blue dots represent ionospheric delay changes. The green line represents the current fixed envelope threshold; the horizontal axis is in hours and the vertical axis is in meters, while the red line illustrates the proposed real-time varying envelope threshold derived from Equation (1), aiming for more effective envelopment. This approach adapts flexibly to ionospheric delay variations.

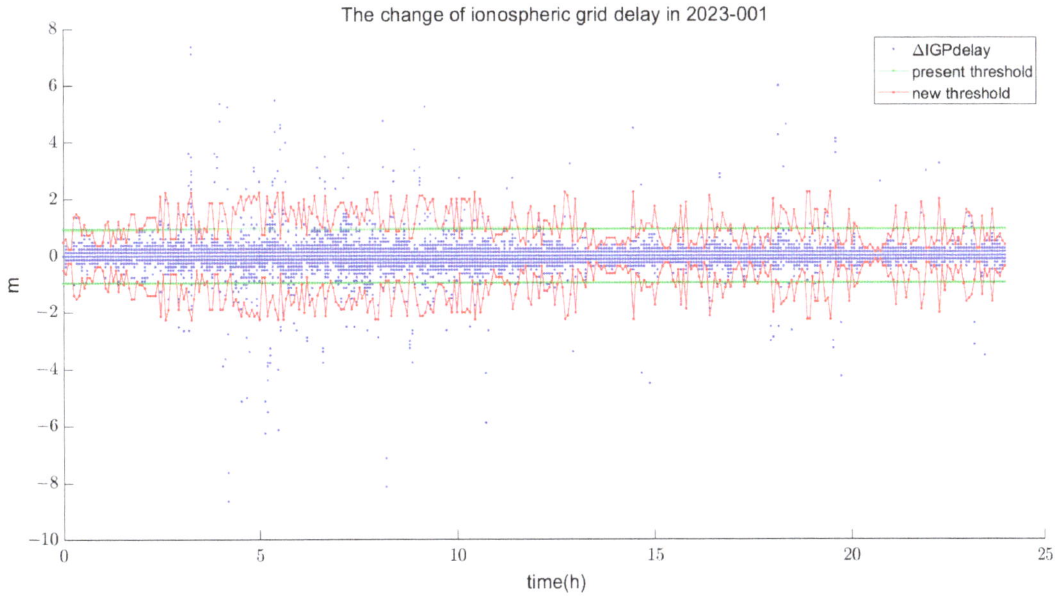

Figure 9. The rate of exceeding threshold probability under different ionospheric degradation parameters in 2023-001.

Applying the new method to the real-time calculation of ionospheric degradation parameters for BDSBAS and MSAS significantly reduces the rate of ionospheric grid delay changes exceeding the threshold. This reduction is especially noticeable in the more active ionospheric years of 2023 and 2024. Table 6 provides the more specific annual probability of exceeding the threshold for the ionospheric grid during the highly active ionospheric period from 2022 to 2024, the BDSBAS probability of exceeding the threshold dropped from 1.46% to 0.58%, while the MSAS probability decreased from 2.37% to 0.4%.

Table 6. The probability of exceeding the threshold of grid ionospheric delay variations under different ionospheric degradation parameters.

	Degradation	2021	2022	2023	2024
BDSBAS	present	0.12%	0.78%	1.64%	1.96%
	new	0.05%	0.33%	0.67%	0.74%
MSAS	present	0.54%	1.44%	2.42%	3.24%
	new	0.05%	0.2%	0.41%	0.59%

The changes in the probability of exceeding the threshold for ionospheric grid delay calculated by Equation (3) from 2021 to 2024, for both BDSBAS and MSAS, are shown in Figure 10, after implementing the new parameters. The horizontal axis represents time in days, while the vertical axis shows the probability of exceeding the threshold. The blue line corresponds to the probability of exceeding the threshold with the new degradation parameter, and the red line represents the probability of exceeding the threshold with the current degradation parameter.

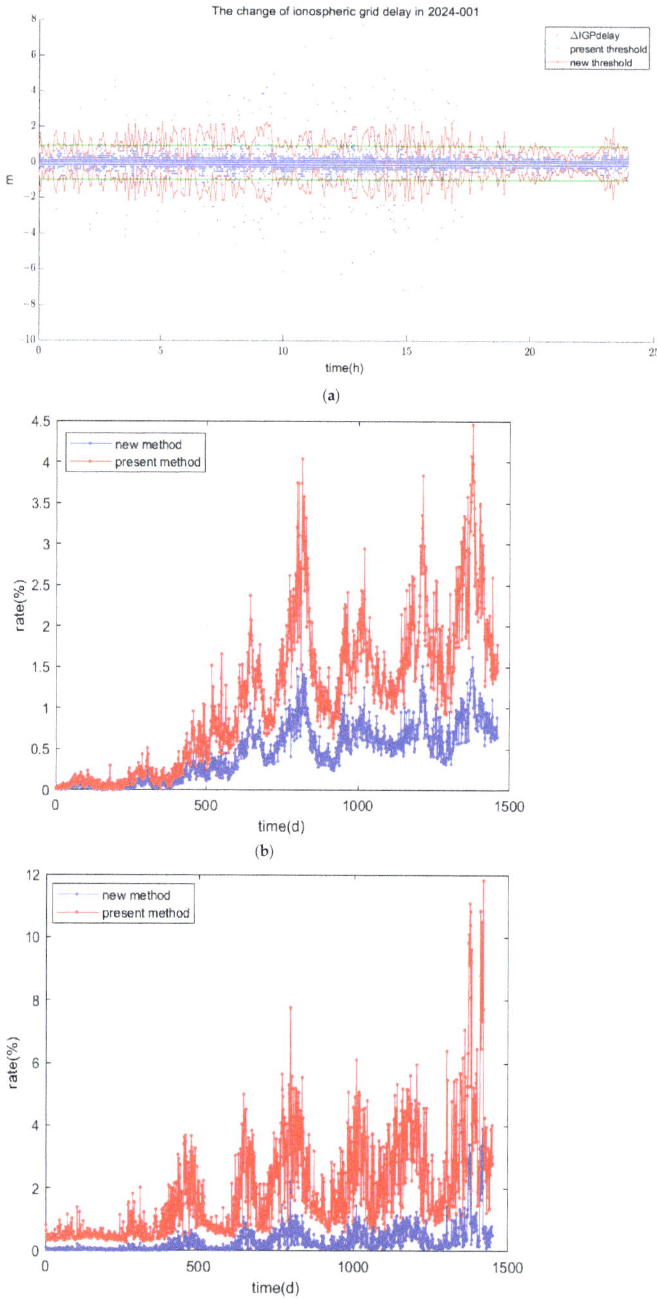

(a)

(b)

(c)

Figure 10. (**a**) The rate of exceeding the threshold probability under different ionospheric degradation parameters in 2024-001; (**b**) The rate of grid ionospheric delay variations exceeding the threshold under different ionospheric degradation parameters for BDSBAS in 2021–2024; (**c**) The rate of grid ionospheric delay variations exceeding the threshold under different ionospheric degradation parameters for MSAS in 2021–2024.

The probability of grid ionospheric delay variations exceeding the threshold decreases overall when the new parameter is used. A higher exceeding threshold rate corresponds to a larger reduction amplitude, with a more significant decrease in the rate of exceeding the threshold for MSAS after the new degradation parameter is applied. Specifically, the maximum rate for exceeding the threshold for MSAS drops from 11.8% to 3.9%, indicating that the proposed new degradation parameter strongly inhibits the growth of the rate for exceeding the threshold during ionospheric activity, effectively buffering the impact of ionospheric anomalies.

4. Discussion

To assess the impact of service availability after updating the efficiency reduction parameter algorithm, this paper employs the BDSBAS single-frequency user performance positioning algorithm and protection level calculation method to evaluate BDSBAS service performance across various dates in 2023. The new ionospheric degradation parameter proposed in this study was applied. The location performance, protection levels, probability of misleading information events, and probability of hazardously misleading information events for several IGS stations were calculated and compared with the results obtained using a fixed degradation parameter.

The data in Table 7 reflect the performance indicators of five IGS stations, including JFNG, URUM, WUH2, HKSL and HKWS, as well as five regional monitoring stations located in Beijing, Chengdu, Sanya, Kashi and Shantou in the same order as the table, during the selected time periods. Table 7 summarizes the positioning accuracy of ten stations across three time periods, highlighting changes in protection levels, availability, and integrity following the implementation of new degradation parameters. Positive values indicate improvements, such as an increase in protection levels, whereas negative values denote a decline, including misleading integrity information. Each statistical period spans 15 days. As shown in the table, during the high-activity ionospheric year of 2023, the updated degradation parameter algorithm enhanced protection levels. However, this improvement led to reduced service availability for stations located in edge-service areas. Despite this, misleading information significantly decreased, and integrity was markedly improved.

The vertical integrity Stanford diagrams for the URUM station on DOY 135 in 2023 using different ionospheric degradation parameters, as shown in Figure 11a–d, display the vertical integrity Stanford diagrams for the WUH2 station on DOY 225 as a more concrete example. The horizontal axis represents the VPL, while the vertical axis denotes the VPE. The Alert Limit is depicted as a horizontal line at 50 m. Each dot represents a specific (VPE, VPL) pair, with its color, based on a logarithmic scale, indicating the frequency of occurrences. The diagrams clearly illustrate a reduction in misleading information. However, during periods of highly active ionospheric conditions, the increased protection level results in fewer available epochs.

The new degradation parameters improve integrity more significantly in regions with higher ionospheric activity, similar to their effect on reducing the ionospheric grid delay exceedance rate. The reduction in misleading integrity information is particularly effective in southern and edge regions of China. In contrast, the probability of misleading integrity events in northern regions is inherently low or even zero, with availability generally remaining stable at 100%. Furthermore, these parameters do not reduce the availability of stations in northern regions. Broadcasting the new degradation parameters decreases the exceedance rate and enhances integrity during periods of high ionospheric activity, while having no adverse effects during low-activity periods. In ionospheric active regions, such as MSAS and BDSBAS, broadcasting these parameters effectively reduces the exceedance rate. Likewise, in ionospheric inactive regions, SBAS systems, such as EGNOS and WAAS, can also feasibly implement this method for broadcasting degradation parameters.

Table 7. Service performance improvement table under new ionospheric degradation parameters (unit: meters).

STATION	DATE	HPE	VPE	HPL	VPL	Availability	PMI	PHMI
JFNG	1-1	0.98	2.11	0.26	0.56	0	−0.19%	0.000%
	5-1	1.14	3.07	0.29	0.58	0	−1.21%	0.000%
	9-1	1.02	2.51	0.29	0.62	0	−0.43%	0.000%
URUM	1-1	0.68	1.36	0.28	0.64	0	−0.01%	0.000%
	5-1	0.85	1.86	0.28	0.60	0	−0.07%	0.000%
	9-1	0.77	1.57	0.29	0.68	0	−0.08%	0.000%
WUH2	1-1	1.01	2.40	0.26	0.57	0	−0.55%	0.000%
	5-1	1.21	3.29	0.29	0.58	0	−1.66%	0.000%
	9-1	1.08	2.81	0.30	0.64	0	−0.86%	0.000%
HKSL	1-1	0.97	2.98	0.25	0.57	−0.01%	−0.68%	0.000%
	5-1	1.29	4.22	0.26	0.59	0	−2.42%	0.000%
	9-1	1.19	4.06	0.27	0.60	0	−2.18%	0.000%
HKWS	1-1	0.97	2.94	0.25	0.57	−0.01%	−0.74%	0.000%
	5-1	1.29	4.17	0.26	0.59	0	−2.34%	0.000%
	9-1	1.19	4.02	0.27	0.60	0	−2.13%	0.000%
BJ01	1-1	0.55	0.98	0.25	0.57	0	0.00%	0.000%
	5-1	0.74	1.36	0.32	0.77	0	0.00%	0.000%
	9-1	0.53	1.36	0.27	0.59	0	0.00%	0.000%
CD01	1-1	0.83	1.90	0.24	0.52	0	−0.13%	0.000%
	5-1	1.32	2.17	0.32	0.74	0	−0.50%	0.000%
	9-1	0.90	2.07	0.25	0.57	0	−0.16%	0.000%
SY01	1-1	1.23	2.23	0.27	0.61	−0.01%	−0.27%	0.000%
	5-1	1.71	2.99	0.36	0.90	−0.01%	−1.29%	0.000%
	9-1	1.69	3.03	0.24	0.60	0	−1.05%	0.000%
KS01	1-1	0.69	1.13	0.10	0.26	−0.01%	0.00%	0.000%
	5-1	0.80	1.12	0.42	1.04	0	0.00%	0.000%
	9-1	0.71	0.94	0.30	0.71	0	0.00%	0.000%
ST01	1-1	2.78	3.40	0.34	0.80	−0.03%	−1.90%	0.000%
	5-1	2.04	2.89	0.30	0.78	−0.01%	−2.21%	0.000%
	9-1	1.96	2.53	0.23	0.53	0	−1.91%	0.000%

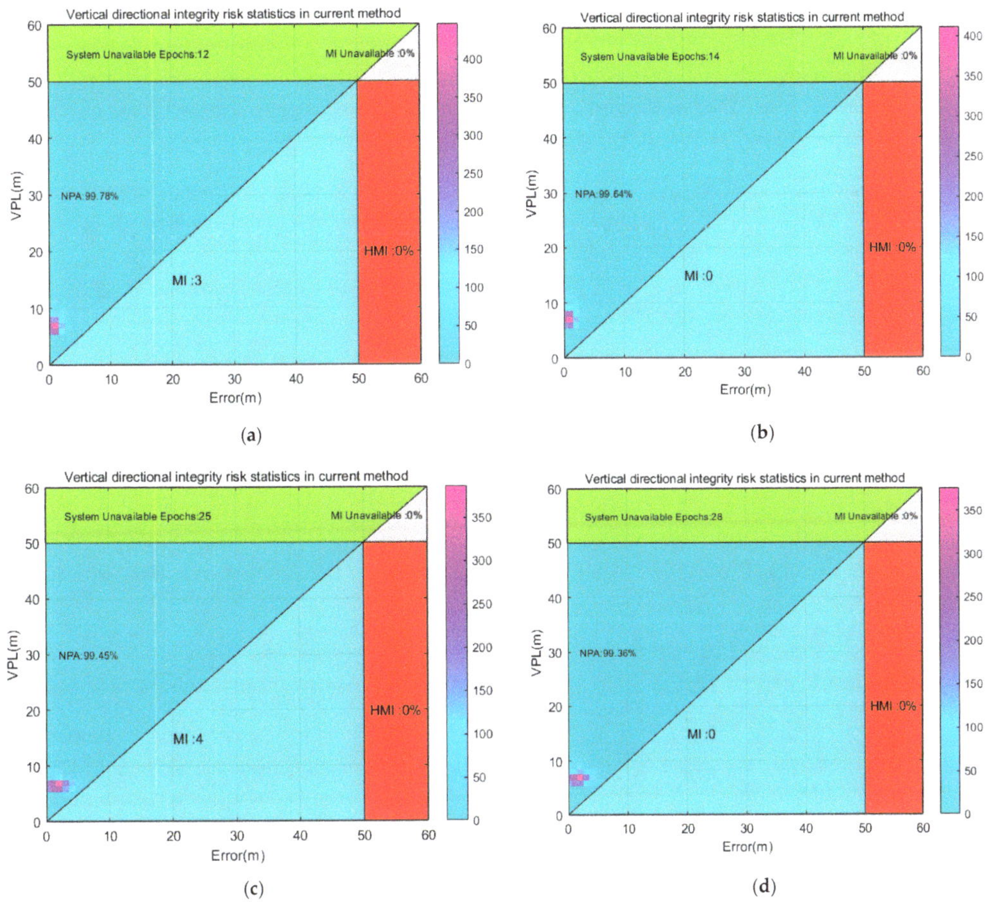

Figure 11. (**a**) Statistical diagram of vertical integrity of URUM on present ionospheric degradation parameters on DOY 135; (**b**) statistical diagram of vertical integrity of URUM on new ionospheric degradation parameters on DOY 135; (**c**) statistical diagram of vertical integrity of WUH2 on present ionospheric degradation parameters on DOY 225 (; and (**d**) statistical diagram of vertical integrity of WUH2 on new ionospheric degradation parameters on DOY 225.

5. Conclusions

This study evaluates the impact of ionospheric delay variations in grid-based ionospheric models and assesses the effect of ionospheric degradation parameters on service availability and integrity by statistically analyzing the packet loss rate of BDSBAS from 2021 to 2024. The results indicate that in 2024, when ionospheric activity was more intense, the rate of grid ionospheric delay variation exceeding the threshold increased by 1.85%, leading to decreased service availability and the emergence of misleading information compared to 2021.

To address the ineffective envelopment of grid-based ionospheric delay variations and its impact on BDSBAS service integrity caused by fixed broadcast values of ionospheric degradation parameters during periods of high ionospheric activity, this study proposes a new method for ionospheric degradation parameters. The proposed method was verified and analyzed using IGS data. The results show that after implementing the new ionospheric degradation parameters, the probability of exceeding the threshold of grid-based ionospheric delay variations during peak ionospheric activity decreased by 1.10%. Misleading information events were decreased by 1.04%, ensuring service integrity.

In summary, the new BDSBAS degradation parameters proposed in this study effectively reduce the exceeding threshold probability of grid-based ionospheric delay variations during high ionospheric activity periods and improve service integrity for users.

Author Contributions: Conceptualization, Z.L.; original draft preparation, Z.L.; methodology, Y.C. and S.Z.; data analysis, Z.L.; data curation, X.H.; project administration, R.L. All authors have read and agreed to the published version of the manuscript.

Funding: This work was supported by the National Natural Science Foundation of China (Grant Nos. 12373077, 41674041, 12173072, and 12273096).

Data Availability Statement: The raw data supporting the conclusions of this article will be made available by the authors on request.

Conflicts of Interest: The authors declare no conflicts of interest.

References

1. Li, R.; Zheng, S.; Wang, E.; Chen, J.; Feng, S.; Wang, D.; Dai, L. Advances in BeiDou Navigation Satellite System (BDS) and satellite navigation augmentation technologies. *Satell. Navig.* **2020**, *1*, 12. [CrossRef]
2. Liu, C.; Gao, W.; Shao, B.; Lu, J.; Wang, W.; Chen, Y.; Su, C.; Xiong, S.; Ding, Q. Development of Satellite-Based Augmentation System. *Navigation* **2021**, *68*, 405–417.
3. Liu, Y. Research on Key Technologies for Improving Performance of BDS-3 Satellite-Based Augmentation Service. Ph.D. Thesis, Shanghai Astronomical Observatory Chinese Academy of Sciences, Shanghai, China, 2023.
4. Xin, J.; Guo, R.; Chen, J.; Tian, Y.; Cao, Y.; Liu, Y.; Li, X.; Wang, D.; Cai, H. Performance analysis of the BDSBAS-B1C message in trial operation stage. *Sci. Rep.* **2023**, *13*, 6043. [CrossRef] [PubMed]
5. Liu, A.; Wang, N.; Li, Z.; Wang, L.; Wang, Z.; Yuan, H. Algorithm Optimization and Terminal Validation of BDSBAS Ionospheric Correction. In Proceedings of the China Satellite Navigation Conference (CSNC 2024), Jinan, China, 24–26 May 2024; Springer: Singapore, 2024; pp. 590–604.
6. Hu, Z.; Liu, X.; Wang, G.; Zhang, Q.; Zhou, R.; Chen, L.; Zhao, Q. Initial performance assessment of the single-frequency (SF) service with the BeiDou satellite-based augmentation system (BDSBAS). *GPS Solut.* **2022**, *27*, 35. [CrossRef]
7. Wang, D.; Zhang, L.; Tian, X.; Zhang, D.; Huang, Y. Research on Terminal Integrity Algorithm Based on BDS Single-Frequency SBAS Service. In Proceedings of the China Satellite Navigation Conference (CSNC 2024), Jinan, China, 24–26 May 2024; p. 9.
8. Zhao, L.; Jin, B.; Dong, Q.; Li, Z.; Liu, N. Pseudorange Bias Correction for BDSBAS Single Frequency Service. In Proceedings of the China Satellite Navigation Conference (CSNC 2022), Beijing, China, 22–25 May 2022; Springer: Singapore, 2022; pp. 129–140.
9. Wang, Y.; Cao, Y.; Hu, X.; Huang, Y.; Tang, C. The Algorithm and Validation of BeiDou System (BDS) Integrity Degradation Parameters with RTCA Protocol. *Chin. Space Sci. Technol.* **2016**, *36*, 25–31.
10. Shao, B.; Ding, Q.; Zhang, J.; Li, P.; Wu, X. Development Status and Trend of International Satellite-Based Augmentation Systems. *GNSS World China* **2024**, *49*, 3–9.
11. Kim, D.; Han, D.H.; Kim, J.-B.; Kim, H.-G.; Kee, C.-D.; Choi, K.-S.; Choi, H.-H.; Lee, E.-S. A Study of SBAS Position Domain Analysis Method: WAAS and EGNOS Performance Evaluation. *J. Position. Navig.* **2016**, *5*, 203–211.
12. Shao, B.; Ding, Q.; Wu, X. Estimation Method of SBAS Dual-Frequency Range Error Integrity Parameter. *Satell. Navig.* **2020**, *1*, 9. [CrossRef]
13. Li, T.; Li, R.; Li, J.; Yang, T. Research on BDSBAS Service Coverage Area Assessment Methodology. In Proceedings of the China Satellite Navigation Conference (CSNC 2022) Proceedings, Beijing, China, 22–25 May 2022; Springer: Singapore, 2024; pp. 215–227.
14. Wang, E.; Chen, Y.; Yu, T.; Zhang, J.; Yang, J.; Xu, S.; Wang, Y. Performance Evaluation Method and Flight Test Analysis of BeiDou Satellite-Based Augmentation Dual-Frequency Service. *GNSS World China* **2024**, *49*, 44–53.
15. Wang, Z.; Wang, L.; Xie, W.; Huang, G.; Yang, W.; Tian, Y. DFMC SBAS Service Performance Analysis of Multi-GNSS Based on BDS-3 in Different Regions. *Meas. Sci. Technol.* **2024**, *35*, 116310. [CrossRef]
16. Bao, J.; Li, R.; Liu, Y.; Liu, Y.; Shao, B. Ionospheric Anomaly Detection to Support the BDSBAS. *IEEE Access* **2020**, *8*, 1691–1704.
17. Wang, H.; Wang, Z.; Fang, K.; Dan, Z.; Zhu, Y. An Airborne Ionospheric Correction Approach for Single-Frequency BDSBAS. *IEEE Trans. Geosci. Remote Sens.* **2023**, *61*, 5802420. [CrossRef]

18. Sahithi, K.; Sridhar, M.; Kotamraju, S.K.; Kavya, K.C.S.; Sivavaraprasad, G.; Ratnam, D.V.; Deepthi, C. Characteristics of ionospheric scintillation climatology over Indian low-latitude region during the 24th solar maximum period. *Geod. Geodyn.* **2019**, *10*, 110–117. [CrossRef]
19. Li, Z.; Lyu, S.; Xiang, Y. Spatial and Temporal Correlation Analysis of Ionosphere in Chinese Regions Based on PPP Method. In Proceedings of the China Satellite Navigation Conference (CSNC 2024) Proceedings, Jinan, China, 24–26 May 2024; p. 8.
20. He, Y.; Yang, L.; Xu, Z. Analysis of Temporal and Spatial Changes of Regional Ionospheric Delay. *Oceanogr. Map.* **2006**, *26*, 19–21+28.
21. Lou, G. Ionospheric Disturbance Monitoring and Its Performance Evaluation for GNSS Positioning During Magnetic Storm. Master's Thesis, Shandong Jianzhu University, Shandong, China, 2024.
22. Singh, P.R.; Singh Kushwaha, U.K. A Comparative Study of Solar Activity Parameters during the Period 2009–2012 and 2020–2023 (Ascending Phase of Solar Cycles 24 and 25). *Astrophys. Space Sci.* **2024**, *369*, 78. [CrossRef]
23. Wang, W.; Xu, Z.; Zhang, R.; Wang, B.; Yang, S. Correlation Analysis of the TEC at Wuhan Station with Solar Activities and Geomagnetic Activities during 23rd Solar Cycle. *Chin. J. Space Sci.* **2012**, *32*, 40–47.
24. Li, Y.; Li, J.; Dai, T.; Pang, P. Influence of Solar Activity on Ionospheric TEC Change. *Chin. J. Space Sci.* **2018**, *38*, 847–854. [CrossRef]
25. Davoudifar, P.; Tabari, K.R.; Shafigh, A.A.E.; Ajabshirizadeh, A.; Bagheri, Z.; Akbarian Tork Abad, F.; Shayan, M. Development of a Local Empirical Model of Ionospheric Total Electron Content (TEC) and Its Application for Studying Solar-Ionospheric Effects. *Sci. Rep.* **2021**, *11*, 15070. [CrossRef] [PubMed]
26. Mansoori, A.; Khan, P.; Ahmad, R.; Atulkar, R.; Aslam, A.; Bhardwaj, S.; Malvi, B.; Purohit, P.; Gwal, A. Evaluation of Long-Term Solar Activity Effects on GPS Derived TEC. *J. Phys. Conf. Ser.* **2016**, *759*, 012069. [CrossRef]
27. Radio Technical Commission for Aeronautic (RTCA). *Minimum Operational Performance Standards (MOPS) for Global Positioning System/Satellite-Based Augmentation System Airborne Equipment (Standard)*; Radio Technical Commission for Aeronautic (RTCA): Washington, DC, USA, 2016.

remote sensing

MDPI

Article

Global Navigation Satellite System Meta-Signals with an Arbitrary Number of Components

Daniele Borio

European Commission, Joint Research Centre (JRC), 21027 Ispra, Italy; daniele.borio@ec.europa.eu

Abstract: Global Navigation Satellite System (GNSS) meta-signals are obtained when components from different frequencies are jointly processed as a single entity. While most research work has focused on GNSS meta-signals made of two side-band components, meta-signal theory has been recently extended to the case where the number of components is a power of two. This condition was dictated by the use of multicomplex numbers for the representation of GNSS meta-signals. Multicomplex numbers are multi-dimensional extensions of complex numbers whose dimension is a power of two. In this paper, the theory is further extended and a procedure for the construction of GNSS meta-signals with an arbitrary number of components is provided. Also in this case, multicomplex numbers are used to effectively represent a GNSS meta-signal. From this representation, multi-dimensional Cross Ambiguity Functions (CAFs) are obtained and used to derive acquisition and tracking algorithms suitable for the joint processing of components from different frequencies. The specific case with three components is analysed. Theoretical results are supported by experimental findings obtained by jointly processing Galileo E5a, E5b and E6 signals collected using three synchronized Software-Defined Radio (SDR) HackRF One front-ends. Experimental results confirm the validity of the developed theory.

Keywords: acquisition; GNSS; meta-signal; multi-frequency; tracking

1. Introduction

Modern Global Navigation Satellite System (GNSS) signal modulations, such as the Binary Offset Carrier (BOC), adopt a subcarrier to split the signal energy away from the signal centre frequency [1]. For this reason, the spectra of BOC modulations are characterized by two main lobes symmetrically placed around the signal Radio Frequency (RF). BOC acquisition and tracking algorithms try to recover all the signal energy by combining the contributions from the BOC side-lobes [2]. BOC modulations have been progressively generalized by considering side-band components carrying different information [3] and with different spectral shapes [4–6]. The concept of the GNSS meta-signal was originally introduced to describe an advanced modulation obtained by combining two signals broadcast on two different frequencies [7]. A meta-signal made of two components can be seen as a generalized BOC modulation where the side-band signals are split on two different RFs using a complex subcarrier [8]. Jointly processing signals from different frequencies allows one to recover and coherently combine the energy from the different components. Moreover, a GNSS meta-signal is characterized by a wide Gabor bandwidth, which can lead to high-accuracy pseudorange measurements [7,9]. Several algorithms have explored approaches for coherently combining signals from different side-bands and GNSS meta-signal approaches have been successfully applied to the Galileo Alternative Binary Offset Carrier (AltBOC) [9], the Beidou Asymmetric Constant-Envelope Binary

Offset Carrier (ACE-BOC) and B1 modulations [4–6]. Combinations involving the Galileo E6 signal have also been considered [9].

A convenient representation of a GNSS meta-signal with two components can be obtained using bicomplex numbers [8], which are two-dimensional extensions of complex numbers [10]. Using bicomplex numbers, it is possible to jointly represent the side-band components and explicitly express the subcarrier as a complex exponential. Bicomplex representation also allows one to derive optimal, in the Maximum Likelihood (ML) sense, acquisition and tracking algorithms [8], effectively combining signal energy from different frequencies.

Meta-signal theory has been recently generalized to the case where the number of components is a power of two [11]. The generalization is based on multicomplex numbers [12,13], which are multi-dimensional extensions of complex and bicomplex numbers. In this case, a set of 2^{M-1} signals, with M as an integer, is jointly represented as a single entity modulated by $2^{M-1}-1$ subcarriers, which split the signal components on 2^{M-1} centre frequencies. While GNSS meta-signals with 2^{M-1} components can be effectively considered as a form of multi-channel modulation [14], they significantly differ from multi-carrier schemes such as the Orthogonal Frequency Division Multiplexing (OFDM) [15,16]. GNSS meta-signals combine a limited number of components irregularly spaced in frequency and with potentially very different spectral characteristics. This is not the case for OFDM modulations, which combine hundreds to thousands of orthogonal components overlapping in frequency.

At the receiver side, the recovery of the meta-signal subcarriers allows one to align in phase the different signal components and effectively combine their energies. The constraint of having a number of components equal to a power of two was dictated by the structure of multicomplex numbers whose dimension is indeed a power of two.

In this paper, the GNSS meta-signal theory is further extended and a procedure for the construction of GNSS meta-signals with an arbitrary number of components is provided. Also in this case, multicomplex numbers are used for the meta-signal construction and representation. The procedure allows one to build a GNSS meta-signal with K components and identify $K-1$ subcarriers that split the different signals on K RFs.

The obtained representations are then used to derive a K-dimensional Cross Ambiguity Function (CAF), which, in turn, is used as basis for the design of multi-dimensional acquisition and tracking algorithms. These algorithms, effectively combining components from several frequencies, are detailed in Section 2.4. Similar to the case with a number of components equal to a power of two, acquisition and tracking algorithms are restated in terms of standard elements from single-component processing. The obtained acquisition approach is a form of weighted linear detector [17], whereas a multi-loop tracking architecture is found. A single Delay Lock Loop (DLL) is used to track a delay assumed to be common to all components, a Phase Lock Loop (PLL) is adopted to track the common meta-signal carrier phase and $K-1$ Subcarrier Phase Lock Loops (SPLLs) are used to recover the subcarrier parameters. In this way, a new class of Multi-Tracking Loop (MTL) approaches [18,19] is found. MTLs from the literature exploit the relationships between the Doppler frequencies of the different signal carriers [18–20]. A Kalman filter is usually adopted to track common carrier variations and estimate residual differences. Alternatively, inter-frequency carrier aiding is adopted [18]. A different approach is considered here: instead of tracking the original carrier components, Doppler shifts and carrier phases are projected at first in a transformed domain where they are expressed in terms of a common meta-signal carrier and subcarriers. In this domain, all the signal components contribute to the estimation of common Doppler and phase parameters. Also in this case, the relationships between the different Doppler frequencies can be exploited [21]. The analysis of this option is, however, left for future work.

Theoretical findings have been supported by experiments involving the Galileo E6, E5b and E5a components, giving emphasis to the case of a GNSS meta-signal with three components. Signals have been processed with a custom Python (version 3.9) Software-Defined Radio (SDR) receiver implementing the developed acquisition and tracking algorithms. Experimental results confirm the validity of the developed theory.

The remainder of this paper is organized as follows: Section 2 describes the methods developed in the paper and the materials used for their experimental validation. These include a brief summary on multicomplex numbers, the extension of meta-signal theory to an arbitrary number of components and the derivation of multi-component acquisition and tracking algorithms. Section 3 describes sample experimental results and Section 4 provides a discussion on the main findings and on future research directions. Finally, Section 5 concludes the paper.

2. Materials and Methods

2.1. Multicomplex Numbers and Multicomplex Signal Representation

Multicomplex numbers are multi-dimensional extensions of complex numbers [12,22] characterized by commutative products. A set of multicomplex numbers is usually denoted as $\mathbb{C}_M$ [12], where M is an integer defining the dimension of the space. $\mathbb{C}_1$ is the set of complex numbers whereas $\mathbb{C}_2$ is the bicomplex number space [10]. A multicomplex number is characterized by 2^M real components or equivalently by 2^{M-1} complex components. It features one real unit, 2^{M-1} imaginary units squaring to -1 and $2^{M-1}-1$ hyperbolic units [23], which square to 1. A multicomplex number can be written as [24]

$$z_M = \mathbf{k}_M^T \mathbf{a}_M + \mathbf{j}_M^T \mathbf{b}_M = \mathbf{k}_M^T \mathbf{z}_M, \tag{1}$$

where $\mathbf{a}_M$ and $\mathbf{b}_M$ are two real vectors of size 2^{M-1}. $\mathbf{k}_M$ and $\mathbf{j}_M$ are the vectors with the hyperbolic and imaginary units, respectively:

$$\begin{aligned} \mathbf{k}_M &= \left[1, k_1, \ldots, k_{2^{M-1}-1}\right] \\ \mathbf{j}_M &= \left[j_0, j_1, \ldots, j_{2^{M-1}-1}\right]. \end{aligned} \tag{2}$$

$\mathbf{z}_M$ is a complex vector obtained as

$$\mathbf{z}_M = \mathbf{a}_M + j_0 \mathbf{D}_M \mathbf{b}_M \tag{3}$$

where $\mathbf{D}_M$ is a square diagonal matrix of size $2^{M-1} \times 2^{M-1}$. The diagonal elements of $\mathbf{D}_M$ are equal to either 1 or to -1. The construction of $\mathbf{D}_M$ can be found in [24]. j_0 is the first imaginary unit common to all multicomplex number spaces.

Since multicomplex numbers feature several imaginary and hyperbolic units, several conjugations can be defined. Only the star-conjugation is considered in the following. It is obtained by negating the sign of the imaginary components. The star-conjugation of (1) is given by

$$z_M^* = \mathbf{k}_M^T \mathbf{a}_M - \mathbf{j}_M^T \mathbf{b}_M = \mathbf{k}_M^T \mathbf{z}_M^* \tag{4}$$

where $\mathbf{z}_M^*$ is the vector whose elements are the complex conjugates of the elements of $\mathbf{z}_M$. Using the star-conjugation, it is possible to define the hyperbolic modulus of a multicomplex number [10,24]:

$$|z_M|_k = \sqrt{z_M \cdot z_M^*} = \mathbf{k}_M \mathbf{r}_M, \tag{5}$$

where $\mathbf{r}_M$ is a vector with 2^{M-1} real components.

A multicomplex number can always be represented in terms of orthogonal components

$$z_M = \mathbf{e}_M^T \mathbf{z}_M^O = \sum_{i=0}^{2^{M-1}-1} z_{O,i} e_{M,i} \tag{6}$$

where $\mathbf{e}_M$ is the vector with idempotent orthogonal units, $e_{M,i}$. $z_{O,i}$ are complex coefficients arranged in the complex vector, $\mathbf{z}_M^O$. The idempotent units are characterized by the following property:

$$e_{M,i} e_{M,j} = \begin{cases} e_{M,i} & \text{for } i = j \\ 0 & \text{for } i \neq j \end{cases}. \tag{7}$$

They can be constructed according to the procedure outlined in [12,24]. Vectors $\mathbf{e}_M$ and $\mathbf{k}_M$ are related through the following transformation [24]:

$$\mathbf{e}_M = \frac{1}{2^{M-1}} (\mathbf{D}_M \mathbf{H}_{2^{M-1}})^T \mathbf{k}_M \tag{8}$$

where $\mathbf{H}_{2^{M-1}}$ is a Hadamard matrix of size 2^{M-1} [25] obtained through the recursive procedure outlined in [24]. Equation (8) provides a simple way to compute $\mathbf{z}_M^O$ from $\mathbf{z}_M$:

$$\mathbf{z}_M^O = (\mathbf{D}_M \mathbf{H}_{2^{M-1}})^T \mathbf{z}_M. \tag{9}$$

Orthogonal decomposition (6) allows one to perform component-wise operations when manipulating multicomplex numbers. For instance, the orthogonal components of the product of two multicomplex numbers are given by the products of the orthogonal components of the factors. In a similar way, the orthogonal components of hyperbolic modulus (5) are obtained as the (complex) moduli of the orthogonal components of z_M.

Finally, multicomplex numbers can be expressed in polar form as [24]:

$$z_M = \mathbf{k}_M^T \mathbf{r}_M \exp\left\{ \mathbf{j}_M^T \mathbf{\Phi}_M \right\} \tag{10}$$

where $\mathbf{r}_M$ and $\mathbf{\Phi}_M$ are real vectors of size 2^{M-1} containing modulus and phase information, respectively. $\mathbf{r}_M$ has been introduced in (5) and $\mathbf{\Phi}_M$ can be computed from the orthogonal components as

$$\mathbf{\Phi}_M = \frac{1}{2^{M-1}} \mathbf{H}_{2^{M-1}} \mathbf{\Phi}_M^O, \tag{11}$$

where

$$\mathbf{\Phi}_{2^{M-1}}^O = \begin{bmatrix} \angle z_{O,0} \\ \angle z_{O,1} \\ \vdots \\ \angle z_{O,2^{M-1}-1} \end{bmatrix}. \tag{12}$$

2.2. Multicomplex Signal Representation

Multicomplex numbers can be used to provide a joint representation of 2^{M-1} RF signals [11,24]. In particular, it is possible to show

$$\begin{aligned} z_{2^{M-1},RF}(t) &= \sum_{i=0}^{2^{M-1}-1} x_{i,RF}(t) = \sum_{i=0}^{2^{M-1}-1} \Re\{x_i(t) \exp\{j_0 2\pi f_{X,i} t\}\} \\ &= \Re\left\{ z_{2^{M-1}}(t) \exp\{2\pi \mathbf{j}_M^T \mathbf{f}_{2^{M-1}} t\} \right\} \end{aligned} \tag{13}$$

where $x_{i,RF}(t)$ is the ith RF signal obtained by modulating the ith baseband component, $x_i(t)$, at frequency $f_{X,i}$. $z_{2^{M-1}}(t)$ is a multicomplex signal obtained as

$$z_{2^{M-1}}(t) = 2^{M-1} \sum_{i=0}^{2^{M-1}-1} x_i(t) e_{M,i}. \tag{14}$$

Equation (14) implies that the orthogonal components of $z_{2^{M-1}}(t)$ are proportional to the baseband signals, $x_i(t)$. In (13), $\mathbf{f}_{2^{M-1}}$ is a real vector obtained from the original RFs as

$$\mathbf{f}_{2^{M-1}} = \frac{1}{2^{M-1}} \mathbf{H}_{2^{M-1}} \begin{bmatrix} f_{X,0} \\ f_{X,1} \\ \vdots \\ f_{X,2^{M-1}-1} \end{bmatrix}. \tag{15}$$

The key point arising from (13) is that a set of 2^{M-1} RF components can be represented as a single baseband multicomplex signal modulated by a set of carrier/subcarriers with frequencies common to all components. The multicomplex exponential in the second line of (13) can be expanded as the product of exponentials, each with a single imaginary unit at exponent. The exponential with respect to the j_0 unit is the carrier, whereas the remaining terms are the subcarriers. The product between the multicomplex signal, $z_{2^{M-1}}(t)$, and the subcarriers defines the GNSS meta-signal with 2^{M-1} components. At the reception side, these carrier/subcarriers are affected by Doppler frequencies and phases introduced by the communication channel. These parameters need to be estimated. In the meta-signal approach, instead of independently tracking 2^{M-1} carriers, the exponential terms in (13) are jointly considered. Since all the signal components in (13) contribute to the estimation of the carrier/subcarrier Doppler frequencies and phases, improved performance is obtained with respect to the individual tracking approach. In order to generalize the theory of GNSS meta-signals, (13) has to be extended to the case where an arbitrary number of components is considered.

2.3. *Meta-Signals with an Arbitrary Number of Components*

In this section, a simple procedure for the construction of a meta-signal with an arbitrary number of components is described. A representation similar to (14) is provided where the missing baseband components required to obtain a number of elements equal to a power of two are set to zero. The process is described in detail for the case with three components. The procedure is then generalized to the case with an arbitrary number of components.

2.3.1. Meta-Signals with Three Components

A meta-signal with three components can be obtained by considering the space of tricomplex numbers, $\mathbb{C}_3$ [12], which can jointly represents up to four (2^{M-1}, with $M = 3$) complex components. The case with four components was illustrated in [11] for the Galileo signal landscape. The space of tricomplex numbers features one real unit, four imaginary units, j_0, j_1, j_2 and j_3, and three hyperbolic units k_1, k_2 and k_3. Moreover, tricomplex numbers are characterized by four orthogonal idempotent units:

$$\begin{aligned} e_{3,0} &= \frac{1-k_2}{2}\frac{1-k_1}{2}, \qquad & e_{3,1} &= \frac{1-k_2}{2}\frac{1+k_1}{2} \\ e_{3,2} &= \frac{1+k_2}{2}\frac{1-k_1}{2}, \qquad & e_{3,3} &= \frac{1+k_2}{2}\frac{1+k_1}{2} \end{aligned}. \tag{16}$$

When four RF signals are present, (13) becomes:

$$\begin{aligned} z_{4,RF}(t) &= \sum_{i=0}^{3} x_{i,RF}(t) = \sum_{i=0}^{3} \Re\{x_i(t)\exp\{j_0 2\pi f_{X,i} t\}\} \\ &= \Re\left\{z_4(t)\exp\{2\pi \mathbf{j}_3^T \mathbf{f}_4 t\}\right\} \end{aligned} \tag{17}$$

where

$$z_4(t) = 4\sum_{i=0}^{3} x_i(t) e_{3,i} \tag{18}$$

is the tricomplex baseband signal obtained by combining the different baseband signals, $x_i(t)$. $\mathbf{j}_3^T = [j_0, j_1, j_2, j_3]$ is the vector with the imaginary units and

$$\mathbf{f}_4 = \begin{bmatrix} f_0 \\ f_1 \\ f_2 \\ f_3 \end{bmatrix} = \frac{1}{4}\mathbf{H}_4 \mathbf{f}_{X,4} = \frac{1}{4}\begin{bmatrix} 1 & 1 & 1 & 1 \\ 1 & -1 & 1 & -1 \\ 1 & 1 & -1 & -1 \\ -1 & 1 & 1 & -1 \end{bmatrix}\begin{bmatrix} f_{X,0} \\ f_{X,1} \\ f_{X,2} \\ f_{X,3} \end{bmatrix}, \tag{19}$$

where $\mathbf{H}_4$ is the Hadamard matrix [25] of size 4. Product $\mathbf{j}_3^T \mathbf{f}_4$ can be explicitly written as a summation of pure imaginary terms. Moreover, exponential $\exp\{2\pi \mathbf{j}_3^T \mathbf{f}_4 t\}$ can be written as a the product of a carrier, with nominal frequency f_0, and three subcarriers.
When three components are present, a representation similar to (18) can be obtained by simply setting one of the orthogonal components of the multicomplex signal to zero. While different choices are possible, $x_3(t) = 0$ is assumed here. By doing so, the last RF, $f_{X,3}$, remains undetermined and can be selected arbitrarily without compromising the validity of (17). This fact can be proved by considering the exponential, $\exp\{2\pi \mathbf{j}_3^T \mathbf{f}_4 t\}$. From the properties of the imaginary vector, $\mathbf{j}_M$, it follows that [24]

$$\mathbf{j}_M = j_0 \mathbf{H}_{2^{M-1}} \mathbf{e}_M. \tag{20}$$

Thus,

$$\exp\{2\pi \mathbf{j}_3^T \mathbf{f}_4 t\} = \exp\{2\pi j_0 \mathbf{e}_3^T \mathbf{H}_4^T \mathbf{f}_4 t\} = \exp\{2\pi j_0 \mathbf{e}_3^T \mathbf{f}_{X,4} t\}, \tag{21}$$

where condition $\mathbf{f}_{X,4} = \mathbf{H}_4^T \mathbf{f}_4$ directly derives from (19) and the fact that the inverse of a Hadamard matrix is equal to its transpose normalized by the matrix dimension [25]. Since the exponential function admits an absolutely convergent Taylor series expansion, it is possible to show that [13]

$$\exp\left\{\sum_{i=0}^{2^{M-1}-1} a_i e_{M,i}\right\} = \sum_{i=0}^{2^{M-1}-1} \exp\{a_i\} e_{M,i}, \tag{22}$$

where a_i are complex numbers expressed with respect to the j_0 unit. Using (22) and (21) becomes:

$$\exp\{2\pi \mathbf{j}_3^T \mathbf{f}_4 t\} = \sum_{i=0}^{3} \exp\{j_0 2\pi f_{X,i} t\} e_{3,i}. \tag{23}$$

Expansion (23) shows that the multiplication with the multicomplex exponential in (17) is equivalent to multiplying each orthogonal component of $z_4(t)$ by the corresponding complex carrier. In (23), the only term depending on $f_{X,3}$ multiplies $e_{3,3}$, which is now multiplied by zero in (18). Thus, the product $z_4(t)\exp\{2\pi \mathbf{j}_3^T \mathbf{f}_4 t\}$ does not depend on $f_{X,3}$, which can be selected arbitrarily. A convenient solution is to fix $f_{X,3}$ such that the last component of $\mathbf{f}_4$ (any other component of $\mathbf{f}_4$ would lead to a similar result) is zero. Using this approach, it follows that

$$\mathbf{f}_{X,4} = \mathbf{H}_4^T \mathbf{f}_4 = \begin{bmatrix} 1 & 1 & 1 & -1 \\ 1 & -1 & 1 & 1 \\ 1 & 1 & -1 & 1 \\ 1 & -1 & -1 & -1 \end{bmatrix} \begin{bmatrix} f_0 \\ f_1 \\ f_2 \\ 0 \end{bmatrix} = \begin{bmatrix} 1 & 1 & 1 \\ 1 & -1 & 1 \\ 1 & 1 & -1 \\ 1 & -1 & -1 \end{bmatrix} \begin{bmatrix} f_0 \\ f_1 \\ f_2 \end{bmatrix}. \tag{24}$$

Since the actual value of $f_{X,3}$ is not of interest, transformation (24) can be written as

$$\mathbf{f}_{X,3} = \begin{bmatrix} f_{X,0} \\ f_{X,1} \\ f_{X,2} \end{bmatrix} = \begin{bmatrix} 1 & 1 & 1 \\ 1 & -1 & 1 \\ 1 & 1 & -1 \end{bmatrix} \begin{bmatrix} f_0 \\ f_1 \\ f_2 \end{bmatrix} = \mathbf{H}_3^{-1} \mathbf{f}_3, \tag{25}$$

where $\mathbf{H}_3^{-1}$ denotes the matrix obtained by removing the last column and row from $\mathbf{H}_4^T$ and $\mathbf{f}_3$ is the vector with frequencies f_i, for $i = 0, 1$ and 2. In this way, a meta-signal with three components is obtained. Representation (17) becomes

$$z_{3,RF}(t) = \Re\left\{ z_3(t) \exp\{2\pi \mathbf{j}_{3,3}^T \mathbf{f}_3 t\} \right\} \tag{26}$$

where $\mathbf{j}_{3,3}$ is the vector with the first 3 imaginary units of $\mathbb{C}_3$ and

$$z_3(t) = 4 \sum_{i=0}^{2} x_i(t) e_{3,i}. \tag{27}$$

The exponential in (26) defines one carrier and two subcarriers. $\mathbf{H}_3^{-1}$ has been obtained by removing the last column and the last row of $\mathbf{H}_4^T$ and is no longer an orthogonal matrix. Its inverse

$$\mathbf{H}_3 = \frac{1}{2} \begin{bmatrix} 0 & 1 & 1 \\ 1 & -1 & 0 \\ 1 & 0 & -1 \end{bmatrix} \tag{28}$$

defines how the meta-signal carrier and subcarrier frequencies are obtained. In this case, the carrier nominal frequency is directly proportional to the narrow-lane combination of the last two signal RFs. The subcarrier frequencies correspond to the wide-lane linear combinations of the other two possible signal pairs.

The result of this construction is illustrated for the Galileo E5a, E5b and E6 signals in Figure 1. The actual allocation of the different components on the leaves of the binary tree displayed in Figure 1 depends on the ordering of the different components and their paring with the orthogonal elements in (27). In this case, the E6 signal is found in isolation whereas the E5a and E5b components are on the same branch and form the AltBOC modulation.

In order to illustrate the fact that alternative triple-component meta-signals can be obtained using a different signal ordering, a second configuration for the Galileo E5a, E5b and E6 signals is provided in Figure 2. In this case, a carrier frequency equal to the Galileo AltBOC RF is found. Moreover, the nominal frequency of the second subcarrier is increased from 15.345 MHz to 35.805 MHz.

In addition to different ordering, alternative triple-component meta-signals can be built using different construction approaches. For instance, a component could be split and and paired with two different orthogonal units in (18). These options are outside the scope of this work.

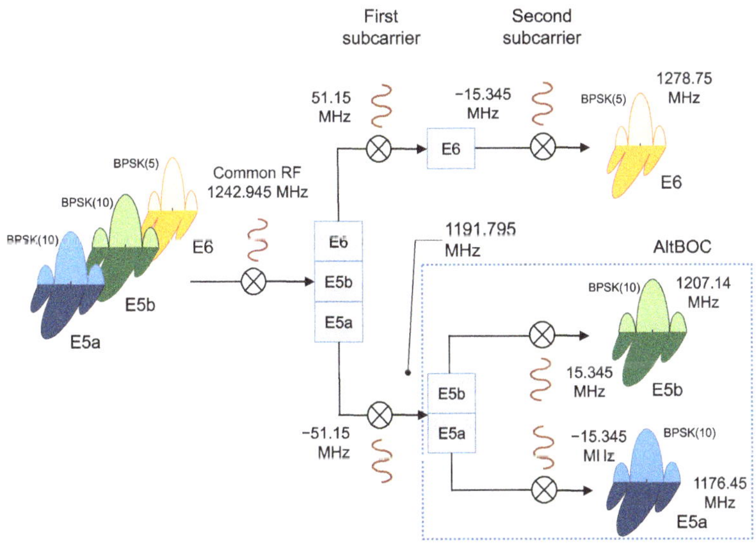

Figure 1. Construction of a GNSS meta-signal with three components. The Galileo E5a, E5b and E6 signals are combined using two subcarriers. With the ordering adopted, the AltBOC modulation is found on the bottom branch of the tree.

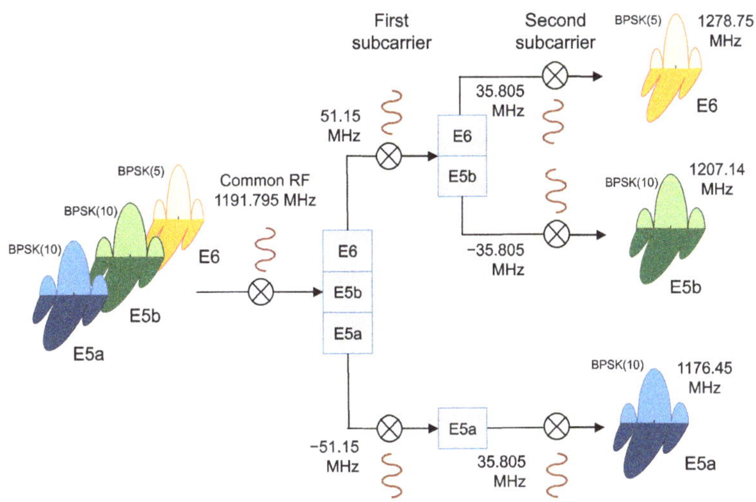

Figure 2. Alternative option for the construction of a GNSS meta-signal based on the combination of the Galileo E5a, E5b and E6 signals. The AltBOC RF is now the nominal meta-signal carrier frequency.

2.3.2. General Case

In the previous section, a procedure for constructing a GNSS meta-signal with three components was provided. This procedure can be easily generalized to an arbitrary number of components by following the steps summarized in Figure 3. Given K RF components, they can be combined using $K-1$ subcarriers and expressed as a single meta-signal using the appropriate multicomplex space. The first step consists in the identification of $\mathbb{C}_M$, a multicomplex space sufficiently large to contain the different components.

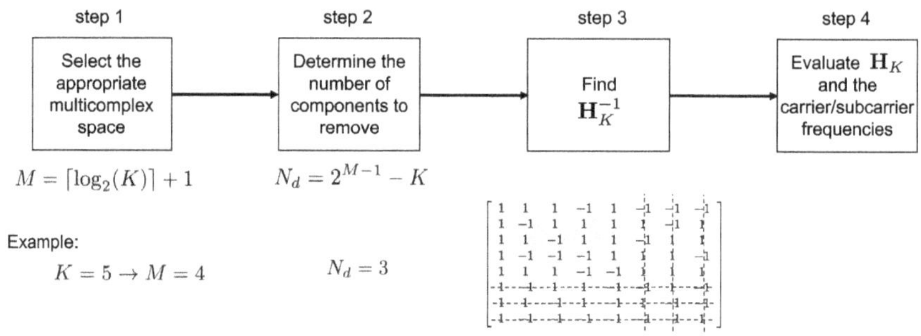

Figure 3. General procedure for the construction of GNSS meta-signals with an arbitrary number of components. K is the number of components considered.

If K is the number of RF signals, then M is found as

$$M = \lceil \log_2(K) \rceil + 1 \tag{29}$$

where $\lceil \cdot \rceil$ denotes the ceiling operator. In Figure 3, the case with $K = 5$ components is considered as example. When five RF signals are present, $M = 4$ and $\mathbb{C}_4$ are considered. In this space, up to eight components can be combined to form a single meta-signal. The second step reported in Figure 3 consists of determining the number of components to remove from the full meta-signal with 2^{M-1} elements. This number, denoted here as N_d, is given by

$$N_d = 2^{M-1} - K. \tag{30}$$

In the example of Figure 3, $N_d = 3$. After these preliminary operations, it is finally possible to determine the matrix $\mathbf{H}_K^{-1}$ connecting the different RFs to the meta-signal carrier and subcarrier frequencies. $\mathbf{H}_K^{-1}$ is obtained by removing the last N_d columns and rows from $\mathbf{H}_{2^{M-1}}^T$, where $\mathbf{H}_{2^{M-1}}$ is the Hadamard matrix characterizing the $\mathbb{C}_M$ space and obtained according to the procedure outlined in [24]. For $K = 5$,

$$\mathbf{H}_5^{-1} = \begin{bmatrix} 1 & 1 & 1 & -1 & 1 \\ 1 & -1 & 1 & 1 & 1 \\ 1 & 1 & -1 & 1 & 1 \\ 1 & -1 & -1 & -1 & 1 \\ 1 & 1 & 1 & -1 & -1 \end{bmatrix}. \tag{31}$$

$\mathbf{H}_K^{-1}$ is finally inverted and the meta-signal carrier and subcarrier nominal frequencies are computed as

$$\mathbf{f}_K = \mathbf{H}_K \mathbf{f}_{X,K} \tag{32}$$

where $\mathbf{f}_{X,K}$ is the vector with the RFs of the original signal components. The resulting multicomplex baseband signal is given by

$$z_K(t) = 2^{M-1} \sum_{i=0}^{K-1} x_i(t) e_{M,i}, \tag{33}$$

and the corresponding RF signal can be represented as

$$z_{K,RF}(t) = \Re\left\{ z_K(t) \exp\left\{ 2\pi \mathbf{j}_{M,K}^T \mathbf{f}_K t \right\} \right\}, \tag{34}$$

where $\mathbf{j}_{M,K}$ is the vector with the first K imaginary units of $\mathbb{C}_M$.

For $K = 5$,

$$\mathbf{H}_5 = \frac{1}{4}\begin{bmatrix} -1 & 1 & 1 & 1 & 2 \\ 1 & -1 & 1 & -1 & 0 \\ 1 & 1 & -1 & -1 & 0 \\ -1 & 1 & 1 & -1 & 0 \\ 2 & -0 & 0 & 0 & -2 \end{bmatrix}, \tag{35}$$

and it involves quad-frequency linear combinations derived from the Hadamard matrix of order 4 and narrow/wide-lane combinations. This case is illustrated in Figure 4, where all the five Beidou Navigation Satellite System (BDS) III signals [26] are combined using four subcarriers.

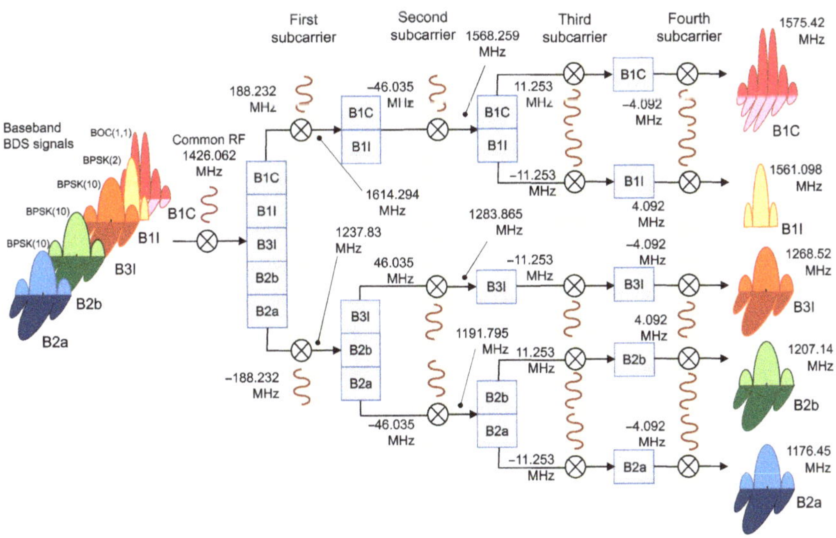

Figure 4. Construction of a GNSS meta-signal with five components. All the five BDS III signals are combined using four subcarriers.

2.4. Baseband Processing

In the previous section, it was shown that a set of K RF signals can be represented as a single meta-signal with a single carrier up-converting all the components to a common RF. $K - 1$ subcarriers are then used to split the different components on their final RFs. K can be an arbitrary integer value. In this section, the obtained signal representation is used to derive meta-signal acquisition and tracking algorithms.

2.4.1. Received Signals

The different RF components, $x_{RF,i}(t)$, are transmitted by a single GNSS satellite and pass through a common communication channel. When received on the ground, a multi-frequency GNSS receiver is used to recover the different components. The receiver front-ends amplify, filter and down-convert the different components. In this way, baseband versions of the received signals are obtained. A simple communication channel is assumed [27]. In particular, each received component is modelled as

$$x_{rx,i}(t) = \alpha_i x_i(t - \tau_0) \exp\{j_0 2\pi f_{X0,i} t + j_0 \varphi_{X0,i}\} + \eta_i(t), \tag{36}$$

where the definitions are as follows:

- τ_0 is a delay assumed common to all signals. While biases among the different components can be present, for instance due to hardware imperfections at both satellite and receiver side, these biases are neglected here. The calibration procedure developed in [11] is adopted to compensate for possible Differential Code Biases (DCBs),
- α_i is the attenuation introduced by the communication channel on the ith signal component,
- $f_{X0,i}$ and $\varphi_{X0,i}$ are the Doppler frequency and carrier phase introduced on the ith signal component,
- $\eta_i(t)$ is Additive White Gaussian Noise (AWGN) with independent real and imaginary parts. Noise terms from different frequencies are assumed independent.

The receiver front-ends digitize the different $x_{rx,i}(t)$, producing the samples

$$x_{rx,i}[n] = x_{rx,i}(nT_s) = \alpha_i x_i(nT_s - \tau_0)\exp\{j_0 2\pi f_{X0,i} nT_s + j_0\varphi_{X0,i}\} + \eta_i(nT_s), \tag{37}$$

where T_s is the sampling interval assumed common to all the receiver front-ends and n is the time index. The inverse of T_s, $f_s = \frac{1}{T_s}$ defines the sampling frequency. The noise components, $\eta_i[n] = \eta_i(nT_s)$, are independent and identically distributed (i.i.d.) random processes with independent real and imaginary parts. The same variance, σ_η^2, is assumed for the real and imaginary part of $\eta_i[n]$.
The received baseband multicomplex signals can be reconstructed by assigning a different component (37) to the orthogonal elements of $\mathbb{C}_M$ as in (33)

$$\begin{aligned} z_{rx,K}[n] &= 2^{M-1}\sum_{i=0}^{K-1} x_{rx,i}[n]e_{M,i} \\ &= A z_K(nT_s - \tau_0)\exp\left\{2\pi \mathbf{j}_{M,K}^T \mathbf{f}_{K,0} nT_s + \mathbf{j}_{M,K}^T \mathbf{P}_{K,0}\right\} + \eta[n] \end{aligned} \tag{38}$$

where

$$A = \sum_{i=0}^{K-1} \alpha_i e_{M,i} \tag{39}$$

is the multicomplex attenuation obtained by combining the α_i terms introduced above. $\mathbf{f}_{K,0}$ and $\mathbf{P}_{K,0}$ are the vectors with the meta-signal carrier and subcarrier Doppler frequencies and phases. They are obtained through the transformation defined by (32):

$$\mathbf{f}_{K,0} = \mathbf{H}_K \begin{bmatrix} f_{X0,0} \\ f_{X0,1} \\ \vdots \\ f_{X0,K-1} \end{bmatrix} \qquad \mathbf{P}_{K,0} = \mathbf{H}_K \begin{bmatrix} \varphi_{X0,0} \\ \varphi_{X0,1} \\ \vdots \\ \varphi_{X0,K-1} \end{bmatrix}. \tag{40}$$

Finally, $\eta[n]$ is a multicomplex noise term obtained as

$$\eta[n] = 2^{M-1}\sum_{i=0}^{K-1} \eta_i[n]e_{M,i}. \tag{41}$$

Equation (38) is analogous to the signal model obtained in [11] for the case with 2^{M-1} components. The main difference here is that the summation in the previous equations now extend to K instead of 2^{M-1}. Moreover, the transformations connecting the parameters of the individual received components to those of the resulting meta-signal are no longer defined by a Hadamard matrix and the procedure outlined in Section 2.3.2 has to be followed. This analogy is exploited for the derivation of acquisition and tracking algorithms jointly processing several components from different frequencies.

In the following, a standard model for the GNSS signals, $x_i(t)$, is assumed. In particular, $x_i(t)$ is modelled as [27]

$$x_i(t) = d_i(t)c_i(t), \tag{42}$$

where $c_i(t)$ is a spreading sequence from a family of quasi-orthogonal codes and $d_i(t)$ represents the navigation message. Note that $x_i(t)$ may include several components such as a data and pilot channel. This case is not considered here and a single component is assumed. The presence of both data and pilot channels can be addressed with an approach similar to that developed in [28]. $d_i(t)$ may be absent if pilot channels are considered. The codes from the different components can be combined into a single multicomplex term

$$\tilde{c}(t) = \sum_{i=0}^{K-1} \tilde{\alpha}_i c_i[n] e_{M,i} \tag{43}$$

and used for the computation of the multicomplex CAF as better described in the next section. In (43), $\tilde{\alpha}_i$ are estimates of the α_i coefficients introduced in (36). Since each signal is received with a different amplitude, the local codes also have to be weighted differently. $\tilde{\alpha}_i$ can be fixed according to previous knowledge on received signals or using a calibration process.

2.4.2. Multicomplex CAF

The goal of the signal processing stage of a GNSS receiver is to estimate the received signal parameters. In this case, these parameters are τ_0, $\mathbf{f}_{K,0}$ and $\mathbf{P}_{K,0}$. When 2^{M-1} components are considered, this is performed through the maximization of a multicomplex CAF [11]. This approach can also be adopted for the general case with K components. In particular, estimates of τ_0, $\mathbf{f}_{K,0}$ and $\mathbf{P}_{K,0}$ are found as

$$\tilde{\tau}, \tilde{\mathbf{f}}_K, \tilde{\mathbf{P}}_K = \arg\max_{\tau, \mathbf{f}_d, \mathbf{P}} \Re\left\{C_z(\tau, \mathbf{f}_d) \exp\left\{-\mathbf{j}_{M,K}^T \mathbf{P}\right\}\right\}, \tag{44}$$

where

$$C_z(\tau, \mathbf{f}_d) = \sum_{n=0}^{N-1} z_{rx,K}[n] \tilde{c}^*(nT_s - \tau) \exp\left\{-2\pi \mathbf{j}_{M,K}^T \mathbf{f}_d n T_s\right\} \tag{45}$$

is a multicomplex CAF obtained by correlating $z_{rx,K}[n]$ with local code (43) delayed by τ and modulated by the frequency vector $\mathbf{f}_d$. In (45), $*$ denotes the star-conjugation introduced in Section 2.1. CAF (45) can be derived following a procedure similar to that developed in [11] for the case with 2^{M-1} components. Following the same approach used in [11], solution (44) can be rewritten as

$$\begin{aligned} \tilde{\tau}, \tilde{\mathbf{f}}_K &= \arg\max_{\tau, \mathbf{f}_d} \Re\{|C_z(\tau, \mathbf{f}_d)|_k\} \\ \tilde{\mathbf{P}}_K &= \mathbf{P}_{CAF} \end{aligned} \tag{46}$$

where $|C_z(\tau, \mathbf{f}_d)|_k$ is the hyperbolic modulus of the CAF and $\mathbf{P}_{CAF}$ is the vector with the phases of the CAF evaluated for $\tau = \tilde{\tau}$ and $\mathbf{f}_d = \tilde{\mathbf{f}}_K$.

2.4.3. Multicomplex Acquisition

Maximization of (46) is performed through a two-step process: acquisition and tracking. Acquisition maximizes $\Re\{|C_z(\tau, \mathbf{f}_d)|_k\}$ though a search on a finite grid of Doppler frequencies and delay values. The process is illustrated in Figure 5 and it is analogous to the case with 2^{M-1} components. The incoming multicomplex signal is multiplied by local code (43) and by local replicas of the carrier and subcarriers. The resulting multicomplex signal is coherently integrated on N samples. $T_c = NT_s$ is the coherent integration time. In this way, the CAF is computed. Different Doppler frequencies and code delays are tested and $\Re\{|C_z(\tau, \mathbf{f}_d)|_k\}$ is finally maximized.

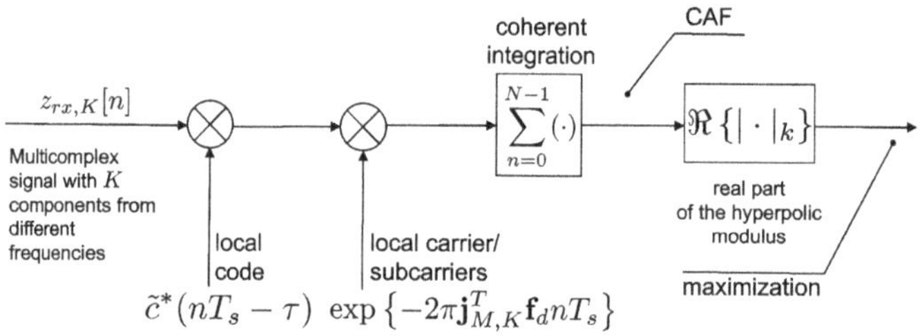

Figure 5. General representation of the joint acquisition process of K components from different frequencies using multicomplex numbers.

Note that while different carrier/subcarriers are present, relationships between the different Doppler frequencies can be exploited to reduce the dimensionality of the search process. For instance, only the carrier Doppler can be considered as a search parameter. The Doppler frequencies of the subcarriers can be obtained by scaling the carrier Doppler by the corresponding nominal frequencies.

As for the case with 2^{M-1} components, $\Re\{|C_z(\tau, \mathbf{f}_d)|_k\}$ can be evaluated in terms of single-frequency signals. Since the proof is analogous to that provided in [11] for the case with 2^{M-1} components, only the final result is reported here. In particular, it is possible to show that

$$\Re\{|C_z(\tau, \mathbf{f}_d)|_k\} = \sum_{i=0}^{K-1} \tilde{\alpha}_i |C_{x,i}(\tau, f_{h,i})| \tag{47}$$

where $C_{x,i}(\tau, f_{h,i}) = \sum_{n=0}^{N-1} x_{rx,i}[n] c_i^*(nT_s - \tau) \exp\{-j_0 2\pi f_{h,i} nT_s\}$ are the standard single-frequency CAFs computed for the different signal components, $x_{rx,i}[n]$. $f_{h,i}$ is the Doppler frequency tested on the ith component. These frequencies are obtained as

$$\mathbf{f}_h = \mathbf{H}_K^{-1}\mathbf{f}_d. \tag{48}$$

As for the case with 2^{M-1}, a form of weighted linear detector is found [17]. The joint acquisition of K components requires a common integration time, T_c. This can be implemented for the Galileo case considering the E5a, E5b and E6 components.

2.4.4. Multicomplex Tracking

A multicomplex tracking scheme can be obtained by adapting the architecture described in [11] to the case where K components are present. The tracking scheme obtained is schematically represented in Figure 6: a single DLL, a single PLL and $K-1$ SPLLs are used to track the code, carrier and subcarrier components present in signal representation (38).

The incoming multicomplex signal, $z_{rx,K}[n]$, is first multiplied by the exponentials generated by the carrier and subcarrier Numerically Controlled Oscillators (NCOs). Three copies of local code (43) are also generated with different delays. The prompt code replica is generated considering the best code delay estimate available at the current processing epoch. Early and late code replicas are obtained by advancing and delaying the prompt replica by fixed amounts, which are the early-minus-late spacings. The same considerations made in [11] on the use of different code spacings for different frequency components apply and are not repeated here. The three code replicas are multiplied by the multicomplex signal obtained after removing the carrier and subcarrier components and finally integrated coherently considering N samples. In this way, three multicomplex correlators P, L and E, are obtained. Code and phase residual errors are extracted using appropriated discriminators operating on the multicomplex correlators. In particular, a generalized non-coherent early-minus-late envelope discriminator [27] can be adopted for the code error:

$$D_c = \frac{\Re\{|E|_k\} - \Re\{|L|_k\}}{\Re\{|E|_k\} + \Re\{|L|_k\}}, \tag{49}$$

where $|\cdot|_k$ denotes the hyperbolic modulus and D_c is the code discriminator output. Carrier and subcarrier phase error estimates are obtained by considering the polar representation of the prompt correlator output that is a multicomplex number:

$$P = |P|_k \exp\left\{\mathbf{j}_{M,K}^T \mathbf{P}_{p,K}\right\} = |P|_k \exp\{j_0 \Delta\varphi_0\} \prod_{i=1}^{K-1} \exp\{j_i \Delta\varphi_i\}. \tag{50}$$

Since P has been obtained from $z_{rx,K}[n]$, it inherits several of its properties. For instance, P has only K non-zero orthogonal components. Moreover, owing to the construction proposed in Section 2.3.2, only its first K phases expressed with respect to the imaginary units, j_i, are different from zero. For this reason, polar representation (50) involves $\mathbf{j}_{M,K}$, which is made of K components. $\mathbf{P}_{p,K}$ is the vector with the K non-zero phases of P. These phases are the residual errors that the different carrier and subcarrier loops are trying to minimize. These phases can be obtained from the polar representation of P. A procedure for extracting them using real operations is detailed in the next section.

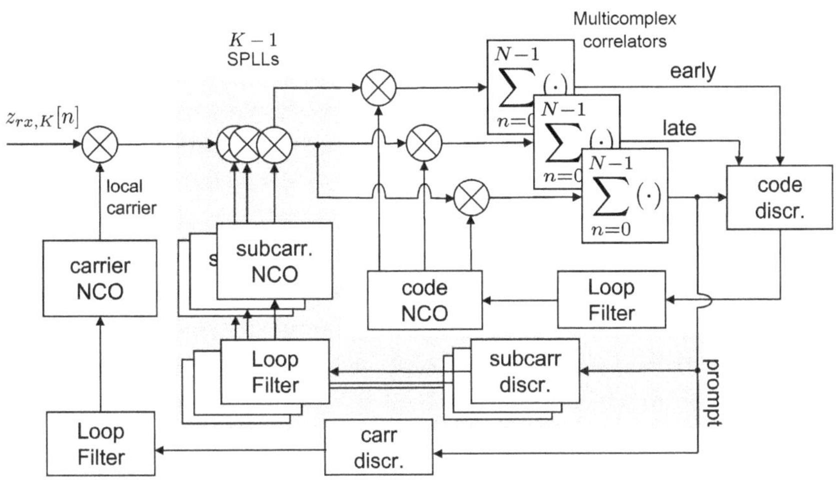

Figure 6. Schematic representation of a multicomplex tracking architecture for jointly processing K signal components from different frequencies. Multicomplex operations and $K-1$ SPLLs are used.

The code, carrier and subcarrier discriminator outputs are filtered using standard loop filters. The filter outputs are estimates of the Doppler frequencies affecting the different components and are used by the NCOs to generate the code, carrier and subcarrier local replicas. In this way, the different loops are closed.

2.4.5. Implementation in Terms of Standard Components

The multicomplex tracking scheme described above can be implemented using the standard components available in single-frequency tracking loops. This is possible using the orthogonal decomposition and the polar representation of multicomplex numbers. In the DLL case, the multicomplex early, late and prompt correlators can be decomposed as follows:

$$
\begin{aligned}
P &= 2^{M-1} \sum_{i=0}^{K-1} \tilde{\alpha}_i P_{x,i} e_{M,i} \\
E &= 2^{M-1} \sum_{i=0}^{K-1} \tilde{\alpha}_i E_{x,i} e_{M,i} \\
L &= 2^{M-1} \sum_{i=0}^{K-1} \tilde{\alpha}_i L_{x,i} e_{M,i}
\end{aligned}
\tag{51}
$$

where $P_{x,i}$, $E_{x,i}$ and $L_{x,i}$ are the complex prompt, early and late correlators computed for the ith signal component, $x_{rx,i}[n]$. Using (51), it follows that

$$
\begin{aligned}
\Re\{|E|_k\} &= \sum_{i=0}^{K-1} \tilde{\alpha}_i |E_{x,i}| \\
\Re\{|L|_k\} &= \sum_{i=0}^{K-1} \tilde{\alpha}_i |L_{x,i}|,
\end{aligned}
\tag{52}
$$

which is analogous of (47) for the early and late correlators. Using (52), it is finally possible to compute (49) in terms of single-component correlator outputs:

$$
D_c = \frac{\sum_{i=0}^{K-1} \tilde{\alpha}_i |E_{x,i}| - \sum_{i=0}^{K-1} \tilde{\alpha}_i |L_{x,i}|}{\sum_{i=0}^{K-1} \tilde{\alpha}_i |E_{x,i}| + \sum_{i=0}^{K-1} \tilde{\alpha}_i |L_{x,i}|}. \tag{53}
$$

Thus, the multi-component DLL can be implemented using standard loop elements. The single-frequency correlators, $\{P_{x,i}, E_{x,i}, L_{x,i}\}_{i=0}^{K-1}$, are at first computed using the code generated by single-component NCOs. These correlators are used for the evaluation of discriminator output (53), which is filtered using a standard loop filter. In this way, a common code Doppler estimate is obtained. Finally, the loop filter output is used to drive the code generation from all the K single-component NCOs. This architecture is similar to that described in [11] for the case with 2^{M-1} components.

For the evaluation of the different correlators, single-component local carriers are needed. These are provided by the multicomplex PLL/SPLLs, which can be also implemented in terms of standard components.

This is illustrated in Figure 7, which shows a joint implementation of the different PLL/SPLLs. The prompt correlators, $P_{x,i}$, are computed using the local codes from the multicomplex DLL and the local carriers generated by single-component carrier NCOs. These correlators are passed to standard carrier discriminators, which provide estimates of the phases of the different $P_{x,i}$. The discriminator outputs are the phases of the orthogonal components of P and are processed using a carrier/subcarrier joint filter. A schematic representation of this joint loop filter is provided in Figure 8: the carrier discriminators are grouped in a vector and multiplied by $\mathbf{H}_K$, the matrix obtained according to the procedure outlined in Section 2.3.2. In this way, the phase vector $\mathbf{P}_{p,K}$ is obtained:

$$
\mathbf{P}_{p,K} = \mathbf{H}_K \begin{bmatrix} \Delta\varphi_{x,0} \\ \Delta\varphi_{x,1} \\ \vdots \\ \Delta\varphi_{x,K-1} \end{bmatrix}, \tag{54}
$$

where $\Delta\varphi_{x,i}$ is the discriminator output from the ith signal component.

The first element of $\mathbf{P}_{p,K}$ is the phase error associated to the meta-signal common carrier, whereas the other elements are the phase errors affecting the different subcarriers. These phase errors are processed using standard loop filters, which produce estimates of the Doppler shifts on the different meta-signal carrier and subcarrier components. Doppler estimates are arranged in a vector and multiplied by the inverse transform defined by $\mathbf{H}_K^{-1}$. In this way, estimates of the Doppler frequencies affecting the individual components, $x_{rx,i}[n]$, are found. These Doppler estimates are finally used to drive the carrier NCOs in Figure 7. This closes the loops in the PLL/SPLLs implementation using standard components.

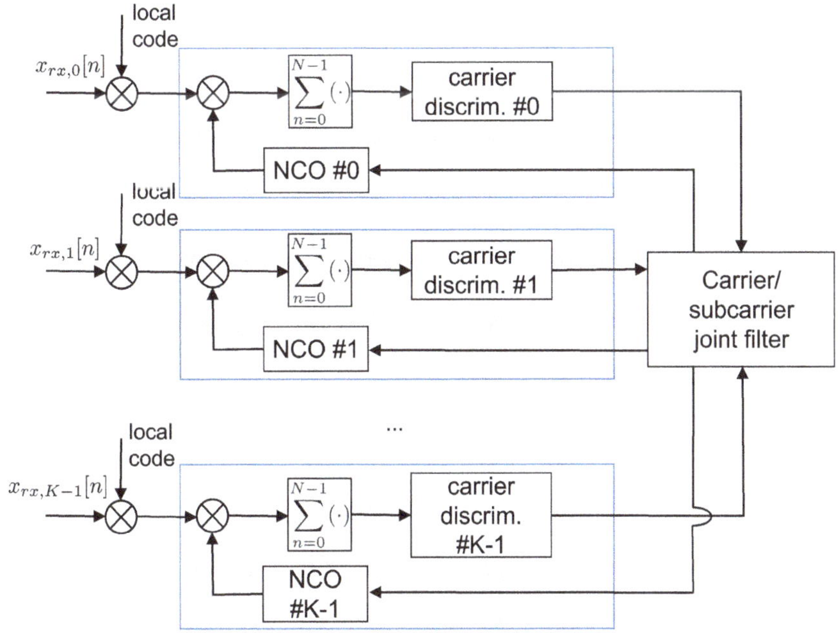

Figure 7. Schematic representation of a possible implementation of multicomplex PLL/SPLLs using standard components. The multicomplex PLL/SPLLs can be implemented using single-frequency standard PLLs sharing a joint multi-dimensional loop filter.

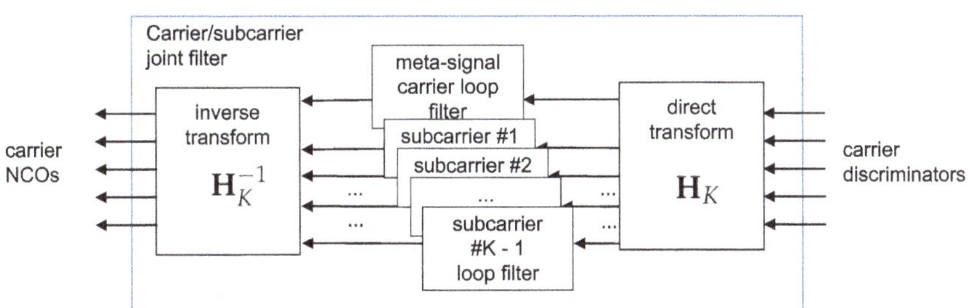

Figure 8. Schematic representation of the multi-dimensional joint loop filter adopted for the PLL/SPLL implementation.

2.5. Experimental Setup

In order to demonstrate the feasibility of the meta-signal approach developed for the joint processing of an arbitrary number of signals, the baseband data used in [11] have been reprocessed considering three components, namely the Galileo E6, E5b and E5a signals. Since a complete description of the experimental setup adopted for the data collection can be found in [11], only a brief summary is provided here. The second part of this section describes the actual implementation in a Python (version 3.9) SDR receiver of the acquisition and tracking algorithms introduced in Section 2.4 considering three components.

The data recording system used in [11] consisted of four HackRF One front-ends (Great Scott Gadgets, Lakewood, CO, USA) [29] sharing the same clock and synchronized according to the procedure detailed in [30]. While four front-ends were used to collect baseband samples from four Galileo frequencies, only three channels are considered in this work. In particular, the In-Phase Quadrature (I/Q) samples from the three front-ends tuned on the E6, E5b and E5b frequencies have been jointly processed. The parameters used for configuring these front-ends are summarized in Table 1.

Table 1. Parameters related to the three data streams from the HackRF One front-ends used for the experimental analysis.

Parameter	Value
Sampling frequency, f_s	20 MHz
Centre frequency, E6 front-end, f_{E6}	1279 MHz
Residual IF, second front-end, $f_{E6,IF}$	0.25 MHz
Centre frequency, E5b front-end, f_{E5b}	1207 MHz
Residual IF, third E5b, $f_{E5b,IF}$	0.14 MHz
Centre frequency, E5a front-end, f_{E5a}	1176 MHz
Residual IF, E5a front-end, $f_{E5a,IF}$	0.45 MHz
Sampling type	Complex I/Q
No. of bits	8

All the front-ends were characterized by the same sampling frequency, $f_s = 20$ MHz. Moreover, each front-end was tuned to a different centre frequency as reported in Table 1. Note that, as analysed in [31], the fractional PLLs used by HackRF One front-ends to synthesize the local carrier for the signal downconversion, are usually not able to completely match the parameters set by the user. This effect leaves small differences in the residual signal Intermediate Frequency (IF), which have been accounted for through a calibration procedure.

The I/Q samples collected using the three HackRF One front-ends have been processed using the custom Python SDR receiver implementing the acquisition and tracking stages described in the previous section. The Galileo E6, E5a and E5b signals are characterized by the same primary code duration, which is equal to 1 ms. Thus, meta-signal acquisition can be implemented directly through (47): individual CAFs are computed on the different components and linearly combined. Appropriate scaling of the different Doppler frequencies has been implemented.

In tracking, the architecture illustrated in Section 2.4.5 has been adopted. The signal configuration depicted in Figure 1 has been considered. This choice led to the implementation of a PLL and two SPLLs. A single DLL was adopted for tracking a common code delay. The tracking loop parameters provided in Table 2 were adopted for the experiments described in Section 2.5.

The Python SDR receiver features several tracking stages, such as the adoption at of Frequency Lock Loop (FLL)/Subcarrier Frequency Lock Loops (SFLLs) for achieving frequency lock conditions. Secondary code synchronization is also implemented on the pilot channels of the three Galileo signals. After secondary code synchronization, the coherent integration time is increased from 1 to 5 ms and pure pilot processing with four-quadrant phase discriminators is used [27]. This type of processing is possible when using the architecture described in Section 2.4.5.

As for the case with four components considered in [11], sub-sample offsets can be present among the data streams collected from the different channels. For this reason and similarly to [11], a calibration procedure has been implemented as part of the DLL operations. For a configurable period of time, the DLL operates in calibration mode using a single component while the other channels are processed in a passive way. The correlators from the passive channels are used to estimate sub-sample delays between components. After estimating such delays, the DLL moves to joint processing using the information from all the channels. In particular, discriminator (53) is used. The DLL also estimates the combining weights, $\tilde{\alpha}_i$. This is performed using an exponential filter on the amplitude of the different prompt correlators.

Table 2. Parameters of the Python SDR receiver used to demonstrate multicomplex tracking using the Galileo E6, E5a and E5b components. The same parameters have been used for the two SPLLs.

Parameter	Value
DLL order	2
DLL bandwidth	2 Hz
PLL order	3
PLL bandwidth	15 Hz
SPLL order	2
SPLL bandwidth	3 Hz
Integration time before secondary code synch.	1 ms
Integration time after secondary code synch.	5 ms
Early-minus-late chip spacing, E6	1 BPSK(10) chip
Early-minus-late chip spacing, E5a and E5b	0.5 BPSK(10) chips

Finally, the computation of the early and late correlators requires the specification of the early-minus-late chip spacings for the different components. These spacings are reported in Table 2 in terms of Binary Phase Shifting Keying (BPSK)(10) chips. This is the modulation of the E5a and E5b signals. The values reported in Table 2 have been selected in order to have a constant product between chip spacing and slope of the correlation function of the corresponding signal.

3. Results

Sample results obtained by processing the data collected using the experimental setup described in Section 2.5 are briefly reported in this section. These results support the feasibility of the approach derived in the paper.

3.1. Acquisition

Since the Galileo E6, E5b and E5a signals share the same primary code duration, they can be directly combined using linear detector (47). Figure 9 shows the four CAFs computed using individual signals (plots (a), (b) and (c)) and by jointly considering the three components. From the figure, it emerges that the three Galileo components are characterized by different Signal-to-Noise Ratio (SNR) conditions. The E6 signal is the noisiest and the corresponding CAF peak is barely visible in Figure 9. In contrast, the E5b component is characterized by the highest SNR. The different signal qualities are due to the different transmit power levels and by different antenna/front-end gains provided at different frequencies.

All the CAFs have been computed using a 1 ms coherent integration time and normalized with respect to the CAF maximum. Non-coherent integrations were not implemented. In acquisition, the different signal gains, $\tilde{\alpha}_i$, are not available. For this reason, all the single-component CAFs have been weighted equally. Moreover, sub-sample delays between components have not been accounted for. While these choices are not optimal, a clear gain is visible in Figure 9. The noise floor of the triple-component CAF in Figure 9d is clearly reduced with respect to the E6 one: the stronger components aid the acquisition of the weakest one. Also, a limited noise floor reduction is observed when comparing the triple component CAF with that obtained using the E5b signal. This is expected given the high quality of the E5b signal.

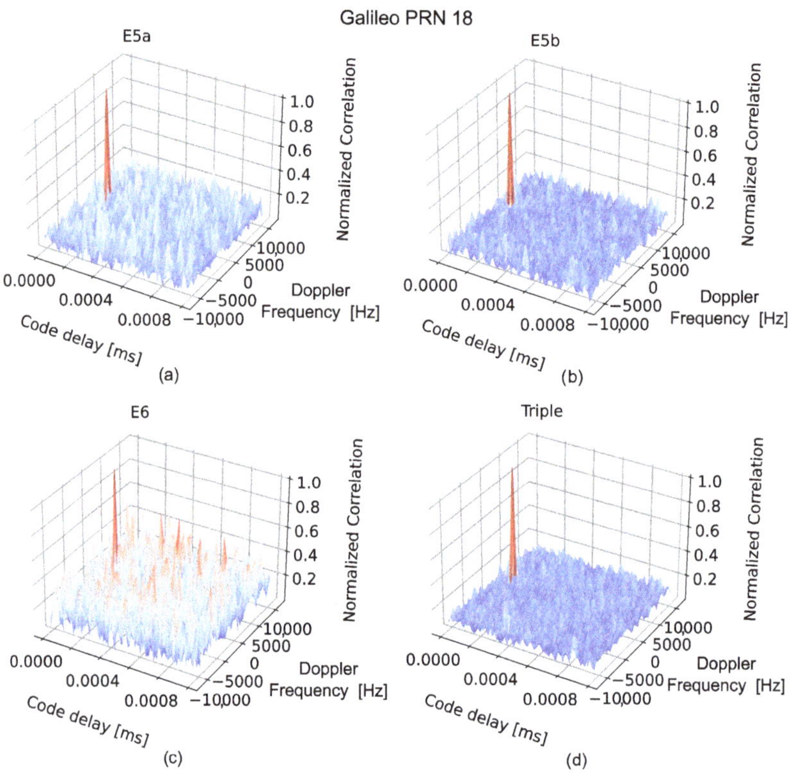

Figure 9. CAFs computed considering the Galileo E6, E5b and E5a components. (**a**) E5a component. (**b**) E5b component. (**c**) E6 component. (**d**) Combined triple-component CAF.

The analysis provided in Figure 9 is qualitative and aims at showing the feasibility of the approach derived in Section 2.4.3. A more thorough analysis of the weighted linear detectors resulting from the meta-signal theory developed in this paper is left for future work.

3.2. Tracking

Results obtained by jointly tracking the E6, E5b and E5a components are provided in the following. The signal representation shown in Figure 1 has been considered with a common carrier, with nominal frequency equal to 1242.945 MHz and two subcarriers with frequencies equal to 51.15 and 15.345 MHz, respectively. Figure 10 shows the single-component prompt pilot correlators obtained using the quad-loop architecture. These are the orthogonal components of the multicomplex prompt correlator obtained by considering the E6, E5b and E5a signals.

Both I/Q branches are provided in Figure 10: after an initial transient period, the meta-signal PLL and SPLLs are able to correctly align all the useful power in the in-phase correlator parts. From the figure, the different operating modes of the implemented tracking architecture are also visible. After less than two seconds, the secondary codes of the pilot components are recovered and removed from the prompt correlators. For this reason, after secondary code removal, only positive values are visible in the in-phase parts of the prompt correlators.

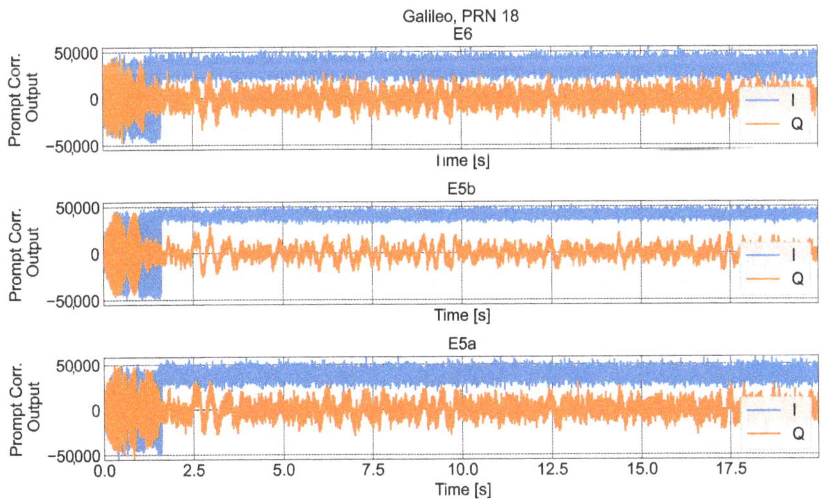

Figure 10. Pilot prompt correlator outputs obtained using the quad-loop architecture jointly tracking the E6 (upper plot), E5b (middle plot) and E5a (lower plot) components.

The proper functioning of the implemented quad-loop tracking architecture is also supported by Figure 11, which shows the normalized code, carrier and subcarrier Doppler estimates obtained at the output of the filters of the DLL, PLL and two SPLLs. Since the architecture depicted in Figure 7 has been implemented, the loop filter outputs correspond to the values obtained before the inverse transform shown in Figure 8. The different Doppler estimates have been normalized by the corresponding nominal frequencies. After normalization, the different components practically overlap, indicating the possibility of aiding between different components. After secondary code removal, the integration time is increased from 1 to 5 ms. In this way, the noise on the different components is reduced. Code estimates are also improved when moving from calibration to standard processing mode. After calibration, all the signal components contribute the estimation of a common delay, thus reducing the associated noise variance.

The right part of Figure 11 provides a zoomed-in section of the Doppler estimates associated to the three carrier/subcarrier components. The curve related to subcarrier C is the noisiest. This is expected since it is characterized by a nominal frequency equal to 15.345 MHz, which is the lowest among the three components. The normalizations used in Figure 11 also scale the noise variations. Noise performance can be improved by reducing the SPLL bandwidth. The transition from 1 to 5 ms is also visible in the right part of Figure 11.

Finally, Figure 12 compares the Carrier-To-Noise Power Spectral Density Ratio (C/N_0) estimates obtained using the meta-signal architecture with three components and single-component tracking loops. As already emerged from the discussion in Section 3.1, the three Galileo components are characterized by significantly different C/N_0 values. The E5b component is about 8 times stronger than the E6 signal. While this significant difference reduces the overall gain of the joint architecture when evaluated with respect to the E5b component, the implemented architecture is still able to effectively combine the different signals. The resulting C/N_0 is close to the maximum theoretical C/N_0 (52.4 dB-Hz), corresponding to the direct sum of the signal energy from the individual channels.

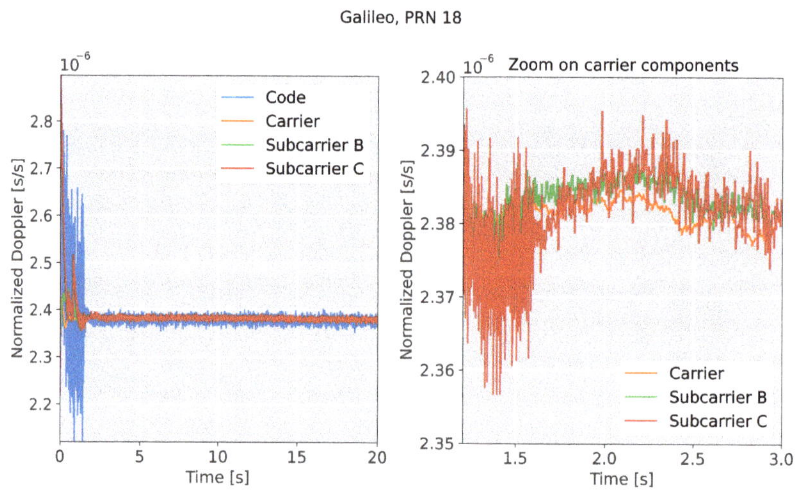

Figure 11. (**Left**) Normalized code, carrier and subcarrier Doppler estimates from the DLL, PLL and two SPLLs used to jointly track the E6, E5b and E5a components. (**Right**) Zoom of the carrier and subcarrier Doppler estimates.

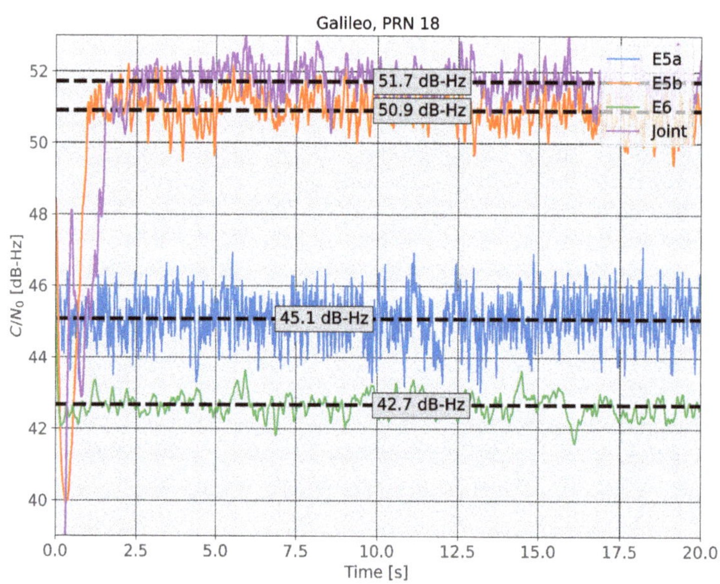

Figure 12. Comparison between C/N_0 estimates obtained using the meta-signal tracking architecture with three components and single-component tracking loops.

4. Discussion

The results illustrated in the previous section confirm the fact that in the proposed acquisition/tracking architectures all signals contribute to the estimation of common parameters. Moreover, stronger components help the processing of the weakest ones. In this case, the E6 signal is the one benefiting the most from meta-signal approaches.

A key result of this paper is the possibility to restate the algorithms derived using multicomplex numbers in terms of standard components. As a result, the cost function optimized at the acquisition stage is a form of weighted linear detector combining the single-frequency CAFs. A single DLL combining all the single-frequency correlators was also designed. Finally, the meta-signal PLL/SPLLs were obtained by modifying standard single-frequency PLLs. The discriminator outputs from different single-frequency PLLs are brought into a transformed domain and processed using carrier and subcarrier loop filters. In this way, the estimates of the carrier/subcarrier Doppler frequencies are found. An inverse transform is finally used to compute the Doppler shifts on the different signal components. The transformation from individual carriers to the meta-signal carrier/subcarriers was obtained from the structure of multicomplex numbers as a truncated Hadamard matrix. The truncation, implemented by removing rows and columns from a square Hadamard matrix of size equal to a power of two, makes the resulting transformation not orthogonal. Additional work is required to better understand the implications of the loss of orthogonality. Moreover, while the use of truncated Hadamard matrices has some appealing properties, future work should consider the adoption of different transformations obtained according to different criteria.

While this paper provides the theoretical foundations for the design of multi-dimensional acquisition/tracking algorithms with an arbitrary number of components, additional tests are required to fully determine the benefits and limitations of meta-signal approaches. The algorithms proposed can also be easily extended by introducing constraints among the different carrier and subcarrier components. For instance, a Kalman filter can be integrated within the multi-dimensional loop filter introduced in the Section 2.4.5.

Finally, the theory developed in this paper has been supported by experiments conducted using the Galileo E5a, E5b and E6 signals. However, the results presented are general and can be applied to signals of opportunity, such those broadcast by Low Earth Orbit (LEO) satellites [14]. Future research directions may include, for instance, the analysis of meta-signal approaches for tracking a limited number of OFDM components, such as pilot tones, sparsely located in an OFDM symbol.

5. Conclusions

In this paper, the general theory for representing and processing a GNSS meta-signal with an arbitrary number of components was derived. The theory, which uses multicomplex numbers, extends previous results, which required a number of components equal to a power of two. It is shown that a general meta-signal can be obtained as the product of a multicomplex baseband term multiplied by $K-1$ subcarriers, which split the meta-signal components on K RFs. Using multicomplex numbers, it was possible to derive a generalized CAF computed using signals from K different RFs. In turn, the CAF was used to derive acquisition and tracking algorithms for the processing of K frequency components, which are jointly used to estimate common parameters such as a single delay and common carrier and subcarrier parameters. While multicomplex numbers proved beneficial for algorithm design and for properly representing the different signal components, the possibility of restating the different processing strategies in terms of standard components allows their efficient implementation. This fact has been proved through the implementation of the derived algorithms in an SDR receiver. Experiments involving the processing of Galileo signals collected using three synchronized SDR front-ends support the theoretical findings derived in this paper.

Funding: This research received no external funding.

Data Availability Statement: Dataset available on request from the authors.

Conflicts of Interest: The author declares no conflicts of interest.

References

1. Betz, J.W. Binary Offset Carrier Modulations for Radionavigation. *Navigation* **2001**, *48*, 227–246. [CrossRef]
2. Zheng, Y.; Yunhan, Q.; Mingquan, L. Subcarrier modulated navigation signal processing in GNSS: A review. *Satell. Navig.* **2024**, *24*, 5.
3. Lestarquit, L.; Artaud, G.; Issler, J.L. AltBOC for Dummies or Everything You Always Wanted To Know About AltBOC. In Proceedings of the 21st International Technical Meeting of the Satellite Division of The Institute of Navigation (ION GNSS 2008), Savannah, GA, USA, 16–19 September 2008; pp. 961–970.
4. Yao, Z.; Zhang, J.; Lu, M. ACE-BOC: Dual-frequency constant envelope multiplexing for satellite navigation. *IEEE Trans. Aerosp. Electron. Syst.* **2016**, *52*, 466–485. [CrossRef]
5. Wang, C.; Cui, X.; Ma, T.; Zhao, S.; Lu, M. Asymmetric Dual-Band Tracking Technique for Optimal Joint Processing of BDS B1I and B1C Signals. *Sensors* **2017**, *17*, 2360. [CrossRef]
6. Tian, Z.; Cui, X.; Liu, G.; Lu, M. LPRA-DBT: Low-Processing-Rate Asymmetrical Dual-Band Tracking Method for BDS-3 B1I and B1C Composite Signal. In Proceedings of the International Technical Meeting of The Institute of Navigation, Long Beach, CA, USA, 25–27 January 2022; pp. 1027–1038.
7. Issler, J.L.; Paonni, M.; Eissfeller, B. Toward centimetric positioning thanks to L- and S-Band GNSS and to meta-GNSS signals. In Proceedings of the ESA Workshop on Satellite Navigation Technologies and European Workshop on GNSS Signals and Signal Processing (NAVITEC), Noordwijk, Netherlands, 8–10 December 2010; pp. 1–8. [CrossRef]
8. Borio, D. Bicomplex Representation and Processing of GNSS Signals. *Navigation* **2023**, *70*, 1–33. [CrossRef]
9. Paonni, M.; Curran, J.; Bavaro, M.; Fortuny-Guasch, J. GNSS Meta Signals: Coherently Composite Processing of Multiple GNSS Signals. In Proceedings of the 27th International Technical Meeting of the Satellite Division of The Institute of Navigation (ION GNSS+), Tampa, FL, USA, 8–12 September 2014; pp. 2592–2601.
10. Alpay, D.; Luna-Elizarrarás, M.E.; Shapiro, M.; Struppa, D.C. *Basics of Functional Analysis with Bicomplex Scalars, and Bicomplex Schur Analysis*; SpringerBriefs in Mathematics; Springer: Cham, Switzerland, 2014. [CrossRef]
11. Borio, D. A General Multi-Dimensional GNSS Signal Processing Scheme Based on Multicomplex Numbers. In Proceedings of the International Technical Meeting (ION GNSS+ 2024), Baltimore, MD, USA, 16–20 September 2024; pp. 2904–2925. [CrossRef]
12. Price, G.B. *An Introduction to Multicomplex Spaces and Functions*, 1st ed.; CRC Press: Boca Raton, FL, USA, 1991. [CrossRef]
13. Nasiruzzaman, M.; Mursaleen, M. The Multicomplex Numbers and Their Properties on Some Elementary Functions. *Thai J. Math.* **2023**, *21*, 77–100.
14. Nardin, A.; Dovis, F.; Zanier, F.; Melman, F. Adaptable Meta-Signal Tracking for Next-Generation Satellite Navigation Systems and LEO PNT. *IEEE Access* **2025**, *13*, 5706–5718. [CrossRef]
15. Wang, X.; Yang, Y.; Deng, L.; Ye, L.; Li, Z.; Xiao, Y.; Dong, W. Design and Performance Analysis of Navigation Signal Based on OFDM. *Appl. Sci.* **2022**, *12*, 9486. [CrossRef]
16. Wang, D.; Fattouche, M. OFDM Transmission for Time-Based Range Estimation. *IEEE Signal Process. Lett.* **2010**, *17*, 571–574. [CrossRef]
17. Marcum, J. A statistical theory of target detection by pulsed radar. *IRE Trans. Inf. Theory* **1960**, *6*, 59–267. [CrossRef]
18. Yang, R.; Xu, D.; Morton, Y.T. Generalized Multifrequency GPS Carrier Tracking Architecture: Design and Performance Analysis. *IEEE Trans. Aerosp. Electron. Syst.* **2020**, *56*, 2548–2563. [CrossRef]
19. Bolla, P.; Nurmi, J.; Won, J.; Lohan, E.S. Joint Tracking of Multiple Frequency Signals from the same GNSS satellite. In Proceedings of the International Conference on Localization and GNSS (ICL-GNSS), Guimaraes, Portugal, 26–28 June 2018; pp. 1–6. [CrossRef]
20. Yang, R.; Morton, Y. An Adaptive Inter-frequency Aiding Carrier Tracking Algorithm for the Mountain-top GPS Radio Occultation Signal. In Proceedings of the 2018 International Technical Meeting of The Institute of Navigation, Reston, VA, USA, 29 January–1 February 2018; pp. 412–419. [CrossRef]
21. Borio, D.; Susi, M. GNSS Meta-Signal Tracking Using a Bicomplex Kalman Filter. *Navigation* **2024**, *71*, 1–32. [CrossRef]
22. Segre, C. Le rappresentazioni reali delle forme complesse e gli enti iperalgebrici. *Math. Ann.* **1891**, *40*, 413–467. [CrossRef]
23. Fjelstad, P.; Gal, S.G. n-Dimensional hyperbolic complex numbers. *Adv. Appl. Clifford Algebr.* **1998**, *8*, 47–68. [CrossRef]
24. Borio, D. A Vector Representation for Multicomplex Numbers and its Application to Radio Frequency Signals. *Axioms* **2024**, *13*, 324. [CrossRef]

25. Yarlagadda, R.K.R.; Hershey, J.E. *Hadamard Matrix Analysis and Synthesis*; The Springer International Series in Engineering and Computer Science; Springer: New York, NY, USA, 1997. [CrossRef]
26. Lu, M.; Li, W.; Yao, Z.; Cui, X. Overview of BDS III new signals. *Navig. J. Inst. Navig.* **2019**, *66*, 19–35. [CrossRef]
27. Kaplan, E.D.; Hegarty, C., Eds. *Understanding GPS/GNSS: Principles and Applications*, 3rd ed.; Artech House Publishers: Norwood, MA, USA, 2017.
28. Borio, D. Meta-Signal Processing with Data/Pilot Combining for Beidou B2 Signals. *Remote Sens.* **2024**, *16*, 2537. [CrossRef]
29. Ossmann, M. HackRF Github Page. 2024. Available online: https://greatscottgadgets.com/hackrf/one/ (accessed on 15 November 2024).
30. Bartolucci, M.; Del Peral-Rosado, J.A.; Estatuet-Castillo, R.; Garcia-Molina, J.A.; Crisci, M.; Corazza, G.E. Synchronisation of low-cost open source SDRs for navigation applications. In Proceedings of the 8th ESA Workshop on Satellite Navigation Technologies and European Workshop on GNSS Signals and Signal Processing (NAVITEC), Noordwijk, The Netherlands, 14–16 December 2016; pp. 1–7. [CrossRef]
31. O'Driscoll, C.; Curran, J.T. Carrier Phase Tracking Considerations for Commodity SDR Hardware. In Proceedings of the 31st International Technical Meeting of the Satellite Division of The Institute of Navigation (ION GNSS+), Miami, FL, USA, 24–28 September 2018; pp. 4182–4196. [CrossRef]

remote sensing

MDPI

Article

Multi-Layer Perceptron Model Integrating Multi-Head Attention and Gating Mechanism for Global Navigation Satellite System Positioning Error Estimation

Xiuxun Liu, Zuping Tang and Jiaolong Wei *

School of Electronic Information and Communications, Huazhong University of Science and Technology, Wuhan 430074, China; hust_lxx@hust.edu.cn (X.L.); tang_zuping@hust.edu.cn (Z.T.)
* Correspondence: jlwei@hust.edu.cn

Abstract: To better understand and evaluate the GNSS positioning performance, it is convenient to adopt corresponding measures to reduce the impact of errors on positioning. A GNSS positioning error estimation scheme based on an improved multi-layer perceptron model is proposed. The multi-head attention mechanism and gating operation are integrated into the multi-layer perceptron model to adaptively select and filter features, enhancing the model's ability to understand input features. First, the original positioning error of the satellite is obtained through the Kalman filter positioning method. The data are then preprocessed to extract available features. Finally, the features are input into the constructed model for training and testing to obtain the estimated positioning error value. Two types of comparative experiments were completed. The performance of the presented model is evaluated by the root mean square error. Experimental results show that the proposed method performs well in terms of performance indicators, and has obvious advantages over other state-of-the-art methods. In particular, the root mean square error of the presented method in the first dataset is 0.239 m, which is 39.2% and 17% lower than the current state-of-the-art long short-term memory network and convolutional neural network, respectively. The presented method can provide higher-precision estimated values for studying the GNSS positioning error estimation problem.

Keywords: global navigation satellite system; positioning error estimation; machine learning; multi-layer perceptron model

1. Introduction

Global navigation satellite system (GNSS) positioning technology has a wide range of applications in aviation, navigation, vehicle navigation, and other fields. GNSS positioning technology is necessary for achieving precise flight trajectory control and target navigation in aviation and navigation [1]. Real-time positioning is required to support autonomous driving and intelligent transportation systems in vehicle navigation [2]. Accurate positioning data are needed to study crustal movement and atmospheric changes in seismic monitoring and climate research [3]. The demand for high-precision and high-reliability positioning is increasing with the continuous development and transformation of GNSS technology. Previously, scholars focused on the research of GNSS high-precision positioning, and less exploration was undertaken on the problem of GNSS positioning error estimation. However, GNSS positioning and GNSS positioning error estimation are equally important [4]. Positioning errors may lead to the inaccuracy and unreliability of GNSS positioning, which has a serious impact on practical applications. Therefore, accurately estimating and controlling the positioning error are the keys to ensuring the accuracy and reliability of GNSS positioning.

GNSS positioning is affected by errors that directly lead to inaccurate positioning results. The error sources in the GNSS positioning process mainly include satellite clock errors, ionospheric delay, tropospheric delay, multipath effect, receiver error, etc. [5]. Early research on GNSS positioning error estimation mainly focused on the estimation of a certain positioning error [6]. To minimize the impact of various factors on positioning results through modeling and analysis, many scholars have explored the problem of single error term estimation [7–9]. Han et al. designed a scheme to achieve high-precision positioning by estimating satellite clock errors [10]. This method has achieved excellent positioning results in both static and dynamic positioning. To address the problem that the dual-frequency ionospheric correction method ignores the high-order ionospheric error, Hoque et al. investigated a different approximation formula to estimate and correct the error caused by the excess path length [11]. Due to the lack of joint estimation of other error terms, the positioning error estimation method for a single error term has the problems of low accuracy and poor robustness. Subsequently, scholars have begun to focus on more refined positioning error modeling methods, and GNSS positioning accuracy has been improved to a certain extent [12]. The focus of related research is gradually shifting to error handling and precision improvement in complex environments [13–15]. Lesouple et al. studied the GNSS positioning problem in a restricted environment, and applied the sparse estimation theory to estimate the GNSS positioning error caused by different environments [16]. Jiang et al. explored the GNSS positioning problem in an environment with severe occlusion and interference, and presented an assisted joint positioning algorithm that exploits near-Earth Doppler and other measurement data to track the target position [17]. Although the research has promoted the advancement of GNSS positioning error estimation methods in complex environments, it has not solved the problem of limited estimation accuracy caused by estimating only a single error term. Nowadays, multisystem fusion has become a popular research topic [18]. Researchers have begun to study how to apply multisystem satellite observation data for error estimation and correction to improve the accuracy and reliability of positioning. Researchers have gradually shifted from estimating individual errors to estimating the overall error in the GNSS positioning process [19–21]. Marais et al. expatiated a GNSS positioning algorithm based on pseudorange error modeling and adaptive filtering [22]. Biswas et al. explored the influence of satellite spatial distribution on GNSS position estimation, and analyzed the GNSS positioning error estimation performance of four Kalman filters in detail [23]. Although estimating the overall error in the positioning process has been proven to be able to effectively mitigate the impact of error factors on positioning results, the method of estimating the overall error has the defect of relying on the assumptions of GNSS observation noise and motion model.

The application of GNSS positioning has introduced many complex and highly dynamic situations. How to solve the problems of error handling in complex environments and reliable positioning in highly dynamic environments urgently needs to be considered. Research on positioning error estimation methods can reduce or eliminate positioning deviations caused by additional positioning error factors. Traditional positioning error estimation methods for single error terms have the problems of low estimation accuracy and the need to make assumptions about GNSS observation noise. Although traditional methods for estimating the overall error can achieve higher estimation accuracy, they also require assumptions about GNSS observation noise. Therefore, a new GNSS positioning error estimation method is urgently needed. In recent years, the GNSS positioning error estimation field has been further developed. Scholars have applied artificial intelligence methods to address the GNSS positioning error estimation problem and obtained good results. The GNSS positioning error estimation method based on artificial intelligence does not rely on prior assumptions, and can exploit GNSS observation data to model complex positioning error factors, providing a new idea for improving the positioning accuracy and robustness of GNSS [24,25]. Closas et al. utilized an iterative approach to deal with multivariate optimization problems, and obtained the maximum likelihood estimator of the position under GNSS [26]. Kuratomi et al. explored a GNSS positioning error estimation method based on the random forest algorithm, and obtained excellent error estimation results [27]. Considering that the least squares method ignores the fact that velocity information is time-dependent, Jiang et al. investigated and implemented a GNSS position estimation method based on graph optimization, using historical states and measured values to estimate the current state [28]. By aiming at the problem that the accuracy and stability of multisource fusion navigation still cannot meet the needs of current practical applications, Bao et al. studied a GNSS position error estimation algorithm based on multitask learning [29]. To obtain more accurate GPS positioning accuracy than existing statistical and linear machine learning methods, Yang et al. utilized the long short-term memory network (LSTM) to dynamically model the factors affecting GPS positioning accuracy within a certain coherent time window and then predict the positioning error in the next second to a few seconds. The effectiveness of the presented method was verified through experiments. Implementing the presented LSTM model on hardware requires further considerations [30]. Liu et al. studied an LSTM model with infinite impulse response (IIR) filtering by constructing a dynamic adaptive filter to remove outliers and interference. This innovative solution enhances the robustness of the LSTM network when processing noisy data and achieves better experimental results [31]. Shen et al. studied the unmodeled error using historical GNSS observation data and proposed a positioning error estimation method based on convolutional neural network (CNN). The effectiveness of the presented method was verified through experiments, especially in the low-frequency component [32]. The research is of great significance to error elimination and reliability estimation in the GNSS positioning process.

Traditional GNSS positioning error estimation methods are mostly based on linear assumptions, and cannot effectively capture complex nonlinear relationships, which limits the accuracy of error estimation. GNSS positioning error estimation methods based on deep learning can comprehensively consider multiple error factors, and have powerful nonlinear modeling capabilities, which can effectively improve the accuracy of error estimation. Therefore, a GNSS positioning error estimation method is studied based on an improved multi-layer perceptron (MLP) model. The multi-head attention (MHA) and gating operation are exploited to selectively weight and transform input features, enabling adaptive feature selection and information filtering. Unlike traditional GNSS positioning error estimation methods that rely on precise mathematical models and assumptions, the proposed method is a data-driven deep learning model. The relationship between error factors is learned through historical GNSS data. The proposed method does not require assumptions about error factors, and can adaptively extract features from GNSS data, thereby providing more accurate error estimates. The main contributions of the paper are as follows:

(1) A MLP model integrating MHA and gating operation is proposed for GNSS positioning error estimation. To our knowledge, no scholars have utilized this method to address the problem of GNSS positioning estimation.
(2) Traditional GNSS positioning error estimation methods are usually based on linear models or empirical rules. The presented method can model nonlinear relationships more flexibly, thereby achieving higher estimation accuracy.

(3) Two types of comparative experiments are performed. The experimental results show that the presented method has higher estimation accuracy than other GNSS positioning error estimation methods.

The remainder of the article is structured as follows. In Section 2, the GNSS positioning error estimation problem is formulated. Section 3 focuses on the presented improved MLP model. In Section 4, the performance of the improved MLP model is evaluated, and the experimental results are analyzed in detail. In Section 5, the proposed method is discussed. Finally, a brief summary is presented in Section 6.

2. Problem Formulation

GNSS positioning adopts the principle of trilateration to determine the position of a receiver by measuring the distance between the receiver and multiple satellites. However, the GNSS positioning process may be affected by various error factors. Therefore, accurately estimating the GNSS positioning error is challenging.

In an Earth-centered, Earth-fixed (ECEF) coordinate system, assuming that there are n satellites, the coordinates of the satellites are $S_i = (x_i, y_i, z_i)$, where the value of i ranges from 1 to n. The coordinates of the receiver are $P = (x, y, z)$. Different positioning methods are utilized in the GNSS positioning process, and different observation equations are obtained. The most common GNSS positioning method is pseudorange measurement positioning, which is achieved by measuring the geometric distance from the receiver to the satellite and various error factors. If each error factor is regarded as an independent random variable, the observation equation of the pseudorange measurement model is

$$P_i = \|P - S_i\| + c\Delta t + T_i + I_i + M_i + \varepsilon_i \quad (1)$$

where $\|P - S_i\|$ denotes the geometric distance from the receiver to satellite i, Δt expresses the receiver clock error, T_i means the tropospheric delay, I_i represents the ionospheric delay, M_i denotes the multipath error, and ε_i expresses the measurement error. Notably, due to space reasons, only several important GNSS positioning error factors are shown in (1). In the GNSS positioning, there are other methods, such as carrier phase measurements, which can be applied to satellite positioning.

An observation equation for each satellite is applied to describe the positional relationship between the receiver and the satellite during the GNSS positioning process. Multiple observation equations are obtained by observing multiple satellites at the same time. Then, by solving the system of simultaneous equations from multiple observation equations, the position coordinates $P = (x, y, z)$ of the receiver in the ECEF coordinate system can be solved. The above process can be realized by utilizing the least squares method or other positioning algorithms.

Assume that the true position of the receiver is P_{true}. Then, the positioning error is $\Delta P = P_{\text{true}} - P$. Therefore, the GNSS positioning error estimation problem is

$$P_{\text{true}} = f(D) + \Delta P_{\text{est}} \quad (2)$$

where $f(D)$ expresses the positioning algorithm based on the observation data D and ΔP_{est} denotes the estimated value of the positioning error.

As shown in Figure 1, the presented method aims to estimate the positioning error ΔP as accurately as possible, and make the difference between ΔP_{est} and ΔP as small as possible, to obtain better GNSS positioning error estimation performance.

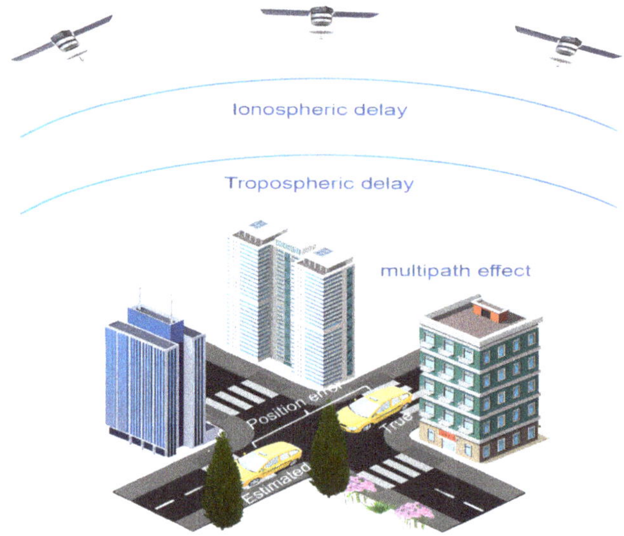

Figure 1. Schematic diagram of positioning error estimation.

The main application scenario of the proposed method is the positioning needs of daily electronic devices, such as smartphones. Although the positioning accuracy of code observation is lower than that of carrier phase observation, the positioning time required for code observation is shorter than that of carrier phase observation, and it is less affected by environmental changes. Therefore, code observation is more suitable for the application scenario of smartphone positioning.

3. Proposed Method

The solution to the GNSS positioning error estimation problem depends on the analysis and processing of historical data. Currently, the common methods of GNSS positioning error estimation tend to ignore the relationships among different types of data, resulting in unsatisfactory estimation accuracy. With the rapid development of artificial intelligence methods, the excellent ability of artificial intelligence methods in data processing and feature learning has been proved. Therefore, after exploring and experimenting with various artificial intelligence methods, a multi-head attention gating unit (MAGU) that combines the MHA and gating operation was designed, and then the MAGU and MLP models were integrated to address the problem of GNSS positioning error estimation. The MHA is beneficial to improving the feature extraction ability of the model. The gating operation can enhance the model's dynamic information selection capability.

The traditional MLP model is a feedforward neural network model. It consists of multiple neural layers. Each MLP neuron is connected to all neurons of the previous layer, and receives input from the previous layer. Weighting calculations and nonlinear transformations are then performed on the input and, finally, the result is passed to the next layer. The explored model represents an enhancement of the traditional MLP model, incorporating a gating operation and a MHA, and updating the network structure and parameters.

According to Figure 2, the overall process of the designed scheme is to first extract relevant parameters from GNSS data and adopt them as feature variables. Next, the positioning error is obtained through the positioning solution, and the positioning error is selected as the target variable. Then, the feature variables and target variables are input into the improved MLP model for multiple iterations of training. Finally, the estimated value of the positioning error is output.

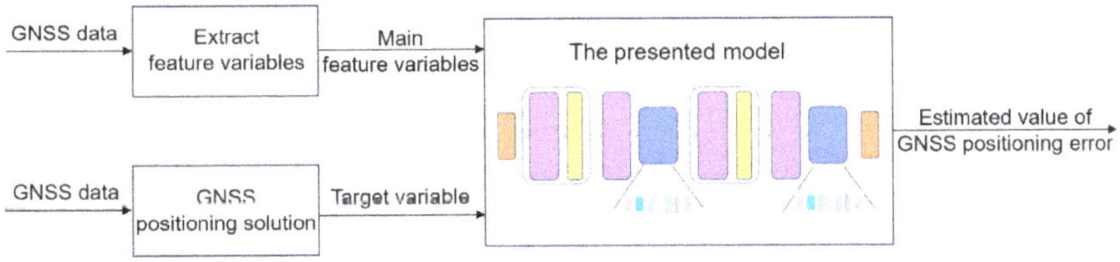

Figure 2. Overall flow chart of the presented scheme.

3.1. Improved MLP Model

Although the traditional MLP model can be used to perform the nonlinear transformation on all input features in each hidden layer, it cannot selectively process different features. Therefore, the presented method integrates the MAGU module into the network structure of the traditional MLP model. This module can dynamically select the input features so that the model can better adapt to the importance and correlation of different features, thus enhancing the feature selection ability of the model.

3.1.1. Model Network Structure

The improved MLP model mainly consists of an input layer, four fully connected layers, two MAGU modules, an output layer, and some activation functions. Four fully connected layers and two MAGU modules constitute the hidden layer of the improved MLP model. The hidden layer mainly performs a series of linear transformations and feature extractions on the input data. The sigmoid activation function is applied in the MAGU module. The fully connected layer uses the rectified linear unit (ReLU) activation function, which can improve the nonlinear ability of the model and speed up its training.

In Figure 3, assuming that the input of the presented model is x, its output after passing through the first fully connected layer and the first ReLU activation function is

$$x_1 = F_1(W_1 x + b_1) \tag{3}$$

where W_1 and b_1 denote the weight matrix and bias vector of the first linear layer, x means the input data, and F_1 expresses the ReLU activation function. The ReLU activation function [33] is

$$F_1(x) = \begin{cases} x & if \;\; x \geq 0 \\ 0 & if \;\; x < 0 \end{cases} \tag{4}$$

The output of the second linear layer can be expressed as

$$x_2 = W_2 x_1 + b_2 \tag{5}$$

After inputting x_2 into the first MAGU, the output is

$$x_3 = g_1 \odot u_1 + (1 - g_1) \odot x_2 \tag{6}$$

where $\odot$ means elementwise multiplication, g_1 represents the output of the gating operation in MAGU, and u_1 denotes the output of the MHA after linear transformation in MAGU.

The output of x_3 after the third linear layer and the second ReLU activation function can be expressed as

$$x_4 = F_1(W_3 x_3 + b_3) \tag{7}$$

The output of the data after the fourth linear layer is

$$x_5 = W_4 x_4 + b_4 \tag{8}$$

After taking x_5 as the input of the second MAGU module, the output can be expressed as

$$x_6 = g_2 \odot u_2 + (1 - g_2) \odot x_5 \tag{9}$$

The final output of the proposed model is

$$y = F_2(W_5 x_6 + b_5) \tag{10}$$

where F_2 denotes the linear activation function. The linear activation function is adopted in the output layer of the improved MLP model. The final estimation result is obtained through the identity mapping of the linear activation function.

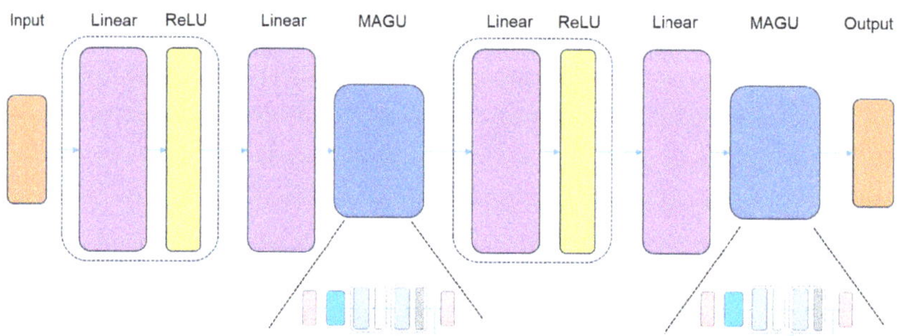

Figure 3. The overall network structure of the improved MLP model.

The number of neurons in the input layer of the presented model is set to be consistent with the number of feature vectors in the dataset. The first and second fully connected layers have 128 neurons. The number of fully connected layer neurons of the first MAGU module is 128. The third and fourth fully connected layers have 64 neurons. The number of fully connected layer neurons of the second MAGU module is 64. The manuscript is studying a regression problem, and the output is a scalar. Therefore, the number of neurons in the output layer is 1.

3.1.2. Model Training and Parameter Setting

The design process of the improved MLP model mainly includes steps such as data preprocessing, model initialization, training parameter setting, loss function definition, and model performance evaluation. First, the input data are standardized and each feature is scaled to a range with a mean of 0 and a standard deviation of 1. The training data and test data in the dataset are divided at a ratio of 7:3. Then, an initialized model, which includes an input layer, a fully connected layer, an MAGU module, and an output layer, is built. Finally, the relevant parameters of the model are set and the positioning error estimation accuracy of the model is evaluated.

The inputs of the presented model in the training phase are feature variables and target variables. Feature variables include carrier-to-noise ratio, satellite clock deviation, etc. The target variable is the positioning accuracy obtained through high-precision positioning solution technology, which is regarded as the true value of the positioning error. The output of the presented model during the training phase is estimated value of the positioning error.

The adaptive moment estimation (Adam) optimizer and $L2$ regularization are adopted in the improved MLP model. Gradients are calculated by the backpropagation algorithm, and the parameters of the model are updated by adopting the optimizer during training. The Adam optimizer combines the characteristics of the momentum method and the adaptive learning rate, effectively updating the parameters and adaptively adjusting the learning rate. *L*2 regularization penalizes model weights, making the weight values tend to a smaller range. It is mainly employed to control model complexity and prevent overfitting to enhance the generalizability of the model. *L*2 regularization is

$$L2_{reg} = L + \lambda \sum_{i=1}^{n} w_i^2 \tag{11}$$

where L denotes the original loss function, λ means the weight decay coefficient, w expresses the weight, n represents the number of data samples, and i is the index.

The improved MLP model adopts the mean square error (MSE) as the loss function, and the MSE is

$$\text{MSE} = L(y, \hat{y}) = \frac{1}{n} \sum_{i=1}^{n} (y_i - \hat{y}_i)^2 \tag{12}$$

where n means the number of samples, y_i and $\hat{y}_i$ represent the true value and its estimated one. The smaller the MSE value, the smaller the difference between estimated value and true value.

3.2. Multi-Head Attention Gating Unit

One defect of the traditional MLP model is that it ignores the interdependencies between different time steps in the sequence data when processing sequence data. The presented model aims to solve this defect by adding MAGU to the MLP model. MAGU mainly consists of three parts, namely MHA, gating operation, and linear transformation. The MHA can extract features from different subspaces in parallel and better capture the dependencies between time series data. The gating operation can selectively control the information flow by ignoring irrelevant information or amplifying important information, enabling the improved MLP model to better process sequence data. A linear transformation can map input data into a new representation space, which usually includes matrix multiplication operations and offset addition operations.

The MAGU network structure in Figure 4 mainly includes an input layer, a multi-head attention layer, a gating subunit, a linear transformation operation, and an output layer. The gating subunit contains a linear layer and a sigmoid activation function, which can generate gating weights through the sigmoid function to achieve gating operations. The linear transformation layer includes a linear transformation layer and a linear activation function. Linear transformation refers to linear transformation of the input data and the hidden state of the previous time step, and then mapping them to a new feature space.

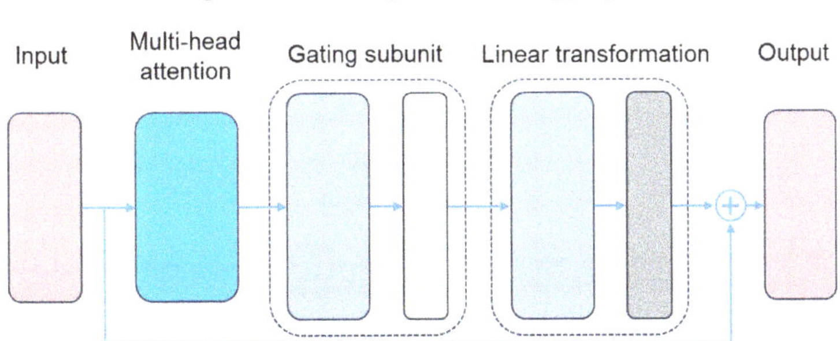

Figure 4. Network structure diagram of the multi-head attention gating unit.

During the forward propagation, the model works as follows: First, an input is received. Then, it is passed to the multi-head attention layer. Subsequently, it is passed to the gating subunit. In the gating subunit, the gating signal is calculated by the sigmoid activation function, and after the calculation is completed, the input is passed to the linear transformation layer. Afterward, the input is linearly transformed in the linear transformation layer. The final output is the result of multiplying the gating signal by the transformation result and adding the input residual.

MAGU is a module that combines MHA and gating operation to enhance the model's ability to process time series data. MHA can capture complex dependencies in the input sequence. Gating operation can improve the nonlinear expression ability of the model by dynamically adjusting the flow of information. Next, the two important components of the MAGU module, the MHA and the gating operation, will be introduced in detail.

3.2.1. Multi-Head Attention Mechanism

The MHA mechanism can be utilized to capture long-term dependencies in the input sequence [34]. Its principle is to adopt multiple attention heads in parallel to capture the features of different subspaces in the input sequence. First, each attention head independently performs a linear transformation on the input. Then, the attention weights are calculated. Finally, the outputs of all heads are concatenated and linearly transformed to obtain the final output.

According to Figure 5, the MHA is mainly composed of four linear layers, a scaled dot product attention layer, and a connection layer. The scaled dot product attention layer includes dot product operation, scaling calculation and softmax operation. The execution process of the MHA is to first pass the input vector through three different linear layers to generate the query matrix, key matrix, and value matrix, respectively. Afterward, the attention score matrix is obtained by calculating the dot product of the query matrix and the key matrix for each attention head. Then, the obtained attention score matrix is divided by a scaling factor to prevent the gradient from disappearing or exploding. Subsequently, a softmax operation is performed on the scaled attention score matrix to obtain the attention weight matrix. Then, the attention weight matrix is multiplied by the value matrix to obtain the weighted value matrix. Repeat the above steps for each attention head. Finally, the outputs of all attention heads are connected and passed through a linear layer to obtain the final output of the MHA mechanism.

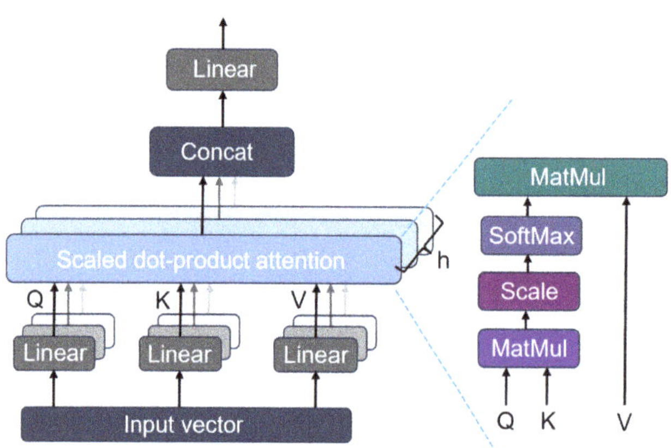

Figure 5. Structure diagram of the multi-head attention mechanism. Q, K, V represent the query, key, and value of the attention head, respectively.

Assume that the input vector passes through a linear layer to obtain the query, key, and value of each attention head.

$$Q_i = xW_i^Q, K_i = xW_i^K, V_i = xW_i^V \tag{13}$$

where W expresses the weight matrix and the subscript i represents the i-th attention head.

The output of each attention head can be expressed as

$$H_i = \text{softmax}\left(\frac{Q_i K_i^T}{\sqrt{d_k}}\right) V_i \tag{14}$$

where d_k represents the dimension of the key vector.

Then, the outputs of all attention heads are concatenated and passed through another linear transformation to obtain the final output.

$$M(Q, K, V) = \text{Concat}(h_1, \ldots, h_n)W^O \tag{15}$$

where W^O means the weight of the last linear transformation layer, and h denotes the total number of attention heads.

The presented model can focus on information at different positions in the input sequence through the MHA, which improves the proposed model's ability to understand and express the input data. It is worth noting that traditional recurrent neural networks, which are commonly used for time series feature extraction, cannot be computed in parallel, while the MHA can be computed in parallel.

3.2.2. Gating Operation

The gating mechanism regulates the flow of information by introducing control signals. The gating weights are generated by the sigmoid function, and the input and the transformed results are weighted and summed to achieve feature screening. Since the sigmoid function has a smooth curve characteristic, it can perform a nonlinear transformation on the input and help the model adapt to nonlinear relationships to learn more complex features.

The gating operation consists of a linear layer and a sigmoid activation function. Its execution process is to first receive the input from the MHA. Then, the input is linearly transformed through the linear layer. Subsequently, the gating weight is generated through the sigmoid activation function. Finally, the gating weight is passed to the next layer of the network.

The output after the gating operation can be expressed as

$$g_i = \sigma(W_g H_i + b_g) \tag{16}$$

where W_g denotes the weight matrix of the gating operation, b_g is the bias vector of the gating unit, H_i expresses the output of the MHA in the corresponding layer, and σ represents the sigmoid activation function.

The sigmoid activation function [35] is

$$f(x) = \frac{1}{1 + e^{-x}} \tag{17}$$

where e means the basis of the natural logarithm. The sigmoid function can map the input x into a continuous value between 0 and 1.

Since the activation function in the gating operation is differentiable, this facilitates the use of the backpropagation algorithm to calculate the gradient and adopt the optimizer to update the model parameters during training to minimize the loss function. The gating operation can promote the stable propagation of gradients, which can alleviate the problem of gradient vanishing or gradient exploding to a certain extent.

The MHA helps the model focus on important features. The gating operation can further filter the features. The adoption of MHA and gating operation enables the MLP model to effectively capture long-term dependencies in the input data, thereby improving the performance of the model.

4. Experimental Process and Result Analysis

To clearly demonstrate the performance of the proposed method, three types of experiments are conducted. First, the estimation accuracy of the presented method is comprehensively demonstrated. Then, the proposed method is compared with the most commonly used methods and the current state-of-the-art methods. Finally, an ablation experiment is performed to verify the effectiveness of the MAGU module.

Two datasets were adopted in the experiments. The datasets were collected on tree-lined streets and roads with wide views. The data were collected dynamically with a collection interval of 1 s. The true position information of the receiver was obtained from the NovAtel SPAN system. The adopted dataset contains code measurements and does not contain carrier phase measurements. The test environment is Python 3.9. The hardware is supported by RTX 3090 GPU, whose total memory is 24 GB. The first dataset was collected and publicly released by Fu et al., which consists of GNSS positioning data collected by multiple low-cost receivers in the San Francisco area of the United States [36]. The second dataset was collected and publicly released by Kuratomi et al. [27]. Both datasets were used for comparative experiments and ablation experiments. Applying the same dataset for comparison can improve the persuasiveness of the comparison experiment to a certain extent.

The experimental process of employing the first dataset is mainly as follows. First, the dataset collected and publicly released by Fu et al. was obtained. Secondly, the original positioning error of the receiver is obtained by the Kalman filter (KF) method. Then, the parameters that may affect the positioning error are extracted from the dataset as feature variables, and the obtained positioning error is adopted as the target variable. Finally, the feature variables and target variables are input into the improved MLP model together, and the estimated results of the positioning error are output after multiple iterations of training. Next, each step will be introduced in detail, and corresponding diagrams will be given.

4.1. Obtain the Original Positioning Error

The KF method is used to obtain the original positioning error in the first dataset. The positioning process not only includes a series of processing and analysis of satellite observation data and navigation message data, but also includes the correction and elimination of various error factors such as satellite clock deviation and ionospheric delay. The second dataset includes original positioning error data. Therefore, the step of calculating the original satellite positioning error can be ignored during the experiment using this dataset, and the subsequent steps can be directly carried out.

According to Figure 6, the calculated positioning error fluctuates greatly, with a minimum error of about 0.1 m and a maximum error of more than 6 m. The average positioning error is calculated to be 2.840 m. Notably, the purpose of the designed method is mainly to explore positioning error estimation, which does not depend too much on the accuracy of the original positioning error. Therefore, not only the KF method can be used to obtain the value of the positioning error, but also other GNSS positioning methods can be utilized to obtain the value of the positioning error.

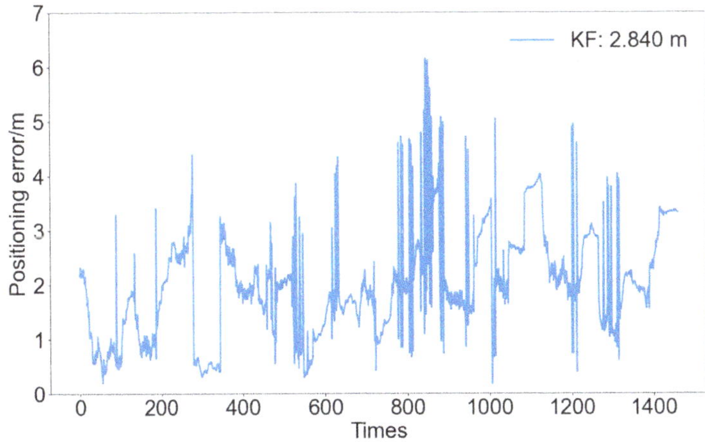

Figure 6. Positioning error results are obtained through the Kalman filter method.

4.2. Preprocessing the Experimental Dataset

The original satellite observation and navigation message files contain many parameters. However, not all of these parameters are directly related to the positioning error estimation. Therefore, the data in these files need to be preprocessed. Data preprocessing includes unifying the data value range and preliminary feature screening. Since the value ranges of different data are different, it is necessary to standardize the data. The standardization operation is to subtract the mean of the same type of data from the data to be standardized and divide it by the standard deviation, so as to achieve the purpose of scaling different data in proportion.

The main features in the dataset and their corresponding descriptions are listed in Table 1. These features are utilized in the training and testing processes of the improved MLP model. The quality of the feature selection strategy will directly affect the GNSS positioning error estimation accuracy of the presented method. The improved MLP model mainly selects features through MHA and gating operation. Then, the weights are updated according to the gradient, thereby gradually optimizing the performance of the model.

Table 1. The main features in the dataset and their corresponding descriptions.

Feature	Description
Satellite clock difference	The difference in time between an atomic clock on a satellite and a reference clock on Earth.
Satellite clock deviation	The offset of an atomic clock on a satellite from its theoretically precise time.
Satellite elevation	The angle between the satellite and the horizon where the receiver is located.
Ionospheric delay	The ionospheric delay means that when a satellite signal passes through the ionosphere, the free electrons in the ionosphere will scatter and reflect the satellite signal, thereby changing the propagation path of the satellite signal. The ionospheric delay is obtained by the Klobuchar model.
Satellite azimuth	The angle between the line from the ground observation point to the satellite and the due north direction.
Tropospheric delay	The tropospheric delay means that when a satellite signal passes through the atmosphere, the water vapor in it will cause the satellite signal to refract, thereby changing the propagation path of the satellite signal. The tropospheric delay is obtained by the Saastamoinen model.
Carrier-to-noise ratio	The ratio of the carrier power of the received signal to the noise power in the surrounding environment.
Satellite coordinate information	The satellite position information in the ECEF coordinate system is calculated using the satellite broadcast ephemeris.

4.3. Performance Analysis of the Improved MLP Model

Next, the improved MLP model is trained and tested with the two datasets. To demonstrate the performance of the presented model more intuitively, the root mean square error (RMSE) is adopted in the experiments as the evaluation index of the estimation accuracy of the model. The smaller the RMSE value is, the closer the estimated value is to the true value, and the higher the accuracy of the model. The formula for calculating the RMSE [37] is

$$\mathrm{RMSE} = \sqrt{\frac{1}{n}\sum_{i=1}^{n}(y_i - \hat{y}_i)^2} \tag{18}$$

where n denotes the number of samples, y_i represents the true value, and $\hat{y}_i$ represents the estimated value.

The x-axis and y-axis in Figure 7 are the training epochs and loss values of the model during the training process, respectively. A total of 500 epochs are iterated during the model training process. In accordance with Figure 7, the loss of the model is continuously reduced during the training process and is ultimately maintained at a low value, demonstrating that the model has converged normally and achieved better performance.

The general idea of GNSS positioning is adopted, and the output results of the model are divided into three components: East, North, and Up. Figure 8 shows the experimental results of the proposed model on the three components of East, North, and Up. It can be seen from Figure 8 that the trained and converged model has a high estimation accuracy.

After the two datasets are input into the model separately, the model gradually converges after several iterations of training. The presented method obtains RMSE values of 0.239 m and 0.029 m for the first and second datasets, respectively. The random forest (RF) algorithm is widely adopted in the field of GNSS positioning error estimation, and it has high estimation accuracy. Therefore, to demonstrate the performance of the presented method more clearly, the experimental results of the presented method and the RF algorithm will be displayed and comparatively analyzed next.

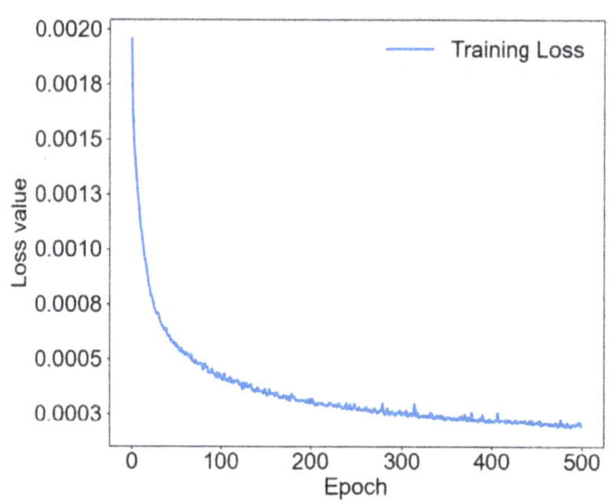

Figure 7. Image of the loss function of the presented model during training.

The gray portion in Figure 9 indicates the confidence interval for the positioning error estimate. The confidence interval is denoted by the area of the estimated value minus two standard deviations and the estimated value plus two standard deviations. The confidence interval is a concept from statistics, which means that the true value has a certain probability of falling within a certain interval around the estimated value. The confidence interval is an interval range and is often used to indicate the credibility of the estimated value. Applying confidence intervals to the problem of positioning error estimation is mainly used to describe the uncertainty range of positioning error estimation, which is conducive to evaluating the reliability and stability of the proposed model.

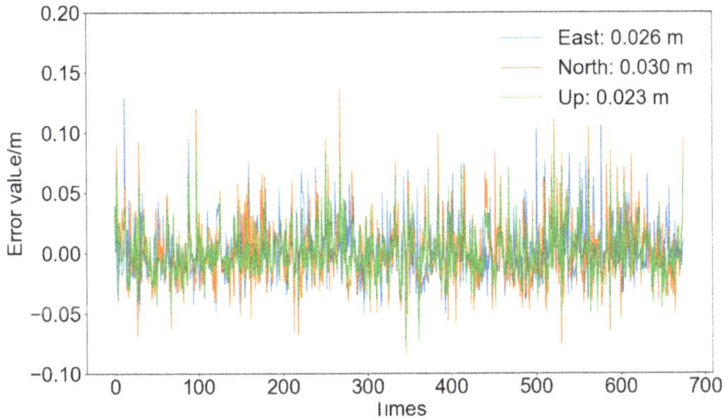

Figure 8. Experimental results of the proposed model on the three components of East, North, and Up.

As shown in Figure 9, the estimated values of the presented method have a large repetition range with respect to the true values, and they are basically within the confidence interval, which shows that the estimation results of the positioning errors of the presented method have high accuracy. However, the RF algorithm is sensitive to noise and outliers in the data, which affects the estimation accuracy of the algorithm. On the whole, the presented method has a larger repetition range of estimated values and true values compared to the RF algorithm. The estimation accuracy of the presented method is significantly better than the RF algorithm.

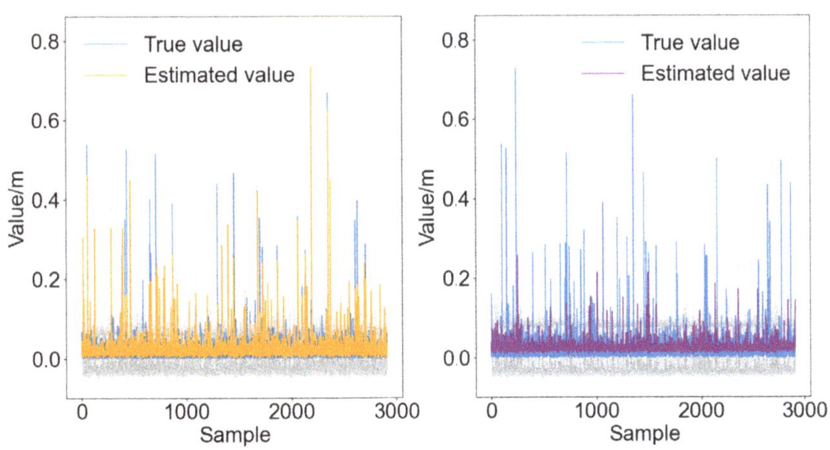

Figure 9. Comparison of the estimated and true values of the presented method (**left**) and the random forest algorithm (**right**).

It can be seen from Figure 10 that most of the estimation errors of the presented method are approximately distributed around 0, and the difference does not generally exceed 0.1. This shows that the estimated values obtained by the presented method are more accurate and have better performance. The RF algorithm is a model based on decision trees that cannot handle the interdependence in time series data well. Compared with the

RF algorithm, the presented method has a more concentrated error distribution and the overall error estimate is smaller.

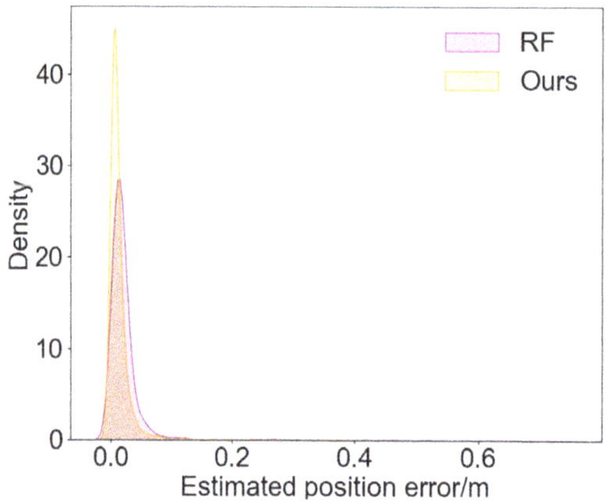

Figure 10. Comparison of the estimation error distributions between the presented method and the RF algorithm.

The Stanford diagram can show how well the estimated positioning error matches the original positioning error. The closer to the original positioning error the estimated positioning error is, the closer the data points are to the diagonal. According to Figure 11, most of the data points of the presented method are in the lower left corner of the graph, and only a very small number of data points are distributed on either side of the diagonal. This shows that the presented method obtains more accurate estimates. However, part of the data in the positioning error estimate obtained by the RF algorithm is biased, and there are a large number of data points where the positioning error estimate is less than the true value of the positioning error.

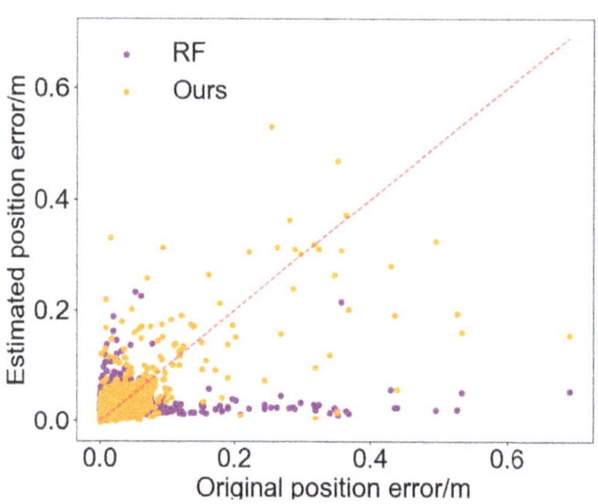

Figure 11. Stanford diagram with the presented method and the RF algorithm. The red dotted line indicates the diagonal line.

A residual is defined as the difference between the true value and an estimated value. The residual histogram clearly shows the residual distribution of the proposed method. The x-axis and y-axis in Figure 12 are the size and frequency of the residual value, respectively. Based on Figure 12, the residual histogram of the presented method has a higher kurtosis than the residual histogram of the RF algorithm, and the peak value is closer to 0. This shows that the residual distribution of the presented method is relatively concentrated, the overall value is small, and it has better estimation performance.

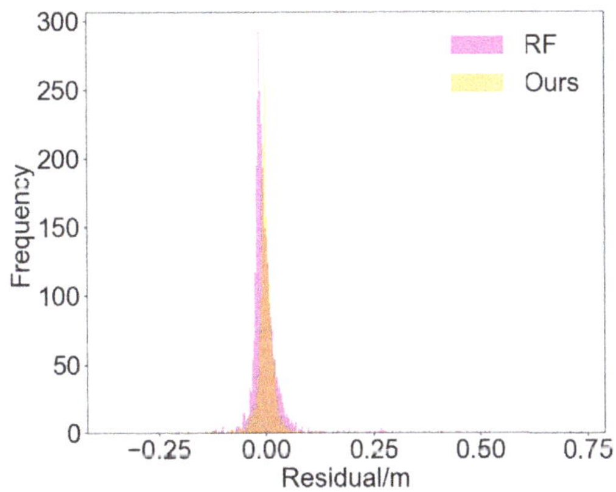

Figure 12. Residual histogram with the presented method and the RF algorithm.

Table 2 records the quantitative analysis results of the proposed method and the random forest algorithm. RMSE, mean absolute error (MAE), and standard deviation are used as evaluation indicators. It can be seen from Table 2 that the performance of the proposed method is better than that of the random forest algorithm in all three evaluation indicators.

Table 2. Quantitative analysis experimental results of the proposed method and random forest algorithm (unit: m).

Methods	RMSE Value	MAE Value	Standard Deviation
RF	0.047	0.022	0.016
Ours	0.029	0.014	0.010

4.4. Comparative Experiments

To better demonstrate the performance of the presented method, two comparative experiments are carried out. The first comparative experiment is between the presented method and other common methods, and the second comparative experiment is between the presented method and methods studied by other scholars.

4.4.1. Comparison of the Presented Method with Other Common Methods

Common methods in the GNSS positioning error estimation research mainly include the extended Kalman filter (EKF), linear regression (LR), support vector machine (SVM), decision tree (DT), and RF. These five common methods and the presented method are compared in the first comparative experiment.

By incorporating the MAGU module into the MLP model, the presented method minimizes the influence of unimportant features on the model, enabling the model to achieve better performance. It can be seen from Table 3 that the presented method obtains smaller RMSE values for both datasets and has obvious advantages over other common methods. This shows that the estimation accuracy of the presented method is higher.

Table 3. Comparative experimental results of the proposed method and five common methods (unit: m).

Methods	RMSE Value on the First, Dataset	RMSE Value on the Second, Dataset
EKF	1.324	0.118
LR	0.864	0.071
SVM	0.618	0.065
DT	0.527	0.060
RF	0.473	0.047
Ours	0.239	0.029

4.4.2. Comparison of the Presented Method with the Current State-of-the-Art Method

To our knowledge, the current state-of-the-art methods are the long short-term memory network model studied by Yang et al. [30] and the convolutional neural network model designed by Shen et al. [32] when dealing with the GNSS positioning error estimation problem. Therefore, the proposed method is compared with two state-of-the-art GNSS positioning error estimation methods to show the superiority of the presented method in the second comparative experiment.

The LSTM model is a method commonly applied to predict time series data. It usually consists of input gate, forget gate, output gate, and cell state. Compared with the MLP model, it requires additional gating units and loop connections. The CNN models are mostly used to process image data, and are not suitable for processing data containing geographic information. It can be seen from Table 4 that the RMSE value of the presented method is significantly smaller than that of the other two methods, especially compared with the LSTM method. The proposed method can adaptively choose which features to retain based on the importance of the input features. This characteristic can improve the representation ability and generalization performance of the model to a certain extent. Since the experiment was conducted with the same dataset, the results of the comparative experiment are convincing to a certain extent.

Table 4. Comparative experimental results of the proposed method and two state-of-the-art methods (unit: m).

Methods	LSTM [30]	CNN [32]	Ours
RMSE value on the first dataset	0.393	0.288	0.239
RMSE value on the second dataset	0.039	0.033	0.029

4.5. Ablation Experiment

To verify the effectiveness of the MHA mechanism and gating operation, ablation experiments are performed. The ablation experiment consists of four groups of experimental objects. The control variable method was adopted in the ablation experiment. Four groups of experimental subjects were tested separately under the premise that other parameters remained unchanged.

The MLP model without the MHA mechanism and the gating operation has difficulty effectively processing different features in the data due to the lack of feature importance screening ability. Therefore, it can be seen from Table 5 that the RMSE values obtained by the MLP model with the MHA mechanism and the gating operation are smaller than those of the MLP model without the MHA mechanism and the gating operation on both datasets. This shows that the proposed model gives full play to the advantages of the MHA mechanism and the gating operation, and can better learn and extract features in the data and obtain more accurate estimation accuracy. In addition, from the experimental results of the MLP model with only the gating operation and the MLP model with only the MHA mechanism, it can be seen that the MHA mechanism improves the model performance more than the gating operation.

Table 5. Ablation experiment results of the proposed model (unit: m).

Gating Operation	MHA	RMSE on the First, Dataset	RMSE on the Second, Dataset
		0.449	0.042
✓		0.427	0.041
	✓	0.398	0.038
✓	✓	0.239	0.029

5. Discussion

In applications that require high positioning accuracy, if the errors of the GNSS positioning system are not promptly warned or corrected, serious consequences may occur. The purpose of studying GNSS positioning error estimation methods is to evaluate and correct errors, ensure that GNSS can still provide high-quality positioning services in complex environments, and avoid large deviations in positioning results. As can be seen from Table 3, the proposed method achieves a lower RMSE value than the traditional method. This is because the traditional GNSS positioning error estimation method has the following shortcomings:

(1) Traditional positioning error estimation methods rely on precise assumptions, and have difficulty accurately reflecting the error size in complex environments. Therefore, the estimation accuracy is usually low.

(2) There are large differences in orbital errors, clock biases, and other factors between different satellite systems when multiple GNSS systems are used together. Traditional GNSS positioning error estimation methods have difficulty compensating for errors between multiple systems at the same time, and are prone to cause accumulation and cross-interference of errors between multiple systems.

However, the GNSS positioning error estimation method based on deep learning has powerful nonlinear modeling capabilities. It can learn the trend of error changes from historical data to effectively handle complex error factors, and has the potential to further improve accuracy with the support of large-scale data. Therefore, a new multi-layer perceptron model is proposed for GNSS positioning error estimation. Multi-head attention mechanisms and gating operations are designed to improve the standard MLP model. The improved MLP model can better capture the relationship between input features by combining multi-head attention mechanism and gating mechanism. It can be seen from the two types of comparative experiments that the proposed model not only has better performance than other commonly used methods, but also has obvious advantages over the current state-of-the-art methods.

Our study aims to demonstrate the potential of the presented method in dealing with real-time position error estimation problems. However, in actual applications, factors such as data transmission delay, data reliability, and application performance requirements may need to be considered. In the face of building occlusion in urban environments and signal interference in natural environments, additional measures need to be taken to deal with data uncertainty to improve the stability and reliability of position estimation.

6. Conclusions

Traditional GNSS positioning error estimation methods rely on prior assumptions about observation noise and motion models, and have difficulty handling the complex nonlinear relationships between different error factors. However, the GNSS positioning error estimation method based on deep learning does not require prior assumptions. It can extract effective information from a large amount of historical data, and capture the complex nonlinear relationship between different error factors, thereby improving the accuracy of positioning error estimation. Therefore, a GNSS positioning error estimation method based on deep learning is explored. The nonlinear modeling ability and feature selection ability of a MLP model are improved by introducing a multi-head attention mechanism and gating operations into the MLP model. Experimental results show that the presented method has better performance than that of the extended Kalman filter, linear regression, decision tree, support vector machine, and other methods for two datasets, and the presented method is also better than the current state-of-the-art long short-term memory network and convolutional neural network considering comparative experiments. This shows that the presented method can reduce the influence of errors on the positioning results to a certain extent, and has more accurate positioning error estimation capabilities. Currently, the proposed model only estimates the positioning error based on GNSS data. However, other sensors, such as IMU and visual sensors, also provide valuable information in practical applications. In the future, we will explore how to fuse data from different sensors to further improve the error estimation accuracy of the model in complex environments.

Author Contributions: Conceptualization, X.L., Z.T. and J.W.; methodology, X.L.; validation, X.L., Z.T. and J.W.; formal analysis, X.L.; investigation, X.L. and Z.T.; writing—original draft preparation, X.L., Z.T. and J.W.; writing—review and editing, X.L. and J.W.; visualization, X.L.; supervision, X.L. and Z.T.; All authors have read and agreed to the published version of the manuscript.

Funding: This work is supported by the National Natural Science Foundation of China (Nos. 62171191 and 62271223).

Data Availability Statement: The raw data supporting the conclusions of this article will be made available by the authors on request.

Acknowledgments: We thank the editor and reviewers for their time and energy in reviewing our manuscript and providing insightful comments, which greatly improved our manuscript.

Conflicts of Interest: The authors declare no competing interests.

References

1. Dautermann, T. Civil air navigation using GNSS enhanced by wide area satellite based augmentation systems. *Prog. Aerosp. Sci.* **2014**, *67*, 51–62. [CrossRef]
2. Li, Z.; Wang, L.; Wang, N.; Li, R.; Liu, A. Real-time GNSS precise point positioning with smartphones for vehicle navigation. *Satell. Navig.* **2022**, *3*, 19. [CrossRef]
3. Melbourne, T.; Szeliga, W.; Santillan, V.; Scrivner, C. Global navigational satellite system seismic monitoring. *Bull. Seismol. Soc. Amer.* **2021**, *111*, 1248–1262. [CrossRef]
4. Siemuri, A.; Selvan, K.; Kuusniemi, H.; Valisuo, P.; Elmusrati, M.S. A systematic review of machine learning techniques for GNSS use cases. *IEEE Trans. Aerosp. Electron. Syst.* **2022**, *58*, 5043–5077. [CrossRef]
5. Jacobsen, K.; Dähnn, M. Statistics of ionospheric disturbances and their correlation with GNSS positioning errors at high latitudes. *J. Space Weather Space Clim.* **2014**, *4*, A27 . [CrossRef]
6. Chen, J.P.; Wang, J.G.; Wang, A.H.; Ding, J.S.; Zhang, Y.Z. A SHAtropE—A regional gridded ZTD model for China and the surrounding areas. *Remote Sens.* **2020**, *11*, 165. [CrossRef]
7. Won, D.; Ahn, J.; Lee, S.; Lee, J.; Sung, S.; Park, H.; Park, J.; Lee, Y. Weighted DOP with consideration on elevation-dependent range errors of GNSS satellites. *IEEE Trans. Instrum. Meas.* **2012**, *61*, 3241–3250. [CrossRef]
8. Park, J.; Veettil, S.; Aquino, M.; Yang, L.; Cesaroni, C. Mitigation of ionospheric effects on GNSS positioning at low latitudes. *Navigation* **2017**, *64*, 67–74. [CrossRef]

9. Ge, H.; Meng, G.; Li, B. Zero-reconvergence PPP for real-time low-Earth satellite orbit determination in case of data interruption. *IEEE J. Sel. Top. Appl. Earth Observ. Remote Sens.* **2024**, *17*, 4705–4715. [CrossRef]
10. Han, S.; Kwon, J.; Jekeli, C. Accurate absolute GPS positioning through satellite clock error estimation. *J. Geod.* **2001**, *75*, 33–43. [CrossRef]
11. Hoque, M.; Jakowski, N. Estimate of higher order ionospheric errors in GNSS positioning. *Radio Sci.* **2008**, *43*, 1–15. [CrossRef]
12. Tang, L.J.; Wang, J.G.; Zhu, H.Z.; Ge, M.R.; Xu, A.G.; Schuh, H. A comparative study on the solar radiation pressure modeling in GPS precise orbit determinatio. *Remote Sens.* **2021**, *13*, 3388. [CrossRef]
13. Pesyna, K.; Humphreys, T.; Heath, R.; Novlan, T.; Zhang, J. Exploiting antenna motion for faster initialization of centimeter-accurate GNSS positioning with low-cost antennas. *IEEE Trans. Aerosp. Electron. Syst.* **2017**, *53*, 1597–1613. [CrossRef]
14. Wang, J.; Han, H.; Liu, F.; Cheng, X. Performance analysis of GNSS/MIMU tight fusion positioning model with complex scene feature constraints. *J. Geodesy Geoinf. Sci.* **2021**, *4*, 1–13.
15. Huang, G.W.; Cui, B.B.; Zhang, Q.; Fu, W.J.; Li, P.L. An improved predicted model for BDS ultra-rapid satellite clock offsets. *Remote Sens.* **2018**, *10*, 60. [CrossRef]
16. Lesouple, J.; Robert, T.; Sahmoudi, M.; Tourneret, J.; Vigneau, W. Multipath mitigation for GNSS positioning in an urban environment using sparse estimation. *IEEE Trans. Intell. Transp. Syst.* **2019**, *20*, 1316–1328. [CrossRef]
17. Jiang, M.; Qin, H.; Zhan, C.; Sun, G. LEO Doppler-aided GNSS position estimation. *GPS Solut.* **2022**, *26*, 31. [CrossRef]
18. Wu, M.J.; Wang, L.; Xie, W.; Yue, F.; Cui, B.B. Performance evaluation and application field analysis of precise point positioning based on different real-time augmentation information. *Remote Sens.* **2024**, *16*, 1349. [CrossRef]
19. Baselga, S.; García-Asenjo, L. GNSS differential positioning by robust estimation. *J. Surv. Eng.* **2008**, *134*, 21–25. [CrossRef]
20. Azarbad, M.; Mosavi, M. A new method to mitigate multipath error in single-frequency GPS receiver with wavelet transform. *GPS Solut.* **2014**, *18*, 189–198. [CrossRef]
21. Niu, X.; Dai, Y.; Liu, T.; Chen, Q.; Zhang, Q. Feature-based GNSS positioning error consistency optimization for GNSS/INS integrated system. *GPS Solut.* **2023**, *27*, 89. [CrossRef]
22. Marais, J.; Nahimana, D.; Viandier, N.; Duflos, E. GNSS accuracy enhancement based on pseudo range error estimation in an urban propagation environment. *Expert Syst. Appl.* **2013**, *40*, 5956–5964. [CrossRef]
23. Biswas, S.; Qiao, L. Effect of PDOP on performance of Kalman Filters for GNSS-based space vehicle position estimation. *GPS Solut.* **2017**, *21*, 1379–1387. [CrossRef]
24. Gao, W.; Li, Z.; Chen, Q.; Jiang, W.; Feng, Y. Modelling and prediction of GNSS time series using GBDT, LSTM and SVM machine learning approaches. *J. Geod.* **2022**, *96*, 71. [CrossRef]
25. Abdolkarimi, E.; Mosavi, M. A modified neuro-fuzzy system for accuracy improvement of low-cost MEMS-based INS/GPS navigation system. *Wirel. Pers. Commun.* **2023**, *129*, 1369–1392. [CrossRef]
26. Closas, P.; Fernandez-Prades, C.; Fernandez-Rubio, J. Maximum likelihood estimation of position in GNSS. *IEEE Signal Process. Lett.* **2007**, *14*, 359–362. [CrossRef]
27. Kuratomi, A.; Lindgren, T.; Papapetrou, P. *Prediction of Global Navigation Satellite System Positioning Errors with Guarantees*; Lecture Notes in Computer Science; Springer: Cham, Switzerland, 2021; Volume 12460.
28. Jiang, C.; Chen, S.; Chen, Y.; Shen, J.; Liu, D.; Bo, Y. Superior position estimation based on optimization in GNSS. *IEEE Commun. Lett.* **2021**, *25*, 479–483. [CrossRef]
29. Bao, L.; Luo, H.; Gao, X.; Ning, B.; Zhao, F.; Zhu, Y. MT-eR: NMEA protocol-assisted high-accuracy navigation algorithm based on GNSS error estimation using multitask learning. *IEEE Trans. Intell. Transp. Syst.* **2022**, *23*, 20464–20475. [CrossRef]
30. Yang, S.; Ben, T.; Xiang, W. Build up a real-time LSTM positioning error prediction model for GPS sensors. In Proceedings of the IEEE 90th Vehicular Technology Conference 2019, Honolulu, HI, USA, 22–25 September 2019; pp. 1–5.
31. Liu, S.; Xiang, W.; Elangovan, V. A GPS distance error forecast model based on IIR filter de-noising and LSTM. *IEEE Trans. Instrum. Meas.* **2023**, *72*, 1–12.
32. Shen, N.; Cheng, L.; Wang, L.; Chen, R. GNSS Site unmodeled error prediction based on machine learning. *GPS Solut.* **2023**, *27*, 77. [CrossRef]
33. Banerjee, C.; Mukherjee, T.; Pasiliao, E. An empirical study on generalizations of the ReLU activation function. In Proceedings of the ACM Southeast Conference, Kennesaw, GA, USA, 18–20 April 2019; pp. 164–167.
34. Vaswani, A. Attention is all you need. In Proceedings of the 31st Conference on Neural Information Processing Systems (NIPS 2017), Long Beach, CA, USA, 4–9 December 2017.
35. Narayan, S. The generalized sigmoid activation function: Competitive supervised learning. *Inform. Sci.* **1997**, *99*, 69–82. [CrossRef]

36. Fu, G.; Khider, M.; Diggelen, F. Android raw GNSS measurement datasets for precise positioning. In Proceedings of the ION GNSS, Online, 22–25 September 2020; pp. 1925–1937.
37. Karunasingha, D.S.K. Root mean square error or mean absolute error? Use their ratio as well. *Inform. Sci.* **2022**, *585*, 609–629. [CrossRef]

remote sensing MDPI

Article

Analysis of Multi-GNSS Multipath for Parameter-Unified Autocorrelation-Based Mitigation and the Impact of Constellation Shifts

Wenhao Xiong [1], Yumiao Tian [1,*], Xiaolei Dai [2], Qichao Zhang [1], Yibing Liang [1] and Xiongwei Ruan [1]

1 Faculty of Geosciences and Engineering, Southwest Jiaotong University, Chengdu 611756, China; xiongwh@my.swjtu.edu.cn (W.X.); zqcswjtu@my.swjtu.edu.cn (Q.Z.); l1ang@my.swjtu.edu.cn (Y.L.); ruanxiongwei@my.swjtu.edu.cn (X.R.)
2 GNSS Research Center, Wuhan University, Wuhan 430079, China; xldai@whu.edu.cn
* Correspondence: yumiao.tian@swjtu.edu.cn

Abstract: Multipath effects can significantly reduce the accuracy of GNSS precise positioning. Traditional methods, such as sidereal filtering and grid-based approaches, attempt to model and mitigate these errors by leveraging the spatial autocorrelation of multipath based on residuals. However, these methods can only approximately handle spatial autocorrelation data, limiting their effectiveness. This study investigates the spatial cross-correlation of residuals between various GNSS frequency bands, analyzes their covariance function parameters, and evaluates the impact of constellation shifts on long-term multipath mitigation. Based on this, a simplified autocorrelation-based approach utilizing unified covariance parameters for multipath mitigation is proposed, with its efficacy assessed for both short- and long-term applications. The study demonstrates the correlation of multipath effects across various GPS and Galileo frequencies, including GPS L1/L2/L5 and Galileo E1/E5a/E5b/E5ab/E6, by analyzing correlation coefficients. A strong correlation (greater than 0.8) is observed between residuals of closely spaced frequencies, such as E5b and E5ab, despite their frequency differences. Additionally, the covariance parameters of the residuals are found to be consistent across all frequencies for a baseline, suggesting that unified parameters can be applied effectively for spatial autocorrelation-based multipath mitigation without sacrificing performance. The orbit shifts of certain GPS satellites, particularly G02, G20, and G21, result in significant changes in orbital parameters and satellite tracks, reducing the effectiveness of long-term multipath mitigation. However, the impact of GPS orbit shifts can be minimized through periodic model updates or by integrating GPS and Galileo modeling. In experiments, the LSC correction strategy based on a GPS/Galileo combination, utilizing unified parameters, outperforms the grid method based on the GPS/Galileo combination, improving the mean residual variance elimination rate by 11.3% for GPS L1 and 11.4% for Galileo E1. These improvements remain consistent, with rates of 11.3% and 15.7%, respectively, even on DOY 365, which is 327 days after the modeling data were collected.

Keywords: multi-GNSS; multipath; least-squares collocation; orbit shift; spatiotemporal correlation

1. Introduction

The signal received by the GNSS receiver antenna is a combination of the direct line-of-sight (LOS) signal and signals reflected and/or diffracted from nearby objects. The distortion of the desired LOS signal, caused by undesired non-LOS signals, results in tracking errors in the receiver and subsequent code and phase measurement errors, with the maximum theoretical magnitude extending to 1/2 of the ranging code chip and 1/4 of the carrier wavelength [1]. While notable advancements have been achieved in mitigating various systematic errors in high-precision GNSS positioning, multipath propagation continues to persist as a prominent error source due to its inherent challenges in parameterization and its resistance to differencing techniques.

The most straightforward way to mitigate multipath effects is by selecting observation environments that minimize reflections near the receiver antenna. However, this approach is often impractical, especially in complex environments. As a result, researchers have explored various methods to address multipath errors, including both hardware enhancements and software processing techniques. Hardware solutions typically focus on improving antenna designs to reduce multipath, such as using right-handed circular polarization antennas [2], choke ring antennas [3], helix antennas [4], and narrow correlator delay-locked loops [5]. While these approaches can lessen the impact of multipath, they are only partially effective and often come with significant costs. Software-based methods involve developing analytical or simulated models for calibration. These approaches include weighting GNSS observation data based on the signal-to-noise ratio [6,7] or carrier-to-noise-power-density ratio [1,8]. Additionally, multipath effects can be extracted or suppressed in the frequency domain using filtering techniques. Common methods include the Vondrak filter [9], wavelet analysis [10,11], and the Kalman filter [12].

Additionally, two methods based on the spatiotemporal repeatability of multipath are commonly used for real-time mitigation in static or quasi-static environments. The first, known as the sidereal filtering (SF) method, exploits the near-daily repetition of GNSS satellite constellations. This repetition leads to recurring multipath effects on almost a daily basis, allowing residuals from post-processing to be applied to correct current measurements [13–15]. Essentially, SF uses constellation repetition to identify the closest residuals from post-processing and applies them to mitigate the multipath error in the current observation. The second method is the grid-based technique, which mitigates multipath effects by stacking residuals within azimuth and elevation grids, usually $1^\circ \times 1^\circ$ [16–20]. Both methods aim to model and calibrate multipath errors by leveraging spatial autocorrelation.

However, these methods handle spatial autocorrelation data only approximately, limiting their effectiveness. Actually, spatial autocorrelation can be more accurately modeled using covariance functions, applied through techniques such as Least-Squares Collocation (LSC). These functions can be homogeneous and isotropic, depending only on distance and exhibiting rotational symmetry, or anisotropic, where spatial orientation affects their values [21–24]. This approach significantly improves the precision of spatial autocorrelation modeling compared to traditional sidereal and grid methods.

Moreover, the growing number of satellites and the increasing frequency overlap among GNSS systems have made signal compatibility and interoperability key trends in high-precision data processing for multi-GNSS applications. These advancements offer new perspectives on improving multipath correction. Research by Lu et al. [25] highlights a strong correlation in the residuals of overlapping frequencies across GPS, BDS-3, and Galileo at the same spatial location. Similarly, Geng et al. [26] demonstrate that the interoperability of GPS, Galileo, and BDS-3 enhances spatial resolution, improves the efficiency of multipath modeling, and boosts correction performance. These findings suggest that using overlapping frequencies can significantly improve multipath mitigation. However, there remains considerable interest in exploring the common multipath characteristics across various GNSS frequencies, including non-overlapping ones. Understanding these characteristics could further enhance the effectiveness of spatial autocorrelation-based multipath mitigation strategies.

Additionally, due to various perturbation factors, all satellite orbits experience shifts [27–29], which result in the residual values of satellites after an orbit drift no longer accurately representing the multipath errors of their pre-drift positions. Consequently, multipath modeling must be reconstructed, and new residuals from post-processing are required to account for these shifts. In contrast, if the constellations remain stable and the surrounding environment of the station remains unchanged, a single model can be used consistently over time without the need for updates. However, it remains unclear whether all satellite constellations drift far enough to significantly impact multipath mitigation. Therefore, investigating the extent of constellation drifts within GNSS systems and their potential effects on multipath mitigation is of particular interest.

To overcome the limitations of traditional multipath mitigation methods and better leverage the spatial autocorrelation of GNSS data, this study proposes a simplified autocorrelation-based multipath mitigation method utilizing unified covariance parameters. Specifically, this method investigates the spatial cross-correlation of residuals across different GPS and Galileo frequencies, analyzing the relationship between the covariance function parameters of the residuals and evaluating the impact of constellation shifts on long-term multipath mitigation. Based on this analysis, the method applies multi-GNSS data for multipath mitigation, simplifying the modeling process while maintaining the performance, thereby improving the short- and long-term positioning accuracy.

This study introduces the multipath modeling techniques and the mathematical tools used for analysis in Section 2. Section 3 explores the cross-correlation of residuals across different GNSS frequencies, the spatial autocorrelation parameters of these residuals, and an analysis of orbit shifts. The study concludes with a summary of the findings in Section 4.

2. Multipath Modelling and Mathematic Tools for Analysis

2.1. Multipath Models

The equation for GNSS carrier phase observations is given by

$$\phi_{r,i}^{s} = \rho_{r}^{s} + c \cdot (\delta t_{r} - \delta t^{s}) - I_{r,i}^{s} + T_{r}^{s} + \lambda_{i} N_{r,i}^{s} + b_{r,\phi_i} - b_{\phi_i}^{s} + M_{r,\phi_i}^{s} + \varepsilon_{r,\phi_i}^{s} \tag{1}$$

where $\phi_{r,i}^{s}$ is the carrier phase observation value on frequency point i; the superscript s represents the GNSS satellite PRN (Pseudo-Random Noise) number; the subscript r represents the receiver identifier; ρ_{r}^{s} is the geometric distance between the satellite and the receiver; c is the speed of light in a vacuum; δt_r and δt^s, respectively, represent the receiver clock errors and satellite clock errors; $I_{r,i}^{s}$ is the ionospheric delay; T_{r}^{s} denotes the tropospheric delay; λ_i is the carrier wavelength and $N_{r,i}^{s}$ is the ambiguity; b_{r,ϕ_i} and $b_{\phi_i}^{s}$ are the receiver phase bias and satellite phase bias, respectively; M_{r,ϕ_i}^{s} is the multipath errors; and $\varepsilon_{r,\phi_i}^{s}$ is the noise term and other unmodeled errors.

Double-difference is a commonly adopted strategy in short baseline positioning, which effectively mitigates correlated errors at both the receiver and satellite ends. Consequently, the carrier phase observation equation is transformed as follows.

$$\nabla\Delta\phi_{r_1,r_2,i}^{s_1,s_2} = \nabla\Delta\rho_{r_1,r_2}^{s_1,s_2} + \lambda_i \nabla\Delta N_{r_1,r_2,i}^{s_1,s_2} + \nabla\Delta M_{r_1,r_2,\phi_i}^{s_1,s_2} + \nabla\Delta\varepsilon_{r_1,r_2,\phi_i}^{s_1,s_2} \tag{2}$$

where $\nabla\Delta$ is the double-difference operator, s_1 and s_2, respectively, represent any two satellites from the same GNSS system, and r_1 and r_2 are the two receivers located at the endpoints of the baseline. If the station coordinates are known with high precision, the baseline vector and integer ambiguity can be solved, leaving only the residual term primarily composed of the multipath effect and measurement noise.

$$\nabla\Delta l_{r_1,r_2}^{s_1,s_2} = \nabla\Delta M_{r_1,r_2,\phi_i}^{s_1,s_2} + \nabla\Delta\varepsilon_{r_1,r_2,\phi_i}^{s_1,s_2} \tag{3}$$

To avoid the impact of reference satellite changes on the residuals, we applied a "zero-mean" constraint to obtain the single-difference carrier-phase residuals from double-difference residuals [26]. We therefore have

$$\begin{bmatrix} w_1 & w_2 & w_3 & \cdots & w_n \\ 1 & -1 & 0 & \cdots & 0 \\ 1 & 0 & -1 & \cdots & 0 \\ \vdots & \vdots & \vdots & \ddots & \vdots \\ 1 & 0 & 0 & \cdots & -1 \end{bmatrix} \begin{bmatrix} \Delta l_{r_1,r_2}^{s_1} \\ \Delta l_{r_1,r_2}^{s_2} \\ \Delta l_{r_1,r_2}^{s_3} \\ \vdots \\ \Delta l_{r_1,r_2}^{s_n} \end{bmatrix} = \begin{bmatrix} 0 \\ \nabla\Delta l_{r_1,r_2}^{s_1,s_2} \\ \nabla\Delta l_{r_1,r_2}^{s_1,s_3} \\ \vdots \\ \nabla\Delta l_{r_1,r_2}^{s_1,s_n} \end{bmatrix} \tag{4}$$

where $w_1, w_2, \ldots, w_n$ refer to the weight of the satellite; and $\Delta l_{r_1,r_2}^{s_1}, \Delta l_{r_1,r_2}^{s_2}, \ldots, \Delta l_{r_1,r_2}^{s_n}$ are the single-difference residuals which will be used to construct the multipath models.

2.2. Interpolation with Grid Method

The grid method for GNSS multipath mitigation models the spatial distribution of multipath errors by partitioning the sky into a grid structure based on azimuth and elevation angles. The method assumes that multipath errors exhibit spatial repeatability, meaning that errors observed at a particular position in the sky can be used to estimate future errors in the same or nearby locations. This approach allows for more precise error correction over time. Mathematically, the grid method can be described as follows.

First, the sky is divided into grids with resolutions determined by azimuth angle θ and elevation angle φ. For example, a $1^\circ \times 1^\circ$ grid resolution means that the sky is divided into cells of size $\Delta\theta = 1^\circ$ and $\Delta\varphi = 1^\circ$, resulting in $N_{az} \times N_{el}$ total grid cells, where N_{az} and N_{el} are the number of divisions in azimuth and elevation, respectively.

Then, over multiple observation days, the post-processing residuals $l(t, \theta, \phi)$, where t represents time, are collected for each grid cell. These residuals capture the differences between the observed and true GNSS signals, with the assumption that multipath errors dominate these residuals. For each grid cell, we compute the average residual:

$$\bar{l}(\theta_i, \phi_i) = \frac{1}{n}\sum_{k=1}^{n} l(t_k, \theta_i, \phi_i) \quad (5)$$

where $l(t_k, \theta_i, \phi_i)$ is the residual at time t_k for the grid cell defined by azimuth θ_i and elevation ϕ_i, and n is the number of observations for that grid.

During subsequent observations, the estimated multipath error for each satellite position is corrected by subtracting the corresponding average residual from the raw GNSS observation. For a satellite positioned at azimuth θ_i and elevation ϕ_i, the corrected observation O_{corr} is

$$O_{corr}(t) = O_{obs}(t) - \bar{l}(\theta_i, \phi_i) \quad (6)$$

where $O_{obs}(t)$ is the raw observation at time t.

In summary, the grid method mitigates GNSS multipath errors by constructing a spatial model of the errors through grid partitioning and residual averaging. This mathematical framework helps to reduce multipath-induced inaccuracies in GNSS observations by using past data to predict and correct future signal distortions.

2.3. Interpolation with Moving-Neighborhood LSC

LSC, widely used in physical geodesy, is a statistical estimation technique employed to separate systematic trends, spatially correlated signals, and random noise in observation data. Originally developed for geoid determination, LSC is based on the spatial autocorrelations of observables, such as gravity, which are described by a covariance function [21–23]. This function can be homogeneous and isotropic [30], meaning its value depends solely on distance and exhibits rotational symmetry, or it can be anisotropic. The primary goal of LSC is to predict unknown values at unmeasured locations (or times) using known observations, accounting for both spatial correlation and random noise. By leveraging the covariance characteristics of the data, LSC provides an optimal estimation of the underlying signal, making it a powerful tool for improving accuracy in geodetic applications.

It decomposes the observed values into three components: the trend, signal, and noise [31]. The observation equation, represented by a vector matrix, can be expressed as

$$l = AX + t + e \quad (7)$$

where l is the measurement vector of the samples, A denotes the design matrix of the trend model, X represents the non-random parameters, t stands for the autocorrelated random parameters at the observation points, and e is the random noise.

Assuming that the other signals at the prediction points are vectors s correlated with the signal t, despite their potential spatial differences, the solution for the unknown signal vector s is derived based on the minimum principle of LSC given as [23]

$$\text{Min}\left(e^T C_{ee}^{-1} e + s^T C_{ss}^{-1} s\right) \tag{8}$$

where C_{ee} is the variance-covariance (VC) matrix of the noise e, and C_{ss} is the VC matrix of the signal s.

If the trend is neglected or has been calibrated, the estimate of the unknown signal s is expressed by [23]

$$\hat{s} = C_{st} C_{ll}^{-1} l \tag{9}$$

where C_{ll} is the VC matrix of the measurements l and has $C_{ll} = C_{tt} + C_{ee}$, and C_{st} refers to the covariance matrix between the signal s at the prediction point and the signal t at the measurement point. Basically, the estimated $\hat{s}$ is predicted based on the spatial correlation between the observed and predicted points. The elements in the matrices C_{ll} and C_{st} are computed using the covariance function, which is determined based on the distances between the sample points and the current satellite. Specifically, the covariance function describes how the spatial correlation between measurements decreases with increasing distance.

LSC considers both random and non-random components in observations. It can be applied when observations are spatially correlated, which is often the case in geodetic and GNSS applications. LSC allows for accurate interpolation of values at unmeasured points by leveraging spatial autocorrelation.

2.4. Covariance Function

The covariance function is a method used to describe the spatial correlation between data points, where the covariance between any two measured values can be represented as the function $C(d)$, and d represents the distance between the two points. Assuming there are two points, x_i and x_j, in the spatial domain with a distance of d_{ij}, their covariance can be expressed as

$$C(x_i, x_j) = C(d_{ij}) \tag{10}$$

The function $C(d)$ decreases as the distance d increases, attaining its maximum value at zero distance and approaching zero as the distance tends to infinity. Commonly used covariance functions that satisfy these conditions include the Gaussian function, Markov's function, the Hirvonen function, and so on [31]. Markov's function can be expressed by

$$C(d) = C_0 e^{-d/d_0} \tag{11}$$

where C_0 represents the covariance at zero distance, d_0 is the coefficient of the covariance function, and d is the distance between two points.

Furthermore, when the observed values are numerical data, the empirical covariance function (ECVF) can be computed and expressed numerically [24], which is commonly employed in practice for LSC. It can be written as

$$\begin{cases} C_0 = \frac{1}{m}\sum_{i=1}^{m} l_i l_i \\ C(d_k) = \frac{1}{m_k}\sum_{i<j} l_i l_j \end{cases} \tag{12}$$

where C_0 is the variance when distance d equals zero, d_k is the distance between two points i and j, m is the number of point pairs at distance d_k, and the terms l_i and l_j represent the multipath values at points i and j, respectively.

In this study, we computed the covariance at various distance intervals using the ECVF. The procedure for obtaining the ECVF can be referenced from Tian et al. [24]. Next, we employed Markov's function to fit the covariance coefficients, and the parameters C_0 and d_0 in Equation (11) are derived.

2.5. Correlation

The Pearson correlation coefficient (ρ) is employed to analyze the correlation between residual sequences of any two frequencies. The formula for ρ is given by [32]

$$\rho_{XY} = \frac{cov(X,Y)}{\sigma_X \sigma_Y} \tag{13}$$

where X and Y represent two sets of residual sequences; and cov and σ denote the covariance and standard deviation operators, respectively. Typically, a ρ value between 0.8 and 1.0 indicates a very strong correlation, 0.6 to 0.8 represents strong correlation, 0.4 to 0.6 is moderate, 0.2 to 0.4 is weak, and 0.0 to 0.2 reflects very weak or no correlation.

Calculating the correlation between residuals for frequencies within the same system is straightforward, as the satellite positions remain consistent across all observations. However, for different systems, this process becomes more complex due to the non-overlapping satellite trajectories.

To solve this, we first generate uniformly distributed positions close to the trajectories of both systems, ensuring a maximum distance of 0.02 rad from each trajectory. Next, the multipath values at these positions are interpolated for each system's residuals using a spatial autocorrelation-based method. Once the multipath values are obtained for both systems at the uniformly distributed positions, the correlation coefficients are calculated.

The positions are illustrated in Figure 1, where the blue and green solid lines represent the satellite trajectories from different systems, and the red-dotted triangles indicate the uniformly distributed sample points located within 0.02 rad of both trajectories.

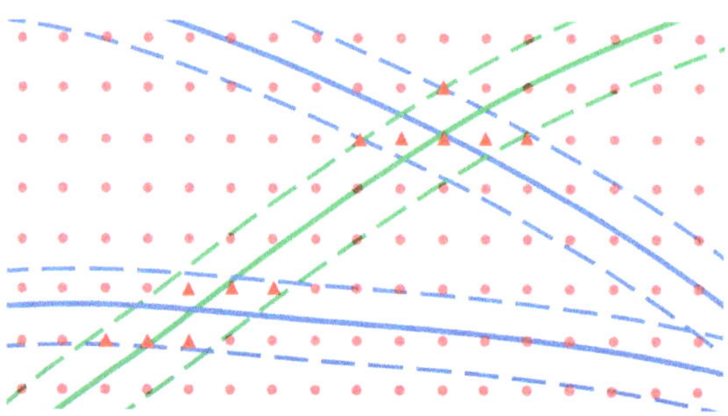

Figure 1. Schematic diagram of the data locations used to calculate the correlation between residuals from different frequencies. The blue and green solid lines represent satellite tracks from two distinct systems, while the blue and green dashed lines indicate the 0.02 radians boundaries to the satellite tracks. All red points, including the red circles and red triangles, represent uniformly distributed samples generated for the correlation analysis. The red triangles are the points that fall within the 0.02 radians overlapping region around the satellite tracks of both systems, i.e., the region where the dashed lines overlap. The red circles represent points that fall outside the overlapping region and are therefore excluded from the analysis.

The spatial correlation characteristics of multipath are analyzed to enhance its mitigation in multi-GNSS, multi-frequency observations. This involves calculating the correlation coefficient between different GNSS frequencies, examining the spatial autocorrelation of residuals, and identifying consistent modeling parameters for the covariance function. Based on these findings, a spatial autocorrelation-based method utilizing a unified covariance function is proposed. Additionally, the impact of orbit shifts on long-term multipath error mitigation is evaluated, and strategies to reduce this impact are assessed. The flowchart of the procedure is plotted in Figure 2.

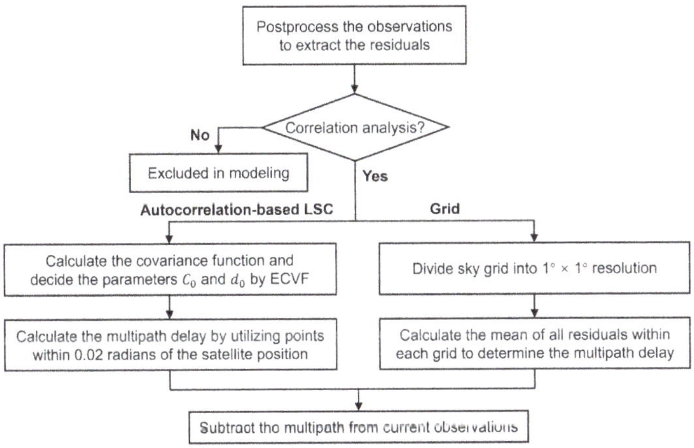

Figure 2. Flowchart of multipath analysis and mitigation.

3. Computation and Analysis

3.1. Employed Data

The observations from the short baselines MATE_MAT1 and KERG_KRGG, provided by the International GNSS Service, are utilized to analyze and validate methods for mitigating multipath effects. The lengths of these baselines are 10.7 m and 8.6 m, respectively, with a sampling interval of 30 s. The antenna photographs of the four stations are shown in Figure 3. Figure 4 illustrates the multipath distribution for both baselines, revealing that the errors are predominantly concentrated in low-elevation areas. The presence of a large metallic structure near station MAT1 leads to significant multipath errors in the MATE_MAT1 baseline, particularly around an elevation angle of 20° toward the southeast. Additionally, the KERG_KRGG baseline shows higher-frequency multipath errors in low-elevation regions, primarily due to the influence of ancillary equipment, such as antenna mounts [33].

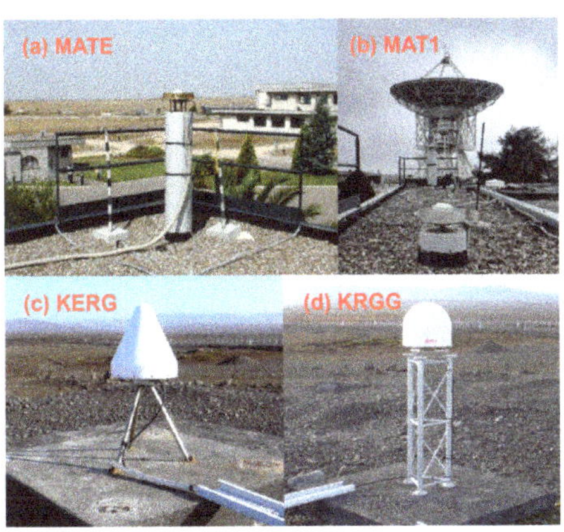

Figure 3. Antenna photographs of MATE_MAT1 (**a**,**b**) and KERG_KRGG (**c**,**d**). Photographs are from www.sonel.org (accessed on 14 October 2024).

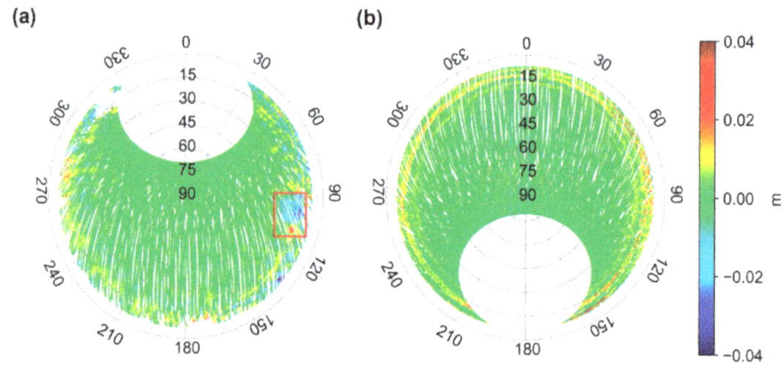

Figure 4. Multipath distributions for the baselines MATE_MAT1 (**a**) and KERG_KRGG (**b**). The red boxed area in (**a**) represents the significant multipath errors of the MATE_MAT1 baseline around an elevation angle of 20° toward the southeast.

Based on the satellite orbit repeat period of GPS and Galileo [13,34,35], we utilize 10 days of observation data from Days of the Year (DOY) 029–038 in 2023 for analysis and modeling, which are then applied to mitigate the multipath effects on DOY 039 and subsequent days. During this period, the GPS and Galileo frequencies available for each baseline are listed in Table 1, with specific frequency values. During data processing, a satellite elevation cutoff angle of 7° is applied, and the igs20_2309.atx antenna file is used to correct the phase center offsets (PCOs) and phase center variations (PCVs) for both satellites and receivers. For the KERG_KRGG baseline, due to the lack of Galileo PCO/PCV correction values, GPS PCO/PCV correction values are used as substitutes.

Table 1. The frequency values of the available open service signals transmitted by GPS and Galileo for the two baselines. "√" means that the baseline has this frequency, and "×" means that the baseline does not.

Frequency (MHz)	Signal		Baselines	
	GPS	Galileo	MATE_MAT1	KERG_KRGG
1575.420	L1	E1	√	√
1278.750		E6	×	√
1227.600	L2		√	√
1207.140		E5b	√	√
1191.795		E5ab *	√	√
1176.450	L5	E5a	√	√

* i.e., E5(E5a + E5b).

3.2. Correlation Analysis Between GPS/Galileo Frequencies

The carrier phase SD residual distribution for all frequencies of the MATE_MAT1 and KERG_KRGG baselines over the 10-day period from DOY 029–038 in 2023 is presented in Figures 5 and 6. The SD residuals for satellites at low elevation angles (below 30°) are relatively large, a common feature associated with the heightened multipath effects in this region [36]. Additionally, it is noted that the number of satellites observed at the E5a frequency for the KERG_KRGG baseline is relatively low when the elevation angle is below 15°. This reduction is primarily due to the lack of low-elevation satellite data at the KERG station for this frequency.

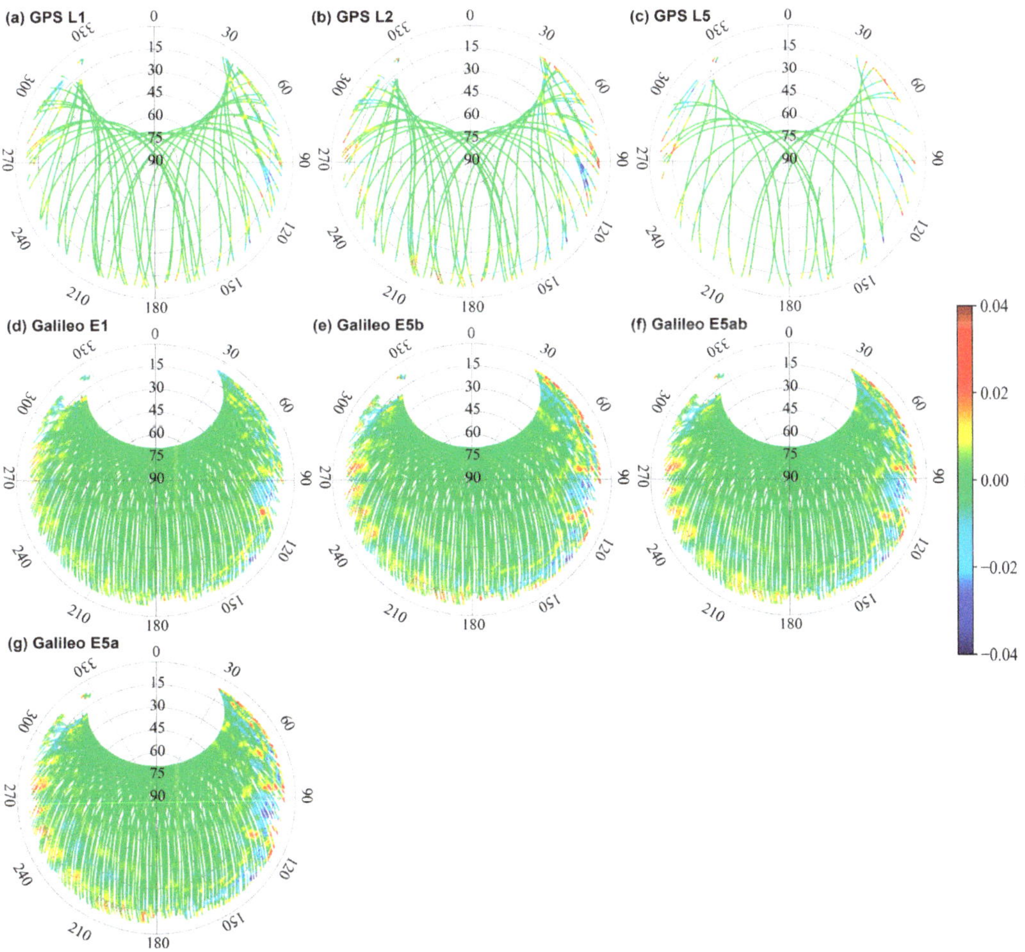

Figure 5. Satellite tracks and SD residuals for all frequencies of the baseline MATE_MAT1 on DOY 029–038 of 2023.

The residuals for the GPS L1/L2/L5 and Galileo E1/E5b/E5ab frequencies over a 10-day period, as the satellites traversed a designated square window, are presented in Figure 7, offering a comparative view of the signal variations across different GNSS frequencies. The windows for the MATE_MAT1 baseline cover an elevation angle of 7° to 17° and an azimuth angle of 90° to 100°, while for the KERG_KRGG baseline, the range is 7° to 17° in elevation and 140° to 150° in azimuth. A consistent pattern is observed for different frequencies at the same spatial locations. Notably, for overlapping or closely spaced frequencies, the satellite trajectory colors appear similar or nearly identical. For instance, this applies to overlapping frequencies such as GPS L1 and Galileo E1, closely spaced frequencies like GPS L2 and Galileo E5b, as well as GPS L5 and Galileo E5ab, all of which follow the same pattern. In contrast, for frequencies with larger differences, such as GPS L1 and Galileo E5b/E5ab, the trajectory colors vary significantly, and in some instances, even exhibit opposite sign values.

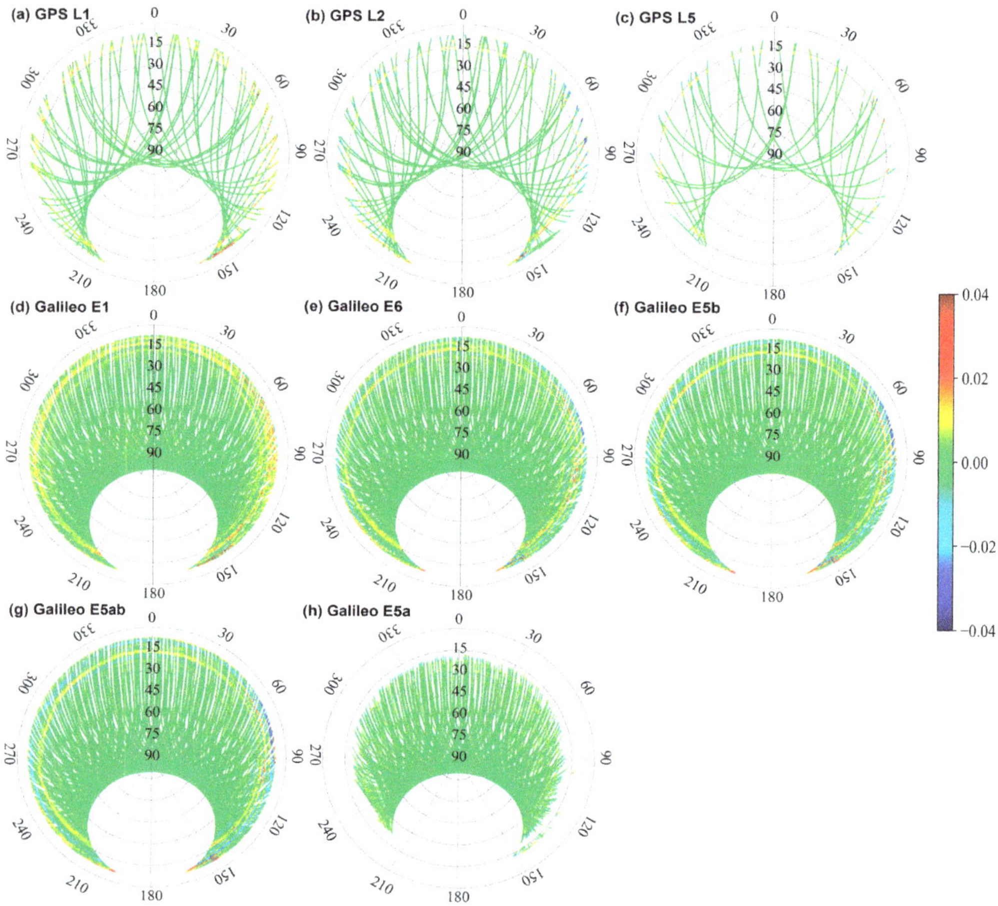

Figure 6. Satellite tracks and SD residuals for all frequencies of the baseline KERG_KRGG on DOY 029–038 of 2023.

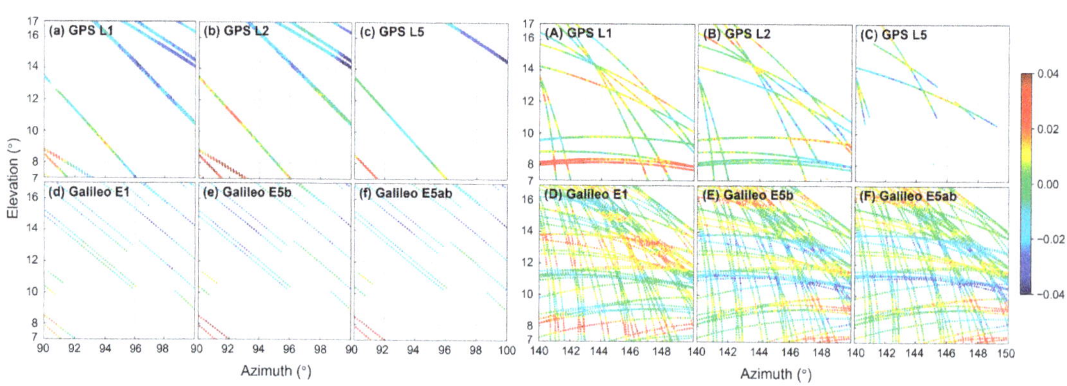

Figure 7. SD residuals for different frequencies of the MATE_MAT1 baseline ((**a**–**f**), 7°–17° in elevation and 90°–100° in azimuth) and the KERG_KRGG baseline ((**A**–**F**), 7°–17° in elevation and 140°–150° in azimuth).

The Pearson correlation coefficient is used to evaluate the correlation between the residual sequences of any two frequencies. The absolute values of the resulting correlation coefficients are displayed in Figure 8, showing diagonal symmetry along the value of 1. To better visualize the strength of multipath error correlations between frequency pairs, a color gradient is applied, ranging from dark (strong correlation) to light (weak correlation). It is clear from the figure that correlations exist between multipath errors for all frequency pairs, though the strength of this correlation decreases as the frequency differences increase. Additionally, the relationship between frequency difference and the absolute value of the correlation coefficient is illustrated in Figure 9, offering a more intuitive visualization of how frequency differences affect correlation strength.

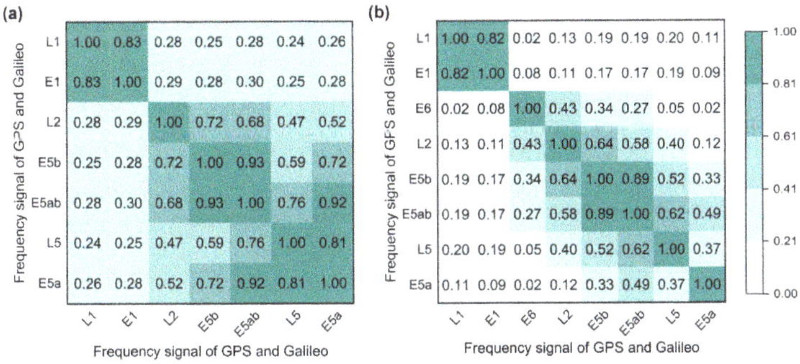

Figure 8. Absolute values of correlation coefficients between all frequencies of the baselines MATE_MAT1 (**a**) and KERG_KRGG (**b**). Utilizing a color gradient ranging from dark to light to represent the hierarchical levels of correlation strength, spanning from very strong to very weak or no correlation.

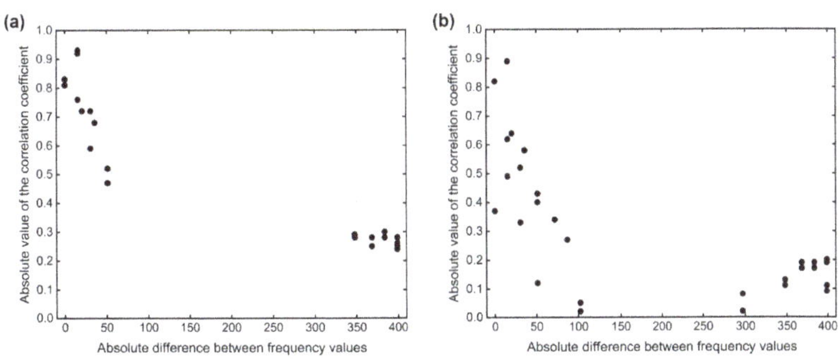

Figure 9. Variation of all correlation coefficients with frequency differences for the baselines MATE_MAT1 (**a**) and KERG_KRGG (**b**).

Due to the lack of low-elevation observations at the Galileo E5a frequency from the KERG station, the low correlations between E5a and the other frequencies in the KERG_KRGG baseline may not accurately reflect the true correlations. Apart from this, the residuals from frequency pairs, such as overlapping ones like L1/E1 and L5/E5a between GPS and Galileo, as well as non-overlapping frequencies like L2 and E5b, and the Galileo frequencies E5a, E5b, and E5ab, exhibit relatively large correlation coefficients, indicating strong correlations. Other frequencies also demonstrate strong to moderate correlations between narrow-spaced frequencies. These strong correlations suggest that their residuals contain similar multipath errors, supporting the potential for combining frequencies to enhance multipath error correction.

3.3. Parameters of Covariance Function

The covariance function used in this study is Markov's function by Equation (11), where C_0 and d_0 determine the function. These parameters for different frequencies are initially calculated by fitting the theoretical function to the empirical covariance function and then compared across the frequencies.

The computed and fitted ECVF curves for the GPS L1 and Galileo E1 frequencies, from the MATE_MAT1 and KERG_KRGG baselines on DOY 038 in 2023, are shown in Figure 10 as an example. The curves of other frequencies are similar to these two frequency signals. Figure 11 presents the computed values of C_0 and d_0 over a 10-day modeling period for both the GPS and Galileo frequencies for the two baselines.

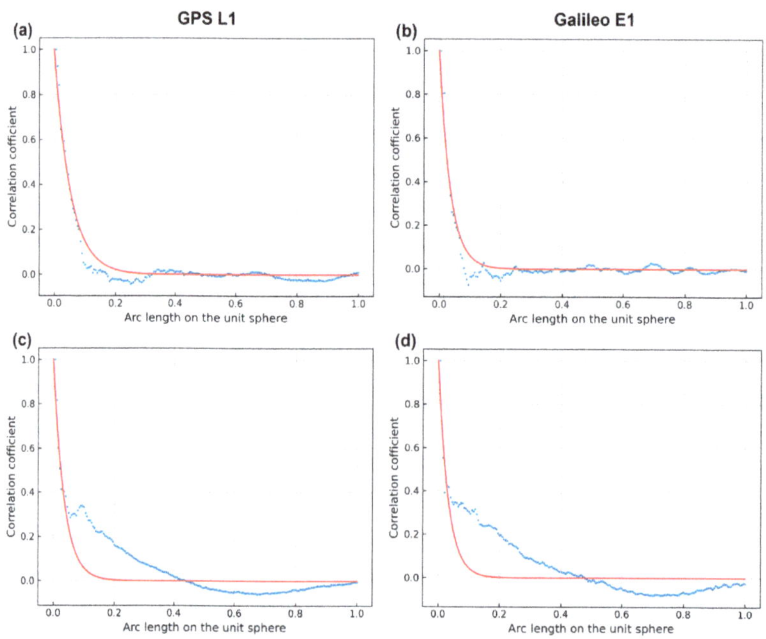

Figure 10. Theoretical covariance functions and ECVFs of GPS L1 and Galileo E1 for MATE_MAT1 (**a**,**b**) and KERG_KRGG (**c**,**d**) on DOY 038 of 2023. The blue dot curve is the ECVF derived with the GPS L1 (**a**,**c**) and Galileo E1 (**b**,**d**) SD residuals. The red curve is the fitted curve.

The C_0 values for the MATE_MAT1 baseline are slightly larger than those for the KERG_KRGG baseline, though the ranges for both baselines overlap significantly. In contrast, the d_0 parameter for the MATE_MAT1 baseline is notably larger than for KERG_KRGG, which is expected due to the higher spatial frequency of multipath observed in the KERG_KRGG baseline.

Additionally, the deviations of parameters relative to the averages of each single baseline and both baselines combined are computed and shown in Figure 12. Due to the absence of low-elevation data with large residuals at the E5a frequency for the KERG_KRGG baseline, the calculated covariance parameters are not representative of the multipath errors, leading to the exclusion of this frequency from the analysis. As anticipated, the differences between the original parameter values and the averages for each baseline are relatively small (Figure 12), while the deviations from the combined averages of both baselines are slightly larger. Nonetheless, the differences in C_0 for this case remain small, around 1×10^{-5}, and the d_0 differences stay within 0.02.

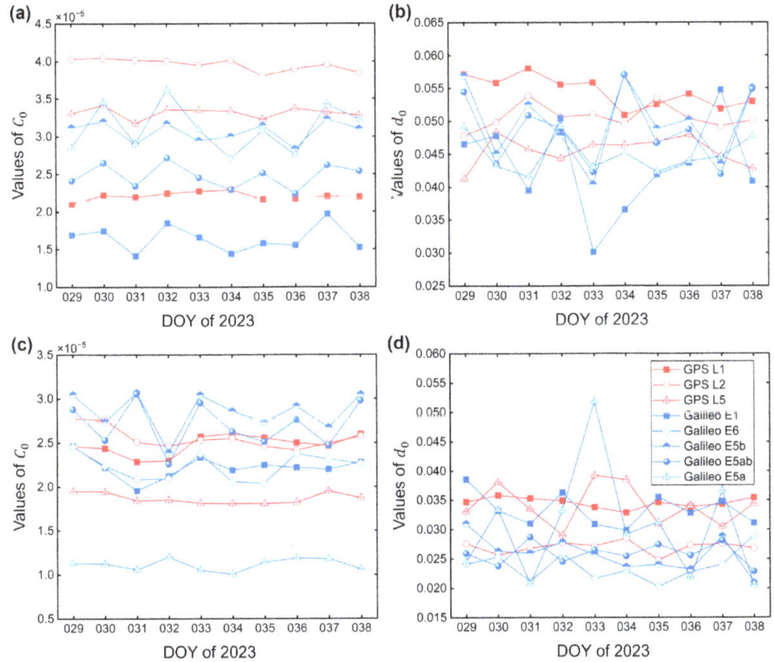

Figure 11. Daily C_0 and d_0 parameters for MATE_MAT1 (**a**,**b**) and KERG_KRGG (**c**,**d**) on DOY 028–039 of 2023.

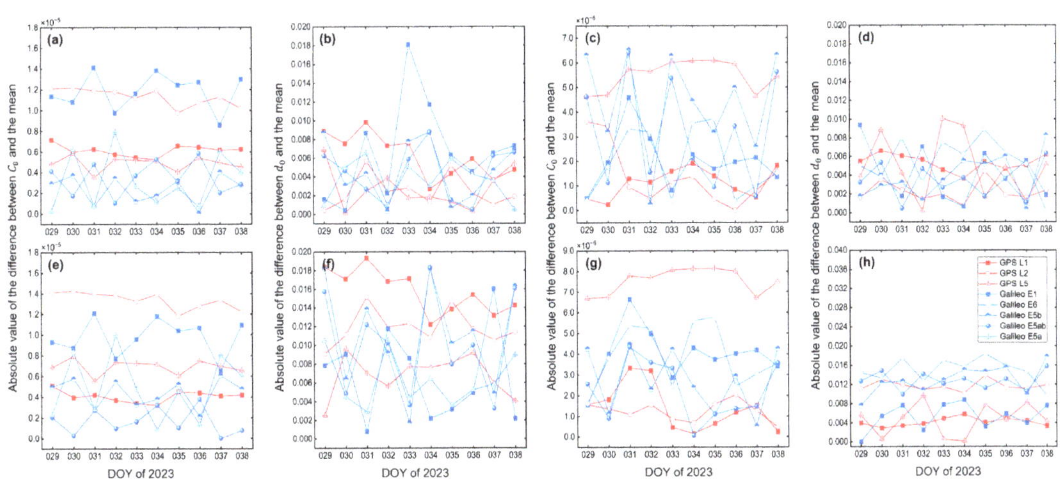

Figure 12. Deviations of the parameters respective to the average of each single baseline (**a–d**) and both baselines (**e–f**) are computed, separately, for MATE_MAT1 (**a**,**b**,**e**,**f**) and KERG_KRGG (**c**,**d**,**g**,**h**).

The parameters C_0 and d_0 of all frequency residuals are unified by averaging all frequency parameters across two baselines. These unified parameters are then applied in a spatial autocorrelation-based method for multipath mitigation. To evaluate performance, GPS L1 frequency observations from both baselines are analyzed. The residual variance reduction rates after applying the unified parameters are displayed in Figure 13, alongside rates using average parameters for each baseline and for each frequency individually.

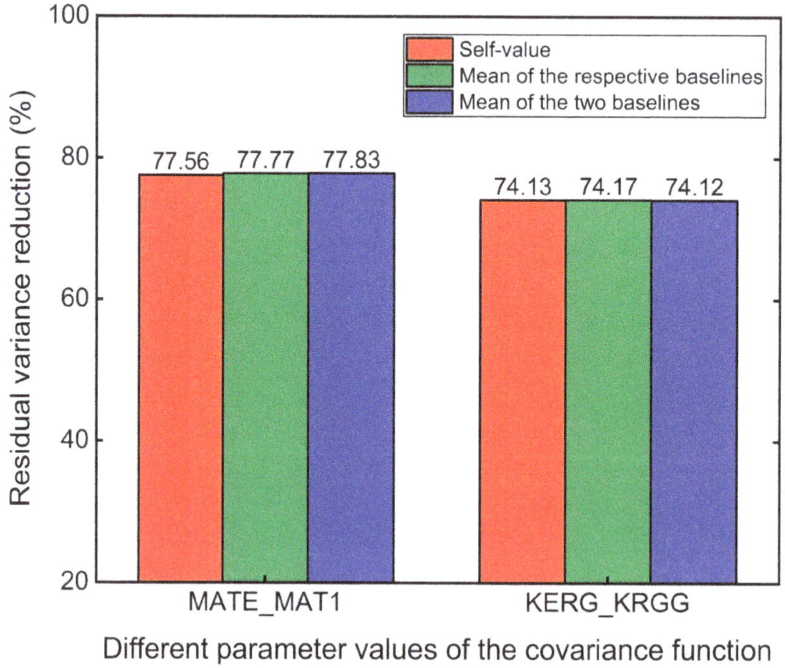

Figure 13. Variance reduction rates of SD residuals after multipath correction using different parameter values. The results of the parameters C_0 and d_0 by averaging all frequency parameters across two baselines (mean of the two baselines), averaging parameters for each baseline (mean of the respective baselines), and for each frequency individually (self-value) are compared.

The results demonstrate that the variance reduction rates achieved with the unified parameters are comparable to those obtained with baseline-specific averages. This suggests that using unified covariance function parameters for different baselines and frequency signals is effective, simplifying the computational process by eliminating the need to estimate the parameters C_0 and d_0 individually. Moving forward, these unified parameters will be employed in subsequent research efforts.

3.4. Orbit Shift over a Long Time Span

Due to various perturbations, satellite orbits exhibit drift, with variations across different satellites and systems. Figures 14 and 15 depict the satellite tracks of all GPS and Galileo satellites observable at the MATE station over a four-year period, from 1 January 2020 to 31 December 2023. In these figures, the satellite tracks for 2020 are shown in red, 2021 in blue, 2022 in orange, and 2023 in purple, visually representing the orbit drift across the years.

Within the Galileo system, satellites E14 and E18 show more pronounced orbit drift compared to others, indicating less stable orbits. Conversely, many GPS satellites exhibit significant orbit drift, notably satellite G02 in 2022 and 2023, G20 in 2021, and G21 in 2020, with their drift far exceeding that observed in the Galileo system. Specifically, substantial orbital drift was observed for G02 during July–November 2022 and April–June 2023, G20 during March–July 2021, and G21 during May–November 2020. Figure 16 illustrates the variation trends of selected orbital parameters—namely eccentricity, Right Ascension of the Ascending Node (RAAN), orbital inclination, argument of perigee, and semi-major axis—for these three satellites during the noted periods. Significant adjustments in these orbital parameters occurred during times of pronounced deviations in satellite trajectories.

The significant changes in the semi-major axis correspond to a substantial variation in the satellite's orbit repeat period. Consequently, if substantial changes occur in the orbital parameters such as the semi-major axis, the residual data from previous periods may not accurately represent the current multipath errors due to significant spatial discrepancies between the expected and actual satellite positions, leading to inaccuracies in multipath error correction.

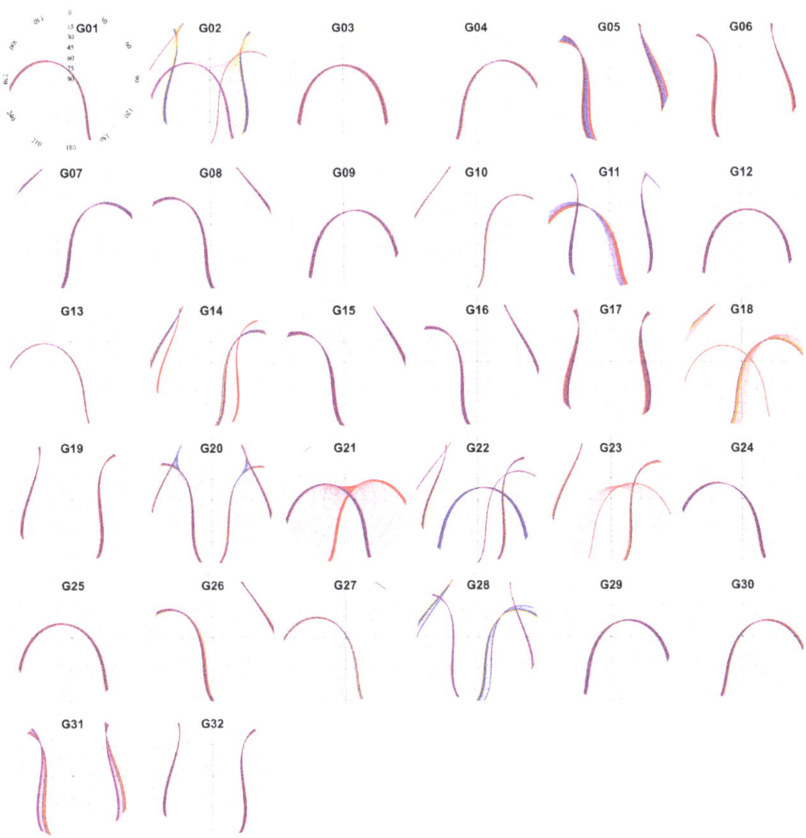

Figure 14. Satellite tracks of all GPS satellites observable at the MATE location over the four-year period from 2020 to 2023. The tracks are color-coded by year: 2020 in red, 2021 in blue, 2022 in orange, and 2023 in purple.

Utilizing orbital data from DOY 029–038 of 2023 as reference orbits for both the GPS and Galileo systems, the average unit spherical distance from the positions of satellites G01–G32 and E01–E36 during DOY 039–365 of 2023 to their nearest reference orbit was calculated. Figure 17a,b demonstrates these deviations. The average deviation distance for Galileo satellites remains relatively stable, peaking at only 0.004 rad, while that for GPS satellites shows a marked increase, with satellite G02 reaching a maximum deviation of 0.029 rad. Interestingly, around DOY 249, the average deviation for GPS satellites stabilizes.

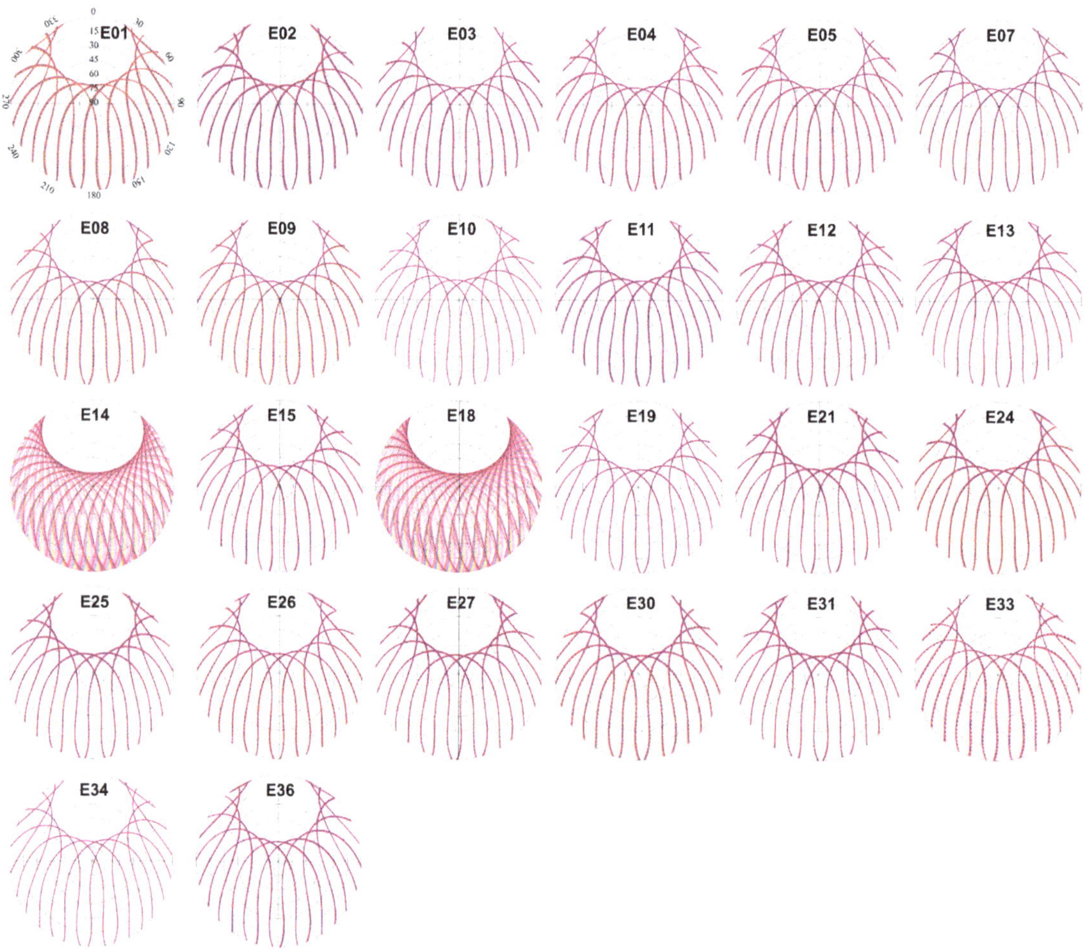

Figure 15. Satellite tracks of all Galileo satellites observable at the MATE location over the four-year period from 2020 to 2023. The tracks are color-coded by year: 2020 in red, 2021 in blue, 2022 in orange, and 2023 in purple.

By recalculating the average unit spherical distance using orbital data from DOY 239–248 as a new reference for the period DOY 249–365, the results, shown in Figure 18, reveal a reduction in the maximum average deviation for GPS satellites to 0.017 rad, which is significantly lower than earlier measurements. This adjustment indicates that using more recent data as reference orbits leads to more accurate multipath error correction, suggesting that outdated historical data may not effectively correct for current GPS multipath errors.

Based on the analysis in Section 3.2, there is a notable correlation in multipath errors between the GPS and Galileo systems, with overlapping or narrowly-spaced frequencies showing very strong correlation. Consequently, integrating these two systems could enhance the density of orbit trajectories, potentially mitigating the negative impacts of GPS orbit shifts on long-term multipath correction.

Using orbital data from both GPS and Galileo systems from DOY 029–038 of 2023, a combined reference orbit was established. We calculated the average unit spherical distance from the positions of satellites G01–G32 and E01–E36 to this combined orbit for the same period. The results, shown in Figure 17c,d, reveal that while the average deviation distance for Galileo satellites from the combined orbit remains stable at 0.004 rad, the deviation for GPS satellites shows a significant reduction to 0.003 rad. Additionally, these distances do not exhibit significant long-term fluctuations.

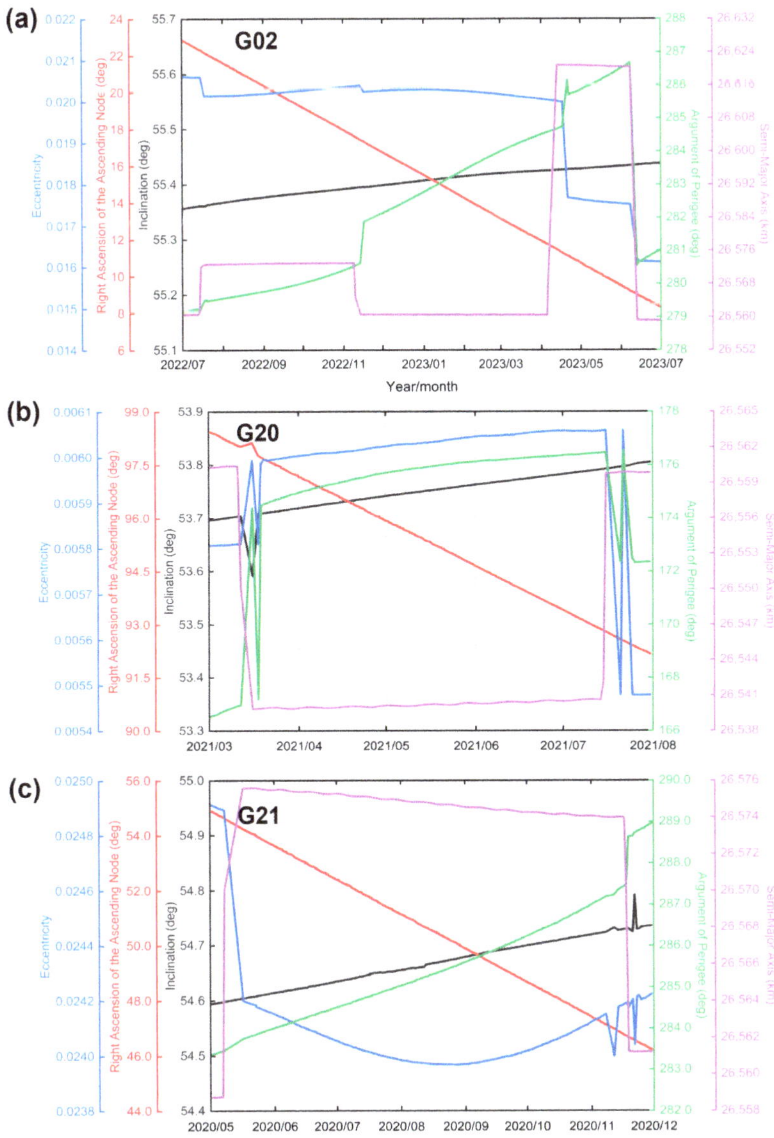

Figure 16. Eccentricity (the blue line), Right Ascension of the Ascending Node (the red line), orbital inclination (the black line), argument of perigee (the green line), and semi-major axis (the purple line) for satellites G02 (**a**), G20 (**b**) and G21 (**c**).

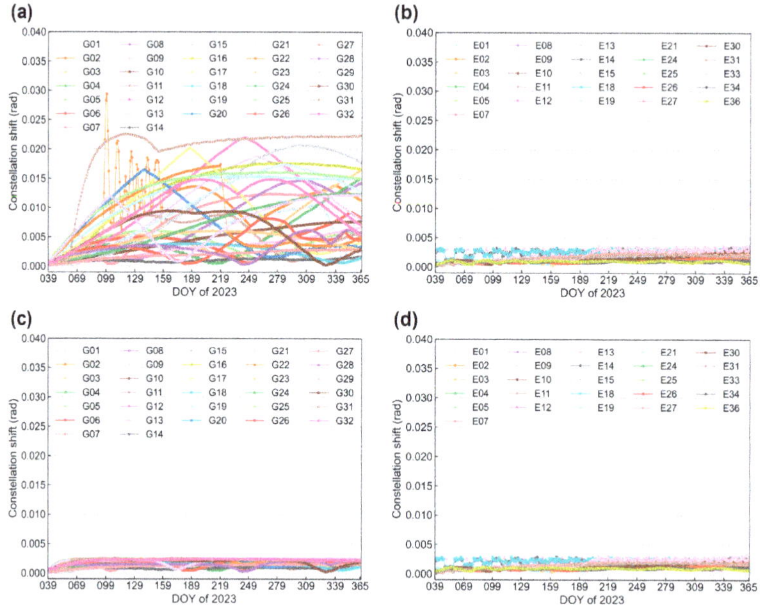

Figure 17. Average unit spherical distances of the GPS and Galileo satellites from reference orbits before and after the GPS/Galileo combination. Panel (**a**) shows the distances of GPS satellites from the GPS reference orbit, panel (**b**) shows the distances of Galileo satellites from the Galileo reference orbit, panel (**c**) shows the distances of GPS satellites from the GPS/Galileo combined reference orbit, and panel (**d**) shows the distances of Galileo satellites from the GPS/Galileo combined reference orbit.

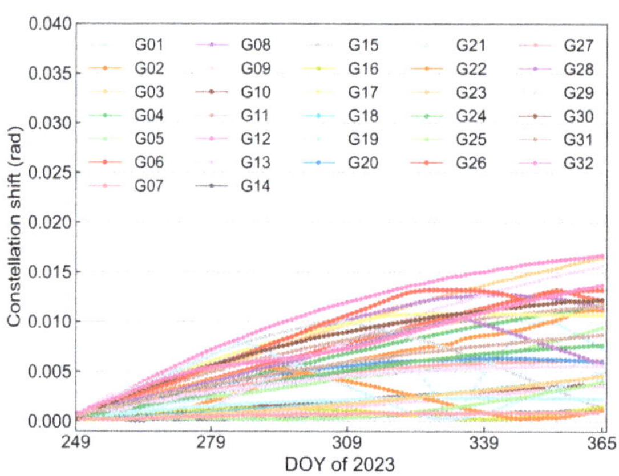

Figure 18. Average spherical distance of GPS satellites from the reference orbit after using the GPS satellite orbits from DOY 239–248 as the new reference.

In Figure 17, the average deviation distances for both GPS and Galileo satellite orbits generally stay within 0.02 rad. Using this threshold, satellite coordinates within this distance from the reference orbit on a unit sphere are considered consistent, indicating an orbit repeat. Orbit repeatability for both systems is then calculated daily from DOY 039 to 365 in 2023 and is presented in Figure 19, with reference to both individual constellations and the combined constellation. When using the GPS satellite orbits from DOY 239 to 248 as the reference, the GPS orbit repeatability gradually declines, dropping to 84.5% after DOY 072.

This reduction negatively impacts the accuracy and stability of the multipath error model. However, when GPS and Galileo satellite orbits are combined, the GPS orbit repeatability remains stable at 100%. Galileo satellites consistently show 100% repeatability throughout the observation period, thanks to their inherent orbital stability. Thus, integrating GPS and Galileo systems for multipath modeling and mitigation is expected to improve correction accuracy and ensure more stable results, particularly for GPS.

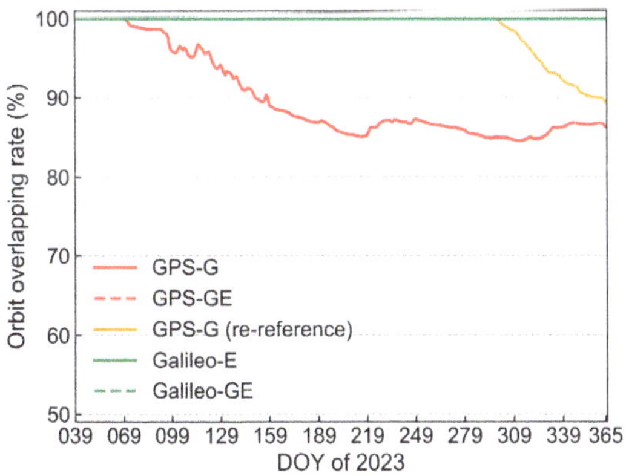

Figure 19. Repetition rate of GPS and Galileo satellite orbits between DOY 039–365 in 2023 before and after the GPS/Galileo combination.

3.5. Unified Covariance Function Parameters for Short and Long-Term Multipath Correction Analysis

Using the MATE_MAT1 baseline's GPS L1 and Galileo E1 frequencies as examples, we applied both the grid method and the autocorrelation-based LSC using unified covariance function parameters in the multipath mitigation. The calibration, carried out from DOY 039 to 365 in 2023, employs a 30-day interval. The moving neighborhood parameter is set at 0.02 rad, and the covariance function parameters C_0 and d_0 are averaged across all frequencies from both baselines. The specific strategies for multipath correction are detailed in Table 2.

Table 2. Multipath correction strategies in this study.

Frequencies to be Corrected	Method	Modeling Date	Frequencies Used for Correction	Abbreviations
GPS L1	Grid	2023 DOY 029–038	GPS L1	Grid_G for L1
			GPS L1/Galileo E1	Grid_GE for L1
		2023 DOY 239–248	GPS L1	Grid_G for L1 (remodel)
			GPS L1/Galileo E1	Grid_GE for L1 (remodel)
	Autocorrelation-based LSC	2023 DOY 029–038	GPS L1	LSC_G for L1
			GPS L1/Galileo E1	LSC_GE for L1
		2023 DOY 239–248	GPS L1	LSC_G for L1 (remodel)
			GPS L1/Galileo E1	LSC_GE for L1 (remodel)
Galileo E1	Grid	2023 DOY 029–038	Galileo E1	Grid_E for E1
			GPS L1/Galileo E1	Grid_GE for E1
		2023 DOY 239–248	Galileo E1	Grid_E for E1 (remodel)
			GPS L1/Galileo E1	Grid_GE for E1 (remodel)
	Autocorrelation-based LSC	2023 DOY 029–038	Galileo E1	LSC_E for E1
			GPS L1/Galileo E1	LSC_GE for E1
		2023 DOY 239–248	Galileo E1	LSC_E for E1 (remodel)
			GPS L1/Galileo E1	LSC_GE for E1 (remodel)

The standard deviations of the SD residuals before and after applying the multipath correction strategies are detailed in Tables 3 and 4. On DOY 039 of 2023, the performance of the grid methods, labeled Grid_G and Grid_GE, are initially comparable. However, Grid_G's performance declines more rapidly over time. In contrast, the LSC method demonstrates robust initial performance, maintaining superiority over Grid_G and showing slightly lesser performance only compared to the combined GPS/Galileo grid method after 120 days. The LSC_GE strategy, specifically modeled on DOY 029–038, consistently offers superior performance compared to all other strategies modeled on later dates (DOY 239–248).

Table 3. STDs of the SD residuals before and after multipath calibration for GPS L1 (unit: mm).

	DOY of 2023											
	039	069	099	129	159	189	219	249	279	309	339	365
Uncalibrated	5.14	5.38	5.31	5.12	5.12	5.17	5.76	5.63	5.75	5.59	5.85	5.66
Grid_G	2.97	3.67	3.99	4.08	4.08	4.15	4.71	4.48	4.61	4.47	4.62	4.66
Grid_GE	2.96	3.34	3.53	3.52	3.55	3.52	4.18	3.91	4.05	3.83	3.88	3.97
LSC_G	2.43	3.02	3.46	3.41	3.54	3.72	4.26	3.99	4.15	3.99	4.07	4.09
LSC_GE	2.43	2.86	3.05	2.99	3.11	3.04	3.70	3.43	3.63	3.31	3.33	3.49
Grid_G (remodel)								3.37	4.05	4.42	4.61	4.59
Grid_GE (remodel)								3.37	3.81	3.77	3.84	3.82
LSC_G (remodel)								2.86	3.50	3.67	4.03	4.14
LSC_GE (remodel)								2.85	3.31	3.18	3.26	3.34

Table 4. STDs of the SD residuals before and after multipath calibration for Galileo E1 (unit: mm).

	DOY of 2023											
	039	069	099	129	159	189	219	249	279	309	339	365
Uncalibrated	4.12	4.10	4.46	4.20	4.09	4.44	4.56	4.45	4.65	4.65	4.89	5.09
Grid_E	2.46	2.45	2.66	2.66	2.51	2.71	2.95	2.76	2.86	2.74	3.04	3.56
Grid_GE	2.49	2.46	2.68	2.64	2.56	2.71	2.95	2.70	2.87	2.75	3.03	3.59
LSC_E	2.05	2.07	2.23	2.28	2.18	2.30	2.59	2.37	2.49	2.33	2.63	3.04
LSC_GE	2.02	2.05	2.18	2.24	2.17	2.27	2.57	2.32	2.45	2.29	2.52	2.97
Grid_E (remodel)								2.51	2.71	2.60	2.96	3.58
Grid_GE (remodel)								2.54	2.74	2.64	2.98	3.51
LSC_E (remodel)								2.07	2.31	2.16	2.48	3.16
LSC_GE (remodel)								2.04	2.27	2.14	2.47	3.00

Regarding the Galileo E1 frequency, the combined GPS/Galileo correction strategy proves to be as effective or slightly more so than the Galileo-only approach. This effectiveness persists even after 210 days of modeling, where the LSC_GE strategy consistently surpasses the grid method's performance. This trend is maintained irrespective of whether the correction strategy involves only Galileo or a GPS/Galileo combination, with the combined LSC method achieving the best results. Thus, the study confirms the advantage of employing unified covariance function parameters for diverse baselines and frequency signals, which simplifies the computational process and enhances the effectiveness of multipath error mitigation.

Then, the residual variance reduction rate calculated by $\left(1 - \sigma_1^2/\sigma_0^2\right) \times 100\%$, where σ_1^2 is the remaining SD residual variance after multipath correction and σ_0^2 denotes the uncorrected SD residual variance, is shown in Figure 20. For GPS L1, the variance reduction rate on DOY 039 of 2023 using the LSC_GE strategy shows improvements of 0.13%, 11.05%, and 11.16% over LSC_G, Grid_GE, and Grid_G, respectively. By DOY 365 of 2023, these improvements markedly increase to 14.24%, 11.25%, and 29.59%, respectively. This suggests that the LSC_GE strategy not only enhances multipath correction immediately after model application but maintains superior correction capabilities over extended periods. Additionally, the LSC_GE's variance reduction consistently exceeds that achieved by the other strategies, even following repeated modeling. For the Galileo E1 frequency, the LSC_GE strategy improves the variance reduction rate by 0.75%, 12.27%, and 11.58% over LSC_E, Grid_GE, and Grid_E respectively on the first day. By DOY 365, these improvements rise to 1.5%, 15.69%, and 14.96%, respectively. Notably, the grid method's performance post-re-modeling remains inferior to the pre-re-modeling performance of the LSC method. On DOY 339, the re-modeled LSC_E strategy matches the pre-re-modeling performance of the LSC_GE strategy, but it falls short after this period. The re-modeled LSC_GE strategy maintains similar performance to the pre-re-modeling performance of the LSC_GE.

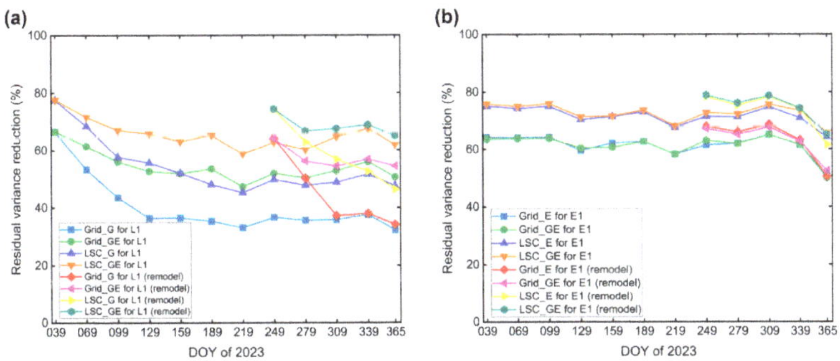

Figure 20. SD residual variance reduction rates of the baseline MAT1_MATE for GPS L1 (**a**) and Galileo E1 (**b**) by the grid method and the autocorrelation-based LSC method.

Kinematic mode processing of GPS L1 and Galileo E1 data was conducted using the outlined strategies, with the results presented in Figure 21. The LAMBDA method was employed for ambiguity estimation in the processing, and the ambiguity fixing rates were over 99%. This figure displays the root mean square (RMS) values for the North, East, and Up components before and after multipath calibration. The combined GPS/Galileo solution for GPS L1 provided better results than the GPS-only solution. For Galileo E1, the combined solution was on par with the Galileo-only solution. Across both frequencies, the autocorrelation-based LSC using unified covariance function parameters yielded lower RMS errors and enhanced positioning accuracy compared to the grid method, indicating its efficacy whether applied to combined or single-system data.

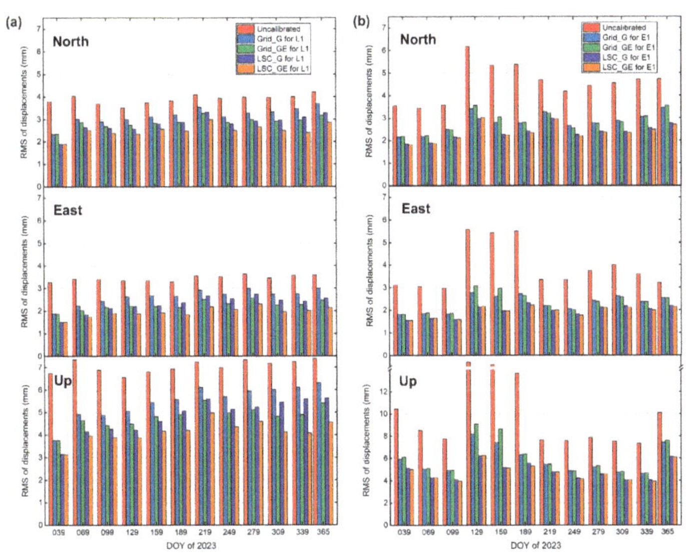

Figure 21. RMS of displacements of the North, East, and Up components of the baseline MAT1_MATE for GPS L1 (**a**) and Galileo E1 (**b**) using different correction strategies.

4. Results

Multipath error significantly affects the accuracy of GNSS precise positioning but can be effectively mitigated using spatial autocorrelation-based methods in static and quasi-static environments. Traditional techniques like sidereal filtering and grid-based approaches only approximate the handling of spatial autocorrelation data. However, the spatiotemporal correlations of multipath can be more accurately modeled by utilizing the covariance function. This study investigates the spatial cross-correlation between different frequency bands across various GNSS systems, along with spatial autocorrelation function parameters and the effects of constellation shifts, to simplify autocorrelation-based methods like Least-Squares Collocation and enhance long-term performance.

The results reveal that overlapping frequencies, such as L1/E1 and L5/E5a, show very strong correlations, while non-overlapping frequencies, such as L2 and E5b, also exhibit strong correlations, with coefficients exceeding 0.8. The covariance function parameters, typically including the residual variance and the range parameter that control the rate of correlation decay with distance, remain relatively consistent across frequencies. This consistency allows autocorrelation-based methods using unified covariance function parameters to achieve a variance reduction rate (VRR) comparable to methods that rely on frequency-specific covariance function parameters. Additionally, the GPS satellites G02, G20, and G21 exhibit noticeable orbit shifts, leading to significant changes in orbital parameters and satellite tracks, which diminishes the effectiveness of long-term multipath mitigation. These satellites display larger average deviations, reaching a maximum deviation of 0.029 rad. In contrast, the Galileo satellites demonstrate more stable orbits, with fluctuations confined to 0.004 rad. When combining GPS and Galileo reference orbits, the maximum deviation for GPS satellites reduces to 0.003 rad, and the orbital repeat rate stabilizes at 100%.

Multipath mitigation experiments using both the grid method and the autocorrelation-based LSC with a unified covariance function for GPS L1 and Galileo E1 frequencies on the MATE_MAT1 baseline were conducted. This analysis considers the GPS, Galileo, and combined GPS/Galileo data separately. For GPS L1, the LSC_GE strategy improves the residual variance reduction rate by 0.13%, 11.05%, and 11.16% compared to LSC_G, Grid_GE, and Grid_G, respectively, on DOY 039, the first day after modeling. These improvements increase to 14.24%, 11.25%, and 29.59% on DOY 365, 327 days from the modeling data. Similarly, for Galileo E1, the LSC_GE strategy shows improvements of 0.75%, 12.27%, and 11.58% on the first day following modeling, rising to 1.5%, 15.69%, and 14.96% by DOY 365. As a result, the accuracy of kinematic positioning is improved.

Overall, the unified covariance function proves effective for autocorrelation-based multipath mitigation, outperforming conventional grid methods. Although GPS orbit drift negatively affects long-term mitigation, integrating GPS and Galileo data consistently produces lower RMS values and more accurate baseline solutions, significantly mitigating the impact of orbit drift. However, the method is based on spatial autocorrelation. Therefore, in the case that the spatial autocorrelation is low, the performance will be limited. This may happen in complex multipath environments. However, this need to be tested when complex multipath environments are encountered in practice.

Author Contributions: Conceptualization, W.X. and Y.T.; methodology, W.X. and Y.T.; software, W.X. and Y.T.; validation, X.D. and Q.Z.; formal analysis, Y.L. and X.D.; investigation, X.R.; writing—original draft preparation, W.X.; writing—review and editing, Y.T. and Q.Z.; visualization, Y.L. and X.R.; supervision, Y.T. All authors have read and agreed to the published version of the manuscript.

Funding: This research was funded by the National Natural Science Foundation of China, grant number 42274032.

Data Availability Statement: The GNSS raw observation data are available at http://www.igs.org/mgex/data-products/#data (accessed on 12 March 2024).

Conflicts of Interest: The authors declare no conflicts of interest.

References

1. Groves, P.D.; Jiang, Z. Height Aiding, Weighting and Consistency Checking for GNSS NLOS and Multipath Mitigation in Urban Areas. *J. Navig.* **2013**, *66*, 653–669. [CrossRef]
2. Manandhar, D.; Shibasaki, R.; Torimoto, H. GPS Reflected Signal Analysis using Software Receiver. *J. Glob. Position. Syst.* **2006**, *5*, 29–34. [CrossRef]
3. Tranquilla, J.M.; Carr, J.P.; Alrizzo, H.M. Analysis of a choke ring groundplane for multipath control in global positioning system (gps) applications. *IEEE Trans. Antennas Propag.* **1994**, *42*, 905–911. [CrossRef]
4. Tatarnikov, D.V.; Stepanenko, A.P.; Astakhov, A.V. Moderately compact helix antennas with cutoff patterns for millimeter RTK positioning. *Gps Solut.* **2016**, *20*, 587–594. [CrossRef]
5. Vandierendonck, A.J.; Fenton, P.; Ford, T.; Inst, N. Theory and performance of narrow correlator spacing in a gps receiver. In Proceedings of the National Technical Meeting of the Inst of Navigation, San Diego, CA, USA, 27–29 January 1992; pp. 115–124.
6. Axelrad, P.; Comp, C.J.; MacDoran, P.F. SNR-based multipath error correction for GPS differential phase. *IEEE Trans. Aerosp. Electron. Syst.* **1996**, *32*, 650–660. [CrossRef]
7. Bilich, A.; Larson, K.M. Mapping the GPS multipath environment using the signal-to-noise ratio (SNR). *Radio Sci.* **2008**, *43*, RS6003. [CrossRef]
8. Zhang, Z.T.; Li, B.F.; Gao, Y.; Shen, Y.Z. Real-time carrier phase multipath detection based on dual-frequency C/N0 data. *Gps Solut.* **2019**, *23*, 7. [CrossRef]
9. Zheng, D.W.; Zhong, P.; Ding, X.L.; Chen, W. Filtering GPS time-series using a Vondrak filter and cross-validation. *J. Geod.* **2005**, *79*, 363–369. [CrossRef]
10. Satirapod, C.; Rizos, C. Multipath mitigation by wavelet analysis for GPS base station applications. *Surv. Rev.* **2005**, *38*, 2–10. [CrossRef]
11. Zhong, P.; Ding, X.L.; Zheng, D.W.; Chen, W.; Huang, D.F. Adaptive wavelet transform based on cross-validation method and its application to GPS multipath mitigation. *Gps Solut.* **2008**, *12*, 109–117. [CrossRef]
12. Bernelli-Zazzera, F.; Campana, R.; Gottifredi, F.; Marradi, L. GPS attitude determination by Kalman filtering: Simulation of multipath rejection. In Proceedings of the 8th Annual Space Flight Mechanics Meeting, Monterey, CA, USA, 9–11 February 1998; pp. 1391–1405.
13. Agnew, D.C.; Larson, K.M. Finding the repeat times of the GPS constellation. *Gps Solut.* **2007**, *11*, 71–76. [CrossRef]
14. Genrich, J.F.; Bock, Y. Rapid resolution of crustal motion at short ranges with the global positioning system. *J. Geophys. Res. Solid. Earth* **1992**, *97*, 3261–3269. [CrossRef]
15. Zhong, P.; Ding, X.L.; Yuan, L.G.; Xu, Y.L.; Kwok, K.; Chen, Y.Q. Sidereal filtering based on single differences for mitigating GPS multipath effects on short baselines. *J. Geod.* **2010**, *84*, 145–158. [CrossRef]
16. Cohen, C.E.; Parkinson, B.W. Mitigating multipath error in gps based attitude determination. In Proceedings of the 14th Annual Rocky Mountain Conf on Guidance and Control, Keystone, CO, USA, 2–6 February 1991; pp. 53–68.
17. Dong, D.; Wang, M.; Chen, W.; Zeng, Z.; Song, L.; Zhang, Q.; Cai, M.; Cheng, Y.; Lv, J. Mitigation of multipath effect in GNSS short baseline positioning by the multipath hemispherical map. *J. Geod.* **2016**, *90*, 255–262. [CrossRef]
18. Fuhrmann, T.; Luo, X.G.; Knöpfler, A.; Mayer, M. Generating statistically robust multipath stacking maps using congruent cells. *Gps Solut.* **2015**, *19*, 83–92. [CrossRef]
19. Wang, Z.R.; Chen, W.; Dong, D.N.; Wang, M.H.; Cai, M.M.; Yu, C.; Zheng, Z.Q.; Liu, M. Multipath mitigation based on trend surface analysis applied to dual-antenna receiver with common clock. *Gps Solut.* **2019**, *23*, 104. [CrossRef]
20. Zheng, K.; Zhang, X.H.; Li, P.; Li, X.X.; Ge, M.R.; Guo, F.; Sang, J.Z.; Schuh, H. Multipath extraction and mitigation for high-rate multi-GNSS precise point positioning. *J. Geod.* **2019**, *93*, 2037–2051. [CrossRef]
21. Hofmann-Wellenhof, B.; Moritz, H. *Physical Geodesy*; Springer: Vienna, Austria, 2005.
22. Lin, M. Regional Gravity Field Recovery Using The Point Mass Method. Ph.D. Thesis, Leibniz Universität Hannover, Hannover, Germany, 2016.
23. Moritz, H. *Advanced Physical Geodesy*; Wichmann: Berlin, Germany, 1980.
24. Tian, Y.; Liu, Z.; Lin, M.; Li, K. Modelling and mitigation of GNSS multipath effects by least-squares collocation considering spatial autocorrelation. *J. Geod.* **2023**, *97*, 37. [CrossRef]
25. Lu, R.; Chen, W.; Li, Z.; Dong, D.A.; Jiang, W.P.; Wang, Z.R.; Huang, L.Y.; Duan, X.L. An improved joint modeling method for multipath mitigation of GPS, BDS-3, and Galileo overlapping frequency signals in typical environments. *J. Geod.* **2023**, *97*, 95. [CrossRef]
26. Geng, J.H.; Li, P.B.; Li, G.C. Combining the GPS/Galileo/BDS-3 signals on overlap frequencies for interoperable multipath hemispherical maps. *J. Geod.* **2024**, *98*, 32. [CrossRef]
27. Krag, H.; Leonard, A.; Ranaudo, P.; Blomenhofer, E. Global Implications of GPS and Galileo RAAN Drifts on GNSS Performance. In Proceedings of the 16th International Technical Meeting of the Satellite Division of The Institute of Navigation (ION GPS/GNSS 2003), Portland, OR, USA, 9–12 September 2003.
28. Pellegrino, M.M.; Scheeres, D.J.; Streetman, B.J. The Feasibility of Targeting Chaotic Regions in the GNSS Regime. *J. Astronaut. Sci.* **2021**, *68*, 553–584. [CrossRef]
29. Sosnica, K.; Bury, G.; Zajdel, R.; Ventura-Traveset, J.; Mendes, L. GPS, GLONASS, and Galileo orbit geometry variations caused by general relativity focusing on Galileo in eccentric orbits. *Gps Solut.* **2022**, *26*, 5. [CrossRef]

30. Moritz, H. *Covariance Functions in Least-Squares Collocation;* Ohio State University Research Foundation: Columbus, OH, USA, 1976; Volume 240.
31. Moritz, H. *Advanced Least-Squares Methods;* Ohio State University Research Foundation: Columbus, OH, USA, 1972; Volume 175.
32. Pearson, W.H. Estimation of a correlation coefficient from an uncertainty measure. *Psychometrika* **1966**, *31*, 421–433. [CrossRef] [PubMed]
33. Liu, Z.F.; Tian, Y.M.; Xiong, W.H.; Liang, Y.B.; Li, K.G.; Tan, S.H.; Yang, X.W.; Li, N. A local filtering approach to mitigating the GNSS multipath effects in relative precise positioning considering the multipath spatial correlation. *Adv. Space Res.* **2024**, *74*, 2709–2727. [CrossRef]
34. Ragheb, A.E.; Clarke, P.J.; Edwards, S.J. GPS sidereal filtering: Coordinate- and carrier-phase-level strategies. *J. Geod.* **2007**, *81*, 325–335. [CrossRef]
35. Wang, M.; Wang, J.; Dong, D.; Li, H.; Han, L.; Chen, W. Comparison of Three Methods for Estimating GPS Multipath Repeat Time. *Remote Sens.* **2018**, *10*, 6. [CrossRef]
36. Larson, K.M.; Bilich, A.; Axelrad, P. Improving the precision of high-rate GPS. *J. Geophys. Res.-Solid. Earth* **2007**, *112*, B05422. [CrossRef]

remote sensing

MDPI

Article

Quantifying Creep on the Laohushan Fault Using Dense Continuous GNSS

Wenquan Zhuang [1], Yuhang Li [1,*], Ming Hao [1], Shangwu Song [1], Baiyun Liu [2] and Lihong Fan [3]

[1] The Second Monitoring and Application Center, China Earthquake Administration, Xi'an 710054, China; zhuangwenquan@smac.ac.cn (W.Z.); haoming@smac.ac.cn (M.H.); songshangwu@smac.ac.cn (S.S.)

[2] Gansu Earthquake Agency, Lanzhou 730000, China; liuby@gsdzj.gov.cn

[3] National Time Service Center, Chinese Academy of Sciences, Shu Yuan East Road, Xi'an 710600, China; fanlihong@ntsc.ac.cn

* Correspondence: liyuhang@smac.ac.cn

Abstract: The interseismic behavior of faults (whether they are locked or creeping) and their quantitative kinematic constraints are critical for assessing the seismic hazards of faults and their surrounding areas. Currently, the creep of the eastern segment of the Laohushan Fault in the Haiyuan Fault Zone at the northeastern margin of the Tibetan Plateau, as revealed by InSAR observations, lacks confirmation from other observational methods, particularly high-precision GNSS studies. In this study, we utilized nearly seven years of observation data from a dense GNSS continuous monitoring profile (with a minimum station spacing of 2 km) that crosses the eastern segment of the Laohushan Fault. This dataset was integrated with GNSS data from regional continuous stations, such as those from the Crustal Movement Observation Network of China, and multiple campaign measurements to calculate GNSS baseline change time series across the Laohushan Fault and to obtain a high spatial resolution horizontal crustal velocity field for the region. A comprehensive analysis of this primary dataset indicates that the Laohushan Fault is currently experiencing left-lateral creep, characterized by a partially locked shallow segment and a deeper locked segment. The fault creep is predominantly concentrated in the shallow crustal region, within a depth range of 0–5.7 ± 3.4 km, exhibiting a creep rate of 1.5 ± 0.7 mm/yr. Conversely, at depths of 5.7 ± 3.4 km to 16.8 ± 4.2 km, the fault remains locked, with a loading rate of 3.9 ± 1.1 mm/yr. The shallow creep is primarily confined within 3 km on either side of the fault. Over the nearly seven-year observation period, the creep movement within approximately 5 km of the fault's near field has shown no significant time-dependent variation, instead demonstrating a steady-state behavior. This steady-state creep appears unaffected by postseismic effects from historical large earthquakes in the adjacent region, although the deeper (far-field) tectonic deformation of the Laohushan Fault may have been influenced by the postseismic effects of the 1920 Haiyuan M8.5 earthquake.

Keywords: northeastern margin of the Tibetan Plateau; Laohushan fault; fault creep; GNSS measurement; present crustal deformation

1. Introduction

In contrast to interseismic locked coupling (continuous strain energy accumulation) and coseismic stick–slip (rapid destabilizing sliding) on earthquake faults, fault creep–slip motions indicate that the fault plane is open/uncoupled, and thus there is no significant accumulation of strain energy or risk of strong earthquakes [1,2]. Research has demonstrated that large earthquakes often nucleate at the transition zones between the locked and creeping segments of a fault [3], and that creeping segments can serve as barriers to the propagation of large ruptures [4]. The effectiveness of these barriers is influenced by the size of the creeping segments and the frictional properties of the fault plane [5,6]. In addition, fault creep may also trigger local small earthquake activity [7]. Therefore, understanding and characterizing fault creep is essential for constraining the seismic hazard assessments of the fault itself, evaluating the potential for cascading ruptures based on fault segmentation, and elucidating the mechanisms driving regional seismicity [8]. To date, numerous major strike–slip faults worldwide, such as the Xianshuihe Fault on the eastern margin of the Tibetan Plateau [9], the San Andreas Fault [2,10], and the North Anatolian Fault [11,12], are believed to exhibit fault creep behavior. With advancements in geodetic technology, GNSS and InSAR observations have become widely employed for characterizing fault creep deformation. However, due to the limited distribution of GNSS stations, studies focusing exclusively on fault creep motion using CNSS data remain relatively scarce. At present, GNSS-based investigations have primarily focused on the Atotsugawa Fault [13], the Alto Tiberina Fault [14], and the Dead Sea Fault [15].

The Haiyuan Fault, located at the northeastern margin of the Tibetan Plateau, is a large left-lateral strike–slip fault zone composed of multiple nearly E–W-trending fault segments (shown as thick black solid lines in Figure 1). Since the Holocene, it has been rapidly slipping at a rate of 3–5 mm/yr [13–18]. Significant seismic events, including the 1920 Haiyuan M8.5 earthquake and the 1927 Gulang M8.3 earthquake, have occurred along this fault (Figure 1). Recent InSAR monitoring studies have revealed a ~35 km long segment of fault creep on the Laohushan Fault within the Haiyuan Fault Zone, with an estimated left-lateral creep rate of 5 ± 1 mm/yr [19]. This rate closely aligns with GNSS-derived far-field displacement rates across the fault [20,21], which is consistent with a velocity-strengthening friction mechanism. This mechanism suggests that the creeping segment may serve as a permanent barrier to seismic rupture propagation and could also potentially trigger moderate seismic activity in the surrounding region [7,19,22]. While recent InSAR results support the presence of creep along the Laohushan segment of the Haiyuan Fault, there remains debate regarding the exact rate of this creep and its associated spatial deformation characteristics [23–26].

It is important to note that while InSAR is a valuable technique for measuring line-of-sight surface deformation, it is not capable of directly and accurately capturing horizontal fault movements. Moreover, the complex vertical crustal deformation present in this region [27] complicates the effective removal of vertical components from the InSAR deformation field [22], potentially leading to an overestimation of the creep rate [23]. Although some studies have incorporated published GNSS horizontal velocity fields as a priori constraints to enhance the accuracy of InSAR-derived deformation results, there remains a lack of GNSS constraints in critical near-field fault zones (<5 km) [23,24]. Accurate near-field fault deformation is crucial for correctly determining fault creep characteristics (such as creep depth, rate, and other kinematic parameters) and understanding its deformation mechanisms [28,29]. This study addresses these challenges by utilizing nearly seven years of observation data from a dense GNSS continuous monitoring profile (with a minimum station spacing of 2 km) across the eastern segment of the Laohushan Fault. Supplemented by data from regional land–state networks, additional GNSS continuous stations, and multiple GNSS survey campaigns (Figure 1), we have calculated GNSS baseline change time series and have generated high-resolution regional crustal horizontal velocity fields across the Laohushan Fault. This comprehensive dataset reveals the dynamic deformation characteristics of both near- and far-field crustal deformation across the fault. It provides precise constraints for determining the current creep rate, its temporal variations, and the depth distribution of creeping and locking segments, thereby offering valuable insights into the fault's deformation mechanisms and contributing to regional seismic hazard assessments.

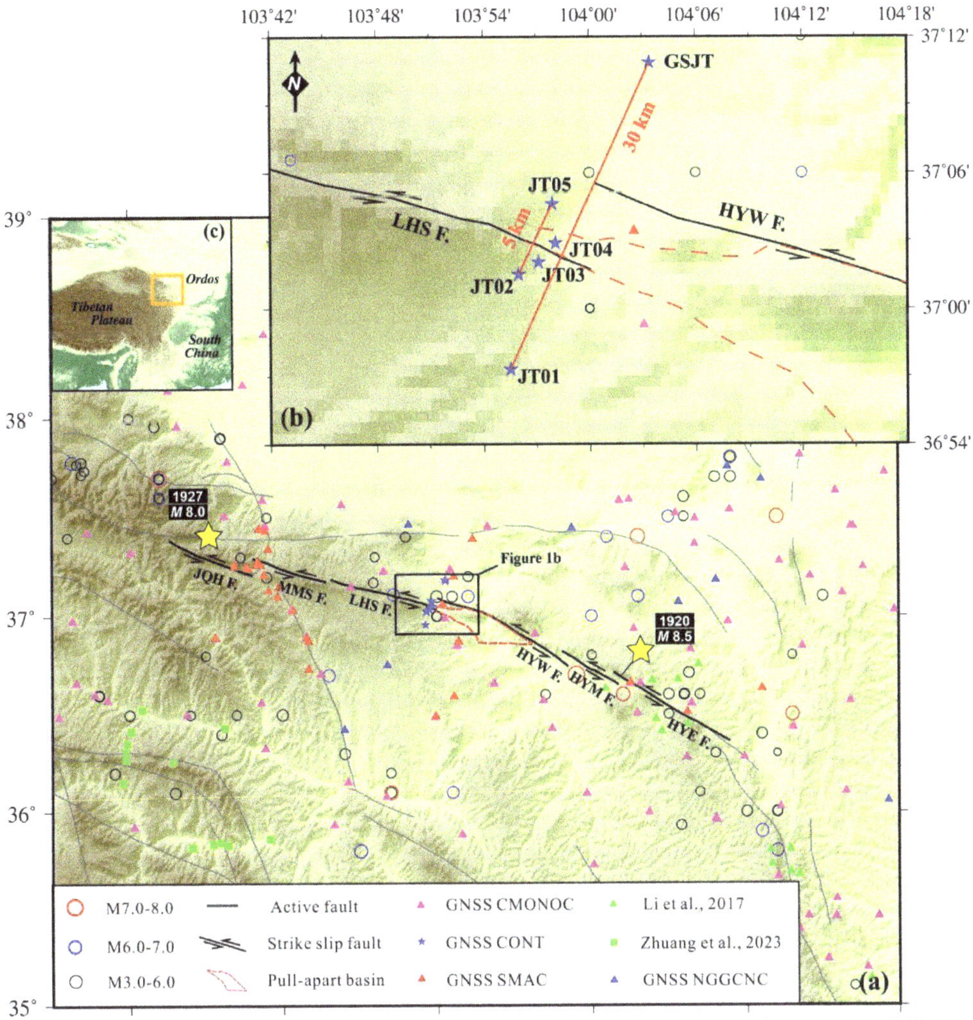

Figure 1. Regional active tectonics map. (**a**) Gray lines represent major active faults [30]. The red dashed line area indicates the Pull-Apart Basin [31]. Hollow circles represent historical earthquakes [32]. Blue pentagons mark the distribution of GNSS continuous observation profile sites established by the Second Monitoring Center of the China Earthquake Administration across the Laohushan Fault. Red triangles represent GNSS survey stations from the Fifteen Digital Seismic Network Construction Project. The pink triangles denote observation stations from the Crustal Movement Observation Network of China. The blue triangles indicate stations from the National GNSS Geodetic Control Network of China. Green triangles and green squares represent published GNSS observation results. Major fault segments of the Haiyuan Fault are shown as thick solid lines, with the following abbreviations: JQH F., the Jinqianghe Fault; MMS F., the Maomaoshan Fault; LHS F., the Laohushan Fault; HYW F., the western segment of the Haiyuan Fault; HYM F., the central segment of the Haiyuan Fault; HYE F., the eastern segment of the Haiyuan Fault. (**b**) Details of the solid black line in (**a**). (**c**) The orange rectangle in (**c**) shows our study area.

2. GNSS Data and Processing

2.1. GNSS Observations and Data Collection

The GNSS continuous observation profile across the eastern segment of the Laohushan Fault in the Haiyuan Fault Zone is shown in Figure 1b. This profile includes a total of 5 GNSS continuous stations. Among them, stations JT02, JT03, JT04, and JT05 are located within approximately 5 km of the fault, while JT01 and GSJT are 30 km apart, forming a combined near- and far-field GNSS continuous deformation observation system consisting of a total of 6 GNSS continuous stations (Figure 1b). We obtained continuous observation data from the above-mentioned GNSS continuous observation system (6 stations) from August 2018 to May 2021. All stations meet the requirements for continuous deformation field studies, with an unobstructed view of more than 10°, a data availability rate of over 90%, and multipath errors MP1 and MP2 both less than 0.5 m. With the exception of JT04, all stations have more than 2.5 years of observation data, which can ensure meeting the need to obtain stable and reliable observation results. In addition, a one-time GNSS survey campaign was conducted at the site in March 2024, with a continuous observation duration of 72 h. The survey met the requirements for unobstructed views of more than 10°, a data availability rate of over 90%, and multipath errors MP1 and MP2 both less than 0.5 m.

This study also collected GNSS survey data from other projects, including the following: (1). Data from the Gulang–Yongdeng GNSS profile and the Jingtai–Baiyin GNSS profile established through key platform projects supported by the Ministry of Science and Technology, National Natural Science Foundation projects, and Sino-French cooperation projects. This includes data from surveys conducted in 1999, 2004, 2005, and 2011. (2). GNSS survey data from the Huazangsi and Shagouhe profiles on the Maomaoshan Fault, established by the Second Monitoring Center of the China Earthquake Administration, with observations in 2005, 2006, 2007, and 2011. (3). Observation stations from the Crustal Movement Observation Network of China from 1999 to 2024. (4). Observation data from the National GNSS Geodetic Control Network of China (NGGCNC), which conducted their first survey in 2005 and 2014; surveys were conducted again twice by the Second Monitoring and Application Center (SMAC) and the China Earthquake Administration in 2012 and 2018–2019, respectively.

2.2. GNSS Data Processing

2.2.1. The Baseline Time Series of GNSS Profile

According to the principles of GNSS relative positioning, short-baseline double-difference solutions are particularly effective in reducing and eliminating regional biases introduced by ionospheric and tropospheric disturbances, as well as inaccuracies in geophysical models, satellite orbit errors, and reference frame instability. With fixed ambiguities, these solutions exhibit rapid convergence [33]. Additionally, noise time series from adjacent continuous observation stations tend to have similar spectral characteristics [34], allowing short-baseline double-difference solutions to cancel out or suppress such noise effectively. Consequently, GNSS short-baseline double-difference solutions offer superior accuracy and reliability in crustal deformation monitoring compared to long-baseline solutions [33,35].

The short-baseline GNSS time series data were processed using the GAMIT/GLOBK software [36,37], applying a free-network solution with fixed satellite orbits (precise ephemerides). To reduce first-order ionospheric effects, a dual-frequency ionosphere-free combination was utilized, and second-order ionospheric corrections were applied based on the Total Electron Content (TEC) data from the IONEX product, provided at two-hour intervals by the Center for Orbit Determination in Europe (CODE). Tropospheric zenith delays were corrected using the GPT2 model, with one zenith delay parameter estimated per station per hour, employing the VMF1 model as the mapping function. Corrections for the Earth's gravity field, solid Earth tides, and pole tides were made in accordance with the latest IERS standards (IERS2003) [38]. Crustal deformation due to ocean tidal loading was accounted for using the FES2004 global ocean tide model, and additional adjustments were made for Earth's center of mass variations induced by ocean tides [39]. Given that this study focuses on monitoring the differential movements across the Laohushan Fault by assessing relative displacement characteristics on either side of the fault, it is unnecessary to constrain the coordinates of the GNSS stations (as shown in Figure 1) within a global reference frame. Instead, a free-baseline solution mode utilizing fixed satellite orbits (precise ephemeris) is employed to obtain GNSS short-baseline time series. This approach mitigates potential errors that could be introduced during the reference frame implementation process.

The GNSS stations selected for calculating the baseline time series in this study include JT01, JT02, JT03, JT04, JT05, and GSJT (Figure 1). The maximum inter-station distance is less than 30 km, with the shortest spacing being approximately 2 km. Due to a receiver malfunction, station JT04 recorded less than one year of intermittent observation data. Additionally, JT04 is located on the periphery of the Laolongwan Pull-Apart Basin (Figure 1a,b), making its observations potentially susceptible to influences from basin activity and non-tectonic factors. Therefore, this station was excluded from the subsequent GNSS baseline and velocity field analyses. Using the aforementioned data processing methods, baseline horizontal vector time series were computed by pairing the five remaining GNSS continuous stations across the Laohushan Fault, resulting in 10 baseline horizontal vector sets (Figure 2). These horizontal vectors were projected both along the fault strike direction (SE102°) and perpendicular to the fault strike (NE12°), with the results presented in Figure 3a,c, respectively.

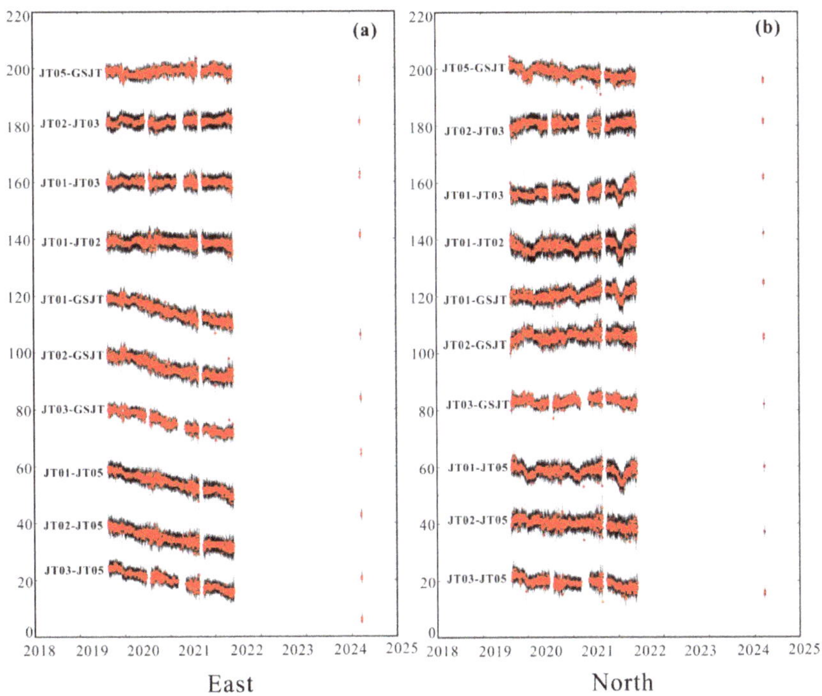

Figure 2. Time series of GNSS short-baseline horizontal vectors. (**a**) Eastward horizontal baseline vector; (**b**) northward horizontal baseline vector.

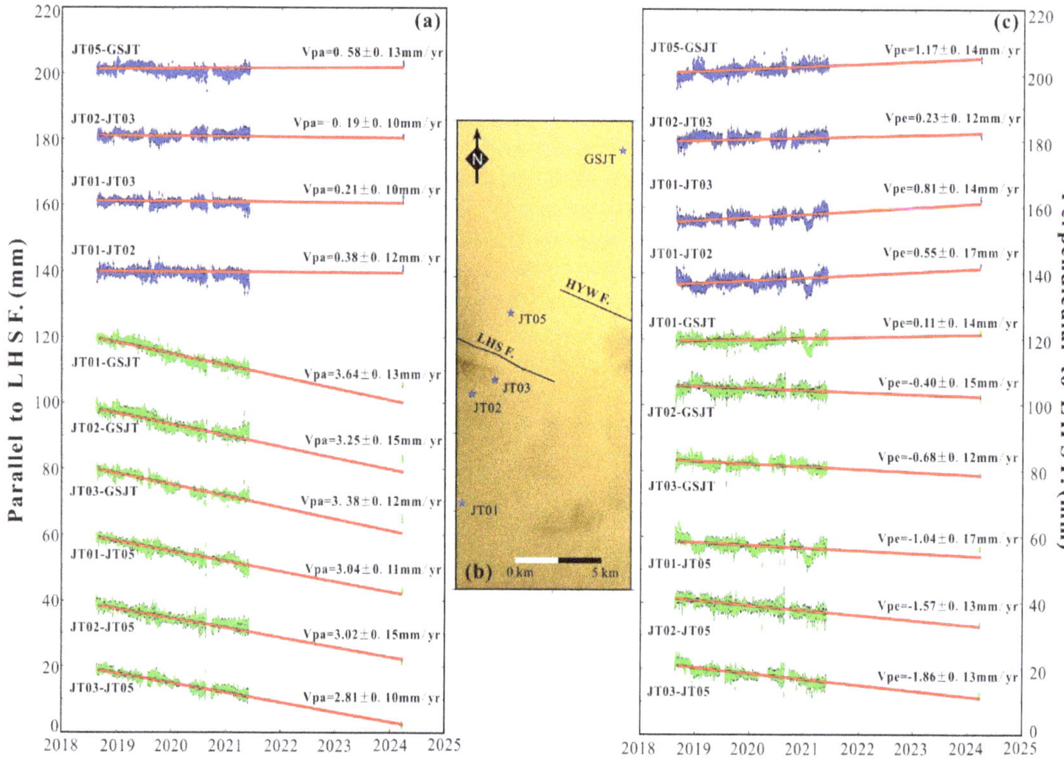

Figure 3. Baseline component variation time series for the Laohushan Fault; (**a**) time series of baseline changes parallel to the Laohushan Fault, where Vpa represents the variation rate of the baseline component parallel to the fault; (**b**) distribution of GNSS stations; (**c**) time series of baseline changes perpendicular to the Laohushan Fault, where Vpe represents the variation rate of the baseline component perpendicular to the fault. Green points represent the baseline time series across the fault. Blue points represent the baseline time series on the same side of the fault. Red solid lines represent the linear rate of best fit from the least-squares method.

2.2.2. Regional GNSS Horizontal Velocity Field

The GAMIT/GLOBK software suite [36,37] was used to process double-differenced carrier phase observations (Figure 1), and daily loosely constrained solutions were obtained. A dual-frequency ionosphere-free combination was employed to mitigate first-order ionospheric effects, while higher-order ionospheric effects were corrected using the 2 h interval Total Electron Content (TEC) IONEX products provided by the Center for Orbit Determination in Europe (CODE). The geophysical models and parameters applied in the processing, such as the FES2004 tidal loading model, the GPT2 global pressure and temperature model, and the VMF1 mapping function, were consistent with those used by [35].

Instead of incorporating global daily solutions provided by the IGS analysis centers, we processed GNSS observation data from 70 core ITRF reference stations globally using the GAMIT package with identical modeling parameters. Daily coordinates and associated uncertainties were estimated by constraining the daily regional solutions with global solutions, and the daily free network solutions were transformed into the ITRF2014 reference frame [40] using the GLOBK package [37]. For campaign GNSS sites, linear velocities relative to ITRF2014 were computed using a weighted least-squares adjustment. Finally, the GNSS horizontal velocity field was transformed into the Eurasian reference frame using the Euler rotation vector for the Eurasian plate provided by [40], as illustrated in Figure 4.

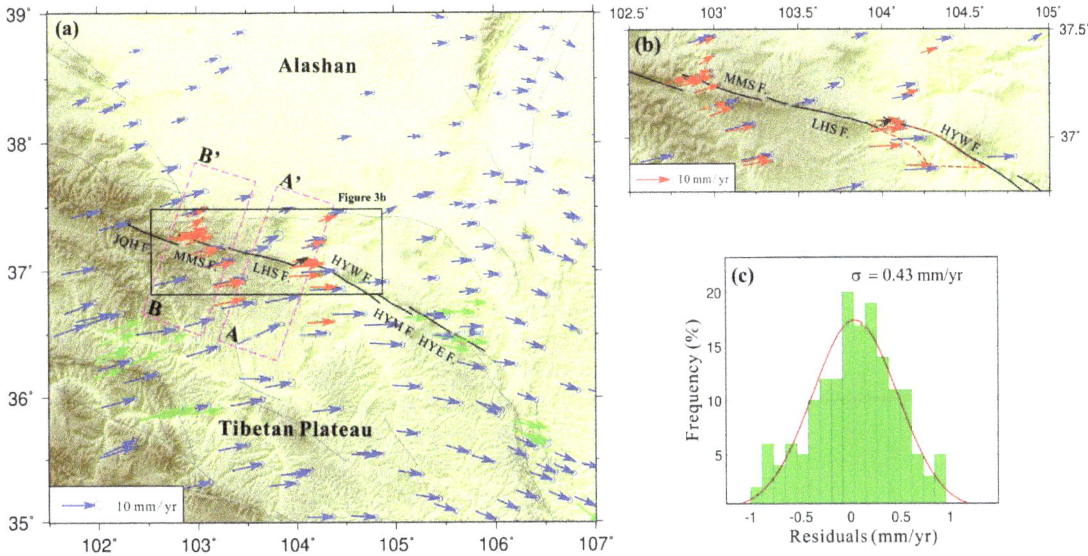

Figure 4. GNSS horizontal velocity field of the Laohushan Fault and surrounding area (relative to the Eurasian reference frame); (**a**) the pink box indicates the location of the GNSS profile shown in Figure 5; (**b**) an enlarged view showing the denser GNSS velocity field around the Maomaoshan Fault and Laohushan Fault; (**c**) the errors associated with common points used in the 200 integration of the published velocity field.

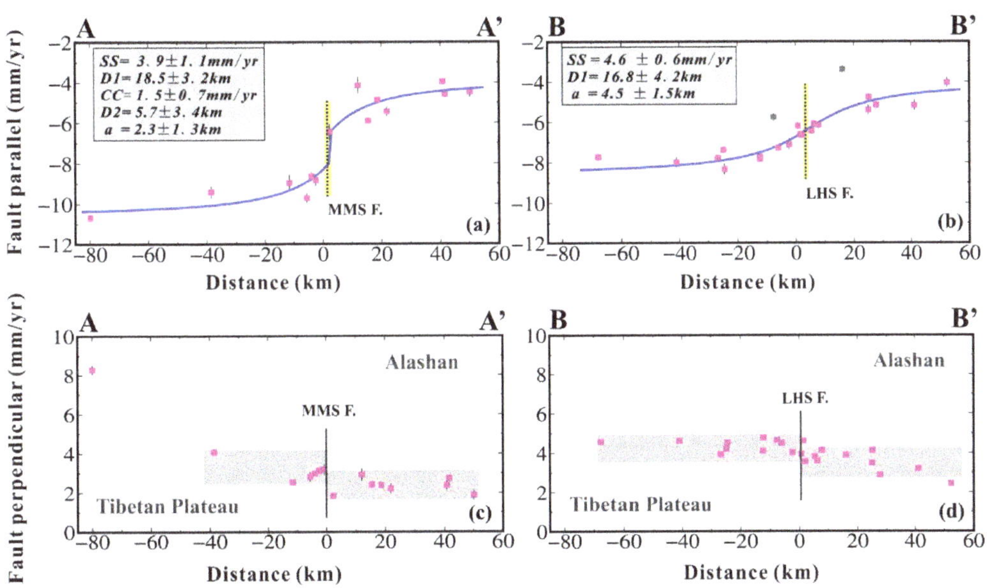

Figure 5. (**a**,**b**) display the GNSS velocity components parallel to the fault, with pink squares indicating the data and gray squares representing outliers (which were not used in the inversion). (**c**,**d**) show the GNSS velocity components perpendicular to the fault, with similar color coding: pink

squares for data and gray squares for outliers. The black dashed lines in (**a**–**d**) indicate the position of the fault as determined by inversion, with yellow shading representing the error margin. The 0 km mark denotes the actual fault position. The blue solid lines in (**a**,**b**) represent the inverse tangent curves calculated from the fault slip model, where SS denotes the slip rate, D1 indicates the locking depth, CC represents the shallow creep rate, and D2 denotes the creep depth. The gray shading in (**c**,**d**) represents the differential motion perpendicular to the fault on both sides. The black solid line shows the actual fault position.

To accurately analyze differential crustal motion between tectonic blocks, we utilized a GNSS velocity field referenced to the stable Eurasian plate [41]. In addition, to precisely model the tectonic characteristics of the Laohushan Fault and enhance our understanding of regional crustal horizontal deformation, we incorporated published high-density GNSS results [42,43], as indicated by the green arrows in Figure 4a. The Helmert parameter transformation method was employed to integrate these published results with the GNSS velocity field derived in this study, ensuring that the residual error at common points in the merged dataset was less than 1 mm/yr (Figure 4c).

3. GNSS Baseline Time Series

Using the GNSS baseline component time series (Figure 3), we applied the least-squares method to fit the rates of change for the baseline components both parallel and perpendicular to the Laohushan Fault. In this analysis, the southernmost station of each baseline served as the reference point, with other stations considered positive if they moved left relative to the reference point and negative if they moved right (as depicted in Figure 3 and Table 1). The results indicate that the baseline components parallel to the Laohushan Fault consistently exhibit a left-lateral strike–slip motion, while the components perpendicular to the fault primarily indicate relatively weak compressional shortening.

Table 1. Fault-parallel and fault-perpendicular movement rates and associated error estimates derived from linear least-squares fitting.

Category	Baseline Name	Parallel Fault Rate (mm/yr)	Parallel Fault Rate Error (mm/yr)	Perpendicular Fault Rate (mm/yr)	Perpendicular Fault Rate Error (mm/yr)
Cross-Fault Baselines	JT03-JT05	2.81	0.10	−1.86	0.13
	JT02-JT05	3.02	0.15	−1.57	0.13
	JT01-JT05	3.04	0.11	−1.04	0.17
	JT03-GSJT	3.38	0.12	−0.68	0.12
	JT02-GSJT	3.25	0.15	−0.40	0.15
	JT01-GSJT	3.64	0.13	0.11	0.14
Same-Side Baselines	JT01-JT02	0.38	0.12	0.55	0.17
	JT01-JT03	0.21	0.10	0.81	0.14
	JT02-JT03	−0.19	0.10	0.23	0.12
	JT05-GSJT	0.58	0.13	1.17	0.14

Notably, all baselines crossing the Laohushan Fault (JT03-JT05, JT02-JT05, JT01-JT05, JT03-GSJT, JT02-GSJT, and JT01-GSJT) display significant variations in fitted parallel fault slip rates (>2.8 mm/yr). Specifically, within a narrow span of approximately 5 km across the fault (JT03-JT05), a left-lateral strike–slip rate of 2.81 ± 0.10 mm/yr is observed (Figure 2), accounting for 77% of the slip rate revealed by the longest baseline, JT01-GSJT (approximately 30 km), which measures 3.64 ± 0.13 mm/yr. In contrast, the baselines on the same side of the fault (not crossing the Laohushan Fault) (JT01-JT02, JT01-JT03, JT02-JT03, and

JT05-GSJT) show minimal variation in their fitted parallel fault slip rates (<0.6 mm/yr) (Table 1). These deformation characteristics collectively suggest that within a 30 km span across the Laohushan Fault, nearly all strike–slip shear deformation is concentrated within a narrow zone of less than 3 km on either side of the fault, with the differential strike–slip motion across the fault inducing negligible deformation on the same side of the fault.

Additionally, the least-squares analysis of GNSS horizontal movement rates perpendicular to the fault strike reveals a degree of horizontal extensional deformation (0.5–1.2 mm/yr) on the same side of the fault (Table 1). Conversely, across the fault, there is consistent but relatively minor horizontal compressional deformation (−0.4 to −1.9 mm/yr) (Figure 3b). These findings suggest that neglecting deformation perpendicular to the fault could introduce uncertainties in accurately estimating the kinematic parameters of the Laohushan Fault's creeping motion when using InSAR data.

4. GNSS Profiles and Inversions

The third section of this paper explores the strain distribution characteristics in the near-field region (within approximately 15 km on either side) of the Laohushan Fault, as revealed by the GNSS short-baseline fault motion components. The analysis identifies significant left-lateral creep along the Laohushan Fault. To further elucidate the detailed variations in strain distribution across both the near- and far-field regions of the fault, and to enhance our understanding of the constraints on fault movement characteristics at different depths, this study utilizes the regional GNSS horizontal velocity field. This field, which includes GNSS creep observation sites along the Laohushan Fault, was used to construct the GNSS velocity profile A–A′ across the fault (Figure 4b).

For a comparative analysis of the transverse differences in interseismic fault motion within the region, an additional GNSS velocity profile, B–B′, was constructed across the Maomaoshan Fault, located approximately 100 km west of the Laohushan Fault, using dense GNSS data along the Maomaoshan Fault (Figure 4b). The GNSS horizontal velocity components within the scope of these profiles were projected along directions parallel and perpendicular to the strikes of the Laohushan Fault (NE102°) and the Maomaoshan Fault (NE103°), respectively, to generate the GNSS velocity profiles for these two faults (Figure 5). A two-dimensional dislocation model incorporating fault creep motion was subsequently employed to invert and determine the geometric parameters (creep depth/locking depth) and kinematic parameters (creep rate/deep loading rate) associated with fault creep and locking.

4.1. Methods and Solution

We assume that plate motion occurs along one or more nearly vertical (upright) strike–slip faults, which are either locked or creeping from the surface to a certain depth. We use a combination of screw dislocation models [44–48] to perform inversion calculations for the Laohushan Fault and the Maomaoshan Fault. This model can simultaneously fit the deformation signals caused by both fault creep and deep locking, which represent two different physical mechanisms. The mathematical expression for this model can be described as follows:

$$V(x) = \frac{S}{\pi}\arctan(\frac{x}{d_1}) + C\left[\frac{1}{\pi}arctan(\frac{x}{d_2}) - H(x)\right] + a \quad (1)$$

V represents the velocity parallel to the fault, x represents the distance perpendicular to the fault, S and d_1 represent the deep slip rate and locking depth of the fault, respectively, C and d_2 represent the shallow creep rate along the fault strike and the creep depth, respectively, $H(x)$ is the Heaviside step function, and a represents the offset between the data and the model. In the actual inversion process, reasonable constraint ranges for the parameters to be estimated were provided based on previous research results [23–25,42]: $0 < S < 20$ mm/yr, $0 < d_1 < 20$ km, $0 < C < 10$ mm/yr, and $0 < d_2 < 10$ km, $-20 < a < 20$ km. These parameters were assumed to follow a uniform prior distribution within the specified ranges, with an additional constraint $d_2 < d_1$. After eliminating some obvious outliers and results with large errors (Figure 5), we used the Levenberg–Marquardt (L-M) method to linearize the nonlinear equations through a Taylor series expansion. The best-fit values and uncertainties for the parameters in the model were then determined through iterative calculations (Figure 5).

4.2. Results of GNSS Profile Inversion

The inversion results show (Figure 5a) that the Laohushan Fault currently exhibits a coupled motion characteristic with shallow creep and deep locking (the fault is not fully open). Specifically, the shallow left-lateral creeping rate is 1.5 ± 0.7 mm/yr, with a creep depth of 5.7 ± 3.4 km. The left-lateral elastic loading rate due to deep tectonic loading is 3.9 ± 1.1 mm/yr, and the elastic locking depth ranges from 5.7 ± 3.4 km to 16.8 ± 3.2 km. In contrast, the same inversion model applied to the Maomaoshan Fault profile (Figure 5b) does not reveal evidence of shallow fault creep. The left-lateral slip rate for the Maomaoshan Fault is determined to be 4.6 ± 0.6 mm/yr, with a locking depth of 16.8 ± 4.2 km, indicating a higher degree of strain accumulation compared to the Laohushan Fault.

Additionally, the horizontal offsets, estimated as free parameters, are 2.3 ± 1.3 km for the Laohushan Fault and 4.5 ± 1.5 km for the Maomaoshan Fault. These offsets correspond closely with the geological fault baselines, providing indirect validation of the observational data and inversion results. Furthermore, the horizontal shortening/extension rates across the faults, inferred from differential motion, suggest varying degrees of crustal shortening at both the Laohushan and Maomaoshan faults, with magnitudes not exceeding 1 mm/yr (Figure 5c,d).

5. Discussion

5.1. Comparison with Existing InSAR Results

In calculating the fault creep rate using InSAR observations, the effects of regional vertical deformation and deformation perpendicular to the fault cannot be neglected [23,25,26]. To account for these effects, we collected precise leveling data for the region from 1970 to 2012. By applying the dynamic least-squares adjustment method [49], we mitigated systematic errors by incorporating the vertical motion rates from continuous GNSS stations as a priori information. This approach enabled us to derive the vertical crustal deformation field for the study area (Figure 6). As depicted in Figure 6, the Haiyuan region exhibits complex vertical deformation, with vertical displacement rates of 1–2 mm/yr observed near the eastern segment of the Laohushan Fault.

We analyzed GNSS profiles and fault dislocation model inversion results for the Laohushan Fault (Figure 5) and found that its left-lateral creep occurs at shallow depths, above 5.7 ± 3.4 km, with a creep rate of 1.5 ± 0.7 mm/yr. These findings were compared with previously published estimates of the Laohushan Fault's creep rate based on InSAR data (Table 2). As shown in Table 2, the GNSS-derived creep rate is lower than the values obtained using InSAR. Even within studies utilizing the same InSAR dataset (InSAR + GNSS (far-field)), discrepancies persist. Specifically, creep rates derived from combined ascending and descending InSAR observations [23,25,26] are lower than those obtained from single-track InSAR data [19,24].

When calculating fault deformation using InSAR technology, single-track InSAR observations require neglecting both vertical deformation and deformation perpendicular to the fault. In contrast, combined ascending and descending track InSAR data only necessitate ignoring deformation perpendicular to the fault [23,50,51]. Existing leveling observations (Figure 6) suggest a vertical deformation of 1–2 mm/yr near the eastern segment of the Laohushan Fault, while near-field GNSS continuous data from this study (Figures 3 and 5) show 1 mm/yr of deformation perpendicular to the fault. In prior InSAR studies, whether using single-track or combined ascending and descending track observations, the inability to accurately separate vertical crustal deformation from fault-perpendicular deformation has likely led to overestimations of the creep rate of the Laohushan Fault. Therefore, to reliably calculate fault deformation characteristics from InSAR data, it is crucial to fully account for both vertical crustal deformation and deformation perpendicular to the fault.

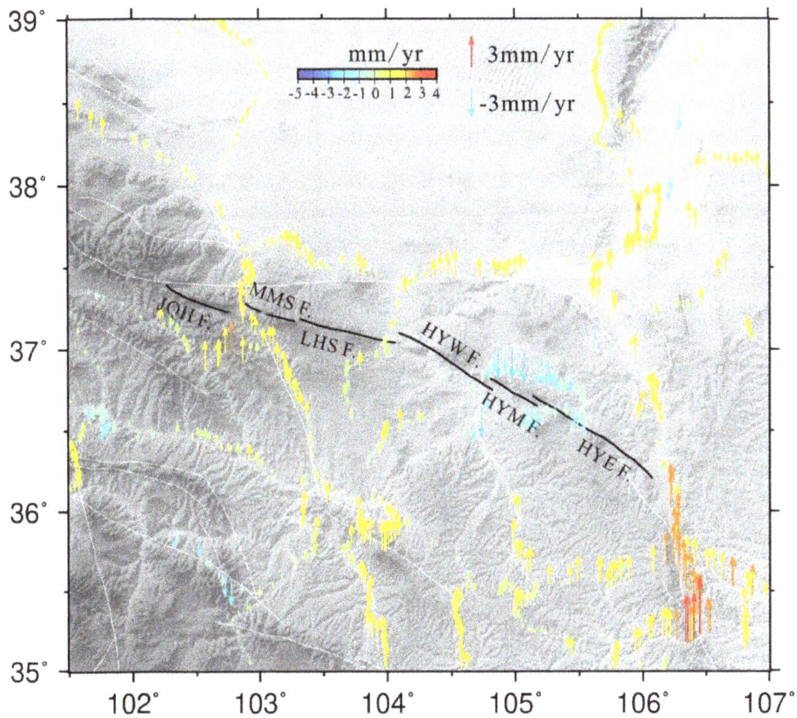

Figure 6. Leveling vertical land motion rates in the Haiyuan Fault and its surrounding regions. Scale of the arrows indicates velocity magnitude, underlined by color. Positive (negative) values denote crustal uplift (subsidence).

Table 2. Summary of creep rates along the Laohushan Fault Zone. (D: descending track InSAR observations; A/D: combined ascending and descending track InSAR observations).

Fault Name	Data Detection	Creep Rate (mm/yr)	Creep Depth (km)	References
LaoHuShan Fault	InSAR (D)	5 ± 1	-	Jolivet et al., 2013 [19]
	InSAR (A/D) + GNSS (Far-field)	2.5 ± 0.4	0~4–8	Li et al., 2021 [23]
	InSAR (D) + GNSS (Far-field)	4.5~4.8	0~0.7 ± 1.3	Qiao et al., 2021 [24]
	InSAR (A/D) + GNSS (Far-field)	2.9 ± 0.4	0~16.9 ± 8.3	Huang et al., 2022 [25]
	InSAR (A/D) + GNSS (Far-field)	2.1 ± 0.1	0~0.3 ± 0.06	Guo et al., 2023 [26]
	GNSS (Near-field)	1.5 ± 0.7	0~5.7 ± 3.4	Our Results

5.2. Temporal and Spatial Variation in Creep Motion on the Laohushan Fault

Considering the lack of observations between June 2021 and March 2024 (Figure 2), and to avoid crustal deformation caused by seasonal changes, we selected observation data from the same season in different years. The slip rates for the two cross-fault GNSS baselines (JT03-JT05 and JT02-JT05) within the near-field of the Laohushan Fault (approximately

5 km) were calculated in segments (Figure 7a). The results show that the creep rate in the near-field of the fault does not exhibit significant changes over different time periods, indicating that the creep motion in the fault's near-field is steady.

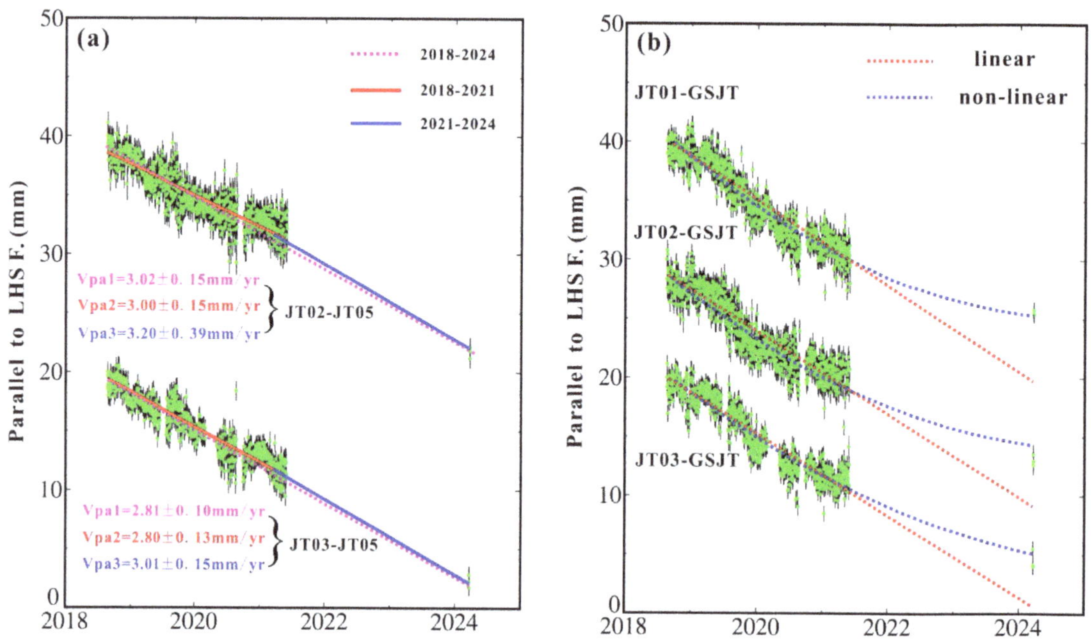

Figure 7. (**a**) Rate fitting of the baseline component parallel to the fault in the near-field across the fault; (**b**) rate fitting of the baseline component parallel to the fault in the far-field across the fault. The red dashed line represents the least-squares linear fit, and the blue dashed line represents the result of the fitting using Equation (2).

In contrast, Figure 3a shows that the cross-fault baseline components parallel to the fault, composed of observation points farther from the fault (greater than 20 km), such as JT03-GSJT, JT02-GSJT, and JT01-GSJT, cannot be accurately fitted with a least-squares linear method for the final observation in 2024. There appears to be significant nonlinear deformation, and the discrepancies in the least-squares linear fit become more pronounced with longer baseline distances. Therefore, we attempt to fit the time series of these three parallel fault baseline components using a postseismic viscoelastic relaxation model (Equation (2)).

$$y = v \times dt + D \times \log(1 + dt/\tau) \quad (2)$$

In this context, y represents the baseline component parallel to the fault, dt denotes the epoch observation time, v indicates the linear rate of change of the baseline component parallel to the fault, and $D \times \log(1 + dt/\tau)$ represents a commonly used postseismic time series fitting function [52–54].

The fitting results presented in this paper (Figure 7b, blue dashed line) demonstrate that the postseismic relaxation model not only closely matches the continuous observations from 2018 to 2021 but also aligns well with the final observations in 2024. Notably, these nonlinear deformation features are absent in the fitting results of the baseline components perpendicular to the fault (Figure 3c). This suggests that if these effects are indeed related to postseismic relaxation from a nearby earthquake, the event in question was likely a left-lateral strike–slip earthquake.

In addition, the nonlinear deformation features observed in the far-field baselines parallel to the fault may be related to postseismic effects from the 1920 Haiyuan earthquake. To explore this possibility, we projected and decomposed the regional high-density GNSS velocity field obtained in this paper (Figure 3) along the direction parallel to the Haiyuan Fault (Figure 8). The results reveal a significant zone of rapid activity south of the 1920 Haiyuan earthquake rupture zone, overlapping with the far-field region of the Laohushan Fault. Additionally, GNSS velocity profiles (Figure 5) indicate that the fault-parallel component on the southern side of the Laohushan Fault is approximately 2 mm/yr faster than that of the Maomaoshan Fault. This observational analysis suggests that the far-field deformation of the Laohushan Fault is probably influenced by postseismic deformation from the 1920 Haiyuan earthquake.

On the other hand, baseline time series and dislocation model inversion results (Figures 3, 5 and 7) indicate that the near-field creep of the Laohushan Fault (within approximately 5 km) does not exhibit the postseismic exponential decay deformation pattern observed in the far-field long baselines. Instead, it shows a stable creep motion. Therefore, the stable shallow creep of the Laohushan Fault is unlikely to be significantly affected by the postseismic deformation of the 1920 Haiyuan earthquake.

5.3. Regional Fault Movement and Earthquake Hazard

Based on the GNSS profiles and dislocation model inversion results for the Laohushan Fault (Figure 5), the left-lateral creep is observed to occur at shallow depths above 5.7 ± 3.4 km, with a rate of 1.5 ± 0.7 mm/yr. This is comparable to the 2–3 mm/yr rate obtained from InSAR data, which accounts for vertical deformation effects [23,25]. Additionally, the results of this study do not support the claim by [22] that the Laohushan Fault is currently experiencing fully open creep, as also suggested by InSAR-based studies [24,42]. Instead, the Laohushan Fault is found to be locked within a depth range of 5.7 ± 3.4 to 18.5 ± 3.2 km (Figure 5).

The characteristics of shallow creep and deep locking observed on the Laohushan Fault are supported by regional seismic activity results [55]. Previous studies indicate that there have been almost no earthquakes of magnitude 3 or greater within the 0–5 km depth range of the Laohushan Fault, which is attributed to the presence of a loose, thick fault gouge in the shallow part of the fault [55]. The collected regional aftershock data from 2008 to 2019 also show that earthquakes of magnitude 3 or greater are primarily concentrated in the 5–20 km depth range of the Laohushan Fault (Figure 9b,c), which aligns closely with the deep fault locking layer position inferred in this study. Additionally, paleoseismic trench studies reveal that the Laohushan Fault has recorded more than two prehistoric earthquakes of magnitude 7.0 or greater since the Holocene [56]. Based on these findings, we conclude that the creep on the Laohushan Fault represents shallow fault activity, while the deep part of the fault is currently locked and experiencing significant strain accumulation, indicating that the fault's potential for a strong earthquake warrants attention.

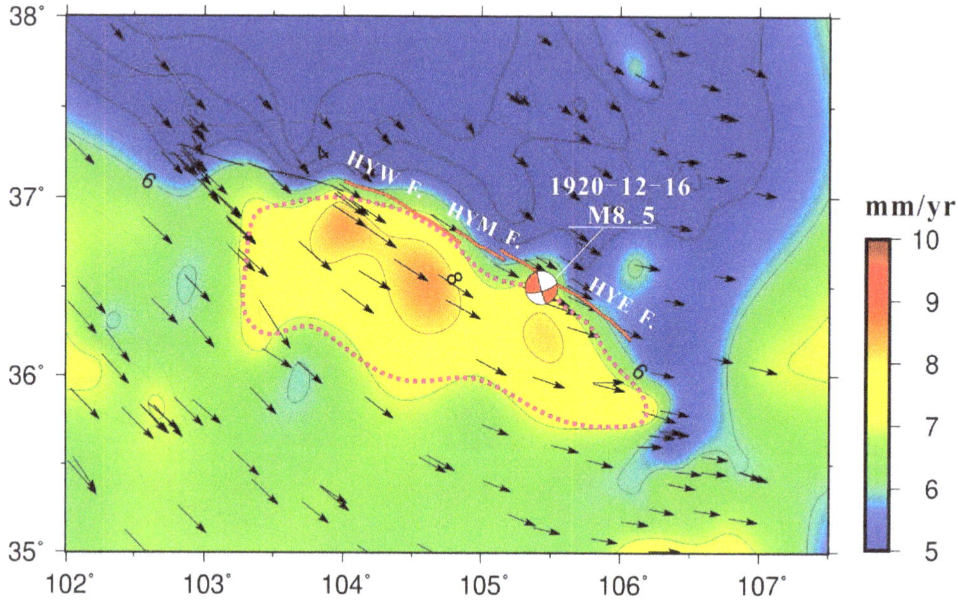

Figure 8. GNSS velocity field parallel to the Haiyuan Fault. The colors in this Figure represent the magnitude of the velocity, while the direction of the arrows indicates the velocity direction. The red beach ball represents the focal mechanism solution [57].

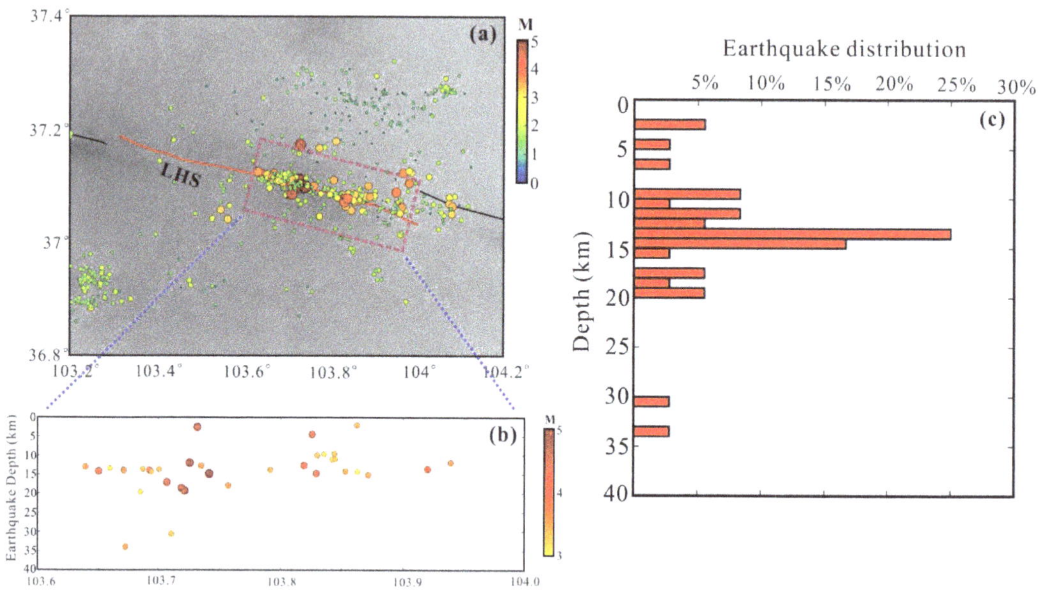

Figure 9. The distribution of small earthquake activity on the Laohushan Fault: (**a**) The spatial distribution of small earthquake activity on the Laohushan Fault. (**b**) Red circles represent the projections of small earthquake activity along the depth of the Laohushan Fault, totaling 36 events. (**c**) The statistical distribution of earthquakes of magnitude 3 or greater with respect to the fault depth.

We compared the inversion results for the Laohushan and Maomaoshan Faults (Figure 4). The Maomaoshan Fault does not exhibit abrupt fault creep signals across the fault but rather shows typical interseismic locking characteristics. The inversion indicates a locking depth of 16.8 ± 4.2 km, which aligns with the 15 km estimate by [42]. The left-lateral slip rate of the Maomaoshan Fault, as determined in this study, is 4.6 ± 0.6 mm/yr, which is consistent with previous geological estimates [13–18] and geodetic estimates of 3 to 5 mm/yr [20,21,58]. Therefore, with its higher slip rate and locking depth, the Maomaoshan Fault currently poses a higher background risk for strong earthquakes.

It is worth noting that the results for the near-field baselines across the Laohushan Fault (approximately 5 km), JT03-JT05 and JT02-JT05 (Figure 3), show a left-lateral slip rate difference across the fault of 2.8–3.0 mm/yr. This is higher than the 1.5 ± 0.7 mm/yr creep rate estimated from the GNSS velocity profile (Figure 5b). This discrepancy may be due to the creep on the Laohushan Fault being concentrated in a smaller region on either side of the fault [23,25], which is smaller than the spacing of the existing GNSS near-field measurements. Therefore, future research should consider integrating dense GNSS and InSAR data to more accurately determine the creep parameters of the Laohushan Fault.

6. Conclusions

We used the latest densely spaced GNSS continuous observations across the Laohushan Fault and regional multi-source GNSS data as constraints to process and obtain precise baseline time series across the Laohushan Fault, as well as a densely monitored regional GNSS horizontal velocity field. The comprehensive analysis of these data reveals that the Laohushan Fault currently exhibits a left-lateral creep motion with characteristics of shallow creep and deep locking. The current fault creep mainly occurs in the shallow crust at depths of 0 to 5.8 ± 3.4 km at a rate of 1.5 ± 0.7 mm/yr. Below the creeping layer, down to a depth range of 16.8 ± 4.2 km, the fault is in a locked state with a locking rate of 3.9 ± 1.1 mm/yr. Additionally, the near-field creep motion (approximately 5 km) of the Laohushan Fault appears stable and seems unaffected by the postseismic effects of historical strong earthquakes in the adjacent area. However, the deep (far-field) tectonic deformation of the Laohushan Fault may be influenced by the postseismic effects of the 1920 Haiyuan M8.5 earthquake.

Author Contributions: Conceptualization, W.Z.; methodology, Y.L.; software, W.Z. and M.H.; validation, W.Z.; formal analysis, W.Z. and Y.L.; investigation, W.Z., S.S., M.H., B.L. and L.F.; data curation, W.Z.; writing, W.Z. and Y.L.; visualization, W.Z.; supervision, M.H. All authors have read and agreed to the published version of the manuscript.

Funding: This work was supported by the National Science Foundation of China (No. 42072243), the Science for Earthquake Resilience (No. XH24059YA), the Natural Science Basic Research Program of Shaanxi (program No. 2023-JC-QN-0309; program No. 2024JC-YBQN-0313), the National Natural Science Foundation of China (No. 12203060), and the Youth Innovation Promotion Association CAS (grant No. 2023426).

Data Availability Statement: All the processed data, models, and computing programs that support the findings of this study are available from the corresponding author upon request.

Acknowledgments: The Figures in this paper are made by GMT tools. We thank the editor, assistant editor, and anonymous reviewers for their constructive comments that improved the manuscript. The authors thank all the staff who participated in the GNSS surveys. We are grateful to the Crustal Movement Observation Network of China and the National Geodetic Control Network of China for providing part of the GNSS data.

Conflicts of Interest: The authors declare no conflicts of interest.

References

1. Burgmann, R.; Schmidt, D.; Nadeau, R.M.; d'Alessio, M.; Fielding, E.; Manaker, D.; Murray, M.H. Earthquake potential along the northern Hayward fault, California. *Science* **2000**, *289*, 1178–1182. [CrossRef]

2. Ryder, I.; Burgmann, R. Spatial variations in slip deficit on the central San Andreas Fault from InSAR. *Geophys. J. Int.* **2008**, *175*, 837–852. [CrossRef]
3. Cattin, R.; Avouac, J.P. Modeling mountain building and the seismic cycle in the Himalaya of Nepal. *J. Geophys. Res. Solid Earth* **2000**, *105*, 13389–13407. [CrossRef]
4. Barbot, S.; Lapusta, N.; Avouac, J.P. Under the hood of the earthquake machine: Toward predictive modeling of the seismiccycle. *Science* **2012**, *336*, 707–710. [CrossRef] [PubMed]
5. Scholz, C.H. Earthquakes and friction laws. *Nature* **1998**, *391*, 37–42. [CrossRef]
6. Kaneko, Y.; Avouac, J.P.; Lapusta, N. Towards inferring earthquake patterns from geodetic observations of interseismic coupling. *Nat. Geosci.* **2010**, *3*, 363–369. [CrossRef]
7. Jolivet, R.; Candela, T.; Lasserre, C.; Renard, F.; Klinger, Y.; Doin, M.P. The burst-like behavior of aseismic slip on a rough fault: The creeping section of the Haiyuan fault, China. *Bull. Seismol. Soc. Am.* **2015**, *105*, 480–488. [CrossRef]
8. Lohman, R.B.; McGuire, J.J. Earthquake swarms driven by aseismic creep in the Salton Trough, California. *J. Geophys. Res.* **2007**, *112*, B04405. [CrossRef]
9. Zhang, J.; Wen, X.Z.; Cao, J.L.; Yan, W.; Yang, Y.L.; Su, Q. Surface creep and slip-behavior segmentation along the northwestern Xianshuihe fault zone of southwestern China determined from decades of fault-crossing short-baseline and short-level surveys. *Tectonophysics* **2018**, *722*, 356–372. [CrossRef]
10. Lisowski, M.; Prescott, W.H. Short-range distance measurements along the San Andreas fault system in central California, 1975 to 1979. *Bull. Seismol. Soc. Am.* **1981**, *71*, 1607–1624. [CrossRef]
11. Ambraseys, N.N. Some characteristic features of the Anatolian fault zone. *Tectonophysics* **1970**, *9*, 143–165. [CrossRef]
12. Cakir, Z.; Akoglu, A.M.; Belabbes, S.; Ergintav, S.; Meghraoui, M. Creeping along the Ismetpasa section of the North Anatolian fault (Western Turkey): Rate and extent from InSAR. *Earth Planet. Sci. Lett.* **2005**, *238*, 225–234. [CrossRef]
13. Ohzono, M.; Sagiya, T.; Hirahara, K.; Hashimoto, M.; Takeuchi, A.; Hoso, Y.; Wada, Y.; Onoue, K.; Ohya, F. Doke Strain accumulation process around the Atotsugawa fault system in the Niigata-Kobe Tectonic Zone, central Japan. *Geophys. J. Int.* **2011**, *184*, 977–990. [CrossRef]
14. Anderlini, L.; Serpelloni, E.; Belardinelli, M.E. Creep and locking of a low-angle normal fault: Insights from the Altotiberina fault in the northern Apennines (Italy). *Geophys. Res. Lett.* **2016**, *43*, 4321–4329. [CrossRef]
15. Hamiel, Y.; Piatibratova, O.; Mizrahi, Y. Creep along the northern Jordan Valley section of the Dead Sea Fault. *Geophys. Res. Lett.* **2016**, *43*, 2494–2501. [CrossRef]
16. Li, C.; Zhang, P.Z.; Yin, J.; Min, W. Late Quaternary left-lateral slip rate of the Haiyuan fault, northeastern margin of the Tibetan Plateau. *Tectonics* **2009**, *28*. [CrossRef]
17. Liu-Zeng, J.; Shao, Y.; Klinger, Y.; Xie, K.; Yuan, D.; Lei, Z. Variability in magnitude of paleoearthquakes revealed by trenching and historical records, along the Haiyuan Fault, China. *J. Geophys. Res. Solid Earth* **2015**, *120*, 8304–8333. [CrossRef]
18. Shao, Y.; Liu-Zeng, J.; Van der Woerd, J.; Klinger, Y.; Oskin, M.E.; Zhang, J.; Yao, W. Late Pleistocene slip rate of the central Haiyuan fault constrained from optically stimulated luminescence, 14C, and cosmogenic isotope dating and high-resolution topography. *GSA Bull.* **2021**, *133*, 1347–1369. [CrossRef]
19. Jolivet, R.; Lasserre, C.; Doin, M.P.; Guillaso, S.; Peltzer, G.; Dailu, R.; Sun, Z.-K.; Shen, X.X. Shallow creep on the Haiyuan fault (Gansu, China) revealed by SAR interferometry. *J. Geophys. Res. Solid Earth* **2012**, *117*, B06401. [CrossRef]
20. Gan, W.; Zhang, P.; Shen, Z.K.; Niu, Z.; Wang, M.; Wan, Y.; Cheng, J. Present-day crustal motion within the Tibetan Plateau inferred from GPS measurements. *J. Geophys. Res. Solid Earth* **2007**, *112*, B08416. [CrossRef]
21. Li, Y.; Liu, M.; Wang, Q.; Cui, D. Present-day crustal deformation and strain transfer in northeastern Tibetan Plateau. *Earth Planet. Sci. Lett.* **2018**, *487*, 179–189. [CrossRef]
22. Jolivet, R.; Lasserre, C.; Doin, M.P.; Peltzer, G.; Avouac, J.P.; Sun, J.; Dailu, R. Spatio-temporal evolution of aseismic slip along the Haiyuan fault, China: Implications for fault frictional properties. *Earth Planet. Sci. Lett.* **2013**, *377*, 23–33. [CrossRef]
23. Li, Y.; Nocquet, J.M.; Shan, X.; Song, X. Geodetic observations of shallow creep on the Laohushan-Haiyuan fault, northeastern Tibet. *J. Geophys. Res. Solid Earth* **2021**, *126*, e2020JB021576. [CrossRef]
24. Qiao, X.; Qu, C.; Shan, X.; Zhao, D.; Liu, L. Interseismic slip and coupling along the Haiyuan fault zone constrained by InSAR and GPS measurements. *Remote Sens.* **2021**, *13*, 3333. [CrossRef]
25. Huang, Z.; Zhou, Y.; Qiao, X.; Zhang, P.; Cheng, X. Kinematics of the ~1000 km Haiyuan fault system in northeastern Tibet from high-resolution Sentinel-1 InSAR velocities: Fault architecture, slip rates, and partitioning. *Earth Planet. Sci. Lett.* **2022**, *583*, 117450. [CrossRef]
26. Guo, N.; Wu, Y.; Su, G. Analysis of the fault slip, creep, and coupling characteristics of the Maomaoshan-Laohushan-Haiyuan Fault using InSAR and GNSS measurements. *Tectonophysics* **2023**, *863*, 229988. [CrossRef]
27. Hao, M.; Wang, Q.; Shen, Z.; Cui, D.; Ji, L.; Li, Y.; Qin, S. Present day crustal vertical movement inferred from precise leveling data in eastern margin of Tibetan Plateau. *Tectonophysics* **2014**, *632*, 281–292. [CrossRef]
28. Xu, X.; Tong, X.; Sandwell, D.T.; Milliner, C.W.; Dolan, J.F.; Hollingsworth, J.; Ayoub, F. Refining the shallow slip deficit. *Geophys. J. Int.* **2016**, *204*, 1867–1886. [CrossRef]
29. Marchandon, M.; Hollingsworth, J.; Radiguet, M. Origin of the shallow slip deficit on a strike slip fault: Influence of elastic structure, topography, data coverage, and noise. *Earth Planet. Sci. Lett.* **2021**, *554*, 116696. [CrossRef]

30. Wu, X.; Xu, X.; Yu, G.; Ren, J.; Yang, X.; Chen, G.; Hao, H. The China Active Faults Database (CAFD) and its web system. *Earth Syst. Sci. Data* **2024**, *16*, 3391–3417. [CrossRef]
31. Pang, Y.J.; Zhang, H.; Cheng, H.H.; Dong, P.Y.; Wang, J.J.; Shi, Y.L. Three-dimensional numerical simulation of pull-apart basins: An example of the Laolongwan basin in the Haiyuan fault zone. *Chin. J. Geophys.* **2015**, *58*, 3615–3626. [CrossRef]
32. Song, Z.; Zhang, G.; Liu, J.; Yin, J.; Xue, Y.; Song, X. *Disaster Information Catalog of Global Earthquakes (9999 B.C. to 2010 A.D.)*; Seismological Press: Beijing, China, 2011. (In Chinese with English Preface)
33. Wang, M.; Shen, Z.K.; Gan, W.J.; Liao, H.; Li, T.; Ren, J.; Qiao, X.; Wang, Q.; Yang, Y.; Li, P.; et al. GPS Continuous Monitoring of Dynamic Evolution of Deformation Field of Xianshuihe Fault. *Sci. China Ser. D* **2008**, *38*, 575–581.
34. Huang, L.R.; Fu, Y. Analysis on the noises from continuously monitoring GPS sites. *Acta Seismol. Sin.* **2007**, *20*, 197–202. [CrossRef]
35. Hao, M.; Li, Y.; Wang, Q.; Zhuang, W.; Qu, W. Present-day crustal deformation within the western Qinling Mountains and its kinematic implications. *Surv. Geophys.* **2021**, *42*, 1–19. [CrossRef]
36. Herring, T.A.; King, R.W.; McClusky, S.C. *GAMIT Reference Manual, GNSS Analysis at MIT, Release 10.6*; Massachusetts Institute of Technology: Cambridge, MA, USA, 2015.
37. Herring, T.A.; King, R.W.; McClusky, S.C. *GAMIT Reference Manual, Global Kalman Filter VLBI and GNSS Analysis Program, Release 10.6*; Massachusetts Institute of Technology: Cambridge, MA, USA, 2015.
38. Petit, G. *IERS Conventions (2010)*; U.S. Naval Observatory Observatoire de Paris: Paris, France, 2010.
39. Lyard, F.; Lefevre, F.; Letellier, T.; Francis, O. Modelling the global ocean tides: Modern insights from FES2004. *Ocean. Dyn.* **2006**, *56*, 394–415. [CrossRef]
40. Altamimi, Z.; Métivier, L.; Rebischung, P.; Rouby, H.; Collilieux, X. ITRF2014 plate motion model. *Geophys. J. Int.* **2017**, *209*, 1906–1912. [CrossRef]
41. Wang, M.; Shen, Z.K. Present-day crustal deformation of continental China derived from GPS and its tectonic implications. *J. Geophys. Res. Solid Earth* **2020**, *125*, e2019JB018774. [CrossRef]
42. Li, Y.; Shan, X.; Qu, C.; Zhang, Y.; Song, X.; Jiang, Y.; Wang, C. Elastic block and strain modeling of GPS data around the Haiyuan-Liupanshan fault, northeastern Tibetan Plateau. *J. Asian Earth Sci.* **2017**, *150*, 87–97. [CrossRef]
43. Zhuang, W.; Cui, D.; Hao, M.; Song, S.; Li, Z. Geodetic constraints on contemporary three-dimensional crustal deformation in the Laji Shan–Jishi Shan tectonic belt. *Geod. Geodyn.* **2023**, *14*, 589–596. [CrossRef]
44. Fattahi, H.; Amelung, F. InSAR observations of strain accumulation and fault creep along the Chaman Fault system, Pakistan and Afghanistan. *Geophys. Res. Lett.* **2016**, *43*, 8399–8406. [CrossRef]
45. Savage, J.C.; Burford, R.O. Geodetic determination of relative plate motion in central California. *J. Geophys. Res.* **1973**, *78*, 832–845. [CrossRef]
46. Savage, J.C. A dislocation model of strain accumulation and release at a subduction zone. *J. Geophys. Res. Solid Earth* **1983**, *88*, 4984–4996. [CrossRef]
47. Savage, J.C. Viscoelastic-coupling model for the earthquake cycle driven from below. *J. Geophys. Res. Solid Earth* **2000**, *105*, 25525–25532. [CrossRef]
48. Harris, R.A. Large earthquakes and creeping faults. *Rev. Geophys.* **2017**, *55*, 169–198. [CrossRef]
49. Li, Z.; Hao, M.; Hammond, W.C.; Cheng, F.; Zhang, G.; Wang, Q.; Liu, L.; Hou, B.; Gan, W. Geodetic constraints on three-component motion of the Ordos block (China) and their implications for lithospheric dynamics. *Geol. Soc. Am. Bull.* **2024**. [CrossRef]
50. Liu, C.; Ji, L.; Zhu, L.; Xu, C.; Qiu, J. Interseismic strain rate distribution model of the Altyn Tagh Fault constrained by InSAR and GPS. *Earth Planet. Sci. Lett.* **2024**, *642*, 118884. [CrossRef]
51. Hao, M.; Wang, Q.L.; Cui, D.X.; Liu, L.W.; Zhou, L. Present-day crustal vertical motion around the Ordos block constrained by precise leveling and GNSS data. *Surv. Geophys.* **2016**, *37*, 923–936. [CrossRef]
52. Marone, C.J.; Scholtz, C.H.; Bilham, R. On the mechanics of earthquake afterslip. *J. Geophys. Res. Solid Earth* **1991**, *96*, 8441–8452. [CrossRef]
53. Tian, Z.; Freymueller, J.T.; Yang, Z. Postseismic Deformation due to the 2012 Mw 7.8 Haida Gwaii and 2013 Mw 7.5 Craig Earthquakes and its Implications for Regional Rheological Structure. *J. Geophys. Res. Solid Earth* **2021**, *126*, e2020JB020197. [CrossRef]
54. Tian, Z.; Freymueller, J.T.; Yang, Z.; Li, Z.; Sun, H. Frictional Properties and Rheological Structure at the Ecuadorian Subduction Zone Revealed by the Postseismic Deformation Due to the 2016 Mw 7.8 Pedernales (Ecuador) Earthquake. *J. Geophys. Res. Solid Earth* **2023**, *128*, e2022JB025043. [CrossRef]
55. Liu, B.Y.; Yin, Z.W.; Yuan, D.Y.; Li, L.; Wang, W. The research on fault plane solution and geometric meaning of the Laohushan fault in the northeastern Tibetan plateau. *Seismol. Geol.* **2023**, *42*, 1354–1369.
56. Liu-Zeng, J.; Klinger, Y.; Xu, X.; Lasserre, C.; Chen, G.; Chen, W.; Zhang, B. Millennial recurrence of large earthquakes on the Haiyuan fault near Songshan, Gansu Province, China. *Bull. Seismol. Soc. Am.* **2007**, *97*, 14–34. [CrossRef]

57. Li, Y.; Shan, X.; Gao, Z.; Qu, C. Interseismic Coupling–Based Stochastic Slip Modeling of the 1920 Ms 8.5 Haiyuan Earthquake. *Seismol. Res. Lett.* **2024**, *95*, 870–878. [CrossRef]
58. Loveless, J.P.; Meade, B.J. Partitioning of localized and diffuse deformation in the Tibetan Plateau from joint inversions of geologic and geodetic observations. *Earth Planet. Sci. Lett.* **2011**, *303*, 11–24. [CrossRef]

remote sensing

MDPI

Article

A Deep-Learning Based GNSS Scene Recognition Method for Detailed Urban Static Positioning Task via Low-Cost Receivers

Yubo Li [1], Zhuojun Jiang [1], Chuang Qian [2,*], Wenjing Huang [1] and Zeen Yang [1]

[1] School of Navigation, Wuhan University of Technology, Wuhan 430063, China; yubo_li@whut.edu.cn (Y.L.); zhuojun_jiang@whut.edu.cn (Z.J.); jennifer@whut.edu.cn (W.H.); jw3194557044@whut.edu.cn (Z.Y.)

[2] Intelligent Transportation Systems Research Center, Wuhan University of Technology, Wuhan 430063, China

* Correspondence: qian_c@whut.edu.cn

Abstract: Global Navigation Satellite Systems (GNSS)-based position service is widely applied in cities, but the precision varies significantly in different obstruction scenes. Scene recognition is critical for developing scene-adaptive GNSS algorithms. However, the complexity of urban environments and the unevenness of received signal especially in low-cost receivers limit the performance of GNSS-based scene recognition models. Therefore, our study aims to construct a scene recognition model suitable for urban static positioning and low-cost GNSS receivers. Firstly, we divide the scenes into five categories according to application requirements, including open area, high urban canyon, unilateral urban canyon, shade of tree and low urban canyon. We then construct feature vectors from original observation data and consider the geometric relationships between satellites and receivers. The different sensitivity to different scenes is discovered through an analysis of the performance of each feature vector in recognition. Therefore, a GNSS positioning scene recognition model based on multi-channel LSTM (MC-LSTM) is proposed. The results of experiments show that an accuracy of 99.14% can be achieved by our model. Meanwhile, only 0.75 s and 1.95 ms are required in model training per epoch and model prediction per data on a CPU, which presents a significant improvement of over 90% compared with existing works. Furthermore, our model can be transferred into different time periods quickly and can maintain robustness in situations where one or two types of observation data are missed. A maximum accuracy of 81.13% can be achieved when two channels are missed, while 96.06% is attainable when one channel is missed. Therefore, our model has the potential for real applications in complex urban environments.

Keywords: GNSS scene recognition; static urban positioning; deep learning; long short-term memory; transfer learning

1. Introduction

Global Navigation Satellite Systems (GNSS) have undergone rapid development and progress in past decades. With the successive establishment of global services such as GPS, BDS, GLONASS, Galileo and regional services such as QZSS, IRNSS, GNSS-based applications have also widely been implemented. Bike-sharing is a typical example among them. Due to its convenience for short-length travel and last-mile connectivity of public transportation, bike-sharing services have expanded to over 300 cities in mainland China and gained widespread popularity especially in super cities [1]. However, the prosperous growth of bike-sharing has brought a series of problems. Over-deployment by operators, false-parking by users and poor scheduling by public management have caused mass disarray of public sharing-bikes, which has become a hot and troubling issue among citizens [2].

A practical solution to this problem is to use electronic fence technology to regulate users' parking behaviors, which means the bike-sharing service is made unable to deactivate unless they are parked in the range of virtual "parking zones". Unfortunately, shared bikes often need to be parked in complex environments, which means GNSS signals could be easily blocked by street trees or affected by strong reflections from buildings on one or both sides [3]. The deteriorated signal quality leads to a reduction in positioning precision, which limits the performance of electronic fence. To address this issue, adjusting the threshold of electronic fence according to the positioning precision is a feasible approach. However, the precision of positioning significantly varies in different obstruction conditions [4], which makes it important to recognize the obstruction scenes before adjusting. On the other hand, improving the precision of GNSS positioning in complex urban environments is also considered as an effective approach and several studies have been proposed in recent years, including anti-multipath [5,6], array signal processing [7,8], high-sensitivity tracking algorithms [9,10] and so on [11]. However, all these scene-adaptive methods require rapid and accurate recognition, or they may cause unexpected errors when scenes are mismatched.

In order to recognize the positioning scenes, researchers have proposed a series of methods, which can be mainly categorized into two technical approaches: multi-sensor fusion and GNSS signal-based methods. From the perspective of multi-sensor fusion, camera, LiDAR [12–14], and signal sensors such as accelerometers and gyroscopes are put into use to detect the environmental features surrounding GNSS receivers in order to recognize contexts [15,16]. Although accurate context recognition could be accessible by multi-sensor fusion, the high computational power and device costs make it economically unappealing for bike-sharing operators. From the perspective of GNSS signal-based methods, different observation data and data-constructed features are selected to recognize GNSS positioning contexts. In the field of indoor and outdoor detection, C/N_0 and the number of visible satellites are widely used as classification features [17–19]. Recently, more and more research has focused on detailed contexts segmentation and the performances of machine learning in this field. Lai et al. [20] use support vector machine (SVM) to divide the environment into open outdoors, occluded outdoors and indoors, which has reached a recognition accuracy of 90.3%. Dai et al. [21] change the GNSS observation data into sky plots and compare the performances of CNN and Conv-LSTM in recognition of open-outdoor, semi-outdoor, shallow indoor and deep indoor. The results show that CNN can reach an accuracy of 98.82% while Conv-LSTM can reach 99.2%. Zhu et al. [22] conducted an analysis of 196 variables related to visible satellite, satellite distribution, signal strength and multipath effect based on statistical methods to find the most important feature elements. They then used those selected features to recognize contexts and compare the performances of eight machine learning models. The results show that LSTM can reach an accuracy of 95.37% in vehicle-mounted dynamic environmental contexts (open area, urban canyon, boulevard, under viaduct and tunnel) awareness.

The above scene recognition methods show that excellent results can be achieved using GNSS observations to recognize scenes in the condition of lacking multi-sensors, which proves that GNSS-based methods are suitable for shared bikes' application. However, there are still several unsolved issues:

(a) Many studies only consider large-scale scenes. With regard to urban static positioning tasks such as shared-bike parking, more detailed scenes should to be taken into consideration.
(b) Research studies about machine learning in GNSS scenes recognition still remain in the stage of directly using existing neural networks or constructing feature inputs that meet the requirements of these networks. Those space-transfer strategies may lead to the degradation of original data features. Being able to make full use of observation data in scene recognition remains a challenging problem.
(c) Multi-systems and complex deep learning networks are used to improve the accuracy of recognition. However, the multi-system observations and computation consumption may not be available in low-cost receivers and microprocessors equipped on public devices such as shared bikes. Moreover, there are relatively few studies on

transfer learning in GNSS scenes recognition. It is crucial to develop methods that can be quickly transferred into different time periods.

Therefore, summarizing the shortcomings of existing works and considering the characteristics of urban static positioning (shared bike parking) application scenes, we propose a deep-learning-based scene recognition method. Our contributions can be summarized as follows:

(a) A more detailed set of scenes that is suitable for urban static positioning is proposed, including open area, shade of tree, high urban canyon, low urban canyon and unilateral urban canyon. A dataset of **15,000 epochs (3000 s for each scene)** is collected for research.
(b) A spatio-temporal correlated method for constructing raw data features is used to analyze the importance of different observation data in recognition based on machine learning. A multi-channel Long Short-Term Memory (MC-LSTM) network for GNSS scene recognition is proposed. The result shows that our method can achieve an accuracy of **99.14%** under the condition of using low-cost GNSS receivers to observe **single satellite navigation system**.
(c) In order to manage the degradation in performances between different time periods, we conduct the transfer learning test of our method's model. The result shows that our pre-trained model can be fine-tuned with a small number of epochs to adapt to different time periods at the same location, which is cost-acceptable for bike-sharing operators.

The remainder of this article is organized as follows: Section 2 introduces the methodology of our proposed model. Section 3 presents the experiments and an analysis of the results. Section 5 provides our conclusions and directions for future works.

2. Methodology

In this section, we first present the feature analysis and construction. Then we propose a deep-learning-based scene recognition model. The overview of our model is shown in Figure 1. The observation data are firstly constructed into different feature vectors. Then those feature vectors are fed into a multi-channel long short-term memory network (MC-LSTM) to identify five scenes (open area, high urban canyon, unilateral urban canyon, shade of tree, low urban canyon).

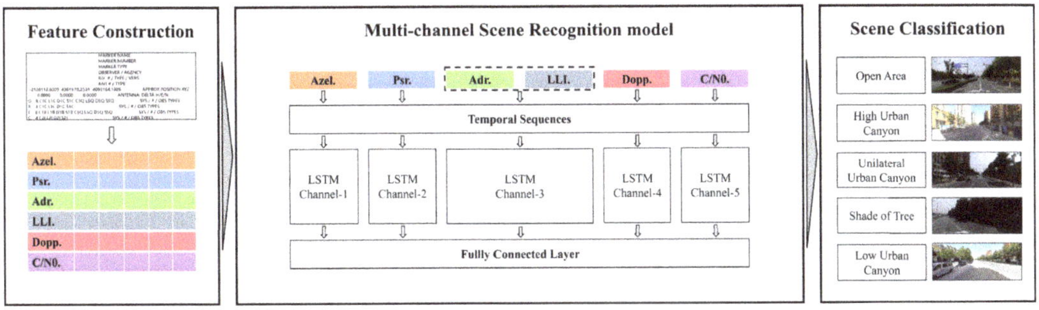

Figure 1. The overview of our proposed model.

2.1. Feature Analysis and Constructions

In this subsection, we first present the observation data and their hidden information. Then we introduce the elevation and azimuth, which shows the geometric relationship between navigation satellites and receivers. Finally, we propose the feature vector definition for our model's input.

2.1.1. GNSS Observations

Global Navigation Satellite Systems (GNSS) typically consist of three segments: the space (space navigation constellation and satellites), the ground control (master control station, monitoring stations and uploading stations), and the user terminals. Users can receive carrier signals transmitted by the space satellites by various of terminals (receivers). The carrier signals are modulated with ranging codes and navigation messages which can be processed by receivers into different observation data, including the Pseudo range [23], Carrier Phase [24], Doppler frequency [25] and C/N_0 [26].

Pseudo range: The pseudo range observation is an absolute measurement of the distance from the satellite to the receiver, obtained by measuring the signal propagation time delay, and is expressed in meters. The pseudo range includes atmospheric delays and clock biases, and its accuracy is typically at the meter level.

Carrier Phase: Carrier phase observations measure the phase difference between the satellite carrier signal and the reference carrier signal produced by the receiver's oscillator, making it a relative observation. It is expressed in cycles and includes atmospheric delays, clock biases and ambiguity of whole cycles.

Doppler frequency: The frequency of the GNSS carrier signal received by the receiver differs from the actual frequency transmitted by the satellite. This difference is known as the Doppler shift, and its magnitude is related to the rate of change in distance between the receiver and the satellite. Specifically, the Doppler observation equation for GNSS is as follows:

$$\lambda D_r^s = \dot{\rho}_r^s - \left(\dot{\delta t}_r - \dot{\delta t}^s\right)c - \delta_{ion}^{\dot{s}} + \delta_{trop}^{s^{\cdot}} \tag{1}$$

where D_r^s is the Doppler frequency and it is expressed in Hz. λ is the wavelength of the carrier wave. $\dot{\rho}_r^s$ is the rate of change of the distance between the receiver (subscript r denotes the receiver) and the satellite (superscript denotes the satellite). $\dot{\delta t}^s$ and $\dot{\delta t}^s$ is the clock drift of the receiver and the satellite. $\delta_{ion}^{\dot{s}}$ and $\delta_{trop}^{s^{\cdot}}$ is the time derivatives of ionospheric and tropospheric delays.

C/N$_0$: The carrier-to-noise density ratio (or CN_0 in more straightforward terms) refers to the ratio of the signal power to the noise power per unit bandwidth in the received GNSS signal. It is expressed in dB and reflects the quality of the RF signal received by the receiver.

Then we consider the characteristics of above observations in urban static positioning scene. Under the conditions of no signal occlusion, the distance between the receiver and the satellite changes in a predictable manner due to the receiver being stationary and the satellite moving in a fixed orbit. In different obstruction scenes, the nature of the obstructions and the characteristics of multipath reflections cause varying rates of change in the distance between the receiver and the satellite. Therefore, we believe that the temporal characteristics of the observations can serve as a basis for scene recognition.

All these observation data can be read directly from Receiver Independent Exchange Format (RINEX format) files or the receivers' output data stream without additional calculations. This is favorable for our application scene which only supports edge computing power. In order to unify the data scale, we normalize the raw observations as follows:

$$d_{norm} = \frac{d - d_{min}}{d_{max} - d_{min}} \tag{2}$$

where d_{max} and d_{min} are the maximum and minimum values of a specific type of observation data within a time step, and d is the original observation data. The normalization only requires simple arithmetic operations, but can ensure network convergence [27,28].

2.1.2. Satellite Elevation and Azimuth

The satellite elevation angle refers to the angle between the vector from the user's location to the satellite's position and its projection onto the Earth ellipsoid tangent plane passing through the user's location. The satellite azimuth angle refers to the angle between this projection and the true North coordinate axis on the tangent plane, with counter-clockwise direction considered positive. The satellite elevation angle and azimuth angle are both related to the position of the user's receiver [29]. They reflect certain aspects of satellite signals, ranging accuracy, and multipath effects. As shown in Figure 2, visible satellites often exhibit different geometric distributions under various obstruction scenes. The calculation methods for satellite elevation angle and azimuth angle are as follows [30]:

$$\begin{bmatrix} N_{zs} \\ E_{zs} \\ U_{zs} \end{bmatrix} = H \times \begin{bmatrix} X_{es} - X_p \\ Y_{es} - Y_p \\ Z_{es} - Z_p \end{bmatrix}, H = \begin{bmatrix} -\sin B_p \cos L_p & -\sin B_p \sin L_p & \cos B_p \\ -\sin L_p & \cos L_p & 0 \\ \cos B_p \cos L_p & \cos B_p \sin L_p & \sin B_p \end{bmatrix} \quad (3)$$

where N_{zs}, E_{zs}, U_{zs} represent the satellite position in the local topocentric coordinate system (ENU). X_{es}, Y_{es}, Z_{es} and X_p, Y_p, Z_p represent the satellite position (calculated from the satellite ephemeris) and station position in the Earth-Centered Earth-Fixed coordinate system (ECEF), respectively. H is the transformation matrix between ENU system and the ECEF system. B_p and L_p are the geodetic latitude and longitude of the receiver. Therefore, the elevation ($El.$) and azimuth ($Az.$) angle is calculated as follows:

$$Az = arctan(\frac{E_{zs}}{N_{zs}}), El = arcsin(\frac{U_{zs}}{\sqrt{E_{zs}^2 + N_{zs}^2 + U_{zs}^2}}) \quad (4)$$

From the physical definitions of the satellite elevation angle and azimuth angle, it can be observed that both the elevation and azimuth angles have upper and lower bounds. The elevation angle ranges from 0 to 90 degrees, and the azimuth angle ranges from −180 degrees to 180 degrees. Similarly to the normalization method used for raw observations, we normalize the satellite elevation and azimuth angles as follows:

$$Az_{norm} = \frac{Az - (-180)}{180 - (-180)}, El_{norm} = \frac{El - 0}{90 - 0} \quad (5)$$

where the elevation and azimuth angles are both expressed in degree.

2.1.3. Feature Vector Definition

The feature vector is used as the input of recognition model. In this subsection, we are going to introduce our feature vector definition, which considers both original observation data and geometric relationships between satellites and receivers. We use fixed-length sequential vectors to represent the constellation of a specific satellite navigation system. For example, a 32-dimensional vector is used to store the normalized pseudo range observations of GPS because the pseudo-random noise (PRN) code of satellites is in range of 1 to 32 (G01∼G32).

Additionally, we regard the satellite elevation and azimuth angles as a combined feature, which represents the geometric distribution of satellites that the receiver can use. Instead of changing visible satellites into sky plots like Figure 2, we use a $2N$-dimensional vector to store the normalized angles(where N is the max PRN of the navigation satellite system). Actually, the different dimensions of observation data is the basic of establishment of our multi-channel model, which will be introduced in Section 3 in detail. Our feature definition can be summarized as Table 1. It is worth noting that these vectors of different dimensions can be used both individually and in combination as the input of the multi-channel model.

Table 1. Brief Feature Definition List.

Feature Information	Feature Dimension
Elevation and azimuth angles	$2N$ [1]
Pseudo range	N
Carrier phase	N
Carrier phase and LLI	$2N$
Doppler frequency	N
C/N_0	N

[1] Where N is the max number of PRN for the navigation satellite system. For example, if the satellite system is GPS, $N = 32$.

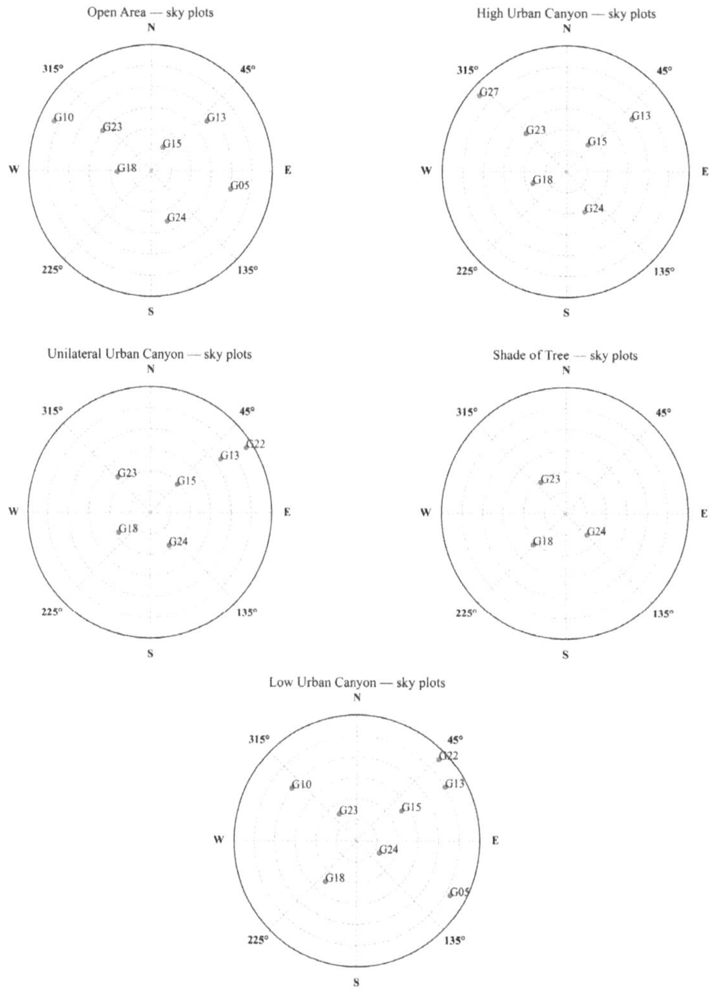

Figure 2. The sky plots of GNSS satellites (GPS system only) in different scenes. This figure shows the elevation and azimuth angles of navigation satellites.

2.2. A Multi-Channel Model for Scene Recognition

In this subsection, we first introduce the Long Short-Term Memory (LSTM) model, which is served as the fundamental building block for the construction of our model. After comparing the performances of different channels (different feature vectors), we propose our multi-channel model to integrate information from different channels. Finally, we consider the transferability of the model and introduce the transfer learning strategy we use in the temporal dimension.

2.2.1. LSTM and Single-Channel Network

Long short-term memory (LSTM) networks are primarily used for learning and predicting temporal features in data sequences. Actually, in the field of GNSS positioning scene recognition, LSTM has been proven to be the most effective network model among numerous machine learning models [21,22,31]. LSTM is composed of numerous LSTM cells Figure 3, which are used to determine whether information is useful. There are three gates in one cell, including a input gate, a forget gate and an output gate. Additionally, a candidate memory cell is set in the LSTM cell in order to process the chosen memory. A hidden state h_{t-1} and the memory passed to the next cell C_t are output by each cell. The key expression of the LSTM cell is as follows [32–34]:

$$I_t = \sigma(X_t W_{xi} + h_{t-1} W_{hi} + b_i) \tag{6}$$

$$F_t = \sigma(X_t W_{xf} + h_{t-1} W_{hf} + b_f) \tag{7}$$

$$O_t = \sigma(X_t W_{xo} + h_{t-1} W_{ho} + b_o) \tag{8}$$

$$\widehat{C}_t = tanh(X_t W_{xc} + h_{t-1} W_{hc} + b_c) \tag{9}$$

$$C_t = F_t \odot C_{t-1} + I_t \odot \widehat{C}_t; \tag{10}$$

where $\odot$ represents the dot product and σ represents the sigmoid activation function. I_t, F_t, O_t are the outputs of the input gate, forget gate and output gate. W_{xi}, $W_h i$; W_{xf}, W_{hf}; W_{xo}, W_{ho} are the weight matrices of input, forget, output gates. W_{xc}, W_{hc} are the weight matrices of the candidate memory cell. b_i, b_f, b_o, b_c are the biases. The hidden state h_t is caculated as follows:

$$h_t = O_t \odot tanh(C_t) \tag{11}$$

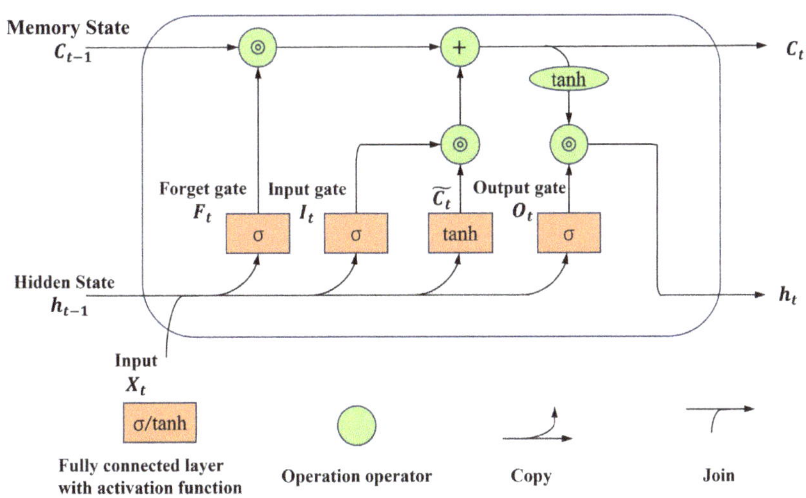

Figure 3. The structure of LSTM cell, including the input gate, forget gate, output gate and the candidate memory cell. The LSTM cell is used to determine whether information is useful.

To change our feature vector to the input of LSTM, the original feature vector sequences should be transmitted into time segments with a sliding window. It means we combine data vectors within a certain time step into a sequence in chronological order and make it the input of single channel. Take pseudo range feature for example, if a N-dimensional vector represents the normalized pseudo range of the time t_0, the input of single LSTM channel is in dimension of (t_s, N), where t_s is the time step of sequence. The overview of our single channel LSTM is shown in Figure 4.

To ensure the real-time performance of the algorithm and consider the practical application requirements, we add only 3 LSTM layers in a single channel which is used to process a single type of feature vector. Meanwhile, the time step is set in 10 s. The sliding window is set in 1 s so that 2990 sequences of each scene can be collected. Similar to the majority of machine learning classification tasks, we use the cross-entropy loss function and optimize the parameters of the network with Adam. The network is trained on training and validation datasets for approximately 500 epochs. The batch size is set to 32 and the learning rate is set to 0.00001.

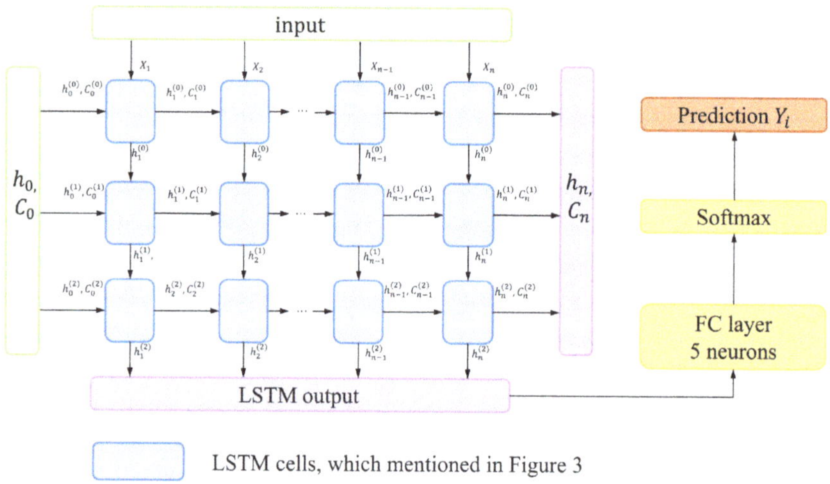

Figure 4. The structure of LSTM network and the full structure of our single channel recognition model. The blue blocks in this figure represent the LSTM cells metioned in Figure 3.

2.2.2. Scene Categories and Single Channel LSTM Performances

Usually, GNSS positioning scenes are divided into two main parts: outdoor and indoor [17–19]. Within urban environments, urban canyons and boulevard are also considered in scene recognition [22,35]. Given the specificity of our application scene (static positioning for shared bikes' parking), we have subdivided urban canyons into high urban canyon, low urban canyon and unilateral urban canyon. So we divide the positioning scenes into five categories: **open area**, **shade of tree**, **high urban canyon**, **low urban canyon** and **unilateral urban canyon**. Due to the bike parking rule, we no longer consider the indoor scenes, which are easily identifiable based on the presence of satellite signals. Figure 5 shows the real-world locations and the street view where we collected data, and Figure 2 shows the distribution of satellites in various scenes.

Figure 5. The street view of real-world locations where we collected data, including open area, high/low/unilateral urban canyon and shade of tree.

In our proposed algorithm, we encode the scene categories as one-hot vectors as Table 2. So the index of the max number of output vector represents the highest possibility of the scene. For example, if the model outputs a vector of $[0.1, 0.1, \mathbf{0.5}, 0.2, 0.1]$, the prediction result of scene recognition is the unilateral urban canyon. Under the training setting mentioned in subsection, we have trained and validated the single channel model and the performances are shown in Figure 6.

Table 2. Scene Categories and Encoding.

Categories	Encoding
open area	[1, 0, 0, 0, 0]
high urban canyon	[0, 1, 0, 0, 0]
unilateral urban canyon	[0, 0, 1, 0, 0]
shade of tree	[0, 0, 0, 1, 0]
low urban canyon	[0, 0, 0, 0, 1]

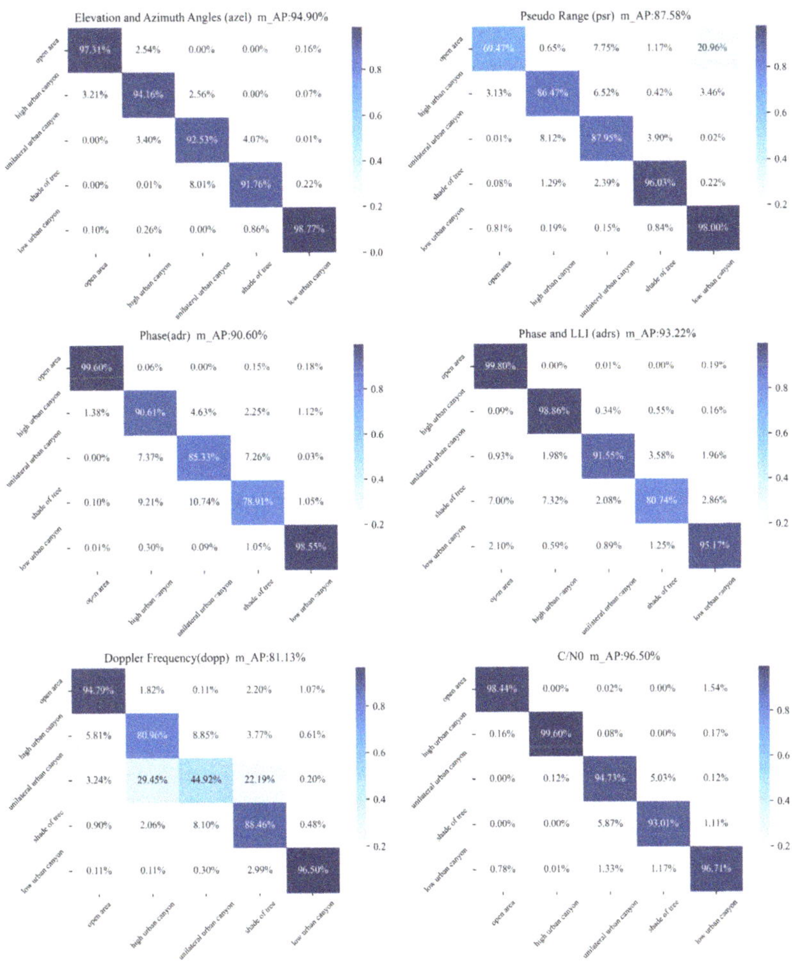

Figure 6. The confusion matrices of different feature vector, which shows the performances of different channels in recognition.

From the confusion matrices shown in Figure 6, we can find that different feature data have varying sensitivities to the recognition of different scenes. The feature of az/el and C/N_0 performs the best in all categories while other feature data performs in specific category (pseudo range in shade of tree, phase + LLI in high urban canyon). Meanwhile, the convergence speeds of the individual single-channel models also show significant differences, as shown in Figure 7. The az/el has the fast convergence speed while the phase has the slowest. Moreover, the performance of phase + LLI is better than single phase no matter the accuracy or the convergence speed, which prompts us to combine the carrier phase and LLI into one feature consideration in the subsequent research.

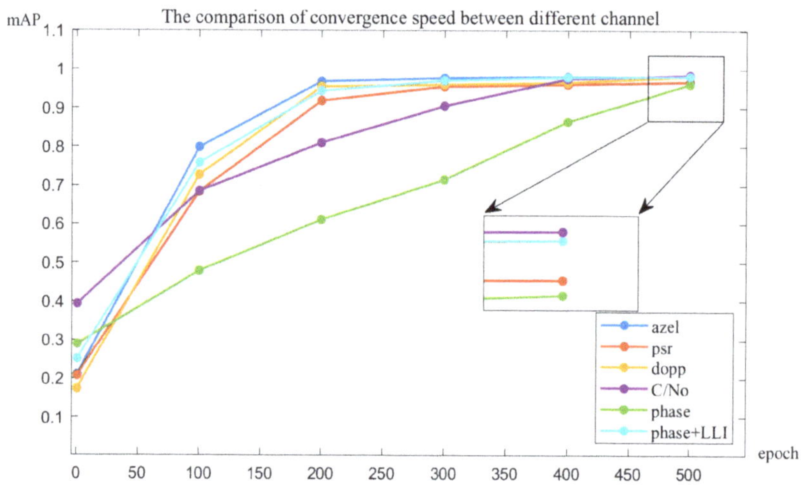

Figure 7. The comparison of convergence speed between different channel, including elevation and azimuth angles (azel), pseudo range (psr), doppler frequency(dopp), C/N_0, phase and phase + LLI. The horizontal axis represents the training epochs, and the vertical axis represents the average accuracy of the model on the validation set.

2.2.3. MC-LSTM Design

In the last subsection, we analysed the performances of different feature vectors. To improve recognition accuracy and convergence speed of our model across all scenes, we integrate information from different feature vectors through a multi-channel parallel design. The overview of our model is shown in Figure 8. The model consists of channel layers and a fusion classification layer. Different feature vectors are input into corresponding LSTM channels. After LSTM processes and outputs hidden states, these are passed through fully connected layers to merge and classify scene recognition results. It is important to note that the dropout layers are used to introduce non-linearity to the model, which helps prevent overfitting and enhances the model's ability to generalize [36,37].

In addition, our proposed multi-channel model allows for flexible combinations of channels based on scenes and requirements. If the receiver cannot output a certain type of observation, the corresponding data vector channel can be excluded.

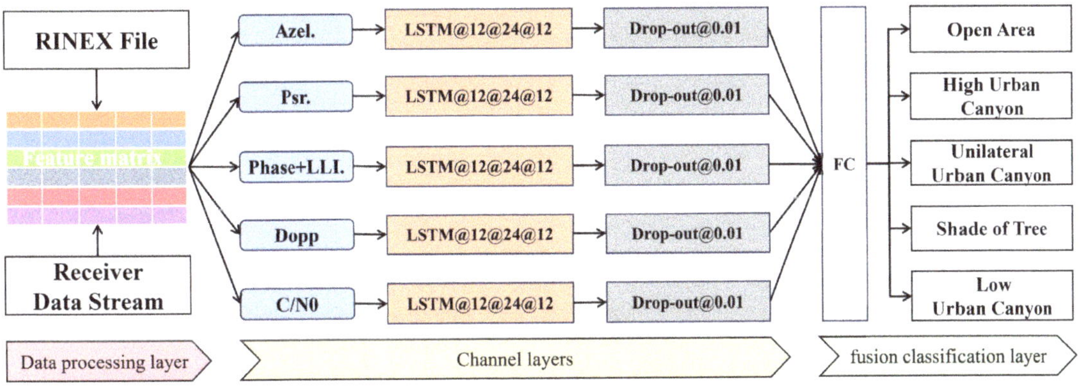

Figure 8. The overview of our proposed multi-channel LSTM model for GNSS scene recognition.

2.2.4. Transfer Learning Settings

A trained model can achieve a high accuracy in training and validation dataset but usually does not perform well in another dataset. Therefore, classification models need to be retrained when changes in the feature space or the feature distribution occur [38]. In our application scene for recognizing shared bicycle parking locations, the absolute position used is generally fixed, but the time when users park is random. It makes the cost exceptionally high for collecting data from all time periods for training. Therefore, the quicker we can execute the transfer learning between time periods, the stronger our model's generalization is. This means the operator of shared bikes can quickly deploy the pre-trained model with simple fine-tuning to different time periods, significantly reducing the time required for model training. To test the temporal transfer performance of our model, we collected data from different time periods and performed full-layer transfer. The distribution of the data over time is shown in Table 3.

Table 3. Transfer Learning Data.

Dataset	Scene	Time of First Observation	Lasting Time
Dataset-1	open area	10:31/13:04/14:16/15:21/16:26	10 min for each
	high urban canyon	10:43/13:16/14:28/15:32/16:38	10 min for each
	unilateral urban canyon	10:57/13:33/14:40/15:45/16:51	10 min for each
	shade of tree	11:16/13:44/14:55/16:00/17:15	10 min for each
	low urban canyon	11:34/14:02/15:07/16:13/17:28	10 min for each
Dataset-2 *	open area	11:09	10 min
	high urban canyon	11:21	10 min
	unilateral urban canyon	11:33	10 min
	shade of tree	11:44	10 min
	low urban canyon	11:57	10 min

* The dataset which is used as the target of transfer learning.

3. Results

In this section, we first introduce the device we used and the dataset we collected. We then analyze the performance of our model and compare it with those from other studies. Finally, we evaluate the real-time performance of our proposed model and present the transfer learning results.

3.1. Device and Dataset

The experimental data were collected using a receiver based on HD8020 series chipset (Allystar Technology, Shenzhen, China), which is a low-cost GNSS receiver chip solution widely used in shared bikes. The appearance of the data collection device is shown in

Figure 9. In order to closely replicate the shared bicycle parking scene, we chose not to alter the structure of the bike's positioning system (GNSS module + antenna). Instead, we directly connected the bike's built-in module to a PC via a serial port to obtain data.

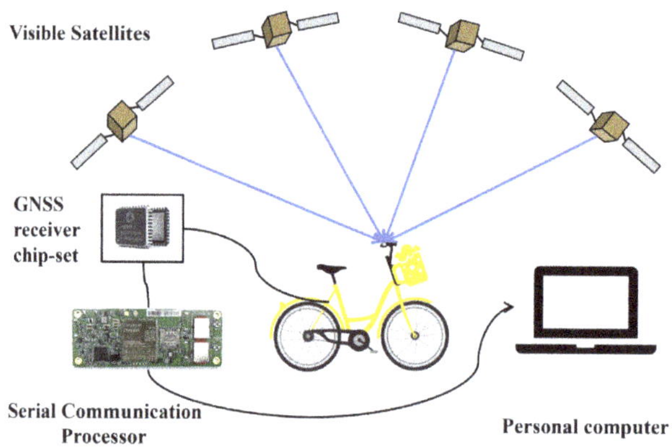

Figure 9. The data collection platform.

To ensure the generalization performance of our proposed model, During each time periods, we collected 10 min's data for each positioning scene (open area, high urban canyon, unilateral urban canyon, shade of tree and low urban canyon). Five such time periods are taken into consideration. Moreover, We shuffled the collected data samples before training and used a cross-validation strategy (70% for training while 30% for validation). As for the dataset for time-transfer-learning, we collected the same distribution data a few days later than training data. The total distribution of the dataset is shown in Table 3.

3.2. Multi-Channel LSTM Model Results

In this subsection, we analyze the performances of our proposed model. First, we introduce the recognition accuracy of our multi-channel LSTM model by confusion matrices and show the changes of accuracy and loss through training epochs by figure. We then compare the results (including accuracy and convergence speeds) of multi-channel model and single channel LSTM mentioned in Section 3. Finally, we compare the performances of our model and other proposed algorithms in recent years.

The confusion matrices and loss and accuracy changes in training epochs are shown in Figure 10. Our model can reach a mean accuracy of 99.14% in all categories and an accuracy of at least 97.5% for each individual category on validation set. From the loss & accuracy changes we can easily find that after only a few training epochs (about 50 epochs) our model is already reaching an accuracy over 90%, which shows it performs well both in recognition as well as training consumption. We observed that the training and validation accuracy of the model initially dropped once during the training process and then quickly increased. This indicates that adding dropout layers in the single channel helps prevent the model from getting stuck in local optima, which aligns with our understanding of nonlinear models [39].

Now we are going to take multi-channel and single-channel into consideration. Compared with the single-channel model mentioned in Section 3, the multi-channel can reach higher accuracy in scene recognition, including mean accuracy of all categories as well as the specific category. The detailed comparison is shown in Table 4.

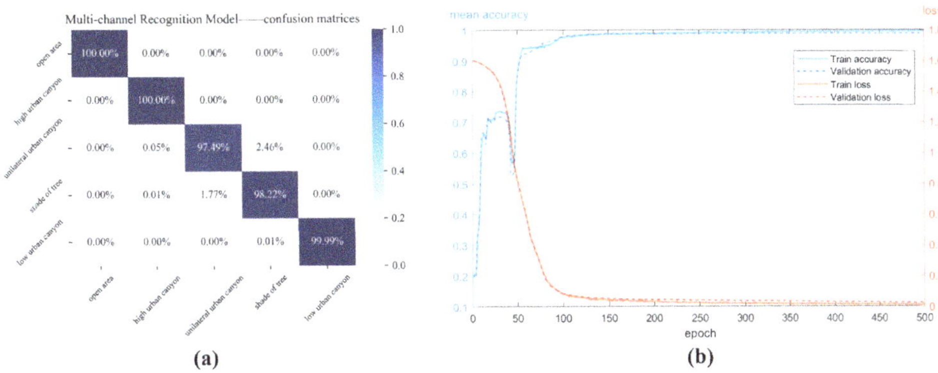

Figure 10. The training results of our proposed multi-channel LSTM GNSS scenes recognition model. (a) is the confusion matrices while (b) shows the changes in loss and accuracy.

Table 4. Comparison between Multi-Channel and Single-Channel.

Channel *	Accuracy in Scenes					Mean Accuracy
	Open Area	High Urban Canyon	Unilateral Urban Canyon	Shade of Tree	Low Urban Canyon	
azel	97.31%	94.16%	92.53%	91.76%	98.77%	94.90%
psr	69.47%	86.47%	87.95%	96.03%	98.00%	87.58%
adr	99.60%	90.61%	85.33%	78.91%	98.55%	90.60%
adrs	99.80%	98.86%	91.55%	80.74%	95.71%	93.22%
dopp	94.79%	80.96%	44.92%	88.46%	96.50%	81.13%
C/N_0	98.44%	99.60%	94.73%	93.01%	96.71%	96.50%
Multi-channel	**100.00%**	**100.00%**	**97.49%**	**98.22%**	**99.99%**	**99.14%**

* Where "azel" represents the azimuth and elevation angles, "psr" represents the pseudo range, "adr" represents the carrier phase, "adrs" represents the carrier phase plus LLI (loss of lock indicator), "dopp" represents the doppler frequency. "multi-channel" represents our proposed recognition model: MC-LSTM.

Despite the accuracy, convergence speed is also an important metric for evaluating model performance. Compared with the single channel model, be similar to accuracy, multi-channel model can also converge faster. Take azimuth and elevation angles channel (the fastest channel in convergence speed) for example, just like Figure 11 shows, the multi-channel model is faster than any other single channel in convergence speed. All results above indicate that our multi-channel design significantly enhances the performance of LSTM in recognizing GNSS positioning scenes, in terms of both accuracy and convergence speed. This demonstrates the benefits of our strategy to integrate multiple channels. It is worth noting that instead directly using existing network, our proposed model design strategy is one of the few specialized network structures specifically designed for GNSS positioning scenes in current related research.

After analyzing the accuracy and convergence speeds, we have already substantiated that our model exhibits superior performance in GNSS positioning scenes recognition. Table 5 shows the comparison between our model and other proposed algorithms in recent years, including accuracy and scene granularity. Our model is the only algorithm which can achieve an accuracy of 99% across five scenes (the scene no GNSS signal received is not included). Indeed, we are also the only researchers who take detailed urban scenes (high, low and unilateral) into consideration.

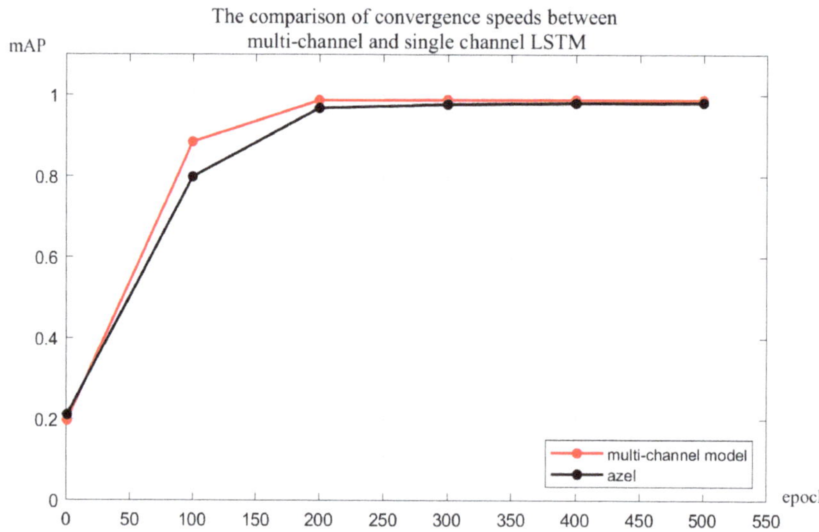

Figure 11. Comparison of convergence speeds between multi-channel and single-channel LSTM.

Table 5. Scene Categories and Encoding.

Authors	Years	Methods	Scenes *	Accuracy
Chen et al. [17]	2017	Threshold judgment	I/O	85.6%
Wang et al. [35]	2019	SVM & temporal filtering	5 urban dynamic scenes	89.30%
Lai et al. [20]	2021	SVM	3 scenes of I/O	90.3%
Dai et al. [21]	2022	CNN & conv-LSTM	4 scenes in I/O	98.82% & 99.92%
Zhu et al. [22]	2024	LSTM	4 urban dynamic scenes	95.39%
Ours	2024	MC-LSTM	**5 urban static scenes**	**99.14%**

* Where the scenes that GNSS signals can not be received is ignored in this table. I/O means the indoor and outdoor scenes.

3.3. Potential Real-Time Ability

In this subsection, we analyze the real-time ability of our proposed model, including the consumption of offline model-training and online model-prediction. We randomly selected 3000 s (2950 time sequences are constructed) of data from the training dataset to evaluate the real-time performance of our model. It is noteworthy that we chose a CPU (Intel (R) Core (TM) i5-12500H, 2.5 GHz) but not a GPU as the computing platform for our real-time performance test and still achieved excellent results.

To evaluate the time consumption of model-training and model-prediction, we first introduce the definition of the metrics: training time per epoch (TPE) and prediction time per data (PPD). Our TPE and PPD is shown in Figure 12. 0.75 s is cost the most in training a dataset of 2950 sequences for one epoch, which makes it possible for our model to do the online fine-tune according to the real-time data. Moreover, only 1.95 ms is needed to perform a full prediction, which makes it possible for our model to be deployed in real-time scenes without any noticeable additional time delay. Compared with existing convolution blocks-relied works, our specially designed multichannel model has a 60% improvement in TPE and 90% in PDD [21,22]. Only 40,817 parameters are included in our model.

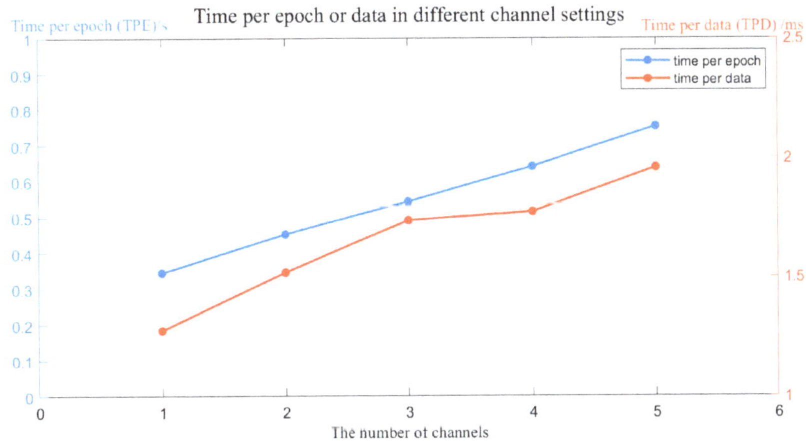

Figure 12. The training time per epoch and prediction time per data of our model. X-axis shows the number of channels. Left y-axis shows the training time per epoch and expresses in seconds. Right y-axis shows the prediction time per data and expresses in milliseconds.

3.4. Time-Transfer Ability

The high precision a pre-trained model can achieve requires that the training data and the test data have similar distributions in feature spaces. Therefore, classification models need to be retrained when changes in the feature space or the feature distribution occur [38]. In our application scenes, we always face that the cost of collecting data of all time periods is unacceptable. Meanwhile, the temporal dimension is a decisive factor in evaluating the generalization capability of our model. So the transfer ability between time periods is taken into consideration in our model. Table 6 shows the details of transfer learning and we can find that our pre-trained model can be fine-tuned with a small number of epochs to adapt to different time periods at the same location. Moreover, the transfer learning performs better than non-transfer (directly) learning in most scenes and training stages. The mean accuracy of transfer learning at every check points (0, 10, 20, 30 and 100 epochs) is higher than non-transfer learning.

Table 6. Comparison between transfer and non-transfer learning.

Scenes *	Accuracy through Epochs									
	0 Epoch		10 Epochs		20 Epochs		30 Epochs		100 Epochs	
	T.	Non-T.	T.	Non-T.	T.	Non-T.	T.	Non-T.	T.	Non-T.
open a.	98.31%	19.84%	100.00%	0.00%	100.00%	85.59%	100.00%	0.00%	100.00%	99.83%
high u.c.	3.39%	19.85%	90.85%	55.76%	94.92%	98.64%	97.63%	100.00%	99.32%	100.00%
unilateral	0.00%	19.68%	98.81%	0.00%	100.00%	0.00%	96.27%	3.90%	96.27%	88.81%
s.o.t.	0.00%	19.97%	0.00%	19.32%	6.61%	88.81%	100.00%	91.53%	100.00%	100.00%
low u.c.	85.93%	20.58%	88.47%	100.00%	95.76%	100.00%	95.93%	99.66%	100.00%	98.64%
mean	37.52%	19.98%	75.62%	35.02%	79.46%	74.61%	97.97%	59.02%	99.12%	97.46%

* Where "open a." represents the open area, "high u.c." represents the high urban canyon, "unilateral" represents the unilateral urban canyon, "s.o.t." represents the shade of tree, "low u.c." represents the low urban canyon. Columns named "T." shows the transfer learning results while "Non-T." shows the non-transfer learning.

3.5. Data-Loss Robustness

Due to the limitations in the number of observation channels of low-cost receivers, the application of our model always faces the situation that some observation data are missing. Therefore, we decide to do an additional experiment to evaluate the robustness of our model in such data-missing situations. As the azimuth and elevation angles are calculated by satellite ephemeris and receiver's position, the pseudo range is the least requirements of calculating the azimuth and elevation angles. So we divide the channels into four groups to test our model's robustness: pseudo range and azimuth & elevation angles (G1), carrier phase and LLI (G2), Doppler frequency (G3), C/N_0 (G4). As the results show in Table 7, our model can achieve an accuracy of over 80% even in the situation that one or two channels' missing. For example, if the carrier phase data and C/N_0 data are missing, our model can also reach an accuracy of 81.13% in scenes recognition.

Table 7. Channel-missing Robustness.

Combinations	Mean Accuracy
G1 & G2	54.80%
G1 & G3	81.13%
G1 & G4	54.24 %
G1 & G2 & G3	90.28%
G1 & G2 & G4	57.17%
G1 & G3 & G4	96.06%

4. Discussion

4.1. Detailed Scenes Recognition in Urban Static GNSS Positioning

Accurate GNSS scenes recognition in urban environments plays an important role in high-precision urban positioning. All the scene-adaptive methods to improve the precision of GNSS positioning require efficient scenes recognition. As shown in Table 5, existing works mainly focus on Indoor and Outdoor recognition [17,20,21] or only consider coarse-grained outdoor scenes [22,35]. Especially for the urban canyon, most research has ignored the different directions of signal blockage. Our proposed model considers the real requirements of urban static positioning application (shared bike parking) and divides the urban canyon scene into high, low and unilateral categories. It achieves an accuracy of over 99% in scenes recognition and is specially designed in multi-channel structure according to the characteristics of GNSS tasks, which represents great progress compared with directly using the existing networks in this field. Detailed scenes and multi-channel network structure design provide a new perspective of GNSS scenes recognition and has the possibility of freely adapting to different tasks according to environments.

4.2. Potential in Real GNSS Scenes Recognition Application

The complexity of urban environment and the limitation of computing power of equipment have always been the main obstacles to the practical application of GNSS scene recognition method based on deep learning. The lightweight network parameter scale, real-time ability and robustness to signal reception failure should be taken into consideration in the real GNSS scenes recognition application. Existing works achieve high precision of recognition by using convolution blocks, which makes the scale of network parameters unacceptable for low-cost computing devices. Our proposed model lightened the hidden layers according to the characteristics of GNSS features and only requires milliseconds of extra time in recognition. Surprisingly, our multi-channel model can also achieve high recognition accuracy even in cases where some observations are lost. All the above findings show that our proposed multi-channel LSTM network structure has a wide potential in real GNSS scenes recognition applications.

5. Conclusions

In this study, a scenes recognition model for urban static GNSS positioning is proposed. The proposed model is able to provide detailed scene information only using a single satellite system's observations via low-cost receivers equipped on the shared bikes. It organizes the original observation data into five feature vectors with different dimensions but without any additional calculation except normalization. After analyzing the performances of different feature vectors, a specially designed multi-channel LSTM model is proposed to further improve the recognition accuracy in different scenes. Experiments are designed to evaluate the proposed model in terms of accuracy, time consumption and robustness. The results show that our proposed model can achieve an average accuracy of 99.14% in all scenes and at least 97.5% in an individual scene. The time delays of model training and prediction on a CPU are 0.75 s per epoch and 1.95 ms per data, which makes it possible for our model to be deployed in real-time applications. Furthermore, our model can be transferred into different time periods with only a few epochs training and maintain a high accuracy even in the presence of missing channel data (96.06% for one channel missed and 81.13% for two channels missed), which shows the robustness of our model in real-world applications. In future works, more detailed and mixed scenes should be taken into consideration. In addition, the contributions from different satellite systems and frequencies require further research.

Author Contributions: Conceptualization, C.Q. and Y.L.; methodology, Y.L.; validation, Y.L., Z.J. and W.H.; investigation, Y.L., Z.J. and Z.Y.; resources, Z.J. and W.H.; writing—original draft preparation, Z.J. and W.H.; writing—review and editing, Y.L. and C.Q.; visualization, Z.J. and Z.Y.; supervision, C.Q.; project administration, C.Q.; funding acquisition, C.Q. All authors have read and agreed to the published version of the manuscript.

Funding: This research was supported by the National Natural Science Foundation of China (42301506).

Data Availability Statement: All data contained within this study are available from the authors for academic purposes on request.

Acknowledgments: The authors would like to thank the hardware and software support provided by Beijing Sankuai Science and Technology Co. (Meituan Co.), Bejing, China in data collection.

Conflicts of Interest: The authors declare no conflicts of interest.

References

1. Sun, Y. Sharing and riding: How the dockless bike sharing scheme in China shapes the city. *Urban Sci.* **2018**, *2*, 68. [CrossRef]
2. Chang, S.; Song, R.; He, S.; Qiu, G. Innovative Bike-Sharing in China: Solving Faulty Bike-Sharing Recycling Problem. *J. Adv. Transp.* **2018**, *1*, 4941029.
3. Yao, H.; Dai, Z.; Chen, W.; Xie, T.; Zhu, X. GNSS Urban Positioning with Vision-Aided NLOS Identification. *Remote Sens.* **2022**, *14*, 5493. [CrossRef]
4. Shytermeja, E.; Paśnikowski, M.J.; Julien, O.; López, M.T. GNSS quality of service in urban environment. In *Multi-Technology Positioning*; Springer: Cham, Switzerland, 2017; pp. 79–105.
5. Closas, P.; Fernández-Prades, C.; Arribas, J. A Bayesian approach to multipath mitigation in GNSS receivers. *IEEE J. Sel. Top. Signal Process.* **2009**, *3*, 695–706. [CrossRef]
6. Zou, X.; Li, Z.; Wang, Y.; Deng, C.; Li, Y.; Tang, W.; Fu, R.; Cui, J.; Liu, J. Multipath error fusion modeling methods for Multi-GNSS. *Remote Sens.* **2021**, *13*, 2925. [CrossRef]
7. Fernández-Prades, C.; Arribas, J.; Closas, P. Robust GNSS receivers by array signal processing: Theory and implementation. *Proc. IEEE* **2016** *104*, 1207–1220. [CrossRef]
8. Sun, Y.; Chen, F.; Lu, Z.; Wang, F. Anti-jamming method and implementation for GNSS receiver based on array antenna rotation. *Remote Sens.* **2022**, *14*, 4774. [CrossRef]
9. Del Peral-Rosado, J.A.; López-Salcedo, J.A.; Seco-Granados, G.; López-Almansa, J.M.; Cosmen, J. Kalman filter-based architecture for robust and high-sensitivity tracking in GNSS receivers. In Proceedings of the 2010 5th ESA Workshop on Satellite Navigation Technologies and European Workshop on GNSS Signals and Signal Processing (NAVITEC), Noordwijk, The Netherlands, 8–10 December 2010; pp. 1–8.
10. Yang, H.; Zhou, B.; Wang, L.; Wei, Q.; Ji, F.; Zhang, R. Performance and evaluation of GNSS receiver vector tracking loop based on adaptive cascade filter. *Remote Sens.* **2021**, *13*, 1477. [CrossRef]
11. Xu, P.; Zhang, G.; Yang, B.; Hsu, L.T. Machine Learning in GNSS Multipath/NLOS Mitigation: Review and Benchmark. *IEEE Aerosp. Electron. Syst. Mag.* **2024**, *1*, 1–17. [CrossRef]
12. Wender, S.; Dietmayer, K. 3D vehicle detection using a laser scanner and a video camera. *IET Intell. Transp. Syst.* **2008**, *2*, 105–112. [CrossRef]
13. Zhu, F.; Shen, Y.; Wang, Y.; Jia, J.; Zhang, X. Fusing GNSS/INS/vision with a priori feature map for high-precision and continuous navigation. *IEEE Sens. J.* **2021**, *21*, 23370–23381. [CrossRef]
14. Cheng, J.; Xiang, Z.; Cao, T.; Liu, J. Robust vehicle detection using 3D Lidar under complex urban environment. In Proceedings of the 2014 IEEE International Conference on Robotics and Automation (ICRA), Hong Kong, China, 31 May–7 June 2014; pp. 691–696.

15. Feriol, F.; Vivet, D.; Watanabe, Y. A review of environmental context detection for navigation based on multiple sensors. *Sensors* **2020**, *20*, 4532. [CrossRef] [PubMed]
16. Gao, H.; Groves, P.D. Context determination for adaptive navigation using multiple sensors on a smartphone. In Proceedings of the 29th International Technical Meeting of The Satellite Division of the Institute of Navigation (ION GNSS+ 2016), Portland, OR, USA, 12–16 September 2016; pp. 742–756.
17. Chen, K.; Tan, G. SatProbe: Low-energy and fast indoor/outdoor detection based on raw GPS processing. In Proceedings of the IEEE INFOCOM 2017-IEEE Conference on Computer Communications, Atlanta, GA, USA, 1–4 May 2017; pp. 1–9.
18. Bui, V.; Le, N.T.; Vu, T.L.; Nguyen, V.H.; Jang, Y.M. GPS-based indoor/outdoor detection scheme using machine learning techniques. *Appl. Sci.* **2020**, *10*, 500. [CrossRef]
19. Bai, Y.B.; Holden, L.; Kealy, A.; Zaminpardaz, S.; Choy, S. A hybrid indoor/outdoor detection approach for smartphone-based seamless positioning. *J. Navig.* **2022**, *75*, 946–965. [CrossRef]
20. Lai, Q.; Yuan, H.; Wei, D.; Li, T. Research on GNSS/INS integrated positioning method for urban environment based on context aware. *Navig. Position. Timing* **2021**, *8*, 151–162. (In Chinese)
21. Dai, Z.; Zhai, C.; Li, F.; Chen, W.; Zhu, X.; Feng, Y. Deep-learning-based scenario recognition with GNSS measurements on smartphones. *IEEE Sens. J.* **2022**, *23*, 3776–3786. [CrossRef]
22. Zhu, F.; Luo, K.; Tao, X.; Zhang, X. Deep Learning Based Vehicle-Mounted Environmental Context Awareness via GNSS Signal. *IEEE Trans. Intell. Transp. Syst.* **2024**, *1*, 1–14. [CrossRef]
23. Chaffee, J.; Abel, J. On the exact solutions of pseudorange equations. *IEEE Trans. Aerosp. Electron. Syst.* **1994**, *30*, 1021–1030. [CrossRef]
24. Forssell, B.; Martin-Neira, M.; Harrisz, R.A. Carrier phase ambiguity resolution in GNSS-2. In Proceedings of the 10th International Technical Meeting of the Satellite Division of the Institute of Navigation (ION GPS 1997), Kansas City, MO, USA, 16–19 September 1997; pp. 1727–1736.
25. Chenggong, Z.; Xi, C.; Zhen, H. A comprehensive analysis on Doppler frequency and Doppler frequency rate characterization for GNSS receivers. In Proceedings of the 2016 2nd IEEE International Conference on Computer and Communications (ICCC), Chengdu, China, 14–17 October 2016; pp. 2606–2610.
26. Falletti, E.; Pini, M.; Presti, L.L. Low complexity carrier-to-noise ratio estimators for GNSS digital receivers. *IEEE Trans. Aerosp. Electron. Syst.* **2011**, *47*, 420–437. [CrossRef]
27. Pei, X.; Zhao, Y.; Chen, L.; Guo, Q.; Duan, Z.; Pan, Y.; Hou, H. Robustness of machine learning to color, size change, normalization, and image enhancement on micrograph datasets with large sample differences. *Mater. Des.* **2023**, *232*, 112086. [CrossRef]
28. Singh, D.; Singh, B. Investigating the impact of data normalization on classification performance. *Appl. Soft Comput.* **2020**, *97*, 105524. [CrossRef]
29. Zhang, X.; He, L.; Li, Y.; Zhang, R. An improved star selection algorithm based on altitude angle and GDOP contribution value. *Softw. Guide* **2016**, *15*, 16–20. (In Chinese)
30. Hu, X.; Liu, F.; Weng, H. Observability analysis of MSINS/GPS complete integrated system. *J. Chin. Inert. Technol.* **2011**, *19*, 38–45. (In Chinese)
31. Zhu, Y.; Luo, H.; Zhao, F.; Chen, R. Indoor/outdoor switching detection using multisensor DenseNet and LSTM. *IEEE Internet Things J.* **2020**, *8*, 1544–1556. [CrossRef]
32. Hochreiter, S.; Schmidhuber, J. Long short-term memory. *Neural Comput.* **1997**, *9*, 1735–1780. [CrossRef] [PubMed]
33. Van Houdt, G.; Mosquera, C.; Nápoles, G. A review on the long short-term memory model. *Artif. Intell. Rev.* **2020**, 53, 5929–5955. [CrossRef]
34. Zhang, A.; Lipton, Z.C.; Li, M.; Smola, A.J. Modern Recurrent Neural Network. In *Dive into Deep Learning*; Cambridge University Press: Cambridge, UK, 2023; pp. 342–348.
35. Wang, Y.; Liu, P.; Liu, Q.; Adeel, M.; Qian, J.; Jin, X.; Ying, R. Urban environment recognition based on the GNSS signal characteristics. *Navigation* **2019**, *66*, 211–225. [CrossRef]
36. Baldi, P.; Sadowski, P.J. Understanding dropout. In Proceedings of the Advances in Neural Information Processing systems, Lake Tahoe, NV, USA, 5–8 December 2013; Volume 26.
37. Gal, Y.; Ghahramani, Z. Dropout as a bayesian approximation: Representing model uncertainty in deep learning. In Proceedings of the International Conference on Machine Learning , New York, NY, USA, 19–24 June 2016; pp. 1050–1059.
38. Ying, W.; Zhang, Y.; Huang, J.; Yang, Q. Transfer learning via learning to transfer. In Proceedings of the International Conference on Machine Learning, Stockholm Sweden, 10–15 July 2018; pp. 5085–5094.
39. Kingma, D.P.; Salimans, T.; Welling, M. Variational dropout and the local reparameterization trick. In Proceedings of the Advances in Neural Information Processing Systems, Montreal, QC, Canada, 7–12 December 2015; Volume 28.

remote sensing

MDPI

Article

Prediction of Deformation in Expansive Soil Landslides Utilizing AMPSO-SVR

Zi Chen [1], Guanwen Huang [1,2,3,*] and Yongzhi Zhang [1]

1 School of Geology Engineering and Geomatics, Chang'an University, Xi'an 710054, China; 2019026017@chd.edu.cn (Z.C.); zhangyz@chd.edu.cn (Y.Z.)
2 Key Laboratory of Ecological Geology and Disaster Prevention, Ministry of Natural Resources, Xi'an 710054, China
3 Smart-BDS Shaanxi University Engineering Research Center, Xi'an 710054, China
* Correspondence: guanwen@chd.edu.cn; Tel.: +86-136-3680-1167

Abstract: A non-periodic "step-like" variation in displacement is exhibited owing to the repeated instability of expansive soil landslides. The dynamic prediction of deformation for expansive soil landslides has become a challenge in actual engineering for disaster prevention and mitigation. Therefore, a support vector regression prediction (AMPSO-SVR) model based on adaptive mutation particle swarm optimization is proposed, which is suitable for small samples of data. The shallow displacement is decomposed into a trend component and fluctuating component by complete ensemble empirical mode decomposition with adaptive noise (CEEMDAN), and the trend displacement is predicted by cubic polynomial fitting. In this paper, the multiple disaster-inducing factors of expansive landslides and the time hysteresis effect between displacement and its influencing factors are fully considered, and the crucial influencing factors which eliminate the time lag effect and state factors are input into the model to predict the fluctuation displacement. Monitoring data in the Ningming area of China are employed for the model validation. The predicted results are compared with those of the traditional model. The model performance is evaluated through indicators such as the goodness of fit R^2 and root mean square error RMSE. The results show that the prediction RMSE of the new model for three monitoring stations can reach 2.6 mm, 6.6 mm, and 2.5 mm, respectively. Compared with the common Grid search support vector regression (GS-SVR), the Particle Swarm Optimization Support Vector Regression (PSO-SVR) and Back Propagation Neural Network (BPNN) models have average improvements of 58.4%, 38.1%, and 25.2% respectively. The goodness of fit R^2 is superior to 0.99 in the new method. The proposed model can effectively be deployed for the displacement prediction of non-periodic stepped expansive soil landslides driven by multiple influencing factors, providing a reference idea for the deformation prediction of expansive soil landslides.

Keywords: expansive soil landslide; displacement prediction; multiple driven factors; AMPSO-SVR; time lag effect

1. Introduction

Expansive soil landslides are a complex clay geological hazard distributed in more than 20 provinces in China, where nearly 400 million people live. Uneven settlements of buildings, instability of engineering slopes, and other detriments are easily attributed to the unstable deformation characteristics of expansive soil due to water absorption expansion deformation and water loss shrinkage cracking [1].

To effectively prevent expansive soil landslide hazards, it is crucial to accurately predict future displacements by utilizing historical landslide data. Expansive soil landslides are unstable nonlinear systems with complex instability mechanisms and special engineering characteristics of repeated expansion and contraction; these special "step-like" characteristics are presented in displacement curves [2]. Compared with traditional loess landslides, the step-like rise in the displacement of expansive soil landslides does not follow strict rainfall periodicity and is often accompanied by strong burstiness [3–8]. This is because expansive soil landslides are jointly affected by multiple internal and external factors. In addition to precipitation, landslide stability is also significantly impacted by earth pressure and soil moisture content [9–12]. Therefore, the dynamic prediction of the displacement of step-like expansive soil landslides driven by multiple factors is an important technical issue that needs to be solved urgently in the current research on landslide prediction.

Solutions for the prediction of step-like landslide displacement have been researched by numerous scholars so far. For example, Zhu Xing proposed an extreme learning machine (ELM) to predict the displacement of the Baijiabao landslide in the Three Gorges [13], Gao Wei fused a grey system (GM) with an evolutionary neural network (ENN) to predict two landslides in the Three Gorges Reservoir Area [14], and Yang Beibei constructed a long short-term memory network (LSTM) for the displacement prediction of the Baishui river landslide [15]. Nevertheless, single factors such as the reservoir water level and rainfall were mainly considered in the above methods, but they are more suitable for landslides with more obvious deformation cycles and a single influencing factor. The internal characteristics of landslides, such as non-periodic soil stress or the internal moisture field of slopes, have not been fully considered. The above neural-network-based prediction models have complex algorithm structures and require high-quality feature data and sample sizes, which is not suitable for predicting the displacement of non-periodic expansive soil landslides under the joint influence of multiple internal and external factors. A strong generalization ability is exhibited by Support Vector Regression (SVR) [16], and the nonlinear relationship between multiple external factors and target displacement can be handled effectively. It does not require complex operation for preprocessing and feature selection in an input dataset, and the model has an excellent robustness. It is very suitable for predicting the deformation of step-like landslides dominated by multiple factors. However, it is easy for the traditional SVR model to fall into local optimal, which can be solved effectively by adaptive mutation particle swarm optimization (AMPSO) [17,18].

Therefore, a machine learning prediction algorithm based on AMPSO-SVR is proposed to predict the displacement of expansive soil landslides. The main idea includes: (1) The total displacement being decomposed into trend and fluctuation terms, respectively, by complete ensemble empirical mode decomposition with adaptive noise (CEEMDAN). (2) The trend displacement being predicted by a fitting method and the relationship between the fluctuation term and the influencing factors being analyzed. (3) The time delay effect between the displacement and multi-factor being considered, the multi-factor wave term component being extracted by CEEMDAN and input into the AMPSO-SVR model to predict the wave term displacement, and the prediction results being compared with grid search support vector regression (GS-SVR), and particle swarm optimization support vector regression (PSO-SVR), verifying that the "step-like" non-periodic displacement can be accurately and dynamically predicted, which has a certain feasibility and scientificity. (4) The total predicted displacement sequence being obtained by superimposing the predicted values of the trend and fluctuation terms.

2. Displacement Prediction Model of Expansive Soil Landslides

2.1. Displacement Time Series Theory

The displacement of expansive soil exhibits nonlinear characteristics under the influence of crack evolution and wet–dry cycles. The total displacement is decomposed into a trend term, wave term displacement, and random term displacement according to the time series theory [19,20].

$$S(t) = T(t) + P(t) + R(t) \tag{1}$$

where $S(t)$ is the total displacement value, $T(t)$ is the trend term displacement, and $P(t)$ is the fluctuation term displacement. The magnitude of $R(t)$ is small and negligible, and the prediction of $T(t)$ and $P(t)$ is mainly focused on in this paper.

2.2. Complete Ensemble Empirical Mode Decomposition with Adaptive Noise

Moving average, wavelet decomposition, and EMD are commonly applied to landslide displacement decomposition [21]. EMD decomposition is a conventional method for decomposing non-stationary data. However, anomalies in displacement decomposition may be caused by the existence of mode aliasing, and the displacement trend term and the fluctuation term can be extracted inaccurately. Moving average has a poor adaptability to noise, and wavelet decomposition has challenges in processing non-stationary signals such as landslide displacement curves. The above problems can be effectively overcome by CEEMDAN decomposition. Adaptive noise is added to landslide displacement curves, which improves the robustness of decomposition and is more suitable for processing non-stationary signals. These decomposed displacement components are more consistent with the actual change trends of landslides. The CEEMDAN decomposition flow is designed as follows in this paper:

(1) The landslide displacement $S(t)$ is added with Gaussian white noise, whose K-th mean is 0 to construct the K sequence $S_i(t)$ to be decomposed. i = 1, 2, 3, ... k, ε is the signal-to-noise ratio, and $\delta_i(t)$ is the white noise sequence added for the i-th time.

$$S_i(t) = S(t) + \varepsilon\delta_i(t) \tag{2}$$

(2) The EMD decomposition of $S_i(t)$ is performed, the first modal component (IMF) is decomposed, and its mean is taken as the first modal component obtained by CEEMDAN decomposition. Each modal component is a different frequency component of the original displacement. $r_1(t)$ is the first residual component, which represents the non-extractable aperiodic component of the displacement sequence.

$$IMF_1(t) = \frac{1}{K}\sum_{i=1}^{K} IMF_1^i(t) \tag{3}$$

$$r_1(t) = S(t) - IMF_1(t) \tag{4}$$

(3) The EMD decomposition is continued after noise is added to the residual component $r_j(t)$ of phase j.

$$IMF_j(t) = \frac{1}{k}\sum_{i=1}^{K} E_1\left(r_{j-1}(t) + \varepsilon_{j-1}E_{j-1}(\delta_i(t))\right) \tag{5}$$

$$r_j(t) = r_{j-1}(t) - IMF_j(t) \tag{6}$$

(4) If the EMD stop condition is satisfied and the displacement residual component $r_n(t)$ of the n-th decomposition is close to a monotone signal, then the CEEMDAN algorithm ends its decomposition.

2.3. Support Vector Regression (SVR)

Expansive soil landslides are driven by various correlation factors [22]. The relationship between displacement and external factors is not a simple linear form, but a more complex curve shape, such as a polynomial or periodic relationship, etc., which can be solved effectively by Support Vector Regression (SVR). SVR began with the research of Drucker et al. (1997) [23]. The principle is to find a nonlinear mapping from the input space to the output space, and the feature data x are mapped to a high-dimensional feature space, and the regression plane with the optimal fitting performance in the high-dimensional space is selected. Its mathematical model is as follows:

$$P\_pre(x) = \omega^T\varphi(x) + b \tag{7}$$

where $P\_pre(x)$ is the output value of the model, namely, the predicted displacement value of the landslide fluctuation term corresponding to the predicted value of item $P(t)$

in Equation (1), ω is the inertia weight, b is the offset term, and $\varphi(x)$ is the mapping function between the landslide correlation factor x and predicted displacement, which includes the input values of the model, namely, the data of multiple influence factors that affect expansive soil landslides and the historical displacement data. The data of multiple influence factors are mapped to a high-dimensional feature space for linear analysis in that space. The nonlinear relationship between displacement and multi-factor features can be modeled accurately by utilizing the kernel function in the SVR model. The key to SVR machine learning is to find the optimal nonlinear mapping relationship so that the output displacement $f(x)$ is closer to the actual value.

2.4. Adaptive Mutation Particle Swarm Optimization (AMPSO)

The particle swarm optimization algorithm (PSO) is an evolutionary algorithm inspired by bird foraging [24]. The solution is regarded as a particle, the particle quality is measured by the particle fitness, and the optimal solution is obtained through multiple iterations. However, the particles are likely to quickly gather near a local optimum in the early stages of the search process, and the search space is not fully explored. It is easy to fall into the local optimum, which makes it difficult to obtain the global optimal solution. The adaptive mutation particle swarm optimization (AMPSO) is an improvement of the standard PSO, and it adjusts the search range of the solution space by dynamically adjusting the weight and adaptive mutation mechanism so that the algorithm has a large global search ability, which can address this issue. The strategy adopted by the algorithm is to adjust the inertia weight adaptively. The larger weight is beneficial for improving the global optimization ability of the particle that can make it skip the local optimum. Therefore, a better optimal position on a large scale can be explored by taking a larger value of ω initially, while a smaller weight will improve the local optimal ability of particles so that the search process can be gradually converged. Later, the value of ω will decrease to improve the convergence accuracy. The weight linear attenuation method is adopted in this paper according to the iterative process, and its mathematical model is as follows:

$$\begin{gathered} V_{id}^{k+1} = \omega V_{id}^{k} + c_1 r_1 \left(P_{id}^{k} - W_{id}^{k} \right) + c_2 r_2 \left(P_{gd}^{k} - W_{gd}^{k} \right) \\ W_{id}^{k+1} = W_{id}^{k} + V_{id}^{k+1} \end{gathered} \tag{8}$$

$$\omega_{iner} = \omega_{start} - \frac{(\omega_{start} - \omega_{end}) \cdot I}{N} \tag{9}$$

where i is the number of particles, d is the dimension of the prediction space, k is the number of iterations, v is the particle velocity, W is the particle position, ω is the inertia weight, Equation (9) is the updating method for weight ω in Equation (8), ω_{iner} is the inertial weight of adaptive change, ω_{start} is the maximum weight (value is 0.9), and ω_{end} is the minimum weight value (value is 0.4). The inertia weight in Equation (8) can be linearly decreased in this way, and the tradeoff between global search and local search can be balanced in the iterative process. This strategy of decreasing weights is aimed at giving the algorithm a greater exploration ability in the initial stage, and gradually increasing the focus on local search in subsequent iterations. Oscillation near the global optimal solution in the late particle search is effectively avoided. I is the current number of iterations, N is the total number of iterations, c_1 is the learning factor 1, and c_2 is the learning factor 2. r_1 and r_2 are random floating point numbers ranging from 0 to 1, P is the optimal position of the particle individual, and g is the global optimal position of the particle swarm.

Simultaneously, random particle mutation should be added to the particle position and velocity search in the iterative process to increase the diversity of the population. The mutation method selected is as follows:

$$if\ R < C_m, \begin{cases} ceil(2 \times R) = 1, v_c = R \times (C_{max} - C_{min}) + C_{min} \\ ceil(2 \times R) = 2, v_g = R \times (g_{max} - g_{min}) + g_{min} \end{cases} \tag{10}$$

In the formula, R is the random number of distributions [0, 1], C_m is the preset mutation probability (constant, set as 0.8 in this paper), v_c is the particle motion velocity of hyperparameter c, v_g is the velocity of the particle with the hyperparameter gamma, and C_{min}, C_{max}, g_{min}, and g_{max} are the search boundaries of the parameters. The adaptive mutation search for the optimal particles is implemented by this method, which enables the SVR model to jump out of the local optimum and find the best parameters c and g. These parameters enable the model to find the best mapping function $\varphi(x)$. The risk of easily falling into overfitting is reduced. The adaptive mutation particle swarm optimization utilizes adaptive optimization to enable the SVR to select parameters more robustly, reducing the problem of model performance degradation caused by improper parameter selection. This is particularly significant for small sample data, as they are more sensitive to parameter selection. The prediction accuracy is further improved.

2.5. The Flow of Prediction Method

The process of the proposed prediction model is shown in Figure 1. The steps are as follows: (1) GNSS (Global Navigation Satellite System) displacement and earth pressure, precipitation, soil moisture content, and other data are collected by multi-source sensors. Among them, the displacement is obtained by self-developed GNSS monitoring equipment with a Ublox ZED-F9P board, and the measurement accuracy is better than 5 mm in both the horizontal and vertical directions. The soil moisture content is obtained by soil moisture meters, with a measurement accuracy of ±3%. The earth pressure is measured by earth pressure boxes, with a nominal measurement accuracy of ±0.5% F.S. The precipitation is obtained by funnel-type rain gauges, with an accuracy of measurement of ±0.2 mm. The multi-source data are processed on a self-built expansive soil landslide monitoring cloud platform; the GNSS displacement is calculated by real-time relative positioning with a solution frequency of 1 s and the GNSS receiver is mainly deployed at the leading edge and the middle of the slope, where the stability of expansive soil landslides is poor. A soil hygrometer determines the soil moisture content by measuring the change in the soil dielectric constant; its measurement frequency is 4 h and it should be deployed 30 cm below the surface of the expansive soil. The earth pressure box converts the mechanical pressure into an electrical signal and converts the electrical signal into the actual pressure value through calibration, with a measurement frequency of 4 h. The box is laid in the middle of the soilbag and reinforced with steel bars. The rain gauge collects precipitation through a tipping bucket structure. Its measurement frequency is 6 h and it is deployed in a stable and open location. (2) The landslide displacement is decomposed into a trend term and wave term, respectively, with CEEMDAN. The trend term displacement represents the long-term trend of landslide development and is predicted by the polynomial fitting method. (3) The correlation and the influence lag time between the fluctuation displacement and external factors are determined. The wave term component of multiple factors is the key part that affects changes in the fluctuation term displacement, which is similarly extracted by CEEMDAN and will be set as the optimal correlation factor to predict the step-like displacement of expansive soil landslides. (4) The AMPSO-SVR model is established by considering the state factors of landslide development. (5) Finally, the performance of the prediction model is evaluated with indicators like R^2, MAPE, and RMSE.

2.6. Evaluation Indicators for Displacement Prediction

Three evaluation indicators were used to evaluate the prediction accuracy of the proposed model R^2 (goodness of fit), *RMSE* (root mean square error), and *MAPE* (mean absolute percentage error). Their mathematical expressions are shown in Equations (11)–(13):

$$R^2 = 1 - \frac{\sum_{i=1}^{n}(y_i - \hat{y}_i)^2}{\sum_{i=1}^{n}(y_i - \overline{y})^2} \tag{11}$$

$$RMSE = \sqrt{\frac{1}{n}\sum_{i=1}^{n}(y_i - \hat{y}_i)^2} \quad (12)$$

$$MAPE = \frac{1}{n}\sum_{i=1}^{n}\left|\frac{\hat{y}_i - y_i}{y_i}\right| \quad (13)$$

where y_i is the measured displacement, $\hat{y}_i$ is the predicted displacement, $\overline{y}$ is the average displacement, which corresponds to the fluctuation term displacement, and i is the monitoring period.

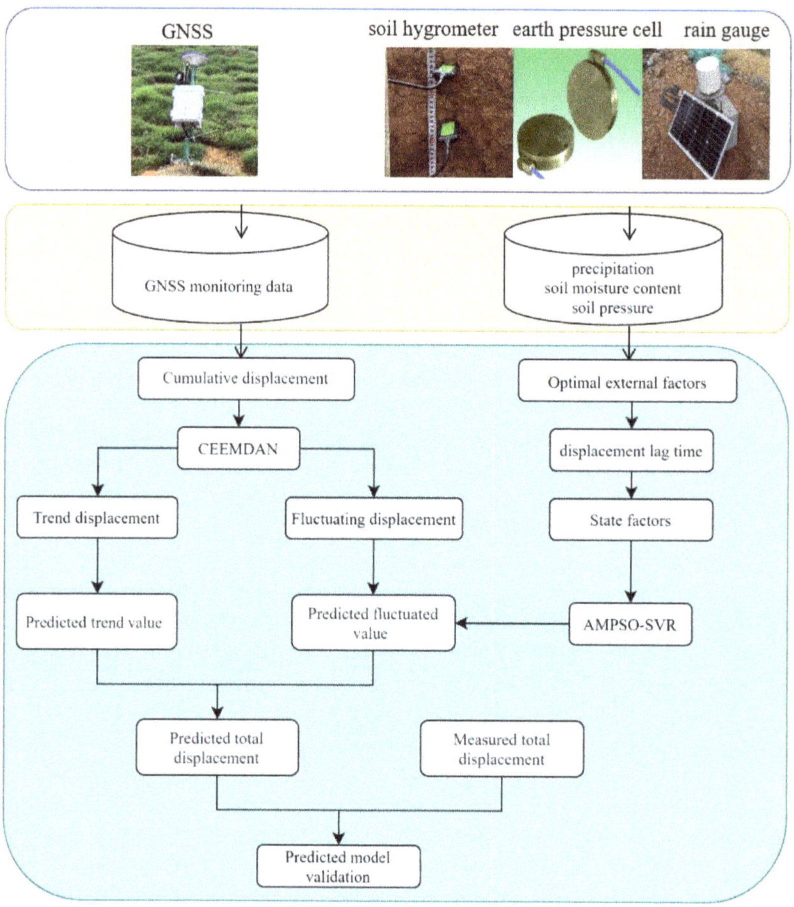

Figure 1. Multi-factor support vector regression displacement prediction flow of expansive soil.

3. Experimental Analysis—Expansive Soil Landslide on Chongai Highway in Ningming

3.1. Landslide Overview

The Ningming expansive soil landslide is located in Ningming Basin, Guangxi Province, China, which has the largest expansive soil deposit in the world and is a typical expansive soil research area. The expansive soil in this area belongs to the type of residual expansive soil [25]. It is a medium expansive soil with a free expansion coefficient of about 58%. The geographical location and overview of the slope are shown in Figure 2. The specific location of the landslide is about 5.2 km on the north bank of the Mingjiang River, a tributary of the Pearl River, 1 km west of Wayao Village, and near the Chongzuo–Aidian highway. Mainly in the shape of a long and narrow fan, it is a highway cutting landslide with a slope of about 1:1.5.

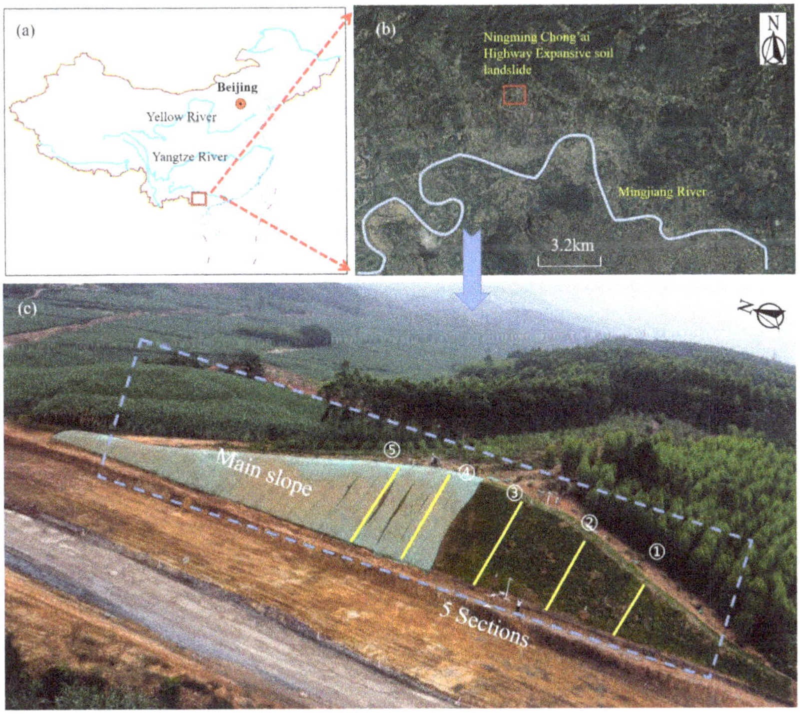

Figure 2. Landslide location and geological overview map. (**a**,**b**) the specific location of the landslide; (**c**) aerial figure of the expansive soil landslide, which is divided into five sections ①–⑤.

The region has a tropical marine monsoon climate with an average annual precipitation of 1200 mm and with spring and summer from April to October and autumn and winter from November to March. A total of eight GNSS monitoring points (NN01–NN08) were deployed to establish a real-time displacement-monitoring system with a monitoring accuracy better than 5 mm due to the obvious displacement of slope unloading. In addition, sensors such as those for earth pressure and soil moisture content were installed to strengthen the monitoring system, and the data transmission interval was 4 h. A rain gauge was arranged at the rear edge, and the data transmission interval was 1 h.

3.2. Deformation Characteristics

The GNSS cumulative displacement curve of the Ningming expansive soil slope is shown in Figure 3, which was calculated by the real-time GNSS landslide cloud platform. Due to the low stability at the three monitoring points on the top position of the slope, there was a significant increase in displacement compared to other points, hence, these three points were selected as the experimental dataset. The red, green, and blue curves represent the displacements of the NN06, NN07, and NN08 monitoring points, showing a step-like characteristic. They are located in sections ④–⑤ of the slope, where the risk of instability is high as a result of excavation unloading, and the slide direction is 78° north by west. The interstitial rise in displacement was mainly due to the occurrence of heavy rainfall events around 25 d and 155 d in the figure. The fissure of expansive soil provided an excellent channel for rainwater infiltration, resulting in a rapid increase in soil water content and a decrease in matrix suction as a result of the decrease in slope shear strength with an increase in saturation. The whole process time was relatively short, so the cumulative displacement increased rapidly with a strong mutability. There was no significant change in displacement during the drying period.

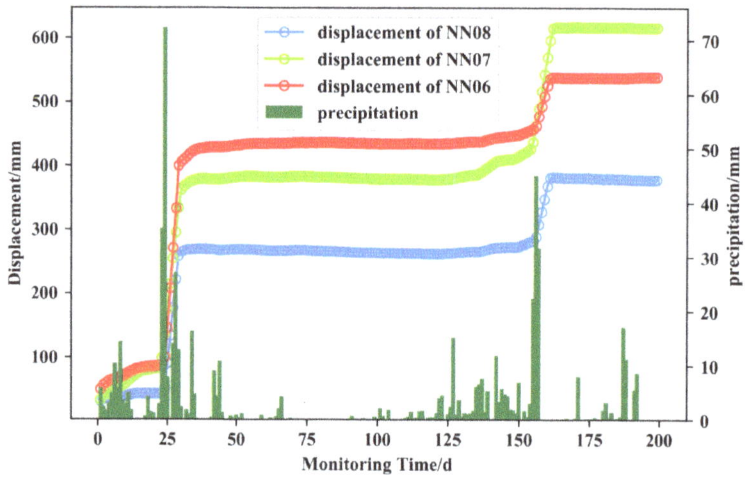

Figure 3. Displacement curve of landslide monitoring points.

The on-site failure form of the slope is shown in Figure 4. The soil near the GNSS monitoring equipment was significantly cracked, and fish-scale shallow failure characteristics appeared after two instability events. Many cracks developed obviously, and these cracks gradually closed and deepened with an increase in instability frequency, finally collapsing in the form of a traction landslide.

Figure 4. Shallow instability characteristics of Ningming expansive soil. (**a–d**) The visible cracks near the monitoring station.

4. Displacement Prediction

A multi-factor support vector regression prediction model was established with the GNSS cumulative displacement, soil moisture content, earth pressure, precipitation, and other data of monitoring points NN06–NN08 for about 200 days. The first 150 days of data were selected as the training set, the last 50 days were selected as the prediction set. The 75% training set ratio helped the model to better learn the pattern of the data, leaving a smaller proportion (25%) of sufficient samples to evaluate the generalization ability of the model, with the problem of overfitting of the training set being effectively avoided. Furthermore, the displacement of expansive soil experienced irregular changes such as a smooth and steep rise in the first 150 days, including the dynamic changes of expansive soil landslides from the safety stage to the instability stage. Hence, the selection of training and prediction sets in this way was more representative.

4.1. Extraction and Prediction of Trend Displacement

The trend term and fluctuation term of the displacement time series were extracted by CEEMDAN decomposition. The decomposition results are shown in Figure 5, from which it can be seen that the trend displacement component was nearly monotonically increasing, while the fluctuation term displacement was characterized by intermittent oscillation.

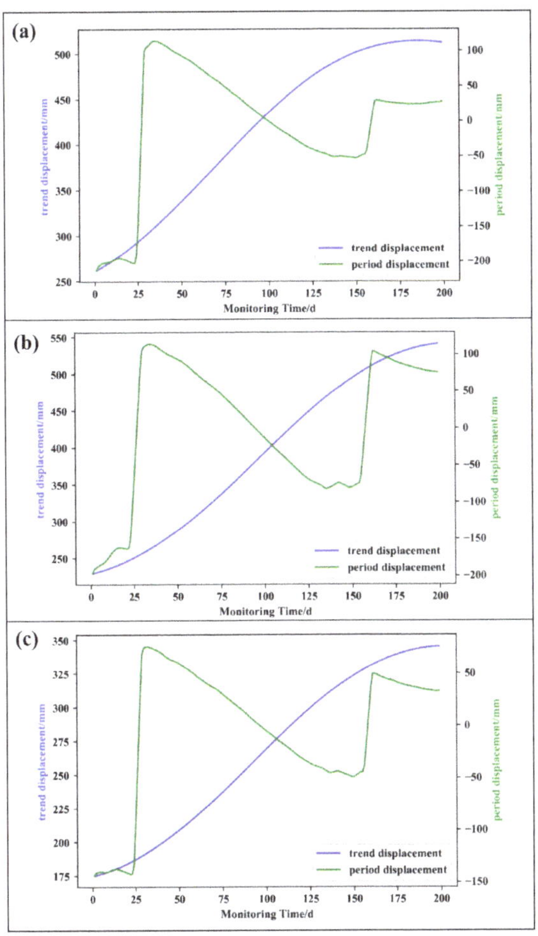

Figure 5. GNSS displacement decomposition results ((**a**) NN06; (**b**) NN07; (**c**) NN08).

The trend displacement was fitted by the least squares cubic polynomial according to the nonlinear curve characteristics of the trend displacement at the monitoring points NN06, NN07, and NN08. The fitting expression is $s(t) = at^3 + bt^2 + ct + d$, where $s(t)$ is the trend term displacement, t is the monitoring time, and a, b, c, and d are the coefficients. The prediction results and accuracy of the trend term displacement are shown in Figure 6. The fitting accuracy RMSEs of the NN06, NN07, and NN08 monitoring points were 0.68 mm, 0.51 mm, and 0.46 mm, respectively, and the goodness of fit R^2 were 0.9989, 0.9959, and 0.9976, respectively.

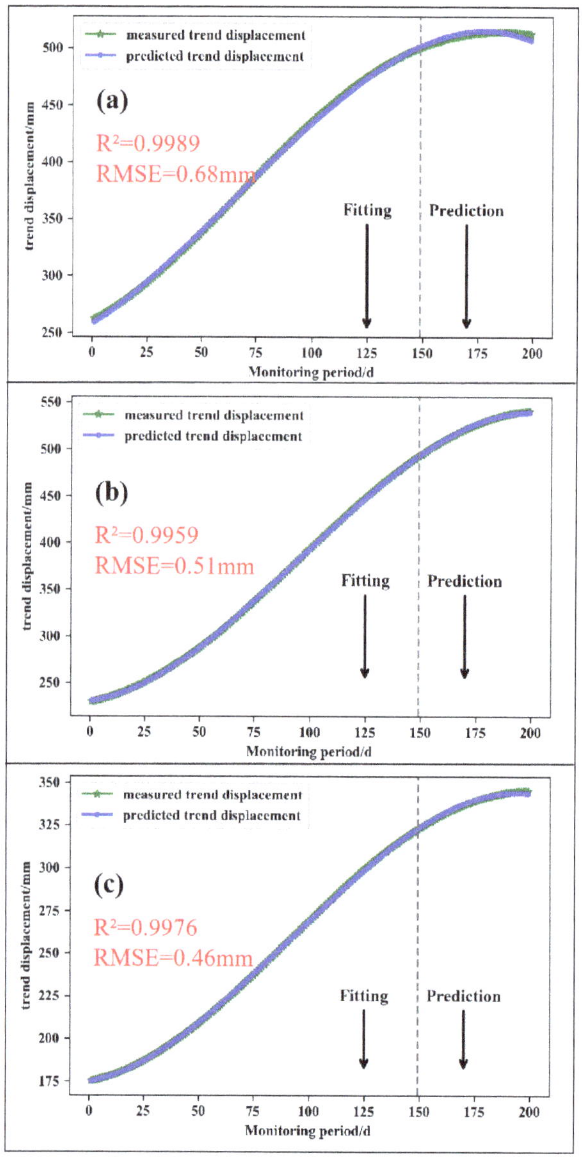

Figure 6. Fitting results of trend terms at monitoring points ((**a**) NN06; (**b**) NN07; (**c**) NN08).

4.2. Extraction and Prediction of Fluctuation Term Displacement

The fluctuation term displacement was obtained by subtracting the trend term displacement from the total displacement, which reflects the landslide displacement changes under the influence of external factors. In this paper, a multi-factor machine learning dynamic prediction model was established for the GNSS fluctuation term displacement at NN06, NN07, and NN08 for 200 d. The displacement data of the first 150 d were selected as the training set, and the last 50 d were selected as the test set.

4.2.1. Determination of Key Disaster-Inducing Factors

The prediction performance of the landslide displacement of expansive soil is directly affected by the selection of disaster-inducing factors [26]. Strength attenuation and swelling deformation are caused by unsaturated seepage under rainfall conditions, which has a significant impact on landslide stability. Moreover, the infiltration effect of rainwater is enhanced by the existence of cracks in expansive soil, and the effective water content of the soil is increased, which leads to the hygroscopic expansion of soil. The volume deformation of expansive soil under the dry–wet cycle has a significant impact on the pressure distribution of the retaining wall, and the lateral earth pressure increases significantly under the action of infiltrating rainwater and the mass of the overlying soil [27–29]. Therefore, precipitation, soil moisture content, and earth pressure are regarded as key factors for displacement prediction.

4.2.2. Time Lag Correlation

During the deformation process of an expansive soil landslide induced by rainfall, the soil moisture content and earth pressure first begin to increase, there is a lag time in the corresponding displacement changes, and the lag time is different at several monitoring points. To find the best correlation factor of the landslide displacement for the model training and avoid obvious deviations in the prediction results, the displacement response lag times of each monitoring point were determined by time delay correlation analysis (TDCA). The results are shown in Figure 7 (Unit: d): for the NN06, NN07, and NN08 monitoring points, the time lag parameters of the fluctuation term displacement and soil moisture content, earth pressure, and cumulative rainfall were (3 d, 1 d, 3 d), (2 d, 0 d, 3 d), and (3 d, 1 d, 4 d), respectively.

4.2.3. Fluctuation Term Displacement Prediction

Compared with the neural network prediction model, the prediction problem of multi-dimensional features can be solved by SVR more efficiently, and the prediction model is built by combining the AMPSO optimization algorithm. Since each non-periodic step change of the expansive soil slope displacement is greatly impacted by the fluctuation term of the external multi-source factor, the fluctuation term of the influencing factor is extracted by CEEMDAN. The mutation information of the influencing factors can be reflected by the fluctuation term component of the multi-factor, which is strongly related to the mutation of the fluctuation term displacement. Hence, it is more suitable for predicting the wave term displacement. As shown in Figure 8, the blue, green, orange, and red circled curves represent the fluctuation term displacement, fluctuation term cumulative precipitation, fluctuation term soil moisture content, and fluctuation term earth pressure at the monitoring points. The correlation between the wave term displacement and the influence factors sequence was more notable when the time lag was eliminated. It can be seen that the fluctuation term cumulative precipitation, soil moisture content, and fluctuation term periodic earth pressure started to increase, and the fluctuation term displacement showed a sudden increase simultaneously. If the cumulative precipitation, soil moisture content, and earth pressure of the fluctuation items continued to decrease overall, although there was a small local rebound, the corresponding fluctuation term displacement decreased synchronously and rose synchronously again during the next instability. Therefore, the fluctuation term displacement was more consistent with the fluctuation term change trend of external influence factors.

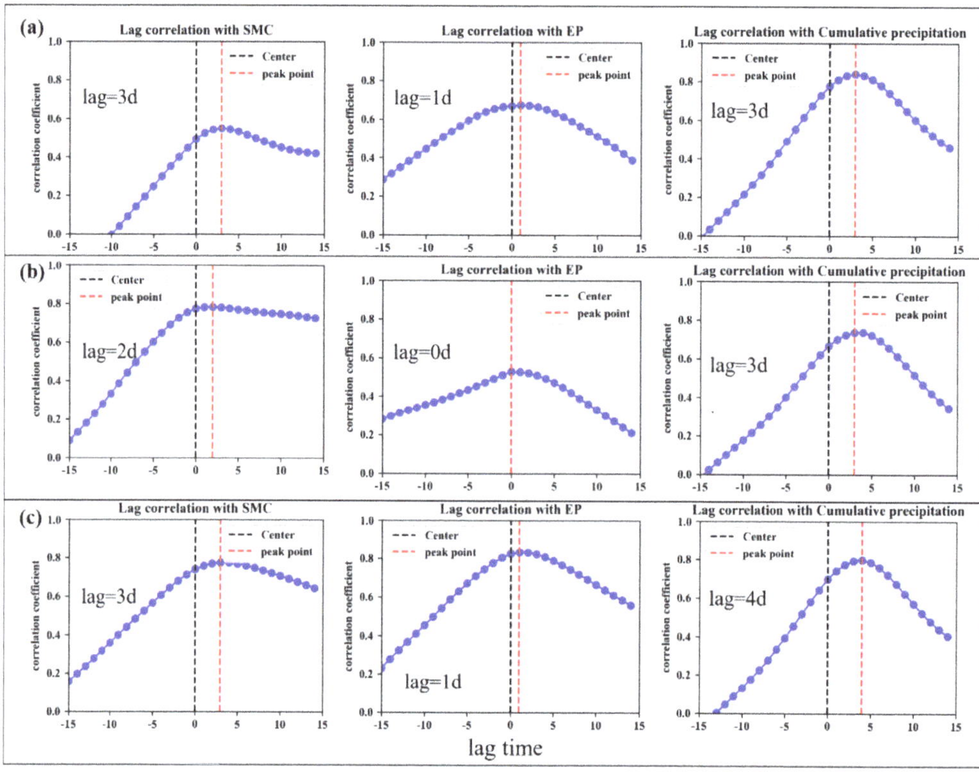

Figure 7. Time lag parameters of fluctuation term displacement and influencing factors ((**a**) NN06; (**b**) NN07; (**c**) NN08).

A self-learning AMPSO-SVR displacement prediction model was established based on the libsvm toolkit in matlab2021b software by normalizing the multi-factor association data and displacement data which eliminated the time lag effect. The model in this section was trained on a Hewlett-Packard computer with an Intel core i7-11800H processor (clocked at 2.3 GHz, eight cores, and sixteen threads), an NVIDIA RTX3050 GPU, and 16 G of memory. The fitness function is defined as the mean square error between the predicted and measured displacement values.

Taking into account the different lag times of the displacements for each monitoring point, multi-factor data that eliminated the time lag effect were added to the training model. The multi-factor data to eliminate the delay effect were added to the training model while considering the different lag times of the displacements at each monitoring point. For the NN06 monitoring point, six types of data, including the fluctuation item soil moisture content (SMC), fluctuation item cumulative precipitation 3 days ago, fluctuation item earth pressure (EP) 1 day ago, and historical displacements in the past 1 d, 2 d, and 3 d, were taken as model inputs. Similarly, for the NN07 monitoring point, the fluctuation term soil moisture content 2 days ago, fluctuation term cumulative precipitation 3 days ago, and fluctuation term earth pressure 1 day ago were combined. For the NN08 point, the fluctuation term soil moisture content 3 days ago, cumulative precipitation of the fluctuation term 4 days ago, and earth pressure data of the fluctuation term 2 days ago were input into the model for data training.

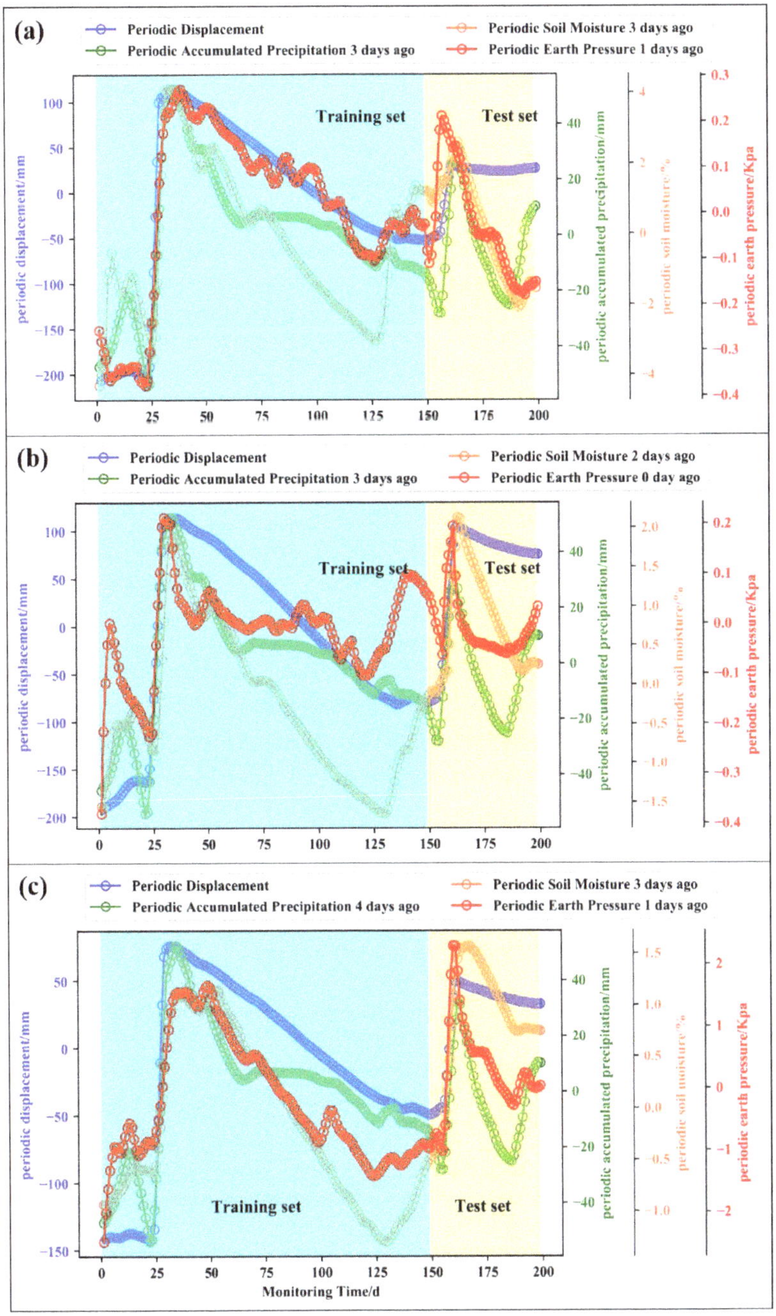

Figure 8. Fluctuation displacement and fluctuation term multi-factor sequence at monitoring point ((**a**) NN06; (**b**) NN07; (**c**) NN08).

However, it is not reliable to predict the displacement by only considering the influencing factors and ignoring the development state of the landslide itself. The historical displacements before 3 d, 2 d, and 1 d were added into the model as predicted state factors.

A total of six sequences were taken as inputs to the AMPSO-SVR model (Figure 9). After 150 iterations of parameter optimization, the final displacement prediction value was obtained.

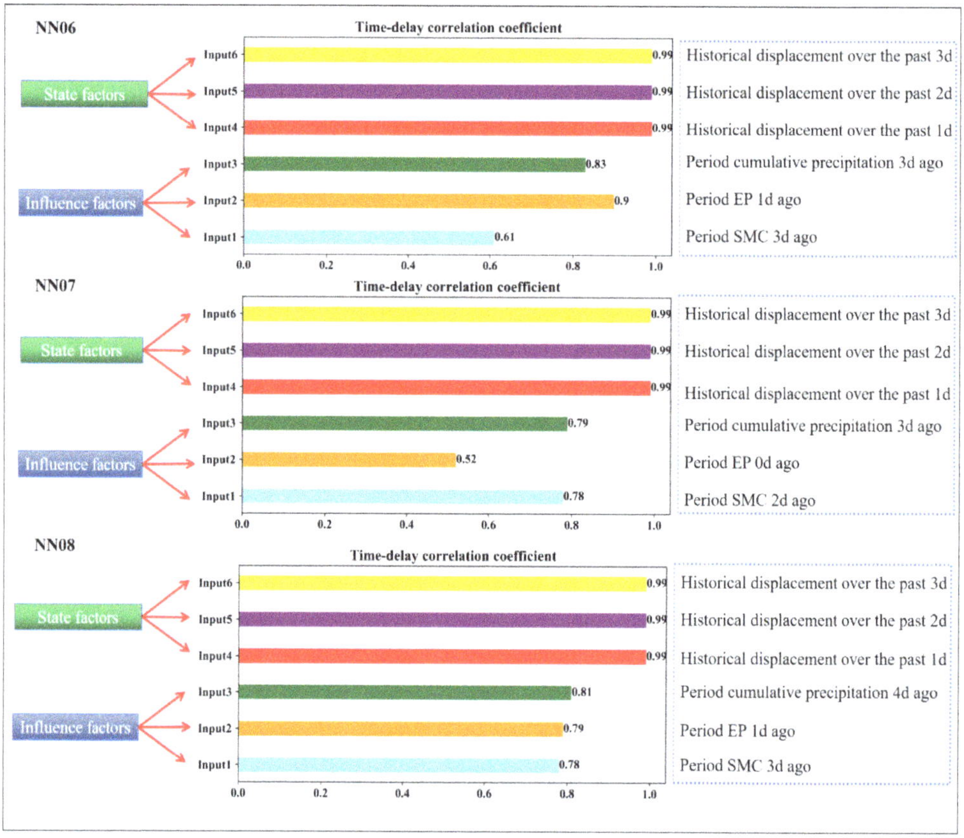

Figure 9. Input items of CEEMDAN-AMPSO-SVR prediction model.

To evaluate the actual performance of the prediction model, the results were compared with the commonly used GS-SVR, PSO-SVR, and BPNN models (see Figure 10). The goodness of fit R^2, root mean square error RMSE, and mean absolute percentage error MAPE were calculated. The calculation formula is as shown in Equation (10). The calculation results of related indicators are shown in Table 1. As can be seen from the prediction results in Figure 10, the green circle curve is the measured displacement data of the landslide, the blue circle curve is the prediction result of the GS-SVR model, the yellow square curve is the prediction result of the PSO-SVR model, the purple triangle curve is the prediction result of the BPNN model, and the red diamond curve is the prediction result of the AMPSO-SVR model. Among them, the predicted results of the GS-SVR model at the monitoring points NN06 and NN08 during the instability period were significantly greater than the measured displacement, and the predicted displacement was not stable in the short time after the slope failure, which is not consistent with the characteristics of the displacement tending to be stable after the actual landslide failure. The goodness of fit and root mean square error at point NN06 were 0.9524 and 5.9 mm, respectively. The goodness of fit and root mean square error at point NN07 were 0.9459 and 13.2 mm, respectively. The goodness of fit and root mean square error at point NN08 were 0.9239 and 8.5 mm, respectively. The prediction error was larger among the four prediction models.

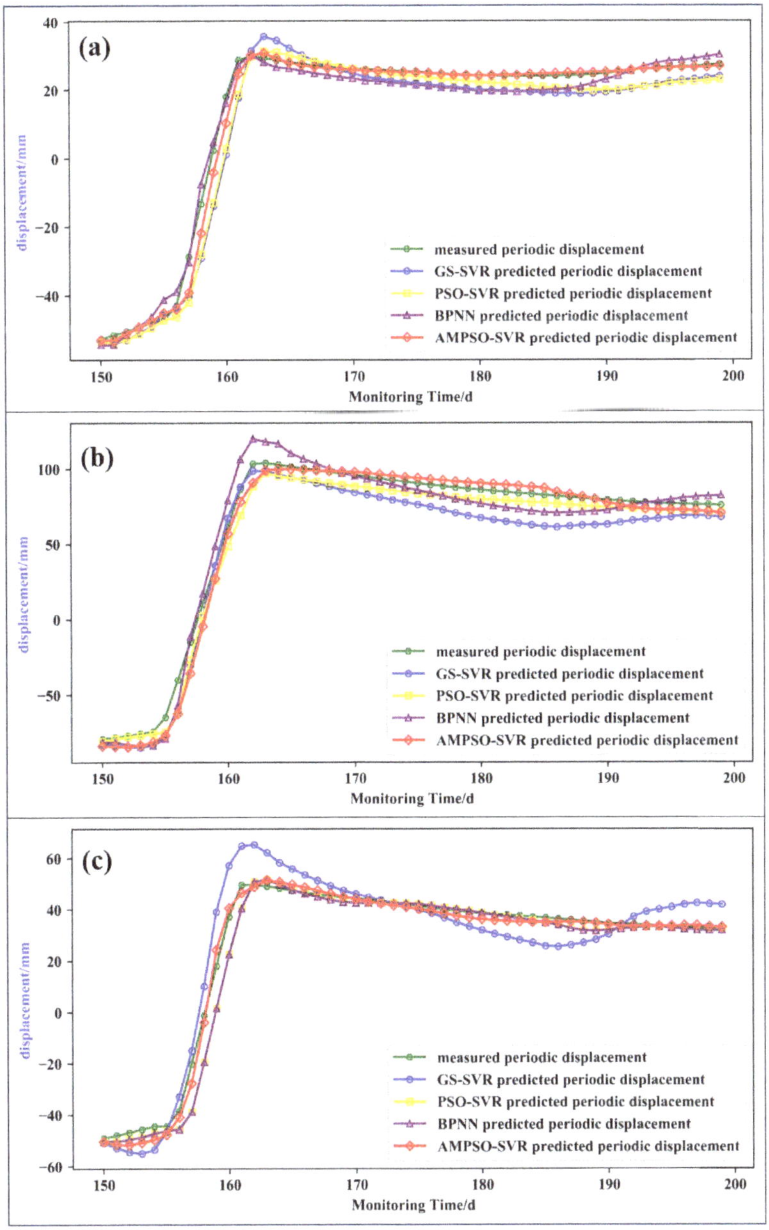

Figure 10. Comparison of prediction performances of different models ((**a**) NN06; (**b**) NN07; (**c**) NN08).

The predicted results of PSO-SVR were similar to the measured results, but the goodness of fit and root mean square error were inferior to the results of the AMPSO-SVR model. The BPNN prediction results showed a good performance at points NN06 and NN08, but a poor performance at point NN07. It can be seen that the predicted value at point NN07 was significantly higher than the measured value during the sliding stage, which had significant errors. Therefore, the generalization ability of this model was insufficient.

Table 1. Evaluation indicators of different prediction models.

Monitoring Points	R^2	*RMSE* (mm)	*MAPE*	Model
NN06	0.9524	5.9	0.3131	GS-SVR
	0.9626	5.2	0.2661	PSO-SVR
	0.9864	3.1	0.1273	BPNN
	0.9907	2.6	0.1017	AMPSO-SVR
NN07	0.9459	13.2	0.1793	GS-SVR
	0.9821	7.6	0.1124	PSO-SVR
	0.9776	8.9	0.1254	BPNN
	0.9857	6.6	0.1181	AMPSO-SVR
NN08	0.9239	8.5	0.3766	GS-SVR
	0.9697	5.3	0.3871	PSO-SVR
	0.9872	3.8	0.2135	BPNN
	0.9934	2.5	0.0964	AMPSO-SVR

The AMPSO-SVR prediction results were most consistent with the measured displacements at the three monitoring points. The goodness of fit and root mean square error at point NN06 were 0.9907 and 2.6 mm, respectively. The goodness of fit and root mean square error at point NN07 were 0.9857 and 6.6 mm, respectively. The goodness of fit and root mean square error at point NN08 were 0.9934 and 2.5 mm, respectively. With the maximum goodness of fit and minimum root mean square error, the comprehensive generalization ability of the model was better than that of the other three models. Hence, the prediction performance of the AMPSO-SVR model was superior to other models.

4.3. Total Displacement Prediction and Accuracy Analysis

The predicted value of the total displacement was obtained by superpositioning the predicted value of the displacement of the fluctuation term and the trend term, which was compared with the measurements. The prediction result curves are shown in Figure 11. The goodness of fit R^2 and RMSE of the predicted and measured displacements of the NN06, NN07, and NN08 monitoring points were 0.99, 2.6 mm, 0.99, 6.6 mm, and 0.99, 2.5 mm, respectively. These prediction results were consistent with the actual measurements, and the accuracy was in the order of millimeters. This indicated that the AMPSO-SVR prediction model can be well applied to multi-factor-driven non-periodic step-like expansive soil landslide displacement prediction.

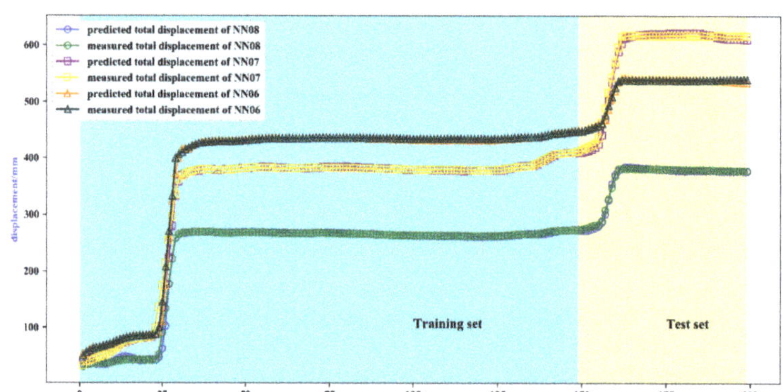

Figure 11. Total displacement prediction results of NN06/NN07/NN08.

5. Conclusions

(1) The characteristics of the "ladder-type" deformation of expansive soil landslides due to their non-periodic repeated instability were recorded and analyzed. The important relationship between key influencing factors such as earth pressure, soil moisture content, and cumulative precipitation with this step displacement were also revealed. Meanwhile, the lag response time of landslide displacement and influencing factors was determined. The GNSS displacement of monitoring points at different parts of the Ningming expansive soil slope was different from the lag time of the influencing factors. The average GNSS displacement lagged behind rainfall, soil moisture content, and earth pressure at 3 d, 2 d, and 1 d, respectively. The GNSS displacement sequence corrected by the lag period was in good agreement with the multi-source influence factor sequence.

(2) The displacements of the trend term and fluctuation term were obtained by CEEMDAN decomposition. The displacements of the trend term were predicted by cubic polynomial fitting. Taking into account the non-periodic step of the fluctuation displacement of the expansive soil landslide that was affected by multiple external factors, a dynamic prediction model driven by multi-factors was established to predict the displacement of the fluctuation term. The prediction results were in good agreement with the obtained measurements. The average RMSE predicted by AMPSO-SVR was 3.94 mm, compared with the results of the GS-SVR, PSO-SVR, and BPNN models, it was increased by 58.3%, 38.1%, and 25.2%, respectively.

(3) The proposed model was feasible and reliable in the prediction of step-like and non-periodic expansive soil landslides, and the stepped deformation and external multiple factors could be modeled efficiently, which gives it the potential to be applied to other expansive soil landslide deformation predictions.

In general, the proposed method had a good performance in predicting the step-wise displacement of expansive soil with a strong suddenness, especially expansive soil landslides driven by multiple internal and external factors. However, it is mainly aimed at the instability of expansive soil landslides with relatively short monitoring periods, and further research is needed for the displacement prediction process of expansive soil landslides with longer terms (e.g., more than 5 years). In addition, more complex prediction algorithms such as deep learning models will be tested and evaluated in subsequent research.

Author Contributions: Conceptualization, G.H.; Methodology, Z.C.; Investigation, Y.Z. All authors have read and agreed to the published version of the manuscript.

Funding: This research was funded by the National Key Research and Development Program of China (Grant No. 2019YFC1509800) and the Special Fund for Basic Scientific Research of Central Colleges (Grant No. CHD300102269305, CHD300102268305, CHD300102263401, Chang'an University).

Data Availability Statement: Data available on request due to restrictions (project data privacy): The data presented in this study are available on request from the corresponding author.

Acknowledgments: Multi-source data in this study were provided by Wuhan University. Xu Yongfu of Shanghai Jiaotong University and Guangxi Jiaotong Technology Group provided guidance and assistance. We are very grateful for the above support.

Conflicts of Interest: The authors declare no conflicts of interest.

References

1. Zheng, J.L.; Zhang, R. Highway Subgrade Construction in Expansive Soil Areas. *J. Mater. Civ. Eng.* **2009**, *21*, 154–162. [CrossRef]
2. Puppala, A.J.; Congress, S.S.C.; Banerjee, A. Research Advancements in Expansive Soil Characterization, Stabilization and Geoinfrastructure Monitoring. In *Frontiers in Geotechnical Engineering: Developments in Geotechnical Engineering*; Springer: Berlin/Heidelberg, Germany, 2019. [CrossRef]
3. Chen, Z.; Huang, G.W.; Xie, W. GNSS Real-Time Warning Technology for Expansive Soil Landslide—A Case in Ningming Demonstration Area. *Remote Sens.* **2023**, *15*, 2772. [CrossRef]

4. Huang, G.W.; Chen, Z.; Xu, Y.F. GNSS Real-time Monitoring Technology of Expansive Soil Slope. *Acta Geod. Cartogr. Sin.* **2023**, *52*, 1873–1882.
5. Chen, Z.; Huang, G.W.; Bai, Z.W.; Zhang, S.C. Monitoring of expansive soil slope based on low-cost millimeter-sized GNSS technology. *J. Cent. South Univ. (Sci. Technol.)* **2022**, *53*, 214–224.
6. Huang, G.W.; Du, S.; Wang, D. GNSS techniques for real-time monitoring of landslides: A review. *Satell. Navig.* **2023**, *4*, 5. [CrossRef]
7. Zhang, Q.; Bai, Z.W.; Huang, G.W.; Du, Y.; Wang, D. Review of GNSS Landslide Monitoring and Early Warning. *Acta Geod. Et Cartogr. Sin.* **2022**, *51*, 1985–2000.
8. Wang, D.; Huang, G.W.; Du, Y. Stability analysis of reference station and compensation for monitoring stations in GNSS landslide monitoring. *Satell. Navig.* **2023**, *4*, 29. [CrossRef]
9. Miao, L.C.; Liu, S.Y.; Lai, Y.M. Research of soil–water characteristics and shear strength features of Nanyang expansive soil. *Eng. Geol.* **2002**, *65*, 261–267. [CrossRef]
10. Ma, Z.J.; Mei, G. Machine learning for landslides prevention: A survey. *Neural Comput. Appl.* **2020**, *33*, 10881–10907. [CrossRef]
11. Shi, B.; Jiang, H.T.; Liu, Z.B. Engineering geological characteristics of expansive soils in China. *Eng. Geol.* **2002**, *67*, 63–71. [CrossRef]
12. Dai, Z.; Chen, S.; Li, J. The failure characteristics and evolution mechanism of the expansive soil trench slope. *2nd Pan-Am. Conf. Unsaturated Soils* **2017**, *2017*, 196–205.
13. Zhu, X. A Hybrid Machine Learning Model Coupling Double Exponential Smoothing and ELM to Predict Multi-Factor Landslide Displacement. *Remote Sens.* **2022**, *14*, 3384. [CrossRef]
14. Gao, W. Landslide prediction based on a combination intelligent method using the GM and ENN: Two cases of landslides in the Three Gorges Reservoir, China. *Landslides* **2020**, *17*, 111–126. [CrossRef]
15. Yang, B.B. Time series analysis and long short-term memory neural network to predict landslide displacement. *Landslides* **2019**, *16*, 677–694. [CrossRef]
16. Miao, F.S.; Wu, Y.P. Prediction of landslide displacement with step-like behavior based on multialgorithm optimization and a support vector regression model. *Landslides* **2017**, *15*, 475–488. [CrossRef]
17. Xun, Z.; Wen, J.Y.; Chao, W. Prediction Study of Wind Energy Based on AMPSO Algorithm and Neural Network. *East China Electric Power.* **2011**, *39*, 0797–0802.
18. Cervantes, A.; Galvan, I.M.; Isasi, P. AMPSO: A New Particle Swarm Method for Nearest Neighborhood Classification. *IEEE Trans. Syst. Man Cybern. Part B (Cybern.)* **2009**, *39*, 1082–1091. [CrossRef]
19. Chae, B.G.; Park, H.J. Landslide prediction, monitoring and early warning: A concise review of state-of-the-art. *Geosci. J.* **2017**, *21*, 1033–1070. [CrossRef]
20. Wei, X.Y.; Gao, W.W. Forecasting the Failure Time of an Expansive Soil Slope Using Digital Image Correlation under Rainfall Infiltration Conditions. *Water* **2023**, *15*, 1328. [CrossRef]
21. Zhang, Z.; Jiang, Z.; Meng, X. Research on prediction method of api based on the enhanced moving average method. In Proceedings of the International Conference on Systems and Informatics (ICSAI2012), Yantai, China, 19–20 May 2012; pp. 2388–2392.
22. Huang, Y.; Zhao, L. Review on landslide susceptibility mapping using support vector machines. *Catena* **2018**, *165*, 520–529. [CrossRef]
23. Smola, A.J.; Schölkopf, B. A tutorial on support vector regression. *Stat. Comput.* **2004**, *14*, 199–222. [CrossRef]
24. Jain, M.; Saihjpal, V.; Singh, N. An overview of variants and advancements of PSO algorithm. *Appl. Sci.* **2022**, *12*, 8392. [CrossRef]
25. XU, Y.F.; Cheng, Y.; Tang, H.H. Failure characteristics of expansive soil slope and standardization of slope slide prevention by geotextile bag. *J. Cent. South Univ. (Sci. Technol.)* **2022**, *53*, 1–20.
26. Zhang, Y.; Chen, X.; Liao, R. Research on displacement prediction of step-type landslide under the influence of various environmental factors based on intelligent WCA-ELM in the Three Gorges Reservoir area. *Nat. Hazards* **2021**, *107*, 1709–1729. [CrossRef]
27. Liu, Y.L.; Li, W.Q.; Zhang, J.W. Research on lateral earth pressure acting on retaining wall in expansive soil considering influences of environmental load. *J. Cent. South Univ. (Sci. Technol.)* **2022**, *53*, 150–159.
28. Li, T.; Kong, L.; Guo, A. The deformation and microstructure characteristics of expansive soil under freeze–thaw cycles with loads. *Cold Reg. Sci. Technol.* **2021**, *192*, 103393. [CrossRef]
29. Wang, Y.; Tang, H.; Huang, J. A comparative study of different machine learning methods for reservoir landslide displacement prediction. *Eng. Geol.* **2022**, *298*, 106544. [CrossRef]

MDPI AG
Grosspeteranlage 5
4052 Basel
Switzerland
Tel.: +41 61 683 77 34

Remote Sensing Editorial Office
E-mail: remotesensing@mdpi.com
www.mdpi.com/journal/remotesensing

www.ingramcontent.com/pod-product-compliance
Lightning Source LLC
Chambersburg PA
CBHW061826210926
51850CB00006B/819
* 9 7 8 3 7 2 5 8 7 5 6 9 6 *